Mountains
of the World

Mountains of the World

A Global Priority

Edited by B. Messerli and J. D. Ives

A CONTRIBUTION TO CHAPTER 13 OF AGENDA 21

Editorial Advisory Committee
Jayanta Bandyopadhyay, Francesca Escher
Lawrence S. Hamilton, Pauline A. Ives,
and Martin F. Price

This publication has been made possible by grants from:
SDC Swiss Agency for Development and Cooperation; UNU Programme: Mountain
Ecology and Sustainable Development; UNESCO International Programmes on
Man and the Biosphere and Geological Correlation; FAO Task Manager for
Chapter 13, Agenda 21; IDRC International Development Research
Centre, and CIP, International Potato Centre

The Parthenon Publishing Group
International Publishers in Medicine, Science & Technology

NEW YORK
LONDON

Library of Congress Cataloging-in-Publication Data

Mountains of the world: a global priority/edited by Bruno Messerli, Jack Ives

    p. cm.

    Includes bibliographical references and index.

    ISBN 1-85070-781-2

    1. Mountains. I. Messerli, Bruno.

II. Ives, Jack D.

GB501.2.M685 1997

551.43'2—dc21   97-7699

                        CIP

British Library Cataloguing in Publication Data

Mountains of the world: a global priority

    1. Mountains 2. Human ecology

    3. Upland conservation

    I. Messerli, Bruno II. Ives, Jack D.

    333.7'84

    ISBN 1850707812

Published in the USA by
The Parthenon Publishing Group Inc.
One Blue Hill Plaza
Pearl River
New York 10965, USA

Published in the UK and Europe by
The Parthenon Publishing Group Ltd.
Casterton Hall, Carnforth
Lancs. LA6 2LA, UK

Copyright 1997© Bruno Messerli and Jack D. Ives

First published 1997

Typeset by AMA Graphics Ltd., Preston, UK
Printed and bound by Butler & Tanner Ltd., Frome and London, UK

Foreword

President of the Swiss Confederation

Mountains and uplands cover about two-thirds of Switzerland. Icy and rocky peaks, steep slopes, wild torrents, and fragile soil and vegetation cover are particular features of our topography. We have learned that areas with these characteristics are especially vulnerable in combination with peripheral social and economic conditions. Over the centuries, the people of Switzerland have experienced both positive and negative aspects of life in the mountains. On the one hand, the negative aspects of fragile mountain ecosystems include natural hazards, degradation due to poverty, lack of sustainable economic opportunities, and inappropriate patterns of development. On the other hand, mountain people have gained a great deal of experience, and have exhibited openness and commitment in creatively adapting strategies for sustainable development to short- and long-term political, economic and climatic change.

It is therefore obvious why Switzerland has emphasised sustainable use of natural resources and sustainable mountain development in relations with its neighbours as well as in international forms of co-operation. One example is Switzerland's co-operation with 7 neighbouring countries in the Alpine Convention. At the international level, Switzerland's experience with technical co-operation and international research has been shared from the outset with many countries in the developing world. This was first done through multipurpose and integrated management projects, and later through programmes concerned with sustainable management of natural resources, often in combination with related research activities. Switzerland has also helped to create and support international centres which focus on different aspects of sustainable mountain development.

At the Rio Conference in 1992, Switzerland stressed the importance of addressing issues pertaining to mountain development and was instrumental in the formulation of Chapter 13 of Agenda 21, 'Managing fragile ecosystems: sustainable mountain development'. Convinced that the momentum for mountain development had to be maintained after the Rio Conference, Switzerland, together with UN, governmental and non-governmental organisations, has been very actively involved in networking, in the creation of synergies, and in taking fresh initiatives in mountain development through new strategies, implementation of projects, and research activities. Today, the concept of sustainable mountain development has taken on new meaning and is receiving more attention at the international level. In the 21st century, humankind will increasingly depend on mountain resources such as water, biodiversity, and recreation. Therefore, the issue of sustainable mountain development deserves even greater international attention and support. In fact, it represents one of the most important challenges we shall face in future. Switzerland is prepared to give top priority to its commitment in this area.

The title *Mountains of the World: A Global Priority*, as well as the messages in this book and in its companion executive summary, *Mountains of the World: Challenges for the 21st Century*, reflect our own conviction and commitment, and correspond to what we have been striving for nationally, regionally and internationally. Our most sincere thanks go to all persons and institutions who contributed to this publication. We warmly recommend it to everyone interested in and committed to sustainable mountain development. We are convinced that this book will stimulate greater world-wide awareness of all aspects of this issue, and we hope that its publication will motivate additional persons and institutions to make commitments of their own. In this sense, it will be valuable in helping to maintain the momentum needed to further the cause of the mountains.

Arnold Koller

Foreword

Rector of the United Nations University

The United Nations Conference on Environment and Development (UNCED), commonly referred to as the *Earth Summit*, was held in Rio de Janeiro in June, 1992. The United Nations University (UNU) had placed special importance on this pivotal world event for many reasons. One was our University's commitment to the mountain issue which centred on Chapter 13 of Agenda 21, UNCED's prime document. This commitment dates back almost to our early years in the mid-1970s. At that time one of our lead projects carried the title: *Highland–Lowland Interactive Systems*, later renamed: *Mountain Ecology and Sustainable Development*. While this project, co-ordinated by Professors Bruno Messerli and Jack D. Ives, drew much intellectual strength from our partners and colleagues, notably UNESCO's Man and the Biosphere (MAB) Programme, the World Conservation Union (IUCN), and several other institutions, the UNU's consistent support, with extensive research and training activities especially in the Himalaya and Southeast Asia, has been crucial as it provided a secure base and sense of continuity for almost two decades.

The very continuity facilitated the expansion of the research and training activities into parts of montane China, Tajikistan, the Andes, East and Northwest Africa and, most recently, Madagascar. It also led to the founding of the International Mountain Society (IMS) in 1981 and the establishment of the UNU/IMS quarterly journal *Mountain Research and Development*, which has since become the leading scholarly journal in the field. These enlarged activities, furthermore, were only possible due to the active and generous participation of the Swiss Development Cooperation together with the emergence of an informal partnership that appropriately has become known as the *Mountain Agenda*.

I do not exaggerate when I claim that without the University's commitment to the mountains, Chapter 13 of Agenda 21 may never have seen the light of day. The University provided the institutional base. The original small group of mountain scholars, now very much enlarged as the reader will see from the number of contributors to this book alone, demonstrated remarkable energy, ingenuity, and courage by insisting that we enter the political arena. This has been an attempt to save the world's mountains. We were well represented in Rio in 1992: Jayanta Bandyopadhyay, Jack Ives, Bruno Messerli, Juha Uitto, and myself. What happened there is now part of the mountain legend. We rather need to pay attention to what happens during the first critical decades of the new millennium. Certainly, the five years following Rio have witnessed an enormous increase in awareness of the mountain issue on the world scene. And I contend that this has been possible, at least in part, because the United Nations University has been able to function as an academic institution with no political mandate and our research scholars and associates can use all the sources of their home institutions and maintain a complete and precious hold on that quality we have come to know as 'academic freedom'.

This book, and its companion executive summary: *Mountains of the World: Challenges for the 21st Century*, will no doubt further increase the much needed growth in world-wide awareness of the mountain issue; I trust they will lead towards providing a base for the first vital steps to new policy; I trust that they will also show that formulation and action are possible. My endorsement of the book constitutes a special personal pleasure; not only have I followed the exciting academic debates and research successes with great interest, I have also witnessed the building of indestructible bonds between scholars, planners, developers, and above all mountain peoples, all over the world. Our project on Mountain Ecology and Sustainable Development has brought together in comradeship sincere and committed people, from every continent and from many walks of life. This alone epitomises the mission of The United Nations University. I would like to thank the very many persons and institutions involved in making this possible.

Prof. Heitor Gurgulino de Souza

Preface

In 1992, during and following the UNCED Earth Summit in Rio de Janeiro, the world's mountains finally received attention at the highest level. This was expressed formally with the inclusion, by acclamation, of Chapter 13 (*Managing Fragile Ecosystems: Sustainable Mountain Development*) in Agenda 21, UNCED's primary product. In this context, mountain ecosystems and mountain peoples were appropriately, and necessarily, afforded the same level of priority for world-wide attention and concern that was being allocated to other endangered ecosystems such as, tropical rain forests, semi-arid lands, wetlands, the oceans, or Antarctica. And collectively, the status of the world's great biomes, or mega-ecosystems, was balanced with the thematic problems of climate change, loss of biodiversity and cultural diversity, water resources, air and water pollution, gender inequity, and the still growing disparity between rich and poor.

In the Preface to *The State of the World's Mountains* (Stone, 1992), one of the *Mountain Agenda's* efforts to ensure the inclusion of Chapter 13 in Agenda 21, we stated: 'One major task is simply to increase awareness of the mountain issue'. This increase in mountain awareness has certainly occurred, although even greater and more detailed awareness, together with understanding, is needed. The present volume, therefore, is designed to carry the original mission further, but also to emphasise the need to move from awareness to initiation of practical steps to achieve sustainable mountain development.

While *The State of the World's Mountains* was the entrepreneurial product of a small group of mountain scholars, with the UNCED political objective as a goad, the present volume is the direct outcome of that 1992 achievement. One of the pragmatic results of the Rio Earth Summit was the establishment of the UN Commission on Sustainable Development (UNCSD). This commission, in turn, designated a series of UN agencies to serve as 'Focal Points' to ensure progress in achieving the various objectives of the individual chapters of Agenda 21. The UN Food and Agricultural Organisation (FAO, Rome) was appointed as Focal Point and Task Manager for the Mountain Chapter

(see Chapter 18). This produced a special meeting in Rome in March, 1994, bringing together representatives of all UN agencies with active mountain interests, along with those of the major mountain NGOs (for example, CIP: International Potato Centre; ICIMOD: International Centre for Integrated Mountain Development; IMS: International Mountain Society; TMI: The Mountain Institute; IUCN: World Conservation Union). From this first meeting a broad political and scholarly agenda was formulated with the following objectives:

(1) To continue the process of augmenting awareness of the critical role played by mountains and mountain peoples in world affairs;
(2) To further mountain research and to accelerate dissemination of applied and scholarly research results;
(3) To create a world-wide forum for mountain NGOs and individuals of all distinctions, together with an effective communication system;
(4) To establish connections amongst mountain peoples, decision-makers, development agencies, and scholars at all levels;
(5) To motivate both inter-governmental and governmental agencies to react in concert to the problems of sustainable mountain development.

Since these objectives required somewhat different sets of responses and competencies, they are being approached along three main, closely interlinked, but separate avenues:

(1) A series of regional/continental intergovernmental meetings have been held: in Asia, South America, Africa, and Europe. These are already beginning to affect policy making at the national and regional levels. A series of meetings are planned for North America during 1997, the first in Washington in April;
(2) A world consultation of mountain NGOs: the first was held in Lima, Peru, in February 1995. On this occasion the International Mountain Forum was established and, under the initial stewardship of The Mountain Institute (Franklin, WV, USA), is rapidly evolving as an enlarged mountain activist network centring

on a series of electronic mail consultations and conferences;

(3) Production of this report on *Mountains of the World: A Global Priority,* together with furtherance of the quarterly journal *Mountain Research and Development.* The editorial and authorship core for this undertaking is derived from the original *Mountain Agenda* group, now significantly expanded to a team of more than one hundred contributors.

All three of these initiatives are responses to FAO's Task Manager remit to report to UNCSD in time for evaluation during the special General Assembly of the United Nations scheduled to be held in New York in June, 1997. This commitment of the UN to review five years of progress (Rio Plus Five) set in motion by the UNCED Earth Summit, is itself one of the more crucial actions ever undertaken by the world body.

As a first step, therefore, it is incumbent upon us to define *sustainable mountain development.* At best, only a first approximation is feasible, considering the manner in which the term *sustainable* has been used this past decade. It appears to have been employed to embrace everything from attainment of a utopian world of health and happiness to a balance between exploitation and conservation of natural and cultural resources. We must take our lead from Gro Harlem Brundtland, and assume that sustainable development means a specific approach to resource use that insures against depletion and reduced options today, protects future generations, and achieves a significant reduction of the enormous imbalance in access to resources. This means a new balance between economic, ecological, and socio-cultural components in mountain development. This imprecise definition, of course, applies to our planet as a whole, its flora and fauna, minerals, hydrosphere and atmosphere, and our own species, *Homo sapiens,* as a *part* of that whole. It is reasonable, therefore, to question why mountains need to be singled out for special attention.

Mountains are indeed critical in this search for sustainability, not only because they account for about a fifth of the world's terrestrial land area and provide the direct life-support base for about a tenth of its human population. We stress that there is also a world-wide linking process to which we have referred as highland–lowland interaction, that cannot be ignored without dire consequences. Mountains not only provide indispensable ingre-dients (including, for instance, half the world's fresh water) for a viable world system, but mismanagement of mountain lands has the potential for overwhelming depredations on the lowlands. These depredations, for example, embrace the physical processes of hydrological disruption, flood, siltation, and landslides acting under gravity, but also mass migration of impoverished peoples who will further add to the pressures on the lowland urban infrastructure and so exacerbate sectoral conflicts. In addition, horrendous problems of brutal acts committed against civilians by organised military forces, together with natural disasters, augment the spread of malnutrition and infectious disease. These negative aspects, nevertheless, are off-set by the many positive aspects of biodiversity and cultural diversity, the untapped resources, but also spirituality and inspiration which have been experienced since time immemorial. Simply put, we believe that we stand at the crossroads; with the sustainment and expansion of achievements deriving from Rio, the necessary and corrective decisions are within reach of World Society.

In the Preface to *The State of the World's Mountains* (1992) we complained of the fact that far too little time had been available for satisfactory completion of that task, given the requirement to ensure distribution of the book in Rio de Janeiro. We took the occasion to plead that we were chasing a political objective and that success in that context would demonstrate the need for a more carefully thought out, reflective, and methodically prepared subsequent volume. This subsequent volume, however, has been prepared under significantly more severe time constraints than those that beset preparation of the first. Once again we have a political objective dictated by the timing of the Special Session of the UN General Assembly. In fact, the planning phase of the book was completed such that requests and instructions to potential authors could only be sent out in April, 1996! Nevertheless, in addition to our UN national delegation and agency readership, we do hope to make a significant contribution to mountain scholarship, regardless of the short-comings, at least partly due to these time constraints. In the present instance, we are taking a fundamentally different approach from that of *The State of the World's Mountains* in that the various chapters deal with mountain themes rather than specific mountain regions.

The current book consists of four parts. *Introduction,* Chapter 1 (Ives, Messerli, and Spiess), together with this Preface, is intended to set the stage and provide an overview. The issue of defining *mountains* is addressed; problems of representativeness in time and space, questions of scale, and especially the need for understanding extreme events and their importance to mountain resource use, are emphasised.

The second section, *The Human Dimension of Mountain Development*, comprises five chapters. Grötzbach and Stadel discuss peoples and cultures and relate them to the extremely diverse physical base, in Chapter 2. Bernbaum, in Chapter 3, highlights the hitherto largely neglected but, we believe, powerfully influential aspect of sacredness and its importance to successful development. In Chapter 4 the issues of comparative inequality and poverty are examined by Ives. Here the topic of marginality is introduced, along with the political necessity of achieving a better balance between the standard of living of mountain peoples and those of lowland and urban centres. Rieder and Wyder, in Chapter 5, confront the issues facing mountain development in an economically integrating world where the international market system is ensuring penetration even into the remotest of mountain communities. They raise the pivotal economic issue of mountain regions in the context of what they describe as their absolute and comparative cost disadvantages. Chapter 6, by Libiszewski and Bächler completes this section with a treatment of the devastating effects of conflict in mountain regions.

Part Three: *Mountain Ecosystems, Resources, and Development*, contains eleven chapters. These range from the major resource issues and their management and potential (water by Bandyopadhyay and colleagues; energy resources for remote highland areas by Schweizer and Preiser; biodiversity by Jenik and Klötzli; mining and mineral extraction, by Fox; protection of nature by Thorsell; montane forests and forestry by Hamilton, Cassells, and Gilmour), to utilisation of mountains for recreation and new forms of tourism and amenity migration (by Price, Moss, and Williams), mountain farming, defined broadly to include agriculture, animal husbandry, and integral use of forest resources (by Jodha), and mountain watersheds (by Hamilton and Bruijnzeel). These chapters are rounded off by a treatment of mountain risk and disasters (by Hewitt) and of climate and climate change (by Price and Barry).

Part Four, *Mountain Agenda for the 21st Century*, represents a summing up; Chapter 18, by El Hadji Sène and McGuire, details the organisational and administrative follow-up to the Rio Earth Summit and sets the framework for our political objectives under the rubric of Rio Plus Five, and beyond; and Chapter 19, by Ives, Messerli and Rhoades draws on the rest of the book to provide a statement on the need for a new approach to policy and politics in-so-far as the prospects for sustainable mountain development during the twenty-first century is concerned. Some suggestions are made on the need for targeted mountain research. There is also a companion volume, more in the context of an executive summary, designed for the decision maker and published separately with the title: *Mountains of the World: Challenges for the 21st Century*.

Each chapter was prepared by one, or several authors who are intimately familiar with its focus. Their statements, overall presentations, and especially their conclusions and recommendations, are personal and reflect their own expertise and convictions. There has been a degree of editorial exchange and discussion, and the editors and members of the editorial board sought to draw up a rational table of contents, in-so-far as that could be matched against availability of contributors under the stress of deadlines. Nevertheless, the book does not, and could not, seek to present a consistent and unanimous 'message'. Had there been more time available, as the editors had ardently hoped would have been so, much more debate between authors and editors, and amongst authors, would have occurred. This is not to deny that, to a certain degree, a common bond, let us say, sense of commitment, exists amongst this enlarged 'mountain agenda' group. Yet there are differences in both interpretation and opinion. Editorial input has entered into the finalisation of every chapter, and into the development of several of them, but has generally been confined to style and efforts to limit repetition. Nevertheless, there is repetition, in some cases by editorial choice – reflecting our desire to ensure that each chapter can more or less stand alone – and in some cases because of dire shortage of time and a complicated communications system regardless of our new-found electronically 'instant' access.

Above all, it is emphasised that this is a book produced by individuals; in no way can it be regarded as an official, governmental, or United Nations, document; the tenets of academic freedom have been adhered to absolutely.

The time and space available for production of this book precludes an exhaustive coverage of the mountain lands and mountain people. Perhaps the time for producing such an exhaustive coverage is long past, due to the great upsurge of interest in mountains and the concomitant and accelerating flood of publication: this includes the news media, popular environmental conservation literature, large format photographic books, academia, and development/policy literature. Much of the latter is what is generally referred to as 'gray' literature and is frequently difficult of access. Beyond our immediate areas of expertise, however, there is a vast mountain literature which ranges from the mountaineering/adventure genre to the extremely erudite technical and physical science literature – an example of the latter is the 81 volumes on tectonics, with several volumes devoted to Himalayan geophysics, published by the Geological Society of London. But within this general setting we would emphasise a number of specific causes, or take-off points. First, the International Biological Programme (IBP) of the late-1960s and early-1970s accelerated interdisciplinary and international collaboration and introduced scientific teamwork to computer modelling. This also provided an important impetus to UNESCO's Man and the Biosphere (MAB) Programme, beginning in 1971. From our point of view, MAB Project 6: *Study of the Impact of Human Activities on Mountain (and Tundra) Ecosystems,* has proved to be seminal (Price, 1995). Next the linkage – MAB-6/IGU Commission on Mountain Geoecology/UNU-Highland–Lowland Interactive Systems – has provided much stimulus and leadership. Regional seminars, field excursions, and volumes of proceedings from mountain ranges all over the world very early demonstrated the need to found the International Mountain Society (IMS) and to publish, with UNU, its quarterly journal *Mountain Research and Development* (from 1981). As part of this complex of activities the establishment of several institutions was encouraged: these include the African Mountain Association, the Andean Mountain Association, and the Laboratory for Mountain Research: this last in Moscow.

These developments, in turn, ensured the means of treating major mountain thematic issues, such as: (i) stability and instability of mountain ecosystems; (ii) hazards and risks, and mountain hazard mapping; (iii) highland–lowland interactions. They also promoted an increasing challenge to populist, and unproven, assumptions that were affecting the

directions taken by much international development aid. From our own point of view, this climaxed with the challenge to the *Theory of Himalayan Environmental Degradation,* which was fundamental to the launch of *Mountain Agenda* and the inclusion of Chapter 13 in Agenda 21 in Rio de Janeiro.

One lesson must be learned, however, from *The Himalayan Dilemma* (Ives and Messerli, 1989) and our first contact with the political arena: the challenge to the *Theory of Himalayan Environmental Degradation* has only been partially successful, in that some aid agencies and national governments still tailor their development policy to the assumption that an 'ignorant' mountain peasantry is responsible for the destruction of mountain forests and that this, in turn, is leading to devastating downstream catastrophe. Despite this, a new generation of more exacting Himalayan research has advanced the debate significantly, expanding to question mountain myths in other regions of the world. This new understanding is a large part of the motivation behind this book. It is perhaps justifiable to introduce some of these topics at the outset.

The primary problem facing formulation of policies to achieve the goal of sustainable mountain development is misinformation. Thompson *et al.* (1987) refer to it within the context of *Uncertainty on a Himalayan Scale*. This misinformation, or misunderstanding, can be divided amongst several categories:

(1) Absolute lack of relevant information (data bases) necessary for any effective policy formulation. In part, this is a reflection of the extreme complexity of the mountain environments;

(2) Misinformation resulting from over-generalisation of data and its analysis in an attempt to encompass much larger regions than studied in the field. This is largely the problem of lack of representativeness in space and time;

(3) Misinformation deriving from the deliberate, or unwitting, assimilation of hypothetical constructs, frequently Eurocentric, or based upon standard Western Science, that is inappropriately extrapolated to developing countries without adequate field check;

(4) Big Science also can be self-serving, so that major societal threats are sometimes conjured and moulded to attract more research funds. This phenomenon, while it has enlarged in

recent decades, essentially produces its own cure through vigorous debate in the scientific literature. The tendency for the popular press to over-develop some of the more dramatic elements that often characterise the early phases of such debates remains a matter for concern.

Many of the items raised in this Preface, of course, will be discussed in greater detail and throughout the remainder of the book. In a sense, we have accepted the delicate task of trying to balance between two stools – to produce documentary and theoretical material for the United Nations Rio Plus Five review process, and to render an addition to contemporary mountain scholarship. We are acutely aware that these two objectives are not necessarily mutually compatible in all aspects. Nevertheless, we hope that we have made at least a partial contribution to each.

Our concern for the principle of sustainable mountain development, however, is the primary motivation prompting this essay. We believe that there are numerous opportunities for considerable progress toward sustainable mountain development. Nevertheless, we must anticipate a general conclusion: sustainable mountain development is feasible; in fact it <u>must</u> be better defined and attained, but some major policy adjustments will be needed. These must be framed to promote rapid progress toward equalising access to natural resources, increasing local autonomy, achieving more effective training and out-reach, obtaining better treatment for women and children, and integrating and optimising socio-cultural, economic, and environmental components in the development process. This will be a fundamental shift in political perspective. It will be attained only in proportion to increased recognition that mountains are distinct from the other major ecosystems of the world, and accelerated accumulation of a much more reliable data base and understanding of the human–physical dynamic of these special three-dimensional landscapes.

JACK D. IVES,   15 March, 1997
    Carleton University,
    Ottawa, Canada
BRUNO MESSERLI,
    University of Bern,
    Bern, Switzerland

References

Ives, J. D. and Messerli, B. (1989). *The Himalayan Dilemma: Reconciling Development and Conservation.* Routledge: London and New York, p. 295

Mountain Research and Development (1981–present). Scholarly quarterly, The United Nations University and the International Mountain Society, Published by University of California Press, Berkeley, CA

Price, M. F. (1995). *Mountain Research in Europe: An Overview of MAB Research from the Pyrenees to Siberia.* UNESCO and Parthenon: Paris and Casterton, UK, p. 230

Stone, P. B. (ed.) (1992). *The State of the World's Mountains: A Global Report.* Zed Books Ltd: London, p. 391

Thompson, M., Warburton, M. and Hatley, T. (1986). *Uncertainty on a Himalayan Scale.* Ethnographia: London, p. 162

World Commission on Environment and Development, (1987). *Our Common Future.* Oxford University Press: Oxford and New York, p. 400

Acknowledgements

The editors of a book such as this obviously owe a great debt to a large number of individuals and institutions. First of all we thank SDC, UNU, UNESCO, FAO, and IDRC for their generous support, without which it would never have been possible to undertake such a difficult and challenging task and to finish it in time, including the policy-oriented summary, *Mountains of the World – Challenges for the 21st Century*.

Second, we thank the authors, co-authors, and box-authors. Without their forbearance, response to appalling deadlines, flexibility, and continuous support, nothing would have been accomplished.

For the editors, this volume is the outgrowth of a life-long commitment to mountain scholarship. It follows that our intellectual and spiritual debts are too far-flung to enumerate. They certainly go back to our personal introduction through Carl Troll's founding of the International Geographical Union's Commission on Mountain Geoecology in 1968. They also include the original UNESCO/MAB-6 Panel of Experts (Salzburg) and International Working Group (Lillehammer) of 1973–75; the formulation of UNU's Project on Highland–Lowland Interactive Systems in 1978, the creation of the International Mountain Society in 1981, and the excitement and camaraderie at Mohonk Mountain House, New York State (The Himalaya–Ganges Problem) in 1986. The formulation of *Mountain Agenda* quickly followed in 1990.

The foregoing statements presuppose contributions and friendships involving some hundreds of individuals; at the risk of unintentional indiscretion, some few must be mentioned by name: Carl Troll, Walther Manshard, Maurice Strong, Melvin G. Marcus, Barry C. Bishop, Ruedi Högger, Peter Stone, Wilhelm Lauer, Walter Moser, Gisbert Glaser, Michael Thompson, David Pitt, Hans Kienholz, Alton Byers, Gerardo Budowski, Rima Zimina, Sun Honglei, Hans Hurni, Carlos A. Baied, He Yaohua, Thomas Schaaf, Juha Uitto, Paul Messerli, Abdelatif Bencheriffa, Hugo Romero, Francis Ojany, Matthias Winiger, Maximina Monasterio, Máximo Liberman, Olivier Chave, Andri Bisaz, Egbert Pelinck, Hubert Zandstra, Suresh Chalise, Sumitra Manandhar Gurung, Kamal Shrestha, Olivia Bennett, and Frank P. Davidson. Alan Hastings, chair, and the entire faculty and staff of the Division of Environmental Studies (DES), University of California, Davis, were most supportive during the preparation of the book. And on the seemingly more mundane level, the second editor would never have completed his many tasks without the patience, care, and commitment of his 'computer guru' Derek Masaki, University of California, Davis.

We were also extremely fortunate in having our Publisher assign Helen Lee as our Scientific Editor. Her performance, under a most stressful time schedule, has been superhuman, a performance backed by the excellent Parthenon staff supporting her. Despite this there will be errors of various kinds for which we are entirely responsible. When parts of the final manuscript reached our publisher two months after the original absolute deadline and with only three months left for the essential publication date, how did we ever succeed?

Our ultimate thanks, nevertheless, go to the hundreds of folk who have entertained us in their small mountain huts, in the Andes, High Asia, Southwest China, East Africa, and Ethiopia, and closer to home, in the Pyrenees, the Caucasus, the Cairngorms, and the Alps. To break bread and take wine and cheese high in the Pays d'Enhaut, to eat chapattis above Darjeeling, and drink butter tea on the Plateau, or to dance at a village wedding in the Pamir; these are the privileges and pleasures beyond compare. We hope that the insights we may have gained, as well as a large sense of the mountain life force, can be repaid, at least in part, by even this modest attempt to move World Society closer to mountain responsibility.

THE EDITORS

Contents

Forewords   v

Preface   vii

Acknowledgements   xii

1   **Mountains of the World – A Global Priority**   1
     Jack D. Ives, Bruno Messerli, and Ernst Spiess

The Human Dimension of Mountain Development

2   **Mountain peoples and cultures**   17
     Erwin Grötzbach and Christoph Stadel

3   **The spiritual and cultural significance of mountains**   39
     Edwin Bernbaum

4   **Comparative inequalities – mountain communities and mountain families**   61
     Jack D. Ives

5   **Economic and political framework for sustainability of mountain areas**   85
     Peter Rieder and Jörg Wyder

6   **Conflicts in mountain areas – a predicament for sustainable development**   103
     Stephan Libiszewski and Günther Bächler

Mountain Ecosystems, Resources and Development

7   **Highland waters – a resource of global significance**   131
     J. Bandyopadhyay, J. C. Rodda, Rick Kattelmann, Z. W. Kundzewicz, and D. Kraemer

8   **Energy resources for remote highland areas**   157
     Petra Schweizer and Klaus Preiser

9   **Mining in mountains**   171
     David J. Fox

10   **The diversity of mountain life**   199
     Jan Jeník

11   **Protection of nature in mountain regions**   237
     Jim Thorsell

12   **Tourism and amenity migration**   249
     Martin F. Price, Laurence A.G. Moss, and Peter W. Williams

13 **Montane forests and forestry**   281
    Lawrence S. Hamilton, Donald A. Gilmour, and David S. Cassells

14 **Mountain agriculture**   313
    Narpat S. Jodha

15 **Mountain watersheds – integrating water, soils, gravity, vegetation, and people**   337
    Lawrence S. Hamilton and L. A. (Sampurno) Bruijnzeel

16 **Risk and disasters in mountain lands**   371
    Kenneth Hewitt

17 **Climate change**   409
    Martin F. Price and Roger G. Barry

Mountain Agenda for the Twenty-first Century

18 **Sustainable mountain development – Chapter 13 in action**   447
    El Hadji Sène and Douglas McGuire

19 **Agenda for sustainable mountain development**   455
    Jack D. Ives, Bruno Messerli, and Robert E. Rhoades

Author biographies   467

Index   473

Mountains of the World – A Global Priority

1

Jack D. Ives, Bruno Messerli, and Ernst Spiess

INTRODUCTION – CHAPTER 13 (1992) AND THE EARTH SUMMIT REVIEW (1997)

In 1992, the Rio Earth Summit (UNCED) not only established the UN Commission on Sustainable Development (CSD), it set in motion the mechanism for review of the progress made within the first five years following the Summit (Rio Plus Five or Earth Summit Review, 1997). In turn, the CSD appointed *Task Managers* for each of the chapters of Agenda 21. Task Managers were related to *Focal Points* located within the existing structures of the relevant UN agencies. As mentioned in the Preface, the UN Food and Agricultural Organisation (FAO) accepted the role of Task Manager for Chapter 13 (*managing fragile ecosystems – sustainable mountain development*) (see also Chapter 18).

The main approach of the Rio Plus Five review has involved each Task Manager unit, through its own committee structure, working in close association with the CSD. The ensuing comprehensive report is to be submitted by the CSD to a Special Session of the UN General Assembly in New York, scheduled for June 1997. This report should contain the following elements:

(1) overall assessment of the economic, social, and environmental situation five years after UNCED, and prospects for the future;
(2) main achievements following UNCED;
(3) most significant problems in Agenda 21 implementation;
(4) major challenges and priorities for the period following 1997;
(5) suggestions for institutional changes following 1997.

This general framework has been applied in standardised fashion to each Task Manager and the space allotted for the individual chapter reports has been tightly constrained. Nevertheless, beyond the precise specifications for the chapter reports, the Task Managers have had considerable freedom of action. In the case of Chapter 13, there has been a virtually unprecedented level of collaboration amongst UN agencies, national governments, international organisations, NGOs, and research institutions. This book, and the companion document: *Mountains of the World: Challenges for the 21st Century* (Mountain Agenda, 1997), of course, is only one element of the FAO Chapter 13 Task Manager initiative. Its preparation, while supported financially by several governmental and UN agencies, especially by the Swiss Agency for Development and Cooperation, has been a strictly open academic and independent process.

There is no doubt that the Rio Earth Summit with Agenda 21 has produced a much heightened awareness of the essential role played by the mountains as an integral part of the global biophysical and socio-economic system. It has also stimulated a large number of initiatives at all governmental and NGO levels. Whether or not this activity has moved from simply further increasing mountain awareness to actually changing conditions for the better, will be part of the task of the Special Session of the UN General Assembly to answer. In practice, we believe that much longer than five years will be required and the question will need a specific answer in a case-by-case evaluation for individual mountain regions. Thus, the present Rio Plus Five exercise, while it will go part way toward a general assessment, should be regarded as a first step in what must be framed as a long-term undertaking. In addition, the review process should aid in the identification of any trends, based on selected case studies, it should point to areas of weakness in the post-Rio process, and lay out, much more precisely, the challenges and priorities that must be faced as we enter the next millennium.

To provide a baseline for future mountain evaluation it should be worthwhile to consider the introductory statement from Chapter 13:

Mountains are an important source of water, energy and biological diversity. Furthermore, they are a source of such key resources as minerals, forest products, and agricultural products, and of recreation. As a major ecosystem representing the complex and interrelated ecology of our planet, mountain environments are essential to the survival of the global ecosystem. Mountain ecosystems are, however, rapidly changing. They are susceptible to accelerated soil erosion, landslides, and rapid loss of habitat and genetic diversity. On the human side, there is widespread poverty among mountain inhabitants, and loss of indigenous knowledge. As a result, most global mountain areas are experiencing environmental degradation. Hence, the proper management of mountain resources and socio-economic development of the people deserves immediate attention. (Agenda 21, Chapter 13, final version, adopted by the Plenary Session in Rio de Janeiro on 14 June, 1992).

We submit that the above quotation is an imperfect statement: it was a compromise that had to be pushed through the UNCED Preparatory Committees in the face of considerable opposition. This opposition challenged the stand that *mountains* constituted a viable unitary system, or formulation, and argued that the mountain issues were already covered in many of the other chapters that were being developed to become Agenda 21 (Table 1.1). This led to the contention that a special chapter for mountains (i.e. Chapter 13) was not necessary.

While it is impressive to see how many chapters of Agenda 21 have elements that relate directly to *mountains*, the same can be said for tropical rainforests, and the other 'fragile ecosystems': and at the moment of potential challenge in Rio, Chapter 13 was approved by acclamation and without controversy. However, we expect that the issue of mountain authenticity will continue to surface during these times of limited resources and international uncertainties. Thus we will contend that, while the mountains do contain elements of all the other major (and more readily defined) ecosystems – tropical rain forests, arid zones, polar-type regions, coastal zones, grasslands, wetlands, and the various other forest ecosystems – their unique characteristic, *verticality*, overwhelmingly warrants their priority ranking. To this argument can be added the warning that many of the past

development projects failed, at least in part, because mountain specificities were either ignored or not fully understood (Chapter 14).

Our argument, however, will only be justified to the extent that an integrated and comprehensive approach to mountain development, and its supporting research structure, is recognised and attained. By this we mean, amongst other requirements, a strong linkage between the human and biophysical elements of the mountain environment must be perfected, and that the mountain problems must be integrated with all related global issues (Figure 1.1).

DEFINITION OF THE MOUNTAIN ZONE

As with many aspects of application of mountain knowledge to policy formulation in the interests of sustainable development, a basic problem remains: how do we define *mountain*? The inability of mountain scholars to produce a rigorous definition

Table 1.1 Agenda 21 chapters with relevance to *mountains*

Chapter 2	International Cooperation
Chapter 3	Combating Poverty
Chapter 6	Protecting and Promoting Human Health
Chapter 7	Sustainable Human Settlements
Chapter 8	Making Decisions for Sustainable Development
Chapter 11	Combating Deforestation
Chapter 12	Combating Desertification
Chapter 14	Sustainable Agriculture and Rural Development
Chapter 15	Conservation of Biological Diversity
Chapter 18	Protecting and Managing Fresh Water
Chapter 24	Women in Sustainable Development
Chapter 26	Strengthening the Role of Indigenous People
Chapter 27	Partnerships with NGOs
Chapter 28	Local Authorities
Chapter 32	Strengthening the Role of Farmers
Chapter 33	Financing Sustainable Development
Chapter 34	Technology Transfer
Chapter 35	Science for Sustainable Development
Chapter 36	Education, Training, and Public Awareness
Chapter 37	Creating Capacity for Sustainable Development
Chapter 39	International Law
Chapter 40	Information for Decision Making (and so on)

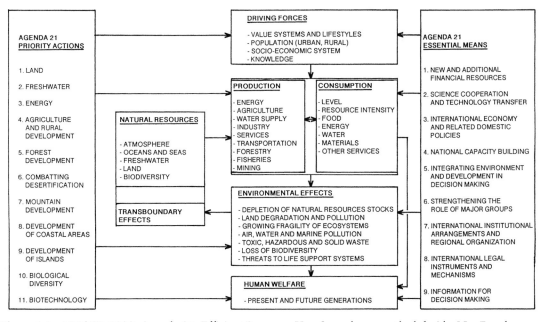

Figure 1.1 UNCED 1992, Agenda 21: Efficient Resource Use. See column on the left side, No. 7 under 'Priority Actions': Mountain Development (UNCED 1992)

that has universal application and acceptance has often led to time-consuming debate with no satisfactory result. We, like several other authors, have tended to side-track the issue and to generalise to such an extent that little rigour is retained. Thus we have relied upon the juxtaposition of 'steep slopes' and 'altitude', facets of mountain landscapes that individually, or in tandem, lead to marginality in the sense of human utilisation and adaptation.

Steep slopes, of course, encompass geomorphologically high-energy environments. Here, weathering processes combine with gravity to induce large mass transfers downslope and hence inhibit the development of mature soils. Together with slope instability, this promotes reduced biomass productivity and increased vulnerability from human intervention. Frequently, those downslope mass movements (rockfalls, landslides, avalanches, dam bursts and flood waves) can be catastrophic to human life and property (Chapter 16).

High altitude, in extreme cases, is also invariably associated with steep slopes. But there are many regions in the world that lie between about 3700 and 5000 m asl that are by no means high-energy environments. These include the Altiplano of the Andes, parts of the Ethiopian Highlands, much of the Tibetan Plateau, and the Eastern Pamirs. In these cases, however, biomass productivity is con-

strained by reduced temperatures and, depending on the climatic zone, where access to irrigation water is restricted, by aridity. It has become traditional, therefore, to bracket steep slopes and high altitudes as essential components of mountain and upland landscapes, separately, or in combination. In this way the extreme high plateaus are included in any definition of *mountain*.

At the risk of appearing overly simplistic, it is pointed out that ambient air temperature is influenced not only by altitude (adiabatic lapse rate), but also by latitude. The phenomena of world upper timberlines and regional snowlines are best described as high altitude–low latitude to low altitude–high latitude continua. It follows, therefore, that as timberlines approach sea level between 56° and 72°N latitude in the Northern Hemisphere, actual northing will depend upon longitude and degree of continentality. According to this approach, the Lofoten Islands in northern Norway, the Torngat Mountains in eastern Canada, and the mountains of southern Greenland, as examples of formerly or presently glaciated areas of high-energy relief, are approximate equivalents of the Alps, but cut off at timberline and reduced to sea level. More problematic, for an acceptable and universal definition, are the Hercynian massifs of Europe – the German Schwarzwald (Black Forest), the French Vosges, and the Scandinavian Calidonides. The

Plate 1.1 Nevado Sajama, 18° South, Bolivia's highest mountain and site of the world's highest forests (*Polylepis* trees close to 5000 m). (Photograph: B. Messerli)

Scottish Cairngorms, and large sections of the mountains in Norway and Sweden, at least rise well above both the climatic and the anthropomorphic timberlines; the *Mittelgebirge* (Middle Mountains) of Central Europe rarely do so. Yet they must all be included as mountains.

We can trace the beginnings of intellectual realisation of the significance of mountains in the 'Western context' to Alexander von Humboldt (1769–1859). His travels in the Andes led him to categorise the three-dimensional nature of mountains and to recognise the influence of latitude on the elevation of the distinctive altitudinal belts (Figure 1.2). These altitudinal belts primarily identify the upslope progression of changes in vegetation cover and landforms. He also recorded the adaptation of the Andeans to these altitudinal 'life zones' in terms of their subsistence and patterns of movement and exchange. Much later, this very

early initiative evolved into John Murra's (1972) concept of the 'vertical archipelago', itself derived from Carl Troll's (1954, 1975, 1978) elaboration of Humboldt's legacy (Bromme, 1851) into his more sophisticated concept of altitudinal belts. Troll went on to attempt a natural science classification of mountains with a clear distinction between *Hochgebirge* (High Mountains, such as the Alps) and *Mittelgebirge* (Middle Mountains, such as the Black Forest), and between the glaciated and non-glaciated mountains.

Nevertheless, since Troll's interests extended to embrace mountains world-wide, he quickly identified the dilemma facing any attempt to produce a universal classification. For instance, he referred to mountain areas in the humid equatorial zone, such as those extending above 3000 m in Indonesia, as 'high mountains without a high mountain landscape'.

This problem facing attempts to produce a classification of mountains becomes even more acute, for instance, when we try to incorporate the high lands of Ethiopia and East Africa. These highland areas traditionally have been much more favourable to human settlement than the surrounding arid and semi-arid lowlands. Further difficulties arise with mountains in the extreme arid zone, such as the Eastern Pamir and the Tibesti and Hoggar of the Sahara that have neither forest belts nor landforms produced by former glaciation. And for mountains in general, the most intensely utilised terrain is that located below the high mountain stage (upper timberline) (Figure 1.3): here are to be found the more serious intensities of human landscape intervention. More recently, these schematic representations have combined physical and human attributes of verticality (Lauer, 1993; Figure 1.4) and have been employed to demonstrate responses to climatic change over thousands of years (Messerli, *et al.*, 1993; Figure 1.5).

Concurrent with, and subsequent to, but also using the Trollean altitudinal belt classification as a starting point, Uhlig (1984, 1995), Grötzbach (1984, 1988), Kreutzmann (1995), and others, developed a tentative cultural geography of mountain regions (see Chapter 2). This depicts a fine line between the impacts of increasing altitude (the verticality issue) and the overlay of culture and, more recently, the penetration of the world market economy with the accelerated spread of modern communication systems. Uhlig, in particular, studied the altitudinal limits of domestic crops and forms of human subsistence, such as *Almwirtschaft* (mixed mountain farming), transhumance, and nomadism (Uhlig, *ed.* Kreutzmann, 1995). Grötzbach (1988) proposed the first cultural geographic typology of mountains.

From the foregoing discussion it should be apparent that the world's mountains do not lend themselves to a unifying definition and classification that goes beyond the simple combination of 'steepness of slope' and 'altitude'. Mountains, obviously, are regions of accentuated relief and altitude, which influence climate, soil fertility, vegetation, slope instability, and accessibility. From the humid sub-tropical and temperate zones polewards, all land-use activities are disadvantaged compared with the subjacent and neighbouring more densely populated lowlands. This comparative mountain disadvantage has been recognised in the recent extant and pending legislation of European countries, such as Switzerland, Germany, Norway, and others (Chapters 4 and 5). But

Plate 1.2 Mountains in high latitude, even of modest elevation, display a 'high mountain' (*Hochgebirge*) landscape, comparable to the Alps cut off at timberline and reduced to sea level. Inugsuin Pinnacles, Baffin Island, Canada, in latitude 70°N. (Photograph: J. D. Ives)

from the sub-tropical arid zones to the tropical humid zones, certain higher altitudinal belts are more beneficial for human land use than lower altitudes. This is due to their more favourable humidity and temperature conditions which, in turn, influence soil formation and fertility, and vegetational growth, and have provided more healthy living environments. Nevertheless, they are also extremely vulnerable to over-use and abuse, again, because of their verticality – the influence of gravity on steep slopes.

Another consideration, but also inconclusive, is cartographic representation. The fold-in map is a new approach to the cartographic portrayal of the world's mountains and highlands. It is based on the latest available electronic mapping capability. Contours are drawn for 500, 1000, 2000, 3000, and 4000 m and the total area above each contour has been electronically computed. Disregarding all other attributes except altitude, we have calculated that 48 percent of the world's total terrestrial surface lies above 500 m; 27 percent exceeds 1000 m;

Figure 1.2 Bromme, T., 1851: Atlas zu Alexander von Humboldt Kosmos, Stuttgart. Dominant vegetation zones in selected mountain systems. The average annual temperature is shown in °C; altitudes are in feet

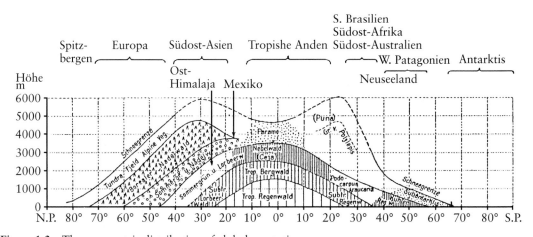

Figure 1.3 The asymmetric distribution of global vegetation
Schneegrenze = snowline; Tundra-Fjeld Alpine Vegetation = Tundra-Fjeld Alpine Vegetation; Borealer Nadelwald = Boreal coniferous forest; Sommergr. U. Nadelw. = Deciduous and coniferous forest; Sommergr. U. Lorbeerw. = Deciduous and laurel forest; Subtr. Lorbeerwald = Sub-tropical laurel forest; Páramo = Paramos; Nebelwald = Cloud forest; Trop. Bergwald = Tropical montane forest; Trop. Regenwald = Tropical rain forest; Puna = Puna; Gr. V. *Polylepis* = *Polylepis* border line; *Podocarpus Araucaria* = *Podocarpus, Araucaria*; Subtr. Regenw. = Sub-tropical rain forest; Schneegrenze = Snowline; Subantarktis = Sub-antarctic; Kühler reg.w. = Cool rain forest. Only humid climates are shown, except for the snowline and the puna region (broken lines and parentheses). Related forms of vegetation in the tropical highlands and the higher latitudes are depicted by the same conventional signs. Global distribution of permanently moist vegetation forms, shown in profile from the North Pole to the South Pole

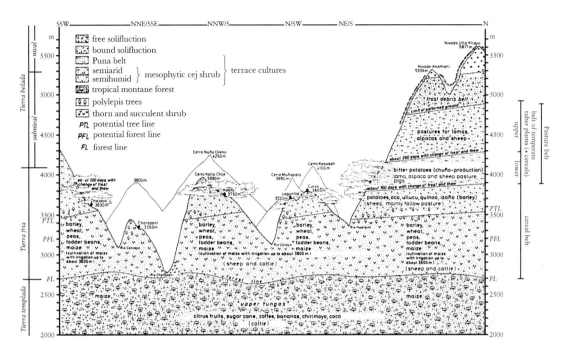

Figure 1.4 Agro-ecological altitudinal belts of the Charazani region, Callawaya, Northern Bolivian Cordillera (Lauer, 1993)

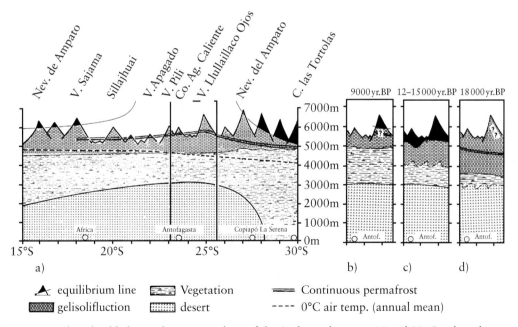

Figure 1.5 Altitudinal belts on the western slope of the Andes: a) between 15 and 30° South at the present time; and between 23 and 25° S at b) 9000 yr BP; c) 12–15,000 yr BP, and d) 18,000 yr BP (Messerli *et al.*, 1993)

11 percent exceeds 2000 m; 5 percent lies above 3000 m; and 2 percent above 4000 m. In certain regions of the world, such as Western Europe, the humid inner tropics, and the Pacific high islands, 500 m is already a significant altitude. However, much of Kansas, Nebraska, and eastern Colorado lies appreciably higher than this and, to the inhabitants of these flat prairie lands, mountains begin much further westward, along the Rocky Mountain front at about 1500 m. Similar comparisons can be made for other parts of the world.

To demonstrate that nearly half of the world's land surface lies above 500 m tends to support our conclusion that the search for a unitary definition of *mountain* is to chase a chimera. It follows that several definitions, which are region-specific, are needed. While we believe it has been necessary to pursue this rather tortuous path in our opening chapter, we also contend that to pursue it further becomes academic. We therefore return to our long-standing guestimate that 'mountains occupy about a fifth of the world's land surface and provide the direct life-support base for about a tenth of humankind'. We will return to this 'truism' after a global survey of mountains and a discussion of the one outstanding aspect – verticality – mountains as high-energy environments.

A BRIEF SURVEY OF THE WORLD'S MOUNTAINS

Mountains are found on every continent, from the equator polewards as far as land persists. Taken together, on account of their three-dimensional nature, as a single great landscape category, or ecosystem in the broadest sense, they encompass the most extensive array of topography, climate, flora and fauna, as well as human cultural differentiation, known to humankind. Furthermore, geologically and tectonically, mountains comprise the most complex elements of the earth's underlying structures.

Mountains and uplands incorporate the inhuman and extremely cold and sterile high ice plateaus of Antarctica and Greenland, and the high, dry, hypoxic, and almost inhospitable ranges of Central Asia and the south-central High Andes. They also include the richly varied, and even luxuriant, ridge and valley systems of the humid subtropics and tropics, such as the Himalaya, the Hengduan Mountains, Mt Cameroon, sections of the Northern Andes, and parts of New Guinea. In East Africa and Ethiopia the flanks of the high

mountains have long been a preferred human habitat compared with the more arid lowland areas that surround them. Additionally, there must be included an enormous compass of other mountains. Thus, there are the high volcanoes, for instance, of Indonesia, the Caribbean, and Hawai'i, where humans have long benefited from access to rich soils and have been exposed to extreme fiery hazards. There are also what are usually called middle mountains (German: *Mittelgebirge*), ranging from Tasmania to South Africa, and from Central and Northern Europe to the Urals and Siberia. While these latter contrast with the Alps (the epitome of the *Hochgebirge*) because of their more subdued relief, their other mountain attributes demand special policies to ensure sustained resource use.

Mountains and uplands contain the largest number of environmentally protected areas, including biosphere reserves, national forests, national and international parks, and World Heritage Sites (see Chapter 11), of any of the world's major landscape categories. Even when we exclude the two big ice sheets, mountains provide more than half the world's fresh water (see Chapter 7), as well as significant proportions of its timber (see Chapter 13), minerals (see Chapter 9), and grazing lands (see Chapter 15). They also serve as the abode of the deities of many of the world's religions (see Chapter 3) and provide an over-arching spirituality, aesthetic, and source of myth, legend, and psychological balm and aspiration for society at large (see Chapter 12).

When the intrinsic and spiritual resources are considered together, we are compelled to assert that mountains, in addition to providing the life-support base for about ten percent of the world population, are vital for the well-being of more than half of humankind. Their priority ranking, as world governance begins to grapple with sustainable development for the twenty-first century, would appear secure from the foregoing introductory remarks alone. Yet we must also take into account the actual and perceived threats to the lowlands if mismanagement of mountain resources continues unabated. In this sense, mountains are not only suppliers of many products, they are protection watersheds for the lowlands. The converse of 'protection' is that they are potential destroyers of the life-support systems of the hundreds of millions of people on the plains. Even when the threats, for instance, of deforestation as a cause of catastrophic flooding of Gangetic India and Bangladesh, are

Plate 1.3 Mount Kasbegh (5047 m), Caucasus, once the place where an angry Zeus chained the miscreant Prometheus for giving fire to mankind. Today, Georgians, Ossetians, Chechens, and Russians devastate both the mountain environment and each others' homelands. (Photograph: J. D. Ives)

ent in the proliferation of trans-border parks-for-peace, inspirational enrichment, recreation, and spiritual and physical challenge, make mountains an indispensable element of the human heritage. These attributes will become increasingly important if the enormous problems of the next century are to be faced and overcome.

Perhaps the only generalisation that can be made about mountains, however, is that, except for the three-dimensional landscape characteristic and the relative marginality and inaccessibility, generalisation should be avoided. In the popular eye, until recently (and still a prevailing trait), mountains have been perceived as vast, rugged, and remote landscapes, seemingly inured to human environmental impacts. Despite this, by the 1960s another, contrasting, popular view emerged. In this view, over-development, winter sports, mass tourism in general, rapid population growth, deforestation, soil erosion, accelerated run-off and devastating downstream effects, have been perceived as leading to a world super-crisis. Yet these viewpoints are overwhelmingly those from centres of population and power from outside the mountains.

At the same time as this great mountain contradiction was emerging (the one viewpoint producing complacency and benign neglect, the other leading to hysteria) bilateral and international development agencies tended to treat mountains as unimportant two-dimensional adjuncts to be accommodated as fringe attachments to the big development projects on the surrounding and much more densely populated plains. It seems that there was (is) a prevailing conviction that, dollar for dollar, better investment value would be obtained, in terms of economic successes, in the lowlands. Here accessibility is superior, infrastructure better developed, or less costly to establish and maintain. Traditionally, therefore, most political and economic interactions between highlands and lowlands have been initiated from the lowlands: policy decisions made in the lowlands and imposed on the highlands. The inhabitants of the mountain regions have been obliged to suffer exploitation, or to defend themselves, or to react to outside pressures; they have had little chance to act out their own destinies.

only perceived rather than proven, there are both serious implications as well as prospects for profligate expenditures to solve problems that have not been correctly analysed (Ives and Messerli, 1989).

In many parts of the world, mountains form international and provincial frontiers, and many of these today are in contention. We need only mention the entire Hindu Kush–Karakorum–Himalayan region; the frontiers of Chile, Argentina, Bolivia, Peru, and Ecuador in the Andes; the Caucasus; and the Balkans. The majority of today's most pernicious and destructive armed conflicts, guerrilla activities (Chapter 6), as well as the devastating drug wars, hinge on our mountains.

But there are also positive aspects that need to be reinforced. Mountains harbour by far the largest number of distinct ethnic groups, varied remnants of cultural traditions, environmental knowledge, and habitat adaptations; they host some of the world's most complex agro-cultural gene pools and traditional management practices (*cf* Peruvian Andes: Zimmerer, 1996). They offer primary challenges to research and scholarship and will likely prove vital field laboratories for early detection of some of the first indications of climate change (Chapter 17). The positive benefits that are inher-

PHYSICAL PROCESSES IN MOUNTAINS

In providing an overview of the dynamic geomorphic processes operating in mountains, we must emphasise that by far the majority of twentieth century research in this field has concentrated on the *Hochgebirge*. The reasons for this are obvious and yet are worthy of emphasis here. First, there is the personal attitude of individuals from 'Western' affluent societies. The high mountains are dramatically appealing and there is no doubt that an adventure element has influenced many mountain researchers. Second, there is also a pragmatic element. With the quantification of geophysical mountain research in the two decades following World War II, steeper and relatively unvegetated slopes greatly facilitated the acquisition of 'precisely' surveyed process data in a relatively short time period, simply because the processes are more dynamic under these types of locale. (Downslope movement proceeds much more rapidly and, therefore, can be measured to a much higher standard of accuracy with the available instruments in unit time.) A third factor could well be that of the relative inaccessibility of the research site, whereby markers and instruments could be left in place over several years with much reduced risk of interference by curious villagers or passers-by.

Today, a vast literature exists on this general field of enquiry. Some of the elements of high mountain process studies are mentioned here because the results must be taken into consideration whenever human interventions are to be perpetrated on mountain slopes at any altitude. The primary English language references are: Rapp's (1960) pioneer study of mountain slope processes; Price's (1981) synthesis of mountain physical geography; Gerrard's (1990) more recent synthesis of research on the same topic; Barsch (1984); and Hewitt's attempts to link our growing knowledge of catastrophic processes and their interactions with humans and their property, in terms of natural hazard research (see Chapter 16).

It can be argued that more geomorphic work can be accomplished during single catastrophic events than from all mass-transfer precisely measured for a decade, or observed over centuries (Ives, 1989). The effects of a hundred-year or even several hundred years' event may also be dwarfed by even more spectacular occurrences: for example, a very large landslide that occurred in the Langtang Himal, Nepal, 25 000 or more years ago. This landslide displaced approximately 10 km^3 of debris through a vertical distance of 2000 m. Something similar occurred at Köfels, in the Ötztal Alps of Austria, about 9500 years ago, when a landslide displaced 3 km^3 of debris through up to 1000 vertical metres (Heuberger *et al.*, 1984).

Some other examples may be mentioned here. First, in 1970, a debris flow, resulting from an earthquake-induced avalanche near the summit of Huascaran (6768 m) in the Peruvian Cordillera Blanca, covered 150 km^2 with debris, annihilated the town of Yungay, and killed 18 000 people (Patzelt, 1983). Second, in October, 1968, rainfall, varying between 600 and 1200 mm fell on the Darjeeling area of the Himalaya, during a three-day period at the end of the summer monsoon period when the ground was already saturated. Some 20 000 people were killed, injured, or displaced (Ives, 1970; Starkel, 1972). Third, in southeast Iceland 155 mm of rain in twenty-four hours on 29 July 1982 caused several dozen debris flows in a small area of Skaftafell National Park. One of us has photographed and observed this area qualitatively from 1952 to 1994 and estimates that (with the exception of the very high rates of glacial erosion, deposition, and associated jökulhlaup or glacial lake outburst flood and fluvio-glacial activity) this single geomorphic event accounted for more 'work' than all other processes combined over the last 100 years. Fourth, on July 19 and 20, 1993, an extraordinary flood event took place in eastern and central Nepal with catastrophic effects: several districts were hit by floods and hundreds of landslides, and many people died or became homeless. Due to the very high level of sedimentation,

Plate 1.4 Unteraargletscher near Grimselpass, Switzerland: frost weathering, glaciated landforms and retreating glaciers. (Photograph: J. D. Ives)

the life-span of the Kulekhani Reservoir, estimated to last for 60–70 years, was reduced by ten years (Schreier and Wymann 1996). These are just a few high magnitude events that can serve as examples.

The regularity, or irregularity, of occurrence of the extraordinarily large events poses a serious problem for any attempt to rank geomorphic processes and deduce long-term denudation rates. Therefore, it is important to have a more accurate understanding of what is happening on contemporary mountain slopes, to dismantle false, if intellectually attractive hypotheses, and to appreciate how and why a wide range of processes are functioning. This leads into the applied aspects of the assessment of what, in the human context, become natural, or mountain hazards (Aulitzky, 1974; Dow *et al.*, 1981; Kienholz *et al.*, 1983; Heuberger and Ives, 1994). It is equally important to appreciate the problem of spatial and temporal lack of representativeness, and to realise that many high mountain areas are rising as rapidly, if not more rapidly, than the processes of erosion are wearing them down. In this respect, the recent Chinese work in Tibet and the Himalaya has demonstrated a very high rate of contemporary uplift (Liu and Sun, 1981). Similarly, the most recent work in the western Himalaya and Karakorum and the Indus Gorge indicates a rate of land uplift that would have been considered impossible even a decade ago. Current progress in plate tectonics and advances in dating techniques, especially application of cosmogenic nuclide techniques to dating rock substrates, are providing an array of powerful tools for future research.

While great strides have been made in our understanding of slope processes, the problems of extreme variability in space and time remain as a challenge for the future. And despite the heavy concentration of research on a very small proportion of the total area of the mountain lands, the more spectacular high mountain belt dominated by cold climate processes, the knowledge so gained probably justifies the bias. However, in our estimation, the future challenge lies in the inhabited lower mountain belts where human activities may constitute a dominant geomorphic process (Ives and Messerli, 1981). Nevertheless, further progress in solving the spatial and temporal variability enigma must be made. Studies of this kind need to be undertaken hand in hand with studies in the human and social sciences, for a holistic approach to mountain landscape dynamics that crosses the

Plate 1.5 Intensive agriculture in the mountains of Cuba: Topes de Collantes 930 m. (Photograph: B. Messerli)

traditional divisions between the so-called human and natural sciences, as proposed by UNESCO's International Man and Biosphere Programme in the 1970s and 1980s.

THE HUMAN DIMENSION

Cultural identity is a multidimensional value by itself and creates the feeling of belonging to a certain community. Only today are we beginning to understand how different mountains are seen by different religions and cultures (Chapters 2 and 3). But the influence of a changing world can be observed even in remote mountain areas. New information influences the existing cultural behaviour, and a search for new balances between internal and external forces, between conservation and development, is beginning everywhere. Normally, indigenous populations have a traditional knowledge and long-term experience relating to the use of mountain resources. External impacts, such as mass-tourism and new traffic systems in industrialised countries, and economic decline and migration in developing countries, have disturbed this balance. The following three examples illustrate how mountain areas under different climatic–ecological, socio-cultural and economic–demographic conditions have been confronted with very different problems and processes on their development path from the past into the future:

(1) Let us first consider Kilimanjaro, a volcanic massif in the tropical zone close to the equator: in addition to its cultural and spiritual significance, this mountain is the primary source of much of the water, food, fuel, and building material for the people of north-central Tanzania. Unfortunately, the capacity of Mount Kilimanjaro to continue to provide these vital products and services is being threatened by inappropriate and, in some cases, over-exploitation of many of its natural resources. Much of the current stress on natural resources is a result of the dramatic increase in human population on the slopes of Mount Kilimanjaro: it has more than tripled during the last forty years, and the mountain area now has the highest rural density of any administrative district in Tanzania (Newmark, 1991).

By contrast with mountain areas in other climatic zones, the tropical mountains offer very favourable ecological conditions with fertile soils, especially on volcanoes, and sufficient precipitation for agricultural production. But will it be possible to find a balance between the new cultural, economic, and ecological forces?

(2) Our second example is the Middle Mountains of Nepal in the subtropical climatic zone with a seasonal monsoon precipitation regime: parts of Nepal are facing a serious problem in being unable to maintain soil fertility in agriculture and forestry. In order to meet the food, animal feed, and fuelwood demands of the rapidly growing population, there has been a gradual transformation from single to multiple annual crop rotation. This practice has developed to the point where three crops are grown annually in all areas of the Middle Mountains where irrigation water is available. Soil fertility has changed and soils have become very acidic. All soil fertility conditions are marginal and put into question the long-term sustainability of current levels of production (Schreier *et al.*, 1994). How can a new balance be struck between ecological conditions and economic needs? Can the needed alteration in cropping intensity and the introduction of nitrogen-fixing trees and crops constitute sufficient responses for promoting sustainability?

(3) Third, let us consider the Alps, situated in the temperate climatic zone, surrounded by highly urbanised and industrialised areas. Recent research projects have found that today about 44 percent of the Alpine population is living in so-called Alpine towns, concentrated in the main valleys along the traffic routes (pers. comm. Paul Messerli; Batzing *et al.*, 1996). These approximately 150 towns could assume new functions or new responsibilities in the future development of Alpine regions. But it is not yet clear, whether these urban populations could preserve a special Alpine-oriented cultural identity, and whether there is the necessary interest and capacity to create a new balance between rural and urban areas within the Alps. Could this represent an opportunity for new forms of co-operation between mountain farming and off-farm employment in the urban areas close by?

These three examples, illustrate three very different situations on the long and difficult path to sustainable development. New questions are being raised everywhere, and new solutions must be found in connection with all the internal and external forces. Solutions could include: economic diversification, appropriate agriculture, adapted forms of tourism, controlled migration, long-term investment in infrastructure, improved social conditions, better education, and more political autonomy.

CONCLUSION

The complexities of the high mountain stage (*Hochgebirge*), in terms of the physical processes that are occurring today and have occurred in the past, have been reviewed briefly above. It is equally apparent, however, that inhabited mountain regions are even more highly complex fields for study and, therefore, for management. Thus, international collaborative efforts are needed, including standardisation of objectives and methods, shared data banks, and identification of minimum needs.

Plate 1.6   The Yulongxue Shan (Jade Dragon Snow Mountains 5596 m) from the black Dragon Park, Lijang City, NW Yunnan, China. (Photograph: J. D. Ives)

In these closing years of the twentieth century that are characterised by increasing scarcity of research funds, this recommendation for expansion of effort, even for the establishment of a world network of mountain research activities, may appear impractical. In this respect it is worth considering the fate of some highly expensive development projects in mountain regions. For instance, the engineer's estimate of the useful life of a large hydroelectric scheme is often proved to have been greatly over-optimistic because no attention was paid to the dynamics of the mountain catchment in which the infrastructure was placed. Reservoirs that rapidly fill with sediment should not be contemplated, let alone constructed; and this occurs so frequently and at great cost to all concerned because of the lack of attention paid to geomorphic and hydrologic conditions. Similarly, poorly designed roads lead to instability of extensive mountain slopes and serious downslope damage. Resource mismanagement in mountain lands in general, often due to ignorance, or to disregard for long-term objectives and concentration on the short term, has the potential for devastating losses in the surrounding and densely populated lowlands. The problem is exacerbated by the extreme unreliability of much of the available data and the tendency to generalisation instead of careful local assessment.

The need for special attention being given to the mountains and their people, therefore, is understandable. If aid and development must take on a capitalist concern with immediate returns on investment then, obviously, it would be rational to neglect the mountains. But this is the short-term view. The long-term implications are becoming increasingly clear and are inextricably entwined in the growing concern that continuing development must be sustainable. This overwhelmingly reflects the remarkable initiatives taken during the Rio Earth Summit (UNCED). The present volume has the intention of aiding the clarification of the long-term viewpoint; this chapter to lay the general framework for that clarification.

13

References

Agenda 21 (1992). *A Guide to Agenda 21 – A Global Partnership*, UNCED: Geneva, p. 115

Aulitzky, H. (1974). *Endangered Alpine Regions and Disaster Prevention Measures*, Nature and Environment Series 6, Council of Europe, Strasbourg, p. 103

Bätzing, W., Perlik, M. and Deklova, M. (1996). Urbanization and depopulation in the Alps (with 3 colored maps). *Mountain Research and Development*, **16**(4): 335–50

Barsch, D. (ed.) (1984). *High Mountain Research*. Special issue of *Mountain Research and Development*, **4** (4): 286–374

Bromme, T. (1851). *Atlas zu Alexander von Humboldt*. Kosmos: Stuttgart

Dow, V., Kienholz, H., Plam, M. and Ives, J. D. (1981). Mountain hazards mapping : Development of a prototype combined hazards map, Monarch Lake Quadrangle, Colorado, USA (with fold-in map). *Mountain Research and Development*, **1**(1): 55–64

Grötzbach, E. F. (1984). Mobility of labour in high mountains and the socio-economic integration of peripheral areas. *Mountain Research and Development*, **4** (3): 229–35

Grötzbach, E. F. (1988). High Mountains as Human Habitat. In Allan, N. J. R., Knapp, G. W. and Stadel, C. (eds.) *Human Impact on Mountains*. Rowman and Littlefield: Totowa, NJ, pp. 24–35

Gerrard, A. J. (1990). *Mountain Environments: An examination of the physical geography of mountains*. MIT Press: Cambridge, MA, p. 317

Heuberger, H., Masch, L., Preuss, E. and Schroecher, A. (1984). Quaternary Landslides and Rock Fusion in Central Nepal and in the Tyrolean Alps. *Mountain Research and Development*, **4** (4): 345–62

Heuberger, H. and Ives, J. D. (eds.) (1994). *Mountain Hazards Geomorphology*. Special issue of *Mountain Research and Development*, **14** (4): 271–363

Ives, J. D. (1970). Himalayan Highway. *Canadian Geographical Journal*, 80: 26–31

Ives, J. D. (1989). Mountain Environments. In Marini-Bettolo, G. B. (ed.) *A Modern Approach to the Protection of the Environment*. Città del Vaticano:

Pontificiae Academiae Scientiarum Scripta Varia, pp. 289–345

Ives, J. D. and Messerli, B. (1981). Mountain hazards mapping in Nepal: Introduction to an applied mountain research project. *Mountain Research and Development*, **1** (3/4): 223–30

Ives, J. D. and Messerli, B. (1989). *The Himalayan Dilemma: Reconciling Development and Conservation*. Routledge: London and New York, p. 295

Kienholz, H., Hafner, H., Schneider, G. and Tamrakar, R. (1983). Mountain hazards mapping in Nepal's Middle Mountains, with maps of land use and geomorphic damages (Kathmandu–Kakani area). *Mountain Research and Development*, **3** (3): 195–220

Kreutzmann, H. (1995). Globalization, spatial integration, and sustainable development in northern Pakistan. *Mountain Research and Development*, **15** (3): 213–27

Lauer, W. (1993). Human development and environment in the Andes: A geoecological overview. *Mountain Research and Development*, **13** (2): 157–66

Liu, Dong-sheng and Sun, Honglie (eds.) (1981). *Geological and Ecological Studies of Qinghai-Xizang Plateau*. Science Press, Gordon and Breach, Science Publishers Inc.: Beijing and New York, 2 vols., p. 2138

Messerli, B., *et al.* (1993). Climate change and natural resource dynamics of the Atacama Altiplano during the last 18,000 years: A preliminary synthesis. *Mountain Research and Development*, **13** (2): 117–27

Mountain Agenda (1997). *Mountains of the World: Challenges for the 21st Century*. Geographical Institute University of Bern: Bern, p. 36

Murra, J. V. (1972). El 'vertical control' de un maximo de pisos ecológicos en la economía de las sociedades Andinas. In Ortiz de Zuniga, I. (ed.), *Visita de la Provincia de Leon de Huanuco (1562)*. Tomo 2, Universidad Hermilio Valdizan: Huanuco, pp. 429–76

Newmark W. D. (ed.) (1991). *The Conservation of Mount Kilimanjaro*. IUCN Tropical Forest Programme: Gland and Cambridge

Patzelt, G. (ed.) (1983). *Die Berg- und Gletscherstürze von Huascaran, Cordillera Blanca, Peru*. Innsbruck, Hochgebirgsforschung, no. 6, p. 110

Price, L. W. (1981). *Mountains and Man.* University of California Press: Berkeley and Los Angeles, p. 506

Rapp, A. (1960). Recent development of mountain slopes in Karkevagge and surroundings, Northern Sweden. *Geografisker Annaler*, **41**: 65–200

Schreier, H. and Wymann von Dach, S. (1996). *Understanding Himalayan Processes: Shedding Light on the Dilemma.* Festschrift B. Messerli, *Jb. Geogr. Ges. Bern*, Bd **59**, 1994–1996:75–84

Schreier, H., Shah, P. B., Lavkulich, L. M. and Brown, S. (1994). Maintaining soil fertility under increasing land use pressure in the Middle Mountains of Nepal. *Soil Use and Management*, **10**:137–42

Starkel, L. (1972). The role of catastrophic rainfall in the shaping of the relief of the Lower Himalaya (Darjeeling Hills). *Geographica Polonica*, **21**: 103–47

Troll, C. (1954). Über das Wesen der Hochgebirgsnatur. *Jahrb. Deutsch. Alpenvereins*, **80**: 142–57

Troll, C. (1975). Vergleichende Geographie der Hochgebirge der Erde in Landschaftsökologischer Sicht. *Geographische Rundschau*, **27**: 185–98

Troll, C. (1978). Der asymmetrische Vegetations- und Landschaftsaufbau der Nord- und Südhalbkugel. *Erdwissenschaftliche Forschung*, Bd. XI, Herausgegeben von Carl Troll und Wilhelm Lauer, F. Steiner Verlag: Wiesbaden, Germany, pp. 10–28

Uhlig, H. (1984). Die Darstellung von Geo-Oekosystemen in Profilen und Diagrammen als Mittel der Vergleichenden Geographie der Hochgebirge. In Grötzbach, E. and Rinschede, G. (eds.), *Beiträge zur Vergleichenden Kulturgeographie der Hochgebirge.* Eichstätter Beiträge, 12. Friedrich Pustet: Regensburg, pp. 93–152

Uhlig, H. (ed., Kreutzmann, H.) (1995). Persistence and change in high mountain agricultural systems. *Mountain Research and Development*, **15** (3): 199–212

Zimmerer, K. S. (1996). *Changing Fortunes. Biodiversity and Peasant Livelihood in the Peruvian Andes.* University of California Press: Berkeley and Los Angeles, p. 308

Mountain peoples and cultures

2

Erwin Grötzbach and Christoph Stadel

MOUNTAIN POPULATION : RESTRICTED RESOURCES AND THE THREAT OF OVER-POPULATION

While a significant proportion of the world's population is living in mountainous regions, it is difficult to provide exact figures as these will vary considerably depending on how 'mountain' is defined (Chapter 1). If a broad definition is used so that 'middle mountains' (*Mittelgebirge)* and 'highlands' are included, then about ten percent of the total human population would be incorporated. This would ensure the inclusion of the very large highland populations of East and Southeast Asia, South and East Africa, and South America. This rough estimate does not include those populations who, although living beyond the actual limits of the mountains, are directly or indirectly influenced by effects of the mountain environment, such as climate (e.g. föhn wind) and floods. Many millions of people are temporarily exposed to long-distance mountain hazards (see Chapter 16) as, for instance, on the plains of northern India and Pakistan and the Po plain of northern Italy. If the term 'mountain' is restricted to 'high mountains' (*Hochgebirge*), however, the proportion of mountain dwellers will drop to about two percent of the world total. (For example: Himalaya 33 million, Andes 26 million, Alps 11 million; see Bätzing *et al.,* 1996, for the most recent demographic data for the Alps.) For the purpose of the current analysis 'mountain' will be defined in the wider sense.

Generally, mountains are considered as areas of low to very low population density, except for some tropical highlands that offer favourable environmental conditions (e.g. Ethiopia, parts of Central America and the Andes). This statement is true with regard to arithmetic density based on the total area. But the total area will frequently include large tracts of land unsuitable for human occupation because of extreme altitude and adverse topographic and ecological conditions, including permanent snow and ice. It is more useful, therefore, to consider the 'physiological density', that is, a

figure based upon the area of agricultural land, with the latter reduced to a standard value according to its productivity. From this it is possible to obtain an understanding of population pressure on the land, at least for those mountain areas primarily dependent on agriculture; and this applies to a large proportion of the total mountain land, however defined. The difference between arithmetic and physiological density is highest in arid mountain environments as in some districts of the Tibetan and High Himalaya or in the driest parts of the Andes. Here the arithmetic density amounts only to a few inhabitants per square kilometre while the physiological density may reach several hundred per square kilometre. Only narrow strips or patches of irrigated land can be utilised intensively, while the remaining land has a very low potential as pasturage with few permanent settlements.

The foregoing discussion emphasises the fact that many mountain areas which seem only sparsely populated are (or were) severely over-populated in relation to their limited agricultural resources. These resources are often over-exploited resulting in serious deterioration of the environment and in the marginalisation of living conditions. This critical situation is further aggravated by a continuing high natural population increase which is characteristic of many mountain regions.

Over-population is manifested by a series of well known phenomena: continued fragmentation of the small holdings culminating, in extreme cases, in collective ownership of individual fruit trees by repeated inheritance (Afghan Hindu Kush); food deficiency despite high labour inputs into agriculture; therefore, a growing necessity to supplement family income, for example from mining, small home industries, itinerant trade, seasonal work outside the mountains, temporary military service and, finally, emigration. These symptoms were characteristic of the Alps during the nineteenth century and prevail today in most high mountain areas in the developing countries. Furthermore, they are aggravated by high rates of under-employment and unemployment, increasing fuel-wood shortages and environmental stress, inadequate infrastructure and services. This critical situation can be overcome only by heavy emigration that pulls off the surplus population and/or, by the development of non-agricultural occupations in the mountains, or by a substantial reduction in the rate of population growth. In many situations, however, out-migration may have a very negative impact in that young male and more highly educated members of depressed mountain communities are often the first to leave, at least in the early stages, and the remaining populations become the enfeebled older segment. Tourism has proved to be an important new source of income for many mountain regions. But the economic benefits from tourism do not reach all segments of the population and such development is frequently accompanied by adverse cultural and social consequences. Industrial growth is usually highly restricted to specialised conditions in high mountain valleys and, therefore, is limited to a very few locations.

The socio-economic and cultural situation across the mountains of the world differs widely in many aspects. A tentative typology of high mountain regions, based mainly on age and density of settlement, and on the dominant economic activities is introduced in Table 2.1 (Grötzbach, 1988).

MOUNTAIN PEOPLES: A CULTURAL MOSAIC

Inhabitants of mountain areas are often reputed to be traditional, conservative, passive, or even backward, in the opinion of people from the plains and cities. This distinction, whether true or not, is sometimes expressed by terms like *Kohistani* (from Persian *Kohistan* = mountain country) in Iran, Afghanistan and Pakistan, or *Pahariya* (from Hindi *pahar* = mountain) in northern India and Nepal. In some cases these terms are used to denote certain ethnic groups as, for example, *Kohistani* for the inhabitants of Indus and Swat Kohistan in Pakistan. The derogatory implications are frequently apparent.

Mountain people may be classified by a number of different parameters. Most frequently mountain population groups are distinguished by ethnic, linguistic, religious, or political criteria. In the Alps, for example, the ethnic and linguistic affiliation is most important and sometimes extends across political borders; for instance, the German speaking group in South Tyrol, Italy, or Italian speaking inhabitants of Swiss Ticino. In the Himalaya and Karakorum, religion is a primary discriminating criterion and populations characterised by a single religion form very large majorities over extensive cultural regions (Figure 2.1). In the Andes religion as a criterion is replaced by race or ethnicity; here the division is between the native Indios and the offspring of European immigrants (Figure 2.2). Another parameter characterising mountain

Table 2.1 Typology of High mountain regions from a human geographical perspective. 'Old' is defined as principally Old World mountain areas that have been settled for centuries; 'young' implies New World areas settled by European immigrants

A. OLD AND RELATIVELY DENSELY SETTLED HIGH MOUNTAINS
1. Largely intact traditional subsistence agriculture and a tendency toward over-population:
   a) Population of mountain peasants
      (*Large parts of the Himalaya–Karakorum–Hindu Kush; Andes*)
   b) A population of mountain peasants, overlain by nomads
      (*High Atlas; mountains of the Middle East; western parts of the Hindu Kush and Himalaya*)
2. Strongly declining traditional agriculture and expanding new activities (tourism, among others)
   (*High mountains of Europe, especially the Alps and Pyrenees*)
3. Largely collectivised, or nationalised, agriculture, in parts with new activities (island-like scattered tourism)
   (*High mountains of the former Soviet Union and China; parts of the Carpathians*)

B. YOUNG AND RELATIVELY SPARSELY SETTLED HIGH MOUNTAINS
   Coincides with areas of European overseas colonisation, with extensive market-oriented agriculture and forestry, and with recent tourism
   (*High mountains of North America and New Zealand*)

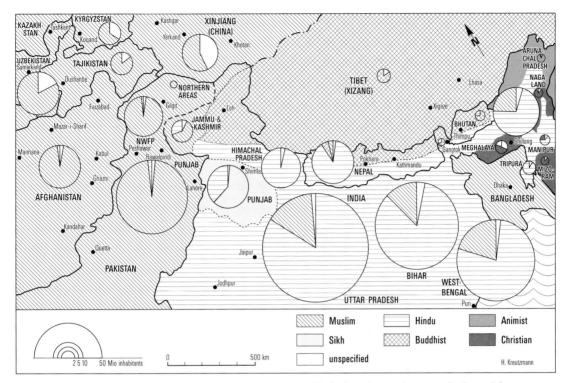

Figure 2.1 Major religious groups of High Asia. Figure published with permission and adapted from Hermann Kreutzmann, 1993

population groups is the degree of spatial mobility; this is highest among the mountain nomads of West and South Asia (Taurus, Zagros, Hindu Kush, Western Himalaya).

Two broad groups of mountain dwellers can also be differentiated in terms of their spatial distribution:

(1) Mountain dwellers who are part of a larger group of people or nation, whose core area is located outside the mountains, for example, the German, French, or Italian speaking populations of the Alps. It should be noted, however, that these primary groups are divided into several sub-groups as, for instance, those who speak German, into Swiss, Tyroleans, Bavarians, and so on, with many sharply distinct dialects, idioms, and local traditions;

(2) Mountain peoples whose living space is (or was) entirely confined to the mountains. Although they tend to be smaller in numbers and spatially more confined, there are exceptions, as shown by the examples of the Tibetans, the Quechua speaking Indios of the Andes, and

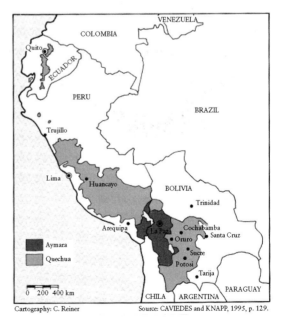

Figure 2.2 Modern native American domains of the central Andes

Box 2.1   Indigenas of the Andes – Mountain people between tradition and modernisation

The indigenous highland peoples of the tropical Andes, over many centuries, have not only adapted physiologically to high altitude conditions, but have proven their ability and resilience to use the numerous ecological niches of the tropical mountain environment. This includes the domestication of animals and a large variety of plants. They have also developed several forms of agriculture that have been referred to as systems of 'verticality' and 'complementarity'; these extend over a considerable range of altitude from the lower Andean slopes to over 5000 m. This ingenious adaptation to, and modification of, the mountain environment (for example, terraces, raised fields, irrigation systems) has sustained considerable populations. At the time of the first European contact, it has been estimated that the total population was as high as 12 million. In addition, a very advanced culture had been developed, including the formation of powerful and extensive empires which have been claimed as the virtual equivalents of European civilisation of the late Middle Ages.

During the colonial and post-colonial periods, the traditional community organisation was threatened or destroyed and the indigenous systems and techniques of agriculture largely collapsed. New crops, domesticated animals, land tenure systems, and agricultural techniques either reflected European tastes and preferences or were geared to maximise profits and were oriented for export according to the prevailing mercantile–capitalist model of the time. This represented not only an ill-fated break with native traditions, but also a growing threat to the mountain environment.

Today, a total of about 17 million people, or approximately 10 percent of the total population of the Central Andean states, are classified as 'native' (Table 2.2). The indigenous Andean population is primarily clustered in specific highland areas of Peru, Bolivia, and Ecuador, with much smaller groups in Chile, Argentina, and Colombia (Figure 2.2).

During the last few decades, the development, fate, and status of the *Indígenas* in the Andean region has revealed a number of patterns and processes, many of them characterised by dualities and dichotomies. While most native people have remained in a marginalised social and economic condition, the political and social influence of the indigenous

Plate 2.1   The village of Parinacota (4392 m) in the dry Andes of Northern Chile. The church was built in the seventeenth century. Today only nine indigenous families live in the village (Photograph: B. Messerli)

Table 2.2 Total population and proportion of native and other peoples in South American countries with a share in the Andean realm

Country	Total population in 1993 (million)	Native population (%)	People with mixed ancestry (%)	'White' population (%)	Black population (%)
Argentina	33.8	0.1	5.0 (Mestizos)	94.9	–
Bolivia	7.1	42.0	31.0 (Mestizos)	27.0	–
Chile	13.8	1.5 (Mapuche)	5.3 (Mestizos, other than Mapuche)	93.2	–
Colombia	35.7	0.06	75.0	20.0	4.0
Ecuador	11.0	20.0	50.0 (Mestizos, Mulattos)	25.0	5.0
Peru	22.9	47.0	32.0 (Mestizos)	12.0	9.0 (black and Mulatto populations, other minorities)
Venezuela	20.9	2.0	69.0 (Mestizos, Mulattos)	20.0	9.0
Total population	145.2	Average: 16.0	Average: 38.2	Average: 41.7	Average: 3.9

Source: Fischer Weltalmanach 1996, 1995

movements and organisations has expanded significantly, especially in Ecuador, Peru, and Chile. They have become the leaders in protest movements for under-privileged peoples – a 'national integration from below'. The growing empowerment of the indigenous peoples is also reflected in a number of bilingual educational programmes and in a greater degree of respect for their cultural traditions. A duality, however, can also be observed between this new move toward national conscience-forming for indigenous rites and traditions and the drive for national integration, modernisation, and economic neo-liberal policies. The latter are geared toward the privatisation of indigenous collective properties and the transformation and neglect, or even dissolution of small-scale subsistence agriculture, in favour of technologically advanced agri-business, intent on maximising profits. Once again, a counter-movement to this trend is observed in the newly acquired pride and the reaction of the *Indígenas* toward their traditions. This becomes, in effect, a 'development' philosophy centred on the following premises: respect for nature, the soil, and life itself, as a basis for native culture and society; the concept of equality and communality; recognition of cultural flexibility and the need to minimise economic risks (Centro de Arte y Acción Popular, 1981). The new pride of the indigenous peoples is best expressed by the concluding statement of an Ecuadorian *Indigena*: 'We have to 'civilise' western society which has not yet recognised our struggle' (Zettl, 1994).

Source: Christoph Stadel, Institute of Geography, Salzburg, Austria

Box 2.2 The Hunza people – survival in a harsh environment by optimising local resources

In the midst of a highly diversified environment of deeply incised arid valleys and Karakorum glaciers, the Hunza people have managed to cultivate isolated patches and to utilise all the available natural resources through a great range of altitude. The former semi-autonomous principality of Hunza occupies an area of about 10 000 km², making it a quarter of the size of Switzerland. Only one percent of the land area is cultivated yet this has permitted the survival of a population that today stands at about 35 000. Despite the small population, there are four different languages belonging to three language families which characterise the ethno-linguistic pattern in Hunza. The lower part of the valley is occupied by Shina speakers (Indic language group) who account for 12.3 percent of the total population. More than two-thirds (67.9 percent) belong to the Burusho group who have settled mainly in the central part of the valley; their isolated language, Burushaski, is only found in Pakistan and continues to pose a problem for linguist scholars. The upper part of the valley is occupied by Wakhi (Eastern Iranian idiom) who replaced Kirghiz (Turkic) nomads in the late eighteenth century (Kreutzmann, 1995). Both groups were attracted to the natural pastures of the Karakorum. The fourth, and smallest, group (1.4 percent of the total) is formed by the Dom (Indic), who represent a professional class, formerly restricted in their activities to music and blacksmithing. While the earliest inhabitants trace their ancestry and settlements back for more than 600 years, most of the high-lying villages were founded as recently as the nineteenth century (Kreutzmann, 1994).

Despite this ethno-linguistic diversity, all inhabitants are involved in practices of high mountain agriculture. The degree of participation and the size of land holdings may vary, but this high mountain valley presently sustains a population which had originally depended to a much greater extent on agriculture. What changes have occurred to bring this about?

The traditional system aimed at utilisation of the natural grazing of different altitudinal belts, by seasonally pasturing flocks of sheep, goats, and yaks outside the settled areas. The village lands, themselves, consist of highly productive terraced fields where irrigated crops of wheat, barley, maize, millets, and buckwheat are cultivated. Fruit orchards supply additional staple foods, such as apricots and mulberries, that can be dried to augment the winter cereals, and processed products from livestock husbandry. Yet this system of high mountain agriculture always failed to provide adequate surpluses for times of need. The oral traditions are full of tales about springtime famine and periods of starvation. In 'olden times' the looting of trade caravans proved to be one solution to the local production deficit.

Since colonial rule entered the Karakorum towards the end of the nineteenth century, however, more and more non-agricultural jobs, in services, were taken up, both inside and outside the valley. Some men joined the army, others became postal runners across the Karakorum passes, or coolies for a variety of employers. This approach of augmenting the household income by earnings from external sources is practised in many remote regions, and in this respect, Hunza is no exception. However, special features of this adaptation for Hunza are new strategies that, in recent years, have been related to migrant labour (Kreutzmann, 1996). Previously, unskilled farmers joined the army and civil service, but now many of the Hunzukuts work as entrepreneurs, teachers, managers, and professionals. The efforts of the Aga Khan network – the great majority of the people belongs to the Ismaili faith, of which the Aga Khan is spiritual leader – have created an educational environment that provides equal opportunities for these rural areas. The literacy rate is now higher than anywhere else in Pakistan. The qualitative improvement in education is supported by better health facilities and rural development programmes which help to overcome the supply deficit. Today, Hunza depends for more than two-thirds of its nutritional needs on external supplies which are imported from down country and largely paid for by the migrants' wages. Locally, additional income is generated from tourism which has expanded, especially since the Karakorum Highway was opened for trans-border travel to China.

Source: Hermann Kreutzmann,
Nurenburg-Erlangen, Germany

the Amhars of the Ethiopian highlands. The speakers of the old Romance languages in Switzerland, the Dolomites, and Friaul/Friuli, in Italy, as well as the Hunzukuts in the Karakorum (see Box 2.2), or the Sherpas of the Mt Everest region in Nepal, occupy much more limited areas.

Many mountain areas exhibit a high ethnic diversity, that is, a variety of groups of type 2: parts of the Himalaya–Karakorum–Hindu Kush (Chitral–Gilgit–Baltistan in Pakistan; Nepal) and the eastern Caucasus (Daghestan). The German ethnologist Karl Jettmar characterised the mountains of northern Pakistan as 'a giant ethnographic museum', an interpretation that might be applied equally to other mountain regions. This ethnic diversity is usually explained by the fact that some mountain peoples lived in social and spatial seclusion for a long time, such as the Nuristani, or former Kafirs, in the Afghan Hindu Kush who were pagans from the viewpoint of their Muslim neighbourhood until the end of the nineteenth century, or the Druzes and the Maronite Christians of Mount Lebanon. These and other examples led to the hypothesis that mountain peoples of type 2 occupy locations of refuge or retreat. The validity of this hypothesis is discussed in the following section (see also Chapter 6).

TRADITIONAL MOUNTAIN CULTURE AND ENVIRONMENT

The mountain refuge, or retreat, hypothesis seeks to explain the variety of mountain peoples as a result of the dissected and fragmented structure of mountains where every valley or basin may form a peculiar world of its own. Similarly another widespread assumption interprets the variety of mountain cultures as the consequence of long-lasting seclusion and isolation of mountain communities caused, or intensified, by restricted accessibility. These hypotheses stress the determining role of the mountain environment on ethnicity and culture. Thus, Jentsch (1984) speaks of a worldwide 'high mountain culture' characterised by an optimal adaptation of people to their natural environment, finding its expression in a 'high mountain cultural landscape'. This perspective is open to the criticism that it is 'environmental determinism', a hypothesis that has been contradicted by modern social science research. There should not be any doubt, however, that the natural environment remains one of the most important factors in shaping mountain cultures. It strongly controls the location of settlements, agricultural land use and the nature of other economic activities, as well as everyday life that must cope with the strain of the vertical dimension; the pronounced mountain environment also greatly influences mobility patterns and transportation. This is particularly evident in traditional mountain peasant societies whose adaptation to their environments has evolved from many generations of experience; the mountains of Asia, the Andes and, in former times, the Alps, provide excellent examples.

HUMAN ADAPTATION TO MOUNTAIN ENVIRONMENTS

Adaptation to the mountain environment means primarily to respond to the potential as well as the problems of highly differentiated ecological conditions arising from the variables of relief, exposure, and altitude. Among these physical factors, altitude, or the 'vertical dimension', is of particular importance as it controls many other aspects resulting from the decrease of temperature with height. It influences agriculture especially in that the latter must contend with shrinking land-use potential with increasing altitude. The highest land-use belt always consists of pastures. Crops grow up to a certain altitude where they find their natural or

economic upper limit. These facts have been demonstrated by a large number of empirical studies and classical models, for example, by Troll (1975, trans. Stadel, 1988), Stadel (1992), and Uhlig (1984, 1995). This is particularly evident in tropical mountains which show a multitude of ecological and land-use belts and limits and a complex pattern of the altitudinal range of crops and human activities. A classic example of such altitudinal belts are those of the tropical Andes where they are named according to their climatic conditions: *tierra caliente, tierra templada, tierra fria, tierra helada*. From these facts, the concept of altitudinal limits and belts, originally confined to natural phenomena, such as snow or vegetation, was transferred to cultural and economic mountain scholarship, especially to land use and settlements.

Adaptation to the natural environment expressed by altitudinal stratification of people, settlements, and land use occurs most distinctly in mountain regions with traditional cultures. Here, agriculture is usually characterised by a great variety of types and products, for instance the combination of tillage and livestock, especially under conditions of subsistence production. The necessity for mountain peasants to utilise the entire range of the environmental potential has led to a vertical arrangement of their land parcels. This is one of the world-wide characteristics of agriculture in mountain regions.

Traditional staggered, or echelon-like, land use and settlement systems

Land utilisation of different altitudes practised by the individual peasant, or a peasant community, is termed by geographers 'staggered' or 'echelon-like' land-use systems, translated from the German *Staffelsystem* (Grötzbach, 1984; Uhlig, 1984, 1995). Common to all these systems is the dominant vertical component from which American social anthropologists derived the concept of 'verticality' or 'vertical control' (Murra, 1975) that describes the same phenomenon. If the parcels of land forming one such system are at a considerable distance from each other, the individual families have often added extra buildings as shelters for men and animals and for storing hay. Therefore, a staggered land-use system also functionally integrates settlements in different altitudes. One peasant family may possess many buildings of different types, according to location and altitude.

Plate 2.2 Kunde and Khumjung villages (3790 m), Sagarmatha (Mt Everest) National Park, Khumbu, Nepal. Ama Dablam (6856 m) and Everest (8848 m) to the right and centre-right, respectively. Both villages have nearly doubled in size over the past 20 years and yet remain in relative harmony with their surroundings (Photograph: A. Byers)

There are two basic types of staggered systems found in tropical and non-tropical mountains:

(1) In the tropics, especially the inner tropics, high-altitude echelons can be used the whole year round as there are no, or only moderately distinct, temperature seasons, as in the tropical Andes and the highlands of Ethiopia. However, the alternation of dry periods (*verano* in the Andes) and wet periods (*invierno*) may still result in seasonally distinct agricultural activities and land uses.

(2) In non-tropical mountains the yearly temperature variation allows only a seasonal use of the higher echelons of a staggered system. From spring to summer, agricultural and pastoral activities ascend step by step up to the highest belts of the ecumene and, conversely, they descend gradually in autumn. During winter the higher echelons are abandoned and usually inaccessible because of snow.

These systems of the second type have been extensively developed in the Alps and in the other European mountains, in the High Atlas of Morocco, and in the mountains of West, South and Central Asia. Although showing a great diversity according to local or regional conditions, they are based on the same requirement: to utilise the full potential of the altitudinal range of the mountain environment.

The most complex staggered systems occur in deep valleys flanked by high mountain slopes where the environment comprises a whole sequence of ecological belts. Classic examples are the Wallis/Valais in Switzerland, long since transformed by modernisation, and some valleys of the Himalaya, Karakorum, Hindu Kush, and the Andes, where they still function today, although even here change is beginning the process of transformation. 'Complex' means that such a system consists of several land use and settlement echelons

Plate 2.3 Highest mountain pasture for yaks at Machherma (4410 m), north of Namche Bazar, Khumbu, Nepal (Photograph: J. D. Ives)

that are alternately or simultaneously used in the course of the year. Figure 2.3 represents a historical example from the Swiss Alps where, over an altitudinal range of more than 2000 m, six distinct echelons were operated, from the vineyards of the Rhône Valley up to the high alp pastures. Figure 2.4 shows a recent example from the Bagrot Valley near Gilgit in northern Pakistan; this consists of four echelons and a similar vertical range of about 2000 m.

These elaborate systems are characteristic of old traditional mountain communities. Maintenance

of the spatially fragmented land-use belts required a high level of mobility of the people and their flocks within staggered, predominantly subsistence, systems. Herding the livestock and processing milk on the summer pastures, tilling and cropping fields in middle and lower altitudes, and transporting the products required numerous movements between the main farmstead, usually permanently inhabited, and the various echelons being operated seasonally. This mobility often has been termed 'vertical nomadism' because it bears some resemblance to true nomadism. Actually true nomadism still exists in the mountains of North Africa (Atlas), and of West and South Asia (Taurus, Zagros, Hindu Kush, Western Himalaya). In contrast to mountain peasants, nomads spend their winters in the low valleys or in the foreland. Moreover they differ completely from peasant communities in social structure and economic behaviour.

Complex staggered land-use systems represent elaborate and specific traditional forms of adaptation to the mountain environment. However, they require very high inputs of labour while yields are comparatively low. They reflect the necessities of a widely self-supporting economy but succumb to contraction or even abandonment when this economy is affected by market influence from the outside. Hunza (see Box 2.2) provides an example of this process.

Other aspects of human adaptation

Mountain peoples developed other forms of adaptation to their natural environment, in addition to the staggered land-use systems. A major feature is exposure, or aspect in relation to the sun. Many mountain societies apply special terms to sun and shadow slopes, a fact that reflects the importance of this locational attribute in the perception of mountain dwellers. While mountains in the inner tropics do not exhibit this difference because the sun is always at or near the zenith, it is extremely

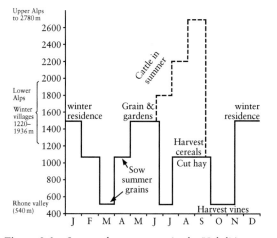

Figure 2.3 Seasonal movements in the Val d'Anniviers, Switzerland (after Peattie 1936)

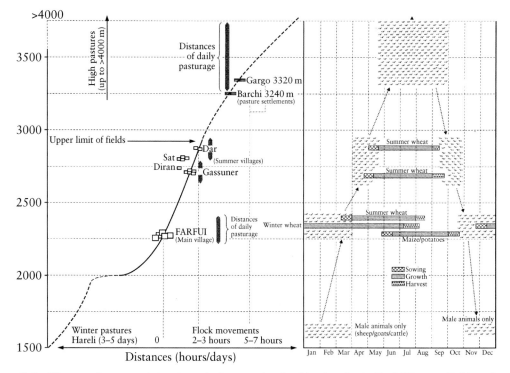

Figure 2.4 The use of space and time in agriculture and animal husbandry in Farfui/Bagrot (Pakistan) (after E. Ehlers 1995)

important in the subtropics, and especially in the more arid parts. Here the shady slopes are preferred for agriculture while the sunny slopes often lack water and soils parch rapidly. In middle latitudes, as in the Alps, these conditions are reversed: sunny slopes are clearly favoured over the shady ones which mostly are covered with forests. However, desiccation may still be a problem on sunny slopes so that crops and meadows often have to be irrigated. This is true, for example, for the upper Etsch/Adige valley (South Tyrol) and in the Wallis/Valais of Switzerland. Here irrigation dates back to Roman time, as the local terminology for irrigation activities and facilities indicates.

Another important aspect is human adaptation to natural hazards. For centuries settlements have been built in locations where, for example, the threat of floods, landslides, rockfall, and avalanches is comparatively low (for the Karakorum see: Kreutzmann, 1994; for general information see Hewitt, 1992 and Chapter 16). In the Alps, forests have been preserved on steep slopes in order to protect villages below from avalanches. Likewise, valley floors exposed to floods were not built on until the nineteenth or twentieth century after the rivers had been canalised. Nevertheless, in many Alpine valleys deforestation had proceeded to such an extent that the protection capabilities of the preserved forest tracts have been dangerously reduced.

Routes of trade and transportation lines in the pre-automobile era were also adjusted to the terrain, partly because of natural hazards, partly due to low technical standards. Roads usually avoided narrow valleys and gorges but took long detours and preferred terraces, slopes, and passes. This pattern of road communication has completely changed with the emergence of railways and, later, of roads for motor vehicles.

MOUNTAINS AS BARRIERS AND REFUGES

As already pointed out (see Chapter 6) mountains are often described as barriers and as refuges or retreats. These attributes are interpreted as an effect of the mountainous terrain that is difficult to penetrate and, therefore, offers shelter to its inhabitants. Although this aspect may still be valid in many cases it must not be generalised as reality was and is much more complex.

Mountain ranges that did act as difficult-to-penetrate frontiers include the Caucasus, the Py-

renees, and the Southern Andes, all of which are crossed by only a few roads or railway lines. The Alps, however, have never been an effective barrier, but were penetrated by a multitude of pass routes of varying importance. In the Himalaya–Karakorum–Hindu Kush ranges sections of impassable precipitous terrain alternate with relatively accessible pass regions. In many cases the lower parts of these pass routes present more obstacles, especially the deep gorges and cliffs, to traffic routes than the higher sections that may have been widened and smoothed by former glacial erosion. It follows, therefore, that even in past centuries, only parts of mountain ranges may have been insurmountable barriers.

The intensity of communication across mountains not only depends on topographic and climatic obstacles but of possibly greater importance is the potential and actual demand for economic and cultural exchanges between the forelands on both sides, or between the forelands and interior mountain regions. If these forelands are well populated core areas with complementary economies, and if they are not separated by rigid political boundaries, the extent of communication and transportation flows is high. This is true for the Alps that connect rather than separate Central Europe and Italy. Traffic and trade across the Himalaya–Karakorum have been much less as a consequence not so much of higher altitude and more difficult terrain but because of much lower demand for exchange and communication between the densely populated plains of northern India and the almost empty Tibetan highland.

Another widely endorsed idea concerning the function of mountains is embodied in the hypothesis that they have been areas of refuge or retreat for their inhabitants. A more detailed survey again reveals that this is not generally valid: confusion has been caused because exceptional cases have been accepted as the rule. History does provide a limited number of examples: in West Asia wooded mountains (Pontic Mountains, Taurus, Lebanon, Elbruz) were for centuries places of refuge for the sedentary peasant population while the nomads dominated the open country of the surrounding arid plains and hills. Pressure on sedentary groups was most intense if they differed from their neighbours in religion, language, or race. The Christians and Druzes of Lebanon and Syria and the Hazara, the descendants of medieval Mongolian invaders in the highland of central Afghanistan, are examples; also, those groups of Indios who fled from the

Spanish conquerors to the most remote parts of the Andes in Peru and Bolivia (Box 2.1). Mountains have also been used as regions of military refuge and resistance as, for example, the Caucasus and the Afghan Hindu Kush (see Box 2.3).

Another attribute ascribed to mountain peoples and societies is their inclination to cling to their cultural heritage. This is usually explained by continuous isolation and seclusion from the outside world as a consequence of difficult access. Again, a thorough investigation shows that this explanation does not reflect the complex reality, even in the past. No doubt, it must be conceded that many mountain peoples preserved old customs, languages, or dialects, religious beliefs and practices, traditional costumes, songs and music, for a long time. But it is debatable whether the cultural identity of these groups is the result of their former isolation, or whether they secluded themselves from neighbouring people in order to preserve their identity or autonomy. The isolating effect of the environment is decisive only under extreme conditions of relief and climate. Such extremes may be found in the rugged mountains of eastern Caucasus, southern Hindu Kush, or in the mountains of New Guinea where Stone Age cultures were literally only discovered during the last 50 years.

Box 2.3   The Afghan Hindu Kush – marginal High Mountains as a stronghold of military resistance

The Afghan Hindu Kush is one of the most traditional and least developed regions of the world. Together with adjacent mountain areas in north-eastern Afghanistan, it covers about 70 000 km^2 and has 0.8 million inhabitants. As the westernmost part of the Central and South Asian high mountain system it has been a region of transit trade between the Indus plain (modern Pakistan) and the lowlands of Turan since prehistoric time.

While only a few relatively easy passes accommodated the transit traffic, numerous secondary passes were used for sumpter trade between neighbouring valleys. But the repeated decline of central political power throughout the region's history impaired this trade. At such times local rulers or communities snatched control of pass routes and levied tributes on merchants, or even robbed them. In order to evade these obstacles transit trade changed its routes several times. The nineteenth century was the last long period of commercial decline and, from 1880 onwards, the new Afghan state more and more guaranteed security. This culminated in 1964 when a modern road and tunnel ensured the concentration of transit traffic through the Salang Pass route. It also attracted much of the local trade that prefers long detours by motor vehicle to direct but difficult pass tracks on foot (Grötzbach, 1972).

A final relapse into seclusion occurred in 1979 when Afghanistan was occupied by Soviet troops and the resistance movement found its main strongholds in the mountains. Large areas of the Hindu Kush were only temporarily, or never, occupied by Soviet forces. This was especially true of the upper valleys that remained under the control of the resistance fighters (*mujahedin*). Minor passes that had been neglected for decades suddenly gained importance and were much used for supplies from Pakistan. Thus intra-mountain communications regained importance, even where they crossed ethnic boundaries. After the Soviet retreat and the fall of the communist Afghan government, the Hindu Kush seems to have relapsed into relative isolation from the rest of the country where civil war has continued to flare up. This has been facilitated by the absence of a central power in Afghanistan.

Modern development had made only very modest progress before 1978/79 and was limited to a few easily accessible valleys. It was these same valleys that were invaded by Soviet troops and devastated; a particularly infamous example is the Panjsher Valley northeast of Kabul, Afghanistan's capital. Over large areas of the Hindu Kush traditional forms of life and economy have survived, modified by the growing pressure on natural resources, such as pastures and irrigated fields. Utilisation of pastures had increased considerably by the end of the 19th century when Pashtun nomads acquired the right to graze their flocks in high altitude valleys where they competed with the interests of mountain peasants. However, elaborate staggered systems of land use and settlements, and specialised crafts (mostly weaving of traditional textiles) in some valleys provided only partial subsistence. Population pressure was (and probably still is) reflected by seasonal or temporal migration of manpower to the oases and towns in the forelands and to several of the valleys; permanent, or semi-permanent, out-migration also accelerated. During the recent war the populations of hard contested valleys fled to Kabul or Pakistan, so that pressure on resources in these areas may even have been reduced in some cases. Although new influences and ideas have penetrated the mountains since 1978, the Afghan Hindu Kush still appears as a region of traditional culture.

Source: Erwin Grötzbach,
Feldafing, Germany

In former times, one of the main components for cultural identity was religion, as already mentioned in the examples of the Nuristani, the Hunzukuts, the Maronites and the Druzes. Other factors are ethnicity and language. But languages spoken by small numbers of people may be easily absorbed by the dominating language of the country, for example, the Romance speakers in Switzerland and the Dolomites (Kraas, 1996) who have more and more abandonend their mother tongue in favour of the German language. In the Andes the original languages have been discriminated against by the Spanish (for instance, the 'Castillanisation' of the indigenous population). Nevertheless, relatively isolated mountain groups have maintained unique languages; perhaps the most extreme case is the Burushaski language of

one of the Hunzukut sub-groups (Kreutzmann, 1996; see Box 2.2).

Although many mountain peoples or communities lived in continuous seclusion they have never been completely cut off from the outside world. There has always been some commercial exchange or other kind of interaction. Even the Kafirs of the Hindu Kush had such economic contacts, despite the fact that they were in a state of protracted hostility with their neighbours before their compulsory conversion to Islam. Therefore, 'seclusion' and 'isolation' must be used as relative terms which imply a low level of external relations. Seclusion, in this sense, depends primarily on socio-cultural factors which may be intensified by the obstacles of the mountain environment, but may not necessarily be the sole cause.

Box 2.4   The Western Alps – influences of the state on mountain culture in France and Switzerland

Studies of traditional mountain communities usually emphasise the dominating importance of ecological conditions specific to high altitude and variable relief and their influence on cultural forms. Despite the world-wide similarities amongst mountain peoples that derive from adaptation to such natural conditions, there are also significant differences that can be related to institutional histories. There is, for example, the particularly striking contrast between the evolution of Swiss Alpine peasants and their French counterparts in the region of the Briançonnais (Wiegandt, 1977, 1980; Rosenberg, 1988). Until the French Revolution the two regions had many features in common. Their production systems combined private land and communally controlled and managed resources, such as pasture, forest, and water. Equal inheritance rules led to the fragmentation of private holdings, while use-rights gave general access to communal resources, precluding the emergence of a class of large landholders. The villages had corporate identity and political autonomy. There was, however, an early and significant difference between the Swiss and the French systems. French mountain communities often had small local elites who had acquired wealth and power from the purchase of offices from the Crown. This social differentiation, based on links to a stronger central power, did not exist in the very centralised Swiss political system. The consequences for community adaptability and sustainability of these early distinctions are felt to the present day.

The French Revolution had a profound influence on the Briançonnais villages. At first the changes were favourable because they introduced cottage industries and trade, bringing relative prosperity.

Politically, however, villagers' rights were sharply circumscribed as local assemblies were abolished and the power to negotiate taxes and military obligations were transferred to the central state. This destruction of community control led to demographic collapse and political destabilisation. The dwindling population became ever more drawn into production for outside markets and, therefore, increasingly dependent on external influences. The state responded by encouraging outside investments, further aggravating domination by non-local financial institutions. Consolidation of economic power was accompanied by centralisation of administrative tasks and their transfer from the village to the state level. Before the Revolution, power brokers had existed to mediate between village and state, but they gradually evolved from being the spokesmen for the communities to manipulators for their own personal and outside interests. Today the communities are so weak politically that there are no longer any intermediaries; decisions are imposed from the centre (see Chapter 4 on the inequality of the economic situation among the Alpine communes).

In the Briançonnais, the inability of communal institutions to withstand powerful pressures of a highly centralised state contributed to the erosion of community economic, political, and social life. The situation in Switzerland, in contrast, is very decentralised precisely because it evolved out of peasant institutions based on equality and solidarity; these, in turn, became the principles on which the central state was based, and even today continue to guarantee local autonomy. The larger institutional context within which Swiss Alpine communities are embedded has meant that the evolution that has taken them from isolated, closed, and subsistence-oriented to cash-based economies has occurred without violence, crises, or impoverishment. Both the timing and the speed of change are part of the explanation: the cash economy was introduced late and gradually, as agriculture continued to nourish substantial proportions of the population well into the twentieth century. The importance of local political institutions cannot be ignored. Powers of taxation and decisions about education, land use, and economic planning are left largely to the communes in Switzerland and have allowed mountain populations to have significant impact on their own destinies.

As important as their unique physical characteristics are in explaining the common vulnerabilities and strengths of the world's mountain communities,

it is also essential to analyse political and economic differences that account for variations amongst them and which can be extremely significant. The two cases presented here, distinguished not by their environments but by their histories, demonstrate the impact of institutions on the contrasts between social systems. Relative autonomy and ensuing local solidarity in the Swiss Alps has allowed villages to maintain vibrant community life despite changing economic livelihoods. Much of the Briançonnais has become a rural backwater. Understanding these contrasts becomes relevant for crafting policy that will facilitate attainment of a sustainable future for many of our mountain populations.

Source: Ellen Wiegandt, CRERI,
Inst. universitaire de hautes études
internationales, Geneva, Switzerland

MOUNTAINS AS PERIPHERAL REGIONS

Another reason why mountains are often regions of cultural persistence is their character as peripheral areas. Periphery not only implies long spatial distances from the centres of political, administrative, and economic power, it also implies a psychological and cultural distance that facilitates the perseverance of a traditional mountain culture. Some authors interpret ethnic and cultural diversity in high mountains as the product more of social marginality than of absolute distance and environment (Kreutzmann, 1996).

Peripheral mountain regions with fully, or largely, independent inhabitants were numerous in the past. They occurred in areas where governmental power was weak or absent, as in the Caucasus before the Russian conquest, the Rif mountains in present-day Morocco, the tribal areas in the mountains along the border between Pakistan and Afghanistan, and in Indus and Swat Kohistan (Pakistan); the latter are under limited control from the centre even today. In other mountain regions, such as the Alps, even the most remote valleys have been controlled for a long time by either a semi-independent government located inside the mountains, or by a superior power from outside (see Box 2.4; Brugger *et al.*, 1984). Nevertheless, some alpine regions have attained a high degree of autonomy from the central government, as shown by the example of the Swiss Confederation that originated in the mountains around the Lake of Lucerne over 700 years ago.

'Periphery', like 'seclusion', is a term that should only be used in a relative sense, applicable only in comparison to a centre, or to the main communication lines between centres. This applies on different spatial scales. In mountain regions both conditions (contact to centre and peripheral) may alternate within short distances. While the lines of transit traffic and commerce in the main valleys and the pass routes have been directly connected with the outside world, side valleys with no egress and settled slopes high above the valley floors also represent the periphery. This contrast reached its peak during the railway era but has almost dissolved with the ubiquitous motor traffic of the modern era.

Peripheral remoteness corresponds to the fact that mountains are obviously rather unsuitable for the development of urban centres with extensive hinterlands. This is true at least for mountains with difficult precipitous relief and narrow valleys between high mountain chains, as in parts of the Alps and the Himalaya. Highlands with a plateau character offer much better conditions for the development of cities at high altitude, especially in the tropics. Here, plateaus and broad valleys are often more accessible and more suitable for the establishment of urban centres than are the adjacent lowlands. In the Andes, important cities have been the foci of power and wealth since pre-colonial times. The Inca Empire is the best historical example, others are Ethiopia and Tibet. Today only a few mountain or highland states exist which are controlled from the mountains – Bolivia, Ecuador, Colombia, Yemen, Bhutan, and, to a lesser degree, Nepal and Afghanistan, are examples.

Plate 2.4 Village in the anti-Atlas Mountains of southern Morocco. With only 200 mm of annual precipitation, the inhabitants have a very carefully managed and terraced agriculture, as well as pastures (Photograph: J. D. Ives)

MODERNISATION AND ITS CONSEQUENCES

Modernisation, or modern development, is a process that has reduced the relative seclusion or isolation of mountain areas and has integrated them into the overall economy, society, and culture of a country. This means that mountain areas have become increasingly opened to the outside world and to intensive interactions with it.

The role of modern transportation and communication

Two of the most important requirements for modernisation are modern transportation and communication systems. This is generally true, but it is of special importance for mountain regions with difficult access. In European and North American mountains modern transportation and communication began with the emergence of the railways, continued with motor transport, and culminated in the ubiquitous availibility of telecommunication. In other mountain regions the railway era was omitted and development proceeded with motor traffic. In the Alps, this process of communication development began about the middle of the nineteenth century; in the Himalaya, and some other non-European mountain regions, it started a hundred years or more later.

The opening up of mountain regions by modern transportation and communication has resulted in a transformation of the mountain economy and society that has been most pronounced in the Alps and other European mountain areas, such as the Pyrenees and Tyrol (Box 2.5). Modern traffic and

MOUNTAINS – A GLOBAL PRIORITY

communication facilitated or even caused the following changes:

(1) It greatly enhanced the mobility of people and ideas between the mountain valleys and the forelands. It stimulated the flow of people in both directions, but also within and across the mountains.

(2) It facilitated the exchange of goods and allowed for adequate provision of the mountains with food where many people previously had suffered from a deficit in food production,

Box 2.5 Tyrol (Austria and Italy)

Tyrol is an example of a mountain region that was thoroughly transformed by a process of modern development. Within less than one century it has changed from a partially poverty stricken area to one of the most modernised and well-to-do regions of the Alps. It emerged in the thirteenth century as one of the semi-independent pass states which controlled both sides of the Brenner Pass (1372 m) that provided the easiest communication between Central Europe and Italy. Tyrol had been a part of the Austrian Empire for more than 500 years when its southern sector (South Tyrol, or the province of Bolzano/Bozen) was ceded to Italy after World War I.

Although Tyrol has been a transit land since prehistoric time, agriculture provided the source of sustenance for the bulk of the population until the beginning of the twentieth century. Nevertheless, the limited agricultural resources were not sufficient to ensure subsistence for the people as a whole. An increasing number, therefore, came to depend upon non-agrarian income. During the fifteenth and sixteenth centuries this was mining, especially of silver and copper ores. The decline of mining in the seventeenth and eighteenth centuries caused local crises and forced the people to look for other occupations. These were various small home industries, in addition to itinerant work and trade. The situation was most critical in the West Tyrol where farm fragmentation by inheritance had been practised for many generations. Seasonal food shortages were common, despite the fact that the peasants utilised all their resources (even mountain meadows that were so steep that the hay could only be cut using crampons and ropes).

This state of affairs began to change with the introduction of railways in about the middle of the ninteenth century. But while the flow of persons and goods was facilitated along the main valleys, remote settlements still suffered from their difficult access that now became more marked than ever before. As a result, the more isolated communities were reduced by out-migration, while the settlements along the railway lines expanded and added new functions.

The spatial inequality was much reduced with the advent of individual motor traffic. This began after World War I, but only penetrated the entire area after 1950. Since then an extensive network of roads has been built, and even single farmsteads on slopes high above the valley floors are now connected. This revolution in the transport system has had two very important effects: (a) it has enabled many mountain dwellers to adopt new professions by commuting to the main valleys without the need to abandon their hereditary environment, and thus has reduced out-migration; (b) it has opened formerly secluded valleys to tourism. Tourism has become the leading economic activity in many formerly poor villages and has transformed them into semi-urban settlements with both summer and winter tourist seasons. As tourist development in Tyrol has been largely under the control of the local people, a great part of the indigenous population has benefited. The Austrian Tyrol, with a population of 655 000 in 1994 recorded 7.5 million tourist arrivals, mostly from Germany, and more than 42 million tourist nights. In South Tyrol the development has been similar, with guests originating mainly from Italy and Germany. In contrast to the tourist areas, the main valleys have attracted modern industry and commerce and have retained their function for transit traffic. In the highly urbanised valleys of the Inn and the Etsch (Adige), the capitals of the North and South Tyrol, Innsbruck and Bozen (Bolzano), have each expanded to over 100 000 inhabitants.

Mountain agriculture has also changed substantially toward market production, with specialisation in cattle breeding, and extensification of land use. Most of the fields that were formerly tilled and devoted to grains for subsistence have been transformed into meadows and pastures. Nevertheless, although the farmers receive considerable subsidies, mountain agriculture has been in a state of recession for more than half a century. One of the major problems facing Tyrol, therefore, is the need to maintain mountain farming; this is of critical importance, in part because it is the best way to ensure the geomorphic stability of the mountain slopes, and in part because the beauty of the cultural landscape is one of the primary attractions for the tourists, at least for the summer visitors (see also, Moser and Moser, 1986).

Source: Erwin Grötzbach,
Feldafing, Germany

especially following cold wet summers when malnutrition and even starvation threatened. Thus it reduced the necessity of the mountain population to strive for self-sufficiency.

(3) It contributed largely to the integration of mountain agriculture into the respective national economy; this has made peasants dependent on national agrarian policies and markets. This process revealed the drawbacks and the problematic competitiveness of mountain agriculture (see Chapters 4 and 5). In more advanced economies this leads to a decline of agricultural land use including the abandonment of marginal land, but also to specialisation in products well adapted to the environment (for example dairy products in the Alps, potatoes in the Andes and the Himalaya).

(4) Modern traffic and communications have encouraged, or enabled, a new functional orientation of mountain regions, particularly in recreation and tourism and in the generation of hydroelectricity for the cities of the forelands.

(5) Modern lines of transportation, especially mountain highways and freeways, also have had negative effects (e.g. air pollution, noise) on the environment and on people living nearby. This has led to massive protest movements, mainly against trans-mountain truck transport as, for example along the Brenner Freeway, the most important transportation axis through the Alps (Stadel, 1993).

(6) The opening up of mountain valleys by modern transport facilities and their socio-economic integration into the national economy and society often result in new forms of dependence on the lowlands, to a loss of traditional values and ways of life, and to erosion of the cultural identity of mountain people.

The new dependence of mountain regions

The new dependence of mountain regions may become evident in various forms. Food deficiency produces a dependence on supplies from outside. They have to be paid for from additional income which mostly comes from work outside the mountains, from national or international support programmes and, usually only to a small extent, from the sale of products. This coincides with the lack of employment opportunities within the mountains

that compels the under- and unemployed to seek work in the forelands, be it seasonal, temporal, or permanent. Part of the ensuing cash income is transferred to relatives in the mountains. In some areas these financial transactions have become essential for sustaining the mountain communities. In many parts of the Himalaya–Karakorum a new economic system has developed called the '*money order economy*' because large groups of the population make their living only with the aid of money from outside. This system is manifested by the spread of bank branches and shops, even to formerly remote places.

In European countries mountain peasants receive subsidies from their governments, or from European institutions in order to compensate for the limited incomes from mountain agriculture (see Chapters 4 and 5). These subsidies that are granted either directly, or indirectly, may constitute considerable monetary flows from the metropoles into the mountain valleys. Without them mountain agriculture would largely have ceased to exist.

Mass tourism has established new forms of dependence (Chapter 12). As tourists come from urban source regions outside the mountains, the local population of major tourist centres is swamped by urban visitors, while becoming economically dependent on this large seasonal influx of guests.

In many less developed mountain regions the indigenous population is not able (or sometimes not interested) to organise and finance adequate tourist facilities because of lack of education, know-how, or capital. In this case, development of tourism is taken over by public authorities or by private entrepreneurs and investors from outside the mountains. This means that, besides the influx of tourists, a flow of capital and know-how comes into formerly remote valleys from the outside. Under such circumstances local people have a limited potential to participate in tourism. Usually they find employment only in poorly paid positions, for example, as unskilled labour in hotels or as guides, porters, and so on. This type of tourist development is most characteristic of many developing countries, but even in the Alps, particularly in parts of the French Alps. Here, in the 1950s and 1960s, heavy investments in tourism were made by private companies from Paris and most of the profits obtained in the mountains flowed back to the centres from where the bulk of the tourists came.

In contrast to regions with this type of development are those mountain regions where the indigenous population has largely developed and controlled tourism. The most striking examples are the Swiss and Austrian Alps where tourism is a widespread source of income for many mountain dwellers who rent rooms or apartments and operate guest-houses or hotels, besides their many other occupations. A similar type of participation in tourism is found in Nepal where tourist lodges along the trekking routes are operated by local people; some of the latter, especially Sherpas, even own modern hotels in Kathmandu.

A third type of tourist development combines the two extremes mentioned before. Here the local population provides cheap accommodation while more comfortable modern hotels are operated by owners and managers from the cities, often by hotel chains who offer international standards. This type occurs in parts of the Alps and of the Himalaya and Karakorum in India and Pakistan.

The type of tourist development has an influence upon mountain culture. In general, total dependence on the world outside the mountains is a menace to mountain cultures that may lose their identity and be modified or absorbed by alien cultures. A more loose dependence, as is often the case with indigenous tourist development, may not erode the cultural identity of a mountain community; it may even revive some local traditions that have become an attraction for tourists, as in the Austrian and German Alps.

Apart from tourism there are other economic interests in mountain areas that may contribute to outside influences and dependence. Among them the commercial exploitation of forests has received much publicity, especially in countries with limited timber reserves. Thanks to new access roads, forestry is frequently practised on a large scale by timber merchants from outside, while the benefits to the local population largely depend on the ownership patterns. If the forests are owned by the local people they may realise a good income from felling timber. But government forests may be exploited without any participation of the local population, especially where the mountain communities have no legal title to protect their traditional forest access. In this case, clear-cutting often results in serious environmental damage that arouses the protest of the mountain dwellers; the most famous example is the *Chipko Movement* in the Himalaya, India (Shiva and Bandyopadhyay, 1986; Chapter 6).

The revaluation of mountain space and environment

A major consequence of the opening-up and integration of mountain areas has been a fundamental revaluation of their space and environment. This process is most advanced in the Alps where not only major valleys, but also many formerly remote and depressed tracts have experienced a new economic and social reappraisal of space.

Until the beginning of the twentieth century, communities with difficult access in many parts of the Alps suffered poor conditions and hardship. Land was cheap as there was virtually no demand for it from outside the mountains. Since then accessibility, the mountain economy, and the settlement system have changed totally. Settlements have expanded in the main valleys to the point of forming continuous ribbons of urbanised and industrialised former villages and small towns along the arteries of modern transportation. In side valleys, on mountain slopes, and even at high elevations, formerly poor peasant villages have developed into tourist resorts, sometimes taking the appearance of quasi-urban centres. This is particularly true for many of the winter resorts that were established on former alpine pastures above the treeline. These pastures, too, have been an integral part of the revaluation process. In former times they were used only during summer and were abandoned in winter; they are now preferred playgrounds for skiers and are equipped with all the facilities for modern winter sports, including roads, cableways, ski lifts, runs, restaurants, and hotels. Only few other environments, such as some seaside beaches, are likely to have changed their value as much as these remote parts of mountains. Nevertheless, as Bätzing *et al.* (1996) have pointed out, the pattern of revaluation, even in the Alps, is by no means uniform, and less favoured communities are facing the prospect of economic and demographic collapse (see also Chapter 4).

A similar process of revaluation can be observed in the mountains of North America, though in a less spectacular way. Here dilapidated former mining towns (ghost towns) have been reactivated by tourism, and new winter resorts have been developed in formerly virgin regions. But here revaluation proceeded in spatial isolation while in the Alps it created continuous patterns. In the Himalaya and in the Andes, this type of development is in its very early stages and is still limited to specific areas and places. The rather sporadic tourist development has contributed to an increase in socio-cultural diversity amongst the mountain population by contrasting the more internally orientated and traditional with the more externally controlled and innovative groups.

To sum up: the process of revaluation has upgraded the mountain environment economically, even in some of the most remote localities, and has made them comparable to the urban centres in the forelands, mainly because of tourist development. Urban influence is demonstrated by architecture, increased land values, a commercial orientation of space and activities, high standards of accommodation and a general change of lifestyles.

OBJECTIVES AND PROSPECTS FOR FUTURE MOUNTAIN DEVELOPMENT

It has been shown that the traditional equilibrium between the natural environment and mountain society has changed completely during the twentieth century, with rapid acceleration since about 1950. This is reflected most clearly by the changes in the cultural landscape. Improved access for modern traffic, integration of mountain areas into the national and international economy, society, and culture, dependence upon centres in the lowlands accompanied by a revaluation of the mountain environment has thoroughly changed the socio-economic functions and settlement patterns of high mountain regions.

These processes have produced rather contrary results: on the one hand, in many regions, the living standard of the mountain peoples has improved considerably, on the other, a reduction of economic and social autonomy, together with cultural alienation, have contributed to a new economic vulnerability and the decline in traditions and cultural identity. Their awareness of the growing

Plate 2.5 Yi women in traditional costume. Lijang Naxi Autonomous County, Yunnan, China (Photograph: J. D. Ives)

domination by and dependence upon outside core areas, in some cases, has brought about frustration and even defensive reactions. These reactions usually are the response to the exploitation of natural resources, often against the interests of the local communities. Protests against excessive tourism and traffic have been reported from many mountain areas: for instance from Tyrol (Austria), but also from upper Swat (Pakistan) where police have had to protect tourists from violence, and from Garhwal (India) where the Tehri dam project, as well as commercial woodcutting, has met strong local opposition (see Chapter 6).

Protest movements sometimes are accompanied by political demands for greater local autonomy and control over access to resources. In highly developed countries mountain dwellers often complain of being exploited or 'colonised' by the outside world and, therefore, demand more, or even decisive, influence on regional development policies. In contrast to this, mountain communities in developing countries may feel neglected by the development priorities of their central governments. They call for more assistance from the outside, but become increasingly aware that they must mobilise local resources (Stadel, 1995).

A much discussed problem concerns the alternatives for sustainable mountain development. As agriculture usually is stagnating, or in recession, and industry finds suitable conditions only in a few highly specialised locations, tourism seems to be the only opportunity for development in many mountain regions. However, as any mono-functional orientation is always prone to crises, tourism itself, as far as possible, should be complemented by other activities.

Sustainable development should not only aim at safeguarding and improving the living conditions of mountain populations and the protection of their vulnerable environment; it should also encourage preservation of their cultural heritage and diversity as the foundation of their cultural identity and self-confidence. In order to achieve these objectives a balance between national and regional (or local) interests is necessary in development policy. This requires a certain degree of political self-determination for mountain people who in many countries are completely dominated from the outside. This means that mountain regions should not be regarded as 'backward' by outsiders, but as complementary areas that have to be developed according to their resources and the needs of their inhabitants.

Development policy must consider the specific socio-cultural context as a basic requirement for acceptance by the people concerned. This is especially true for mountain regions of great cultural diversity. Moreover, it has to be stressed that development models and methods already approved in highly developed countries should not automatically be transferred to mountain areas in the developing world. Response by the people and success of the measures may be unsatisfactory if the socio-cultural background has not been taken into account. This suggests that the dialogue between development experts who are rooted in 'western' ways of thinking and representatives of the people concerned in developing countries must be intensified.

References

Batzing, W., Perlik, M. and Dekleva, M. (1996). Urbanization and depopulation in the Alps (with 3 colored maps). *Mountain Research and Development*, **16**(4): 335–50

Brugger, E. A., Fürrer, G., Messerli, B. and Messerli, P. (eds.) (1984). *The Transformation of Swiss Mountain Regions*. Paul Haupt: Bern, p. 699

Caviedes, C. and Knapp, G. W. (1995). *South America*. Prentice Hall: Englewood Cliffs, NJ

Ehlers, E. (1995). Die Organisation von Raum und Zeit – Bevölkerungswachstum, Ressourcenmanagement und angepaßte Landnutzung im Bagrot/Karakorum. *Petermanns Geograph. Mitteilungen*, **139**(2): 105–20

Grötzbach, E. (1972). *Kulturgeographischer Wandel in Nordost-Afghanistan seit dem 19 Jahrhundert*. Afghanische Studien, 4, Verlag Anton Hain: Meisenheim, p. 302

Grötzbach, E. (1984). Mobility of labour in high mountains and the socio-economic integration of peripheral areas. *Mountain Research and Development*, **4**(3): 229–335

Grötzbach, E. (1988). High Mountains as Human Habitat. In Allan, N. J. R., Knapp, G. W. and Stadel, C. (eds.), *Human Impact on Mountains*. Rowman and Littlefield: Totowa, NJ, pp. 24–35

Hewitt, K. (1992). Mountain Hazards. *GeoJournal*, **27**(1): 47–60

Jentsch, Ch. (1984). Methodische Ansätze in der vergleichenden Geographie der Hochgebirge. In Grötzbach, E. and Rinschede, G. (eds.) *Beiträge zur vergleichenden Kulturgeographie der Hochgebirge*. Eichstätter Beiträge, **12**. Verlag Friederick Puste: Regensburg, pp. 57–72

Jettmar, K. (1979) Foreword: In Müller-Stellrechl, I., *Materialen zur Ethnographie von Dardistan (Pakistan)*. Bergvölker im Hindukush und Karakorum, Vol. 3, Part 1. Akademische Druch und Verlagenstadt: Graz

Karan, P. P. (1987). Population characteristics of the Himalayan region. *Mountain Research and Development*, **7**(3): 271–4

Kraas, F. (1996). The decline of ethnodiversity in high mountain regions: the Rhaetoromansch minority in Grisons, Switzerland. *Mountain Research and Development*, **16**(1): 41–50

Kreutzmann, H. (1993). Entwicklungstendenzen in den Hochgebirgsregion en des indischen Subkontinents. *Die Erde*, **124**:1–18. [Translation published as: Development trends in the high mountain regions of the Indian subcontinent. *Applied Geography and Development*, **42**: 39–59]

Kreutzmann, H. (1994). Habitat conditions and settlement processes in the Hindukush–Karakoram. *Petermanns Geograph. Mitteilungen*, **138**(6): 337–356

Kreutzmann, H. (1995). Sprachenvielfalt und regionale Differenzierung von Glaubensgemeinschaften im Hindukusch–Karakorum. *Erdkunde*, **49**(2): 106–21

Kreutzmann, H. (1996). *Ethnizität im Entwicklungsprozeß. Die Wakhis in Hochasien*. Dietrich Reimer Verlag: Berlin, p. 488

Moser, P. and Moser, W. (1986). Reflections on the MAB-6 Obergurgl Project and tourism in an alpine environment. *Mountain Research and Development*, **6**(2): 101–18

Murra, J. V. (1975). El Control Vertical de un Máximo de Pisos Ecológicos en la Economía de las Sociedades Andinos, In *Formaciones Económicas y Políticas del Mundo Andino*. Instituto de Estudios Peruanos: Lima, pp. 59–115

Peattie, R. (1936). *Mountain Geography, A Critique and Field Study*. Harvard University Press: Cambridge, MA, p. 257

Rosenberg, H. (1988). *A Negotiated World: Three Centuries of Change in a French Alpine Community*. Toronto

Shiva, V. and Bandyopadhyay, J. (1986). The Evolution, structure and impact of the Chipko movement. *Mountain Research and Development*, **6**(2): 133–42

Stadel, Ch. (1992). Altitudinal Belts in the Tropical Andes: Their Ecology and Human Utilization. In Martinson, T. L. (ed.), *Bench Mark 1990. Conference of Latin American Geographers*, **17/18**: 45–60

Stadel, Ch. (1993). The Brenner Freeway (Austria–Italy): mountain highway of controversy. *Mountain Research and Development*, **13**(1): 1–17

Stadel, Ch. (1995). Development Needs and the Mobilization of Rural Resources in Highland Bolivia, In Robinson, D. J. (ed.), *Yearbook 1995, Conference of Latin Americanist Geographers*, **21**: 37–48

Troll, C. (1975). Vergleichende Geographie der Hochgebirge der Erde in landschaftökologischer Sicht, *Geographische Rundschau*, **27**:185–198. (Translator C. Stadel, 1988). Comparative Geography of High Mountains of the World in the View of Landscape Ecology. In Allan, N. J. R., Knapp, G. W. and Stadel, C. (eds.), *Human Impact on Mountains*, Rowman and Littlefield: Totowa, NJ, pp. 36–56

Uhlig, H. (1984). Die Darstellung von Geo-Ökosystemen in Profilen und Diagrammen als Mittel der vergleichenden Geographie der Hochgebirge. In Grötzbach, E. and Rinschede, G. (eds.), *Beiträge zur vergleichenden Kulturgeographie der Hochgebirge*. Eichstätter Beiträge, 12, Regensburg, pp. 93–152

Uhlig, H. (1995). Persistence and change in high mountain agricultural systems. *Mountain Research and Development*, **15**(3): 199–212

Von Baretta, M. (ed.) (1994 and 1995). Der Fischer Weltalmanach, 1995 and 1996. Fischer: Frankfurt

Wiegandt, E. (1977). Inheritance and Demography in the Swiss Alps. *Ethnohistory*, **24**(2): 133–148

Wiegandt, E. (1980). Un village en transition. *Ethnologica Helvetica*, **4**: 68–93

Zettl, C. (1994). Die westliche Gesellschaft zivilisieren: Der Kampf der indianischen Bewegungen in Ecuador. *Aufbrüche*, **4**: 9–10

The spiritual and cultural significance of mountains

3

Edwin Bernbaum

INTRODUCTION

As the highest and most impressive features of the landscape, mountains have a natural tendency to evoke a sense of wonder and awe. The storms that swirl around their peaks, the wandering shafts of light that illuminate their summits, the eerie mists that veil their ridges, the towering heights that reduce the viewer to insignificance, all contribute to creating impressions of overwhelming power, splendour, and mystery. Moved by such impressions, people around the world – even from modern, secular societies – experience something in mountains that imbues them with an aura of sanctity. Ansel Adams, one of the most influential landscape photographers of the twentieth century, wrote:

> No matter how sophisticated you may be, a large granite mountain cannot be denied – it speaks in silence to the very core of your being. There are some that care not to listen, but the disciples are drawn to the high altar with magnetic certainty, knowing that a great Presence hovers over the ranges. . . . (Adams and Alinder, 1985: p. 143).

Due in large part to their evocative nature, mountains throughout the world have come to reflect the highest and most central values and beliefs of many different cultures and traditions. Mount Sinai occupies a special place in the Bible as the imposing site where Moses received the essential teachings of Judaism and the Ten Commandments, the basis of law and ethics for much of Western civilisation. The remote Himalayan peak of Mount Kailas, rising aloof above the Tibetan Plateau, directs the minds of millions of Hindus and Buddhists toward the utmost attainments of their spiritual traditions. The graceful cone of Mount Fuji has come to represent the quest for beauty and harmony that lies at the heart of Japanese culture. For many in the modern world, the summit of Mount Everest symbolises the highest goal they may strive to attain, whether their pursuit be material or spiritual.

The sacredness of mountains highlights cultural and spiritual factors that profoundly influence how people view and treat the environment. The values and beliefs associated with major sacred peaks like Sinai and Kailas underlie many of the ways in which members of different societies see the world and their place in it. These values and beliefs determine to a great extent which natural resources and features of the environment people are willing to exploit and which ones they feel deeply motivated to protect. For any assurance of long-term success, policies directed toward environmental preservation and sustainable development need to take such cultural and spiritual factors into account; otherwise they will not win the support of communities and individuals who have the greatest stake in the areas and projects under consideration.

The sacredness of mountains manifests itself in three major ways. First, mountains in general commonly awaken a sense of wonder and awe that sets them apart as places imbued with evocative power and significance. Many people in modern, secular societies view ranges like the Alps and the Sierra Nevada in this light, often going to them for aesthetic and spiritual inspiration or regarding them as embodiments of important cultural values. William O. Douglas, a United States Supreme Court Justice, extolled the value of mountains in building national character:

> A people who climb the ridges and sleep under the stars in high mountain meadows, who enter the forest and scale peaks, who explore glaciers and walk ridges buried deep in snow – these people will give their country some of the indomitable spirit of the mountains (Douglas, 1951: p. 328).

Plate 3.1   Mount Kailas, Tibet. The symmetrical dome of Mount Kailas, the most sacred mountain in the world for nearly one billion people in Asia, appears to rise up in the shape of a Hindu temple or a Buddhist monument looming over a remote region of western Tibet (Photograph: Edwin Bernbaum)

Second, certain mountains are singled out by particular cultures and traditions as places of special sanctity directly linked to their deepest values and aspirations. These mountains – the ones traditionally known as 'sacred mountains' – have well-established networks of myths, beliefs, and religious practices such as pilgrimage, meditation, and sacrifice. Examples include such peaks as Mount Sinai in Egypt, T'ai Shan in China, Tongariro in New Zealand, and the San Francisco Peaks in the southwestern United States. Illustrating the kind of religious beliefs associated with traditional sacred mountains, a major Navajo or Diné myth describes the creation of Doko'o'slid or the San Francisco Peaks by First Man and First Woman:

> The mountain of the west, they fastened to the earth with a sunbeam. They adorned it with abalone shell, with black clouds,

he-rain, yellow corn, and all sorts of wild animals. They placed a dish of abalone shell on the top, and laid in this two eggs of the Yellow Warbler, covering them with sacred buckskins. Over all they spread a blanket of yellow evening light, and they sent White Corn Boy and Yellow Corn Girl to dwell there (Matthews, 1897: p. 78–9).

Finally, mountains that may or may not be revered in themselves frequently contain sacred sites and objects such as temples, monasteries, hermitages, stones, springs, and groves. Great numbers of people, for example, visit pilgrimage shrines located in mountainous regions, such as the Christian site of Montserrat in Spain and the Hindu shrine of Badrinath in the Indian Himalaya. Singling out this role of mountains, a passage in the *Mahabharata*, the religious epic of ancient India, describes a Himalayan peak as a 'refuge of hermits,

Plate 3.2 Badrinath Temple, Indian Himalaya. Pilgrims climb the steps to the temple, the major Hindu pilgrimage place in the Indian Himalaya and the site of ritual ceremonies to distribute and plant tree seedlings in a programme to reforest the region (Photograph: Edwin Bernbaum)

treasury of sacred places' (Van Buitenen, 1975, 2: p. 308).

In general, whatever people regard as sacred, such as a mountain, possesses for them some degree of ultimate reality and value. As a manifestation of ultimate reality, it acquires a special status that puts it beyond the reach of mundane manipulation. In addition, what is revered as sacred has a value that makes it worth protecting at all costs – a value which often transcends all cost. If cultivated and taken seriously, such beliefs and attitudes can function as powerful forces helping to preserve the integrity of natural environments.

Of all the different kinds of natural sacred sites, mountains form the most diverse and complete environments and ecosystems. They include shrub lands, forests, meadows, deserts, tundras, glaciers,

rivers, lakes, ranging from tropical to arctic environments. In addition, mountains are the largest features of the natural landscape that can be directly perceived as wholes – rivers are much longer, but their entire lengths from source to mouth cannot be seen from a single viewpoint. Mountains function, in fact, as microcosms of the environment perceived as a whole. Sacred mountains can, therefore, give us some of the best pictures of what aspects of nature various societies revere and feel deeply motivated to protect.

People of different cultures generally experience the sacredness of mountains through the views they have of them, such as the mountain as centre of the universe, place of the dead, or source of life. The various expressions of these views or themes differentiates the experience of the sacred in nature and provides articulation for doing research and formulating policy for environmental and cultural preservation.

Each view or theme brings together various ideas, images, and associations to evoke the experience of some deeper reality in a mountain revered as sacred. A Tibetan pilgrim, for example, views the peak of Mount Kailas as the temple of a deity. The two images fuse in his or her mind so that they become indistinguishable: as far as the pilgrim is concerned, the mountain *is* the temple. The fusion of these two images awakens the experience of something that suffuses Mount Kailas with an aura of sanctity. The pilgrim feels a divine power and presence radiating from the mountain.

The process works a little like the fusion of two slightly different photographs in a stereoscopic viewer to trigger a vivid perception of the third dimension represented in each two-dimensional picture. The scene that looks flat suddenly seems to pop open with depth – in the case of a mountain like Kailas, a depth full of meaning and significance for the pilgrim who reveres it as sacred. Something similar happens when certain impressions click together in our minds and the peak we are climbing or seeing appears to come momentarily alive with a peculiar depth and intensity that we have difficulty putting into words.

Many of the ideas and images brought together in traditional views of sacred mountains are found in myths that express a culture or tradition's most basic assumptions about the nature of reality – the givens that make action possible and life meaningful. These myths provide valuable keys for identifying different views of mountains needed to guide research and develop policies. They also function as sources of data and background information, as well as means of understanding different cultures and traditions and their attitudes toward the environment.

MAJOR THEMES

Using specific examples, we will examine some of the most important and widely distributed themes associated with mountains. These themes have been selected for their potential to suggest guidelines for doing research and formulating policy. Others could, and should, be chosen for further exploration. Many of the themes overlap, suggesting that we not treat them as distinct, isolated views of mountains to be analysed apart from each other.

Centre

An extremely important and widespread theme is that of the mountain as centre – of the cosmos, the world, or a local region. A number of Asian mountains, such as Mount Kailas in Tibet and Gunung Agung in Bali, are patterned on the mythical Mount Meru or Sumeru, which stands as a cosmic axis around which the universe is organised in Hindu and Buddhist cosmology. As a piece of Meru transported there by the gods, Gunung Agung provides the Balinese with their sense of geographical and psychological orientation – everything on the island exists and finds its place in relation to the volcano (Covarrubias, 1946). Elsewhere in Southeast Asia, in countries like Burma (Myanmar), Thailand, and Cambodia, rulers reinforced the centrality of their political

power by associating their capital cities and palaces with the cosmic axes of Mounts Meru and Kailas (Heine-Geldern, 1956).

Many mountains in East Asia, Japan in particular, are viewed as centres of *mandalas* – circular spaces and arrangements of deities used in Buddhist meditation. Some, such as Mounts Koya and Omine, are treated as *mandalas* themselves, their peaks, trees, and other natural features regarded as parts of a sacred circle representing the divine space of a visualised deity with whom a meditator identifies himself or herself (Bernbaum, 1990). In the period leading up to World War II such *mandalas* spread out from sacred mountains to engulf the surrounding landscape and transform the entire country of Japan into a divine nation in the eyes of the Japanese people (Grapard, 1982).

In North America the highest peak of the Black Hills, known to the Lakota as Paha Sapa, appears as the centre of the world in vision quests by religious leaders such as Black Elk. Delphi on the slopes of Mount Parnassus played a central role as the site of the Omphalos or navel of the ancient Greek world. On a more local level, Sherpa artists commonly place Khumbila, the sacred mountain of the Sherpas living near Mount Everest, in the centre of paintings depicting their homeland of Khumbu (Bernbaum, 1990).

Power

Many sacred mountains, including those viewed as centres, are places of awe-inspiring power. This power is often regarded as dangerous, a source of fear to be treated with the greatest care and respect.

Box 3.1   Mount Olympus – the power of a cultural symbol

Western culture recognises the sacred character of a number of mountains. Among them, Mount Olympus is the most celebrated – a 'pillar of the sky', rich in a profound symbolism that belongs to the interior geography of multitudes of human beings. The sacredness of the mountain, however, no longer derives from a persistent belief in myths and divinities; nevertheless, Olympus preserves intact its cultural fascination, just as it preserves the natural aspects of a mountain extolled by the ancient Greeks. Abundantly celebrated in universal poetry inspired by Homer, it is without doubt a *locus major*, a site hallowed by the traditions and history of Greece as a precious legacy for all of humanity.

At 2917 m, Mount Olympus is the highest peak in Greece. Its sheer cliffs and soaring summits rise up

from forests of Balkan pine creased with beautiful gorges and watered by pristine streams. The first national park in Greece was established here in 1937 to acknowledge the mountain's cultural importance and protect its natural environment.

Despite its status as a national park, the spiritual and physical integrity of Mount Olympus has come under repeated attack. Since 1985 a series of proposals by public agencies and private interests have sought to develop a large-scale ski resort on its slopes and to build a Homeric theme park complete with imitation ancient temples, statues of the twelve Olympian deities, and even fast-food restaurants. Plans to construct a cable car to carry thousands of people to the top of the highest summit led to a campaign of protest by Mountain

Wilderness International, an organisation dedicated to the preservation of unspoiled mountain environments.

The importance of Mount Olympus as a transnational cultural symbol rallied people from all over Europe to the mountain's defence. In June 1989 some 250 mountaineers, hikers, journalists, and ecologists from Greece, France, Italy, and other countries climbed Mount Olympus and held a widely publicised demonstration on its summit. Their number included Reinhold Messner, perhaps the best-known climber in the world today. That action, plus numerous letters of protest sent to the Greek Minister of Culture and the President of the European Parliament, forced the temporary cancellation of development plans.

In 1995 a petition with 100 signatures of well-known figures from all over the world, including noted authors and Nobel Laureates, such as Günter Grass, Umberto Eco, and Claude Simon, was presented to the Greek Ministry of the Environment.

The signatories asked for special protection for Olympus based on a recognition of its cultural significance. They called on the Greek government to forbid the building of a ski centre and any other intervention that could alter the nature of the national park.

Efforts continue to forestall future attempts to exploit Mount Olympus in inappropriate and unsustainable ways. In 1996 an international congress was convened in Trikala, Greece, for the purpose of protecting Mount Olympus and the Pindos Mountains and developing culturally and ecologically sensitive programmes that will benefit the mountains and the communities near them. These programmes include such projects as renovating abandoned monasteries, improving already established refuges, developing educational materials, and encouraging tourist activities that have minimal impact on the environment.

Source: Rita Charitakis and Constantin Tsipiras, Mountain Wilderness Greece

The way to the awesome summit of Mount Olympus was blocked by doors of clouds and darkness controlled by the Horae, goddesses of time entrusted with protecting the sacred mountain. In the past Maori warriors crossing the plateau beneath Mount Tongariro in New Zealand would avert their eyes from its summit for fear of provoking a blinding snowstorm. Maori conceptions of the *mana* of *tapu* mountains such as Tongariro emphasise the sheer power of a sacred mountain that makes it forbidden to ordinary humans (Orbell, 1985; Yoon, 1986).

The power of mountains can assume various forms. The five principal sacred mountains of China, T'ai Shan in particular, enshrined the political authority of the emperor to rule over the four quarters of his empire, supported by the mandate of Heaven. The earliest annals of Chinese history say that the legendary first emperors of China performed sacrifices on a mountain in each quarter of the empire in order to establish their sovereignty over the princes of the realm (Chavannes, 1910). Later historical emperors climbed T'ai Shan, the eastern peak, to thank Heaven and Earth for the success of their dynasties. Frequent storms made the jagged peaks of Olympus a dramatic setting for vividly displaying the power of Zeus, king of the gods and deity of thunder and lightning. The molten lava of Kilauea in Hawai'i embodies for many Hawai'ians today the fiery energy, both de-structive and creative, of the volcano goddess Pele (Taswell, 1986; Bernbaum, 1990). As the examples of Olympus and Kilauea demonstrate, the power of a sacred mountain can be both natural and supernatural.

Deity or abode of deity

In many traditions the power of mountains derives from the presence of deities. People may venerate a mountain itself as a supernatural being or as the abode of a deity. A poem from the earliest collection of Japanese poetery refers to Mount Fuji as 'a god mysterious' (Manyoshu, 1969: p. 215). For many native Hawai'ians today, Kilauea represents the physical body of the goddess Pele: they see drilling for geothermal energy on the volcano as though someone were jabbing a spear into her side (Taswell, 1986). In other traditions a mountain may be regarded as the seat of a divine power. Traditional Kikuyu of Kenya revere Mount Kenya or Kere-Nyaga as the resting place in this world of Ngai or God. Sand paintings used by Navajo singers in healing rituals will sometimes depict the four sacred mountains of the Navajo in the form of hogans, traditional homes of the Navajo people. Present-day descendants of the Incas believe that *apus* or mountain deities have their palaces inside sacred peaks of the Andes such as Ausangate near Cuzco (Bernbaum, 1990).

As deities or abodes of deities, mountains frequently play important roles as divine guardians or protectors. People invoke their power to protect livestock, crops, religion, the local community or even the state. Like the gods of many other peaks in Tibet and Himalayan border areas, the 'country' god of Mount Khumbila is revered as a warrior deity who watches over the Sherpa region of Khumbu, warding off the forces of evil. Herds of yaks, goats, and sheep fall under his protection (Stevens, 1993). In the Andes mountain gods are considered protectors of llamas and alpacas as well as wild vicuñas (Bernbaum, 1990).

Temple or place of worship

Deities are commonly associated with other themes connected with the sacredness of mountains: in particular, the widespread view of the mountain as a temple or place of worship. Members of many traditional societies revere sacred peaks as temples in which their deities reside. Tibetan Buddhists, for example, view Mount Kailas as the pagoda palace of Demchog, the One of Supreme Bliss, a tantric deity embodying the ultimate Buddhist goal of enlightenment. Many actual temples in India are modelled on the idealised form of the sacred peak and bear the name of Kailasanatha, Lord of Kailas – a reference to Shiva, the Hindu form of the supreme deity who dwells on the mountain (Bernbaum, 1990).

Mountains may also take the form of places of worship, viewed or imagined as altars, shrines, churches, and cathedrals. Here a deity does not necessarily reside in or on the peak, but rather the mountain provides a special setting for making contact with a divine presence through prayer, ritual, or contemplation. Just as Christians revere churches as sacred places of worship but do not regard them as objects of worship in themselves, so the same applies to mountains for many followers of monotheistic religions.

The view of mountains as places of worship plays an important role even in modern secular societies and is reflected in the names of numerous groves, valleys, and peaks, such as Cathedral Peak in the Sierra Nevada of California. John Muir, the founder of the Sierra Club and a key figure in the genesis of the modern-day environmental movement, wrote of Cathedral Peak that he would climb it 'to say my prayers and hear the stone sermons' (Muir, 1987: p. 198). He also spoke of nearby Yosemite Valley as a temple far finer than any made by human hands.

Garden or paradise

Modern societies also share with traditional cultures the view of mountains as gardens and paradises. The Muslim Kirghiz of western China, for example, believe that the snows on the summit of Muztagh Ata, one of the highest peaks in the Pamir, conceal an earthly paradise that goes back to the time of the Garden of Eden. The Buddhist Dai people of southwestern China regard the flora of their Holy Hills as the gardens of the gods (see Box 3.2 and Pei, 1993). Mount Athos, a mountainous peninsula in northern Greece that serves as the monastic centre of Orthodox Christianity, is known as the 'Garden of the Mother of God' – a reference to the role of the Virgin Mary as the patron saint of the holy mountain.

Many in the modern world view the untrammelled environment of mountains as a pristine paradise that preserves the purity of creation. Mountain meadows and wildflowers tend to evoke such views: one of the most popular destinations for visitors in Mount Rainier National Park in the United States, for example, is Paradise, an idyllic meadow close to the glaciers. Even barren regions of rock and snow can evoke images of a primordial Garden of Eden. Gaston Rébuffat, a noted French alpinist, wrote of his first view of the Himalaya:

Plate 3.3 Dai Holy Hill, China. A stand of trees on a Nong or Holy Hill above Man-Jing village shows how Dai religious beliefs and practices have protected forests and maintained islands of biodiversity on holy hills in the Xishuangbanna region of Yunnan Province in China (Photograph: Zhu Hua, courtesy of Dr Pei Shengji)

It is one of those places marked in ochre and white in the atlas, high, sterile and good for nothing; nothing marketable grows there, and higher still nothing can exist at all. It is one of those spots made solely for the happiness of men, in order that in this changing world, grown every day more artificial, they might yet find a few gardens still unspoiled in their silence of forgetfulness, a few gardens full of primal colours that are good for the eyes and for the heart (Rébuffat, 1956: p. 11–12).

Ancestors and the dead

A major theme frequently linked to conceptions of paradise as the other world is the view of mountains as divine ancestors and places of the dead, often involved in origin myths. Traditional Chinese beliefs hold that the spirits of the dead go to the foot of T'ai Shan. Mount Koya, the centre of Shingon Buddhism, has one of the most impressive graveyards in Japan. East Africans traditionally bury their dead facing sacred peaks like Kilimanjaro and Mount Kenya (Bernbaum, 1990).

Box 3.2   The Holy Hills of the Dai – Culture and biodiversity conservation

An indigenous ethnic group in southwest China, the Dai (T'ai) people who inhabit Xishuangbanna in Yunnan have a long tradition of conservation practices. These practices are characterised by the management of Holy Hills for conserving biological diversity and habitats, through formal or informal norms and rules of their ethics and religious beliefs.

The Dai practise a predominantly Buddhist religion. In their traditional concepts a Holy Hill or *Nong* is a forested hill where the gods reside. All the plants and animals that inhabit the Holy Hills are either companions of the gods or sacred living things in the gods' garden. In addition, the Dai believe that the spirits of great and revered chieftains go to the Holy Hills to live, following their departure from the world of the living.

Holy Hills can be found wherever one encounters a hill of virgin forest near a Dai village and are a major component of the traditional Dai land management ecosystem. In Xishuangbanna approximately 400 of these hills occupy a total area of roughly 30 000 to 50 000 ha, or 1.5 to 2.5 percent of the total area of the prefecture. There appear to be two types of Holy Hill. The first, *Nong Man* (or *Nong Ban*), refers to a naturally forested hill, usually 10 to 100 ha, that is worshipped by the inhabitants of a nearby village. Where several villages form a single, larger community, another type called *Nong Meng* is frequently found. Forested hills of this second type occupy a much larger area, often hundreds of hectares, and they belong to all the villages in the community.

Traditionally the Holy Hills constitute a kind of natural conservation area with great biological diversity. Gathering, hunting, wood-chopping, and cultivation are strictly prohibited. The Dai people believe that such activities on the hills would make the gods angry and bring misfortune and disaster as punishment. A Dai text warns: 'The Trees on the *Nong* mountains (Holy Hills) cannot be cut. In these forests you cannot cut down trees and construct houses. You cannot build houses on the *Nong* mountains, you must not antagonise the spirits, the gods, or the Buddha'.

Despite modern development interventions that have covered some of these hills with cash crops, the Holy Hill concept has made a significant contribution to the conservation of biological diversity in Xishuangbanna. First, it has contributed to ecosystem conservation within the ecotone region where there are consequently hundreds of well-preserved dry, seasonal rain forest localities characterised by species of *Anitiaris*, *Pouteria*, *Canarium*, and others. Second, a large number of endemic, old, or relict species of the local flora have been protected, including about 100 species of medicinal plants and more than 150 species of economically useful plants. Third, the large number of forested Holy Hills distributed throughout the region forms hundreds of 'green islands'. This pattern could help the natural reserves, which were established by the state government in recent years, by exchanging genes and playing the role of 'stepping stones' for the flow of genetic materials. The natural reserves are separated into five large sections and seven locations totalling 334 576 ha and are usually surrounded by larger and smaller Holy Hills.

The Dai's traditional practices and Holy Hill concept demonstrate the co-existence of biological and cultural diversity and suggest that the principle of biodiversity conservation and the conservation of cultural diversity should be considered as a concomitant process and as integral factors in overall conservation.

Source: Dr Pei Shengji, International Centre for Integrated Mountain Development (ICIMOD)

Most of the major peaks in New Zealand, including Aoraki, or Mount Cook, are revered as ancestors of the Maori who came to the North and South Islands in migration canoes (Orbell, 1985). According to the origin myth of the Korean people, they are descended from the union of a sky god and a bear woman on the sacred mountain of Paekdu (Henthorn, 1971). The natural features of Uluru, or Ayers Rock, in Australia record the formative activities of the dreamtime ancestors of the Pitjantjara and Yankunjatjara tribes (Mountford, 1965).

Communal and personal identity

As divine ancestors and places of origin, a number of mountains provide different societies with their sense of communal identity. One of the most dramatic examples of a sacred mountain holding a people together is Mount Kaata in the Bolivian Andes. Joseph Bastien (1978) has shown how the metaphor of Kaata as a human body unites the various communities who live on its slopes into an organic unit that has been able to resist all efforts to break it apart into smaller administrative subdivisions.

At intra-tribal gatherings Maori ritually identify themselves by first stating their tribal mountain, followed by their river, or lake, and then their chief (Yoon, 1986). One clan of the Yakutat Tlingit has Mount St Elias, or Washetaca, the second highest peak in Alaska, as its crest or totem. They stitch symbolic representations of the mountain on ceremonial blankets and shirts (Laguna, 1972). The Armenian people regard Mount Ararat, a volcano in eastern Turkey believed to be the site of Noah's ark in the Bible, as the symbol of their national and cultural identity (Lang, 1980).

Source

People throughout the world look up to mountains as sources of blessings – most notably, water, life, fertility, healing, and general well-being. Mountains such as Ausangate in Peru, Tlaloc in Mexico, the Himalaya, and the San Francisco Peaks in Arizona are revered as abodes of weather deities, places of springs, and sacred reservoirs of waters on which societies depend for their very existence (Reinhard ,1985). In times of drought, modern-day Kikuyu still fall back on pre-Christian traditions and face Mount Kenya, or Kere-Nyaga, to ask Ngai, or God, for rain. Traditionally each village in China had a temple dedicated to the local mountain deity responsible for clouds and rain. As sources of life-giving water, sacred mountains also provide the blessing of fertility. Today great numbers of elderly women climb T'ai Shan to make offerings to have grandchildren if their daughters or daughters-in-law have been infertile (Bernbaum, 1990).

Showing the close relationship between fertility and healing, two divine attendants of the goddess who grants grandchildren on T'ai Shan cure childhood diseases and restore sight to the blind (Chavannes, 1910). Female shamans in Japan and Korea routinely climb sacred mountains to charge themselves with healing powers and conduct rituals for their patients (Blacker, 1975; Kendall, 1985). Flora Jones, a leading spiritual doctor of the Wintun tribe in northern California, feels that she gets her power to heal from Mount Shasta or Bohem Puyuik – in particular, from a sacred spring in a pristine meadow threatened by the proposed development of a ski area. Traditional singers or medicine men go to the four sacred mountains of the Navajo in the southwestern United States to collect medicinal herbs and pebbles; they also invoke these mountains in rituals and sand paintings intended to restore a sick person to health and harmony. In Europe and North America, there is a tradition of putting sanatoria, particularly for tuberculosis, on mountains, based on the idea that the mountain air and environment have special curative properties. Thomas Mann's Nobel Prize winning novel about such a sanatorium in the Alps, *The Magic Mountain*, reflects the power of this view in Western societies (Bernbaum, 1990).

Much of the modern appreciation of mountains, in fact, derives from the perception of them as sources of spiritual and physical well-being. Jean-Jacques Rousseau, the French philosopher who played a major role in transforming European attitudes towards the Alps in the eighteenth century, wrote: 'In effect, it is a general impression experienced by all men, even though they do not all observe it, that on high mountains, where the air is pure and subtle, one feels greater ease in breathing, more lightness in the body, greater serenity in the spirit' (Rousseau, 1967: p. 45). A century later John Muir exhorted Americans to:

> Climb the mountains and get their good tidings. Nature's peace will flow into you as sunshine flows into trees. The winds will blow their own freshness into you, and the

storms their energy, while cares will drop off like autumn leaves (Muir, 1901: p. 56).

Inspiration, revelation, and transformation

The final theme to consider is that of the mountain as place of inspiration, revelation, and transformation. Mount Sinai plays a prominent role in the Old Testament as the imposing site where God reveals the Torah and the Ten Commandments to Moses. In the New Testament Jesus is transfigured on a mountain – usually identified as Mount Tabor – and revealed there as the Son of God. Muhammad received his first revelation of the Koran, the basic text and holiest scripture of Islam, in a cave on Mount Hira. Buddhist pilgrims to Wutai Shan in China expect to see visionary manifestations of Manjushri, the Bodhisattva of Wisdom, who may appear to them in the form of a dragon, a prince seated on a lion, or as a ball of fire (Birnbaum, 1986).

In China and Japan mountains are regarded as such ideal places for meditation and spiritual transformation that the Chinese expression for embarking on the practice of religion means literally 'to enter the mountains' (Demiéville, 1965). Before the Communist Revolution, Huang Shan, the most spectacular of the five principal sacred peaks of China, was a favourite haunt of Taoist hermits intent on transforming themselves into immortals; traces of this practice have returned to the mountain now that religion is permitted again. North American Plains Indians, such as the Lakota and the Crow, seek out high places for vision quests that give them spiritual power and determine the course of their lives (Bernbaum, 1990).

The mountain monasteries and hermitages of Mount Athos in Greece have the primary function of providing an environment conducive to pursuing the Orthodox path of spiritual purification and illumination. The monk or hermit who follows this path to its goal is transformed, in the words of Saint Syméon, the New Theologian, into:

> . . . one who is pure and free of the world
> and converses continually with God alone;
> He sees Him and is seen, loves Him and is
> loved, and becomes light, brilliant beyond
> words (Syméon, 1969, 1:Hymn 3; Bernbaum, 1990: p. 114).

Many people, both traditional and modern, seek out mountains as places of artistic inspiration and spiritual renewal. Peaks and ranges such as Huang Shan, Mount Fuji, and the Diamond Mountains have long inspired works by artists and poets in East Asia. Kuo Hsi, one of China's greatest landscape painters, wrote of the reason for painting landscapes (Kuo Hsi, 1935: p. 30–1):

> The din of the dusty world and the locked-in-ness of human habitations are what human nature habitually abhors; while, on the contrary, haze, mist, and the haunting spirits of the mountains are what human nature seeks.

Contemporary artists and actors in Tokyo routinely climb Mount Ontake, seeking visionary experiences to inspire them in their artistic work (Blacker, 1975).

Interest in the Alps for scientific, artistic, and mountaineering reasons developed when philosophers and poets such as Rousseau, Goethe, and Shelley began to extol their spiritually uplifting qualities. Inspired by the ideas and sentiments of such writers and their successors, many Europeans today flock to mountains like the Alps and the Pyrenees, seeking spiritual nourishment and inspiration from views imbued with transcendent power and mystery lacking in their everyday lives. In the United States John Muir, a major figure in the genesis of the modern-day environmental movement, founded the Sierra Club primarily to preserve the natural environment of mountains, such as the Sierra Nevada, as places where people could go for spiritual and physical renewal. Such motivations continue to inspire and energise the environmental movement today (Bernbaum, 1990, 1996).

RESEARCH AND POLICY IMPLICATIONS

Because they cut across many different cultures and regions of the world, the themes associated with the sacredness of mountains have a number of practical implications for doing research and formulating policies regarding environmental and cultural preservation and sustainable development. This section will present general guidelines for working out these implications, giving illustrative examples suggesting how to apply the approach in specific situations.

Before proceeding, we should note the following important point made in the Summary Report and Recommendations to the United Nations Commission on Sustainable Development of the

International NGO Consultation on the Mountain Agenda held in Lima, Peru, in 1995:

> The sacred values of mountains are of utmost sensitivity to people from these cultural and religious traditions. It should be absolutely clear that the study and understanding of religions or sacred sites is not universally an acceptable activity. Therefore, neither governments nor NGOs should presume to initiate or support any activities without first ascertaining that these are welcomed by local people and faith keepers (International NGO Consultation, 1995: p. 14).

Some cultures, such as the Zuni in North America, place a great value on keeping knowledge of sacred places, beliefs, and practices restricted to a few people entrusted with guarding them. The secrecy of sites in such cases should be respected: they should not be researched or publicly documented, nor should their implications for policy be developed or otherwise implemented.

Many societies draw vitality and cohesion from their relationship to mountains and other sacred features of the landscape. Destroying what makes such a site sacred may undermine a culture, resulting in adverse social, economic, and environmental impacts as the society falls apart and traditional controls are lost.

This last point brings out an important role for researchers and policy makers. Under the impacts of modernisation, the influx of outside forces, and population growth, many traditional beliefs and practices that have been extremely effective in preserving the environment are now being overwhelmed. Policy makers need to work with local communities to support and revitalize these controls and adapt them to new circumstances. Local and national governments can help by acknowledging and reinforcing the legal authority that traditional leaders and caretakers have had to protect and maintain sacred sites under their jurisdiction. An awareness of the themes associated with the sacredness of mountains and their implications for conservation and sustainable development can help to guide this process.

In order to work out the implications of these themes in a specific case, we need first to identify the groups and individuals for whom the mountain or sacred site under consideration is sacred. Stakeholders and other concerned parties should be involved from the beginning as full participants in the process. Their views, needs, and wishes must take precedence in doing research and formulating policy.

The second step is to identify the major themes associated with a particular mountain or mountain site. An examination of pertinent myths, stories, rituals, and other practices can provide important keys for making this identification. The third step is to ascertain the meaning and relevance of the themes at a specific site. The same theme can have different meanings and significance in different cultural contexts, even on the same mountain. Consultation with local people entrusted with traditional knowledge about a particular site is essential.

Once the relevant themes and their significance have been determined, their implications for environmental conservation and sustainable development at a particular site can be worked out. Most of these themes take the form of metaphorical views, such as the mountain as a temple or the mountain as place of the dead. Many of these views link mountains and their environments to features of places where people live and work, such as houses, gardens, temples, and cemeteries. This gives us a general approach that we can use with different themes. We can identify practices and attitudes that people have toward features of their daily lives metaphorically linked to mountains and determine ways of encouraging beneficial applications of those practices and attitudes to mountain sites. If they are already in place, they can be reinforced in people's minds.

Centre: Implications

Societies need central symbols to hold them together – key ideas, attitudes, and values around which their people can unite. Such symbols have an important role to play, even in the modern world. This is illustrated by some of the most widely quoted lines of twentieth century poetry: 'Things fall apart; the centre cannot hold; Mere anarchy is loosed upon the world'. These lines, from *The Second Coming* by William Butler Yeats, are frequently cited by political commentators. One important consequence of such anarchy 'loosed upon the world' is environmental degradation as the institutions upholding controls of the environment weaken and vanish.

In many cultures mountains function as concrete expressions of this kind of unifying symbolism – or are closely associated with the ideas and values central to a given society. If this is the case, to keep things from falling apart, measures may be needed to preserve and reinforce natural and man-made features that enhance the view of a particular mountain as a centre with the power to orient people in the world. Such measures might include setting aside a circle of undeveloped land around a peak or making sure that nothing obscures views of it from important places, such as temples and shrines. It may be critical to situate settlements in sites or in configurations that highlight the central location and prominence of a sacred mountain. If local people orient their lives with respect to a central peak, as they do in Bali with Gunung Agung, it may prove counter-productive to try to replace traditional with more scientific or 'rational' systems of orientation. A better approach is to treat the two systems as complementary.

Researchers and policy makers need to identify other, non-spatial ways in which mountains are associated with central values and beliefs of a society and determine how conservation and development activities on a mountain may impinge on the people who hold those values and beliefs. The Hopi, for example, believe that the correct practice of rituals on the San Francisco Peaks – known to them as Nuvatukya'ovi – is central to maintaining their right to their land, even though they live more than a 100 km away from the mountain itself. The expansion of ski runs, in their view, has scarred the slopes of the sacred mountain and impaired the performance of these rituals, undermining a basic underpinning of their society and leaving them feeling insecure in their homes (Bernbaum, 1990).

Power: Implications

The perceived power of a mountain frequently underlies other views of mountains and may sustain a community that holds that particular mountain sacred. It may also provide people with a sense of protection that is essential to their individual and communal well-being. If development or conservation measures appear to weaken that power, they may have negative psychological and social consequences that work against the intended benefits of a particular project. On the other hand, programmes that enhance the perceived power of the mountain will bolster local communities and stand a better chance of succeeding.

Many societies, both traditional and modern, revere mountains as places of dangerous power. The Bible, for example, warns that if anyone other than Moses sets foot on the slopes of Mount Sinai he or she will die. Other traditions and cultures share similar views of their sacred peaks and restrict access to them. Many people in the modern world value certain mountains for their aura of danger and inaccessibility as embodiments of the wild power of nature itself. If these qualities form an essential part of a particular mountain's mystique, policies for programmes of environmental protection or sustainable development should avoid making it safe or easy to visit. This may involve establishing a zone above which trails and huts are not allowed or where access is otherwise limited or entirely curtailed. Climbers can also be informed that they are on their own and should expect no rescue. Such policies need not conflict with tourism: they can actually make a mountain more attractive, setting it aside as a place of special status to be appreciated from a safe distance.

The power of a mountain may be expressed through various features of the natural environment. People may see abundance of vegetation as a sign of overflowing power. Or the very barrenness of a peak ravaged by storms or natural disasters may bear witness to a divine force so powerful that it makes life impossible. These features need to be ascertained in order to formulate policies that respect the power a mountain has for the people who revere it. Where fecundity reveals this kind of power, programmes of reforestation and agriculture may find considerable support among local communities. In other cases, where the opposite is true, policy makers may be well advised to leave untouched barren areas that reveal, through lightning or fire, the presence of some kind of natural or supernatural force.

Park and protected area managers can work with local communities to set mountains aside as places of special power. In the latter part of the nineteenth century, Europeans in New Zealand were starting to buy up parcels of land for sheep farms on Tongariro, the sacred mountain of the Ngati Tuwharetoa. This provoked great concern because the *mana* or power of the chief – and through him, that of his entire people – depended on keeping Tongariro whole and intact. In order to preserve the *mana* of the mountain, a European advisor counselled the Paramount Chief, Horonuku Te Heuheu Tukino IV, to give the mountain to the Crown as a park for the benefit of everyone. This he did in 1887, and Tongariro National Park became the first national park in

New Zealand and the fourth in the world (Lucas, 1993).

Deity or abode of deity: Implications

For many people the power of mountains derives traditionally from the presence of deities. The iconography and mythology of such deities can provide important clues for doing research and formulating policy. For example, if a god appears in paintings or myths in association with certain domestic animals, these will probably be the kind of livestock permitted on that deity's sacred mountain. The Hindu deity Shiva rides a bull suggesting that programmes of cattle grazing would be appropriate and likely to succeed in mountainous areas sacred to him. The goddess Durga, on the other hand, is known for destroying an evil buffalo demon: allowing buffalo on one of her mountains may offend the local community.

Not taking such considerations into account can alienate people and cause unnecessary problems. Sherpas in the Sagarmatha National Park of Nepal believe that goats, yaks, and sheep are under the special protection of Khumbu Yul Lha, the major local deity of the Khumbu region. They were outraged when in 1983 the park banned the grazing of goats on the slopes of the god's sacred mountain, Khumbila. Some of them believed that the action offended the deity and had brought misfortune on their herds of yaks, on which they depended for much of their livelihood (Stevens, 1993). Greater sensitivity on the part of park managers, perhaps working out an arrangement to allow goats to graze on certain parts of the mountain where they would do the least damage, would have avoided creating suspicion and resentments that made other park policies much more difficult to implement.

Many mountain deities appear with particular wild animals, often as their protectors: reinforcing this kind of association in people's minds can help to protect wildlife. Vicuñas are considered the domestic livestock of *apus* or mountain deities of the Peruvian Andes. National laws protecting these animals have helped to reinforce older prohibitions on hunting them based on fears of offending

Plate 3.4   Tesi Lapcha Pass, Nepal. Two Sherpas walk above cloud-filled valleys near Mount Everest, crossing the Tesi Lapcha Pass, nearly 5800 m high. Many hikers and climbers seek experiences of transcendence in the mountains (Photograph: Edwin Bernbaum)

their divine protectors (Mishkin, 1940). This illustrates an important way in which policy makers can use modern legal systems to complement and revive traditional controls that have lost their effectiveness under the impact of change and modernisation.

Temple or place of worship: Implications

The widespread view of mountains as temples or places or worship can provide powerful motivation for protecting mountain areas and developing them in responsible ways. This is especially true for Western societies that tend to give sacred status primarily to man-made structures like churches and cathedrals. John Muir galvanised a great deal of popular support for preserving Yosemite Valley in the Sierra Nevada with arguments of this kind. His passionate appeal to prevent the flooding of nearby Hetch Hetchy Valley shows how deeply felt attitudes toward man-made places of worship can be transferred to features of the natural landscape: 'Dam Hetch Hetchy! As well dam for water tanks the people's cathedrals and churches, for no holier temple has ever been consecrated by the heart of man' (Muir, 1988: p. 196–7).

Parks can draw on the implications of the theme of mountains as places of worship to create educational materials that make visitors think of how they behave in the sacred space of churches and act accordingly in corresponding places outdoors. Informational material of this kind can also convey attitudes across cultures so that people can more easily understand and appreciate the beliefs and practices of traditional societies. Chief John Snow of the Assiniboine or Stoney in Canada very effectively uses this approach in his book *These Mountains Are Our Sacred Places* to explain the significance of the Rocky Mountains for his tribe:

> These mountains are our temples, our sanctuaries, and our resting places. They are a place of hope, a place of vision, a very special and holy place where the Great Spirit speaks with us. Therefore, these mountains are our sacred places (Snow, 1977: p. 13).

Rather than detract from a wilderness setting, the construction of temples and shrines in appropriate places may enhance the spiritual experience of a mountainous area and help to preserve its natural environment. In the Khumbu region of Nepal near Mount Everest, sacred groves around monasteries and temples have been better preserved than forests protected by the Nepali government and the Sagarmatha National Park (Stevens, 1993). In places with a similar kind of history, an effective strategy for protected area managers would be to identify sacred sites in areas that need protection and to sponsor the building of new shrines and the renovation of old ones near or at those sites, encouraging local religious leaders to declare the surrounding forests sacred as well.

Garden or paradise: Implications

Different conceptions of gardens and paradises can provide the basis for the protection and sustainable use of mountain environments. Policies can draw on specific notions of gardening and stewardship in a given society to encourage its members to nurture and care for a mountain forest as they would their household garden. This approach will be especially effective where such notions play major roles in religious texts and teachings. Orthodox monks on Mount Athos, for example, can look to conceptions of the Garden of Eden and ideas of stewardship in the Bible as added incentives for preserving the flora and fauna of the mountain peninsula they already regard as the 'Garden of the Mother of God'. A policy of encouraging this kind of thinking would help to mitigate the environmentally destructive effects of unrestrained road building and logging on Mount Athos (Oikonomou *et al.*, 1993).

Cultivating the theme of mountains as gardens and paradises can also encourage people to view and treat wilderness in positive ways. Instead of regarding wild mountain areas as useless and hostile places to avoid or subjugate, they can begin to appreciate the beauty and value of nature as it is. In particular, they may start to look for signs of underlying order in the environment since order characterises most cultures' conceptions of gardens and paradises. People may also develop an interest in ecology and science as a means of discerning this kind of order in what may appear to be chaos.

Ancestors and the dead: Implications

If a community views a mountain as a place of the dead, policies can be formulated to encourage its members to treat it as they would a cemetery containing the remains of their relatives. Just as cutting down trees or grazing goats would be, for some cultures, a desecration of a graveyard, so doing the same things on the mountain might offend the spirits of ancestors on whom people depend for their well-being and blessings. The presence of a kilometer and a half long cemetery on Mount Koya in Japan, for example, has succeeded in preserving a magnificent forest of giant cedars (Bernbaum, 1990). If people avoid contact with signs of death, as they do in many cultures, such as the Navajo or Diné in North America, reinforcing views of mountains as places of the dead, or spirits of the dead, can help to keep intruders out of sensitive core areas of biological nature preserves, many of which are in inaccessible mountain regions that have remained relatively intact.

Communal and personal identity: Implications

If a mountain functions as a symbol of personal or communal identity, people may feel that what happens on the sacred peak happens to them – even if they live nowhere near it. Where such a view has influence, projects that may appear relevant only to the immediate environment also can have profound effects on distant communities. Environmental impact assessments will need to take into account these kinds of psychological and social aspects. On the other hand, policy makers can discourage people from environmentally harmful activities on their sacred mountain by pointing out that doing those things will harm them as well. Just as some governments prohibit burning a flag that

stands for the nation, so communities may want to ban destructive practices of burning forests and grasslands on mountains with which they identify themselves.

As Box 3.3 on Taranaki shows, the name of a mountain that symbolises the identity of a people can have major significance, influencing how they see many other matters. In order to assure the goodwill and co-operation of local communities, policy makers need to take special care with the names of mountains that they use in official and unofficial communications. In many cases they will need to go back to earlier, indigenous names that do not appear on contemporary maps. This can become a particularly sensitive problem when a mountain represents several groups and each group has its own name for the peak. A partial solution may be to make separate documents with different names for use with different communities.

Policies of multiple use also need to be implemented with particular sensitivity on mountains that function as symbols of communal identity. It is essential to make sure that proposed uses do not undermine or conflict with this symbolism. Policy makers need to consult with the people who derive their sense of identity from a particular mountain and work out with them the measures and activities they would consider appropriate there. In the case of Tongariro in New Zealand, recreational use was more acceptable to the Ngati Tuwharetoa than cutting the sacred mountain up into parcels and

Box 3.3   'Who is that mountain standing there? It's Taranaki . . .'

Taranaki, the ancestral mountain of the Taranaki tribes, lies in the province that bears its name on the West Coast of New Zealand's North Island. Taranaki Maori take their name and their identity from it, and imbue it with a romantic past. Anciently, Taranaki lived in the centre of the North Island with a rival mountain, Tongariro, and competed with him for the love of a female mountain, Pihanga. Taranaki was defeated, and ploughed his way to the west. In settling into his present position he obliterated a village named Karaka Tonga, found a new love in the eroded volcano, Pouakai, who threw out a spur to capture him, and was ritually 'fixed' in place by tribal elders. Over time, his lower slopes provided us with food and shelter, and his high, sacred places received the bones of our dead.

The spiritual waters that fall upon him as rain flow to meet the ancestral waters of the sea, over which our founding canoes came to this country

centuries ago. At one time one of our rivers, Pungaereere, was deep enough to dive into, for an ancestor named Tamakaha did just that, giving the name of Upokoruru ('the shaking of the head upon surfacing') to the spot where it happened. Pungaereere exemplifies many rivers in Taranaki that have been diverted or infilled to meet the needs of local farmers. Today, no one would think of diving into its shallow waters.

Ironically, Mount Taranaki was constituted a National Park in 1900 because dairy farmers were concerned about downstream flooding if its forests were logged. The mountain formed part of lands confiscated by the Crown during the colonising wars of the 1860s, when settlers forced the issue of sovereignty and land ownership on Taranaki. In an assertion of their new-found rights they called the mountain Egmont, the name given by Captain Cook in 1770 in honour of the Earl of Egmont, whose link with New Zealand is tenuous.

Over the past century, the Taranaki Maori tribes have pressed for the return of their mountain and the reinstatement of its name. In 1978, the Mount Egmont Vesting Act provided for its symbolic return and immediate gifting to the Crown as a National Park for the people of New Zealand. But as a prominent tribal leader, Titi Tihu, noted in a petition to government in 1979, this was no just way to remedy its confiscation, for the Crown had ordered that the gift to remedy the injustice should be made by those who had suffered the injustice.

In 1985, the Taranaki tribes applied to the New Zealand Geographic Board for a change of name from Egmont to Taranaki. Calls for objections under Section 13 of the New Zealand Geographic Board Act 1946 resulted in an overwhelming, negative response from the largely non-Maori population of Taranaki. In the face of such opposition, the Board's decision to approve the name change was a courageous one.

The reinstatement of Taranaki's name represents a major triumph for the Taranaki tribes, although it is only part of a dynamic process. Negotiations are currently underway with government for the resolution of long standing land grievances. Ideally, the mountain will form part of the settlement package, although its high conservation and recreation value is acknowledged and respected. Nevertheless, in an ongoing struggle between belief systems and political systems, the need to restore integrity to the mountain and hope to its people transcends all else.

Source: Ailsa Smith, lecturer in the Centre for Maori Studies and Research at Lincoln University, New Zealand, is of Taranaki Maori descent

---

grazing sheep on its slopes. The latter course, they felt, would destroy the unity of the place that represented them as individuals and as a tribe (Lucas, 1993).

Source: Implications

Views of mountains as sources frequently evoke closely associated notions of purity based on the observation that the closer one comes to the source of a river high on a mountain the purer the water that flows from it. Policy makers and researchers can ascertain and develop the implications of concepts of purity in a particular culture and work with the local community to apply them to protecting watersheds and other features of mountain environments from which blessings are perceived to flow.

Policy measures should build on motivations and practices people already have for preserving mountains as sacred sources of water, fertility, life, and healing. Sherpas, for example, avoid polluting mountain springs from fear that the *lu* or serpent deities residing in the water will inflict leprosy on local inhabitants. Rather than try to persuade them that bacteria instead of gods or spirits cause illness, a more respectful and effective approach would encourage them to hold both views, applying each where it is appropriate to the end of safeguarding the water supply. This does not have to be contradictory: in many cultures traditional beliefs may lend support to arguments that spiritual entities can act through bacteria as their material agents in the physical world.

Members of many cultures feel naturally motivated to protect and develop mountains as sustainable sources of medicinal plants. Researchers and policy makers can work with traditional healers to determine how best to preserve and pass on their knowledge of these plants and how to use them for the benefit of their communities. In many societies not only the plants themselves, but the ways they are harvested and where they come from, play critical roles in whether or not they will have the power to cure. These traditional ideas and practices deserve acknowledgement for both cultural and medical reasons. Recent scientific studies have shown that strongly held beliefs can physiologically effect the immune system and stimulate the body's resources for healing. Moreover, traditional attention paid to the provenance of medicinal plants and the ways they should be harvested frequently provide a sound basis for protecting the environments in which they grow.

The modern Western tradition of putting sanatoria and spas in mountains can also provide motivation for protecting and developing mountain environments in responsible ways. People who go to ranges like the Alps seeking health and fitness will respond favourably to arguments that, like hospitals and spas, mountainous areas need to be kept clean and sanitary if they are to have their salutary effects. In particular, they will tend to support the conservation of natural features, such as pristine forests, clear air, and sparkling streams, that contribute to the sense of well-being they derive from the mountains and that also have beneficial effects downstream.

Inspiration, revelation, and transformation: Implications

The theme of mountains as places of inspiration, revelation, and transformation is the one most closely linked to issues of tourism and pilgrimage. People throughout the world are going to mountains in increasing numbers to seek some form of spiritual renewal. Will their presence interfere with the practice of religious rituals and meditation by indigenous practitioners, as well as the quality of their own experience? On mountains where inspiration and transformation come from quiet, solitary contemplation in a pristine environment the answer is clearly yes. Either visitors must be encouraged to walk quietly in small parties and leave little trace of their passage, or they must be restricted in where they can go. On other mountains where inspiration and renewal come traditionally from large numbers of people engaged in group activities, such as religious ceremonies and pilgrimage festivals, this may not be such a problem.

Effective policies must take into account the various ways in which different kinds of sites function as places of inspiration, revelation, and transformation. Religious ceremonies performed at sacred places in mountains may require secrecy for their intended effects. Non-participants need to be barred from these sites, especially on ceremonial occasions. The natural setting of a grove of trees or a rocky grotto in a national park may create an intimate atmosphere that quiets the mind and makes one aware of subtle feelings and insights. Such a site calls for regulations that minimise talking and limit the number of visitors to individuals or very small groups. Other places, such as high peaks and ridges, offer expansive views that open people to a sense of something far vaster and greater than themselves – something that can so overwhelm them that the presence of others, even in large numbers, may not appreciably detract from the quality of their experience. Educational materials and signs can prepare visitors for experiencing different kinds of sites and acting in ways appropriate to each.

Many parks and wilderness areas in mountain regions have been set aside for protection because of the cultural and spiritual values placed on them by modern societies – values that often outweigh economic, political, and scientific considerations. People see them as places of inspiration and transcendence where they can temporarily leave behind the materialistic, competitive concerns of the urban world and immerse themselves in a primordial reality that renews them spiritually and physically. The inspirational value of mountains, in particular, has played a vital role in the establishment of national parks and is one of the most effective tools for galvanizing support for the conservation of wilderness areas.

In order to retain and cultivate this important source of support, park and protected area managers need to adopt policies that reinforce, rather than undermine, such spiritual and cultural values. In the United States, for example, it is becoming a common practice for national and state parks to have private companies handle campsite reservations and even issue wilderness permits. Although it may not be as cost-effective in the short term, it may be preferable for rangers to perform this function themselves. Contracting such services out to commercial ticket agencies may alienate the public by appearing to reduce parks to theatres or amusement parks that no longer embody the natural qualities of mountain environments that inspire a sense of renewal. Visitors who deal directly with rangers will feel that their money is actually going to the park they are visiting and will have a greater sense of personal involvement that will make them more inclined to support park systems economically and politically.

GENERAL ISSUES AND RECOMMENDATIONS

Pilgrimage

Most of the themes associated with mountains come together in the widespread practice of religious pilgrimage. Hindu and Buddhist pilgrims approach Mount Kailas as the centre of the universe. They also revere it as a place of power – the abode of their most important deities. Orthodox Christians come to Mount Athos to wander in the Garden of the Mother of God and receive her blessings. The Hopi go on group pilgrimages to the San Francisco Peaks, or Nuvatukya'ovi, to invite the *katsinas* to bring summer clouds and rain to their homes. Christians from all over the world seek cures for their ailments at shrines such as Lourdes, situated at the base of the Pyrenees. Chinese pilgrims seek visions of Buddhas and Bodhisattvas on the heights of sacred mountains such as Wutai Shan and Emei Shan (Bernbaum, 1990).

As Chapter 12 points out in greater detail, pilgrimage involves millions of people and can have

Plate 3.5 Tree planting at Hanumanchatti, Indian Himalaya. Hindu swamis and holy men carry seedlings to plant in a ritual ceremony at Hanumanchatti, an important shrine near Badrinath, the major Hindu pilgrimage site in the Indian Himalaya (Photograph: Edwin Bernbaum)

major impacts on mountain cultures and environments. Many more pilgrims than trekkers visit the Himalaya, for example, and heavily frequented shrines like Gangotri and Badrinath are being severely degraded with deforestation, litter, sewerage, overcrowding, and other problems (Kaur, 1985; Academy for Mountain Environics, 1995). Although few people stay long at their pilgrimage destinations, they have more than a passing interest in them since they hold them sacred. Pilgrims, therefore, should be treated as non-resident stakeholders whose interests must be taken into account in environmental and sustainable development plans. Sometimes their interests come into conflict with those of local and indigenous peoples, who may view the same mountains and pilgrimage sites very differently – even from a religious point of view. For example, many Hindu pilgrims come to Muktinath in Nepal to worship the Hindu deity Vishnu, but the people living in the immediate vicinity of the shrine are nearly all Tibetan Buddhists.

The motivation that pilgrims have for coming on pilgrimage can be a powerful force for protecting the environment and improving conditions at a particular site. The accompanying Box (3.4) shows how this kind of motivation has been employed at Badrinath to inspire pilgrims to replant deforested areas as an act of religious devotion. Using the themes and their implications sketched out above, this approach can be extended to other sites, cultures, and conservation activities. It is critical, however, that programmes of this nature be implemented in such a way that the religious experience of the participants is deepened and enhanced at the

Box 3.4 Badrinath – Pilgrimage as a basis for conservation in the Himalaya

Badrinath is the major Hindu pilgrimage place in the Indian Himalaya. The northernmost of four shrines situated at the four points of the compass in India, it lies at 3122 m in a valley surrounded by sacred mountains. The site has been a focal point of religious devotion and spiritual practices for thousands of years.

In 1993 scientists from the G. B. Pant Institute of Himalayan Environment and Development visited Badrinath and noticed how the area had been stripped bare of forest. Every year 450 000 pilgrims come to the shrine from all over India, arriving on roads built in the early 1960s. Their influx has had a substantial impact on the local environment. At the suggestion of the scientists, the Chief Priest of the

temple at Badrinath agreed to use his religious authority to encourage pilgrims to plant trees for restoration of the site.

In September of that year, about 20 000 seedlings were brought to Badrinath and blessed by the Chief Priest in a special ceremony. He gave an inspirational talk to pilgrims and local people highlighting religious beliefs and myths about the spiritual importance of trees. He concluded by encouraging them to plant the seedlings as an act of religious devotion. There was a great rush and all the seedlings were planted.

A sign was posted asking pilgrims for donations to care for the seedlings. In no time a substantial amount of money was collected. The numerous beggars at Badrinath were offered an amount in cash equal to their daily take from begging, plus food, if they would care for the trees instead of begging. All of them participated in the planting and expressed eagerness to take up the offer, but controversy over who would collect and distribute the funds prevented full implementation of the idea.

Harsh winter conditions with heavy snowfall killed off many of the seedlings planted in 1993. The G. B. Pant Institute has done further research and established a nursery near Badrinath to select and prepare native trees acclimatised to the weather and altitude. Subsequent ritual distributions and plantings, incorporating specially designed protection measures, have increased the survival rate of the seedlings. A ceremony conducted in 1996 at Hanumanchatti, an important shrine at the lower end of the Badrinath Valley, represents the first step in extending the programme to other sites.

The G. B. Pant Institute, in collaboration with villagers, pilgrims, priests, local stakeholders, and other institutes, such as the High Altitude Plant Physiology Research Centre of H. N. B. Garhwal University and The Mountain Institute in the United States, plans to expand reforestation to shrines elsewhere in the region and to include additional environmental measures, such as cleaning up litter and disseminating seeds for pilgrims to plant back in their home communities throughout India. The extended programme will be used to develop and test guidelines for replicating the approach initiated at Badrinath at other pilgrimage sites in the Himalaya and, where appropriate, in other traditions and parts of the world.

The reforestation at Badrinath shows how science and religion can work together for the benefit of the environment and the preservation of cultural and spiritual values. The collaborative efforts of the different parties involved provide a promising model for developing ways of involving people in environmental conservation measures for reasons that are culturally motivated and sustainable over the long term.

Source: Prof A. N. Purohit, Director of the High Altitude Plant Physiology Research Centre of H. N. B. Garhwal University, formerly Director of the G. B. Pant Institute of Himalayan Environment and Development

same time as the environment is benefited. Otherwise people will feel exploited and resentments will arise.

The reforestation programme at Badrinath also presents an instructive example of how myths and religious beliefs can be used in a practical way to explain and motivate environmental conservation. As part of his inspirational sermon to pilgrims, the Chief Priest of Badrinath related an important myth about the origins of the sacred Ganges in the Himalaya. A sage prayed to the goddess of the river to come down from Heaven. Not wanting to leave her blissful abode, she protested that the force of her fall would shatter the earth. Shiva, the form of the supreme deity most closely associated with the Himalaya, offered to break her descent with the hair on his head. Classic art and contemporary religious posters depict Shiva seated on the summit of Mount Kailas with the river winding down through his matted locks.

Indian environmentalists have noted that Hindu texts regard the trees of the Himalaya as Shiva's hair. In the summer the Ganges does indeed fall from Heaven in the form of monsoon rains, and when the Himalayan forests are cut down, the earth literally shatters under the monsoon's impact with landslides and floods (Nanda, 1990). The Chief Priest also pointed out how tree roots hold the soil in place and added that if the pilgrims were seeking the blessings of Shiva they would do well to replant trees and restore his hair.

In order to draw on this kind of motivation, we must first determine why pilgrims go on a particular pilgrimage, what blessings they seek, and how they obtain them. We need to know why because people may not be seeking anything from their efforts other than doing the practice for its own sake. If that is the case, then we can look into the question of how the pilgrimage might be extended to include conservation efforts as selfless activities. If pilgrims seek blessings, we need to know their nature and how they receive them. Do they obtain them directly from the site, from a ritual performed there, and/or from the dispensation of priests or other religious figures? This information will show who needs to be involved, what kind of influence and stake they have, and how best to enlist their efforts in programmes of environmental conservation and sustainable development – provided, of course, that they approve of the programmes and their involvement in them.

Tourism

Pilgrimages and sacred sites in mountains frequently draw tourists and trekkers seeking colourful sights and interesting experiences. Not only the beauty of many of these sites, but festivals and rituals associated with them can be major attractions – attractions in danger of being reduced to staged performances done for the sole benefit of paying visitors. In addition, sacred peaks, like Kangchenjunga in Asia and Devil's Tower, or Mato Tipila, in North America, attract mountain climbers who may engage in climbing activities considered sacrilegious by local people.

Tourists coming in great numbers or lacking respect for religious and local traditions can have extremely negative impacts. Mass tourism with an endless stream of visitors in large buses has made the traditional practice of monasticism nearly impossible at the scenic monasteries of Meteora perched on rock pinnacles in Greece: as a consequence, most of the monks have left for Mount Athos. Such an impact will eventually destroy much of what makes a sacred site attractive to tourists in the first place. Visitors can also stimulate theft and illegal trade in religious objects and art – a serious and widespread problem of great concern to local people.

To avoid such consequences, tourism needs to be adapted to the needs of sacred sites. The measures taken will depend on what makes a particular site significant: here, once again, an understanding of themes can be extremely helpful in formulating policies. If, for example, a mountain is viewed as a divine centre off-limits to climbing, visitors could be encouraged to circumambulate it as a substitute activity – provided that pilgrims and other stakeholders have no objections. Literature on the site could point out advantages of such substitute activities, such as sharing a pilgrimage experience and seeing the peak from all sides.

To preserve the integrity of a sacred site, tourism may have to be limited in various ways. If the presence of tourists interferes with traditional practices, making them difficult or impossible to perform, restrictions must be placed on the number of visitors, and on where and when they can go. The religious and secular government of the Himalayan kingdom of Bhutan adopted a policy of granting only a limited number of tourist visas at a high daily fee – thereby maintaining income from tourism – and closing all but a few of the country's monasteries to foreigners. In the United States national monument officials worked out a promising compromise that would have banned climbing on Devil's Tower, or Mato Tipila, during a month of particular religious significance for the Lakota and other indigenous groups: a commercial guiding company, however, contested this policy in court, and the judge in the case ruled against the government's plan (*New York Times*, July 1, 1996). When a mountaineering expedition wanted to make the first ascent of Kangchenjunga, one of the most sacred peaks in the Himalaya, people in India, Nepal, and Sikkim feared that the deity of Kangchenjunga would be offended and would visit disasters on them. Authorities persuaded the expedition to stop just short of the summit, thereby allaying local fears and preserving the sanctity of the mountain while giving the climbers the satisfaction of having overcome all the obstacles of the climb (Bernbaum, 1990).

The international furore provoked by plans to construct a cable car up Mount Sinai and open a night club on the summit illustrates the perils of tourist development that ignores the cultural and spiritual significance of a sacred site. When word of the scheme got out in 1990, people from all over the world expressed outrage at what they considered crass exploitation and desecration of a mountain that functions as a prime symbol of revelation and ethical values in Judaism, Christianity, and Islam. *Time* Magazine published a full-page essay by Lance Murrow titled 'Trashing Mount Sinai' that decried the proposed development and ended with the following sarcastic comment:

> Perhaps they will make the cable cars in the shape of calves and gild them. The golden calves can slide up and down Mount Sinai and show God who won (*Time*, March 19, 1990: p. 92).

Embarrassed by this kind of criticism and concerned about its impact on tourism elsewhere in Egypt, the government suspended its plans. Similar negative reactions to inappropriate development on Mount Olympus show that sacred mountains and sites can have a powerful force as cultural symbols for people in the modern secular world, even if the religious traditions that originally established their sanctity no longer exist.

Mountains as models for other environments

As a number of preceding examples have shown, sacred mountains and sites often preserve ways of life and traditional practices that conserve the environment and use natural resources in sustainable ways. These ways of life and traditional practices

Plate 3.6 Monastery of St Paul, Mount Athos, Greece. The Byzantine monastery of St Paul stands like a fortress beneath the imposing heights of Mount Athos, the monastic centre of Orthodox Christianity (Photograph: Edwin Bernbaum)

can serve as models for people living in non-mountainous environments. Orthodox Christians throughout the world tend to look to monastic life on Mount Athos as a source of spiritual inspiration for their own lives, even if they do not live as monks. Educational and other religious materials can build on this pre-existing tendency to point out how highly respected monks and hermits have also protected flora and fauna on the Holy Mountain. Priests in the outside world who play active roles in lay communities can promote the idea that the kind of spiritual life highlighted by exemplary figures on Mount Athos reduces the kind of self-centred greed that destroys the environment and encourages the conservation of natural resources for the benefit of all.

Mountain pilgrimage places that draw visitors from widespread regions provide ideal dissemina-tion points for spiritually inspired environmental ideas and practices. The hands-on experience of participating in a conservation activity at a pilgrimage site, such as planting seedlings at Badrinath, can reinforce such ideas and practices and motivate people to implement them back in their own communities. Where feasible and appropriate, priests and other religious figures can also distribute seeds suited to the pilgrims' home environments. If these seeds and other distributed materials are blessed, there will be added incentive to plant and use them. Pilgrimage sites in mountains have the great advantage over many other pilgrimage shrines of being more closely associated with features of the natural landscape so that pilgrims can more clearly see and appreciate connections between religious beliefs and practices, on the one hand, and environmental restoration and conservation, on the other.

CONCLUSIONS

For many people, both modern and traditional, the environment is not just the natural environment. It includes cultural and spiritual aspects that make it meaningful – a source of life in its deepest and broadest sense. People who do rituals to draw water from a sacred mountain, for example, do not view the water and mountain simply as physical parts of the ecosystem needed to grow their crops. They see them as essential components of a larger system of meaning, expressions of a deeper reality that sustains them spiritually and culturally, as well as physically.

An examination of the role and significance of the sacredness of mountains suggests that we adopt broader and deeper views of environmental conservation and sustainable use. Researchers and policy makers need to take into account issues of cultural and spiritual, as well as ecological and economic, sustainability. If trees are replanted, clear-cut logging can be sustained from an economic point of view, but the biodiversity of the forest will be lost.

In a similar vein, measures taken only to preserve the natural environment without regard for what it enshrines may kill the spirit of a place, destroying its value for the people for whom it is important.

Just as we need to maintain biodiversity for its own sake and for the benefits it can bring, such as new pharmaceuticals and other products, so we need to preserve cultural diversity for its intrinsic, human value and for the riches it can provide. Different cultures have a right to exist in their own ways and have much to contribute in their intimate knowledge of the environment and in their instructive examples of diverse ways of living in harmony with nature. Like unwise dependence on a single species of plant for food crops, reliance on a single world culture, no matter how scientific and practical it may seem, could endanger the future of the planet and jeopardise the survival of the human race. The views of diverse cultures that revere mountains remind us that there are many ways of seeing the world and many reasons for valuing and protecting the environment.

References

Academy for Mountain Environics (1995). *Mountain Tourism for Local Community Development: A Report on Case Studies in Kinnaur District HP and the Badrinath Tourist Zone*, MEI Series No. 95/10. ICIMOD: Kathmandu

Adams, A. and Alinder, M. S. (1985). *Ansel Adams, an Autobiography*. New York Graphic Society Books, Little, Brown: New York and Boston

Bastien, J. W. (1978). *Mountain of the Condor: Metaphor and Ritual in an Andean Ayllu*. West Publishing Co.: St Paul

Bernbaum, E. (1990). *Sacred Mountains of the World*. Sierra Club Books: San Francisco

Bernbaum, E. (1996). Sacred Mountains: Implications for Protected Area Management. *Parks*, 6(1): 41–8

Bhardwaj, S. M. (1973). *Hindu Places of Pilgrimage in India: A Study in Cultural Geography*. University of California Press: Berkeley

Birnbaum, R. (1986). The Manifestation of a Monastery: Shen-ying's Experiences on Mount Wu-T'ai in a T'ang Context. *Journal of the American Oriental Society*, **106**(1):119–37

Blacker, C. (1975). *The Catalpa Bow: A Study of Shamanistic Practices in Japan*. Allen and Unwin: London

Chavannes, E. (1910). *Le T'ai Chan: Essai de monographie d'un culte Chinois*. Ernest Leroux: Paris

Covarrubias, M. (1946). *Island of Bali*. Alfred A. Knopf: New York

Demiéville, P. (1965). La montagne dans l'art littéraire chinois. *France-Asie/Asia*, 20:7–32

Douglas, W. O. (1951). *Of Men and Mountains*. Victor Gollancz: London

Grapard, A. G. (1982). Flying Mountains and Walkers of Emptiness: Toward a Definition of Sacred Space in Japanese Religions. *History of Religions*, **21**(3):195–221

Heine-Geldern, R. (1956). *Conceptions of State and Kingship in Southeast Asia*, Data Papers 13. Southeast Asia Program, Dept. of Far Eastern Studies, Cornell University: Ithaca, NY

Henthorn, W. E. (1971). *A History of Korea*. Free Press: New York

International NGO Consultation on the Mountain Agenda, Lima, Peru April, 1995: *Summary Report and Recommendations to the United Nations Commission on Sustainable Development*. (Copies available from The Mountain Institute, Franklin, WV)

Kaur, J. (1985). *Himalayan Pilgrimages and the New Tourism*. Himalayan Books: New Delhi

Kemf, E. (ed.) (1993). *The Law of the Mother: Protecting Indigenous Peoples in Protected Areas*. Sierra Club Books in association with WWF and the IUCN: San Francisco

Kendall, L. (1985). *Shamans, Housewives, and Other Restless Spirits: Women in Korean Ritual Life*. University of Hawaii Press: Honolulu

Kuo Hsi (1935). *An Essay on Landscape Painting*. John Murray: London

Laguna, F. D. (1972). *Under Mount Saint Elias: The History and Culture of the Yakutat Tlingit*. Smithsonian Institution Press: Washington, DC

Lang, D. M. (1980). *Armenia: Cradle of Civlization*. George Allen and Unwin: London

Lucas, P. H. C. (1993). History and Rationale for Mountain Parks as Exemplified by Four Mountain Areas of Aotearoa (New Zealand). In Hamilton, L. S. *et al.*, (ed.) *Parks, Peaks, and People*, pp. 24–8. East-West Center Press: Honolulu

Matthews, W. (1897). *Navajo Legends*. American Folklore Society, 5: Boston

Mishkin, B. (1940). Cosmological Ideas among the Indians of the Southern Andes. *Journal of American Folklore*, 53(210):225–41

Mountford, C. P. (1965). *Ayers Rock: Its People, Their Beliefs and Their Art*. East-West Center Press: Honolulu

Muir, J. (1901). *Our National Parks*. Houghton Mifflin: Boston/New York

Muir, J. (1987) (1911 reprint). *My First Summer in the Sierra*. Penguin: New York

Muir, J. (1988) (1914 reprint). *The Yosemite*. Sierra Club Books: San Francisco

Nanda, N. (1990). Who is Destroying the Himalayan Forests? In Rusomji, N. K. and Ramble, C. (eds.) *Himalayan Environment and Culture*. Indian Institute of Advanced Study and Indus Publishing Company: Shimla/New Delhi pp. 48–60

Oikonomou, D., Oswald P. and Palmer, M. (1993). *Final Report of the Ecumenical Patriarchate, WWF International and WWF Greece Team on the Ecological Status of the Monasteries of Mount Athos*, (draft)

Orbell, M. (1985). *The Natural World of the Maori*. Collins/Bateman: Auckland

Pei Shengji (1993). Managing for Biological Diversity Conservation in Temple Yards and Holy Hills: The Traditional Practices of the Xishuangbanna Dai Community, Southwest China. In Hamilton, L. S. (ed.), *Ethics, Religion and Biodiversity*. The White Horse Press: Knapwell, Cambridge pp. 118–32

Rébuffat, G. (1956). *Mont Blanc to Everest*. Thames and Hudson: London

Reinhard, J. (1985). Sacred Mountains: An Ethno-archaeological Study of High Andean Ruins. *Mountain Research and Development*, 5(4):299–317

Rousseau, J.-J. (1967 reprint). *Julie ou la Nouvelle Héloïse*. Garnier-Flammarion: Paris

Snow, Chief John (1977). *These Mountains are our Sacred Places: The Story of the Stoney People*. Samuel Stevens: Toronto/Sarasota

Stevens, S. F. (1993). *Claiming the High Ground: Sherpas, Subsistence, and Environmental Change in the Highest Himalaya*. University of California Press: Berkeley/Los Angeles

Syméon, the New Theologian, trans. J. Paramelle (1969). *Hymnes*. Les Éditions du Cerf: Paris

Taswell, R. (1986). Geothermal Development in Hawaii Threatens Religion and Environment, *Cultural Survival Quarterly*, 10(1):54–6

The Manyoshu: The Nippon Gakujutsu Shinkokai Translation of One Thousand Poems (1969). Columbia University Press: New York

Van Buitenen, J. A. B., trans. and ed. (1975). *The Mahabharata*, 3 vols. University of Chicago Press: Chicago

Yoon, H.-K. (1986). *Maori Mind, Maori Land: Essays on the Cultural Geography of the Maori People from an Outsider's Perspective*. Peter Lang: Berne, Frankfurt am Main, New York, Paris

Comparative inequalities – mountain communities and mountain families

4

Jack D. Ives

INTRODUCTION

As emphasised in Chapter 1, any attempt to define *mountains* will involve some combination of altitude and steepness of slope. We have reasoned that these two factors relate to either reduced rates of soil profile development or increased instability of soils on slopes, and hence to lower levels of available nutrients, when compared with adjacent lowlands. This, in turn, implies comparative limits to biomass productivity. This reduction in biomass productivity, of course, is further accentuated by the temperature lapse rate typical of all mountains. To this must be added the topographic factor that reduces accessibility and augments the cost of infrastructure emplacement and/or its maintenance. Taken together, this characterisation of *mountains* implies that, in terms of human exploitation and/or adaptation, mountainous regions can be described as marginal areas for achievement of specific levels of human welfare or affluence. This is particularly relevant to the late-twentieth century setting of a materialistic global culture and widespread penetration of the capitalistic world market economy. This line of reasoning can be broadened by incorporation of the fact that much of the on-going warfare, guerrilla activity, or general human conflict (Chapter 6) is concentrated in mountain regions. Mountainous areas are also especially prone to natural catastrophic events, such as earthquakes, landslides, avalanches, and floods; these phenomena also add to the costs of living and resource use, whether in terms of cash expenditures or input of human subsistence energy (Chapter 16). Some inner tropical arid zone mountains, however, such as the Ethiopian Highlands and the mountains of East Africa, stand in contrast to most other mountain areas since the greater abundance of precipitation with increasing altitude, compared with their surrounding lower lands and plains, historically have ensured them a preferred status.

To conclude that, with the stated exception of the arid zone examples, mountains, relative to their surrounding lowlands, are necessarily regions of poverty, is taking a facile generalisation to the point of absurdity. Other chapters in this book emphasise the enormous assets that many mountain areas control, from most of the world's fresh water and hydroelectric power, to forests, minerals, and recreational facilities. Nevertheless, the Government of the People's Republic of China explained at an international seminar in Beijing in March, 1993, that the great majority of its 80 million citizens designated as existing below the poverty line, are mountain minority people. Some of the Least Developed, and hence, presumably, poorest countries are land-locked and/or mountainous. These include Nepal, Bhutan, Afghanistan, Ethiopia, Rwanda, Burundi, Papua-New Guinea, Ecuador, and Bolivia, to name only a few. Furthermore, their peoples are partially invisible because of the tendency to aggregate census data, and the mountain sections of Uttar Pradesh in India, Yunnan and Xizang in China, northern Pakistan, Tajikistan, Myanmar, Thailand, Peru, Argentina, and Chile, amongst others, would presumably rank with Nepal, Bhutan, and Ethiopia.

In comparison with the inhabitants of these mountain areas of extreme poverty, the turn-of-the-century descendants of poor mountain villagers in the Ötztal, Tyrol, or the canton of Valais, Switzerland, would be classed as wealthy beyond measure. Yet both Austria and Switzerland have post-World War II legislative histories with the goal of subsidising mountain communities in an attempt to reduce the considerable inequity of living standards and opportunity between them and the non-mountain regions within their national borders. Germany, France, the United Kingdom, and Norway, for instance, all have in place or under consideration legislation to address the undeniable inequities between their highland and lowland areas. Moreover, during two inter-governmental conferences on Chapter 13 of Agenda 21 (UNCED, 1992) held at Aviemore, Scotland, and Trento, Italy, in 1996, participants endorsed the need for a European mountain convention that would recognise the ubiquitous problem of inequity in wealth and health. And in the United States, Appalachia has been characterised as a 'Least Developed Country' **within** one of the most advanced of the industrialised nations (see Chapter 8, Schnelling *et al.*, in Stone, 1992).

To tackle, on a world mountain scale, this issue of inequity in health and wealth in any detailed and analytical manner is not feasible in the space of a single short chapter. Nevertheless, an attempt is needed, even if restricted to an iteration of principles and problems, with the emphasis on relativity. Furthermore, while there are enormous differences in absolute standard of living between the mountain inhabitants of the industrialised countries and those of least developed mountain countries, the relative position in comparison to their neighbouring lowland and urban dwellers is possibly of greater importance. Another topic that requires urgent attention is the relative position of women and children within mountain communities. In this case the situation in the least developed countries would appear to be much the more critical. Thus a section of this chapter will be devoted to women, children, and poverty in mountain regions. Furthermore, the efforts to reduce inequities between mountains and lowlands in Europe provide a number of examples and lessons for transfer of experience with the resolution of mountain problems to less developed countries and regions. Nevertheless, the foregoing discussion is constrained by the notion of poverty and how it is defined (see Ives and Messerli, Chapter 7).

THE HUMAN SITUATION IN MOUNTAINS OF INDUSTRIALISED COUNTRIES

While a clear distinction must be made between North America, New Zealand, and Australia, on the one hand, and Europe, on the other, emphasis will be placed here on the latter, and especially on the Alps.

In a very general sense, the position of Alpine inhabitants prior to about 1870 has been compared to that of those in developing mountain countries of today. While there are obvious differences, the Alps, during the Little Ice Age (1500 to 1850 AD), constituted a region of relative poverty, periodic and systemic out-migration, malnutrition, and even outright starvation. Netting (1981) has produced a classic study of one small village (Törbel, in the canton of Valais) which he aptly entitled 'balancing on an alp'. His hypothesis that the finely balanced socio-economic controls on population level were adjusted upward with the introduction of the potato in the early 1800s, clearly indicates the tenuous nature of the situation (for an excellent account of potato introduction and population increase amongst the Sherpas of Khumbu, see Stevens, 1993; for the classic analysis of the potato in terms of mountain agriculture, see Rhoades, 1985). The Törbel situation, and presumably that of many other villages in the Alps, was a very delicate adjustment between private field and property ownership and closed village corporate community management of forest resources, water for irrigation, and alp grazing. It was reinforced by endorsement of celibacy, late marriages, and out-migration of young males, principally as mercenary soldiers who supplied a crucial return of cash remittances to their families. As the nineteenth century progressed the pattern of out-migration became one of movement of entire families to the New World as colonists, and to the nearby cities. There was virtually no in-migration.

The over-riding pattern of subsistence, moreover, involved dependence on mixed mountain farming (*alpwirtschaft*) whereby maximum use was made of resource extraction from multiple altitudinal belts. Netting, somewhat obliquely, does raise the issue of carrying capacity. Moser and Moser (1986) draw more forceful attention to the well perceived threat of uncontrolled population growth in the case of the Ötztal, Austrian Tyrol, whereby marriage was forbidden for a period in the mid-nineteenth century on village authority; young women who had the misfortune to become pregnant were exiled from the valley. (This anecdote is relevant to the later section on 'women, children, and poverty'; too often women, but rarely men, are 'blamed' for pregnancies, so that child-bearing comes to be regarded as women's primary function next to securing household subsistence.)

Despite these attempts to 'balance on an alp', over the centuries the upper timberline was lowered catastrophically, in places by as much as 400 m. This, in turn, augmented losses from mountain torrents, landslides, and avalanches (Plate 4.1, Chapter 16). While this is only one example of the

Plate 4.1 The view from the Furka Pass, Central Swiss Alps, eastwards in the direction of the Andermatt, tells a dramatic tale. Centuries of forest clearing, expeically on the adrêt (sunny) slopes, to increase available grazing land, has lowered the upper timberline by as much as 400 m in places. The only remaining trees in the scene depicted here serve the small village as a tenous defence against avalanches (Photograph: J.D. Ives)

impacts of human activities on the Alpine landscape prior to industrialisation, it does reinforce the notion of marginality – costs of mountain living both in terms of loss of life and the necessary repairs to, and replacement of, private property and infrastructure.

The potato, railway development after the mid-nineteenth century, and the beginning of tourism *(Belle Epoque)*, set in motion fundamental changes that are still unfolding today. The ability to import cereals from the lowlands, together with the potato, put an end to actual starvation and malnutrition early in the century; and a warming climate, later in the century, undoubtedly produced a less marginal environment. Tourism, and the assumption that high altitudes were beneficial to victims of respiratory ailments (and especially tuberculosis), began to influence a few favoured central places, such as Davos and St Moritz, as resorts developed. The initiation of mass tourism,

and especially large-scale winter sports, after about 1960 led to apparent Alpine-wide major transformation and affluence.

This view of region-wide alpine affluence has been challenged recently by Bätzing *et al.* (1996). They conducted the first demographic assessment at the level of the commune and have demonstrated the fallacy of this assumption of near-uniform affluence generated by mass tourism. They show both cartographically and by analysis of census data (1870 to 1990) that large sections of the Alps face serious socio-economic difficulties reflected in severe loss of population. Their approach and conclusions are reviewed here because they are important *per se* but also have a relevance to other mountain regions in the industrialised countries; they may also be important to some of the newly industrialising countries with mountainous areas (e.g. Taiwan, South Korea, Thailand).

As Bätzing et al. (1996) emphasise, the signing of the Alpine Convention in 1991 has moved the current type of discussion from the category of interesting academic debate to one of vital practical and political concern. The Convention's definition of 'Alpine commune' is given in Bätzing et al. (1993). The factors considered in their analysis include: (1) population change in alpine communes between 1870 and 1990; (2) distribution of the least populated and most densely populated alpine communes; and (3) categorisation of all alpine communes according to four altitudinal belts.

Their study embraces 5814 communes with a total area of just over 180 000 km^2 and a total population slightly in excess of 11 million in 1990. The comparison date of 1870 allows assessment of the pre-industrial population distribution (with a few minor exceptions).

During this period of about a century the overall alpine population increased by 57 percent (from 7 to 11 million) which is considerably below the average for Europe as a whole. Within this aggregate, however, 43 percent of all communes have lost population while 47 percent have experienced marked growth. This allows a classification of the Alps into two types of area of approximately equal total size: (1) losing population; and (2) marked growth. Declining communes are concentrated in the southwestern Alps – especially in the south-western and northern French Alps where 600 communes have lost more than two-thirds of their 1870 populations and 28 communes have less than 50 inhabitants – and in the southwestern Italian Alps. Other areas of decline occur in the central and eastern Italian Alps, eastern Austria, and Slovenia, although in these areas the pattern is a complex mosaic of gainers and losers. Switzerland also shows this mosaic of declining and robust communes.

In contrast, strong growth has occurred in the western half of the eastern Alps, especially between the upper Rhine and Salzburg, including Bavaria, Vorarlberg, Tyrol, Salzburg, Leichtenstein, Alto Adige, and Trento. The situation in the Swiss Alps is perhaps surprising because this region has been the beneficiary of the most comprehensive mountain subsidies. Despite this, however, the policies and measures designed to promote development appear not to have been strong enough to counterbalance the declining trend in many of the Swiss Alpine communes (see Chapter 5 – Pays d'Enhaut). While there are several reasons for this, the patterns of agricultural development over centuries carry a significant part of the explanation. For instance, agriculture in the cantons of Valais, Tessin, and Grisons, developed to a pattern with small individual fields and fragmented family holdings. In contrast the pattern of land holding of the northern and eastern Alps has been much more conducive to consolidation of holdings and to a more viable agricultural development in the present day.

Bätzing et al. (1996) used a population minimum of 300 inhabitants to define an effectively functioning commune. According to this criterion, 157 of 201 communes that constitute the French departement of Drôme have less than 300 inhabitants. This implies economic breakdown.

The classification of communes according to altitude is especially revealing in this context. Altitude is determined at the point in the commune where population density is highest. According to this the following pattern emerges:

(1) below 500 m – overall growth rate of 89 percent (1870 to 1990);
(2) 500–1000 m – this group embraces half of the Alps in terms of area and shows a growth rate of 40 percent;
(3) 1000–1500 m – stagnation and decline predominates;
(4) over 1500 m – growth rate of 27 percent but with a few tourist centres exerting heavy influence; without these the trend would be one of decline.

The below-500 m category shows that the main areas of strong growth occur along the borders of the Alps and in the broad inner longitudinal valleys. Communes located entirely within the mountains have generally lost population and this trend has been significantly reversed by tourism only in western Austria, Bavaria, parts of Switzerland, and the Tyrol. Because these findings contradict the widespread popular image, a special investigation was made of communes located between 1500 and 2042 m. Of the 109 communes in this category, only 36 have experienced growth during the 1870 to 1990 period.

The results of this study were then compared with those of an investigation carried out under the authority of the European Union (European governments and inter-government studies, 1994) because of the great contrast in conclusions. The EU report predicts an excellent economic future for the alpine region. However, the foundation for this investigation is aggregated data at the level of 'statistical units' (i.e. canton, departement, *provincie, bundeslander*) and then further aggregation to the regional level. This approach accentuates the appearance of region-wide homogeneity; thus there can be no distinction of problem areas within the mountains. This failing is compounded when we consider how the 'alpine region' was defined in the first place. In contradistinction to the area considered by Bätzing *et al.* (1996) – 180 000 km^2 – the EU study embraces an area of 436 000 km^2. Furthermore, this includes the cities of Stuttgart, Munich, Zurich, Vienna, Milan and Lyon, giving a total 1990 population of 67 million. This leaves the Bätzing *et al.* Alpine population of 11 million as a small minority. The central argument of the EU study is that the big cities provide the main driving force for economic development and, therefore, they need further improvement of their transport linkages. This supports the long-term EU interests of 'improving' high capacity north–south highways through the Alps (Stadel, 1993).

The Bätzing *et al.* study shows that current developmental trends indicate a growing concentration of almost all available jobs and the bulk of the population into a few favourable locations (transportation corridors and nodes) while the real Alpine zone, with a few exceptions, is losing its productive potential. Thus it appears that a 'European policy' would accentuate this process and lead to highly productive locations within comparatively small areas and a potential breakdown of economic activity in all other areas. This raises the critical question: would this lead to serious ecological, social, cultural, and economic problems throughout much of the Alps? Early in the Swiss MAB project it was noted that disparities between the Swiss Alps and the Central Plateau had been decreasing since about 1970; disparities amongst the alpine communes, however, had been increasing (Brugger *et al.* 1984). The Bätzing *et al.* study has carried this understanding much further so that it now embraces the entire Alpine region.

Even without the impact of policy based on the EU study, Bätzing *et al.* predict that unless present trends are arrested a total collapse will occur, together with the effective disappearance of several hundred mountain communes, accompanied by vigorous growth in the more favoured areas.

The major lesson to be learned from the two studies relates to the specific research approaches that are adopted for mountain regions, especially as they reflect selection of scale. The issue of choice of 'statistical unit' becomes especially critical; the

EU study neglects a basic, if simplistic, characteristic of mountain regions – their extreme horizontal and vertical variability. Aggregation of data, in the current instance, blinds us to the actual geoecological dynamics of the Alps and, by extension, of all mountain regions. This is one of the salient justifications for our insistence that mountain regions must continue to receive special treatment, and hence for the continuation of their priority established in 1992 in the context of Chapter 13 (Agenda 21, UNCED, 1992). The need for region-specific solutions to be incorporated into mountain policy formulation at the national and trans-national levels is crucial. Fortunately, this major issue is being addressed within the context of the Alpine Convention.

This discussion of demographic trends at the communal level in the Alps only provides indirectly for any assessment of the position of the individual person or family; nor is it directly applicable to other mountain areas of the industrialised countries. However, there will likely be many similarities, although large differences must be assumed. These will relate both to the physical geoecological variability between mountain regions, but also to the very great differences in cultural background and to historical and political evolution. The Pyrenees will display many similarities, for instance, but the much later impact of modern developments on Pyrenean land use, and hence landscape and demography, must be considered (Garcia-Ruiz and Lasanta-Martinez, 1993). Nevertheless, for our next case study we will focus on the situation of the *Mittelgebirge* of Germany.

THE GERMAN MIDDLE MOUNTAINS (MITTELGEBIRGE): POLITICAL RECOGNITION OF INEQUALITIES

The traditional restriction of the term *Mittelgebirge* (middle mountains) to Europe is followed in this discussion. Stadelbauer's (1991, 1996) categorisation of the German *Mittelgebirge* as 'naturally less favoured and therefore backward areas' anticipates his description of a wide range of Federal and State policies and management tools intended to offset growing post-World War II inequities between lowland and urban areas and mountains.

The *Mittelgebirge* of Germany fall into two broad types: (1) hill and low plateau regions ranging in elevation between 400 and 900–1000 m; and (2) uplands between 900 and 1500 m. The former show striking contrasts between upland plateaus with open fields and wooded slopes and, in climatically favoured regions, vineyards up to altitudes of 350 m. Examples include the Swabian–Franconian Alb and the extensively forested areas of the Weser–Leine uplands, Hessen, and the Palatinate. The latter include the Harz, the Black Forest, the Bohemian Forest, and the Bavarian Forest of southern Germany.

As Stadelbauer (1991) maintains, although Germany contains extensive mountain regions, it cannot be described as a mountain country. He underlines several critical physiographic factors that serve to place the *Mittelgebirge* in a marginal category compared with the rest of the highly developed and very rich lowlands and urban areas:

- regular increase in annual precipitation totals with increasing altitude, and potential for increased surface run-off;
- a significant temperature lapse rate and, therefore, a reduction in biomass productivity;
- higher silicate and carbonate contents of the underlying rocks with their influence on soil development;
- slope gradients that may restrict agricultural use, but will accentuate differences in radiation receipts between northerly and southerly aspects;
- natural or artificial vegetation based upon location, with predominance of closed forest formations;
- and in a more positive sense, variations in slope, and hence augmented landscape aesthetics which, in recent decades, has become a 'scenic resource' for recreation and tourism.

In general, agriculture has become progressively more marginal and forestry dominates all other production activities. According to the data base used by Stadelbauer (data for 1990, pre-reunification), German forests cover nearly 7.5 million ha, of which 30 percent are state forests, 24 percent public forests, and 46 percent privately owned. These support more than 100 000 forestry enterprises, while nearly a quarter of a million farms manage 1.55 million ha of woodland. Over the last 200 years forest utilisation has produced a substantial change in tree species composition from deciduous and mixed forests to conifers (Chapter 13).

The slow but progressive marginalisation of this considerable area of upland and mountains has been partially off-set in recent decades by the growth of tourism. Today, the total contribution of the tourist sector to gross productivity of the mountain regions (including the Bavarian Alps) considerably exceeds the proportionate contribution of tourism in all other areas of the country. Even more recently, however, the environmental importance of the mountain areas has been recognised. Thus the former emphasis on resource evaluation has shifted from productivity and economics to the ecological and cultural functions and to the environmental balance they are expected to guarantee in a very densely populated country.

To preserve and further develop this environmental contribution, governmental policies at all levels have been, and are being, introduced in an effort to reduce inequities in living standards between urban agglomerations and remote rural (principally mountain) areas. Central to this is the planning of transportation and traffic systems in a strategy to improve connections between urban centres and the rural periphery. Related to this is location of new industries to take advantage of the relatively cheap labour force following progressive reduction in agricultural activity. In some states programmes are specifically devoted to mountain regions, while in others the more generally designed programmes predominantly affect mountain regions.

The various programmes and subsidies differ from state to state and from region to region. Nevertheless, many of the specific goals are strictly comparable. They include: protection of meadows in small valleys; maintenance of fallow for several years in the context of landscape preservation; moist areas must not be drained; no conversion is allowed from grassland and pasture to cultivated land or to forestry. There are direct subsidies to individual farms in a variety of forms: additional quotas of milk production are allowed above EU quota regulations, in order to support dairying which is the most effective form of farming at these higher elevations. Individual farms, for instance, in North Rhine–Westphalia are subsidised by up to 300 DM/ha annually with amounts up to 1600 DM/ha for areas decreed of particular importance for nature conservation.

Other programmes include support for rebuilding or repairing traditional farm and village buildings and to encourage conservation of, and reintroduction of, traditional types of land use. These measures are aimed at firmly linking the environmental and cultural heritage of the mountain regions. In Hessen, for instance, so-called equalisation payments are made to individual farms ranging from 55 to 286 DM annually per livestock unit. In this example payment is made for a maximum of 20 cows and is applicable only where the annual income of the farmer and his wife does not exceed 65 000 DM.

An example of a somewhat different type of programme is provided by Lower Saxony which contains the Harz (971 m). While there is no special programme for mountain regions in Lower Saxony *per se*, except for general assistance to agriculture or economic activities, the towns of Seesen and Herzberg, which are oriented toward the Harz, receive a subsidy of 15 percent of total costs for establishment of new enterprises.

These several examples of regional planning within Germany are based on the recognition that it is necessary to develop a more equitable standard of living throughout the nation in rural and urban areas alike. While many of the individual programmes are not mountain-specific, the fact that the majority are applicable to such regions demonstrates the marginal nature of the mountains. Even so, it is believed that the economic situation in some peripheral mountain regions will remain precarious unless tourism can be accelerated. Nevertheless, acceleration of tourism in many localities has its own negative aspects (Chapter 12). In the Black Forest region, for example, Stadelbauer (1996) believes that tourism has reached its upper limit. It generates about 3000 million DM and almost 30 million overnight stays each year. About 100 000 persons are employed and 4200 different enterprises depend directly upon the tourist sector. There is an urgent current trend, however, to develop other forms of management to mitigate such impacts. It is ironic that these include efforts to reduce accessibility, at least by private vehicle, in some of the more remote areas.

The experience from Germany, in addition to the Alpine situation as analysed by Brugger *et al.* (1984), and especially by Bätzing *et al.* (1996), demonstrates the enormous structural problems facing the people of mountain regions even in some of the most affluent of the industrialised countries. The expenditures of vast sums of money and considerable political will to ameliorate the built-in mountain inequities still fall far short of general solutions. This, however, does provide an important object lesson for any approach to sustainable mountain development, both for other industrialised countries and for Least Developed mountain regions. In the latter regions, the problems facing sustainable development are much more severe. Here again, however, control of mountain resources by power centres on the plains or lowland urban nodes, whether directly, or indirectly through the manner in which foreign bi-lateral and multi-lateral aid is applied, demonstrates the marginalisation of mountain regions. Nevertheless, we shall next turn to one of the more daunting problems facing these least developed regions – the inequitable and highly vulnerable position of women and children, and especially the girl child, a problem that is superimposed on an already desperate situation in mountain regions (Chapter 5).

WOMEN, CHILDREN, AND POVERTY IN MOUNTAINOUS DEVELOPING COUNTRIES

Problems facing mountain environments and mountain peoples failed to receive any special attention from the international community until recently. Foreign aid and development projects have tended to treat mountain areas as adjuncts to areas and issues of primary concern on the plains below (Chapter 1). Despite earlier efforts by a number of agencies, the Pontifical Academy of Sciences Study Week of November, 1987 (Marini-Bettolo, 1989) was the first occasion when mountain issues world-wide were accorded equal status alongside tropical rain forests, deserts, the oceans, and such thematic mega-issues as climate change, loss of biodiversity, and atmospheric pollution. But it was not until the Rio de Janeiro (UNCED, 1992) Earth Summit that *mountains* received a primary priority mandate at the global level.

As with *mountains*, women's issues received only belated attention on the world agenda. The United Nations Decade for Women: Equality, Development, and Peace (1976–1985), brought the profound gender inequalities, and concomitant restrictions on sustainable development, to the general public's awareness, as well as to national governments. Byers and Sainju (1994) and Bajracharya *et al.* ((1993), following a long period of concern with the status of women in Nepal, firmly linked women's issues with the mountain problem.

Furthermore, it has been customary to associate children and poverty with women's problems. This is generally done because of the relationship between mothers and children and their marginalisation in society. Recognition of the adverse conditions women often experience has been reinforced by the growing realisation that, amongst the poorest of humankind, women suffer most in times of increasing stress. There is ample evidence to support the disproportionate effect this has on children in their care. Conversely, much of the comparatively enormous expenditure on foreign aid and international development has benefited men disproportionately, in part because of a general poor understanding of women's role in subsistence agriculture, and in industry, in the developing world. The statistical invisibility of women has also served to reinforce this. Yet there is a significant drawback to dealing with the problems facing women and children within a single context. This tends to depict them **both** as dependent, whereas women are as resourceful and as capable of work of equal importance as their men (personal communication, October, 1996, Farida Hewitt). This drawback is emphasised here and should be kept in mind throughout the remainder of this chapter.

The Chinese government's categorisation of the vast majority of its 80 million citizens below the poverty line as mountain ethnic minorities has already been mentioned (page 61 above). In India, almost half of all deaths occur by the age of five; deaths of young girls exceed those of young boys by almost 350 000 every year. One infant death in six is specifically credited to gender discrimination. Yet it is extremely difficult to disaggregate these data to show the situation in mountain communities. Nevertheless, in Himachal Pradesh, West Bengal, Jammu and Kashmir, and Uttar Pradesh (all states with significant areas of mountain territory), males outnumber females by almost 10 percent (Chatterjee, 1990). The maternal mortality rate of over 400 per 100 000 live births for India, is 40 times the rate for Singapore and 150 times the rate for Norway (State of the World's Children Report, 1994, UNICEF).

Box 4.1   Women of the high pastures – Karakorum–Himalaya

In the Hushe Valley women go up with the livestock to the high pastures. They use their considerable skills and intimate knowledge of the environment to produce and process the resources of the mountains. Men join them occasionally from the main village to help with the initial movement, or later to assist with moving to a different pasture as the grasses of the first are depleted. They will also repair the houses and cut timber for house-building in the valley below. But it is about the experiences of the women that I will write.

Women's work here can be separated into three categories:

(1)   Women *produce* for subsistence. This is essential work that forms the main economy;
(2)   Women *enable* their men to live and work outside the village for long periods at a time while they maintain the traditional way of living;
(3)   Women *replace* men in tasks that are traditionally performed by men, thereby increasing the burden of their own workload.

In all of this work, the status of the women is subordinated to that of their men *because* they are 'left behind' – both physically, in the valley, and ideologically, through non-participation in development. They inhabit the domestic sphere, which shrinks as the dominant, male, public sphere expands. Thus the women, children, and old people inhabit a 'domestic' sphere, in a sense hidden as the household economy, while the young men move between this and the development sphere, or live

entirely in the latter. The young men are beginning to look elsewhere for work for wages and to abandon their traditional roles, being increasingly drawn to the towns and cities of the plains, or to work as porters with foreign mountaineering expeditions. Most of the mountaineers are from northwest Europe and Japan; they pay handsomely and outfit their porters with expensive and practical gear.

In Hushe village, which carries the same name as the valley, this work as porters and guides has increased in importance over the last five years, and the opening up of the Ghondoghoro Pass has permitted access to the Baltoro Glacier. The Government is now encouraging mountaineering and trekking into regions previously barred to foreigners as politically 'sensitive'. Porters and guides earn between Rs 100 and 300 per day, plus food and clothes; most of all they gain prestige in the eyes of their community. Some may even leave for Islamabad to work for trekking agencies. It is not surprising, therefore, that young men and youths cluster around camp grounds all summer in hopes of gaining employment from one of the expeditions which pass through Hushe, the final village before the mountain wilderness; thus the women are left to work on the land.

This situation is extremely unfortunate for people who still believe in family values rather than individual rights and for whom an arranged marriage is an alliance between clans. The husband can depend on his wife for subsistence while regarding her as 'backward' and 'dirty', or treat her as inferior.

Language is also a barrier that separates the elderly and females from the younger men. While men and women speak the local dialects with each other, *only* boys go to school, and so learn to speak Urdu, the official language of Pakistan. These are government schools which were opened in the northern areas two decades ago. Because of the conservative, Islamic sanctions, girls cannot go out in public spaces to attend school in most Balti villages.

There are three main areas of high pasture, or *bluq*, for the village of Hushe. Herds and humans migrate to the higher pastures as the snow melts and the vegetation regenerates after winter. The work the women do in the *bluq* follows an unchanging pattern that must have developed many generations ago. They produce *mar*, or clarified butter, the main resource of the *bluq*. While barley, grown in the main village, is the staple food, *mar*, at Rs 150/kg, is especially valuable because it can be exchanged for cash or other goods in the villages or the main town of Skardu.

It should be noted that there are differences *within* Baltistan in the gendered division of labour, but there always *is* such a division. For example, at the northern end of the Shigar-Basha valley, in the village of Arindu, it is the women, accompanied by small children and nursing babies, who care for the flocks of yak, dzo (a hybrid of cow and yak), goats, and sheep. In Chutrun, 25 km to the south in the same valley, only the men go up to tend the herds in the summer pastures, due to a belief that women are 'impure' because of their physical functions of menstruating and child-bearing. Mountains are believed to be the 'abode of the gods', therefore 'pure'. In Askole, in the Braldu valley, in contrast, entire families accompany their livestock for the summer.

The cash income for the *mar*, which is derived *solely* from women's work, is central to the changing, increasingly cash-based economy. Cash is replacing barter both within the community and between villages. The *mar* produces a steady,

Plate 4.2 Two girls in a 'Brangsa' (high pasture settlement) in Hushe valley. The girls are in a typical 'nang' (house) of stone, with wooden supports. Cracks and crannies in the walls are utilised as shelves. One girls is churning 'mar' (butter) with a leather thong wrapped around the churn. July 1993 (Photograph: Farida Hewitt)

Plate 4.3 View of Chutrun across River Basha. The bath-house is visible on the extreme left (arrow). Behind it is the winding jeep road. Fields and fruit trees fill the view. June 1993 (Photograph: Farida Hewitt)

dependable source of income, especially as goods from 'outside' become increasingly desirable. So when women make their daily, or twice weekly trips down to the main village they carry with them all they have produced or collected, necessary to the well-being of the community. While in the village they carry out other women's tasks – weeding and irrigating the fields of barley, peas, and vegetables, adding the manure which they have carried down, cooking, and processing cereals, fruits, and nuts. Thus they work and care for children constantly, in contrast to their men.

While women do almost all the work of food production in both village and high pasture, it is reasonable to ask what the men do. They are traditionally responsible for ploughing, terracing, timber cutting, construction, and other heavy work. Men also market livestock and other products which women may have produced; they also weave and sell their crafts. However, as society changes, older men and the women are now having to handle more of the workload. The younger men, if not working with

expeditions, are in the army, or working as teachers, road builders, or on contract with the Government; some go to the Middle East. Thus the older men and women not only have to take on more and more work, they also risk that they will gain little benefit from it. The earned money is generally used by the males for luxury consumer goods, yet it is the work performed by the women that frees them to take on the more glamorous, 'outside' jobs. The women, moreover, are forbidden by patriarchal codes and conservative interpretation of Islam, from being seen in public places by strange men; and this happens to be the political–economic arena. It is *while* they are working themselves to death in the hidden economy, that the women are also being marginalised. Sadly, their world will disappear for the following reasons: they live in a high mountain valley, on the edges of the 'modern' world; the work they do is labour-intensive; and half the economically active population is leaving.

This, therefore, is the problem of women: their forced inclusion into a capitalist system of labour –

Plate 4.4 Chutrun: Women and girls weeding a field of wheat. The older women have turned away from the camera. From May to July, when harvest begins, this is womens' most time-consuming task. June 1993 (Photograph: Farida Hewitt)

first by their own men due to the patriarchal organisation of their society which turns partner and co-worker in the household into master and slave; by the political–religious elites and leaders of the village, who maintain and confirm this division, because it enables them to become richer; by middle-men and entrepreneurs, who profit from women's low, or unpaid labour; then by the State and the global economy, into which the women are inserted, without their knowledge, consent, or control. Mean-while, their husbands are 'who they are', producing for benefits which they can directly enjoy.

Source: Dr Farida Hewitt, Department of Geography, Wilfrid Laurier University, Ontario N2L 3G9, Canada.
Based on a paper presented to the Canadian Association of Geographers, Annual Meeting, May, 1996.

On another related topic, a study in Nepal (CWD, 1988) showed that mountain communities had an out-migration rate of 20 percent, compared with a 2 percent rate for lowland communities. In both cases, however, 85 percent of the migrants were male. Another study in the mountains of western Nepal showed that 73 percent of men and 35 percent of women who left their villages did so because of reduced availability of land per capita, falling agricultural productivity, and depletion of natural resources (Schuler, 1981). All these data for Nepal indicate a rapid growth in the proportion of female heads of households and the increasing pressure such a trend is producing in its effect on poverty and, presumably, on the status of mountain children.

So far, only a few random examples have been introduced. It is important, however, that the problem of data reliability be considered. There is not only the question of aggregation of data and the general statistical invisibility of women, there is the more basic issue of the accuracy and completeness of the data that is available. This issue has been elaborated in detail by Ives and Messerli (1989) for the Himalaya, but it is without doubt a prevailing problem throughout the mountains of the developing world. Some generalisations, therefore, are introduced below in an effort to better focus on the position of women and children in mountain communities.

It is now increasingly acknowledged that poverty weighs much more heavily on women than on men and, through women, impacts severely on children, whether in mountain communities or society at large. A number of trends and problems have become apparent over the last 20 years, and it is only for the period, 1970–1990, that there is even rudimentary data available (UN, 1992). Nevertheless, it is important for this discussion to review what is known.

First, in most societies, but in some much more than in others, the wife or mother in a household has a much reduced access to resources, control over her life and effective decision-making role

Box 4.2 Women's life in the Rwenzori (Uganda)

According to the tradition of the Bukonzo tribe, which lives in a mountainous area of western Uganda at about 1400 m, fathers pass their land to their sons. Nothing is left to the female children as they will be 'sold' to their husband's family and will lose all rights and responsibilities towards their birth family.

The dowry mostly consists of 3 to 12 goats, depending on the beauty, behaviour, and education of the bride. The price is discussed for hours or even days among the male relatives of the bride and the groom and the marriage is only declared official when the entire dowry has been paid. Only then will the couple be allowed to marry in a Christian church in European style, the dream of every Bukonzo woman.

Even though women are no longer forced to marry a man selected by their family, they have no great choice, because unmarried women greatly out-number unmarried men. Women have to marry to be respected in the tribe, especially older women who often have to accept an already married man and become his second, third, or even fourth wife. As children, especially boys, are still considered to be the fortune and the security for a secure old age, one man may have as many as 50 children.

Even today, women and children of the Bukonzo tribe are regarded as the property of their husbands. If a man dies, his widow will become the property of one of his brothers, as an additional wife. If there are too many children, they might be distributed among the family members of the deceased husband.

The illiteracy in Uganda is as high as 52 percent (women 65 percent) and the average schooling is about 1.1 year. But in this rural and mountainous area of Rwenzori, illiteracy is certainly much higher. Lack of capital for investment, and low levels of education and training, hinder development. The local government of the Bukonzo people is slowly beginning to realise the important role of education, and primary school projects in different villages have been undertaken and supported by local contributions. Senior secondary schools are still very rare and most are located in remote areas so that very few children, whose parents can afford it, obtain a higher education.

The living standard of the Bukonzo tribe is very elementary. Their cottages are made of clay and roofed with dry grass and banana leaves; the beds are simple and hard. Mattresses are stuffed with banana leaves and are usually covered with home-woven mats made of coloured fibres of palm tree leaves.

The morning meal consists of a porridge made of cassava flour, although the Bukonzo people mainly survive on beans, which they grow themselves. These beans are eaten together with cooked bananas, sweet potatoes, or a thick, doughy paste made from cassava roots, called *Obundu*. Fish and meat are luxuries!

The farmers can harvest twice a year, but during the dry season, between the two harvests, shortages of food often occur and the children, therefore, are undernourished. A few women plant cabbage, maize, and tomatoes as supplements. Most families own only 2–5 acres of good land and coffee is the main cash crop. The money they sometimes earn from coffee trading is used for the purchase of

Plate 4.5 Women working in the fields of beans and maize in the Rwenzori area of Uganda, close to the Zaire border (Photograph: E. Wismer-Strebel)

corrugated iron sheets, a wooden family bed, or a precious blanket. The work is hard, as the fields are situated on the slopes of the deforested hills and the women have to labour with their babies on the backs.

The most important problem they face is the heavy rainfall in November that can create landslides and thus wash away their precious land; it can also dissolve their clay huts and leave them with nothing.

In this kind of situation the obstacles facing attainment of any form of sustainable mountain development are particularly severe.

Source: Mrs Eleonore Wismer-Strebel,
Allenwinden Switzerland

than her husband or other male head of household. This is exacerbated by the strong preference for male children in many societies. For instance, in much of South and Southeast Asia, sex-selective abortions and female infanticide, neglect of young female children, and related problems, have resulted in the elimination of many millions of girls. These problems are especially apparent from demographic statistics for India, but also prevail in other countries. In Nepal, there is a saying that 'boys are brought up; girls are starved up' (Messerschmidt, personal communication, In Ives and Messerli, 1989: Chapter 7). This demographic trend alone demonstrates the status of girls and women. And, where there are vigorous government policies to limit family size, there is already the beginning of a shortage of young women of marriageable age. A

somewhat comparable situation in the Alps is the difficulty young male farmers have in finding wives who are willing to undergo the heavy workload involved in maintaining higher altitude farms. However, this indication of the liberation of women from former traditions is also in striking contrast to the position of women in less developed mountain regions. Furthermore, it should be noted also that in traditional, subsistence-based societies women's access to resources is (was) much greater than in transitional economies, or cash-based households. This distinction is important because it affects the status of women who become more dependent on their men with the introduction of cash. They may work as hard, but they are confined more to 'private' spaces and lose their freedom and command over valuable commodities (personal communication, October, 1996, Farida Hewitt).

Second, as economic pressures have mounted, accentuated by the world debt crisis and reductions in government-subsidised social programmes in the indebted developing countries, more and more young males migrate from rural to urban areas and overseas. This migration may be seasonal, for a year or so, or be semi-permanent. The result is that women must function as *de facto* or actual heads of households, with the accompanying increase in pressure on their time, since they often must operate the family farm with little or no assistance (UNICEF, 1994).

Third, until very recently, international and bilateral aid and development agencies targeted their projects at males. A few examples will demonstrate the difficulties that have resulted from this tendency:

(1) Many development projects have emphasised the introduction of cash crops. Regardless of where any increased burden of work may fall, in many societies, males control any cash income and much of this is spent on their personal consumption, such as alcohol and cigarettes, rather than the family.

(2) Cottage industries have been a favourite with development agencies over the years. This does the woman the double disservice by providing the male with access to cash income and by adding to the already disproportionately heavy workload on the female.

(3) Even where development agencies attempt great sensitivity by seeking to determine the 'felt needs' at the village level prior to taking development initiatives, in many societies women either do not attend village councils or, if they do, remain silent observers. Thus, even with the best of intentions, development agencies often have no contact with women in terms of selecting development projects to be pursued, types of crops, species of trees to be planted, and so on.

(4) Especially in rural areas, where women's work is unpaid (for instance, on the family subsistence farm), there is a fundamental lack of data. Development agencies, therefore, may have no way of determining possible negative impacts. As the workload for women increases, so does their dependency on female children to help with household and other chores. This is considered a major factor in

educational disparities between boys and girls and contributes to the statistical invisibility of women. In many countries, census data are not disaggregated by sex. Thus, there is great need for trained female census enumerators, especially in societies where women are partially or completely secluded. Also, seasonal activities, and activities that fall outside the reference period of the censuses, are not recorded, further distorting the apparent workload women are required to carry.

(5) It is becoming increasingly apparent that domestic violence, the hidden or secretive element of female inequality, also takes an enormous toll on women, children, and families. This is especially insidious since such violence is often viewed as a social ill rather than a crime.

(6) The result of unbalanced male out-migration means that motherhood is increasingly unsupported by marriage and that the elderly are more often unsupported by children.

These trends increase the burden on women, and hence the pressure on children in their care. While the issues are at last becoming better understood and documented, many are intractable while others will only be alleviated after a considerable time lag required for agency adjustments to take effect. Some data are available, and Tables 4.1 through 4.5 (pages 74–78) provide a selection of indicators that have been extracted from *The State of the World's Children Report, 1994* (UNICEF). Data related to thirteen mountain countries are displayed. The countries selected include tiny Bhutan with a population estimated at 1.6 million in 1992; Colombia, with a 1992 population of 33.4 million; and India, which is only partially mountainous. Examination of the tabulated data allows a number of alarming generalisations:

(1) Mortality under age 5 ranges from 222 per 1000 live births for Rwanda (prior to the recent strife), 201 for Bhutan, to only 20 for Colombia;

(2) Other basic indicators, such as life expectancy and total adult literacy, show the same pattern among the 13 countries ranked in order of improving conditions, with Rwanda 'worst' and Colombia 'best'. There are, however, a number of off-sets to this ranking;

(3) GNP per capita data are in reverse order, as would be expected;

Table 4.1 *Basic Indicators – Quality of life for 13 mountain countries or countries with mountainous territory

	Under 5 mortality	Infant mortality	Total population (millions)	Annual no. of births (thousands)	Annual no. of deaths under 5 (thousands)	GNP per capita (US$)	Life expectancy at birth	Total adult literacy	% of age-group in primary school
Rwanda	222	131	7.5	396	88	270	46	50	69
Ethiopia	208	123	53	2627	547	120	47	24	38
Bhutan	201	131	1.6	65	13	180	48	38	26
Nepal	128	90	20.6	778	100	180	53	26	86
India	124	83	879.5	25900	3212	330	60	48	97
Bolivia	118	80	7.5	261	31	650	61	78	82
Tajikistan	85	65	5.7	229	19	1050	69	–	–
Papau N.G.	77	54	4.1	136	11	830	56	52	71
Guatemala	76	55	9.7	380	29	930	64	55	77
Kenya	74	51	25.2	111	82	340	59	69	94
Peru	65	46	22.5	658	43	1070	64	85	126
Ecuador	59	47	11.1	332	20	1000	66	86	118
Colombia	20	17	33.4	809	16	1260	69	87	110

*Derived from *State of the World's Children Report*, UNICEF, 1994

Table 4.2 *Basic Indicators – Health/water and sanitation

| | % of population with safe water 1988–1991 | | | % population with access to adequate sanitation 1988–1991 | | | % population with access to health services 1985–1992 | | | % fully immunized 1990–1992 | | | | pregnant women | ORT use rate 1987–92 |
| | | | | | | | | | | 1-year-old children | | | | | |
	Total	Urban	Rural	Total	Urban	Rural	Total	Urban	Rural	TB	DPT	Polio	Measles	Tetanus	
Rwanda	66	75	62	58	77	56	80	–	–	94	85	85	81	88	26
Ethiopia	25	91	19	19	97	7	46	–	–	21	13	13	10	7	68
Bhutan	34	60	30	13	50	7	65	–	–	81	79	77	82	43	85
Nepal	42	67	39	6	52	3	–	–	–	82	72	72	64	18	14
India	85	87	85	16	53	2	63	90	36	96	89	89	85	77	37
Bolivia	52	77	27	26	40	13	–	–	–	86	77	84	80	52	63
Tajikistan	–	–	–	–	–	–	–	–	–	–	92	–	–	–	–
Papau N.G.	33	94	20	–	57	–	–	96	–	67	61	61	66	52	46
Guatemala	62	92	43	60	72	52	34	47	25	56	65	69	58	18	24
Kenya	49	74	43	43	69	35	77	–	40	93	85	85	81	27	27
Peru	56	77	10	57	77	20	75	–	–	82	80	81	80	27	31
Ecuador	55	63	43	48	56	38	88	–	–	99	83	83	66	5	70
Colombia	86	87	82	64	84	18	60	–	–	86	77	84	74	40	40

*Derived from *State of the World's Children Report*, UNICEF, 1994
ORT: Oral rehydration therapy
TB: Tuberculosis
DPT: Diptheria

Table 4.3 Basic indicators – *Nutrition

| | % Infants with low birth weight | 1986–1992 % of children who are: | | | 1980–1992 % of children suffering: | | | | | Total goitre rate (6–11 yrs) % | Daily per cap. cal (% req.) | % share of household consumption |
| | | Exclusively breast fed | Breast fed & sup. food | Still breast feeding | Underweight (0–4 years) | | Wasting (12–23 mo.) | Stunting (25–59 mo.) | | | | |
	1990	0–3 mo.	6–9 mo.	20–23 mo.	moderate	severe	mod. & sev.	mod. & sev.	1980–92	1988–90	1980–85
Rwanda	17	90	75	–	29	6	9	58	49	82	29
Ethiopia	16	74	–	35	48	16	12	63	22	73	49
Bhutan	–	–	–	–	38	–	4	56	25	128	–
Nepal	–	–	–	–	–	–	–	–	44	100	57
India	33	–	–	–	63	27	–	65	9	101	52
Bolivia	12	59	57	30	13	3	2	51	21	84	33
Tajikistan	–	–	–	–	–	–	–	–	20	–	–
Papau N.G.	23	–	–	–	35	–	–	–	30	114	–
Guatemala	14	–	–	44	34	8	3	68	20	103	36
Kenya	16	24	87	46	14	3	5	32	7	89	38
Peru	11	40	62	36	11	2	3	46	36	87	35
Ecuador	11	31	31	23	17	0	4	39	10	105	30
Colombia	10	17	48	24	10	2	5	5	18	106	29

*Derived from *State of the World's Children Report*, UNICEF, 1994

Table 4.4 Basic indicators – *Education

| | Adult literacy 1990 | | No. of sets per 1000 pop. 1990 | | Primary school enrolment rate 1986–1991 net | | % of grade 1 enrolment reaching final primary grade | Secondary school gross enrolment 1986–1991 | |
	Male	Female	Radio	TV	Male	Female	1988	Male	Female
Rwanda	64	37	62	–	65	65	36	9	6
Ethiopia	33	16	191	2	32	24	44	17	12
Bhutan	51	25	16	–	31	20	26	7	2
Nepal	38	13	34	2	84	43	–	42	17
India	62	34	79	32	109	83	53	54	33
Bolivia	85	71	599	163	83	75	50	37	31
Tajikistan	–	–	–	–	–	–	–	–	–
Papau N.G.	65	38	72	2	79	67	61	15	10
Guatemala	63	47	65	52	82	70	36	20	17
Kenya	80	59	125	9	92	89	62	27	19
Peru	91	79	253	97	125	120	70	66	60
Ecuador	88	84	315	83	118	117	63	55	57
Colombia	87	86	170	115	109	111	56	48	57

*Derived from *State of the World's Children Report*, UNICEF, 1994

Table 4.5 Basic indicators – *Women

	Life expectancy females as % of males	Adult literacy rate, females as % of males	Enrolment ratios females as % of males 1986–1991		Contraceptive prevalence	Pregnant, immunized for tetanus	% of births attended by trained health personnel	Maternal mortality rate	Total life expectancy
	1992	1990	prim. school	second. school	% 1980–93	1990–1992	1983–1992	1980–1991	
Rwanda	107	58	99	67	21	88	29	210	46
Ethiopia	107	48	65	71	2	7	14	560	47
Bhutan	103	49	65	29	2	43	7	1310	48
Nepal	98	34	51	40	14	18	6	830	53
India	101	55	76	61	43	77	33	460	60
Bolivia	108	84	90	84	30	52	55	600	61
Tajikistan	–	–	–	–	–	–	–	–	69
Papau N.G.	103	58	84	67	4	52	20	900	56
Guatemala	108	75	85	85	23	18	51	200	64
Kenya	107	74	96	70	27	37	50	170	59
Peru	106	87	96	91	59	27	52	300	64
Ecuador	107	95	99	104	53	5	84	170	66
Colombia	109	99	102	119	66	40	94	200	69

*Derived from *State of the World's Children Report*, UNICEF, 1994

(4) Another factor is missing data. There are no available data, for instance, for Tajikistan. This very mountainous country, with an estimated population of five to seven million in 1992, is facing one of the highest birth rates in the world (Badenkov, personal communication, 1992; Cunha, 1994; personal field notes) and has been in the throes of serious civil disturbance, if not actual warfare. Since the author's several visits (1987–1991) it has been estimated that more than 20 000 casualties have been sustained;

(5) The high prevalence of goiter in all mountain countries is especially distressing, running as high as 44 percent of children aged 6–11 in Nepal, 49 percent in Rwanda, and 36 percent in Peru;

(6) Tables 4.4 (education) and 4.5 (women) enable a degree of differentiation along gender lines for women and children. Secondary school education for girls in Bhutan is as low as 2 percent, compared with 7 percent for boys; Rwanda, Ethiopia, and Papua New Guinea are nearly as bad, while Nepal records 42 percent for boys against 17 percent for girls. The situation has been improving in recent years so that even the data incorporated here are becoming out-of-date. Nevertheless, they present a realistic picture;

(7) Maternal mortality, introduced earlier in this chapter, is amongst the worst in the world. According to 1992 figures for India, the infant mortality rate was 460 per 100 000 live births, but was much worse for other countries: Bhutan – 1310 per 100 000 live births; Papua New Guinea – 900; Nepal – 830; Bolivia – 600; and Ethiopia – 560;

(8) The question of data availability and reliability needs to be raised again. Nepal, for instance, claims, for 1988–1991, that 67 percent of urban dwellers had access to safe drinking water. The accuracy of this figure can be justifiably questioned based on the widespread susceptibility of tourists to dysentery in Kathmandu. No doubt data from rural areas in mountain countries are even less reliable.

These data, while alarming by the standards of the industrialised countries, do not provide any significant statistical separation of mountain countries from non-mountain countries. With the exception of Colombia, which ranks 106th out of a total of 145 countries in descending order of their 1992 under 5 mortality rates, the other twelve countries listed fall within the upper 69 of all countries (that is, the 69 worst-case examples using this single criterion). The data, however, do point to the need for comparative studies between plains villages and mountain villages.

An example of how many of the problems discussed above interrelate and compound each other is drawn from Pakistan's northern areas.

Dr Farida Hewitt (1989) undertook a study of a group of women in Hopar, Karakorum Himalaya. Born in Pakistan, Hewitt expected to be able to conduct detailed interviews in the five Hopar settlements of Nagyr, a district of Gilgit Agency. She found, however, that the women of Hopar incurred a double separation from adjacent 'outside' societies. First, as women, their traditional way of life ensures that they are secluded; second, they speak only one or two of four local languages that belong to the Dardic group, unrelated to Indo-Aryan languages. One of these languages, Boroshaski, is the only one in the world not known to be connected to any other language. Urdu is the mainstream language of Pakistan and is taught to the male children of Hopar. Due to religious restrictions, until very recently girls were not sent to school so, in addition to the seclusion, the women could not speak directly with Hewitt, a female Urdu-speaking interviewer whose collected data had to be filtered through a male interpreter.

In making a general statement about the modernisation process in this hitherto remote mountain fastness, Hewitt explains that: 'the life based on a local vernacular construct of gender is changing to one based on the competitive values of the market economy. When this happens, women truly become the second sex'. She itemises the following changes:

(1) With the introduction of new technology for growing cash crops, women are losing control and responsibility over resources and thus over their own activities;

(2) As men leave for employment in the outside world (as porters, guides), or down to the plains to join the army or to work in towns and cities, the pressure of work for women increases;

(3) With fewer young men able or willing to go to the high pastures, livestock will diminish. This will decrease the availability of manure and make food growing more difficult for women. Government-subsidised chemical fertilisers,

which must be paid for from the cash income of the villagers, are not an adequate substitute for organic fertilisers on the poor morainic soils;

(4) Every year land is lost to cultivation due to landslides. Unless men build irrigation channels, no additional land can be added for cultivation to make up for the annual losses. The burden of intensification, necessary to achieve the same level of food production with reduced resources, falls on the woman;

(5) Women are left as temporary heads of households with all the responsibilities and hardships involved;

(6) Remittances sent, or brought back by the men are often not invested in family property to raise the standard of living, but appear to be acquired for luxury purchases for men's personal use.

While Hewitt does not discuss the situation of children in her study, it can be assumed that this projected increase in pressure on women will have a disproportionately negative effect on children, and especially on the girl child.

Both the Food and Agricultural Organisation (FAO) and the Aga Khan Rural Support Programme (AKRSP) have tended to direct household technology toward women and agricultural technology toward men. This defines women's sphere as private and domestic and, therefore, subordinate, and men's as public, dominant, and powerful. The approach must be compared with the traditional subsistence level, where a gendered division of labour gave women much better equity and much more security and respect.

However, it must be pointed out that in the last few years, the AKRSP has taken vigorous steps to arrest this deterioration in the position of women. The Programme's 10th Annual Report (1992), and subsequent reports, shows that serious attention is now being paid to the problem of women in development. Already, in Pakistan's northern areas approximately 570 women's organisations have been established. In addition, there has been a great increase in educational facilities for girls, as well as in actual school attendance. Nevertheless, the loss of equity incurred by women will take substantial time and effort before it is corrected, or even to return to the initial *status quo*, which was already inequitable. Moreover, there also remains a serious short-coming with the AKRSP women's organisations. They have only been initiated in the Ismaili

Muslim provinces; in the much more conservative Shia Muslim provinces of Nagyr and Baltistan, women are not allowed to participate in 'public' activities. Education for girls, even at the primary level, is not available in most villages. In Chutrun, Baltistan, for instance, the process began only in 1992. Furthermore, women's organisations are joined only by elite women – that is, the wives of elite men. Poorer 'peasant' women are omitted, or marginalised, because they cannot afford the nominal cash savings that are obligatory for membership, or their heavy workload in securing the family's subsistence leaves them no time to work on AKRSP projects. Neither are they visible to Programme officials who conduct the formative group dialogues (the above comments on Nagyr and Baltistan are virtually verbatim: Farida Hewitt, personal communication, October, 1996). Thus it is necessary to contrast the Shia districts with the Ismaili districts. This contrast was especially evident to the author during his visit to Hunza in September, 1995.

The foregoing discussion about the status of women in a single small mountain community reads as a prototype of any synthesis that may be drawn from the two world-wide surveys of the position of women in general (UN, 1989 and 1992; UNICEF, 1994). To this must be added the conclusion that female-headed households are amongst the poorest in all countries for which data are available, and this is increasingly the situation in mountain communities.

But there are also positive signs. For instance, progress has been made in some areas, such as introduction of appropriate tools and technologies, especially those for irrigation and processing, to improve the productivity and working conditions of rural women. In China, for instance, rice transplanters, fertiliser appliers, water pumps and grain-drying equipment have been introduced. These all save women's labour but do not displace them in agricultural work (Satterthwaite, n d: p. 96).

Recent fieldwork in Yunnan Province, China, supported by the Ford Foundation and the United Nations University (Ives and He, 1996) has uncovered some interesting insights. The study involved village/household level questionnaires among Yi and Naxi mountain villages, both close to and remote from road access. The initial findings indicate:

(1) In many instances there was much greater equity between men and women than was

originally expected. Among individual families, labour usually classified as women's work, such as fetching water and fuelwood, and caring for the children, was often shared between women and men on ratios of 30 : 70, 40 : 60, and 50 : 50;

(2) Some women at times, but by no means in the majority of cases, had control or equal say in deciding how the small amounts of earned cash wages should be spent;

(3) Village meetings often had active female participation. It should be noted, however, that the Naxi have traditionally been a matrilineal society, although these first three points, introduced here, are also applicable to the Yi;

(4) In very poor villages (both Naxi and Yi) young men have difficulty finding wives, since girls from neighbouring villages would not marry into poor villages;

(5) It was relatively easy to discuss birth control and use of contraceptives with village women, even involving a male outsider. We were told that, even in a very poor Yi village, most women practised artificial contraception and appeared to approve the government's policy to minimise family size (three children per family for rural national minority people and for Han people dependent on subsistent agriculture; this was reduced to two in 1993). This is counterbalanced by an experience from a previous visit by one Yi woman (1991) who explained that the prospect of the new government regulation would force her into a pattern of virtual annual abortion since she already had two small daughters and could not afford to risk a third girl. She feared that another girl, the last chance of legally adding to the family, would induce her husband to abandon her.

What can be gained from the above discussion is that there is a severe lack of data and that women's (and children's) position will likely vary enormously from family to family, from ethnic group to ethnic group, from region to region, and from country to country. But in terms of the gender division of labour, the position in northwestern Yunnan is at least flexible (Swope *et al.*, 1997).

A general conclusion, nevertheless, is in order. As it is widely assumed that women in developing countries experience the greatest pressures during periods of economic recession, or any other kind of difficulty, such as periods of poor harvest,

floods, and so on, and as they are already in a disadvantaged position, 'bad times' can be especially severe. It seems reasonable to extend this general 'truth' to the mountain situation in which all of these disadvantages will be magnified by varying degrees. The very small size and geographic fragmentation of mountain societies inhibits development of any effective political leverage for correction of these circumstances. Warfare, 'defence', independence movements, guerrilla activities (Chapter 6) will further exacerbate the inequity conditions.

So far, mainly negative aspects have been discussed. There are positive aspects, or situations where much greater gender equality within communities prevail, and where the standard of living has improved considerably over the past several decades. Perhaps the most striking example is that of the Sherpas of Khumbu, Nepal. Here, tourism is in the process of transforming the traditional life style of these high mountain dwellers. Stevens (1993) has demonstrated, however, that, given the Sherpa characteristics of adaptability and entrepreneurial skill, very considerable improvements have been won. In this instance, also, the position of the Sherpani is much closer to one of equity with their menfolk. While life style is changing, resilience of the communities, so far, has ensured preservation of local traditions, increased support for religious elements, and has furthered the development of ethnic pride and sense of identity. This is not to say that there are **no** negative aspects, and the continuing struggle of Sherpa life within a National Park and World Heritage Site needs constant attention.

Poverty and mountain children

The general conclusion that children in extremely difficult circumstances are especially vulnerable is inescapable. This is largely due to their dependence on women, who themselves are frequently in the position of second class members of society with reduced access to resources and restricted power over their own activities and well-being. Among children, the girl child is the most disadvantaged; it is worth reviewing the general assumptions that support this claim:

(1) When the mother is under pressure of increased workload, sickness, or limited access to resources, she must depend on the actual working support of her children;

(2) Since much of this heavy workload is regarded at the village level traditionally as 'women's work', such as fetching water, collecting fuelwood, care of small children, weeding, transplanting, it follows that the boy child is frequently 'protected' from being incorporated into these numerous, laborious, and time-consuming household chores;

(3) As women appropriate their daughters' assistance, their school attendance is much reduced compared to that of boys;

(4) Disproportionately low school attendance by girls, in general, relates to the lower value placed upon them compared with their male counterparts;

(5) Limited, or no education, for girls brings further disadvantages to them in adult life. Their prospects for advancement are reduced by a lower educational level in general and by illiteracy in particular. This results in severe curtailment of their access to general health care, especially birth control information, as well as other opportunities outside the village.

Bajracharya (1994) in his report, *UNICEF and the Challenge of Sustainable Livelihood: An Overview,* makes an insightful statement in reference to women and children as part of a discussion on the need for Primary Environmental Care (PEC):

> The implication of poverty and environmental deterioration are evident most dramatically in the increasingly greater time spent by women and children in the agricultural fields for meagre production of food grains and other crops and in the collection of fuelwood, fodder and drinking water. Less time is thus available for child care or self-development, or for seeking out alternate livelihood opportunities. Implementing PEC becomes, therefore, even more challenging. Innovative and imaginative measures are indeed essential. More importantly, finding sustainable solutions is possible only through proper assessment of local resource potentials and through dialogue, as well as active participation.

This statement also can be more narrowly focused and made highly relevant to the mountain-poverty nexus.

The exacerbating problems of mountain poverty in developing countries, leading to the statement that the mountain girl child is one of the most vulnerable members of humankind, can be explained in much greater detail than space permits. Nevertheless, as already mentioned, the statistical invisibility of women in many societies is especially insidious in poor mountain areas and affects any discussion of the status of children in such societies. Lack of data, compounded by the extreme variability of mountain environments, and the numerous ethnic groups that depend on them, leave us in a critical position. This position is characterised by an inability to make generalisations, and by unreliability and uncertainty regarding much of the available data. The prevailing general assumption that the mountain regions of the developing countries are universally experiencing population growth also needs re-examination. Certainly some specific areas do not fit this general assumption, which once again emphasises the danger of aggregated data (MacDonald, 1996).

It is argued, nevertheless, that the general situation of mountain children, and of females in particular, is so serious that this overall problem must be identified as demanding special treatment at the world level.

CONCLUSIONS

This chapter has introduced a considerable range of issues centred on inequalities both within mountain regions, between mountain regions, and between mountains and adjacent lowlands. The north–south disparities tend to submerge distinctly mountain problems at all levels. But inequality (or poverty) must be viewed as relative. Thus, in terms of sustainable mountain development, direct comparisons between, for instance, the Alps and the Himalaya or Andes, are not particularly helpful. Moreover, it would appear that, within Europe, the development of tourism (in all its various forms) is not a process whereby homogeneous increase in affluence within mountain regions is being attained. Bätzing *et al.* (1996) rather demonstrate that it is one factor, along with general urbanisation along communication corridors and nodes, that will tend to accentuate inequities. The German Federal and States efforts (Stadelbauer, 1991) to ameliorate inequities through extensive subsidies have set a pattern that may well spread to all European upland and mountain regions. Indeed, similar moves are being initiated, or at least recognised as necessary, in many other European countries. This may well provide an effective instruction for policy modification to address structural problems in less developed mountain countries.

So far, however, exploitation of the natural resources of these regions has tended to be initiated from the power centres of the lowlands for the benefit of the lowlanders. This tendency increases inequities between the mountains and the lowlands, and between rural and urban communities. The situation of the Daniel Palacios dam at 1855 m on the Paute River in the eastern Andes of Ecuador, development of the Arun III project (now at least temporarily suspended) in Nepal, or the socio-economic and political crisis over the Tehri Dam (and others) in the Indian Himalaya, all indicate the degree of marginalisation of mountain peoples. The very large resource development projects, whether they be mining, hydroelectricity, irrigation, flood control, could be restructured so that mountain peoples who inhabit the areas from which these resources are being extracted, can benefit. Good examples of this type of management are available from the Alps and could provide a format for transfer to other mountain regions. It is also necessary to be aware of the dynamic situation in the Alps and other mountain regions of Europe. For instance, with additional entries into the European Union, especially following the GATT Uraguay Round, there has been a move away from subsidising mountain products in preference for extending subsidies to livestock units and actual land units.

This chapter has also attempted to show that, in most of the less developed mountain regions, the vulnerable position of women, and through them, dependent children, further exacerbates the problem of how to approach a policy of sustainable development. While this problem has been progressively highlighted over the last decade, an acceptable degree of resolution is still far removed from reality. A solution, or solutions, however, is imperative. Perhaps the most damaging situation that will hinder the attainment of a degree of sustainable development in mountain regions, and especially in those regions located in developing countries, is the continuing tendency toward widening the gender inequality gap. Nevertheless, this gender gap must be viewed as part of the overall inequality characteristic of mountain regions. Provision of better infrastructure, including facilities for health care, education, and banking, will be needed as a clearly defined effort to compensate mountain communities for the services and goods they provide for the large populations of the lowlands and for the world at large.

Yet it is emphasised again that the data base upon which assessments of poverty are prepared is fraught with uncertainty. The same situation prevails with respect to literacy, GNP, and the whole range of the Physical Quality of Life Indices. There remains the equally difficult question of mountain perspective since conclusions will shift according to whose perspective dominates the assessment.

References

AKRSP Aga Khan Rural Support Programme (1992)

Bajracharya, D. (1994). *UNICEF and the Challenge of Sustainable Development: An Overview*. UNICEF: New York

Bajracharya, D., Banskota, M., Cecelski, E., Denholm, J. and Gurung, S. M. (1993). *Women and the Management of Energy, Forests, and Other Resources*. MPE Series No. 3. ICIMOD: Kathmandu

Bätzing, W., Perlik, M. and Dekleva, M. (1996). Urbanization and depopulation in the Alps (with 3 colored maps). *Mountain Research and Development*, **16**(4): 335–50

Brugger, E. A., Furrer, G., Messerli, B. and Messerli, P. (eds.) (1984). *The Transformation of Swiss Mountain Regions*. Verlag Paul Haupt: Bern/Stuttgart, p. 699

Byers, E. and Sainju, M. (1994). Mountain ecosystems and women: Opportunities for sustainability, development, and conservation. *Mountain Research and Development*, **14**(3): 213–28

Chatterji, M. (1990). *Indian women: Their health and economic productivity*. World Bank Discussion Paper, 109. The World Bank: Washington DC

Cunha, S. (1994). *Applicability of the Biosphere Reserve Model to the Pamir Mountains, Tajikistan*. Unpublished doctoral dissertation presented to University of California, Davis

CWD (Center for Women and Development) (1988). *Women's Work and Family Strategies under Conditions of Agricultural Modernization.* Unpublished Report, Kathmandu

Garcia-Ruiz, J. M. and Lasanta-Martinez, T. (1993). Land-use conflicts as a result of land-use change in the Central Spanish Pyrenees: A review. *Mountain Research and Development,* **13** (3): 295–304

Hewitt, F. (1989). Women's work, women's place: The gendered life-world of a high-mountain community in Northern Pakistan. *Mountain Research and Development,* **9** (4): 335–52

Ives, J. D. and He Yaohua (1996). Environmental and cultural change in the Yulong Xue Shan, Lijiang District, NW Yunnan, China. In Rerkasem, B. (ed.), *Transformation of Montane Mainland Southeast Asia,* Proceedings of a Conference, Chiang Mai University, Thailand, 1–16

Ives, J. D. and Messerli, B. (1989). *The Himalayan Dilemma: Reconciling Development and Conservation.* Routledge: London/New York, p. 296

MacDonald, K. I. (1996). Population change in the Upper Braldu Valley, Baltistan, Pakistan, 1900–1990: All is not as it seems. *Mountain Research and Development,* **16**(4): 351–66

Marini-Bettolo, G. B. (ed.) (1989). *Study Week on: A Modern Approach to the Protection of the Environment.* November 2–7, 1987, Pontificiae Academiae Scientiarum Scripta Varia, 75, Citta del Vaticano, p. 606

Moser, P. and Moser, W. (1986). Reflections on the MAB-6 Obergurgl Project and Tourism in an Alpine Environment. *Mountain Research and Development,* 6(2): 101–18

Netting, R. McC. (1981). *Balancing on an Alp.* Cambridge University Press: Cambridge, p. 278

Rhoades, R. E. (1985). *Traditional Potato Production and the Farmer Selection of Varieties in Eastern Nepal.* Food Systems Research Report No. 2. International Potato Centre: Lima, p. 52

Satterthwaite, D. (no date). *Children, Environment and Sustainable Development in the Third World.* Report submitted to UNICEF, manuscript, p. 126

Schuler, S. (1981). The Women of Baragoan. In *The Status of Women in Nepal,* Vol. 2, Pt. 5. CEDA: Kathmandu

Stadel, C. (1993). The Brenner Freeway (Austria–Italy): Mountain highway of controversy. *Mountain Research and Development,* 13(1): 1–17

Stadelbauer, J. (1991). Utilization and management of resources in mountain regions of the (former) Federal Republic of Germany. *Mountain Research and Development,* 11(3): 231–8

Stadelbauer, J. (1996). Resource use and environmental stress in the Central European Mittelgebirge. *Mountain Research and Development,* 16(1): 17–25

Stevens, S. F. (1993). *Claiming the High Ground. Sherpas, Subsistence, and Environmental Change in the Highest Himalaya.* University of California Press: Berkeley/Los Angeles, p. 537

Stone, P. B. (ed.) (1992). *The State of the World's Mountains: A Global Report.* Zed Books: London/New Jersey, p. 391

Swope, L., Swain, M. B., Yang, F. and Ives, J. D. (1997). Trees and tourists in SW China: Sustainable development or cultural and environmental exploitation? In Johnston, B. R. (ed.), *Life and Death Matters: Human Rights and the Environment at the End of the Millennium.* Alta Mira Press: Thousand Oaks, CA (in press)

UNCED United Nations Conference on Environment and Development (1992). *Earth Summit 1992.* Regency Press: London

United Nations (1989). *World Survey on the Role of Women in Development.* U.N.E. 89 IV 2, p. 397

United Nations (1992). *The World's Women, 1970–1990.* U.N.E. 90 XVII 3, p. 120

UNICEF (1994). *State of the World's Children Report.* Oxford University Press: Oxford/New York, p. 87

Economic and political framework for sustainability of mountain areas

5

Peter Rieder and Jörg Wyder

INTRODUCTION

Mountains have a strong influence on human living spaces world-wide. According to UN estimates, about one tenth of the world population lives in mountainous areas, even though large parts of them are subjected to ecological, economic, and social problems. A tendency toward out-migration demonstrates the marginalisation of these areas. If they are to be preserved as human living space in the long run, the demands of the Rio 1992 Earth Summit (UNCED) declarations must generate long-term, sustainable development.

This chapter will show the ways in which mountain areas are influenced by their natural features. Economic influences, especially the liberalisation of the world markets, and the implications of the political relationship of mountain regions to a particular nation or to a union of states will be considered. The principle of sustainability, emphasised within the declaration of Rio 92, will be evaluated from an economic point of view and discussed in terms of the specific conditions prevailing in mountain areas. Finally, case studies from the three major world mountain areas – the Andes, the Alps, and the Himalaya – as well as two from smaller massifs – the Caucasus and the Balkans – are presented.

GENERAL DEVELOPMENT TENDENCIES IN MOUNTAIN AREAS

Today, the mountain regions of Asia, Europe, and Latin America are on the periphery of economic development. Prospering centres of trade and industry are located at the base of these mountains, such as Lima in relation to the Andes, Lombardia, Lyon, and Munich on the edge of the Alps. Economic centres that lie further away from the mountains, for instance the Ruhr and Paris conglomerations in Europe, have no direct mountain connections. This results mainly from the course of economic development during the last centuries.

Earlier, many ancient centres were locate in mountain regions because the lower altitudes were less accessible or more dangerous at that time. For instance, the coast of Peru was a desert, the coast of Ecuador a jungle, and the lower regions of the Alps were marked by inundations and malaria (Magadino). The conquerors of Latin America settled amongst the Incas because of their secure settlements, such as Cusco, whereas the aborigines of the coastal areas were mostly hunters and gatherers.

As time went by, the mountain regions lost their importance within their own economic systems. Today, the interrelationships between different centres, and also between different countries, are more pronounced than the relationships between peripheral regions within a single country. This divergence of economic centres and peripheral regions has accelerated during the last forty years, especially in industrialised countries.

The reasons for this are the increasing importance of international trade and continuing industrialisation. Economically advantageous locations are determined by the purchasing advantages of production factors. The growing necessity of human capital for production and refinement processes promoted the development of industrial agglomerations and thus economic centres. This concentration reinforced the already weak economic position of the mountain areas.

Nevertheless, recently some mountain areas have shown a more positive trend of development. Because of the strengthened tourism economy, these regions have become places of recreation and relaxation for the urban, lowland agglomerations. Every year, the Alps, the Andes, and the Himalaya are visited by millions of tourists, especially Americans, Europeans, and Japanese, but also increasingly by peoples from the developing countries. In this way, some regions with attractive scenery or culture have become major tourist centres, and this has led to a direct economic dependency on the lowlands. A decline in prosperity in these lowland centres could diminish purchasing power and so threaten the tourist economy of the mountain regions (see Chapter 12).

Differences in history, culture, and the visual landscape have affected the mode of development in each mountain region. Where there is no tourism, agriculture and forestry have remained dominant. These peripheral mountain areas become progressively more disadvantaged as industry is attracted to locations that are more economically favourable This tendency toward development, or lack of development, in mountain areas has economic roots which will now be assessed in more detail.

EFFECT OF ECONOMIC FACTORS ON MOUNTAIN AREAS

Land, labour, and capital are the most important factors of production. They entail costs that are determined by scarcity of the factor, and the absolute and relative scarcity of these factors usually varies from one country to the other, and also between urban agglomerations and mountain areas within the same country. These differences can be discussed using the concepts of absolute and comparative cost advantages.

The concept of the absolute cost advantage compares the total costs of two regions and two products. This concept is best explained by assuming that each product can be produced at lower cost in one region than in another. In the Alpine region, for example, this is the case if grain is grown more cheaply in the lowlands and milk is produced in the mountain areas. Thus, producers in each area can develop co-operative agreements and compete successfully on the market with their products.

Such a situation can develop if the production costs for some items are lower in the mountains than in the lowlands. Nevertheless, goods with an absolute cost advantage produced in mountain areas are rarely found on the international market today, with the exception of those tied to specific mountain locations, such as winter tourism, the mining industry, collection of special spices, and so on. Therefore, the increasingly poor economic situation of mountain areas is related to short-term exploitation of the resources which has caused soil degradation and loss of biodiversity. This problem

has been recognised internationally and is noted in Agenda 21 of UNCED.

The concept of comparative cost advantage can be utilised when two goods are produced in two regions and, although there is an absolute cost disadvantage for both goods in one of them, they are not equally significant. Let us assume that the production costs in the lowland agglomeration (a region with absolute cost advantage) are 100 units for both products A and B, and in the mountain area (region with absolute cost disadvantage) the production costs are 150 units for product A and 200 units for product B. If a company decides to produce in a mountain area despite absolute cost disadvantages – for whatever reason – then it will choose product A. The absolute cost disadvantage in comparison to the competitors in the agglomeration is 50 units; for product B the absolute disadvantage is 100 units. The comparative cost advantage for the mountain area results from the comparison of the cost disadvantages between products A and B. In this case, product A has the comparative cost advantage. This situation may have bitter consequences for workers in the mountain areas, since the producing enterprises can survive only if they lower the production costs and thus reduce the salaries so they are significantly lower than in the lowland agglomeration areas.

The concept of comparative cost advantages is built upon the assumption that the production factors, especially labour, are barely or not at all mobile. If this assumption is investigated in different mountain areas of the world over a period of time, two different migration patterns can be described:

- In the mountains of Europe, especially the Alps, over the past hundred years there has been a strong out-migration from the mountains to the agglomeration areas, both inside and outside the home country. At the end of the nineteenth century thousands of mountain people emigrated to America. Therefore, population pressure and poverty were reduced. However, while the residual mountain population benefited from this migration, especially by gaining access to more land, the emigrants had a very insecure destiny. In contrast, many of those who left the mountains after World War II found adequate job opportunities in the lowland agglomerations of their own country, where they mostly occupied unskilled positions in the industrial or service sectors.

- Many Latin Americans and Asians migrated to the lowland agglomerations of their own countries only after decolonialisation; the beginning of industrialisation after World War II induced many people to move to these economic centres. This migration was not caused solely by differences in opportunities between the mountains and the lowland, but also by the general contrast between amenities available in rural compared with urban areas. Some people from mountain areas with weak national economies migrated to the economic centres of other industrial countries. In some countries this has led to social conflicts (see Chapter 6).

The migration of people across national frontiers will probably become more difficult in the future due to economic and political factors. Furthermore, in many cases, the expectations of past generations to find opportunities for a new successful existence were not fulfilled. This fact is illustrated impressively by the world-wide growth of urban slums.

THE EFFECT OF THE WORLD MARKET DEVELOPMENT ON MOUNTAIN AREAS

Since the Uruguay Round of General Agreement on Tariffs and Trade (GATT) and the founding of the World Trade Organization (WTO) have liberalised world markets in a decisive way, it is important to investigate how the mountain areas will be affected.

The liberalisation of the world market has a significant impact because of the large number of existing tariffs and other obstacles to trade, with which single states protected their national economic sectors from foreign competitors. These manipulations of the economy to the advantage of special interest groups, for example producers, resulted in a loss of prosperity for other groups, mostly for consumers. High prices for protected goods and production factors and low, distorted prices for export goods were the consequence.

Therefore, the liberalisation of the markets, as invoked by the Uruguay agreements, is encouraging new price levels which will lead to a gain in prosperity for the national economies. On the one hand, products with an absolute cost disadvantage

inside a country can be purchased more cheaply from abroad; on the other hand, products with an absolute cost advantage can be exported more easily. Since all countries are reducing their customs and excise duties, each national economy will be able to import and export more goods according to absolute cost advantages and disadvantages. Thus, the volume of trade on the world market will increase. The effects of this liberalisation of world trade are of special interest and will determine whether agriculture will remain an important support base of mountain economies. The following factors play a central role:

(1) Mountain regions seldom have any absolute cost advantages, as outlined above, either for industrial or for agricultural products; if goods are produced at an absolute cost disadvantage, local workers will have lower salaries than those equally qualified workers (including emigrants) in more favourable lowlands.

(2) With a lower income, the population of such mountain regions will not generate extensive savings; imported goods will be expensive in proportion to income. This lack of savings will severely limit the necessary net investments in industry and agriculture and the economy in the mountains will stagnate, or even decline.

(3) Because of the stagnating economy, mountain areas will remain mainly agrarian. Following the liberalisation policies of the Uruguay Round as applied to the agrarian sector, these areas are especially vulnerable. Theoretically, prices on the world agrarian markets should rise since the export subventions in industrial countries are being reduced (Alexandratos, 1995); nevertheless, the trends of agricultural prices suggests that they will decline even further. The reasons for this are the return to higher rates of production that were unrealised or under-realised until now in the agrarian regions of North America and Western Europe, and also because of a rise in production, especially in Eastern European countries. These price expectations have ominous implications for mountain areas.

(4) From an international perspective, the mountain areas will not be able to compete because of their natural disadvantages. Due to lower import prices, some states will have to provide more subsidies to maintain the structure of the disadvantaged mountain regions; otherwise,

out-migration from mountains to industrial areas will definitely increase.

The factors discussed above will differ from one mountain region to another, but population growth is a predominant factor. For example, in the Western Alps the consequences of GATT will be of less importance because population growth is low and a tourism industry has developed together with accelerating amenity migration (Chapter 12). In other mountain areas of the world, these problems will be much more severe, because agricultural workers cannot migrate so easily to other economic sectors; this is the case for mountain areas of the former Eastern Block countries and for large parts of the Andes and the Himalaya. Subsistence economy will continue to play a central role in these regions.

INFLUENCE OF POLITICAL AFFILIATION ON MOUNTAIN AREAS

Mountain regions world-wide are tied to different political unities. This is of decisive importance in light of trends towards liberalisation. The following three categories are distinguished:

(1) Mountain regions that are a part of strong national economies: for example, Switzerland, Austria, France, Italy – the last three are also members of the European Union.
(2) Mountain regions that belong to weak national economies: for example, Peru, Ecuador, Guatemala, Mexico, Russia, Poland, Pakistan, India.
(3) Mountain regions that constitute an independent national economy in themselves, for example, Bhutan, Nepal, and Andorra.

States in the first category are already involved in manifold and specific policy developments that favour their disadvantaged mountain areas. Their populations are willing to maintain the special support to mountain regions and there is no reason to assume that this process will be eliminated in the future. It is significant that the European Union (EU) has established extended programmes to promote the economy of disadvantaged regions during recent years (see Chapter 4). Because of a persistent demand for sports and recreation, summer and winter tourism in these mountain areas is likely to dominate; also, the exploitation of their hydropower resources will certainly gain even more importance. Nevertheless,

at the scale of the mountain commune, significant economic divergence should be anticipated (Bätzing *et al.*, 1996).

States in the second category face severe problems concerning their mountain regions. Economic support funds are used mainly in the principle national development centres – the lowlands and industrialised agglomerations. The danger is that the mountain areas will be marginalised even further since they stand in the shadow of the economic centres. If these regions cannot avert the threat by utilising their resources as, for example, Western Europe did with the increased development of tourism and waterpower, then out-migration will be inevitable, especially where there is a high population growth rate. Where economic and social problems in such mountain regions cannot be alleviated ethnic problems often arise, such as the Kurdish problem in Turkey and Iraq, and the former Shining Path uprising in Peru (see Chapter 6).

There are only a very few countries belonging to the third category and their national economies are relatively small and their governments strongly support the mountain populations, either directly or through foreign aid. Because of their absolute cost disadvantages, these countries are amongst the poorest of the world and, therefore, rely primarily on international development funds. These nations will not become internationally competitive by their own means (Kinlay, 1995).

ON THE SUSTAINABILITY CLAIM FOR MOUNTAIN AREAS ACCORDING TO RIO 1992

As documented so far, the future economic prospects for mountain areas world-wide are not very favourable, with some exceptions in Western Europe and North America. This situation is now brought into sharp focus by the demand for long-term sustainable exploitation of mountain regions (UNCED). Research and politics are challenged to elaborate and realise long-term solutions. If this does not happen, mountain areas are in danger of short-term exploitation as a first step, and abandonment as a second.

The sustainability concept of the Rio 1992 declaration contains the following three components:

(1) Ecology;
(2) Economy;
(3) Social issues.

The underlying values of these three components must be realised, according to Agenda 21, to the highest measure possible.

(1) The **ecological** component represents the natural (i.e. physical) state of the environment which should not be degraded. This implies that erosion, increased slope instability, loss of soil fertility, increase in chemical residues, loss of biodiversity, excessive forest cutting, overgrazing, soil, water, and air pollution, and over-use, should be prevented.

(2) The **economic** component represents a productive economy aided by technological know-how, a modern infrastructure, and other factors. Local industries must be competitive and offer employees a satisfactory salary. Such business structures must ensure that unemployment will not increase and that the economy will strengthen.

(3) The **social** component comprises the different societal factors as, for example, well-balanced demographic structures, appropriate real incomes with sufficient purchasing power, secure working and living conditions, social institutions and, finally, a sense of value in the daily life of the population – a life without war, oppression, and need, and also a viable local culture and language.

These requirements are to be measured by means of indicators over a long period of time, patterned after the concept of The World Bank (1995). In this way, information on the specific changes in sustainability can be recorded. The values of the three registered indicator groups can be related to each other in a triangle as illustrated in Figure 5.1. The shape of the triangle expresses the respective condition of a qualitatively evaluated sustainability.

A change in degree of sustainability can be determined using this theoretical construct. The triangle ABC, in Figure 5.1, shows such a change, in which the values of the three indicators rotate on the three axes so that an increase in one indicator causes the other indicators to decrease. The sustainability, expressed by the surface of the triangle, hardly improves because of this trade-off. Thus strategies must be found that improve all three sustainability components simultaneously. Such a win-win situation is illustrated by the triangle DEF in Figure 5.1. If one attempts continually to influence the three components in a positive way, a progressive improvement of the overall sustainability evolves.

Even in this simplified form the concept is valuable since the degree of sustainability can be compared over time, e.g. every ten years. The following five case studies are introduced according to this concept of sustainability. In this way, the relative situations of different regions can be compared. Also, the main problems, as well as the specific measures which are to be introduced in each mountain region, become clearly visible. Ecological, economic, and social obstacles that hinder any improvement in sustainability are illustrated.

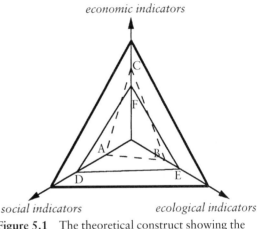

Figure 5.1 The theoretical construct showing the change of sustainability due to social, economic, and ecological indicators

Plate 5.1 Typical mountain village in the region of Puka, in the north of Albania

Plate 5.2 and 5.3 Mountain valley in the north of Albania: small structured fields with irrigation in between steep slopes with a high erosional potential

Box 5.1 Case study in Bhutan

Plate 5.4 The Dzong Tongsa in an ecologically over-used landscape (Photograph: P. Rieder)

Bhutan is a nation located in the Himalayan mountains, with the exception of a narrow strip of land along its southern border – the Terai. It consists largely of steep valleys and ridges between 300 and 6000 m asl. The total area of 40 000 km^2 contains land that ranges from tropical lowland to high alpine and polar-type uplands and mountains. About 60 percent of the surface is covered by forests and much of the remainder is agriculturally unproductive. Thus, with a population of between 800 000 and 1.4 million people the population density is relatively high in terms of fertile agricultural land available, especially in the east of the country.

About 85 percent of the Bhutanese people still depend upon subsistence agriculture and produce about 42 percent of the GDP, according to official statistics based upon monetary goods and services. Industry produces about 9 percent, electricity about 8.3 percent, construction and trade 6.3 percent each, transport and communications 8 percent, and the entire service sector 19 percent (Dorjee, 1995). For years, exports have amounted to only two thirds of the cost of imports, so that every year the country has shown a strongly negative trade balance. This balance, therefore, is equalised each year with funds from outside (e.g. foreign financial support). Nevertheless, there is a real, and large, potential for the sale of hydroelectricity, although there are serious associated problems, common to those of any small country with its only market provided by a powerful neighbour, in this case, India (Dhakal, 1990).

Outline of economic, ecological, and social indicators

Economic indicators

Bhutan is in an early phase of general economic development. The still dominant agrarian sector shows low work productivity (due to very poor mechanisation). It follows that salaries are low, or virtually non-existent in the overwhelming subsistence sector, and savings that could be used for investments in the economy are negligible. Therefore, the development of the country heavily depends on financial support from abroad (e.g. The World Bank, International Monetary Fund – IMF). Because of the mountainous location the infrastructure is weakly developed. Transport and transaction costs for involvement in international trade are relatively high. Due to the high percentage of labour in the agricultural sector, unemployment is officially very low. In reality, however, the official figures disguise relative under-employment in the primary sector, as in many other developing countries, namely because there is no urban development to facilitate out-migration from the rural mountains. The purchasing power on the national market is low, and thus the demand for new workers in the industrial and service sectors is also low. Tourism is beginning to develop, but it is wisely kept to a level which impedes any cultural exploitation of the country. The location offers few possibilities abroad and apart from a few emigrants to the countries of the Arabian Peninsula, work in neighbouring countries, such as China or India, is not available.

Ecological indicators

The extensive area of forest has not yet been effectively exploited. Forest resources in mountainous areas are exceedingly expensive to extract and in addition, about 50 percent of the timber which is cut cannot be used for industrial purposes. Shifting cultivation still occurs near settlements and this results in erosion on the steep slopes. Agricultural practices also lead to soil degradation and much good land is over-grazed.

Social indicators

The social indicators in Bhutan have relatively low values. The literacy rate is beginning to rise, however, although it is still only 51 percent for adult males and 25 percent for adult females. Under 5 mortality and infant mortality are amongst the worst in the world (see Chapter 4, Tables 4.1 and 4.4). In many places, nutrition is highly unsatisfactory, so that the effects of deficiencies are very common and general health is poor. The ratio of physicians to the number of people is low and other regional medical services can be reached only on foot from the scattered mountain settlements. The country has a high birth rate, estimated at 2 percent per annum.

In conclusion, it is apparent that Bhutan suffers the disadvantages of a mountain country in relation to all three groups of indicators. In contrast to the Alps, the natural disadvantages of high altitude are combined with a low standard of development of the national economy. The entire country lacks resources and there are no regions that have a viable economy to attract out-migration from the mountains. These negative factors introduce many difficulties. It would be an international cultural calamity if the traditional values of this unique country, with its many monasteries, distinctive architectural style, and spectacular scenery, were to be sacrificed. One of the most recent problems has been the acceleration of serious ethnic tensions resulting in the movement of large numbers of refugees from Bhutan to eastern Nepal.

Economic and political constraints for sustainability

In the section above the constraints facing Bhutan's economy were outlined. The most binding constraint, however, lies in the lack of opportunity for farmers to migrate in order to find waged work in the secondary sector, such as business and small industries. Therefore, the number of households in rural areas steadily increases and leads to further over-exploitation of the fertile soils and increasing rivalry between inhabitants. Pressures on all three factors are culminating in widespread unrest at the village level. Sustainability is now becoming increasingly at risk in terms of the ecological, economic, and social components of development.

Attention must be given to the very severe constraints on mobility of the population. For political reasons, there are few opportunities for employment in other neighbouring countries such as China, India, and Pakistan. Therefore, solutions must be found based on purely national concepts. All historical experiences, however, suggest that the one solution to overcome this early phase of economic development, is to induce off-farm migration. This may be attained by enhancing opportunities in the secondary sector: by introducing businesses and small-scale industries linked to agricultural production. Governments and external donors should focus their efforts on the promotion of semi-commercialisation, which will draw people away from the agrarian sector while increasing the value of local agricultural raw products and local trade. Investment funds must be introduced to establish installations for processing, refinement, storage, conservation, vehicle maintenance, and infrastructure in rural areas. Only through a reduction in the over-exploitation of the fertile agricultural areas can all three components of sustainability gradually become positive.

This case study of Bhutan, a small mountain country, demonstrates how important it is to precisely analyse the perceived problems in order to draw the correct conclusions for action, always considering the need to improve all components of sustainability simultaneously, and to prevent short-term improvement in one component at the expense of others.

Source: Thomas Bernet and Peter Rieder,
Institut für Agrarwirtschaft, ETH Zurich

Box 5.2 Case study of Encānada in the Peruvian Andes

The Andean eco-region is one of the most extensive mountain regions in the world and its 200 million ha stretch across the territory of seven different South American countries, from Venezuela to Argentina and Chile.

In Peru, the Andes extend from north to south, a distance of 2300 km, and occupy about 39 million ha of land above 1500 m asl. More than 7 million people live in the Peruvian Andes and their livelihood depends mainly upon agriculture.

This mountain sub-region is considered representative of the Central Andes, and it is also the centre of origin of the most important civilisations in prehispanic times, including the Tiahuanaco and Inca empires. The domestication of many important crops: tubers, such as potatoes, oca, olluco, and mashua; grains, such as quiñoa and amaranth; legumes such as lupins; fruits and medicinal plants; took place here and today these are produced throughout the world. The prehispanic Andean inhabitants also domesticated animals, such as camelids (alpacas and llamas) for transport and wool, and guinea pigs and ducks for human consumption. They also initiated the use of products made of refined fibres and leather.

An explanation for this widespread adaptation and use of natural resources for domestic purposes was the existence of many agro-ecological altitudinal belts (niches) and the requirement to satisfy basic human needs in an environment where a single niche would prove extremely marginal for community survival. It is assumed that, at the time of the Spanish incursion, the level of nutrition was quite satisfactory due to the highly successful exploitation of the wide range of resources available. The high level of culture under both the Tiahuanaco and Inca systems was, in many ways, comparable to that of Europe from the twelfth to the fifteenth century.

Outline of the economic, ecological, and social indicators

Economic and ecological indicators

Today, Andean peoples have one of the lowest standards of living in the entire continent; malnutrition dramatically affects more than half of all children. The ecological problems, such as soil erosion, decreasing yields in agriculture, as well as the presence of heavily over-grazed areas, are serious but vary considerably from place to place. On the one hand, there are still areas with potential for more intensified use of resources; on the other, there are areas where drastic measures to alleviate environmental degradation are urgently required.

To understand these variable conditions, it is useful to examine the actual land-use pattern and the socio-economic characteristics of the different units of production. However, to achieve an adequate evaluation, it will be helpful to define the conditions prevailing in one representative area, such as a single watershed. Each of the numerous watersheds within the Peruvian mountain region displays a great variety of cultivars, including more than 100 varieties of potatoes and other crops, as well as diverse species of livestock, such as llama and alpaca, and also cattle, sheep, goats, horses, and donkeys, depending on the altitude as well as the available soil and water resources.

It is important, therefore, to understand the highly heterogeneous environments that comprise this high altitude mountain region if a programme of sustainable development is to be introduced. Most development projects in this area so far have focused on the use of a few high yielding varieties, and on livestock that requires a quality of forage not likely to be obtained under local conditions.

Social indicators

Economic conditions in the Andean region in Peru are also variable. Most people have no access to basic facilities such as safe drinking water, electricity, or health services. The Peruvian government is making a great effort to improve the level of education as a basic first step. Yet, education is lacking in quantity and quality. The salaries of teachers are low and educational programmes, being very similar to those in urban communities, are inappropriate and inadequate. Thus, much more remains to be accomplished.

LA ENCAÑADA

La Encañada is selected here as a case study to assist in understanding problems of the much wider Andean region. It embraces a watershed with considerable biological diversity, variability in climatic and soil conditions, as well as the range of traditional knowledge in use of natural resources. It is in the northern Andes of Peru and is one of the more than 300 micro-watersheds distributed over an area that is eight times the size of Switzerland.

The Encañada is a district located in the *Departement* of Cajamarca; it is 12 000 ha in extent and populated by 1100 families of whom about 90 percent are dependent on agricultural production. It consists mainly of small and middle size farms located in four different agro-ecological belts: the first, called locally the 'quechua zone', from 2500 to 3100 m asl and characterised by pastures for dairy cattle, and horticultural crops; the second, the 'low hillside belt' with maize and low altitude potato varieties as the main source of income; the third, the 'high hillside belt', with cultivation of potatoes, other tubers such as oca and olluco, and also lupin, quiñoa, and cereals, as well as annual forage species; and fourth, the 'jalca belt', located above 3400 m and characterised by extensive pastures, production of native potatoes, together with medicinal plants from the very highest areas which do not have a complete vegetation cover.

The size of the farms also varies from less than one to more than 25 ha. Thus, income varies not only with farm size, but also with location (i.e. altitude), access to water, and soil quality. At present, milk production has become the main source of income for the majority of families and the area dedicated to forage has steadily increased. Milk is the only agricultural product that guarantees a secure annual income due to the presence of a milk processing company (a Nestlé subsidiary) located in the area. A detailed study of the household economy shows that the total annual income of the farmers varies from US$ 380 to US$ 5000, with the exception of one case where the income was more than US$ 8000 due to off-farm activities, such as handicrafts and the production of wool blankets.

CONCLUSION

Rural development projects usually do not consider the scale of diversity exhibited by La Encañada; they tend to propose few alternatives that satisfy local requirements. Thus, any integrated development proposal should not only focus on the need to augment agricultural production, it must also emphasise activities which meet the overall needs of mountainous regions in the Andes and ideally also those of other mountain regions of the world.

Based on three years' experience with a project undertaken in La Encañada, some conclusions can be presented:

(1) Full participation of the local people is required if sustainable goals are to be met.
(2) Local biodiversity and resources must be identified and taken into account so that their effective utilisation can be anticipated.
(3) Factors such as the local climatic conditions and fluctuations in agricultural prices are external factors that must be considered before adequate programme goals can be defined.
(4) Better levels of human nutrition, as well as the availability of training facilities would more likely ensure significant acceptance, involvement, and participation of the local people.

Source: Mario E. Tapia, Centro Internacional de la Papa, Lima

Box 5.3   Case study of Pays d'Enhaut, Switzerland

Switzerland is an inland state in Western Europe with an area of 41 000 km² and ranging in altitude from 200 to 4600 m. Seventy percent of the area lies in the alpine and sub-alpine zone and in the Jura – all hill and mountain land. About 25 percent of the land area is unproductive (lakes, rivers, infrastructure, rocks, permanent ice and snow, and badlands), 25 percent is occupied by forest, and a further 25 percent by natural pastures; the area devoted to agriculture is 25 percent of the total.

The country is situated in the humid temperate climatic zone and, because of its mountain character and geographical location, it comes under the influence of several different air mass types: Mediterranean in the south; Atlantic oceanic in the midlands; and continental in the alpine valleys, progressing to arctic conditions at higher altitudes.

Despite a lack of raw material resources, Switzerland has attained a position amongst the highest developed countries of the world. About half of the seven million inhabitants are employed: about 4 percent in the primary sector; 29 percent in the secondary sector; and 67 percent in the tertiary sector. Domestic agricultural production satisfies about 60 percent of Swiss food requirements.

Plate 5.5 The multi functional landscape of Pays d'Enhaut, Switzerland, has developed over centuries of use (Photograph: B. Messerli)

Regardless of its traditionally negative trade situation, the Swiss national balance is mostly positive due to profits from banking and tourism. Because of the absolute cost disadvantages, especially in agriculture, Switzerland has adopted tight net of tariff and non-tariff measures to protect its domestic farmers. Otherwise, Switzerland is a very liberal country with insignificant customs duties applied to non-agrarian products and services. By signing the Uruguay agreement (GATT) and by joining the WTO, Switzerland moved to extend its liberalisation into the realm of agriculture.

About one quarter of the Swiss population lives in mountain areas. Already by the 1970s Switzerland had introduced legislation to favour these areas (Federal law on investment aid for mountain areas – LIM). Presently, funds are provided for burden–benefit equalisation in mountain areas and there are measures for the promotion of tourism and development infrastructure (transportation, communication, construction), and also for agriculture.

Pays d'Enhaut is used here as a case study to demonstrate the problems of sustainable development faced by many of the Swiss mountain districts.

PAYS D'ENHAUT

The Pays d'Enhaut region lies in the lower Alps in the western part of Switzerland and has an area of 182 km². It extends between 800 and 2500 m asl; the climate is rather mild with about 1400 mm of precipitation in the form of rain and snow, equally distributed over the year; about a third is occupied by forest, of which 46 percent can be exploited for agriculture, including alpine pastures and meadows, with pastures in the valleys. Eighteen percent is infertile and the remainder is occupied by settlements.

The region comprises three municipalities with a total population of 4520. A local initiative within the scope of the Federal Aid for Mountain Areas (LIM), founded the 'Association for the Development of the Pays d'Enhaut' (ADPE) in 1975, with the aim of developing the region in a sustainable and harmonious way. Its members come from different economic and social backgrounds of the three municipalities. The specific goals were:

- stabilisation of the population, i.e. to discourage out-migration;
- maintenance of agricultural activities and creation of additional employment opportunities in trade and tourism;
- improvement of infrastructure and living conditions;
- reduction of environmental pollution to the lowest practical minimum, preservation of the traditional landscape, and appropriate utilisation of the available land surface.

The ADPE became a forum that today co-ordinates, or initiates, all efforts to reduce out-migration and to raise the living standard of the region. The Pays d'Enhaut was also one of the survey areas of the UNESCO Man and the Biosphere (MAB) Programme between 1979 and 1984 (see Messerli, P. 1989; Price, 1995).

Economic, ecological, and social indicators

Economic indicators

Pays d'Enhaut is a typical agro-touristic mountain area of Switzerland. There are about 1950 classifications of employment which are distributed as follows: 17 percent agriculture and forestry; 23 percent construction and trade; 21 percent tourism and transport; 15 percent business and repairs; 7 percent banking, insurance, and other services; and 17 percent education/health, and administration. About 10 percent of the employed population work outside the region and about 5 percent of the jobs in the region are occupied by workers from outside. Unemployment is 4 percent, the Swiss national average.

The region therefore, demonstrates a well-balanced distribution of the different economic sectors. The primary sector has experienced a reduction in the number of jobs (27 percent in 1980), whereas positive development occurred in construction, trade, and non-tourist service sectors. Employment in tourism has increased only slightly. Tourism is a rather traditional economic activity of the region, beginning at the turn of this century when the railways, and subsequently, the first hotels were built;

overall 121 jobs have been added to this sector since 1980.

The prerequisites for this positive development were, on the one hand, financial credits and monetary funds from the state and, on the other, the intensive efforts directed by ADPE. Nevertheless, the positive developments and the relatively favourable economic structure do not fully compensate for the weak points that include: the small percentage of activities with a high added value; the strong dependency on the secondary sector (construction); and the weak development of tourism. Until now the potential for processing agricultural products has not been fully realised and, through the liberalisation of the market, the pressure to lower the prices of agricultural products will increase. Moreover, there are few private entrepreneurs willing to take risks. Another obstacle to economic innovation is the lack of available industrial land.

Ecological indicators

Since the 1970s, Switzerland has systematically extended and tightened its legislation concerning protection of the environment. Soil, air, water, species, and biotopes have been protected by Federal and cantonal funding programmes. In settlements, polluted water is usually purified, garbage is separated before collection and, in part, recycled. In terms of agricultural exploitation, there are concerns about the extent of fertiliser use, and also subsidy for farming on slopes. In a global sense, the landscape has retained its typical and harmonious character but this impression cannot mask a pronounced loss of ecological diversity. Tall fruit trees, hedges, stone walls, and natural river beds are rapidly disappearing. The open landscape of the valley has been reduced by the expansion of forest cover, settlements, and tourist infrastructure. The need to conserve the natural space is recognised but it is becoming more difficult to reconcile this with regional policy and private interests.

Social indicators

Infrastructure for improving the living standards in the Pays d'Enhaut has been extensively developed over the last 15 years. Good transportation facilities (to larger economic centres, and also to remote farms), a satisfactory educational infrastructure (at least for lower school levels) and a favourable supply of social and medical services, all ensure a high level of social sustainability. This can be guaranteed only because of inter-regional financial transfers and Federal, as well as cantonal, financial support. The parity income of the region, about 70 percent, is clearly below the national and cantonal averages. The income of the municipalities from taxes is relatively low. A steady increase in expenses, including needs for investment, has greatly reduced the self-reliance of the municipalities during recent years.

In relation to the Swiss average, the quality of life in the Pays d'Enhaut has remained attractive, as shown by the record of demographic change. Following 50 years of out-migration up to 1987, the population has now started to increase and today it equals that of the early 1970s. A detailed analysis of the demographic structure shows a conspicuous growth in the above-60 age group, while the below-20 age group has declined. This partly results from more general trends, such as the baby-boom of the 1960s and fewer children born in the 1970s due to changes in family structure, and partly from out-migration. The benefits of a good infrastructure, especially improved medical facilities, have induced older people to migrate into the region and compensate for the loss of young people. However, this has led to a distortion in the overall demographic structure and thus to a weakening of the vitality of the region in the longer term.

The broad activities of private organisations in cultural and social affairs, and involvement of all social and economic groups in decision making for regional development, indicate an encouraging integration of the population of Pays d'Enhaut.

CONCLUSIONS

Due to intensive promotion and financial support for regional development during the last 15 years, the Pays d'Enhaut has attained a high level of sustainability in all sectors. Demographic changes, the weak points in the economy (especially the unsatisfactory conditions for new businesses in terms of access to national or international markets, which contributes to a high added value), and growing conflict of interests regarding conservation of nature, together demonstrate the narrow margin for action and the delicate regional balance that prevails. In the future, further efforts will be necessary to enhance the region's attractiveness, and continued financial support will be essential to maintain the vitality of the region within the framework of constantly changing conditions.

Source: Jörg Wyder and Jeannine Bossert,
Schweizerische Arbeitsgemeinschaft für
die Berggebiete (SAB), Brugg

Box 5.4   Case study of North Ossetia–Alania (Caucasus)

The Republic of North Ossetia–Alania is situated on the northeastern macro-slope of the Central Caucasus. Compared to other mountain republics of the Northern Caucasus – members of the Russian Federation – it is relatively small (8000 km²), has a high population density (82.3/km²), and an unusual geopolitical border situation. Two highways cross this region connecting Russia and Trans-Caucasia: the Military Georgian highway (closed in winter) and the Military Ossetian highway (year-round).

Despite its small dimensions the Republic is characterised by an exceptional environmental, ethnic, and cultural variety. Eighty-eight percent of the area is covered by uplands and foothills of the Caucasus, while almost 30 percent is located above 2000 m asl. Altitude ranges from 131 m in the north to 4870 m in the south. The natural landscapes include a full spectrum from lowland steppes and forest-prairies in the north to alpine grasslands and mountain glaciers (covering 1.9 percent of the area) in the south. Sub-alpine forests occupy 22 percent of the area.

Natural resources

North Ossetia is rich in mineral resources; the most valuable are polymetallic ores which have been developed since the end of the nineteenth century and are located in the southern (upland) zone of the Republic. The foothill zone has small oil fields which have been developed since 1960. Granite, marble, and limestone are exploited to a limited extent. A resource of especially high value are the hot water springs with various mineral properties which are actively utilised as health spas and resorts.

Population

More than half of the population are Ossetians; a third are Russians; and other nationalities include Armenians, Georgians, Ukrainians, and Kumyks. Ossetia is one of the most ancient nations of the Caucasus. It was founded by the aboriginal peoples of the Northern Caucasus and invaders who belonged to the Persian language group of Scythians, Sarmatians, and Alans, who arrived in the first century AD. The powerful Alanian state existed until the thirteenth century. It was eventually destroyed by the Mongols led by Genghiz Khan, and the Ossetian population was driven from the fertile Cis-Caucasian plains into the mountain valleys of the Central Caucasus. As a result, four large mountain communities were formed on the northern slopes: the Digori, Alaghir, Kurtatin, and Tagaur communities. Smaller groups settled behind the main Caucasian ridge in present-day Georgia.

Since the middle of the nineteenth century, Ossetia has been actively negotiating contacts with Russia, promoted by a massive migration of Ossetians to the northern foothills and the lowlands. In addition, Christian influence increased significantly, originally spreading from Byzantia in the sixth and seventh centuries, and Georgia. An Islamic influence became significant during the seventeenth and eighteenth centuries from neighbouring Kabarda. It is the minority religion there.

At present the Republic of North Ossetia–Alania has a population of 658 000, of whom 69 percent live in urban areas. The employment structure is as follows: 12.2 percent in agriculture; 28.8 percent in the industrial sector; and 28 percent in services, over half of whom are in culture and science. The recorded level of unemployment was 4.1 percent in 1995, which is almost twice as high as Russia's claimed average of 2.4 percent.

Economy

The leading industries are engineering (production of instruments and agricultural tools), mining, non-ferrous metallurgy, food processing, textiles, glassware and porcelain, timber processing, and construction materials. Nearly all the largest enterprises are located in the northern foothills and lowlands. Similarly, agricultural production is also concentrated in the northern lowlands. Major crops are cereals (wheat, barley, maize), potatoes, fruit, and grapes; animal husbandry is mostly sheep-breeding and production of pork. Cattle are pastured in the upland areas.

The Republic produces a rather modest percentage of Russia's total output: 0.11 percent of its industrial production and 0.54 percent of its agricultural output. Fifty-one percent of the Republic's budget is supported by Federal grants, compared with the Russian average of 9 percent; 78 percent of the taxes collected are allocated to the budget of the Republic, whereas 61 percent is the average for Russia.

Compared with other regions of Russia and the Northern Caucasus, the Republic of North Ossetia has a relatively high density of highways: 289 km/1000 km² while the average for Russia is 26 km.

Table 5.1 The population structure of North Ossetia

Altitudinal zone	area		population		density pers./km²
	km²	%	1000 pers.	%	
Lowlands up to 200 m	770	9.6	74.3	11.7	96.4
Foothills 200–500 m	1372	17.1	70.2	11.1	51.2
Low mountains 500–1000 m	2607	32.6	476.1	75.3	182.8
Middle mountains 1000–2000 m	1473	18.4	11.1	1.8	7.5
High mountains above 2000 m	1771	22.3	0.7	0.1	0.4

Problems of development – economic and social indicators

One of the most acute problems in North Ossetia is depopulation of the mountain areas and the related problem of spatial polarisation along the 'highland-lowland' axis. The process of out-migration of the mountain people to the foothill and lowland areas has occurred in several historical stages. The most active surge took place in the second half of the nineteenth century after the Caucasus was incorporated into Russia. By 1900, two-thirds of the Ossetian upland population had migrated to the lowlands. The major reasons were: shortage of land, industrial development and new job opportunities in the urban centres of the lowland such as Vladikavkaz and Mozdok, and the development of large-scale commercial agriculture and better social infrastructure in the centres. During the Soviet period these tendencies persisted, supported by the state in accordance with the doctrine of centralised planning and management that emphasised industrial development and collectivisation of farms. The modern settlement structure in North Ossetia is given in Table 5.1 (Badov, 1993).

The recent migration of the population to the lowlands in the 1990s resulted in a concentration of about 75 percent of the Republic's population on 32 percent of its territory. The high mountain zone, which accounts for about 20 percent of the territory, was inhabited by only 0.1 percent of the total population. The existing settlement system has several features producing a negative effect for sustainable development:

- marginalisation of the mountain communities and destruction of the ethnic and cultural traditions and values;
- aggravation of the peripheral features of mountain regions due to the 'allocation' of the economic resources to the existing settlement structure;
- deterioration of the natural resources and the environment; this process is taking place both in the lowlands which are 'overloaded' by pollutants, soil depletion, and so on, and in the uplands which are 'inadequately managed' and most sensitive to inappropriate legislation for there are no land-use laws nor local self-government.

Environmental problems – ecological indicators

In North Ossetia industrial development and a high population density have led to a typical spectrum of the problems of a mountain country:

(1) **Atmospheric and water pollution:** over 700 enterprises in the Republic and 7500 point sources annually emit about 40 000 tonnes of pollutants to the atmosphere. Most severe is the situation around Vladikavkaz and in the lowland Mozdok region. Cars add about 75 000 tonnes of pollutants each year. Water pollution is mostly produced by industrial enterprises and by inadequately processed domestic sewage; annually the rivers receive about 40 000 tonnes of pollutants. There is a threat that the major rivers – the Terek, Ardon, and Kambeleevka – will be turned into collectors of industrial and municipal sewage.

(2) **Storage and utilisation of industrial wastes:** in 1995 the amount of waste in North Ossetia was more than 12 000 000 tonnes, or about 20 tonnes per capita (Vaghin, 1995), including highly toxic heavy metals. Also, over 7 tonnes of radioactive wastes are stored within the territory of the city of Vladikavkaz.

(3) **Conflict between those who exploit the forests:** two mutually exclusive users have equal rights of use of the forested areas: forestry farms and agricultural farms (collective state farms, or co-operative societies). This problem is typical for all ex-Soviet states and regions where land is still in state ownership, and no resolution has been found.

(4) **Soil erosion, and the loss of fine earth** that provides the substratum for soil formation, is

critical in the mountain regions of North Ossetia (Ilyichiov, 1995). If the current high rates of erosion persist over most of the mountain rangelands, the Republic will lose from 25 to 50 percent of its loose cover in the next 50 years. This cannot be restored within the foreseeable future and the traditional land use will become increasingly unsustainable.

Social problems – social indicators

The Republic of North Ossetia–Alania has a complete spectrum of the social problems which are found in most mountain regions: underemployment, inadequate medical services, poor education, communication, and transport facilities. However, there is at present a specific problem for the Northern Caucasus, namely the problem of inter-ethnic conflicts and territorial claims. There is severe hostility between the Ossetians and the Ingush, their eastern neighbours. In 1944 the Ingush were deported to Central Asia, and their lands and settlements were partly occupied by the Ossetians. After the Ingush were rehabilitated by the Russian Parliament in 1992, they took up their territorial claims and this action later degenerated into armed ethnic conflict. The war in Chechnya and the Georgian *versus* Ossetian conflict in Trans-Caucasia since the early 1990s together create a real threat to peace and stability in the Caucasus. At the moment the conflicting parties have concluded a peace agreement, but because the Ossetian agreement requires the presence of federal troops as guarantors the conflict is not really settled.

Potential and strategy for sustainable development

The state of conflict limits the potential for North Ossetia to take advantage of its geographical position as a transportation corridor leading to Georgia and Trans-Caucasia. This route is an important element of the economic and geopolitical stability in the Caucasus. In 1994 the Government of North Ossetia approved 'The Strategy of the North Ossetian State for Protection of the Environment and Supporting Sustainable Development'. This sets priorities for social and economic development, taking into account the key role of environmental issues as a basis for political and economic decision making.

One of the important elements of this strategy is the Programme of Rehabilitation of the Mountain Regions of Ossetia. The first phase of the programme is planned for the period 1995–2005 and includes the following points:

(1) research and legally protected management of the mountain economy;
(2) upgrading of the living standard in the mountain zone;
(3) integrated use of the natural resources by developing mountain economic agglomerations.

The Government of the Republic is fully aware of the benefits of using the European experience in mountain development. During the European Intergovernmental Consultations on Sustainable Development of Mountain Regions (Aviemore, Scotland and Trento, Italy, 1996), an initiative was taken to establish a European Network of demonstration projects. This will evaluate models of sustainable development in different environments, and various ethnic, cultural, and political systems of the European mountain countries. It is proposed that North Ossetia–Alania be incorporated in this process as a model case of an economy in transition.

Source: Yuri Badenkov,
International Mountain Laboratory,
Moscow

Box 5.5 Case study of Puka, Albania

Albania is situated on the western side of the Balkan Peninsula. It has a land area of 28 750 km² and 3.2 million inhabitants, 45 percent of whom live in the mountain areas. Mountains and upland occupy 60 percent of the total area, a third of which is above 1000 m asl with the highest point of Mount Korab at 2751 m. Albania is dominated by a Mediterranean climate with hot, dry summers and cool, moist winters; continental influence increases toward the northeast. Precipitation is highest in the northern and southern mountain areas (up to 3500 mm/annum in the Northern Alps); this signifies a large potential for erosion.

About 39 percent of the land in Albania is used for agricultural production (cultivable area and pastures); 36 percent is forested; and 25 percent is occupied by lakes, rivers, infrastructure (roads, cities, villages), and areas regarded as unproductive (rock outcrops, badlands). About half the cultivated land is arable, but because of steep slopes only 64 percent, and in mountain areas only 50 percent, can be cultivated by tractor. In spite of an impressive increase in area of cultivated land since 1950, there has been a dramatic decrease in the amount of arable land per capita – from 0.28 ha in 1950 to 0.18 ha in 1992. This is due to the rapid population growth during this period. In 1990, 47 percent of the active population was employed in agriculture. The most important resources of Albania are chromium, copper, nickel, oil, gas, water power, and also timber and decorative rocks such as marble.

For over forty years Albania was isolated economically and socially, so that the development of the country stagnated. Because of obsolete technologies the industrial and agricultural production decreased during the 1980s and came to a total standstill with the political change that occurred in 1990/91. A major part of the state-owned businesses were shut down; marketing, transportation, and storage infrastructure disintegrated. Between 1989 and 1992, the GDP sank by 50 percent, in 1992 inflation reached 230 percent, and only massive humanitarian aid from abroad prevented famine in 1991/92. According to statistical records, Albania has achieved a positive economic development since 1993, with support from the International Monetary Fund (IMF), The World Bank, and the European Bank for Reconstruction and Development (EBRD). This positive development, however, includes only the service sector, where the high interest applied to loans can be recovered[1]. Until now, very few investments have been made in the secondary sector. Agricultural undertakings have been privatised and consequently about 380 000 families are cultivating an average of 0.3–0.4 ha. Due to a lack of capital and of availability of inputs, such as seeds and tools, productivity is low. The absence of marketing opportunities and the unstable administrative structure have reduced production to bare subsistence. A large part of the processed food is imported.

In addition, the Albanian national economy depends strongly on money transfers (remittances) from Albanians working abroad. This so-called 'post office economy' is based mainly on illegal work in Macedonia, Greece, and Italy, and is estimated to about US$ 400 million per year, that is about three times as high as the entire Albanian export earnings. Both the out-migration rate and the internal migration rate from the mountains to the urban areas are high. Unemployment is estimated at 30–80 percent, depending on the information source and sector.

With the collapse of the economy, the public services also floundered; a large part of the railway system is out of commission and the water supply and irrigation system is disintegrating. An enormous obstacle to restructuring the country is the lack of functioning institutional installations and organisations at all levels of government, administration, and private economy.

PUKA

The mountain region of Puka lies in the north–northeastern part of Albania and extends southward from the valley of the River Drin. Its land area is about 1304 km^2; 80 percent is forest and only 8 percent is cultivated, of which 37 percent is pasture; the rest is unproductive. The most important resources are timber and water power; there are two hydroelectric power stations. The copper mines were closed in 1992. The dominant Mediterranean climate provides 500 mm of precipitation in summer and 1500 mm in winter.

Outline of economic, ecological, and social indicators

Economic indicators

According to the official statistics, the employment breakdown structure in the district of Puka before the political overthrow was 54 percent workers, 30 percent members of farm co-operatives, and 16 percent employees. The workers group was involved mainly in ore mining and timber processing. After the collapse, all mines and industrial enterprises were closed. Only in the lumber industry have small firms survived or new ones been established. The two regional hydropower stations are still operating and supply the entire country with electricity, but since they are state-owned their impact on the region's economy is not significant. Most workers returned to their families and are now considered to be part of the agricultural population but, naturally, cannot find adequate employment in this sector.

Most of the former members of the farm co-operatives have now become private farmers and own only about 0.2 ha per family. The limiting factor is not so much a lack of available land as deterioration of the irrigation systems. After the centralist government disintegrated there were no organisations that could maintain the water supply system in the new decentralised and private economy. Nevertheless, there are already villages that are reorganising their own infrastructural maintenance which enables them to increase the production of vegetables, fruits, and berries.

A traditional agricultural activity which is adapted to the oak forest vegetation is goat breeding.

---

[1]As of the beginning of 1997, there has been a collapse in local investment which has resulted in public unrest in the capital

As elsewhere, there is a lack of efficient organisation of the pasture economy. The consequence is low productivity of meat and milk, and a heavy pressure on the environment. There is a high potential for development of the region's timber economy and agriculture – milk, meat and vegetables – even though this would still require small business structures for regional self-sufficiency. Potential by-products, such as medicinal herbs and mushrooms, should not be neglected. The recommendation is to process the products as far as possible locally, thus gaining a value-added advantage, and to build up good marketing structures.

Compared to other regions, Puka lies relatively close to the lowlands, especially to Shkodra, the main urban centre of northern Albania, and is fairly well developed with 50 percent of the population living along main roads. Nevertheless, access to remote villages remains very difficult.

Ecological indicators

The low standard of development and the total neglect of the environment in the past has caused serious damage. The most impending ecological problems today are the over-exploitation of the forest and pastures, erosion and loss of soil fertility.

The forest is under stress from both felling of wood and over-grazing by cattle. After the communist system disintegrated, foreign corporations began a process of unrestricted clear cutting of the forests for export. Until 1995 entire tree trunks were exported, but since 1996, timber is at least partly processed before being exported. The concepts of sustainable forest cultivation and effective control have not been developed and there is no efficient pasturing system that would improve the productivity of agriculture and animal husbandry (mainly goats) yet still protect the forest.

Before 1992 the agricultural area was exploited to the highest possible limit, aided by irrigation and a high input of auxilliary products. Today, a large part of the fields lie fallow, the so-called wastelands. Often they are situated far from the nearest village, the necessary machinery and materials to cultivate them are unavailable, and the irrigation system has been destroyed. Erosion is causing serious soil deterioration in a region never favoured with highly fertile soils. Currently, the use of wastelands by seeding for pastures or for reforestation is being considered. There is no infrastructure for protection of the environment in much of the settled area. There are no restrictions on garbage and sewage disposal and, since the political overthrow in 1992, plastics have been used increasingly as package materials, making the garbage problem even more visible.

Social indicators

In early 1993, 47 621 people lived in Puka – 2.8 percent less than in 1989. Whilst before the overthrow the mining industry attracted people, there is now a heavy out-migration of young people. The average family size is five; 34 percent are under 15 years of age. The local infrastructure including roads, water, sewage, telecommunications, schools, medical facilities, and public services, is very inadequate.

The collapse that beset Albania at the end of the communist regime is one of the most serious that a society has ever experienced. It can very well be compared to the defeat after a war. In addition to economic disorder, there is a deep distrust by the population of state or community institutions that extend beyond the family. The 45 years of centrist government have left the population with a critical lack of initiative and sense of responsibility. The basic foundational cell remains the family and village cohesion, based on family groups, is becoming effective for reconstruction. In some cases farmers' associations have been established, but they are normally built on family and local ties, or on former co-operative structures. Much remains to be done in order to erect new economic and social structures and to define the role of the individual *versus* the community.

CONCLUSIONS

The economic and social collapse due to the political overthrow underlines the total lack of any sustainable development in communist Albania, economically, ecologically, and socially. In effect, sustainable development began at zero in 1992. There are two crucial points if development of the mountain areas of Albania is to occur. On the one hand, credit possibilities must be provided, with manageable interest rates, to enable development of roads, communications, health and school services and also to support the exploitation of resources and sustainable agriculture, new processing companies, and small industries. A major task is the awakening and support of the self-initiative of the population; it is also very important to create private and public institutions and organisations that have the full trust and support of the population. Finally, the mountain areas of Albania as a whole will also depend on the transfer of knowledge and financial help from abroad.

Source: Jörg Wyder and Jeannine Bossert,
Schweizerische Arbeitsgemeinschaft
für die Berggebiete (SAB),
Brugg

CONCLUSION

From the economic point of view mountain regions only have cost advantages compared with lower regions if they can benefit from specific resources associated with their mountain locations – e.g. in the fields of winter tourism, the mining industry, or the production of special mountain herbs. However, these resources are seldom sufficient for independent economic development; often they are even completely lacking. In all other industrial and agrarian sectors, mountain areas are at great disadvantage in terms of production costs, compared to the industrial centres in the lowlands. In addition, the liberalisation of the world markets, as promoted by the Uruguay Agreement (GATT), has increased the difference in production costs between mountain regions and lowlands. This enforces a strong focus on agricultural production in mountain regions, low salaries, and weak or stagnant economic development. Depending on the given conditions in the specific region – especially on population growth – this economic pressure manifests itself over the short or medium term in over-exploitation of natural resources, out-migration, and social conflicts, which again harm the prospects for regional development. The extent of these economic problems is heavily dependent on the political affiliation of the mountain region under consideration.

Taking in account the world-wide importance of mountain regions as human living spaces, the UNCED requirements in the 1992 Rio declaration

Table 5.2 Classification of levels of sustainability in the five mountain regions used as case studies

| Mountain region | Sustainability | | |
	economic	ecological	social
Bhutan	low	rather low	low
Encañada (Peru)	very low	low	low
Pays d'Enhaut (Switzerland)	high	rather high	high
North Ossetia (Russian Federation)	rather low	very low	low
Puka (Albania)	very low	very low	very low

on the sustainable utilisation of mountain areas warrant close attention. In response to this declaration, the authors have tried to describe the situation of five different mountain regions in terms of their sustainability in the economic, ecological, and social spheres. A rough classification (Table 5.2) of these regions shows how strongly the economic, ecological, and social components are interlinked. Neglecting any one of the three components will have a negative impact on the other two over the short or medium term. Thus, whatever the given conditions, concepts of sustainable development must be used to create a balance between all three components. The priorities of the three components and the concrete measures required, however, must be adapted to the specific conditions of the individual mountain region.

References

(not all references are cited in text)

ADPE (1995). *Révision du programme de développement régional; Rapport préliminaire*. Bilan 19978-1993; Objectifs de déveleppement du Pays-d'Enhaut. Château-d'oex

AGRISWISS, IUED (1994). A Development Strategy for the Mountain Areas of Albania. Final Report 2. Analysis of Potentials and Constraints. Report to the Cooperation Office for Eastern Europe (BZO) of the Swiss Federal Department of Foreign Affairs

AGRISWISS, IUED (1994). Several reports of the 'project Puka' to the Cooperation Office for Eastern Europe (BZO) of the Swiss Federal Department of Foreign Affairs, May 1994, Genf

Alexandratos, N. (ed.) (1995). *World Agriculture: Towards 2010*. FAO Study, FAO: Rome

Badov, A. D. (1993). *Formation, Development and Functioning of Settlement in North Ossetia*. Vladikavkaz. p. 174 [in Russian]

Bätzing, W., Perlik, M. and Dekleva, M. (1996) Landwirtschaft im Alpenraum – unverzichtbar, aber zukunftslos? Blackwell: Berlin

Brando and Martin (1993). Implications of agricultural trade liberalization for the developing countries. *Agricultural Economics*, **8**

Braun, J. von (ed.) (1995). Agricultural commercialization: Impacts on income and nutrition and implications for policy. *Food Policy*, **20**(3)

Civici, A. *Albanie – Revival in Pain. Analysis of the Economical and Agricultural Policies between 1945 and 1994*. International Centre of High Agronomical Mediteranean Studies: Montpellier

Dhakal, D. N. S. (1990). Hydropower in Bhutan: A long-term development perspective. *Mountain Research and Development,* **10**(4): 291–300

Dorjee, K. (1995). An Analysis of Comparative Advantage and Development Policy Options in Bhutanese Agriculture. Dissertation, ETH 11081, Zürich

Edwards, C., *et al.* (1990). *Sustainable Agricultural Systems*. The Soil and Water Conservation Society: Iowa

FAO (1995). *Impact of the Uruguay Round on Agriculture*. FAO: Rome

Federal Department for Statistics (1996). Employment Statistics, first quarter of 1996. Bern

Heks (1995). *Albania. Country Outline*. Zürich

Honan M. (1995). *Switzerland: Travel Survival Kit.* Lonely Planet: NY

Ilyichiov, B. À. (1995). *Loose Cover in the Northern Caucasus in the Erosional Trends: the Material Balance and the Ecological Prognosis*. Abstract of the Conference on Safety and the Mountain Environments. Vladikavkaz, p. 265

Jodha, N. *et al.* (1992). *Sustainable Mountain Agriculture*. ICIMOD: Kathmandu

Keating, M. (1993). *Summit 1992: Agenda for a Sustainable Development*. Centre for Our Common Future: Geneva

Kinlay, D. (1995). *An Analysis of Comparative Advantage and Development Policy Options in Bhutanese Agriculture*

Messerli, P. (1989). *Mensch und Natur im alpinen Lebensraum: Risiken, Chancen, Perspektiven*. Verlag Paul Haupt: Bern/Stuttgart, p. 368

OECD (1995). *Agricultural policies, markets and trade in the central eastern european countries, selected new independent states, Mongolia and China*

Pingali, P. L. and Rosegrant, M. (1995). Agricultural commercialization and diversification: Processes and policies. *Food Policy*, **20**(3)

Price, M. F. (1995). *Mountain Research In Europe*. The Parthenon Publishing Group: Casterton and New York, p. 230

Siebter Landwirtschaftsbericht (1992). EDMZ: Bern

Soto, H. de, (1992). New Rules for Development: a Liberal Order as Way Out of the Poverty Crisis. *Neue Zürcher Zeitung*, **96**

Stone, P. (ed.) (1992). *State of the World's Mountains: A Global Report*. Zed Books: London

Vaghin, V. S. (1995). *Environment and Safety Issues in the Republic of North Ossetia-Alania. Proceedings of the Conference on Safety and the Mountain Environments*. Vladikavkaz, p. 45 [in Russian]

Vosti, St. (ed.) (1991). *Agricultural Sustainability, Growth and Poverty Alleviation: Issues and Policies, German Charity for International Development (DSE)*. Fedafing

WCED (1987). *Our Common Future*. World Commission on Environment and Development, Oxford University Press: New York

World Bank, The (1994). Making Development Sustainable. From Concepts to Action. ESD paper series 2, Washington DC

World Bank, The (1995). *World Development Report 1995: Employees in the Worldwide Integration Process*. The World Bank: Washington DC

Conflicts in mountain areas – a predicament for sustainable development

6

Stephan Libiszewski and Günther Bächler

INTRODUCTION: THE MOUNTAINS AS CONFLICT 'HOT SPOTS'

Throughout history, mountain areas have been arenas of political unrest, of competition over resources, and of violent conflicts. Since World War II, many of the most bloody and destructive wars have occurred in mountains and highlands. The Caucasus, the Balkans, northern Iraq, the Hindu Kush, Kashmir, Tibet, the Peruvian and Colombian Andes, the Vietnamese and Laotian mountains, and the Ethiopian Highlands, to name only a few, have experienced extensive military disturbance. In 1995, according to the war register maintained at the Unit for the Study of Wars, Armaments, and Development at the University of Hamburg (AKUF), 35 wars and 13 armed conflicts (both internal and international) were ongoing, involving 43 countries in total. Nineteen of these wars and seven of the armed conflicts were taking place in mountains, or included areas defined as mountainous (see also map on next page, Figure 6.1)[1]. Moreover, several of the world's major trouble spots with a high risk for escalation, such as the Indo-Pakistani Himalayan border areas, including Kashmir, or the contested Golan Heights between Israel and Syria, involve mountain regions. The proportion of mountain land affected by armed conflicts, in relation to total world land area or total population, is significantly above average.

In recent years, scholars have ascertained an intrinsic causal interrelationship between environmental transformation and many of the present wars and armed conflicts, especially in the poorest regions of the developing world. Environmental change and subsequent shortages of renewable resources, such as fertile land, fresh water, forests, and fisheries are deepening poverty and widening social divides. This generates large-scale and destabilising population movements, aggravates tensions along various societal boundaries, and debilitates political as well as social institutions. These phenomena, in turn, appear to be the main causes of many, if not most, of the current armed conflicts and wars. Thus, environmental hardship either triggers or contributes to violence in many parts of the developing world (Homer-Dixon, 1994; Bächler *et al.*, 1996).

This causal relationship between current environmental transformation and violent conflict is especially relevant to mountain regions. Mountains and highlands are exposed to war and conflict due to their particular ecological and socio-economic vulnerability and to the fact that they contain some of the last remaining natural reserves. Mountain ecosystems are generally less likely to recuperate from substantial perturbations, such as widespread soil erosion or loss of vegetation, than are lowlands. Moreover, due to their relative impenetrability and the difficult conditions that prevail, highlands have been frequently excluded from national development programmes and thus have remained marginal or even isolated, both economically and politically (Omara-Ojungu, 1990: p. 143; Mountain Agenda, 1992). Mountain

---

[1] War is defined by the Unit for the Study of Wars, Armaments, and Development at the University of Hamburg, 'as a massive violent conflict with three constitutive qualitative criteria: 1) it must have a minimum of continuity; 2) both war parties must have a central organisation leading their operations; 3) at least one of the war parties has to be a government with regular or at least government associated troops'. Massive violent conflicts matching these criteria only in part are referred to as 'armed conflicts'. The figures on wars and armed conflicts in mountain areas are based on assessment by the authors

Figure 6.1

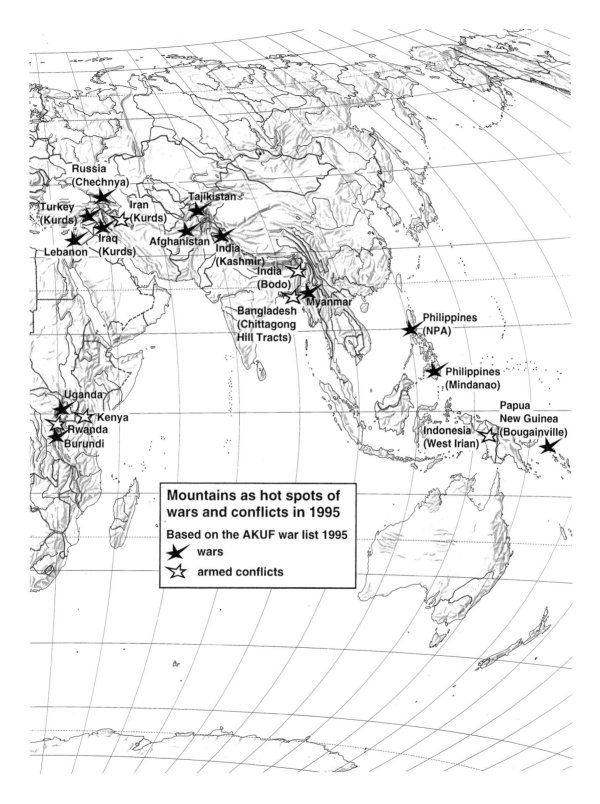

Russia
(Chechnya)

Turkey
(Kurds)

Iran
(Kurds)

Tajikistan

Iraq
(Kurds)

Afghanistan

Lebanon

India
(Kashmir)

India
(Bodo)

Bangladesh
(Chittagong
Hill Tracts)

Myanmar

Philippines
(NPA)

Philippines
(Mindanao)

Uganda

Kenya

Rwanda

Burundi

Papua
New Guinea
(Bougainville)

Indonesia
(West Irian)

**Mountains as hot spots of
wars and conflicts in 1995**

Based on the AKUF war list 1995

★   wars

☆   armed conflicts

communities in most parts of the developing world still rely heavily on agriculture for their subsistence and there are few off-farm alternatives. They have limited means to cope with environmental challenge and, thus, they may rapidly precipitate into political crisis when the traditional socio-ecological balance is disturbed by internal population growth, external pressure, or global climatic change.

On the other hand, many mountain regions have been excluded from the market economy because of their inaccessibility and long-lasting marginalisation. Therefore, they have remained relatively unspoiled and their natural resource potential is largely intact, until recently. Access to mountain areas, however, has improved significantly during the last decades and, as natural resources have become increasingly scarce and depleted in the lowlands, these highland reserves become more attractive for exploitation. Timber, fuelwood, water power, minerals, virgin soils, biodiversity, and the landscape itself as a tourist resource, all encourage the intrusion of people and capital from outside. The political implications of these processes, such as greater control by the central state, changes in the ethnic or national composition of the population, and often simply the disruption of the way of life of the indigenous people, trigger hostilities in many mountain regions of the world.

Environmental conflicts in mountains almost always stem from interaction with lowland areas. Although difficult to access, the mountains have always interrelated with the surrounding foothills and plains; for instance, through the water cycle, or in socio-economic respects concerning the use of pastures, as a reservoir of labour, or as transportation routes. The various forms of highland–lowland interaction imply specific social and political organisations which are severely challenged by the ecological consequences of population growth and of the global modernisation process. Social tensions and conflicts which arise from these challenges escalate into violence and war if and when the environmental threats are perceived as existential, the traditional mechanisms of political regulation are over-strained, and the parties in question have the capacity to organise and arm themselves, or eventually to recruit allies to support their claims (Bächler *et al.*, 1996, Vol. I: p. 308). In the following pages, typical causal patterns of environmentally influenced violent conflict in mountain areas are identified. The patterns are not intended to strictly categorise the empirical cases, rather, they are presented as representative, or

idealised, types of conflict development. The examples of the Euphrates–Tigris water dispute, the Rwandian civil war, and the armed conflict in the Chittagong Hill Tracts of Bangladesh (Boxes 6.1, 6.2, and 6.4), and two examples of violent warfare in the mountain regions of the former Soviet Union (Box 6.3) illustrate the interrelations between the individual types. In the second part of this chapter, some cultural aspects of violence in mountains and highlands are introduced and aspects of environmental conflict management are considered. Above all, the interrelationships of sustainable development and conflict management are highlighted and specific recommendations for policy and research are formulated.

CONFLICT PATTERNS IN MOUNTAIN AREAS

Mountains as resource divides: conflicts over the Earth's fresh water

One of the most significant characteristics of mountains derives from their effect on the control and distribution of natural phenomena and resources on earth. Their function as weather-divides has a decisive influence on the climatic conditions and geomorphological processes that also influence the surrounding regions. Mountains function as the world's water towers by attracting much of its precipitation, thereby providing a major proportion of the world's water supply (Chapter 7). Of particular significance, is that much of the precipitation at high altitudes falls as snow which is

Plate 6.1 Compared with other rivers, the River Jordan is an extremely tiny stream. The snow-melt of Mt Hermon controls the River Jordan's yearly runoff which, therefore, shows a high variability. The distribution of its water has been an issue of conflict for decades (Photograph: S. Libiszewski)

preserved through the winter and acts as a form of reservoir extending stream run-off into the dry season. Most of the world's rivers originate in the mountains so that the upper portions of the watersheds are extremely important for environmental and geopolitical security. At least half of humanity relies (for drinking, domestic use, irrigation, hydropower, industry, and transport) on water that accumulates in the mountains (Denniston, 1995: p. 41).

The specific highland–lowland interaction relating to the water cycle is at the root of many acute conflicts, both international and internal. A primary source of inter-state conflict derives from the fact that most of the large rivers of the world flow through the territory of two or more countries. There are about 240 of these river systems worldwide that cross national boundaries. Competition over the allocation of shared waters is widespread, especially in arid and semi-arid regions such as the Middle East and Central Asia, where hydrological resources are scarce and many countries rely entirely on rivers for their water supply. When politically strained relationships already exist, downstream states often fear that their upstream neighbours might use control over the water as a means of coercion (see Box 6.1 on the water disputes in the Euphrates–Tigris basin). In some

Box 6.1   Upstream-downstream conflicts in the Euphrates–Tigris basin

Of the world's river systems, few have provoked as many conflicts as the Euphrates–Tigris basin. Three nations share most of it: Turkey, controlling the head waters, and Syria and Iraq, the downstream riparians. With their generally very dry climate, Syria depends on exogene water sources for 79 percent and Iraq, for 66 percent of its supply. All three nations have above-average birth rates, and all three have been pursuing aggressively ambitious plans to use the rivers more intensively. However, since Turkey controls the head waters, it has strategic advantage. Over the past 20 years, the Turks have planned and partly carried out an enormous project on the upper reaches of the Euphrates and Tigris in southeastern Anatolia without taking into account the interests of the countries downstream.

The Turkish project, called the Günedogu Anadolu Projesi (or GAP), includes 21 dams and 19 hydroelectric plants. At its heart is the huge Ataturk dam, completed in 1992. Furthermore, the GAP encompasses industries and irrigation projects. It is the most important single development project in the history of Turkey and is supposed to transform southeastern Anatolia into a region of economic growth. By heavily affecting the amount and quality of the water being discharged into Syria and Iraq, however, the GAP is causing major international repercussions. If all the plans are completed, of the 30 billion cubic metres of water per year originally flowing down the Euphrates across the Turkish–Syrian border, only 11 billion will remain. And because the project is intended to provide water for both agriculture and industry, the quality of the residue entering Syria will be impaired.

Tension has risen dramatically among the riparians whenever water has been used to fill the reservoirs behind the various dams. In January 1990, Turkey completely dammed the Euphrates for a month, damaging crops in Syria and interrupting its hydropower supply. A year later, during the Gulf War, Ankara cut the flow again – officially for technical reasons. But the decision may also be explained as political support for the Western allies in their war against Iraq. These incidents are a foretaste of the power Turkey will obtain by being able to control the flow of water once the GAP is completed. The water that Turkey will allow over the spillways will appear as a concession for which, directly or indirectly, Turkey will expect consideration.

The disputes over water have also exacerbated other political conflicts. Most significant are the problems with the Kurds who make up a significant minority in all three countries, and who are demanding greater autonomy and even independence. In the case of Turkey, the Kurdish settlements are located in the mountainous region of southeastern Anatolia and partly coincide with the area of the GAP. The bloody war which the secessionist Kurdish Workers Party (PKK) has been waging against the central state since 1984 has retarded work on the project and discouraged many of the foreign investors who were expected to participate.

At the international level, the GAP and the Kurdish question are interwoven in many ways. Although the Syrian and Iraqi governments have always brutally suppressed Kurdish demands for autonomy within their own borders, the same states have been providing refuge and active logistical support for the PKK. This was their approach to retarding progress on the GAP and to acquiring a card to play against Turkey in the diplomatic dispute over water allocation. Turkey, in turn, uses the reverse strategy, withholding Euphrates water to pressure its neighbours to stop supporting Kurdish rebels.

Although there have been attempts to negotiate a water-sharing system, the situation in the region remains extremely strained. The dirty game being played with the water and the Kurds could escalate into regional warfare at any time; it also creates serious problems within Turkey. The economic and political interests involved in the GAP project partially explain the government's uncompromising attitude toward the Kurds' desire for autonomy. This intransigence, in turn, is the main reason for the popularity of the fanatical PKK and thus for the inexorable escalation of the civil war in southeastern Anatolia. Finally, Turkey's stubborn insistence on complete control of the water rising on its territory gives Syria and Iraq a reason to support any rebel movement that may weaken their upstream neighbour.

Source: Stephan Libiszewski,
Center for Security Studies and Conflict Research,
Swiss Federal Institute of Technology, Zurich

instances lower riparian states have resorted to military power to rectify a situation perceived as dangerous.

The Jordan Basin, the Golan Heights and the mountains of the West Bank have long been areas of contention as part of the Arab–Israeli conflict. When Israel captured these territories in the Six-Days'-War in 1967, the importance of the highlands was emphasised for reasons of defence. However, one of the causes for escalation of the conflict was an Arab project on the Golan which aimed to divert the source of the Jordan River. Although control over water resources was not the prime reason for occupation of the West Bank, in view of the existing water shortages affecting the country, the rich aquifer that originates in these hills has become one of the incentives for Israel to retain the area in the current Middle East peace negotiations (Libiszewski, 1996).

Similarly, Egypt is using its political and military superiority to compensate for its unfavourable geographical location as lowest riparian of the Nile Valley. Dependent almost entirely on water that originates from outside its own territory, Egypt has always viewed with great suspicion the development of water projects in the eight countries upstream. The country reacted with heavy verbal attacks and overt threats to any plan by the upstream riparians that involved damming the Nile or diverting any of its water before it reached the Egyptian border. During recent years, Ethiopia, which encompasses the source of the water-rich Blue Nile, has repeatedly been the target of such intimidation.

The Syr Darya and Amu Darya in the Aral basin of Central Asia, were formerly internal rivers during Soviet times, but are now divided amongst five individual nations following the collapse of the Soviet Union in 1991. As there was no international legislation on water-sharing in the basin, disputes over water allocation have broken out. Uzbekistan, the most populous republic in the region, now depends on external flow for 91 percent of its water needs. Uzbekistan has repeatedly accused the upstream riparians of retaining too much water during summer when the need for agricultural irrigation is most acute. It is interesting to note that Kyrgyzstan, which maintains the Toktogul dam near the source of the Syr Darya, was also blamed for releasing *too much* water during winter 1993. According to a petition from the Aral Committee, the extra water did not reach the Aral Sea because of winter freezing, and not only was lost into the Ayarkul depression, but caused environmental damage there (Klötzli, 1996: p. 292).

Moreover, competition over water distribution is not the only trigger of upstream–downstream disputes. Because steep slopes and pronounced changes in altitude increase the available water energy potential, most of the world's hydropower projects are located in mountains. As the industrialised countries did before them, many developing countries with major hydropower potential are implementing large-scale projects to meet growing demands for electricity. Very often these huge constructions are seen as prestige monuments for national self-awareness. Although the damming of rivers for hydropower production does not substantially reduce their total discharge, it may cause severe ecological and economic impacts downstream by affecting the seasonal availability of the water. The Ganges, Mekong, and Salween rivers in Asia, and the Paraná in South America, among the great international river systems, are most at risk from this kind of dispute. Recently in Europe, also, a hydropower project on the Danube was the cause of a crisis between Slovakia and Hungary which exacerbated the political tensions already existing among the national minorities within each country.

Upstream–downstream conflicts can also arise from land-use practices in mountainous areas which indirectly affect stream development and related geomorphological processes. In the Ganges–Brahmaputra basin, for instance, Bangladesh has long been accusing the upstream riparians, Nepal, India, and Bhutan, of responsibility for the disastrous floods that regularly affect the delta country. Although the causal relationships and the extent of the phenomenon are controversial, Bangladesh claims that deforestation in the Himalayan watershed has caused an increased rate of erosion and thereby siltation of the rivers. In the dry season, when the water discharge is further diminished by Indian diversions, it is purported that the silt is deposited on the river beds thereby decreasing their depth and raising them above the general level of their floodplains. It is further argued that, during the rainy season, these shallow rivers cannot contain the huge volume of monsoon water, both from the upper watersheds of the rivers themselves and from the heavy rainfall on the plains; as a consequence destructive floods occur. The speculations about the responsibilities for these environmental disasters have aggravated the already strained political relations of Bangladesh towards the neighbouring states and are also at the root of violent internal conflicts (see Box 6.4 on the strife in the Chittagong Hill Tracts). This situation is particularly ironic because recent research on the Himalaya–Ganges (and Brahmaputra) problem (Ives and Messerli, 1989) has clearly indicated that the assumed effects of deforestation in the Himalaya have been grossly exaggerated, if not completely misplaced. Nevertheless, the perception of cause of disaster remains as a politically dangerous phenomenon.

Finally, although under different conditions, the upstream–downstream dilemma manifests itself between lowland and upland areas within the same country. While on the international level conflicts over resource use occur – at least formally – between equal sovereign entities, on the intra-state level the role of the central government is clearly dominant. Within states, upstream mountain communities cannot use their geographical advantage over downstream communities in the same way that governments of independent upstream countries can respond to the lower riparians. Rather, they are at the mercy of the political and economic power, which usually resides on the plains.

Internal water conflicts often arise from development projects aimed at exploiting the water resources of the mountains for the benefit of the lowlands without the appropriate participation of the mountain people. Hydropower projects, for instance, require the inundation of extensive areas, the clearing of forests, and the relocation of the local population. Therefore, they become the source of tensions and conflicts. Examples of mountain people struggling against dam construction on their lands are reported from all over the world. Depending on the conflict resolution mechanisms that exist in the respective countries, on the fighting capabilities of the concerned populations, and eventually on the support from external powers, or international public opinion, such conflicts may be arbitrated peacefully, as in the case of the Narmada dam in India, or result in violence, as exemplified by the struggle of the local people against the dam project on the Chico River in the Philippines. (A survey on internal conflicts over dam projects is provided in Bächler *et al.*, 1996, Vol. I: p. 206ff)

Mountains as a niche of minority peoples: ethnicity and conflict

A second factor leading to conflict in mountain regions is their cultural and ethnic diversity. This heterogeneity is not a historical accident. Rather, it is the direct outcome of the local people learning to live in harmony with the surrounding extraordinary biological and morphological diversity. In extensive mountain regions with deep and narrow valleys, separated by precipitous alpine ridges, people throughout history have tended to settle in individual valleys and to have little contact with their neighbours. This physical segregation has enabled those groups to preserve their distinctive traditions and customs until today. In other areas, for instance in the Great Lakes region of Central Africa, diverse peoples have been settling in the same mountain area and have maintained their ethnic and cultural identity, which is heavily dependent on different patterns of resource use. Moreover, because of their relative inaccessibility, many mountain regions have been long-lasting areas of retreat for indigenous peoples and ethnic minorities, or sites of refuge for communities which were forced to move because of political or economic reasons, or both.

This built-in individuality, or distinctiveness, of groups contributes to the heterogeneity of populations often encountered in mountains. The Northern Caucasus, for instance, is home to at least two dozen different nationalities. The autonomous republic of Dagestan alone has twelve official national languages, not counting the various dialects. Estimates of the number of remaining indigenous (or tribal or native) peoples world-wide vary from 200 million to 600 million. Although no figures exist for the total number of indigenous people living in mountains, the proportion is clearly high since mountains account for a substantial portion of the landscape that has not been transformed by modern economies. For instance, more than 10 million Quechua, descendants of the Incas, reside in the Central Andes, and more than 16 million indigenous people live throughout the 19 major mountains ranges of the former Soviet Union (Denniston, 1995: p. 42).

Ethnic and cultural diversity *per se* does not trigger conflict or violence. On the contrary, in many regions of the world, people with different ethnic and cultural affiliations have been living side-by-side peacefully, often developing special forms of interaction which are to their mutual benefit. But such diversity provides a base of cleavages which can be politicised easily if other factors of social and economic stress appear on the scene. The globalised transformation of the environment and the resulting scarcity of renewable resources that have been the source of increasing concern in recent years has challenged the interrelations between different ethnic groups in many eco-regions of the world, including highlands and mountains. The current revival of ethno-political conflicts, for example, although in many cases a response to long-lasting oppression by centralised or external authority, often also reflects profound crises in the relationship between people and their environment.

In Rwanda and Burundi the decreasing capacity of the land to sustain life, especially on steep slopes, coupled with increasing population pressure, has exacerbated the ethnic divisions that existed in pre-colonial times. The depletion of renewable resources has repeatedly triggered violent inter-group clashes and has contributed decisively to the hostilities of the 1990s. This situation has led to the war crimes, massacres, and the genocide of today (see Box 6.2). In the highlands of Kenya, deforestation, over-grazing, soil erosion, and drought have resulted in an absolute decline in total food production since the late 1980s. For the same reasons the production of coffee, Kenya's most important cash crop, declined by 30 percent between 1989 and 1992. The resulting economic losses, which recently led to severe food shortages, and even famine in parts of the country, are seriously challenging the sensitive relationship between the 40 different ethnic groups in Kenya.

Box 6.2 Land use and land tenure conflicts in the Rwandan Hills

Until the war which began in April 1994, 92 percent of the Rwandan population lived and worked in rural areas. The mountainous country of the Great Lakes region was always considered to be relatively rich in terms of the natural capital upon which Rwanda so heavily depends. However, both major ethnically and socially defined groups, the farming Bahutu and the livestock breeding Batutsi (in fact, there was often a mix of both forms of livelihood), have extensively over-used this so-called 'thousand hills' landscape.

Geographically, Rwanda can be divided into favourable and unfavourable areas. The central highlands are an under-developed mountainous region with steep slopes and deeply weathered acid soils of limited fertility. The fertile volcanic area in the northwest, and the previously almost unusable swamp and savanna region in the south and the east were being used up to the limits of their capacity. On the western boundary, up to the Central African Rift Valley, even the most extreme slopes were cultivated. Today, however, the once fertile soils of these areas are degraded and geographical alternatives are rapidly diminishing. The annual loss of humus is high, averaging 10.1 t/ha. While 50 percent of the soils in the relatively favourable areas are sustaining a moderate loss of 3.7 t/ha, it is above all the marginal soils of the unfavourable areas that are incurring extreme losses: on 5 percent of the land more than 36 t/ha and on 1 percent more than 68 t/ha (Becker, 1993 p. 114; see also Bart, 1993).

According to oral history – the only sources we have – during a long period from the sixteenth to the eighteenth centuries the Bahutu peasants and the Batutsi pastoralists are assumed to have lived relatively harmoniously side-by-side. The Bahutu used the fertile areas on the hills, whereas the Batutsi prefered the grasslands in the eastern parts of the region. However, with population growth and the evolution of small kingdoms, each group began to overlap more and more onto the favourable eco-regions. The Batutsi adapted the Bahutu culture (language, clan structures, and kingdoms), but became politically dominant at the same time. They competed with the Bahutu over the use of the land and, above all, over the land tenure system. The peasants had established clear property rights of lineage groups and, later, larger clans in the hills, whereas the pastoralists neglected to acquire any such security; they endorsed the herder's freedom of movement, claiming that the land belonged to the cow (Lachenmann, 1990 pp. 59–119).

During the eighteenth and nineteenth centuries, the respective territorial and political–administrative claims, derived from the extensive livestock economy, on the one hand, and from subsistence farming, on the other, became incompatible with the establishment of geographically expansive dynasties and kingdoms. These encompassed several Bahutu as well as Batutsi clans. The territorial separation of the two groups was gradually replaced by ethnic intermingling and social differentiation. Social stratification and political hierarchy, combined with power struggles between clans and the evolution of a centralised Batutsi dynasty in central Rwanda, led to numerous ethnopolitically motivated riots and minor wars. These have culminated in the genocide and massacres of 1994 to the present (Guichaoua, 1989; Newbury, 1988).

During the twentieth century, the *socio-ecological* indicators worsened in every respect, especially over the last few decades. Large areas of Rwanda, a country once half-covered by trees, were deforested over the centuries and transformed into arable land. For example, the 56 percent expansion of acreage that occurred between 1970 and 1986, pushed the rapidly growing population of cultivators onto the poor soils of the marginal mountain regions. This caused the lower limit of the mountain rain forests, previously at about 1800 m asl, to be driven upwards because of deforestation to as high as 3000 m. At the same time, possibilities for economic restructuring and off-farm opportunities were very limited. A shift from subsistence production, partly destroyed by political interventions, to trade and industry is severely restricted by the lack of purchasing power and by the precarious resource situation. The ore deposits are nearly exhausted. There is not enough wood available for the production of bricks and charcoal. Prices for the main export commodity, coffee, have collapsed and tea production has structural problems. Finally, the extensive state-owned, or co-operative, plantations have transformed those farmers, who were tempted to abandon their subsistence farms, into dependents and have exposed them to the risks of the world market.

The system of self-sufficiency, which in the recent past had provided the basic supplies of a large proportion of the rural people, cannot cope with the rapidly growing population pressure and has become an obstacle to urgently needed economic and social restructuring. The population growth rate of 3.3 percent per year, compared to a yearly increase in total agricultural production of only 0.3 percent between 1980–1988. The tradition of partible inheritance has resulted in continually smaller farm plots, as low as 0.5 ha per capita in the more densely settled parts of the country. This means that fallow cycles can no longer be maintained. The younger generation is abandoning the exhausted land, a trend which will intensify. Since it is difficult to connect the isolated farms and scattered settlements on the hills (in Rwanda there are no compact or linear villages) with road and energy systems, problems of infrastructure reinforce the inclination to migrate to the few towns. Unequal distribution of resources seems even more relevant than the simple lack of agricultural land. About 16 percent of the farms, those at any one time more than 2 ha, together control more than 42.9 percent of the agriculturally productive land. Put another way, 182 000 farms, of a total of 1 112 000 in 1984, accounted for about half of the productive land. The plight of the Rwandans became tied increasingly to the diminishing productivity of progressively more and more degraded land.

In the past, the increased pressure on the natural resource base in Rwanda led to a politically organised caste, or class, society. The power struggle between the lineages was entwined with a conflict between semi-nomadic livestock herders and settled farmers. This manifested itself periodically in land use and land distribution conflicts in the hills. The struggle over the distribution of scarce resources has further exacerbated the tendency to use land, forest, and water beyond their capacity.

The above discussion characterises the situation in Rwanda that reached breaking point in 1994, precipitating into catastrophic collapse. It is postulated that the first Rwandan republic (1960–1994) failed to eliminate the socio-economically and environmentally harmful agrarian structures, to overcome the spatial fragmentation and political weakness of the farmers, and to induce sustainable development on a broad scale by establishment of a rational agrarian constitution. The long-lasting conflicts over land use and land tenure, the excessive use of limited renewable natural resources in the mountains, the lack of off-farm opportunities, and the high dependence on coffee as a cash crop, are all background conditions to the historical power struggles between clans, dynasties, and regions. The distinct ethnicity, which was exaggerated by the former colonial powers and misused as a political tool for ruling the country, together with the unsuccessful management of high rural population growth, low agricultural productivity, and over-use of the hills, combined to trigger the most deadly power struggle in history; the prospects are virtual extermination (see also Bächler, 1996: pp. 461–503).

Source: Günther Bächler,
Swiss Peace Foundation, Bern

In the north of the country, tensions between ethnic groups related to land tenure and land use degenerated into armed fighting in 1991 that led to several thousand deaths. More recently, in 1994, the clashes spread to areas that formerly were calm, such as parts of the coast and western Pokot. Here, up to 10 000 people are reported to have died. The exact figures are not known, since emergency measures put in place by the government hinder the flow of information. However, Kenya's reputation as one of the most stable countries in Africa has been demolished.

A primary event that affects peoples in mountain areas and often leads to apparently 'ethnic' conflicts between them and the central government is the introduction of large-scale development projects into their settlement areas. Internal conflicts arising from the damming of rivers and inundation of settlements and farming land have already been mentioned. Of all the economic activities in the world's mountains, however, nothing rivals the destructive power of mining (Chapter 9). For geological reasons, mountains are the source areas of many of the world's mineral deposits required for industrial production. Thus, they have always attracted the attention of multi-national mining companies and national governments which have a major interest in exploiting this natural wealth.

The environmental impact of the various forms of mining activities includes habitat destruction, increased erosion, air pollution, acid drainage, and metal contamination of water bodies. Currently, the greatest mining projects are open-cast mines for the extraction of copper, iron ore, and coal. These activities involve the dismantling of great tracts of the landscape. The morphology, hydrology, and atmosphere are severely affected, as well as the vegetative cover and, consequently, the fauna. This overall destruction undermines the habitat and, as a consequence, the livelihood and security of indigenous peoples. Violent conflicts arise, proportionate to the ability of the local people to organise themselves and to orchestrate resistance.

Such conflicts are widespread in the mountain regions of the developing world, for instance, in the vicinity of Mount Nimba in West Africa. In the mining region of 'Jharkhand' in the southern part of the Indian state of Bihar, and in the region of the Ok Tedi mine in the Star mountains of Papua New Guinea. But there are also examples in the industrialised world, involving, for instance, the indigenous people of the arctic regions of Canada and Alaska, or the Navajo on the Colorado Plateau in the southwest of the USA. Although not attaining the same level of violence, these latter cases show a similar pattern of conflict.

Among the most evident cases of armed conflict triggered by the environmental impact of mining activities on mountain areas is the current guerrilla war between the Bougainville Revolutionary Army (BRA) and government forces on the island of Bougainville in Papua New Guinea. The conflict was provoked by the environmental destruction arising from the Panguna mine in the mountainous interior of the island. This is one of the largest open-pit copper mines in the world; it has produced a huge excavation 7 km^2 in extent and 500 m deep, which is regarded as the 'second greatest artificial hole' in the world. The overlay shelf is also very extensive, and additional areas were lost through the construction of roads and residences for the mine workers, most of whom originate from outside the island. Millions of tons of tailings laced with heavy metals have been dumped into the water system and have contaminated the Kawerong and Jaba rivers; the rate of run-off has accelerated and their silt content has increased enormously. Lands along the rivers, as well as the surrounding forests, have been destroyed by the massive deposition of silt. The indigenous population was forced to transfer their settlements from the formerly fertile river banks to marginal land on the surrounding mountain slopes. Fishing has virtually collapsed as a source of subsistence.

This severe environmental disturbance triggered a political movement that at first demanded only appropriate compensation. With the government's refusal to respond to these demands, however, the conflict further escalated to violent sabotage against the mine. This forced the temporary closure of the mine in 1989, prompting the government to proclaim a state of emergency and to send in troops to ensure continuing copper production. The guerrilla war which developed has continued to the present day. The conflict that began as an opposition movement against the destructive effects of the copper mine has therefore been transformed into a secessionist war. Because the inhabitants of the island belong to a different ethnic group from the main population of Papua New Guinea, the BRA now demands secession from PNG and unification with the neighbouring Solomon Islands, the inhabitants of which share a common language and ethnic identity with the Bourgainvilleans.

Bougainville is a concrete example of a war caused by the environmental consequences of mining activity in a mountain area. However, many indigenous peoples simply do not possess comparable capabilities and/or the external support necessary to trigger an armed conflict. Small in number, and hopelessly inferior in terms of military strength, many affected groups must rely on weaker or non-violent forms of protest. Hence, when large-scale projects are undertaken by unscrupulous states, and when the affected indigenous population is unable to find external allies or mobilise international public opinion, such conflicts very often do not reach the stage of organised armed conflict. Rather, the lack of opportunities, coupled with an oppressive regime, may result in the silent marginalisation of the indigenous people; in a worst case scenario such situations have led to ethnocides.

Mountains as poverty 'reserves': marginalisation and conflict

Resource depletion and environmental degradation often affect economic productivity in poor countries which tend to be highly dependent on their natural capital. All mountains share the common physical attributes of steep slopes, instability, and ecological complexity that make them extremely vulnerable to environmental disturbance. Above all, the destruction of the vegetation cover and widespread soil degradation seriously endanger subsistence production and livelihood security of mountain populations in most countries of the developing world. Growing populations, on the one hand, and inequitable distribution of, and difficult access to, land resources, on the other, compel farmers to extend their activities onto

Box 6.3   Violent conflicts in the mountains of the former Soviet Union

THE INTER-ETHNIC OSSETIAN–INGUSH CONFLICT IN THE NORTHERN CAUCASUS

An extreme example of cultural and ethnic heterogeneity of mountain regions is the Northern Caucasus. Dagestan alone recognises 12 official national languages; this does not include the various dialects. The tangle of inter-ethnic and socio-political relationships is most complex. Moreover, the region looks back to a rather turbulent history, with arbitrary political borders drawn by hegemonial powers and a long tradition of resistance to external rule. After the region was absorbed into Russia in 1813, it took 80 years for the Czarist troops to gain full control. During the Soviet period the northern Caucasus was subjected to repeated boundary adjustments and coercive movement of people; this has led to a series of almost insoluble violent disputes.

One of these is the inter-ethnic struggle between Ossetians and Ingush. Like the Chechens, the Ingush belonged to the Caucasian people that were most severely affected by Stalin's deportation policies. In 1944, under the pretext of 'punishment for pro-German sympathies', the autonomous status of the Ingush republic was abolished and the entire population was deported to Central Asia. As a result of this policy, the neighbouring republic of North Ossetia, which had traditionally endorsed Russian influence in the Caucasus and had always maintained close connections with Moscow, occupied the frontier district of Prigorodny. After 1944 this area was populated by Ossetian and Russian immigrants who took over the confiscated lands and settlements of the Ingush. In 1957, however, the exiled Ingush were

allowed to return to their homeland and their republic was re-established. Regardless, the Prigorodny region was left under the control of the Republic of North Ossetia, thus opening a serious territorial issue.

Today, the conflict between the Ingush and Ossetians is occurring within the Russian Federation, of which both republics are now a part. The current escalation began in 1992 after the Federation Parliament passed a law which rehabilitated the 'nationalities subject to compulsion in 1944' and restored their rights to the confiscated lands and dwellings. A large number of the Ingush entered the Prigorodny district and demanded restitution of their former properties and territorial rights. The violent demonstrations and clashes which followed resulted in about 600 deaths; some 30 000 Ingush had their citizenship rights suspended and were compelled to leave North Ossetia.

Currently, this conflict is in a 'frozen' condition, the Federal Government having proclaimed a state of emergency in both republics. The Russian army and the national guard of North Ossetia have separated the warring parties and established a militarised buffer zone. Despite the governmental commissions that were set up to manage the conflict, the negotiation process has been complicated and inefficient. The low level of economic development, combined with a high migration pressure from the most mountainous areas to the lowlands, is exacerbating this historical struggle and impeding a territorial compromise. Moreover, the strategic importance of North Ossetia, which carries two important highways connecting Russia and Transcaucasia, makes it difficult for the central government to take a neutral position in its attempts to resolve the conflict.

WAR IN THE PAMIRS REGION OF TAJIKISTAN

Since 1992, a bloody civil war has been taking place between the communist government and the Islamic and national liberal opposition in Tajikistan, one of the five Central Asian republics that gained independence after the breakdown of the Soviet Union. In 1994, the sparsely populated Pamir mountain region of Tavil Dara (upper Obi Hingo Valley) became the main battlefield of the conflict.

The roots of the conflict in Tajikistan and its escalation in the Tavil Dara region must be sought in the overall status of the upland areas during the Soviet period. Tajikistan was the poorest republic in the Union. Moreover, large internal socio-economic disparities characterised the relationship between the different eco-regions. While some industry had developed in the northern province of Chodschent

and vast cotton plantations were established in the southwest, the isolated mountainous regions in the east remained economically marginalised. Beginning in the 1930s, the government of the Republic consistently implemented a policy of moving the population from the highly under-developed mountains to the lowlands. This policy had three goals: (a) liquidation of foci of opposition to Soviet rule in the conservative mountain communities; (b) reduction of expenditure on infrastructure in the mountain areas; and (c) supply of cheap labour to the cotton plantations.

These migrations, which in part were performed as deportations, led to a mixing of people with different cultural and ethnic roots. They had been accustomed to maintaining close clan cohesion in relation to their respective homelands. This also influenced their political and ideological affiliations. Tensions between these regional clans have been exacerbated by the environmental and economic crises in the fertile lowlands. The long-lasting cotton monoculture has severely affected the soils and has led to reduced yields. The very high rate of population growth, especially in the rural areas where it is estimated to exceed 4.0 percent per annum, has markedly reduced the availability of arable land per capita. Unemployment rose to 26 percent of the labour force by the end of Soviet rule. After achievement of national independence in 1991, the political rivalries between the regional factions escalated into open warfare. The clashes began with rival demonstrations in the capital in the spring of 1992; they resumed in the southern province of Kulyab and Kurgan-Tyube, after people originating from the eastern mountains had attempted to occupy lands owned by the local Kulyabes.

During the fight for control of the capital, Dushanbe, in 1992, and in the following 'purges' carried out by the current regime, at least 50 000 people have lost their lives and 250 000 have been forced to flee. Although negotiations, under the auspices of the UN, began in 1994, fighting has continued in the Tavil Dara region between Islamic rebels advancing from the southeast and the Tajik national guard. Tavil Dara is part of the historical province of Vahio, connecting the western part of the country, which is under government control, with Gorno-Badakhshan, a mountainous and isolated area in the east, mostly controlled by the Islamic opposition. Both its strategic position at the intersection of the conflicting parties, and its religious importance as the location of sacred Islamic sites, ensure that this mountain region of Tavil Dara remains a focal point of the conflict.

Source: Yuri Badenkov, Institute of Geography, Academy of Sciences, Moscow

Plate 6.2 Main road from Dushanbe to Nurek (1987) – differences in accessibility in isolated mountain regions (Photograph: B. Messerli)

marginal lands which are far more vulnerable to erosion. An example of cassava cultivation in Nigeria may illustrate this: on a field with a gradient of 1 percent, the annual loss of soil is 3 t/ha; on fields with a gradient of 5 percent, this increases to 87 t/ha, equivalent to the loss of 10 cm of humus in one generation; with a gradient of 25 percent, erosion could destroy all topsoil within a decade (Brown and Wolf, 1984: p. 10).

Although statistics for comparing mountainous regions to the lowlands within a single country are scarce, those that have been compiled demonstrate these serious negative impacts on the natural conditions. For instance, more than 60 percent of the rural Andean population lives in extreme poverty. In the Himalaya, the inhabitants of the Indian state of Himachal Pradesh achieve a GNP which is less than half that of the national average (Denniston, 1995: p. 39). The Philippines provide another flagrant example for the marginalisation of mountain areas; the National Capital Region has a high Human Development Index of 0.871, whereas the mountainous region of West Mindanao is at the bottom of the rating, with only 0.410; it also has the lowest life expectancy (55 years) and more than 55 percent of its people live in poverty, compared with 15 percent in the National Capital Region (United Nations Development Programme 1996: p. 31). These disaggregations indicate disparities in human–nature relationships that can provoke violent conflicts. Economic decline caused by resource degradation disrupts major social and political institutions, including the state itself. Poor state performance in general, and in its mountains in particular, aggravates grievances between the mountain inhabitants and the central government, triggers rural protests, and increases rivalry among different factions belonging to the political elite. All these factors together contribute to the erosion of the state's authority.

The long-lasting insurgency in the Philippines, as well as most of the past armed conflicts along the Central American mountain chain, and the present conflict in the Peruvian Andes, provide powerful evidence of the links between environmental depletion in mountain areas, economic deprivation, and civil strife. In the rural uplands of the Philippines, the current war waged by the New People's Army against the government is motivated by the poverty of landless workers in the rural sector and displaced farmers settling in remote hill areas, where the central government is weak. The absence of property rights governing upland areas has encouraged in-migration. Yet most peasants find themselves under the authority of concessionaries and absentee landlords who have claimed the land. Thus, they have no incentive to conserve the land. Peasants' small-scale logging, production of charcoal for the cities, and slash-and-burn farming cause severe environmental damage. This environmental stress, in turn, often causes a cycle of falling food production, clearing of new plots, and further land degradation. During the 1970s and 1980s, the communist insurgents found the poverty-stricken peasants of the uplands highly receptive to revolutionary ideology, especially where coercive landlords and local governments left them little choice between rebellion and starvation (Homer-Dixon, 1994: p. 28ff).

Similarly, the rise of the *Sendero Luminoso* (Shining Path) rebellion in Peru can be attributed to a subsistence crisis caused, at least in part, by ecological marginalisation. The country's mountainous southern highlands are not suitable for agriculture. The hills are steep, the soils thin and dry. Nonetheless, during the colonial period, indigenous peoples in the region were displaced onto the hillsides when Spanish settlers seized the rich lands in the valleys for their haciendas. In the 1970s, the Velasco government undertook a sweeping land redistribution programme. But people in the highlands benefited little because the government was reluctant to break up large agricultural enterprises that generated much of the country's export earnings. By 1980, cropland availability had dropped below 0.2 ha per capita. These rural population densities exceed sustainable limits, given the inherent fragility of the region's land and prevailing agricultural practices. Cropland, therefore, has been badly degraded by erosion and nutrient depletion. Family incomes, which were already among the lowest in Peru, dropped further in real terms. In 1980, per capita income in the Peruvian highlands had fallen to 82 percent compared to the 1972 level. There is a strong correlation between areas suffering extreme poverty, on the one hand, and the *Sendero Luminoso* strongholds, on the other (Homer-Dixon, 1994: p. 29ff). While the central government has recently weakened the power of the *Sendero Luminoso*, unless the lot of the extremely poor can be radically improved, it is only a matter of time before violence will reassert itself.

Another phenomenon typical of some mountain areas that is highly conducive to the provocation of conflict is drug production. By coincidence of favourable soils, climate conditions, and the poverty of peasant farmers, virtually all the world's cocaine and heroin production is concentrated in three relatively small mountain regions. Mountain farmers in the Golden Triangle of Southeast Asia (northern Thailand, Myanmar, Laos, and south-western China) and Southwest Asia's Golden Crescent (northern Pakistan and Afghanistan) grow nearly all the poppies used in opium and heroin production. Similarly, virtually all coca leaves for cocaine and crack production are grown in the White Triangle of the Andean regions of Bolivia, Colombia, and Peru. Drug trade provides a major source of income for the farmers. However, drug production causes environmental as well as cultural devastation: deforestation of hillsides, decline in soil fertility, and water pollution, not to mention drug addiction, AIDS, and violence among competing drug lords (Denniston, 1995: p. 44). In Colombia, for instance, an endemic war is taking place between the traditional leftist guerrillas, the *Drogue Mafia*, and other paramilitary groups. In 1993, according to *American Watch*, an average of 11 people were killed daily for political reasons, and death by firearms had advanced to become the most common cause of death among the young Colombian male population.

Mountains as last frontiers: migration and conflict

Environmentally exacerbated poverty not only causes social unrest and violent conflicts, it also induces the movement of people. When available resources are inadequate to support a particular population density, the local population must find alternative ways to sustain themselves, or they will be forced to leave a degraded homeland to find new land that will support them. The movement of people and livestock has always been a characteristic of mountain regions. Due to the vertical distribution of different environments and the seasonal variation in conditions at each level, resources are stratified and require a staggered schedule for their effective exploitation (Chapter 2). Nomadic pastoralism, transhumance, and mixed forms of grazing and farming cultures that include different uses for various altitudes have been the prevailing human response to these conditions in the mountains of the subtropics and those of middle and high latitudes; furthermore, migration into the plains historically has been a means of coping with the surplus population which could

not be sustained by these various forms of altitudinal adaptation.

In the tropics, in contrast, sedentary agriculture has been widely practised in mountain areas. In these latitudes, weather conditions at any given altitude remain much the same throughout the year; seasonal temperature variations are entirely masked by the diurnal cycle, so there is no need for migration or movement within the system such as occurs in higher latitudes. In addition, because of the prevalence of malaria and other diseases in the lowlands, the higher elevations have been healthier places for human habitation. As a result, highlands in the humid tropics have tended to support denser populations than many lowland areas (Price, 1981: p. 393). But here, too, the enormous pressure on the landscape, accelerating through the last half century, has degraded the natural resource base, in many places to such a degree that people from these originally favoured highlands have migrated either to the urban areas or to other, more fertile, rural areas.

Traditionally, migration has been a means to solve the problem of population surplus and inadequate environmental resources. In many cases, migration benefited both the place of origin and the destination. However, especially when people from different countries, or with different ethnic affiliation, were involved, migration almost always required complex forms of social regulation and clearly defined limits. Environmental change and the increase in shortage of basic renewable resources over the last several decades have severely challenged the sensitive institutions regulating these movements. Ethnic groups which had developed and practised complementary modes of living may now discover their distinct identity and become engaged in violent conflict, when they are forced to compete with each other for their own existence on the shrinking resource base.

In most regions of the world, migration movements almost never ascend, they usually descend: from degraded mountains to the urban areas of the plains (see also Chapter 2). However, in some circumstances, fertile uphill oases surrounded by a harsh lowland environment can attract an inverse flow of people. In Darfur in Western Sudan, for instance, the mainly Arab pastoralists, the Dars, and the sedentary Fur farmers were interlocked for centuries in a complex solidarity relationship. Both groups have depended heavily on the region's natural capital: the Arab pastoralists on the pastures in the arid and semi-arid Sahel, the Fur on the agricultural resources of the humid Jebel Marra massif. The Arab pastoralists used to withdraw with their cattle into the well-watered Jebel Marra mountains during the dry season and in times of hardship. In turn, they provided the crop and fruit producing Furs with trade goods and dung from their cattle. Conflicts occurred from time to time, but usually they could be solved by peaceful means. In this type of situation, settled farmers and pastoral nomads belonging to distinct ethnic groups were able to live more or less peacefully alongside each other for centuries by developing interactions that were mutually beneficial.

This relatively tranquil situation was profoundly disrupted during the 1980s, when hardship in the Sudanese plains and foothills became chronic due to a combination of the prolonged drought, the limitation of the nomadic movement by large-scale mechanised farming, and the increase in human and cattle populations. The nomads tried to avoid the consequences of the drought by searching for permanent settlements in the foothills and mountains. Thus, they became increasingly competitive with the sedentary Fur farmers. The traditional solidarity, based on seasonal trade-offs, turned into mortal combat throughout the year, since human survival was at stake. The skirmishes between the Arab Zaghawa and Maharia tribes and the Furs,

which have been occurring since 1983, escalated in 1987 into an organised military campaign. Twenty-seven Arab tribes repeatedly attacked the non-Arab groups of the region which, to some extent, retaliated with measures such as poisoning the water wells for the cattle.

This is a typical environmentally-caused conflict between highland and lowland communities across distinctive socio-ecological borders, a conflict which was categorised as the outcome of the drought driven 'desert-*versus*-oasis-syndrome' (Suliman, 1996). These types of conflict can be found in similar circumstances throughout the world; they occur where traditional patterns of seasonal migration and transhumance have been disrupted, or new movements of people have been enforced by environmental transformation.

CULTURAL ASPECTS OF VIOLENCE IN MOUNTAINS AND HIGHLANDS

The significance of violent conflicts occurring in mountains and highlands and the four causal patterns are highlighted, with special attention given to the environmental dimension. There is, indeed, ample evidence for a specific conflict history in mountains and highlands which is illustrated by the four case studies discussed in Boxes 6.1–4.

Box 6.4   The Guerrilla war in the Chittagong Hill Tracts of Bangladesh

A guerrilla war between the 'Shanti Bahini' insurgents and the government troops of Bangladesh has been underway in the Chittagong Hill Tracts (CHT) since 1976. This region is situated in the southeastern part of Bangladesh, near the Indian and Myanmar frontiers. It comprises the only upland area of the entire country. The CHT is inhabited by a number of indigenous tribes whose languages and religion (Buddhist) differ from the Muslim Bengali mainstream population. During British rule, due to its peculiar ethnic structure and its relative inaccessibility, the region enjoyed an 'excluded area' status. This was lifted after the amendment of the Pakistani constitution in 1964 and the door was opened to settlers from outside. The tribals developed a politicised ethnic awareness after the secession of Bangladesh from Pakistan in the early 1970s and demanded autonomy for the Hill Tracts and imposition of a ban on the influx of non-tribals. The government's negative reaction radicalised the autonomy movement and led to the formation of a military wing which began an armed struggle.

While this struggle is often referred to as an ethnic conflict related to state formation, its strong environ-

mental dimension must be emphasised. The initiation of tension between the indigenous people and the government coincided with the construction of the Kaptai Hydropower Project which caused the forced evacuation of large numbers of tribal people. While the dam submerged a vast area and inflicted great harm on the directly affected people, it did not improve the level of prosperity of the wider region. The hydropower generated was for the benefit of the urban areas as well as to provide power for the industries set up in the region. Moreover, the job opportunities created by the industrial growth did not assist the local people because of their lack of skills and education. Rather it opened the door wider for the influx of outsiders and further marginalised the tribals.

The conflict was escalated by the declining carrying capacity of the land across the entire country. The cyclones which devastated Bangladesh in 1970 and 1991, along with other environmental disasters such as floods, river bank erosion, declining soil fertility, exacerbated the long-standing Bangladeshi problems stemming from landlessness due to high birth rates; all these problems combined to produce millions of environmental refugees. Moreover, many of them were attracted by the comparatively better environment in the CHT, still further increasing the influx of outsiders. The government allowed, and sometimes actively encouraged, such migration and settlement. Thus additional pressure was placed on the limited resources and opportunities for employment.

The tribals are totally dependent on the land and could not adapt to any alternate forms of subsistence livelihood. Clashes over land distribution, therefore, became a regular phenomenon, but in most cases the tribals were the losers. This further fueled the years of deprivation and intensified the insurgency. The conflict even acquired an international dimension, since India has provided logistic support for the rebels, as a means of increasing pressure on Bangladesh to aid its overall political struggle with that country. Although it was hoped that the District Council Election in 1989, whereby limited autonomy was vested in the tribals within their own administrative units of the CHT, would pacify the ethnic turmoil, peace was short-lived and the violence has continued to the present.

Source: M. Abdul Hafiz,
Mohakhali, Dhaka, Bangladesh

Nonetheless, from a cultural ecological point of view there remain problems of defining the term 'mountain'. Is it correct to generalise that mountains imply scarcity, harshness, backwardness, and a scattered population? This statement can be tested by referring to the plains: do they, in contrast to mountains, suggest abundance, ease, wealth, and good living in comfortable cities? As F. Braudel pointed out in his colourful and rich study of the Mediterranean during the age of Philip II, they were not 'the thickly populated plains which today are the image of prosperity the culmination of centuries of painful collective effort', originally a land of 'stagnant waters and malaria, or zones through which the unstable river beds passed'? (1995: p. 52).

Shortcomings of the Sanctuary Theory

Against this background, the assumptions of the *sanctuary theory* must be questioned. Its protagonists maintain that mountains are generally less suitable and, therefore, less desirable for human settlement than plains. This is considered to be so because of several environmental constraints, such as lower temperatures, scarcity of arable land, transportation difficulties, and the like. The crux of the matter is that, due to their assumed disadvantages, mountains theoretically should be abandoned eco-zones. As a consequence, whenever a large population is found to settle in mountains, one might conclude it has been forced to do so – displaced either by war, central governments, or destiny (see Bencherifa, 1990: p. 370). But why then are there so many conflicts in supposedly deserted and unfavourable places?

It might be true that mountains belong to the poorest regions throughout the world, but there are also extremely poor regions on the plains. Indeed, while some mountains are quite deserted, mainly where low rainfall renders the land unfit to support even pastoral life, there are also relatively rich mountain areas. Equitable precipitation, generally favourable climate, or occurrence of abundant mineral resources ensure the attractiveness of some mountain areas. Historically, other mountain areas have been densely populated and still are, because of their agricultural and forest resources, and not because of lowland peoples having been driven up from the plain by force.

There may be clear-cut geographical boundaries between highland and lowland: 500 m, or so, as an arbitrary contour. Thus, it can be inferred that mountains and highlands share a set of common characteristics determined by specific natural conditions – altitude, aridity, rainfall patterns, strong relief, soils, and so on – that differentiates them from lowlands. Producers, households, and communities in the mountains do have comparable problems, and likewise, producers in the world's plains, deserts, and deltas. However, the boundaries between highlands and lowlands may be blurred, indicating inter-dependence rather than complete separation. Therefore, the effects of indistinct cultural geographical boundaries on mountain conflict will be examined. These boundaries cannot easily be shown on a map, because they shift over time depending on changes in the human–nature relationship: a concept of cultural ecology that deals with the transformation of this relationship. Cultural ecology encompasses both a horizontal and a vertical dimension when applied to highland–lowland interactions.

The horizontal dimension

The *horizontal dimension* is important for comparative reasons, for instance, in order to stress correlations between environmental change and social behaviour (see the organisation of material in previous section). The horizontal or comparative approach highlights the rich variety of cultural, and mainly *agri*-cultural, systems throughout the world's mountain ranges; culture is defined here as the process of transforming the societal relations with natural resources by production methods, resource-use techniques, labour applications, and ways of life, over a certain period of time. The Mediterranean mountains – even though marginalised in comparison with the plains and plateaus of the wider bi-continental region – are perhaps comparable, to a certain degree, to the Andes in Latin America, but in no way do they resemble the relatively impenetrable mountain areas of Central Asia, China, Indochina, or India. In the Atlas, for instance, agro-pastoralism has been, and still is, a vital if not *the* dominant resource-use system even though the major urban and economic activity is located along the Mediterranean littoral. Because it involves both farming and livestock-raising, agro-pastoralism has survived as a risk-aversion strategy which represents a response of traditional communities to the inherent environmental mountain hazards (Bencherifa, 1990: p. 373). On the other hand, the Ethiopian Highlands, in contrast to most mountains outside Africa, are considered not to have been originally hostile to human settlement. They not only hosted the famous ancient empire of Aksum, but they also provided an excellent human habitat, at least before the recent degradation of forests and soils (Hurni, 1990: p. 51). For centuries, the Ethiopian Highlands were the most favourable areas of the country for both settlement and farming, and this was due to their temperature and rainfall patterns. The Highland Provinces comprise slightly more than 29 percent of the total area of the country, but they contain over 66 percent of the total cultivated land and nearly 45 percent of the grazing land. More than 63 percent of the urban, and more than 69 percent of the rural population live in the Highlands (Wolde-Mariam, 1991: p. 293). These figures for the Horn of Africa alone, disrupt the sanctuary hypothesis almost completely. Nevertheless, it is also evident that the transformation of the Highland's environment has placed enormous pressure on the landscape: settlement, crops, grazing, fuelwood consumption, construction, have all degraded the renewable resource base to such a degree that urban areas and lowlands are attracting more and more farmers and agro-pastoralists from the degraded Highlands.

The vertical dimension

The *vertical dimension* focuses on the highland–lowland interaction in single conflict regions. The crucial point can be illustrated by reference, once more, to F. Braudel who suggested that the degree of the contrast between plain and mountain is, first of all, a 'question of historical period' (1995: p. 53). He concentrated on cultural and socio-ecological disparities – thus, on an approach which is still valid today. In his historical analysis it is the disparities between *old* and *new* soils, that is, between exhausted and newly developed landscapes, that are the key factors responsible for all existing forms of vertical migration. To differentiate this from the premise of sanctuary theory, it can be postulated that, if a population is relatively dense in one area, then the resources available necessarily make this concentration possible (Bencherifa, 1990: p. 370). If available resources do not support a certain level of population density, then alternative ways to sustain life must be found, or there will be a forced evacuation of at least part of the population, in an attempt to locate more productive land. The usual response to local overuse of old soils is migration. The direction of such movement, however, whether uphill or downhill, will depend, at least in principle, on where the new lands are located. In reality, regular migration, such as nomadism or transhumance, occasional migration between villages, for instance for marriage, and forced migration such as flight or displacement, break the concept of eco-geographical determinism repeatedly, and in various ways.

Another important factor is the means of communication, especially roads, for transportation and travel. Roads have significant impact on the intensity and frequency of highland–lowland interaction; they serve as a kind of 'extension of the plain and its power through the hill country' (Braudel, 1995: p. 41). Thus, the changes in the relationship between highland and lowland are mainly induced by modernisation processes following the construction of roads or railways, however slow and imperfect these processes. The pros and cons of opening up mountains by new transportation lines, of themselves, are nothing less than an expression of the collective will of a specific culture in a region. If politically and economically dominant groups concentrate on the lowlands for whatever reason, the marginalisation of the high-lands becomes absolute. Nevertheless, drought and scarcity of resources in the lowlands can lead to a new understanding of the relatively favourable conditions of the mountains (see the discussion on Jebel Marra, above).

Some general remarks on the conflicts in mountains and highlands can now be made. They are based on the following hypothesis: the Janus-headed process of modernisation *versus* marginal-isation (see Bächler, *et al.*, 1996) of mountain and highland arenas promotes the rapid transformation of all forms of the complex migration patterns which have evolved over many centuries. The arena consists of individual and collective actors who mobilise their cultural, social, and economic capacities in order to participate in policy design and to receive a significant share of the benefits (Sottas, 1996). Among the factors contributing to ecological conflicts, structural change or even disruption of downhill–uphill traffic has proved to be a major trigger for conflict in the vertical dimension: the penetration of the mountain peoples' control of resources by the lowland power centres becomes the main force in generating conflict. The real, or the perceived, closure of geographical boundaries between highlands and lowlands leads to clusters of separate populations, or to pockets of poverty, in remote mountain areas – cultural phenomena that were explained by the sanctuary theory *ex post facto*. Ethnic, regional, and religious problems attributed to those conflicts may play a causal role, but in general less as a trigger than as a channel that amplifies inter-group confrontation or provokes exaggeration of group identity. It is always important to differentiate carefully between the causes of hostility and what actually induces the people to fight.

Impenetrability as a myth *versus* interdependence as a reality

Conflicting highland–lowland migration patterns focus on two poles, described here by the notion of *impenetrability,* for the one, and *interdependence* for the other.

The myth of the impenetrable mountain

Mountains always have been regarded as a world apart from civilisation. At least for some mountain regions this may be not very far from the truth. Because access has always been difficult, and still

is, people originating from the mountains are deprived of important exchanges necessary for them to keep pace with urban civilisations that are developing in the foothills and the plains. In 'impenetrable' mountain areas, society, culture, and economy all bear the mark of backwardness and poverty. There is a lack of mutual exchange through movements while inter-dependence and functional solidarity are weak or absent. Thus social anachronisms have persisted, one of which is the Mediterranean Vendetta and its system of honour and shame. Where there is a Vendetta, there are mountains, and usually those that had not been penetrated by feudal justice or by modern laws of the nation state (be it communist or democratic). The Berber countries, Corsica, Sardinia, and Albania, amongst others, provide examples. But at least in Europe, the mountains historically were the lands of the free in the double sense of the word. This implies that people lived beyond reach of pressure and tyranny, with 'no landed nobility with strong and powerful roots, no well-fed clergy to be envied and mocked' (Braudel, 1995: p. 39). The historical 'mistake' of Syrian despotism, for instance, was to stop at the first rocks of the mountains; and Turkish despotism served as a ruler of the roads, passes, and towns, but was rather weak in eastern Anatolia and in the Balkan high-lands. Quite often mountains were considered as zones of religious dissidence, sometimes as real, or perceived, strongholds of foreign religious influence. When circumstances did permit, new religions were able to make substantial, though unstable, conquests in mountainous regions. In the Balkans in the fifteenth century, as F. Braudel (1995) states, whole areas of the mountains converted to Islam in Albania, as in the Her-zegovina, around Sarajevo. For the observer of the present geographical map those evangelisation attempts, half-hearted invasions, and unstable con-quests, or, in other words, those *imperfect penetra-tions* of 'impenetrable' mountains turn out to be the distant trigger, or reason for on-going struggles, social conflicts, and all-out wars. At the present time, Syria is struggling for water controlled by the upstream ruler (Turkey) over the headstreams of the Euphrates and Tigris. Simultaneously, the war between the central government of Turkey and the Kurds (officially the 'Mountain Turks') is still volatile, whereas the Balkans encompass various fragments of ethno-political groups, accentuated by the division between traditional rural mountain areas and modern urban cultures that can be traced back to the early history of conflicting highland–lowland interactions (Albanians *versus* Serbs; Serbs *versus* Muslims in Bosnia-Herzegovina).

The concept of *imperfect penetration* can be supported by further examples. Where the population was so thinly distributed, widely dispersed, and separated by deep valleys that it prevented the establishment of states, of dominant languages, and of important civilisations, the mountains were both the birthplace and the battlefield of the struggle for liberty, democracy, and peasant 're-publics'. The rather belligerent history of the Swiss Cantons adds weight to this line of reasoning. In Rwanda, where the dispersed habitat on the 'im-penetrable' hills had been part of a decentralised culture for centuries this very structure was perceived by the colonial power as 'un obstacle à la division du travail, à l'échange, à l'économie de marché' (AESED: p. 27). The structure of the complex Rwandan society 'qui a résisté des siècles durant aux incursions des Etats voisins et des Arabes' (Guichaoua, 1989: p. 44), therefore, was radically changed by the Belgians, a change that is an important factor in explaining the deep roots of the present war and genocide (Box 6.2).

Finally, 'impenetrability' is a one-way process; it always entails downhill migration. It is the harsh life, the widespread poverty, the hope of an easier existence, or the attraction of good wages, that encourages mountain people to move downwards to more favourable eco-zones. In fact, throughout the world large numbers of mountain dwellers have become indispensable to the life of towns, low-lands, and the sea coast. The disparities manifest themselves in social and cultural barriers that are raised to replace the imperfect geographical barriers which are always broken in one way or another. This is the case in Central Asia where a strong regionalism among former mountain dwellers creates group identity conflicts in the plains and plateaus (Klötzli, 1996: p. 277ff). However, the latter issue relates more particularly to the second premise and embraces the concept of conflict emanating from *interdependence*, with mountains serving as regions of vertical human migration.

Highland–lowland interdependence

Highland–lowland interdependence is founded upon a more or less dense network of routes along which both downhill and uphill movements occur. Transhumance is by far the most important of these migrations from the hills to the plains, but, al-though it is both a temporary as well as a round-trip journey, it is on a larger scale and is more regular than other forms of movement. Mountain dwellers may be hired in the lowlands at harvest time; landless peasants, unemployed artisans, and casual agricultural workers may decide to seek cash income in the lowlands, be it in the formal or the informal economy. Market places in the foothills or plains periodically attract producers from steep valleys and remote upland areas. In addition, resource scarcity, special occasions (for instance, ceremonies in the plains), or military migration (mercenaries for regular armies or war lords), all induce downhill movements. Mountains always have provided a reservoir of labour for other people's use, and have contributed to the development of regions beyond their boundaries almost everywhere in the world and throughout history.

Conflicts generated in the context of migration are related to various kinds of interventions that induce a change in the pattern of movement; hinder or force migration artificially; create new forms of movement; and interrupt or eliminate traditional forms, such as nomadism and transhumance. *External intervention in highland–lowland interdependence and in functional solidarity established between people in the mountains and on the plains force the system into conflictual dependency and disfunctional partiality.* To explain this in more detail: most mountain conflicts are associated with transhumance and nomadism, the regular movement of men and their flocks, which is one of the most distinctive characteristics of the mountain world. Changes in domestic consumption, household mobility, and the degree of commercialisation are obviously important variables for all producer groups depending on mountain resources. However, today those changes seem to be less critical for the more settled farmers and other mountain dwellers, whereas it is the pastoralists, almost universally, who are challenged. They must compete with capital-intensive commercial ranching or dairying, on the one hand, and small-holding peasant husbandry, on the other. This latter indirectly reduces the role of herding and natural pastures by integration of animal husbandry and crop production. In addition, the pressure on pastoralists, caused by both governmental attempts at sedentarisation and by environmental transformation, is accelerating (see Galaty and Johnson, 1990).

Today the phenomenon of *transhumance* is even more complex than in the past, and it involves many factors, physical, human, and historical (Galaty and Johnson, 1990). It has become exceedingly difficult to classify the many different forms of transhumance; there is 'normal' transhumance, which is under pressure because of the scarcity of renewable resources. Cattle farmers, sheep farmers, and shepherds, in the early summer, leave the lowlands where they usually live because the season is unfavourable for stock raising. This vertical movement from the lowland winter pastures to the high quality summer pastures in the mountains was once a sustainable way of combining two geographically and climatically distinct altitudinal belts. However, the mountain pasture may either be the property of other peasant farmers from the plains, or be rented out to the mountain dwellers; it may also belong to settlers and farmers living in the mountains. Problems of ownership, rental, and use of mountain grazing rights are by no means new. Braudel (1995: p. 91) reports on a notorious conflict between the Grisons farmers in the southeastern Swiss Alps and people from Venice; the first group had had to drive their stock to the southern Alps and towards Venice, to a mountain area rented by the Vicentini for their own herders who needed to pasture huge flocks. Over time, population pressure continued to grow, resource degradation accelerated, and productivity declined, so that the struggle for one and the same resource is far more serious today than in past centuries. However, generally speaking, the problem has shifted from Europe to areas where the dependence on natural resources is much higher than in the successfully modernised economies. The different phases of the highland–lowland conflict in Jebel Marra massif (see above) clearly demonstrate the pattern of intensified potential for conflict between farmers and pastoralists arising from the last two decades of unusual drought in the Sahel (Suliman, 1996, Vol. I: pp. 109–145).

Inverse transhumance, in contrast, involved shepherds and their flocks descending from the highlands, mainly for marketing purposes. Herders and animals rapidly left the high altitude pastures at the beginning of the cold season to escape the harsh conditions and rushed into the lower regions 'like an invading army', according to reports on seventeenth century transhumance and related conflicts in the Mediterranean region and: 'All doors were padlocked against these unwelcome visitors, and every year saw a renewal of the eternal war between shepherd and peasant, first on the way down, until the flocks reached the open plains or the wide grazing lands . . .' (Braudel, 1995: p. 86). When, because of land scarcity, the same grazing areas were used during the same season by the practitioners of both inverse and normal transhumance, then conflict became virtually inevitable. The traditional trade-offs, after the principle: 'our flocks may go to your land in summer, but your flocks come to ours in winter' fell by the wayside. Thus, when the resource base became degraded, the different land-use interests determined the separation between peasants and (agro-)pastoralists.

All transhumance resulted from an agricultural situation that was unable to support the pastoral economy in its entirety; this, therefore, induced the off-loading of its burdens either onto the lowland or the mountain pastures. This has persisted until the present day. Governments, however, tend to disregard the advantages of pastoralist resource use; furthermore, in recent years pastoralism has also been blamed for dryland degradation. Although this critical view of pastoralism has been challenged, climate change, population growth, advances in agriculture, and giant projects (mountain dams and land irrigation schemes on the plains) are still threatening to break-up the formerly pastoral ways of life.

Transhumance is only one form of pastoralism which concerns a specialised population, namely the shepherds; it implies a division of labour, and also involves, in combination, a settled form of agriculture with crops, fixed dwellings and villages. *Nomadism,* in contrast, embraces the entire community which moves over long distances: people, animals, and even dwellings. The policies of present-day governments, aimed to foster sedentary life styles, have converted the traditional nomadism to a modified pastoral way of life, a division of labour, that increasingly resembles transhumance. Nomads have to defend themselves against the multiple challenges to their way of life as such – be it on the plains or in the mountains. Thus, nomadism is confronted with similar problems to those faced by transhumance: the pressure on, and the interruption of, seasonal vertical and horizontal movements.

Summary

Conflicting patterns of migration in mountains are focused around the two poles: *impenetrability* and *interdependence*. It is the imperfect, and therefore unstable, penetration of 'impenetrable' mountains that has become a distant trigger of on-going struggle, social conflict, and war. The concept of an impenetrable mountain area high above the plains, with some rather wild-looking poor people is misleading in many respects, even in regions that have remained on the fringe of the great waves of civilisation. On the contrary, it is the unsustainable, weak, and temporary penetration of the mountains that has caused, and still causes, conflict of interests amongst the different actors involved. The marginalisation of mountain communities – which is a result of the unsustainable penetration – forces, at best, the separation of highland and lowland cultures through various interventions, such as the establishment of clear-cut political or economic boundaries. In the worst case, this marginalisation of mountain communities provokes the formation of organised identity groups which are ready to defend their interests with military force.

On the other side, there is a trend towards conflict-prone dependence and dysfunctional partiality. Since the asymmetric structure in most cases favours the lowlands, all forms of vertical migration are challenged. Highland–lowland interactions which, in principle, have resulted from inter-dependent developments, collapse. Changes in the essential environmental factors around which all agricultural and most pastoral economies are organised lead to painful, and sometimes violent, adaptation processes: the (re-)distribution of pasture, the seasonal over-use of available land, the struggle for control of water resources, the nature of animal and household mobility, the division of labour, and the interaction between the highland and lowland economies.

SUSTAINABLE MOUNTAIN DEVELOPMENT AND CONFLICT MANAGEMENT

Governments tend to perceive mountains as unfavourable places for regional development. In self-fulfilling prophecies, mountains become ever more marginalised regions, zones of dispersed habitat where hamlets are islands set in the middle of uncultivable space. As a consequence, travel remains difficult with long-distance transport on inferior roads. People here are forced to be self-sufficient for their basic needs, to produce everything as best they can, even if the soil and the climate are unsuitable. Finally, migrants have to accept the influence of modernity which hinders all traditional forms of movement and creates new, and often more precarious, modes of living.

Environmentally-induced conflicts in the mountains have an overall similarity yet an infinite variety in detail, so that it is impossible to provide one coherent and encompassing strategy to address all such problems together. Nevertheless, it is vital that a way be found to deal more effectively with both potential and existing mountain conflicts. It is not surprising that current and past attempts to resolve uphill–downhill conflicts have not included the more equitable sharing of resources – government performance in mountains tends to be inadequate. Reconciliation attempts are frequently biased by political power struggles among regional or national elites which, in light of their motivation to develop remote areas, 'sets the fox to keep the geese'.

Management of mountain conflicts, therefore, should be related to the transformation of the societal relations with natural resources in their very specific geographical settings. It must address the relationships among the various actors caught up in this syndrome, here described as *imperfect penetration* and *conflict-prone dependence*. In other words, conflict management has to focus on the economic and environmental roots which cause multiple stress upon the vertical highland–lowland interactions. It is the participatory approach, based on grounded theory, that produces adequate, sustainable, and realistic therapies. In the years ahead, therefore, it will be necessary to highlight the participatory aspects of the traditional mechanisms that regulate access to resources; these have been greatly under-estimated so far. In order to effect a geographically-adapted and politically-integrated conflict and resource management strategy, the following questions may serve as guidelines:

(1) **What methods of traditional conflict management still exist in highland and mountain areas?** To answer this, it will be crucial to undertake an inventory of the methods in specific areas. Data will be needed to facilitate the categorisation of regions, communities, and villages, on the one hand, and to identify the actors (groups, leaders, chiefs, individuals), on the other. The two dimensions together shape what is called the arena. An evaluation of such data should lead to a typology of those traditional methods that are still viable.

(2) **What kind of data are required to answer the previous question?** In principle, disaggregated data are needed, that is, data at the local level,

or stemming from interviews with focal groups in particular communities. These data will be mainly qualitative, that is, on behaviour, perceptions, interests, tools/methods, procedures, concerns, difficulties, successes, and failures.

(3) **Are there any viable traditional methods?** The following problem areas must be identified: who has the knowledge; who is involved in conflict management; at what level; what are the environmental/resource problems that people in a given arena have to face?

(4) **How do the viable traditional methods function?** Here, one has to find out where they work and when they work (stage of conflict). Is there intervention by regional governments, by the central government, by experts? It is essential to know if intervention by third parties (external) is positive in conflict management or not. Additionally, the role that cultural relationships between kinship groups play, and also, those between kinship groups and other actors must be determined.

(5) **Are traditional methods under severe stress because of accelerating environmental degradation and associated social or political responses, or for any other reason?** What role do environmental constraints play concerning perceptions of adequacy of farm land, productivity, resource competition (livestock raising *versus* farming), or human security?

(6) **Why do traditional methods work in some places and not in others?** Is there a need for modified or new methods of environmental conflict and resource management? How can traditional (local) and modern (state) methods be combined?

(7) **Is third party intervention a panacea, or must most conflicts be regulated by the actors directly involved?** Do third parties play a major role or not, should their role be minimised; should it be strengthened; or should it focus only on certain phases or aspects of a given conflict?

The issues addressed here indicate the direction in which practical steps should lead. The primary approach is decentralisation of political decision-making, subsidiarity in conflict resolution attempts, participation of traditional groups, and new approaches to non-discriminatory resource sharing. It should be acknowledged that the continued degradation of mountains and highlands is undoubtedly one of the main factors in the increas-

ingly serious problem of food security in several countries (this correlates with data on food crises provided by the FAO). Nevertheless, caution is needed so as not to raise the false hope that the lowlands may be able to relieve the pressure on the highlands; nowadays, they are also under heavy pressure. The most decisive factor, therefore, will be management of all available resources, including land, water, and vegetation, to ensure the improvement of human security. This must be, primarily, the improvement of self-help capacity, the diversification of the economy by providing off-farm opportunities, the development of indigenous resource and conflict management strategies, and local political empowerment.

The transformation of the guidelines into practical solutions should not only enhance the awareness of both the favourable, as well as the unfavourable aspects of mountain life, it should also contribute to a shift in political power from urban centres to regional and local authorities. It is the local communities who are most aware of their own ecological and social fragility, as well as of the advantages of being a mountain dweller.

POLICY STRATEGIES FOR MOUNTAIN ENVIRONMENTAL CONFLICT MANAGEMENT

Conflicts in mountain areas have been presented here in terms of the dichotomy: *impenetrability* versus *interdependence*. It is only logical, therefore, to develop policy recommendations within the aegis of these two constructs. Impenetrability has a positive aspect and is related to certain attributes of mountain systems; these may be described as *resource diversity*. Inter-dependence can be linked to the idea of *freedom of movement*. The problem is how to achieve sustainable mountain development within the dual framework of conservation of resource diversity and freedom of movement.

First, it must be realised that resource diversity does not necessarily mean easy access to, or best use of, resources, and that freedom of movement *per se* offers no panacea. The latter can easily lead to disastrous migration to higher and more sensitive eco-zones. Examples of this include the adverse effects of winter tourism in parts of the Alps, and the dispute between the European Union and Switzerland over the admission of heavy trucks across the high mountain chain that bisects the European free market. Sustainable use of diverse resources and freedom of movement, therefore, are designed as recommendations which must be qualified according to specific conditions. Otherwise they will tend to transform common 'goods' into common 'bads'. Thus, labour intensive development should be seen as a link between the use of the large variety of resources and freedom of movement.

Labour intensive resource use

One of the advantages of a mountain region is that, because of the considerable environmental variation due to the range of altitude, it offers a wide variety of resources: from vineyards, arable fields, and fruit trees on the lower slopes, to the forests and pastureland above. Almost every mountain has some arable land, for instance in the valleys, or on the terraces. Cultivation of crops can be supplemented by the produce from stock raising, mainly sheep and goats, but also cattle. On slopes which are unsuitable for animals, stones must be cleared from the fields by hand, soil prevented from slipping downslope, and, if necessary, carried back upslope and secured with dry stone walls. This is never-ending work; if it stops, the mountains revert to a wilderness and the induced instability will possibly also damage the lowlands. This, for instance, has been experienced in Eritrea; after a thirty-year struggle for independence, which forced the neglect of the land, renewed terracing has become crucial, once more, to sustainable farming. Similarly, in Rwanda, at the end of both the monarchy and the colonial period in 1959, peasants destroyed their own terraces because their original construction had been forced by the ruling Batutsi elites and Belgian colonial power. A major benefit of labour intensive production in mountains is that it provides a sustainable and environmentally sound mechanism to absorb surplus rural labour. Whenever labour intensive projects, such as terracing, can be funded by governments, international agencies, rich farmers, or entrepreneurs, local income can be increased and local economic growth assured; such an approach will also improve the infrastructure, provide rural employment, and reduce rural–urban migration.

Strengthening local responsibility

Labour intensive resource use should encompass another component; the welfare of the mountain inhabitants and their environment should not be

sacrificed to dominant national interests, whereby revenues and profits are collected solely to benefit central governments. Regional resources should be used, not only to improve the local and regional allocation of income, but also to strengthen, or establish, regional political structures. Self-determination, federal resource and power sharing, subsidiarity in decision making, and autonomy can help mitigate potential vertical conflict, on the one hand, and ensure appropriate development, on the other. In many mountain areas, small is not only beautiful, it is also powerful, because the diversity of conditions, as well as the local interests, are recognised. The demarcation of boundaries is also very important; in some instances, it might be beneficial to combine sections of highland and lowland into a single administrative unit (province, canton) in order to facilitate communication, as well as vertical economic activities and solidarity. In other situations, such action might not be appropriate; for instance, in Eritrea the recent change in provincial boundaries has led to new conflicts between farmers in the western lowlands and migrants moving down from the highlands in search of new land (Spangenberg, 1996).

Multiple resource-use systems

The establishment of transboundary mountain eco-regions would be a complementary strategy to the establishment of regional political structures. Why not introduce a multiple resource-use systems approach to ensure the most effective treatment of the different altitudinal belts? A 'mountain charter' could encompass conservation measures for the upper belts, together with protection of primary vegetation and habitats as gene pools; it could allocate forest areas for agroforestry and other purposes, and delineate areas for cattle grazing, intensive recreational use, and so on. The previously discussed inventory of resources would need to be taken into account and monitoring systems could be established; this would also create new jobs.

Labour intensive forms of movement

Policy directed to labour intensive forms of movement should entail improvement in the status of the pastoralists. Pastoralism represents a particular activity within regions where more diverse economic activities are pursued; and it is often a strategy for exploiting areas that are too marginal for other uses. In more fertile areas pastoralism is a means of strengthening, or diversifying, a peasant economy. At the same time, pastoralism may guarantee the autonomy of domestic groups, or it may be pursued as a regional specialisation. The negative image of pastoralism could be off-set by conceptualising it as a strategy for providing animal products to others within a system of reciprocal dependence and functional solidarity (Galaty and Johnson, 1990: p. 20).

Sustainable pastoralism

Three systems of animal production are emerging throughout the world: commercial ranching; modern dairying; and mixed farming. Only recently has it been acknowledged that most pastoral systems achieve better returns on pastoral labour than do other systems, such as subsistence, or semi-commercialised mixed farming, although the returns are far less than those of fully commercialised systems. The dilemma is that market systems require greatly reduced rural populations on the plains; for instance, in the industrialised countries, but not where 70 to 95 percent of the population is rural. Only pastoral systems support an optimal rural population on relatively marginal land; they are relatively low-density, low-intensity systems, based on extensive use of land and, in the absence of capital improvements, relatively intensive use of labour (Galaty and Johnson, 1990: p. 16). Livestock densities reflect land potential: lower densities occur in drier regions, linked to a high degree of movement, while greater rainfall makes higher densities possible, associated with more sedentary husbandry – closely interconnected with cropping (see also the analyses of the Pastoral and Environmental Network on the Horn of Africa: PENHA).

Intensification of agriculture can occur principally by increasing the productivity of land (density) or by intensifying labour. Where the first is not possible in the short term, the latter should be targeted by local leadership, central governments, and international agencies. Ethiopia provides a good example for a labour intensive strategy and leads Africa in virtually every category of livestock holding, with animals serving multiple functions: for milk, meat, manure, transport, and traction. Here is located the largest national herd on the continent, occupying highland and lowland pastures in combination. Also, with the largest agricultural population on the continent, Ethiopian

per capita holdings are 0.85 cattle/person (with a tendency to decrease).

Coherent policy aims for integrating the following elements:

Risk reduction This is an essential pastoral strategy, an attempt to decrease uncertainty by anticipation. Human security can be increased by creating alliances across ecological zones, distributing livestock among co-operative partners, securing rights in dry season pastures, and increasing herds in anticipation of future losses.

Variation of tenure systems There is great variation among different pastoral systems in the nature of land control, tenure, and use. Where herders form part of larger, more complex regional societies, they rarely exercise exclusive control over a pastoral domain, but rather retain, or periodically negotiate, seasonal rights for grazing. For instance, in the Hindu Kush, Afghani herders may own land in the valleys, but must negotiate the rents for distant winter pastures from permanent residents (Balikçi, 1990). Many pastoralists need access to two or more ecological zones because the maintenance of animals in dry season, or winter, refuges is a necessary condition for effective exploitation of the more extensive wet season, or summer, pastures.

Adaptation to the environment If pastoral mobility represents an adaptation to environments with sparse vegetation, the timing and direction of movements are also shaped by the availability of markets, political control of pastures, labour availability, and sociability. With population growth and enclosure, the nature of land tenure and mobility tends to change in tandem. Sedentary occupation comes to be recognised as 'ownership', with seasonal grazing rights easily extinguished or transformed into relationships of leasehold or rent.

Diversification The effective assimilation of pastoral domains into increasingly restricted territories by the states – through the creation of ranches, associations, villages, game parks and reserves, or gazetted forests – has artificially stimulated a process of land intensification. Pastoralists have to respond by adding labour, be it in more intensive animal care or by seeking a wider base for economic security. The latter could encompass activities such as involvement in wage labour and petty commerce, animal marketing because of higher demands from a growing urban population, in real estate, and other off-farm jobs.

New programmes in resource management Many programmes of pastoral development seek to initiate change through altering the vital relationship between herder and pasture resources, by creating co-operative ranches, villages, or groups, or simply by subdividing and enclosing the rangeland. One element could be to secure rights in fixed land resources while attempting to minimise herd mobility, as far as it is practical both economically and ecologically.

Labour intensity in agriculture and other sectors Many highlands and mountains have the potential for agriculture and forestry development, and other economic activities (crafts, small industries, services, tourism), but these can only be achieved if the resources are managed and utilised on a sustainable basis.

CONCLUSIONS

In conclusion some crucial points are listed with examples of successes in brackets:

(1) constructing a network of furrows and irrigation channels in order to manage and distribute water for food crops and domestic use (hills of Sicily);

(2) terracing steep slopes in order to use marginal land without substantially degrading it; terracing catchment areas of small dams that serve a village without jeopardising water use of downstream users (Areza dam, in Eritrea);

(3) establishing mixed farming systems where there is adequate water (the Alps); local and regional marketing of agricultural products;

(4) planting of drought-tolerant food crops and cash crops for more food security and for accumulating some cash (Tanzania);

(5) shift to stall-fed cattle where grazing land is scarce or where the land tenure systems favour dairying and mixed farming (partially, in Rwanda);

(6) electrification to promote crafts and small industries for landless people, mainly young men and women; diversifying the production base of an area by developing labour-intensive manufacturing and handicrafts, cottage industries, with the purpose of absorbing both surplus peasant labour and produce, and the livestock herders, to cope with the 'too-many-people-doing-the-same-thing-syndrome' (European mountains; Swiss and French Jura);

(7) realisation of land reform to return land to its original owners, roll back mechanised farming and nullify concessions of large tracts to absentee landlords; clarify competing land tenure systems. The latifundia system, and its imitations, are serious inhibiting factors in terms of the empowerment of rural populations in developing countries and, thus, of the efforts to establish viable political institutions (Lane and Ersson, 1994: p. 189).

The long-term success of such strategies will largely depend on the will of the actors involved to share available resources, on central governments to

Box 6.5 Chipko – a unique movement of mountain people for sustainability

The Chipko movement emerged in the early 1970s in the Himalayan areas of the state of Uttar Pradesh, India, as an expression of resistance of mountain communities to commercial logging in the local forests by rich private contractors. With increasing population, declining job opportunities, and dwindling open forest resources in the mountains, the profiteering from government reserved forests by private contractors from the plains already provided the objective framework for a plains *versus* mountains conflict. Commercial forestry in the reserved forests of Uttarakhand, as declared under the British Raj, had always led to a deep sense of marginalisation and anguish among the mountain people. Forest policy in post-colonial India did not alter this situation and the potential for the conflict remained alive. In 1973, in the Mandal forest area of the district of Chamoli, this long-standing local anguish expressed itself through spontaneous action; the local people forced forest contractors to flee. They had come into the area to cut ash trees for a sports goods company located in the plains. The news that felling by the highly influential forest contractors, who had seemed politically invincible to the small and isolated mountain villages, could be stopped by collective community action, soon spread throughout the mountain region. Various social activist groups, ranging from non-violent Gandhians, organised under the Uttarakhand Sarvodaya Mandal, to Marxists organised by various communist parties, immediately seized the opportunity; in the short span of a few months, determined local resistance spread and commercial forest fellings were stopped throughout this region, known historically as Uttarakhand (Guha, 1989). In 1974, the first formal victory of the movement came when the state government was forced to agree to demands to discontinue the private contract system of forest felling. A state-run organisation, the Forest Development Corporation, replaced it and worked with co-operatives formed by people from the nearby villages.

The movement, at this stage, was aimed solely at resolving the conflict over access to the resources of the local forests and was eminently successful. It derived its now famous name 'Chipko', meaning to embrace, from speeches made by activists such as Bhatt, and from a poem written by the famous folk poet of the movement, Shailani; they called for the protection of the local forests from the onslaught of felling by the private contractors, endorsing the physical act of embracing the trees, if necessary. Contrary to popular perception created by ill-informed popular writings, this strategy of embracing the trees was never actually used, with a single exception. In 1977, large groups of women led by the courageous Bachhni Devi, herself the wife of a forest contractor and local village head, physically prevented trees being felled in the Adwani forests by

Plate 6.3 Sunderlal Bahuguna, Ghandian activist and philosopher – the most successful campaigner of the movement (Photograph courtesy of *Mountain Research and Development*, 1986, 6(2))

axemen under contract. It was on this occasion that the ecological slogan 'What do the forests bear? Soil, water, and pure air' was born. (Shiva and Bandyopadhyay, 1986). In the other parts of Uttarakhand, leading activists, such as Bhatt, went on to consolidate their gains by strengthening local initiatives in forest protection and afforestation. Another leading activist, Bahuguna, used the Chipko lessons to articulate the deeper conflict over the question of sustainable development of mountain communities. Metamorphosis of 'Chipko', from a peasant movement tied to control of forest resources in Uttarakhand, into a movement for sustainable development of the mountain communities, was thus launched (Bandyopadhyay, 1992).

Mountain development, as practised, was seen in the Chipko view, as nothing but selfish exploitation of the rich natural mountain resources by industrial economies of the plains. To gain a broader understanding of this conflict, and to generate wider public attention to the destructive nature of development projects in the mountains, Bahuguna embarked on a remarkable 4870 km long trek in 1981, along the entire length of the Himalaya, from Kashmir to Kohima, including the mountain kingdoms of Nepal and Bhutan. His concepts of conflicting developmental interests between the mountains and the plains has led Bahuguna to protest the plans for exploitation of the rich water resources of the Himalaya through construction of high dams; the Tehri high dam, on the Bhagirathi River, a Himalayan tributary of the Ganges, has become a major flash point and has attracted widespread international attention.

In recent years, the Chipko experience has became more widely known and incidences of non-violent protests against forest felling in various parts of the world have drawn strength from the Chipko successes. As a non-violent mode of collective social action for sustainable development, the Chipko experience has become a remarkable and valuable source of inspiration at the global level. The tiny mountain communities of Uttarakhand have shown clearly that humankind's move towards sustainability can be initiated from the remotest villages; progress can be made, while high level participants of innumerable conferences and negotiations in the megacities of the world continue to spin wheels.

Source: Jayanta Bandyopadhyay, International Academy of The Environment, Geneva

perform as active partners, and on the explicit targeting of development programmes in order to mitigate social turbulence induced by environmental factors. As Mohamed Suliman pointed out:

'Where ecology is fragile, social peace is also fragile and armed conflict can only be avoided through some form of equitable sharing of available resources.' (Suliman, 1996: p. 176).

References

AESED (Association Europeenne des Societes d'Etudes pour le Developpement) (1961). *Etude globale de développement du Rwanda et du Burundi.* Rapport général, rapport analytique: Bruxelles

Bächler, G., Böge, V., Klötzli, S., Libiszewski, S. and Spillmann, K. R. (1996). *Kriegsursache Umweltzer-störung. Ökologische Konflikte in der Dritten Welt und Wege ihrer friedlichen Bearbeitung*, Bd. I. Verlag Rüegger: Zürich, p. 401

Bächler, G. (1996). Rwanda: The Roots of Tragedy. Battle for Elimination on an Ethno-Political and Eco-logical Basis. In Bächler, G. and Spillmann, K. R., *Environmental Degradation as a Cause of War*, Vol II. Verlag Rüegger: Zürich, pp. 461–501

Bandyopadhyay, B. J. (1992). The Himalaya: Pro-spects for and constraints on sustainable development. In Stone, P. B. (ed.), *The State of the World's Moun-tains*. Zed Books: London, pp. 93–126

Bart, F. (1993). *Montagnes d'Afrique. Terres Pay-sannes. Le cas du Rwanda.* Presses Universitaires: Bordeaux

Balikçi, A. (1990). Tenure and Transhumance: Strati-fication and Pastoralism among the Lankenhel. In Galaty, J. G. and Johnson, D. L. (eds.), *The World of Pastoralism. Herding Systems in Comparative Perspective*. Guilford Press: New York/London, pp. 216–55

Becker, Peter (1993). Rwanda. In: Nohlen, D. and Nuscheler, F. (eds), *Handbuch der Dritten Welt* Bd. 5, Ostafrika und Südafrika. Bonn, pp. 114–133.

Bencherifa, A. (1988). Demography and Cultural Ecology of the Atlas Mountains of Morocco: Some New Hypotheses. *Mountain Research and Develop-ment*, 8(4): 309–13

Braudel, F. (1995). *The Mediterranean World in the Age of Philip II.* Volume I, University of California Press: Berkeley, Los Angeles, London, [first published in France under the title: *La Méditerranée et la Monde Méditerranéen à l'Epoque de Philippe II,* 3 Volumes, Paris 1949]

Brown, L. R. and Wolf, E. C. (1984). *Soil Erosion: Quiet Crisis in the World Economy.* WorldWatch Paper Nr. 60. WorldWatch Institute: Washington DC

Denniston, D. (1995). Sustaining Mountain Peoples and Environments. In *State of the World*. World-Watch Institute Report on Progress Toward a Sustain-able Society. W. W. Norton and Company: New York, pp. 38–57

Galaty, J. G. and Johnson, D. L. (1990). Introduction: Pastoral Systems in Global Perspective. In *ibid.* (eds.), *The World of Pastoralism. Herding Systems in Comparative Perspective.* Guilford Press: New York/London, pp. 1–33

Guha, R. (1989). *The Unquiet Woods: Ecological Change and Peasant Resistance in the Himalaya*. Uni-versity of California Press: Berkeley, Los Angeles and Oxford, p. 214

Guichaoua, A. (1989). *Destins paysans et politiques agraires en Afrique Centrale*, Tome 1. L'édition Har-mattan: Paris

Homer-Dixon, T. (1994). Across the Threshold: Empirical Evidence on Environmental Scarcities as Causes of Violent Conflict. *International Security*, 19(1): 5–40

Hurni, H. (1990). Degradation and Conservation of Soil Resources in the Ethiopian Highlands. In Messerli, B. and Hurni, H. (eds.), *African Mountains and Highlands. Problems and Perspectives*. Wals-worth Press: Marceline, pp. 51–65

Ives, J. D. and Messerli, B. (1989). *The Himalayan Dilemma: Reconciling Development and Conserva-tion.* Routledge: London/New York, p. 295

Klötzli, S. (1996). The Water and Soil Crisis in Central Asia – a Source for Future Conflicts?, In Bächler, G. and Spillmann, K. R. (eds.), *Environmental Degrada-tion as a Cause of War*, Vol II. Verlag Rüegger: Zürich, pp. 247–336

Lachenmann, G. (1990). *Ökologische Krise und sozialer Wandel in afrikanischen Ländern*. Verlag Breitenbach: Saarbrücken/Fort Lauderdale

Lane, J.-E. and Svante, E. (1994). *Comparative Politics. An Introduction and New Approach*. Black-well: Cambridge, MA

Libiszewski, S. (1996). Water Disputes in the Jordan Basin Region and their Role in the Resolution of the

Arab–Israeli Conflict. In Bächler, G. and Spillmann, K. R. (eds.), *Environmental Degradation as a Cause of War*, Vol II. Verlag Rüegger: Zürich, pp. 337–460

Newbury, C. (1988). *The Cohesion of Oppression. Clientship and Ethnicity in Rwanda, 1860–1960*. Columbia University Press: New York

Wolde-Mariam, M. (1991). *Suffering under God's Environment. A Vertical Study of the Predicament of Peasants in North-Central Ethiopia*. Walsworth Press: Marceline

Mountain Agenda (1992). *Fragile Ecosytems and Mountains*. United Nations Earth Summit 1992, Research Paper Nr. 57

Omara-Ojungu, P. H. (1990). Resource Management Implications of Ecological Dynamics in Mountainous Regions. In Messerli, B. and Hurni, H. (eds.), *African Mountains and Highlands. Problems and Perspectives*. Walsworth Press: Marceline

Price, L. W. (1981). *Mountains and Man: A Study of Process and Environment*. University of California Press: Berkeley

Shiva, V. and Bandyopadhyay, B. J. (1986). The evaluation, structure and impact of the Chipko Move-ment. *Mountain Research and Development*, 6(2): 133–42

Sottas, B. (1996). Die Ewaso Ng'iro Arena. Legitimationen und Diskurse bei der Nutzung knapper Ressourcen nördlich des Mt. Kenya. In Müller, H.-P. (ed.), *Weltgesellschaft und kulturelles Erbe. Gliederung und Dynamik der Entwicklungsländer*. Reimers Verlag: Berlin

Spangenberg, U. (1996). *Ökologische Probleme und innenpolitischer Machtkampf. Eine Untersuchung am Beispiel von Eritrea*. Hamburg (Diplomarbeit im Rahmen des Environmental Conflict Management Projekts ECOMAN im Horn von Afrika)

Suliman, M. (1996). War in Darfur or the Desert *versus* the Oasis Syndrome. In Bächler, G. and Spillmann, K. R. (eds.), *Environmental Degradation as a Cause of War,* Vol II. Verlag Rüegger: Zürich, pp. 145–80

United Nations Development Programme, UNDP (1996). *Human Development Report 1996*. OUP: New York/Oxford

(An extensive bibliography on environmentally induced conflicts can be found in: Bachler *et al.*, 1996: Vol. 1)

Highland waters – a resource of global significance

7

J. Bandyopadhyay, J. C. Rodda, Rick Kattelmann, Z. W. Kundzewicz, and D. Kraemer

INTRODUCTION

Background

Almost all the world's major rivers, and many of the minor ones, begin in mountainous regions and supply a large percentage of the water resources of the entire globe. Countries that contain significant areas of mountains are usually well blessed with water resources, often much of it initially in the form of snow and ice. Rivers draining the mountains convey a share of these resources to neighbouring nations through the basins that they occupy jointly. In the arid and semi-arid areas of western Asia and Latin America, the juxtaposition of well-watered mountain ranges and dry plains crossed by rivers draining from the mountains, highlights this point. Even in temperate parts of the globe, downstream countries depend on mountain areas in upstream countries for much of their water resources. The earliest civilisations, Egypt, Mesopotamia, and China, for example, owed their existence to copious water resources provided by the rivers flowing from the mountains upstream, particularly in the form of annual floods. In these and later civilisations, water was esteemed because it was thought to be endowed with special life-giving powers. For example, in the case of the Celts, who inhabited much of Europe prior to the Romans, water was basic to their rituals and practices, a position it occupies today in a number of religions.

This chapter examines the role of mountains with respect to water resources. It discusses how these resources are assessed and some of the problems involved. It briefly mentions floods and droughts in the context of mountain communities. The health of aquatic systems in terms of human impacts and prospects for sustainable development of mountain water resources are discussed in the following sections. The chapter concludes with a research agenda and recommendations for watershed management as an approach to addressing problems of mountain water resources.

Hydrological characteristics

From the hydrological point of view, mountains present a paradox. Although they are the source of the greatest part of the world's water resources, knowledge of the hydrology of mountains is generally much less extensive, reliable, and precise than that of other physiographic regions. And it is this knowledge on which the assessments of water resources are based. Indeed, some authorities consider that mountain regions represent, in practical terms, 'the blackest of black boxes in the hydrological cycle'. Only their output is known to any degree of exactitude. Yet mountain hydrology is a challenge to researchers and practitioners wishing to improve their understanding. Hydrological studies in mountain areas are often pursued under the various programmes of representative and experimental catchments. Significant failures of hydraulic engineering projects in mountain areas are relatively common when the projects have been designed with only the sketchy data that are normally available. Within a few years, or decades, the annual runoff or flood magnitudes may turn out to be much different than those anticipated.

The particular form and structure of mountain areas provide most of them with three important hydrological attributes:

(1) **Temporary storage** in the form of snow and ice that brings about a delay in runoff. Snow cover built up in the higher regions during the winter months melts in the spring and contributes the largest volume of water on a seasonal basis to river flows and, in particular areas, to ground water recharge.

131

Figure 7.1 Global distribution of precipitation in relation to mountain ranges

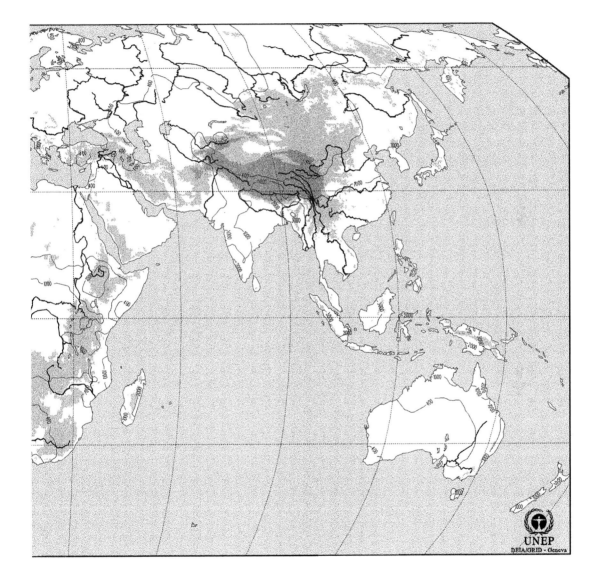

(2) **Natural lakes and man-made storage reservoirs** in mountain areas may be used for a number of purposes, but especially for water supply, irrigation, power generation, and flood control. The runoff retained to reduce flooding downstream can be employed later and beneficially.

(3) **Potential energy** that can be utilised to generate hydroelectricity. A number of countries meet much of their demand for electricity in this way: Norway, Switzerland, and Canada, for example.

The importance that mountains enjoy, in terms of water resources, derives principally from the enhanced precipitation that they engender due to the uplift and ascent of moist air over them. This orographic effect results from the operation of several meteorological mechanisms, which may act singly or together, according to the circumstances. Part of this enhanced precipitation falls as snow at higher altitudes and latitudes, resulting in glaciers and ice sheets at locations where conditions favour the process of firnification. The exact relation between precipitation and altitude varies considerably in different parts of the world, depending on the volume of moisture in the air and its temperature, the steepness of the ascent, the mechanism (or mechanisms) operating, and several other factors such as time of year. However, within the extensive literature on this topic, most results fall in the range between 0.05 mm and 7.5 mm per metre for the uniform increase in annual precipitation totals with altitude on windward slopes. There is also a range of altitudes in some mountain ranges within which maximum precipitation occurs; this is usually between 1500 and 4000 m, with totals decreasing above these levels, although this topic is subject to considerable debate and the inference may be a construct of the distribution of observing sites. On leeward sides of mountains precipitation gradients are usually less steep, distribution patterns are more variable, and significant areas can be found with low totals in 'rain shadows'. Of course, there are some mountains where there are not clear-cut windward and leeward sides and where the patterns of precipitation do not exhibit such obvious differences. Figure 7.1, showing the global distribution of annual precipitation totals, illustrates these points on the macro-scale. It also demonstrates that mountains have a significant effect on the distribution of the world's climatic zones, not only in terms of precipitation but also for other variables. On the micro-scale, mountains can also cause large differences in climate, and these effects can be of considerable economic importance.

A unique feature of mountain regions is the large amount of cloud. In some semi-arid areas, substantial amounts of water are harvested from these clouds, but their role is probably more significant in humid mountain areas, particularly where forests are present. There the forest canopy (Chapters 13 and 15), continually wetted by moisture intercepted from the cloud, can alter net precipitation and evaporation rates much more than short stature vegetation and, in turn, it alters the water balance. The presence of clouds also reduces the incoming radiation and consequently, depresses rates of evaporation. In contrast to precipitation, the literature is sparse on the relation between evaporation and altitude. However, it seems logical to assume that evaporation decreases with altitude, although all the main controls do not respond in the same way (Lang, 1981). For the Alps, evaporation/altitude relationships show gradients from 0.07 mm to 0.36 mm per metre for annual totals.

The precipitation that reaches the ground, liquid or solid, is partitioned over time between

Box 7.1 Switzerland – The water tower of Europe

Switzerland, a small mountainous country in south-central Europe, is sometimes called the water tower of the continent. The country receives on average 1450 mm of precipitation annually, which amounts to some 60 km^3 of water. It supplies four major European rivers with much of their flow – the Rhine, the Rhone, the Danube, and the Po. These four rivers and a number of smaller ones, carry 67 percent of this water across the borders of Switzerland into the neighbouring countries, namely France, Germany, Austria, and Italy, and to the other countries below them in the basins of the Rhine and Danube. Within Switzerland, 136 km^3 of water is stored in lakes and reservoirs and a further 74 km^3 in glaciers, sufficient water to maintain river flows for almost 5 years in the absence of further precipitation. There are similarly-placed countries and regions in other parts of the world, but the data from them are not sufficiently reliable to enable similar assessments to be made to the same precision.

Source: John Rodda,
President IAHS, Yngslas, UK

evaporation, saturated and unsaturated subsurface flow, and overland flow, the relative magnitude of each component being largely determined by the nature of the particular surface, including factors such as gradient and soil depth. When typically thin mountain soils are saturated, frozen, or replaced by bare rock, overland flow is enhanced. Vegetation type, areal extent, and canopy density can affect hydrologic response. Coniferous trees, for example, can intercept precipitation, modify microclimate and local energy balance, alter evaporation rates, and delay snow melt. Trees alter soil conditions via their litter, roots, and soil moisture uptake and affect the pattern of overland and subsurface flow and the time taken for the concentration of these flows into the network of stream channels (Chapter 15). Except where deep percolation and ground water recharge occur, mountains ensure a high specific runoff, the basis of their importance in water resources terms. Runoff efficiency (steamflow as a proportion of total precipitation) generally increases with increasing altitude and increasing precipitation. Mountain catchments, even when undisturbed, are also effective producers of sediment, the other principle constituent of streamflow. Some runoff from mountain areas may result from glacier melt at higher locations where the configuration of the topography allows this type of storage, but seasonal snow melt is a feature of more widespread significance. These factors, together with others such as aspect, exposure, the presence of peat and natural pipes, endow mountain regions with a variety which other physiographic regions may lack. This hydrological heterogeneity provides a basis for many of the features that attract mankind to mountain regions: it is also the source of most of the hazards which endanger human occupancy of mountains (Chapter 16).

Human impacts

Mountains may appear to offer a pristine environment, one free from the problems of lowlands and other regions. However, this impression is not true. In most parts of the world, mountains are presently subjected to intense pressures from human activity, pressures that will rise as the population of the Earth explodes and mobility increases so that remoteness is no longer a safeguard. Most of these pressures act on the hydrological cycle to change the quantity and quality of water yield and these changes, in turn, have impacts on other sectors. One major consequence is that in the future mountains will be less able to meet the demands for water resources than in the past.

Pressures on water resources in mountains were initially associated with development of natural resources, such as, minerals, forest products, agriculture, and water itself. More recently, they have resulted from continued expansion of settlements, agriculture, transportation corridors, and also from the growth of leisure pursuits, particularly winter sports (Chapter 12). These activities occur within mountain regions, but in recent years, mountains have been subject to increasing pressures generated from outside, with air pollution and acid rain being prime problems. Cloud droplets charged with emission products, such as sulfur compounds, are intercepted by the forest canopy, particularly conifers. Deposition of these products occurs in large volumes due to the increased precipitation, while the process of acidification is further enhanced because of the lack of soil buffering in many mountainous regions and the existence of acidic granitic bedrock over wide areas.

Within mountain regions, production of minerals causes water quality problems through removal of vegetation, soil disturbance, mine drainage, washing of products, and leaching from the heaps of waste. These difficulties can continue long after the mining has ceased (Chapter 9). Clearing of forest for roads, agriculture, residential development, or ski runs, can also change the quality of mountain rivers and streams, especially by increasing the volume of sediment transported (Chapter 15). In addition, forest felling changes the volume and distribution of runoff. The most noticeable effect is that storm runoff is more rapid and concentrated within a shorter period of time, while over longer periods the total volume of runoff is increased, until the vegetation cover regrows. When forests are not allowed to regenerate and when conversion to other land uses is semi-permanent, hydrologic impacts persist. Roads are a particularly significant conversion because vegetation is eliminated, soils are severely compacted and effectively sealed, natural slope drainage is interrupted, routes are often located in the riparian zone, and the road network may occupy a substantial proportion of a catchment. Building and construction can cause similar problems locally, while water pollution can result from the discharge of untreated waste from settlements and drainage from paved surfaces. Human activities that occur within and near streams have much greater hydrologic impact than those located well away from

Plate 7.1 Dam and Lake Oberaar (2303 m) with the calving glacier. This lake is part of a complex system of altitudinal lakes and power plants in the high mountain region of the Grimsel Pass, Switzerland (Photograph: KWO, Bern)

watercourses. Because the dimensions of the water balance for mountains are generally larger than for other physiographic regions, and gradients are much steeper, the loads of materials transported are usually greater, causing the degradation of basins to proceed more rapidly. Leaching of nutrients takes place at a greater pace, impoverishing the soil and polluting the receiving waters. These conditions have obvious consequences for the sustainability of mountain ecosystems.

ASSESSMENT OF MOUNTAIN WATER RESOURCES

Methodology

Water resources, world-wide, can neither be protected nor managed rationally without the continuing assessment of their quantity and quality, and what is available for use; mountain water resources are no exception. Water resources assessment programmes consist of a number of components: collection of hydrological data; collection of data about the basins where the assessment is being made, such as data on geology, vegetation, and soils; and the application of scientific methods for using these data in the assessment (WMO/UN-ESCO, 1991). Education, training, and research are often parts of such programmes. Networks of instruments have to be established to collect the hydrological data and they must be operated continuously to record the main variables, namely: precipitation, evaporation, river flow, soil moisture, ground water content, and storage in lakes and reservoirs and as snow and ice. These networks and allied observations can also collect data on water quality and the transport and deposition of sediment. Remotely-sensed data from satellites and

Box 7.2 High mountain regions and the water resources of the continents

Precipitation is the atmospheric source of our water resources on the Earth's land surface. However, the transport of water vapor in atmospheric general circulation and the generation of precipitation produces a very uneven distribution of precipitation and therefore of water resources throughout all continental areas outside the inner part of the tropical belt. This extremely uneven areal distribution of precipitation is caused by the uneven distribution of lifting processes of moist air masses in the atmospheric circulation. Besides the generation of precipitation in frontal systems the orographic forcing of advective and of convective precipitation in mountain regions is obviously a key process in the large-scale distribution of water resources origin.

As a consequence mountains represent the main 'water supply towers' for a great part of the continental areas in the subtropical, middle and high latitudinal zones. This is of vital importance in some arid and semiarid areas of the world where the water resources supply depends completely on river flow which may be fed to more than 90 percent from precipitation in the mountain source areas of such rivers.

Many regions in the world and their populations, economy and ecology can be identified in their strong dependence on water resources originating from high mountain regions. Typical examples are:

(1) Great parts of Central Asia from the Caspian Sea, Lake Aral and the Turan plain to the far east, with extreme continental climatic conditions, including the large desertic areas where average precipitation (P) can be less than 100 mm per year, and the greatest part of the whole continental area with P < 500 mm, most of which is lost by evaporation. The extended high mountain chains to the south of these central continental dry areas have precipitation of approximately 500–2000 mm. Here, many rivers have their main source areas, for example: the Tarek river (Caucasus); Amu Darya (Pamir) and Syr Darya (Tianshan) which feed Lake Aral; Tarim (Nanshan/Kuenlun) which drains into the Tarim basin; the Urumqi river (Tianshan), providing the only water supply for the whole region of Urumqi.

(2) Arabian Desert Peninsula with river flow from the Jemen and Oman mountain chains.

(3) Pakistan arid and semi-arid plains with the Indus river flow from the Karakorum/Himalaya.

(4) Rivers descending from the West Ghats in India.

(5) The Nile River, particulary the tributaries from the Ethiopian high mountain regions.

(6) North America West and Mid-West regions with main river flows from the Rocky Mountains, the Cascades and the Sierra Nevada, both to the dry west and to the dry mid-west areas.

(7) South America with rivers from the Andes feeding the dry regions of southern Argentina and Northern Chile and Peru.

(8) Australia with its water resources originating from the mountain regions in the east of the continent.

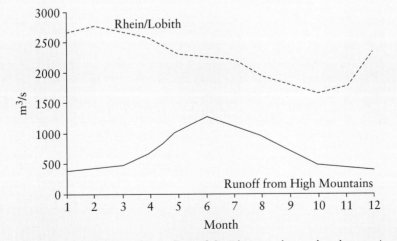

Figure 7.2 Seasonal distribution of river flow of the Rhein as observed at the margin of the Alps (18 035 km²) and in The Netherlands (160 800 km²), close to the North Sea (Rhein/Lobith)

(9) Europe and Asia Minor where most of the large rivers are fed to a great extent by high mountain regions: Ebro, Rhone, Po, Rhine, Danube/Inn, Tiber, Scandinavian Rivers, Euphrates and Tigris.

There is a particular aspect of many mountain rivers in their seasonal distribution of streamflow: a great part of the precipitation is accumulated as snow, and meltwater flow maximum occurs in spring and summer. This is very favourable in regions where at the same time evaporation is at a maximum, or in some places, for example, in the Mediterranean region and in the west of North America, where a summer minimum of precipitation is observed.

The figure (Figure 7.2) showing the seasonal flow conditions of the River Rhine illustrates the great importance of the alpine contribution to the total river flow downstream through France/Germany and The Netherlands. The alpine drainage basin, only 11 percent of the total basin area of 160 800 km² at Lobith, contributes 31 percent of the annual average streamflow, and more than 50 percent in June and July, when runoff from the lowland parts of the basin is strongly reduced because of maximum evaporation losses. This situation is much more extreme in more arid areas of the continents.

There is no doubt that the climatic and hydrologic conditions in the high mountain areas of our continents play a dominant role in the water supply of a large part of mankind. Therefore, high mountain areas deserve our greatest attention with a view to monitoring and research of their climatic, hydrologic and ecological variations.

Source: H. Lang and M. Rohmann,
ETH Zürich

Plate 7.2 View from Bangladesh towards the Meghalaya Hills, an area of high flood relevance (Photograph: T. Hofer, September 1994)

radar are also valuable for the information they can produce. Images from satellites, maps, and field surveys provide the data that are used to characterise the basins concerned, but much of these data, with the exception of land-use data, change only slowly with time. Measurement programmes should be designed to address specific questions relevant to local water-resource development and impact avoidance or mitigation. Hydrological data are processed by computer to convert them into areal averages of the particular variable, to calculate the volume of river flow from a series of point values, the concentration of sediment, or the concentration of a particular chemical. The combination of flows with concentrations will give the load being discharged. Various methods have been evolved to undertake the necessary processing of these data, and these methods have been incorporated in models to simulate the water balance of a basin. Careful interpetation of model output and recognition of the uncertainty of both data and model results are critical to an objective water resources assessment.

From indirect observations and models it is quite clear that the mountains account for a very large share of the total annual flow in rivers and streams. In order to establish the crucial place of the mountains in the supply of useful freshwater, it would be important, therefore, to obtain precise quantification of their annual fresh water output. A rough estimate can be obtained by adding the discharge data at all suitable points on the world's rivers as they emerge from the mountains onto the plains. The Global Runoff Data Centre (GRDC) at Koblenz in Germany was established in 1988 by the WMO as a storehouse for such data. It is not possible, however, to make a precise assessment because long-term flow data from points near the mountains are not available. Furthermore, the zone of influence of the mountains, in terms of precipitation distribution, does not have a sharp boundary. The GRDC, nevertheless, collects data from national institutions responsible for hydrology and then examines their quality, archives them, and makes them generally available. In this manner, data from more than 140 countries are collected covering the discharge of more than 2900 of the major rivers of the world. In the future this will permit a much more accurate assessment.

Measurement problems in mountain regions

Making reliable measurements of the hydrological variables in mountainous regions is widely acknowledged to be far from simple. Indeed, the combination of access difficulties, extremes of climate and topography, the wide range in the magnitude of the variables to be measured, and uncertainties in the performance of the instruments and methods of observation, is a severe test of human ingenuity and persistence. Difficulties of access are a great hindrance, roads frequently end some distance away from instrument sites, loads are heavy, gradients are steep, and the wet and cold are deterrents to field activities. Helicopters offer one alternative where the high cost can be accommodated. Telemetry is another, using telephone or radio, sometimes through a satellite, but again the high cost of the installation is a drawback. High levels of humidity and the severe cold cause the mean time between failures to be short in most mechanical equipment and in many electrical and electronic devices. Ice accretion affects many sensors unless precautions are taken to clear surfaces, such as by using vibration. Heat is one solution to instrument problems, but heat requires storage of electricity in batteries. Batteries are heavy and need efficient solar panels to keep them properly charged. Large-capacity measuring devices are needed to sense some of the variables, flow for example. Some rivers and streams may be dry at one moment then subject to flows of 1000 m^3 or more the next, due to a flash flood. Most of the widely-used types of weir and flume and several other methods of discharge measurement are not able to cope under these conditions and with the large volumes of sediment being transported, the steep slopes, and high water velocities. High wind speeds are a problem for the measurement of precipitation; gauges can lose up to 80 percent of the catch, particularly where snow is prevalent. Wind also redistributes the fallen snow, making snow depth measurements unrepresentative. Images from satellites offer alternatives to some methods of measurement at the most remote sites, providing these images can be calibrated at sites where access is easy. The scientific literature contains examples of where some of these problems have been overcome but there are many circumstances in mountains where even the most modern methods do not provide realistic measurements. The World Meteorological Organization's *Guide to Hydrological Practices* (WMO, 1994) contains advice on the best instrument practice in mountains (and in other regions), advice assembled over many years from the experiences of professional hydrologists.

Networks

The *Guide to Hydrological Practices* (WMO, 1994) recommends densities of instrument networks appropriate to each physiographic region. Because of the hydrological heterogeneity of mountains, recommended densities are higher there than for other regions. However, in practice, hydrological networks in mountains tend to be less dense than elsewhere. Table 7.1 provides a comparison of the area of the globe rising above specified bands of altitude and the number of precipitation gauges in each band. For altitudes above 500 m, except the highest band, precipitation networks world-wide are less dense than would be expected from the simple pro-rating of the area.

Of course this picture may be different for certain countries and for specified regions of the globe, and there are likely to be differences for other hydrological variables. For example, the comparison of the number of observations of water level in rivers in different parts of the world shown in Table 7.2 indicates that in no region is the density of observations in proportion to the area of mountains in that region. The evidence is that for most other variables the situation is the same. In other words, the densities of networks are inadequate.

Further studies of mountain regions conducted by the WMO, using results from 50 countries, have compared the actual densities of networks for the different variables with the minimum densities recommended in the *Guide to Hydrological Practices* (WMO, 1994). This study found that for most of the variables the actual network densities did not match the minimum recommended. Taking the

Table 7.1 Proportion of the land area and the global number of non-recording and recording precipitation gauges at different altitudes

Elevation (m asl)	% Area	% Non-recording gauges	% Recording gauges
< 500	70.0	77.0	76.6
500–1000	19.0	15.0	15.9
1000–1500	6.0	4.5	5.3
1500–2000	2.0	1.5	1.5
2000–2500	1.0	0.7	0.5
> 2500	1.0	1.0	0.2

Table 7.2 Proportion of mountainous area and the proportion of the total number of observations of river water level in mountainous areas for the regions of the globe

Region	% Mountainous area	% Water level observations in mountains
Africa	16	13
Asia	39	34
South America	23	8
North and Central America	21	10
South-west Pacific	26	0.3
Europe	22	8
World	22	10

individual basins being examined, the WMO recommended minimum density was not reached in 74 percent of basins for non-recording precipitation, 52 percent for recording precipitation, 65 percent for evaporation, 65 percent for discharge, 85 percent for sediment, and 8 percent for water quality.

Recent trends in hydrological research and assessments in the mountains

In recent years more attention has been focused on hydrological research and assessment in the world's mountains. Significant progress has been achieved in improving hydrological instrumentation leading to a better understanding of the physics, chemistry, and biology of the movement and storage of mountain waters. Some of these projects represent advanced scientific research, while in other cases they include innovation and adaptation techniques for measurements more suitable for the mountain situation. The examples of the use of trace elements for flow determination in steep rivers and streams and the use of diatoms for the investigation of habitats can be cited. A broad sample of recent research in mountain hydrology can be found in several publications of the International Association of Hydrological Sciences (IAHS): Lang and Musy (1990), Bergmann et al. (1991), Molnar (1992), and Young (1993).

Remote-sensing data from aircraft and satellites have been used in the assessment of snow and ice cover for the estimation of surface runoff since the 1940s (e.g. Rango, 1993; Thapa, 1993). Information about the status of glaciers is available in volumes of the Satellite Image Atlas of Glaciers (e.g. Williams and Feringo, 1991), publications of the World Glacier Monitoring Service (Haberli and Herren, 1991), and the World Atlas of Snow and Ice Resources (Kotlyakov, 1996) provides archival information. There have been several studies on the application of weather radar in mountain areas (Andreu et al., 1990; Joss and Lee, 1993). Another important trend is the development of models that are designed to handle missing data, a frequent problem in harsh mountain conditions.

In the past few years many studies have considered the possible effects of global climate change on mountain areas (Barry, 1992; Chapter 17). Several of these studies have focused on potential hydrologic responses to different scenarios of future climate (e.g. Gleick, 1989). The use of large-scale hydrologic models (e.g. Becker and Nemec, 1987) helps to identify how streamflow, soil moisture, and evaporation might be altered if precipitation amount and type and energy balances change. Such models can also be useful in describing feedback mechanisms. Modelling efforts applied to mountain ranges in western North America have suggested that winter season runoff could increase and summer season runoff could decline if air temperatures increase consistently (Gleick, 1989). Unfortunately, because there are so many uncertainties, unknowns, and modelling difficulties, we can do little more than recognise the extent of possible effects at the scale of a mountain range. Nevertheless, climate change and its consequences in mountain systems continues to be an active area of research (Becker and Bugmann, 1996).

Implications

Inadequate water resources assessments can lead to project failure as well as environmental damage. Water resources planners must recognise limitations of short-term hydrologic data and associated inferences as well as consider potential extremes that are physically possible but have not been encountered during the short period of record. Uncertainty about hydrology and natural hazards requires careful evaluation of the full range of risks inherent in developing water resources in mountain areas. In many cases, lack of knowledge may be sufficient cause to delay a project until there is a more objective basis for decisions – the risks of proceeding in ignorance simply may be too great (Kattelmann, 1994). Water resource assessments must also include evaluations of the current state of the aquatic system, potential degradation caused by water projects and independent changes in land use, and opportunities for mitigation and restoration of impacts generated by human activities.

At a time when the demands for water are escalating globally and most of the resource to meet this demand will continue to be derived from mountains, it is salutary to find that in the majority of cases the human ability to assess this resource is so poor. This situation means that mountain water resources are likely to be exploited in an unplanned manner rather than safeguarded and developed in a rational way. The consequences for mountain ecosystems will be serious.

HYDROLOGIC EXTREMES

Variability of precipitation and streamflow is a basic characteristic of most mountain ranges. Although most of us like to think in terms of averages

and what is typical, such generalities often obscure the broad range of possibilities.

Floods are simply events of high streamflow that result from storms or other large inputs of water to a stream. The water level required to be considered a 'flood' depends upon the characteristics of a particular channel and the local hydrology. When stream discharge exceeds the capacity of the channels to convey the flow, floods may become hazards to people and structures located nearby. Floods are a critical geomorphic process and an important disturbance factor for aquatic biota and riparian vegetation. In mountain areas, floods may be generated by a variety of mechanisms, including intense rainfall of short duration, prolonged rainfall of moderate intensity, seasonal snowmelt, rainfall combined with snowmelt, and outbursts of water from storage. These mechanisms and their variants differ in relative importance between mountain ranges, by elevation within a particular

---

Box 7.3 The inhabitants of the Himalaya are not responsible for the floods in Bangladesh

Every year during the monsoon season the overall Himalyan region appears in the headlines because of disastrous flooding on the vast plains of the Ganges and the Brahmaputra. The question as to to what extent these processes are natural and to what extent influenced by human activities has resulted in passionate discussions among scientists and politicians and forms a basis for flood management, or even flood control. These issues are particularly sensitive because the Ganges and the Brahmaputra are international rivers. The arguments of politicians, journalists, or even scientists, regarding the effects of human activities in the Himalaya on the ecological processes in the lowlands can be summarised by the following, indeed rather convincing, chain of assumed mechanisms: population growth in the mountains → increasing demand for fuelwood, fodder, and timber → uncontrolled and increasing forest removal in more and more marginal areas → intensified erosion and higher peak flows in the rivers → severe flooding and siltation on the densely populated and cultivated plains of the Ganges and Brahmaputra. Such statements are highly explosive politically, but are not at all based on solid scientific foundations.

In this context, and as a consequent follow-up of the long-term research tradition in the overall Himalayan region (Ives and Messerli, 1989), the Department of Geography of the University of Bern has, since 1992, undertaken a project to achieve a

better understanding of the large-scale processes leading to flooding in Bangladesh. The question, to what extent the floods in Bangladesh are influenced by climatological and hydrological processes outside Bangladesh and to what extent by factors inside the country, is a particular focus of the study. The following primary conclusions are based on the comparative investigation of the hydrological contribution from 13 defined sub-catchments in the Ganges–Brahmaputra–Meghna basin for selected years over the period 1950 to 1990 (these 13 sub-catchments include Bangladesh, the Meghalaya Hills, the Indian Ganges Plain, Assam, and the Himalaya, excepting Arunachal Pradesh):

• The intensity of the monsoons in any particular year is highly variable according to region: it is almost a rule that rather dry conditions in the west (Ganges catchment) coincide with above average monsoon intensity in the east (Brahmaputra and Meghna catchments), and *vice versa*. Against this background it was possible to document that the Brahmaputra and the Meghna catchments are much more important for the flooding conditions in Bangladesh than the Ganges catchment: in extraordinary flood years, such as 1974, 1987, and 1988, the rainfall in these areas was well above the average, whereas in the Ganges system negative rainfall anomalies predominated.

- The rainfall patterns in the Meghalaya Hills, part of the Meghna catchment, are particularly relevant: during spring and in June in the Meghalaya Hills, an average of 20–25 percent of the total input into the 13 sub-catchments is recorded, although this area represents barely 2 percent of the total surface area. This very high average input continues throughout the monsoon period. Moreover, in the Meghalaya Hills, the rainfall was much above normal in the years of extraordinary flooding in Bangladesh. This means a remarkable additional input on top of the already very high average figures!
- The rainfall patterns in the Himalaya have almost no impact on flood processes in Bangladesh. With several case-studies it could be shown that heavy rainfall events over the first Himalayan ranges and in the foothills may be the decisive factor for flooding in the adjacent plains (that is, Terai areas), owing to a sharp rise in the hydrographs of the rivers of the affected areas. As the distance from the Himalayan foothills increases, the discharge peaks are reduced, and further downstream the flooding disappears altogether. This means that the hydrological input into Bangladesh originating from the Himalaya is very limited and only one of several relevant elements considering the size of the river catchments and the huge amount of water involved in that country's flood processes. Much more important are the hydrological and climatological processes in Bangladesh itself, such as heavy rainfall, high groundwater tables, short-term discharge peaks, or even drainage congestion due to road and river embankments. The inflow into Bangladesh through the Ganges and the Brahmaputra, however, can be very important when the discharge peaks of the two rivers coincide.

This brief insight into the complexity of the processes leading to floods in Bangladesh, especially into some aspects of the 'highland–lowland interactions', highlights the fact that for sustainable management of water resources in the Ganges–Brahmaputra–Meghna system new thinking is needed. This thinking must be guided by a thorough understanding of the physical processes involved and not by simplistic arguments, or even political disputes. Moreover, only through co-operation among the states sharing the Himalayan waters will future approaches to sustainable watershed management be successful.

Source: Thomas Hofer,
Geographical Institute,
University of Bern, Switzerland

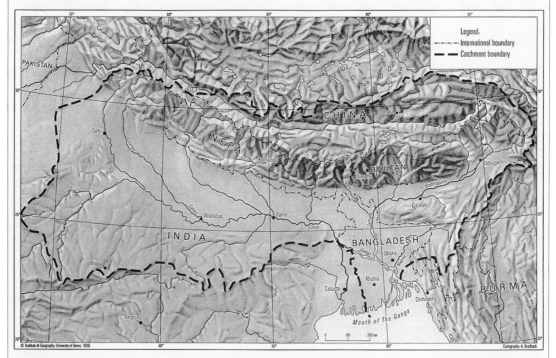

Figure 7.3 Catchment area of the Ganges, Brahmaputra and Meghna rivers (after Brammer, 1990)

range, and over time. Outbursts of glacial lakes (e.g. Watanabe *et al.,* 1994) have recently gained attention as a serious hazard in some high mountain areas. Human activities can both suffer from floods as well as modify floods. Flood damage occurs from occupancy of flood-prone lands, which are easily identified in most cases. Unfortunately, such lands are often the flattest properties in the mountains and are highly attractive for a variety of economic activities when they are not part of a river. Construction and operation of dams, large and small, are the primary human influences on mountain flood waters. Large flood-protection reservoirs can contain rare, large-magnitude events and reduce the downstream discharge to a tiny fraction of the incoming flow. Because of the loss of high flows of water and sediment during floods, aquatic ecosystems below larger dams can be totally different from their pre-dam condition.

Droughts represent the other extreme of hydrologic potential from floods. Similar to floods, droughts are defined on the basis of the local conditions. When precipitation and runoff are significantly (defined locally) below average for an extended period (defined locally), we can say we have a drought. The combination of the magnitude of the deficit and the duration taken together determine the severity of the drought. Dry periods lasting several years have occurred in many mountain ranges in the past century, and severe droughts of several decades have been inferred from palaeobotanical evidence. Both human and natural communities are stressed by the relative lack of water. People living in the comparatively water-rich source areas of the mountains usually fare better than those in the plains during drought conditions. If droughts persist long enough, the distribution of vegetation may change. The altitudinal zonation of vegetation found in most mountain ranges can shift dramatically with changes in water availability.

Most of the water resources infrastructure in mountains around the world was designed for conditions observed in this century. Many of these structures have proven, or will prove, to be inadequate to handle extremes outside the expected range. Some climatologists believe that circulation patterns are changing such that extreme events will become more common and variability will be enhanced in coming decades in some parts of the world. Chapter 16 provides additional treatment of these forms of natural hazards and discusses many other types of hazard that are especially characteristic of mountain regions.

HEALTH OF AQUATIC SYSTEMS

Sustainable development of mountain water resources implies maintenance of the physical and ecological processes that influence the generation of high quality streamflow. If society wishes to enjoy the benefits of water produced in the uplands, then our development practices must respect the integrity of mountain watersheds and streams. The ideal attributes of long-term, dependable water supply are: unimpaired flows of water produced by minimally-modified hydrologic processes; sediment loads that have not been augmented by human activities; naturally low concentrations of chemical constituents and pathogenic organisms; and fully-functioning aquatic and riparian communities. Natural systems can only be pushed so far before the ecological services that we take for granted begin to break down and imperil our highly valued water sources. Besides maintaining high water quality and supplying harvestable food resources, aquatic biota can serve as valuable indicators of the overall health of the water resource system. If conditions permit native aquatic life to flourish, then the stream system is likely to be a reliable water source. When native aquatic communities are in decline, the water resource is likely to have been compromised.

Unfortunately, our knowledge about aquatic systems in mountain areas is poor even under the best of circumstances. Very few assessments of biological integrity (Karr, 1991) in mountain rivers have been attempted. If anything was studied in a stream, fish were the most common target (Kottelat and Whitten, 1996). Knowledge about amphibians, invertebrates, aquatic plants, and ecological interactions is virtually non-existent in most mountain ranges. A comprehensive evaluation of the state of the environment of the Sierra Nevada in California, USA, included assessments of water resources and aquatic ecology (Kattelmann, 1996; Moyle *et al.,* 1996) that are illustrative of generic problems in many mountain ranges. The combined aquatic and riparian system is usually the most productive part of the mountain landscape with a great diversity of life-forms compared to the drier uplands. Unfortunately, the components of this system are easily disturbed, leading to ecological changes. Although minimal information about aquatic invertebrates is available, even in the comparatively well studied rivers of western North America, these small creatures tend to be quite sensitive to changing conditions. Human impacts can cause dramatic shifts in the composition of invertebrate communities, which can have further consequences higher in the food chain. Amphibians appear to be declining throughout the world, including those mountain areas that have been surveyed. The decrease in amphibian populations seems to be caused by the combined effects of many documented and hypothesised factors. Habitat alteration and predation by introduced species are two of the prime hazards. Fish are usually better known and have greater economic value than other aquatic organisms. Nevertheless, native fishes also appear to be declining where streams have been degraded and exotic species have been introduced.

Resource exploitation that damages riparian and aquatic habitat is a principle cause of the noted declines in the distribution and abundance of native aquatic organisms. Direct modifications of stream-flow, channels, and riparian areas have much

Box 7.4 Titicaca – A sacred Andean lake in danger

Lake Titicaca is located at an altitude of 3810 m asl on the Altiplano (high plain) of the Andes. This second largest lake of South America has an area of 8301 km^2 and exceeds 300 m in depth. It is an international lake shared by Peru and Bolivia. More than two dozen rivers drain into this vast high altitude water body. It is a major water transport route with boats and ships criss-crossing it with passengers and cargo; sudden storms have caused big waves which have led to loss of life of people sailing on the lake.

Titicaca has an important place in the mythology of the local people. It is believed that their God, after creating the Sun, the Moon, the Stars and the Inca Empire, had descended into the lake. The once pristine lake with crystal clear water is now a matter of the past. Pollution from all over the basin, in the form of agricultural chemicals, urban wastes, industrial and mining effluents, have now endangered not only the habitat of God, but also the well-being of the common people. The quality of the input to the Chimu pumping station, which transfers potable water from the lake, is now under threat. Anthropogenic pollution has led to strong eutrophication causing proliferation of algal blooms, severe reduction of dissolved oxygen, and foul odours. The ecosystem of this unique high altitude lake is in jeopardy; so are many rare endemic species, such as the Titi duck.

Lake Titicaca was declared as a National Reserve in 1978 and forms the refuge for many native species and areas of extensive reed beds. International interest became focused and several efforts have been made to assess the situation, as well as to take remedial measures. Through bilateral efforts, a network of hydrological stations was established in order to assess the inflow and outflow of the lake. Several national and international agencies have tried to measure the water quality and identify pollutants. In some analyses, the presence of 10 000 faecal coliform bacteria per litre of water was detected. The people living around the lake have taken concerted steps for the collection and removal of algae. In Puno, a lakeside city of 100 000, some 1500 m^3 of algae were collected and removed. A sustainable solution, however, requires better management of the upstream areas which lie at the root of the polluting economic activities.

Plate 7.3 Lake Titicaca (3810 m) from the Sun Island, Bolivia (Photograph: B. Messerli)

Source: Dieter Kraemer, WMO, Geneva, Switzerland

Box 7.5 Without mountains there is no life in the Aral Sea basin

The Aral Sea basin of Central Asia, including parts of Afghanistan, Kazakhstan, Kyrgyzstan, Tajikistan, Turkmenistan, and Uzbekistan, has an area of about 690 000 km^2, with a population of 32 million. Most of them live in the foothills and valleys where the middle-reaches of the rivers emerge from the mountains, and soil and temperature conditions are favourable to agriculture; only 3 million live around and near the Sea. The basin has been the home of many civilisations since 6000 BC; irrigation has been practised for millennia. Today, 8 million hectares are irrigated.

The basin has three distinct ecological zones: the mountains, the deserts, and the Aral Sea with its deltas. The Tien Shan and Pamir mountains, situated in Kyrgyzstan and Tajikistan in the eastern part of the basin, rise to more than 7000 m. In these high mountains, the annual precipitation ranges from 600 to over 2000 mm, with 30 percent falling as snow, and 60 percent from December to May. However, because of snow and glacier melt, the flows of the two main rivers – the Amu Darya in the south and the Syr Darya in the north – are higher in summer. The annual runoff of the rivers in the basin is about 120 km^3. Of this, the Amu Darya brings approximately 77 km^3 from the mountains, and the Syr Darya contributes 39 km^3. Thus, the mountains provide more than 95 percent of the basin's water.

Lowland deserts cover most of the basin and are characterised by low rainfall (less than 100 mm/yr) and high evaporation: potential evaporation can be as much as 1500 mm/yr. The Aral Sea itself is situated in a vast geological depression in the Kyzzyl-kum and Karakum Deserts, astride the border between Kazakhstan and Uzbekistan. In the historic past, oscillations of the meteorological and hydrological conditions in the basin were quite large, causing pronounced changes in the state of the Sea; its water level varied over a range of 25 m. From the mid-nineteenth century until 1960, about half the runoff of the Amu Darya and Syr Darya was used for irrigation and thereby lost for natural evapotranspiration in the middle and lower reaches of the rivers; the remainder flowed into the Sea. In 1960, the Aral Sea was the world's fourth largest inland lake, with a surface area of 69 500 km^2 and a volume of 1040 km^3. However, with little inflow from the Amu Darya and Syr Darya, and relatively little local rainfall or recharge from groundwater, the Sea was losing 5 km^3/yr. By 1992, it had shrunk to a surface area of 32 985 km^2 and a volume of 231 km^3. Together, the Amu Darya and Syr Darya provided an input of only 5 km^3. More recent accurate data are not available, but the Sea appears still to be decreasing in area and volume.

The desiccation of the Sea has resulted in the destruction of its ecosystem and deltas, the loss of the fishing industry, the blowing of salts from the exposed sea bed which are toxic to humans and deleterious to crops, and a depressed regional economy. The indiscriminate use of water for non-agricultural purposes, inefficient irrigation practices, excessive use of chemicals for growing cotton and rice, and inadequate drainage have caused extensive waterlogging, salinity and pollution of the ground-water and drainage inflows to the rivers and the Sea. Urban and industrial water pollution has further aggravated these problems.

The states of the Aral Sea basin have considerable economic potential. They are endowed with huge water and land resources, extensive irrigation systems, large hydropower bases, important minerals, and a considerable labour force. Yet, in these desert countries, water is essential for sustaining life and all economic activities. The sustainable development and realisation of the full potential of the region's diverse resources depends largely on the water from the mountains. Although the water resources are considerable, the demands are also very large. Acute water scarcity and seasonal shortages adversely affecting agricultural and energy production are unavoidable. Moreover, water in most of the region is so polluted that it can no longer be used for human consumption or for agricultural production. There is an urgent need for action to reverse those changes which are not irreversible, and to use the mountain waters wisely.

Source: M. Spreafico,
Swiss National Hydrological and
Geological Survey

Plate 7.4 Water capture for irrigation in the Garm valley, Tadjikistan (Photograph: B. Messerli)

greater impacts on aquatic systems than disturbances occurring well away from streams. As development activities have changed attributes of streams (such as water volume, seasonal timing, peak flows, minimum flows, sediment transport, dissolved oxygen, nutrient cycling, and water temperature), aquatic and riparian ecosystems have had to change in response. Other ecological impacts have been deliberate (such as channelisation, conversion of riparian areas to roads, introduction of exotic species, and conversion of streams to lakes). The various impacts tend to interact and produce combined stresses on aquatic ecosystems that are much more effective agents of change than individual impacts.

Water resource development is usually the main human influence on aquatic systems. Large-scale hydraulic engineering works regulate about three-quarters of the flow of large rivers in the northern third of the world (Dynesius and Nilsson, 1994). Dams and diversions affect biotic communities by changing flow regimes, reducing water availability, converting running water habitats to reservoir environments, limiting sediment transport, changing temperatures, blocking movement of biota up and down the channel, and isolating populations. The intentional modification of the hydrologic regime via water management practices often produces residual streamflows far outside the natural range of variability to which aquatic communities have adapted. Creation of artificial barriers to migration at the mountain front have eliminated vast amounts of river habitat for anadromous fishes. The combination of altered habitat and introduction of exotic fishes has allowed many mountain streams to become dominated by the non-natives (Moyle *et al.*, 1996). When native fish assemblages become isolated by dams, natural recolonisation is no longer possible and extinction becomes a greater threat.

Degradation of mountain riverine habitats occurs through many other human activities besides construction and operation of water storage and diversion facilities. Other damaging actions in the stream channel itself include mining of minerals and gravel; construction of bridges, culverts, fords, bank stabilisation structures, and artificial channels; removal of large woody debris and aquatic plants; and disposal of waste. Human actions in the adjacent riparian area can be equally damaging to the riverine system and may be more widespread (e.g. Gregory *et al.*, 1991; Kattelmann and Embury, 1996). Removal of stream-side vegetation and conversion to roads, structures, recreational facilities, or agriculture destroys ecological values for terrestrial biota dependent on the riparian area and can have dire consequences for aquatic communities. Riparian vegetation both controls the radiant energy reaching streams by providing shade and adds chemical energy and nutrients derived from insects and plant materials. Stream-side soils and vegetation influence delivery of surface runoff, groundwater, sediment, organic matter, and dissolved nutrients and pollutants to the adjacent stream. Riparian vegetation and woody debris slow the movement of flood water and dissipate stream energy. Riparian areas, and especially wetlands, moderate the adverse effects of increased transport of water, sediment, nutrients, and pollutants generated by upslope disturbances. Destruction or degradation of riparian areas impairs all these important functions that maintain high quality streamflow. Because streams act as corridors for wildlife migration and move water, energy, particulates, chemicals, and biota from upstream to downstream, artificial breaks in the linear nature of river systems interferes with the natural movement of materials and life. This fragmentation of aquatic and riparian habitat is perhaps the most pervasive human impact on rivers resulting in segmentation of communities and interruption of flows.

Maintenance of the naturally high quality of mountain water is an integral part of sustainable development of water resources. Water supplies containing low amounts of sediment, chemical contaminants, nutrients, and pathogens reduces difficulties and costs for human uses. Relatively pure water minimises reservoir siltation, turbine wear in hydroelectric generating facilities, clogging of irrigation canals and urban supply pipelines, and treatment for domestic and industrial uses. Fortunately for human society, most mountain areas have relatively few naturally-occurring water quality problems, and opportunities for contamination by human activities are much less than in the lowlands. Nevertheless, there are a wide range of water quality issues in mountain areas. Excessive sediment tends to be the most common pollutant in mountain regions because its production can be increased above natural background levels by almost any human activity that disturbs soil or reduces vegetative cover. Pathogenic organisms derived from human and livestock waste that enter drinking water supplies are a widespread human health concern. The rapid growth of pesticide and fertiliser use in mountain agriculture caused by falling prices of the chemicals could imperil water supplies that have been free from such contamination because of economic factors. In developed countries, a variety of chemicals are used in intensive forest management that can ultimately end up in streams. Compared to the lowlands, there are relatively few urban and industrial (mainly mining and mineral processing; see Chapter 9) pollution sources in the mountains. For those highland point-source problems that do exist, technologies are readily available to treat such effluents where there is sufficient political will to avoid this type of water pollution. In mountains downwind from urban centres, there is potential for water contamination from precipitated air pollutants. Some surface waters in poorly buffered headwater catchments are at risk of acidification from these airborne materials. The various sources of water pollution mentioned above have combined effects for human uses and aquatic communities and are further influenced by the quantity of water in mountain lakes and streams available for dilution and natural treatment. Artificially reduced water quantity is often a primary water quality problem in managed mountain rivers.

On a hopeful note, aquatic systems can be highly resilient. The critical factors are ending the disturbance and providing adequate water for geomorphic and ecological processes to reassert themselves. For example, most rivers of the western slope of the Sierra Nevada of California were devastated by intensive gold mining during the mid-1800s. In the past century, the hydrological and ecological functions of these highly degraded rivers appear to have largely recovered, where not disturbed by more modern impacts (Kattelmann, 1996). Although recovery appears to have been to a less ecologically complex state than existed before mining, this response illustrates an inherent long-term resiliency of mountain fluvial and aquatic systems. When we are impatient with nature's schedule, active restoration can greatly reduce recovery time (US National Research Council, 1992).

MANAGEMENT FOR SUSTAINABLE WATER SUPPLIES

If society accepts the maintenance and enhancement of mountain water resources as a worthwhile goal, a broad range of management strategies are available. Ideally, these strategies should be organised and applied on a river basin and watershed basis (see Chapter 15). Despite the inherent logic of designing water resource policies and plans from a basin perspective, this approach has not been widely used. France and Spain are rare examples of countries that have used river basins as the basic unit of analysis in water resources administration. The use of watersheds as a planning unit, especially in water quality control, is suddenly gaining popularity in the United States. Assessment of water resources and planning for water development within river basins allows more coherent management than occurs with traditional political jurisdictions. Cumulative effects of development activities that are ignored in piecemeal evaluations of individual projects can be identified and opportunities for mitigation of such effects can be studied in the basin context. With a watershed perspective, water resource professionals, land managers, property owners, water users, planners, politicians, and the public have better opportunities to visualise interactions between upstream and downstream, sources and sinks, land and water, slopes and valleys, waste disposal and water intakes, and human actions and natural processes. Though hardly a panacea, a watershed approach to water problems can serve a valuable environmental education role as well as form the basis of rational management of water resources.

In attempting to manage water in ways that do not threaten its future availability or impair ecological processes, the current status of a river basin must be evaluated. In addition to describing the basic hydrologic characteristics of, and natural influences on, runoff production within the basin of interest, the existing human uses and impacts should be identified. The chief problems that currently impair water quantity and quality should be investigated. Opportunities for solving those problems can then be studied. In many cases, there are simple and direct solutions to water problems through common sense once the problem is identified. Compared with some of the immense water issues of densely-populated lowland areas, many of the water problems in the relatively undeveloped mountains seem straightforward. In the absence of integrated watershed-based planning, such problems are simply ignored when they could be readily solved. For example, many hydroelectric facilities are operated with archaic specifications that do not account for current electricity demand or ecological needs. New operating rules can sometimes be developed that both increase net revenue and improve downstream aquatic ecosystems. Other

Box 7.6 Debate over big dams in the mountains – The case of the Tehri Dam

The Himalaya receives enormous amounts of monsoon precipitation during July, August, and September each year. Combined with the topography, the runoff from this large input of water adds up to a hydroelectric potential that has been estimated to exceed 200 000 MW. This spectacular figure has long attracted the attention of engineers and policy makers. Most of this potential is concentrated in the three major rivers – Ganges, Indus, and Brahmaputra – and their tributaries. They are perceived as the answer to the rapidly growing demands for water and power in one of the world's most densely populated regions – the South Asian sub-continent. Seen from a purely hydrological viewpoint, the idea is laudable and, in principle, the water wealth in the Himalaya could be a vital instrument for economic development of the region which contains a significant proportion of the world's poor.

Over the last four decades, several major dams for irrigation, hydropower, and water supply have been planned in the Himalayan areas of the sub-continent. Some have been constructed, some are under construction, and several are in the stage of negotiation and planning. However, dams in the Himalaya, as in many other mountain regions, have become objects of public protest questioning their efficacy as instruments of equitable and sustainable development. A high dam under construction on the Himalayan river Bhagirathi near the town of Tehri in India has emerged over the last twenty years as an important representative case in this respect. The administrative clearance for the 260.5 m high rock-filled dam at Tehri, with a impoundment volume of 3539 million m^3 of water, was granted by the state government in 1976. The proposed dam was calculated to have a power potential of 2200 MW; irrigation, urban water supply, and flood control, were also claimed as additional benefits.

Following the administrative clearance for the dam in 1976, the people living in the area scheduled to be submerged when the reservoir is filled formed the Tehri Dam Opposition Committee to protest what they regarded as unacceptable consequences of the project. These included displacement of a large number of people, inadequate compensation, and improper rehabilitation. Environmental issues were added subsequently as a major component of the opposition to the proposed dam after a massive landslide dam break and flood occurred on 6 August, 1978 in the upstream vicinity of the dam site. The tenacity of the Tehri Dam Opposition Committee has ensured that a plethora of environmental impact studies were undertaken following the airing of the perceived dangers. Serious questions about the longevity of the proposed dam have also been raised as more complete information on the rate of sedimentation became available. Valdiya (1993) has estimated that the original calculation of the economic life of the dam would have to be reduced from 100 to 40 years if realistic figures for sediment load were taken into consideration. In addition, the geodynamic sensitivity of the Himalaya in general and, in particular, a major earthquake on 20 October 1991 in the vicinity of the proposed dam site triggered an intense debate on the risks associated with the design of the proposed dam and the costs associated with redesign to higher standards. This issue is fraught with controversy and has underlined the serious loss of public confidence in the various levels of government. Because the Tehri Dam is located in the so-called seismic gap of the Himalayan front the question of the near-term prospects for a 'great earthquake' (above 8.0 on the Richter scale) has intensified the dispute (Gaur, 1993). The accelerating environmental and social criticisms have further raised the question of who gains at whose cost in the utilisation of the rich water resources of the Himalaya; certainly it would appear that the Terhi Dam is another example of development for the benefit of the plains at the expense of the mountain peoples and their environment. This widening of the debate, which has now extended around the world, and increasing public protest, however, have not come about without pain and sacrifice. In order to force a comprehensive impact assessment of the proposed dam, Sunderlal Bahuguna, social activist of Chipko Movement fame, has had to undergo two periods of indefinite prayerful fast, extending over periods of 49 and 46 weeks, together with threats and arrests.

At a time when major international financial institutions are distancing themselves from investment in large dams, the debate on the costs, risks, and benefits from the Tehri dam, now far into its construction phase, is of immense political significance; it can be taken as a forerunner to decision making on a series of large dams all along the 2500 km-long southern slope of the Himalaya (Bandyopadhyay, 1995). It is probably the most complex issue for decision making in the context of sustainable development of the water-rich slopes of the Himalaya.

Source: J. Bandyopadhyay,
International Academy of the Environment,
Geneva, Switzerland

problems can seem intractable, but they must be identified explicitly as a first step toward potential improvement of the situation. Evaluations of the current situation can also identify opportunities for improved efficiency in the water system. Modifications of existing hydro-technical structures can have much lower financial and environmental costs than new construction.

As new water projects are considered in mountain areas, there is a wealth of experience available that can be used to improve project design and minimise adverse environmental effects. Despite a general human tendency to repeat the mistakes of the past, there are some hopeful signs. The era of easy water projects is largely gone because most of the ideal sites have already been developed and people are more aware of the potential pitfalls

of ill-considered water development. Therefore, almost all projects are subject to much greater internal study and public scrutiny than occurred in the recent past. Water projects are no longer casually considered by governments and donor agencies because the financial costs of construction and awareness of environmental consequences have risen dramatically.

Changing attitudes and practices with respect to development of water resources have been described in a policy paper of The World Bank (1993). If these new policies are thoroughly implemented, future projects funded by The World Bank will be greatly constrained and much more sensitive to social, economic, and environmental concerns. The new policy framework uses river basins as the planning unit, treats water as an economic good,

Box 7.7 Water management by rural mountain communities – The case of Cusco in the Peruvian Andes

The communities of the southern Andes of Peru, like many mountain communities of the world, depend upon an agricultural economy at altitudes ranging from 3000 to 4200 m. Their agro-economic livelihoods are closely connected with the management of local water resources. Management strategy has evolved at two levels: first, the satisfaction of basic water needs of the individual families and, second, the institutions through which the control of the community has been affected in its management. Both strategies have evolved over time and have been adjusted in response to changing demands.

The communities mainly practise a single cropping system, with maize and some vegetables being grown in irrigated areas. In the rain-fed areas the main crop is the potato. Irrigation is chiefly used for the preparation of the soil and the seedlings. The irrigation system involves diversion and gravity-feeding from small streams. The water transport process resulting from this system may appear inefficient because of the rustic loam channels, but the water losses from the channels actually recharge many springs emerging on the slopes below.

During the period of Inca administration a completely different situation prevailed. Authority and control was maintained within a vertical management structure and a horizontal social structure. Both were characteristic of Inca organisation as a whole. This allowed for optimal resource management and for the realisation of important engineering works, such as canal construction. Today, the most critical water issue is the provision of domestic drinking water. For many villages the situation is so

difficult that the collection of drinking water necessitates walking considerable distances each day. In most cases the women have to walk for about 30 minutes three-times a day in order to supply the basic domestic needs. Due to shortage of time, or unavailability of family members to help with this task, many families are reduced to using poor quality water from nearby open channels and streams.

In recent years, drinking water systems have been placed under the management of local councils and private parties. The entrusted parties constructed collectors, storage tanks, and delivery pipelines. Such systems, however, rapidly became ineffective because of poor maintenance, since the local people did not feel responsible. Tanks began to leak and the delivery pipes broke, leading to an efficiency far below that attained during the Inca period. In addition, in those mountain villages where water is in very short supply, new conflicts have arisen between the needs for domestic use and agriculture.

Local people have now decided to take water management into their own hands. In order to ensure positive use of the scarce resources, local people's organisations are being reinforced. They have assumed responsibility for control of water quality and maintenance of the storage and delivery systems. Domestic distribution to the individual households is being arranged, with health education as back-up. This will ensure the optimal distribution and utilisation of all the water that is available, a basic necessity for sustainable mountain development.

Source: Jorge Legoas Pena,
Andea Qullana, Cusco, Peru

encourages active public participation in project planning, recognises water needs for maintenance of biodiversity and ecological processes, and attempts to establish a more holistic view of water resources administration. These changing attitudes at The World Bank may have played an important role in the Bank's decision against providing funding for the Arun III hydroelectric project in eastern Nepal. Given this shift at The World Bank, similar policies may be adopted by other international donors and individual governments. The private sector has generally avoided the more grandiose schemes, simply because of the level of investment required. In the developing world, privately supported water development tends to have fewer environmental impacts than larger public projects because of the typically smaller scale and the investors' pursuit of economic efficiencies rather than as a result of explicit environmental goals. However, with the growing availability of private capital, larger projects are being considered. For example, a private consortium is contemplating financing of the Arun III project.

Integrated river basin planning must also consider watershed disturbance as an important influence on water resources. Watershed assessments can determine the level of degradation in aquatic systems, identify contributing land-use practices, and suggest potential remedies (e.g. Montgomery et al., 1995). The type, location, extent, and intensity of watershed disturbances determine hydrological and ecological consequences (Chapter 15). Careful watershed assessments can identify potential trade-offs in rehabilitation by determining appropriate actions to address particular problems that will result in the greatest benefits for the least cost. Riparian protection can be a cost-efficient and ecologically-effective means of stabilising and restoring the health of aquatic systems. The broad range of ecological services provided by riparian corridors makes protection of these areas a critical part of watershed management (e.g. Kattelmann and Embury, 1996). Many negative impacts to water production and water quality can be avoided simply by staying out of streams and riparian areas. Most of the human activities that occur in riparian areas do not require close proximity to streams and could be located elsewhere. Land management agencies in the United States have recently recognised the importance of riparian corridors and regard stream-side zones as areas requiring careful management.

At the international level, The World Bank has adopted watershed management as an integral part of their water resources management programme. The Bank now recognises that combining land-use practices with water resources administration can provide benefits for both water users and aquatic ecosystems by maintaining natural processes of runoff generation and thereby enhancing water quality. The water policy paper of The World Bank (1993) also endorses economic incentives for improving land management practices that could degrade water supplies and stream ecosystems.

Active involvement of mountain communities in water management can have benefits throughout a river system. Historically, residents of the headwaters have been ignored in planning of water projects designed to benefit population centres in the lowlands. People living in the mountains typically have little political power compared to the far more numerous inhabitants of the valleys and plains. Nevertheless, both social equity and practical arguments can be advanced in support of mountain communities with respect to water. Many projects have crippled highland economies when export of water left the rural area without its own reliable source. Alternatively, meeting the basic needs of people living in the highlands through development of village water supplies is a crucial step in aiding impoverished mountain areas. Improving water supply and sanitation for rural communities can have great benefits for human health and well-being at relatively low cost. In more developed regions, providing economic opportunities to mountain communities based on water management can link the upstream and downstream economies to mutual advantage. Past water development in both rich and poor nations has been largely exploitive of the source areas. In the context of integrated river basin management, new opportunities for reinvestment in the headwaters need to be explored. Through taxes on delivered water and fees for new projects, funds could be generated to finance riparian restoration, pollution control, and rehabilitation of degraded watersheds. Such programmes could be the basis of employment for some mountain residents who would act as watershed stewards. Inclusion of people throughout a river basin in the water planning process can identify overlooked costs, suggest optimal solutions to allocation issues, and reduce the risk of future conflict (e.g. Mosley, 1996).

RESEARCH AGENDAS

The scientific basis for sound water resources management is not yet thoroughly established (US National Research Council, 1982). Many important questions and uncertainties remain in the nature of the hydrologic cycle and its interactions with life. To develop water resources in a sustainable manner, we need to understand the abilities of natural processes and aquatic ecosystems to respond to human impacts. In this regard, a detailed proposal for accelerated scientific research relating to fresh water has recently been prepared (Naiman et al., 1995). Although it was directed to government agencies in the United States, issues raised in this proposal have broad applicability throughout the world. Its general themes are: water availability; ecological impoverishment; and issues relating to human health and quality of life. Three broad questions were posed that integrate human needs with water resources:

(1) 'What are the ecological effects of changes in the amount and routing of water and waterborne materials along the hydrologic flow path, from precipitation falling on land to the ocean, under natural and altered conditions?'
(2) 'What are the effects of human activities on freshwater ecosystems, and how do they influence the sustainability of inland aquatic resources?'
(3) 'Are there key features of freshwater systems that can be used to evaluate and predict the effects of human influences at regional to continental scales?' (Naiman et al., 1995: p. 52).

These questions succinctly frame the principle unknowns facing water resource managers in any environment, including mountains. The book describing the details of this proposed programme, The Freshwater Imperative: A Research Agenda (Naiman et al., 1995), contains a thorough range of research topics applicable to mountain waters.

In the past decade, many scientific questions have been raised regarding human impacts on Himalayan hydrology (e.g. Dixit and Gyawali, 1994; Kattelmann, 1994). These questions relate to water resources availability and ecological degradation and could be applied to almost any mountain range. Although the questions come easily, there has been little response from funding agencies or the scientific community. Perhaps deliberations inspired by this book will lead to implementation of the research strategy proposed by Ives and Messerli (1989) and its successor (Chapter 19) and some progress will be made toward better understanding of human influences on mountain hydrology.

CONSERVATION OF HIGHLAND WATER RESOURCES

A variety of pronouncements regarding the water wealth of mountains have been made at high-level

Box 7.8 Mountains and plains – political relations with hydropower: experiences from the Swiss Alps

Hydropower is a very important source of energy in mountainous countries, such as Switzerland, Norway, and Canada. Switzerland's total power generation capacity is 15 500 MW, of which 75 percent is produced from hydro sources. About 70 percent of this hydroelectricity is generated in the 'mountain cantons' of Glarus, Graubünden, Obwalden, Schwyz, Ticino, Uri, and Valais. These cantons form part of a commission whose aim is to protect the economic interests of the 'mountain cantons' in their dealings with the Federal government and, hence, with the rest of the country.

The installations for generation of hydropower in the mountain cantons are of special importance to the non-mountain cantons in the lower land north of the Alps where the main cities and industries are located. The supply of hydroelectricity from the mountains more or less covers the major demand during the critical periods of cold and dry winters.

The returns on the investment in hydropower installations are generally high since they were completed when construction costs were much lower than today. More recently, additional capacity has been added by increasing the storage volumes of the reservoirs.

Management of mountain watercourses is generally under the control of local authorities. In the framework of Swiss law, the right of utilisation of the watercourses lies with the cantons; the Federal government, however, establishes legal controls that the cantons must follow. In the event of any party wanting to develop watercourses for hydroelectric generation, the potential developer must obtain permission from the local cantonal authorities, or from any other agent to whom the canton has entrusted the relevant responsibility. Permission is usually granted for a period of less than 80 years, with the rights and responsibilities of the generating station clearly defined. The utility company must pay a royalty to the canton, the amount of which is adjudicated by the Federal government. In spite of such measures to protect the interests of the mountain people, conflict situations may arise between the mountain cantons and the potential developer or user. The interesting case of the mountain canton of Ticino is introduced here as an example.

The canton of Ticino wanted to profit directly from the rich hydro-potential within its territory; therefore, in 1958 it established its own electric utility company. Before that, a powerful utility company from the plains canton of Solothurn had been the main generator of hydroelectricity in the canton. Nevertheless, despite the 1958 arrangements, the plains influence on the hydro-potential of the mountain canton of Ticino persisted. The hydro-potential of the Maggia and Blenio valleys was developed by a consortium with an 80 percent share owned by the plains cantons and only 20 percent by the canton of Ticino. The mountain inhabitants have become dissatisfied because they now believe that greater economic benefits should have accrued to their canton. It is almost impossible, however, to change the contractual agreements without mutual concurrence since the existing contract will remain valid till 2040 AD. Nevertheless, in 1995 the mountain cantons requested an upward revision of the royalties for storage reservoirs in the Alps. However, the plains cantons succeeded in derailing this effort at the federal level. This demonstrates that even the mountain areas of an industrialised country with a well developed decentralised administration and a tradition for legalised compensation cannot totally avoid neglect and marginalisation.

Source: Franco Romeiro,
CUEPE, University of Geneva,
Switzerland

forums from the Mar del Plata Water Conference of 1977 to the present. For example, 'The Commission recognises the importance of mountains as the predominant and most dependable source of fresh water presently used by humanity, and therefore stresses the importance of adequate protection for both quality and quantity of water resources from mountainous regions' – Commission on Sustainable Development, 1995. Although such declarations are a strong signal that water problems are gaining attention of policy makers and high government officials, action is urgently required in many mountain ranges where demands on watersheds and water sources are accelerating. Rapid population growth already outpaces development of water supplies in many countries where little thought is given to sustainability (e.g. Engelman and LeRoy, 1993; Postel *et al.*, 1996). The alarming prospect is addressing water demand for billions of additional people when current needs are poorly met. The disjunct distribution of water and people implies strong regional differences in future water availability for human needs. Severe future shortages are likely in some regions and will probably lead to deprivation and violent conflict (Gleick, 1993; Chapter 6).

In regions where the current situation is not so dismal, there are better prospects for developing mountain water resources in a sustainable manner. Where shortages do not already exist, water planners have the comparative luxury of being able to determine how much water is likely to be

Plate 7.5 High Atlas of Morocco (Toubkal, 4165 m). Water storage for water supply and irrigation from the high mountain valleys down to the plain of Marrakech (Photograph: B. Messerli)

available under different scenarios of climate variability and change. With knowledge of basic water availability and environmental needs, planners can then identify what level of water demand can be met and attempt to guide growth under those constraints of supply. Having some information about water resources allows decision-makers to explore the opportunity costs of additional water development – what is given up (usually some aspect of environmental quality) to supply more water? Monitoring of water conditions and environmental indicators can allow detection of warning signs before problems progress too far for effective treatment. Given changing attitudes on the part of financial institutions, such as The World Bank, and water development agencies, such as the US Bureau of Reclamation, there are brighter prospects for incorporation of environmental values in water resources administration. The adoption of holistic river basin planning by water management agencies should allow more thorough evaluation of cumulative effects of resource development.

Designation of mountain catchments as water source protection areas could accomplish a variety of conservation and social goals. If destruction and fragmentation of riparian and aquatic habitat can be avoided, water production and quality will be maintained. If degraded watersheds and streams can be restored, water values will be enhanced. Creative institutional arrangements are needed to reinvest some portion of the benefits derived by water users back into the water source areas. Programmes of financial incentives funded by water users to encourage sound watershed management could also dramatically improve the economies of some headwater areas. Watershed management, restoration, and monitoring activities could provide employment opportunities for mountain inhabitants. Because water often has the highest value of any product of the mountains, reinvestment of some of that value in the lands that generated the water is a rational means of maintaining the water wealth of the highlands: the lowlands will be beneficiaries over the longer term.

References

Agenda 21: Programme of Action for Sustainable Development (1993). The final text of agreements negotiated by Governments at the United Nations Conference on Environment and Development (UNCED), 3–14 June 1992, Rio de Janeiro, Brazil. United Nations Publications: New York

Andreu, H., Creutin, J. D., Leoussoff, J. and Pointin, Y. (1990). In Molnar, L. (ed.), *Hydrology of Mountainous Areas*. IAHS Publication 190: Wallingford, pp. 67–85

Bandyopadhyay, J. (1995). Water management in the Ganges-Brahmaputra Basin: emerging challenges for the 21st century. *Water Resource Development*, **11**(4):411–42

Barry, R. (1992). Climate change in the mountains. In Stone, P. B. (ed.), *The State of the World's Mountains: A Global Report*. Zed Books: London, pp. 361–80

Brammer, J. (1990). Floods in Bangladesh I. Geographical background to the 1987 and 1988 floods. *The Geographical Journal*, **156**(2):12–22

Becker, A. and Bugmann, H. (eds.) (1996). *Predicting Global Change Impacts on Mountain Hydrology and Ecology: Integrated Catchment Hydrology/Altitudinal Gradient Studies*. IGBP–BAHC core project office: Potsdam

Becker, A. and Nemec, J. (1987). Macroscale hydrologic models in support to climate research. In Solomon, S. I., Beran, M. and Hogg, W. (eds.), *The Influence of Climate Change and Climatic Variability on the Hydrologic Regime and Water Resources*. IAHS Publication 168: Wallingford

Bergmann, H., Lang, H., Frey, W., Issler, D. and Salm, B. (eds.) (1991). *Snow Hydrology and Forests in High Alpine Areas*. IAHS Publication 205: Wallingford

Commission on Sustainable Development (1995). *Decisions and recommendations adopted by the third session on sustainable development*. United Nations: New York

Conway, D. and Hulme, M. (1996). The impact of climate variability and future climate change in the Nile Basin on water resources of Egypt. *Water Resource Development*, **12** (3):277–96

Dixit, A. and Gyawali, D. (1994). Understanding the Himalaya–Ganga: Widening the research horizon and deepening cooperation. *Water Nepal*, **4**(1): 307–24

Dynesius, M. and Nilsson, C. (1994). Fragmentation and flow regulation of river systems in the northern third of the world. *Science*, **266**: 753–62

Engelman, R. and LeRoy, P. (1993). *Sustaining Water: Population and the Future of Renewable Water Supplies*. Population Action International: Washington DC, p. 56

Guar, V. K. (ed.) (1993). *Earthquake Hazards and Large Dams in the Himalaya*. Indian National Trust for Art and Cultural Heritage: New Delhi

Gleick, P. H. (1989). Climate change, hydrology, and water resources. *Reviews of Geophysics*, **27**: 329–44

Gleick, P. H. (1993). Water in the 21st century. In Gleick, P. H. (ed.), *Water in Crisis: A Guide to the World's Freshwater Resources*. Oxford University Press: New York, pp. 105–13

Gregory, S. V., Swanson, F. J., McKee, W. A. and Cummins, K. W. (1991). An ecosystem perspective of riparian zones. *BioScience*, **41**(8): 540–51

Haberli, W. and Herren, E. (1991). *Glacial Mass Balance Bulletin* No.1. IAHS/UNEP/UNESCO: Wallingford

Higham, S. (1996). Soiling the Sacred Lake. *Water and Environment*, September, pp. 26–7

Hofer, T. (1997). Floods in Bangladesh: A highland-lowland interaction? *Geographica Bernensia* G48. Institute of Geography, University of Bern

Ives, J. D. and Messerli, B. (1989). *The Himalayan Dilemma: Reconciling Development and Conservation*. Routledge: London, p. 295

Joss, J. and Lee, R. (1993). Weather radar: Operational processing for nowcasting and precipitation estimation. In Moore, R. (ed.) Proceedings of WMO Regional Association IV (Europe) Workshop on Requirements and Application of Weather Rader Data in Hydrology and Water Resources. WHO: Geneva

Karr, J. R. (1991). Biological integrity: a long-neglected aspect of water resource management. *Ecological Applications*, **1**: 66–84

Kattelmann, R. and Embury, M. (1996). Riparian areas and wetlands. *Sierra Nevada Ecosystem Project: Final Report to Congress*. University of California–Davis, Centers for Water and Wildland Resources, vol. III, chapter 5, pp. 201–73

Kattelmann, R. (1996). Hydrology and water resources. *Sierra Nevada Ecosystem Project: Final Report to Congress*. University of California–Davis, Centers for Water and Wildland Resources., vol. II, chapter 30, pp. 855–920

Kattelmann, R. (1994). Improving the knowledge base for Himalayan water development. *Water Nepal*, **4**(1): 89–97

Kotlyakov, V. M. (ed.) (1996). *World Atlas of Snow and Ice Resources*. Institute of Geography, Russian Academy of Sciences: Moscow

Kottelat, M. and Whitten, T. (1996). *Freshwater Biodiversity in Asia with Special Reference to Fish*. Technical Paper 343, The World Bank, Washington DC, p. 59

Lang, H. and Musy, A. (eds.) (1990). *Hydrology in Mountainous Regions I – Hydrological Measurements: The Water Cycle*. IAHS Publication 193: Wallingford

Lang, H. (1981). Is evaporation an important component in high alpine hydrology? *Nordic Hydrology*, **12**:217–24

Molnar, L. (ed.) (1992). *Hydrology of Mountain Areas*. IAHS Publication 190: Wallingford

Montgomery, D. R., Grant, G. E. and Sullivan, K. (1995). Watershed analysis as a framework for implementing ecosystem management. *Water Resources Bulletin*, **31**(3): 369–86

Mosley, M. P. (1996). A participatory process approach to development of a comprehensive water resources management plan for Sri Lanka. *Water International*, **21**(4): 191–7

Moyle, P., Yoshiyama, R. M. and Knapp, R. A. (1996). Status of fish and fisheries. *Sierra Nevada Ecosystem Project: Final Report to Congress*. University of California–Davis, Centers for Water and Wildland Resources., vol. II, chapter 33, pp. 953–75

Naiman, R. J., Magnuson, J. J., McKnight, D. M. and Stanford, J. A. (eds.) (1995). *The Freshwater Imperative: A Research Agenda*. Island Press: Washington DC, p. 165

Postel, S. L., Daily, G. C. and Ehrlich, P. R. (1996). Human appropriation of renewable fresh water. *Science*, **271**: 785–8

Rango, A. (1993). Snow hydrology processes and remote sensing. *Hydrological Processes*, **7**: 121–38

Thapa, K. B. (1993). Estimation of snowmelt runoff in Himalayan catchments incorporating remote sensing data. In Young, G. J. (ed.), *Snow and Glacier Hydrology*. IAHS Publication 218: Wallingford, pp. 69–74

US National Research Council (1982). *Scientific Basis of Water-Resource Management*. National Academy of Sciences: Washington DC, p. 127

US National Research Council (1992). *Restoration of Aquatic Ecosystems*. National Academy of Sciences: Washington DC, p. 552

US Bureau of Reclamation (1972). *Colorado River Water Quality Improvement Programme*. US Government Press: Washington DC, pp. 43–5

Valdiya, K. S. (1993). *High Dams in the Himalaya*. Pahar Publications: Nainital, p. 23

Watanabe, T., Ives, J. D. and Hammond, J. E. (1994). Rapid growth of a glacial lake in Khumbu Himal, Nepal: Prospects for a Catastrophic Flood. *Mountain Research and Development*, **14**(4):329–40

Williams, R. S. and Ferringo, J. G. (1991). *Satellite Image Atlas of Glaciers of the World: Middle East and Africa*. USGS Professional Paper 1386-G: Reston, VA

The World Bank (1993). *Water Resources Management: A World Bank Policy Paper*. International Bank for Reconstruction and Development: Washington DC, p. 140

World Meteorological Organization (1994). *Guide to Hydrological Practices*. WMO-164; WMO: Geneva

World Meteorological Organization/UNESCO (1991). *Water Resources Assessment*. World Meteorological Organization: Geneva, p. 64

Young, G. J. (1980). Monitoring glacier outburst floods. *Nordic Hydrology*, **11**: 285–300

Young, G. J. (1993). *Snow and Glacier Hydrology*. IAHS Publication 218: Wallingford

Energy resources for remote highland areas

8

Petra Schweizer and Klaus Preiser

INTRODUCTION

For centuries the energy demand and supply options of mountain people have developed within the pre-existing mountain ecosystems. Under extreme environmental conditions an energy-use system evolved depending strongly on the locally available human and natural resources. With the spread of twentieth century industrialisation and, accordingly, the escalating energy demands of the lowlands, new challenges have arisen for the mountains. Exploitation of the mountain energy potential, with the extensive utilisation of hydropower, has had a massive impact on the environmental and social structures which have been established for centuries. The hydropower installations have required the construction of roads, dams, and reservoirs and, in turn, have caused landslides, due to their often destabilising effects on steep slopes, flooding of agricultural land, and forced resettlement of mountain villages. Furthermore, the incidence of collapse of dams and associated structures, due to catastrophic natural events, sometimes augmented by either unwise location or unsatisfactory engineering, leaves people and property below them seriously at risk.

In parallel, the increasing demands of the growing rural population has put additional pressure on the local energy resources. Influenced by widespread development in the cities and rapid spread of telecommunications and tourist contacts, mountain people have come to desire a share of the materialistic culture of the lowlands. New energy sources have had to be imported, for instance kerosene and dry cell batteries, that represent a new potential for contamination. Today we know that the combined effects of all these rapid changes endanger the fragile equilibrium of the older established energy-use system in mountainous regions.

This energy-use system is characterised by a weak infrastructure, dispersed settlement, and long distances between villages and from urban centres, as well as a relatively low energy consumption per capita and in total. Household energy needs, besides agriculture, handicrafts, tourism, and small-scale industries, are the main users. If there is no vehicular access, everything has to be carried by human and animal power. If these needs are to be met by external energy sources, costs will be very high and distribution highly inefficient, compared with the situation in the lowlands. Use of locally available energy resources, therefore, usually takes precedence.

TRADITIONAL ENERGY SOURCES AND ENERGY DEMANDS IN REMOTE HIGHLAND AREAS

All over the world, biomass and human or animal power have been the main energy sources for centuries. Due to the remoteness of mountainous regions, this has not changed until very recently, and in many remote areas it still persists. However, the 'new' energy options, such as fossil fuels or hydropower, are assuming a growing share of the total energy budget. They complement or displace traditional sources and also cover the quantitatively and qualitatively increasing energy demand.

Table 8.1 shows different types of energy demands and the most important energy sources used today. Biomass, including firewood, as well as agricultural and animal residues, is mainly used for cooking, lighting, and space heating. These energy sources are limited and often their availability is restricted by ownership, for example governmental or private forests. Additionally, there have been environmental effects (see also Chapters 13 and 15) and, due to smoke, the use of biomass also has had a negative impact on health.

Human and animal power – for transportation, driving machines, and collecting firewood, for instance – is naturally limited. On the one hand, this

Table 8.1 Energy demands and main energy sources in the mountains ◆ main energy source, □ very commonly used energy source

	Locally available energy sources					Imported energy sources		
	Biomass	Charcoal	Hydro-power	Solar energy	Human/animal power	Fossil fuels	Candles	Dry cell batteries
Households								
cooking	◆	□				□		
lighting	□					◆	□	□
water pumping					◆			
water heating	◆							
space heating	◆							
radio/TV								◆
Agriculture								
fruit/crop drying	□			◆				
husking			◆		◆	□		
oil pressing			◆		◆	□		
grain grinding			□		◆	□		
irrigation					◆	□		
soil management					◆			
Forestry					◆			
Handicrafts		◆	□		◆			
Transport					◆	□		
Tourism	□				□	◆	□	□

restricts the available power capacity but, on the other hand, it leads to the over-exploitation of local resources. When human power is bound to the supply of basic energy demands, it will not be available for other creative or productive activities.

Imported forms of energy, such as kerosene, candles, or dry cell batteries, have several undesirable effects in remote mountain areas. Their supply from the urban lowlands creates a link with all the inherent advantages and disadvantages (transportation network, cash flow, subsidies). The new services that they provide also include detrimental effects (CO_2 emission, smoke, contamination of soil and water with chemicals[1]).

Recently, a new energy demand has arisen for communication devices such as radio, TV, and telephone, household appliances, and small-scale industries. Most of it can be satisfied only by electricity. Due to the difficulties of extending the public electricity grid to mountainous areas, the only option is to provide decentralised electricity on the basis of locally available resources. The installation of hydroelectric power stations and diesel generators is increasing but there are other suitable energy options, described in the next section.

LOCALLY AVAILABLE RENEWABLE ENERGY RESOURCES AND THEIR POSSIBLE FIELDS OF APPLICATION

In contrast to the centrally supplied lowland areas, mountain regions tend to be decentralised. Therefore, a decentralised energy supply structure can best meet this specific demand pattern. Local and renewable resources facilitate the creation of an environmentally benign, locally adaptable and manageable supply for mountain societies. With the variety of different energy options, reliability can be increased, for example by using hydropower during the monsoon season and solar power during the dry season.

Since the utilisation of renewable resources strongly depends on local conditions, emphasis must be placed on the energy options with the most appropriate social, environmental, and technical

---

[1]One simple dry cell battery can contain 8 mg Hg, 32 mg Pb and 16 mg Zn. The mercury alone in one battery is sufficient to contaminate 800 m^3 of air, 160 l of drainage water, 8000 l of drinking water or 27 kg of soil.

potential. Figure 8.1 indicates how site and altitude determine the suitability of different renewable energy sources, taking the Himalayan region as an example.

In the Terai, the population density is higher than in the mountains, temperatures are higher, and roads, as well as other infrastructure, are easier to develop; thus, biogas, diesel, and the national electricity grid are the most suitable energy options. This changes dramatically in the mountain regions because the population is dispersed, few roads are available, fuel has to be transported long distances, and the climate is colder. It is often difficult to extend the national electricity grid into these regions (long distances and rugged terrain) and it cannot reach all settlements. Where there are rivers close to roads, decentralised hydropower stations can supply the inhabitants with mechanical power and electricity. It is evident that in the most remote and dry plateaus the use of different solar energy technologies becomes the most suitable option.

Biomass

In many mountain regions biomass provides for more than 90 percent of the total energy consumption. Wood, the traditional, locally available fuel, fulfils the household demand for cooking, heat, and

light and requires no, or little, community collaboration outside the immediate family. When the need for natural replacement is respected, forests remain the principal renewable resource (Plate 8.1). This fragile equilibrium, however, is endangered in most mountain systems, and this can lead, or has already led, to a shortage of firewood.

The people already affected by this shortage are responding by: (a) saving half-burnt firewood, for example by extinguishing embers directly after cooking; (b) using other forms of biomass, such as

Plate 8.1 Nepalese villagers harvesting firewood in their community forest

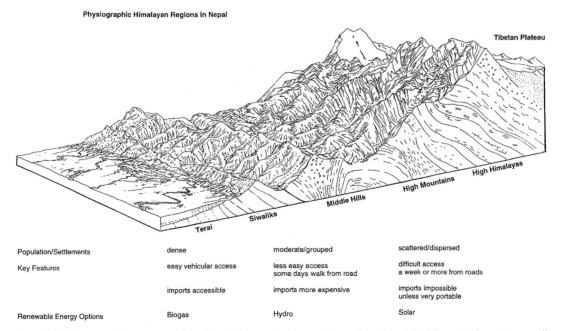

Figure 8.1 Renewable energy options for different physiographic regions (modified from Aitken, Cromwell and Wishart, 1991 p. 10, and Ramsey, 1986)

Plate 8.2 Sherpa women besides a locally built firewood saving stove in the Bamti/Bhandar Village, Ramechhap District, Nepal

dried animal dung or agricultural residues; and (c) changing the construction of traditional fireplaces.

In mountain societies, the fireplace is the social centre of the house. Not only eating, but also sitting, talking, and even sleeping take place around the fire and it has a special significance for different purposes, including communication, religion, and hygiene (Plate 8.2). Often open fireplaces are used and pots are placed on three stones or on an iron tripod. The fire is regulated by moving the firewood in and out. More advanced versions include closed and more efficient clay or iron stoves. These accommodate larger quantities of wood than are needed for stoves in the warmer lowlands. With the clay stoves it is also possible to integrate iron pipes to heat water.

Wood is the raw material for the production of another local fuel, namely charcoal. It supplies very

Box 8.1 The forest as a main source of energy in Nepal

The forests continue to be the major source of domestic energy, especially in the mountains where alternative sources are usually not available.

The indigenous management and use of forests has been quite effective in Nepal from time immemorial. Traditionally the forests were used for cultural (birth to death rites) and domestic purposes. Local tax systems were applied to forest resource use and to grazing in the forests. However, after nationalisation of forests in 1957, the traditional forest management system of Nepal was put under severe pressure. Consequently, Nepal faced a high rate of deforestation, which the community forestry programme was intended to counteract. A forest sector master plan study was submitted to the government in 1988 and the Forest Bill was drafted to implement this plan. It was improved in the Forest Act of 1993 to recognise and authorise forest user groups which undertook responsibility for activities such as conservation, management, and proper utilisation of forest products (Plate 8.3). The forest user groups spread the message that well-managed forests are to everybody's benefit. In this context, the chairperson of Budhikhoria forest user's group, Mr Ganesh Das

Plate 8.3 Nepalese women carrying heavy loads of firewood collected in the nearby forest

Shrestha, (Ramechhap District – Middle hills of Nepal) says: 'Four years ago, the forest was handed over to our group. By that time, its condition was hopeless. We completely protected the forest for two years, except the ground grass, as per the group's decision. In the third year our assembly decided to do the pruning and thinning of one tenth of the forest. We were able to share some firewood. Now we have developed a quite simple but detailed forest management plan. According to that plan we will have enough firewood and some fodder for our group (121 households)'.

This author has been working with about 150 forest user groups (continually increasing). In his experience, the need for forest resources differs substantially among the different strata of society and between men and women. Women see the forest as a source for fulfilling basic domestic needs, whereas men consider the forest more for its commercial possibilities. However, almost all forest user groups in Ramechhap and Dolakha districts have given highest priority in their forest management plans to fuel requirements. In this context, the secretary of Bhange Hodumba forest user group, Mr Sunuwar, said: 'The basic needs of people will receive highest priority. Firewood, fodder, poles, timber in that order, are our priorities'.

Likewise in the five-day community forestry workshop for women in Ramechhap (1995), the participants showed great concern and interest in protecting, managing and utilising forests: 'Women are responsible for domestic work. We need 60–70 bhari (1 bhari = 50 kg) of firewood in a year for cooking alone, depending on the family size (5 to 10 members). So, the women go to the forest to collect firewood, fodder, and leaf litter (for animal bedding). If there is no forest the women will have a harder time than men. Therefore, the women love the forest more than men'.

Source: Damber Tembe, Forest Adviser with the Swiss Development Co-operation in Nepal

intense and durable heat that is suited for industrial purposes and especially for the traditional work of the blacksmith. The manufacture of charcoal, however, consumes large quantities of wood.

Biogas processed by fermentation of animal residues has a long tradition in several rural societies. For the activation and operation of the bacterial process, temperatures of about 25 to 35 °C are essential. From this it is evident that the use of biogas in mountain regions is almost impossible, except under very favourable conditions.

Hydropower

In hilly areas, water wheels have been used for centuries, mostly for milling grain. There are different technologies available, ranging from the large hydropower stations with a capacity of several hundreds of megawatts that supply the plains with electricity (Chapter 7) to directly used, decentralised, small-scale applications which range from a few kilowatts up to tens of kilowatts. The latter are very often used in mountainous areas for mechani-

cal operations, such as grinding, husking, and pressing. Today, installation of micro-hydropower plants is increasing. With single-phase or triple-phase alternating current (AC), any conventional, commercially available electrical appliance can be used if the maximum power output is observed.

A typical hydropower installation is constructed as shown in Figure 8.2: water is taken from the river and routed in a canal through a storage basin into the penstock, to the powerhouse. The higher the head and flow rate, the more power can be gained. Certain types of turbines are suited for certain characteristics of head and flow rate; for instance water wheels are suitable only for low heads, while Pelton wheels are used mainly for higher heads of more than 15 m.

Although utilisation of renewable water power is essentially environmentally benign, there are some constraints that have to be taken into account. Only a certain amount of water can be taken from a river, to ensure that fish can continue to swim upstream or downstream at all times of year. This means that in dry seasons reductions must be

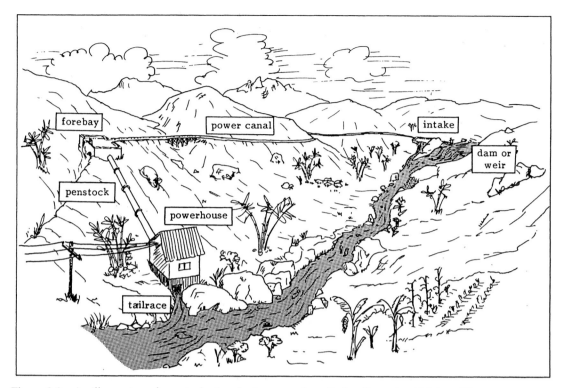

Figure 8.2 An illustration of a typical micro-hydropower installation (Inversin, 1986, p. 63)

made in energy production for the benefit of the surrounding natural habitats. Impacts resulting from construction work should be kept as low as possible. There may be a danger of initiating landslides, for example by digging the water canal.

The power stations in the range between 1 kW and about 50 kW have the following characteristics:

(1) are easily manageable for local entrepreneurs,
(2) easily co-ordinated with other families or even villages,
(3) locally manufactured generators are cheaper and easier to organise,
(4) operation of the power station is comparable to traditional tasks in water management,
(5) the electricity needs of the dispersed mountain population can be met with small units.

The installation of larger water turbines, favoured by state-owned utilities because they are more efficient to administer, cannot rely only on traditional knowledge. New levels of competence in construction, operation, and maintenance are needed when a power station is installed in a mountain village. Furthermore, the involvement of local people and

organisations is necessary to encourage their identification with, and responsibility for, the power supply and its use; this is a key factor for integrated and sustainable water energy management based on local structures.

Solar energy

Solar energy is locally obtainable everywhere. Different technologies – from simple to very complex systems – are available for the conversion of sunlight into thermal and electrical energy. Their range from very small to large systems makes it possible to adapt technical units to the specific energy demand of rural households and communities. The power production depends directly on solar radiation. There are technologies, such as concentrators, that need direct sunlight, while others are also capable of operating under diffuse irradiation conditions. In most solar applications, an energy storage unit is needed to ensure reliable service for 24 hours a day and throughout all seasons of the year.

Figure 8.3 presents the monthly distribution of solar radiation in mountains using the Chilean

Box 8.2 Micro-hydropower in Nepal

In rural Nepal, the traditional water wheel (less than 1 kW) has been in use for centuries. It is estimated that there are over 25 000 such units in operation, meeting the agro-processing energy needs of the scattered villages in the hilly and mountainous regions.

The types of micro-hydropower turbines manufactured in the country are determined by various factors such as power requirements, future load growth, cost, head, flow, local management capability, local availability of technically trained manpower, availability of construction materials, and the remoteness of the site. Those in popular use are:

- Traditional Nepalese water wheels, Ghatta < 1 kW for agro-processing (grinding only);
- Improved Ghatta (< 2.5 kW) for agro-processing (grinding and rice-hulling);
- Multi-purpose power units (MPPU) (5–30 kW) for agro-processing (grinding, rice-hulling and oil-pressing; sometimes with add-on electricity generation < 500 W);
- Pelton turbines (5–30 kW) for electricity generation only;
- Peltric sets (a small Pelton turbine with an induction generator; < 1 kW) for electricity generation only;
- Cross-flow turbines (10–50 kW) for agro-processing (grinding, rice-hulling, and oil-pressing; some have supplementary electricity generators and some are for stand-alone electricity generation only).

The locally manufactured micro-hydropower turbines have developed from the traditional water wheel (Ghatta) to the Pelton turbine. They evolved through different stages (from Ghatta to cross-flow turbine) in an effort to bring the modern turbine technology closer to the traditional water wheel with which the rural people are already familiar. Micro-hydropower installations in Nepal are most commonly in the 1–25 kW range, and equipped with MPPU, cross-flow turbines, or Peltric sets.

Presently, there are more than 900 micro-hydropower installations with a total capacity of about 5 MW. They are used mainly for agro-processing purposes. About 20 percent of them produce electricity in the form of supplementary power. Of the total, only about 100 units are stand-alone electricity-generating units.

Development of micro-hydropower in Nepal presents a good example of private sector initiative, with the government and donor agency's role confined to that of facilitator and promoter. The policy-level initiative with regard to subsidies (up to 75 percent of the equipment costs in the mountains) and loans through banks, the de-licensing of plant of up to 100 kW capacity in 1984, and the recent de-licensing of plants up to 1000 kW capacity, have all been instrumental in making this programme sustainable.

Source: Dr Kamal Rijal, Energy Specialist at the ICIMOD, Kathmandu, Nepal

Box 8.3 Rural electrification in Muktinath, Nepal

Purang village, located in Mustang District adjacent to the Muktinath temple, has been electrified following an initiative taken by local people. This village is situated at an altitude of 4200 m asl and lies on the tourist trail of the Annapurna conservation area.

The area surrounding Muktinath is especially scenic and exceptionally dry with an annual rainfall below 100 mm. Forest resources are minimal to non-existent. Firewood is used to satisfy the demand for domestic energy. This wood is collected from the nearest forest, which is several days walk. There is not even a primary school or government office in the area. Many of the villagers depend on income generated from the tourist trade. Potatoes and buckwheat are harvested once a year. The land which can be used for cultivation is limited. The villagers, though illiterate and totally isolated from the development activities of the rest of the world, are

intelligent enough to think of alternatives to ensure their own survival. The only natural resource available in the area is the water that flows downhill from the shrine at Muktinath.

Following the customs of their people, the villagers came together to discuss details of a project to use hydroelectric power to fulfil the energy needs of their community and replace kerosene for lighting and reduce the consumption of firewood for cooking. The ideas of the villagers were formulated into a practical plan with the help of the district chairman. The necessary funds were raised by the village people with some small support from external sources.

The villagers hired a consultant contractor from Butwal, an industrial centre in southwestern Nepal. He was responsible for the survey, design, and installation of the project. The village was electrified by a micro-hydroelectric plant with a 9 kW capacity. The availability of electricity brought about a realisation among the villagers of its other potential uses. Low-wattage cookers were introduced to heat water, using excess energy at off-peak times. This changed the cooking habits of the villagers to a remarkable extent for they understood that the technology could be used to improve their living standards. The increasing demand, however, could not be met by the 9 kW plant. As the project area is sited on the tourist trail, the need for a further increase in power was recognised by the villagers and hotel owners alike.

A detailed survey and studies were carried out for expansion of the original plant. Replacement of the cross-flow turbine with a Pelton type and certain changes in alignment were suggested. The existing head was increased to give 25 kW from the same water flowing down from Muktinath. The necessary funds were raised with a bank loan, a subsidy from the donor, and contributions from the local community.

The approach the villagers have taken to development seems to be sustainable. The changes introduced have not disturbed the existing system of the society but have raised the local standard of living. The micro-hydroelectric project is cost-effective, sustainable, and locally manageable.

Source: Devendra Adhikari, ADBN,
Nepal (Adhikari, 1993)

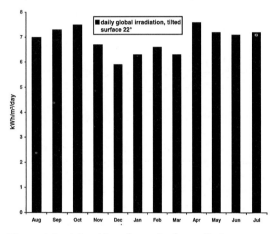

Figure 8.3 Monthly values of solar radiation on a plane tilted at 22°. Data from Visviri (18°N and about 4200 m asl) on the Altiplano of Chile, taken by the Universidad de Tarapacá, Chile

Altiplano as an example. Solar radiation is very high and constant throughout the year, especially near the equator on high plateaus with minimal precipitation.

Thermal use of solar energy

In high altitude, cold desert areas, such as the northwestern Himalaya, a long tradition of energy-conscious, earth-building practices exists. Settlements are built on south-facing hillsides, walls are insulated with mud, and windows of the traditional buildings face south. Since it is now possible to transport glass into these areas, increasing numbers of passive solar buildings, including winter gardens, have been constructed (Norberg-Hodge,

1991). Nevertheless, governmental or newly built tourist residences often lack the integration of traditional passive heating systems, such as mud walls and especially insulated roofs made from local materials; the traditional structures have been replaced with thin sheet roofing.

The newest innovations, including transparent insulation material (TIM), reduce heat losses from houses (as with conventional opaque material) and additionally allow solar energy to contribute to house heating. This can greatly reduce the amount of fuel needed for use in traditional iron stoves, especially in the tourist centres. Applications of this technology can be found more and more in European buildings; its transferability to mountainous regions in the so-called developing countries has yet to be proven.

The drying of fruits, vegetables, herbs and other crops in direct sunlight is one of the oldest methods for preservation of agricultural products. With simple technical means, for example wooden constructions with black metal absorbers and glass covers, the drying process can be improved considerably. Different approaches can be found all over the world, more or less at an experimental stage, although the widespread dissemination of small units for decentralised application has not yet occurred.

For decades, scientists, development agencies, and local NGOs have been experimenting with different technologies to cook with solar power. Most of these solar cookers can only be used outdoors and thus require a certain change in cooking habits. This may be the reason why solar cookers are not as successful or as popular as their experimental performance promised. To make the cooking process more independent of the sunshine hours, further efforts are concentrating on units with integrated energy storage. For mountain regions, the requirements are even more stringent due to the cold climate. There are three major solar cooker options:

(1) The solar cooking box is the most simple way to cook with the sun. It consists of an insulated box that is painted black inside and covered with glass. The lid can be used as a reflector to increase the active solar aperture. This type of cooker can be produced locally, easily, and relatively cheaply. Its disadvantage is that it cannot attain very high temperatures, above 90–120 °C. Compared to a conventional fireplace, the cooking needs more time: stirring and spicing during the cooking process are not possible, because opening the box would result in an enormous loss of heat.

(2) Concentrating reflector cookers may be suitable, especially for regions with a high proportion of direct sunlight. In Tibet for example, they are used to prepare tea. This cooker is manufactured mainly in Europe and North America, but can easily be shipped and re-assembled. It attains very high temperatures (up to 200 °C) relatively quickly when exposed to direct sunlight. The risk of accidents is higher than for other solar cookers due to concentration of sunlight at the focal point of the parabolic form. Their light-weight construction makes them easy to move but also more sensitive to disturbance by wind.

(3) Storage cookers are characterised by a large storage volume and complex construction. The heat produced by a solar collector is transferred to a separate or integrated storage and cooking place. Storage media may be stones or sand. Plant oils provide an excellent liquid to transfer heat. The advantage with this form is that it is possible to cook during the evenings and inside the house; however, it is expensive and more complicated.

Another application of thermal solar energy is the heating of water. Water is led through a black painted pipe into an insulated storage tank. Under cold environmental conditions freezing must be prevented. Hot water is used mainly in the areas where tourism is a common source of income; hot showers are one form of simple application.

Electrical use of solar energy

Photovoltaics (PV), the direct conversion of sunlight into electricity, is one of the most environmentally benign ways to produce electricity. The solar cells used are characterised by a theoretically unlimited lifetime and low maintenance. To obtain higher power values, the individual solar cells are connected together to make modules (typically between 50 and 70 Wp[2]), and these modules can be combined for the construction of PV generators. With appropriate encapsulation, the modules can be used even under extreme environmental conditions. The efficiency of commercially manufactured solar cells is now in the range of 12 to 17 percent, and there is still potential for improvements; for example, in research laboratories, cells with efficiencies of over 20 percent have been developed. To make use of PV, other system components are necessary. In mountain regions, a decentralised, grid-independent power supply is the focus of interest. An overview of the different concepts for suitable PV systems is given in Figure 8.4.

The supply to several houses by way of a decentralised grid in a village power system is characterised by a central PV generator, a large storage battery bank with a charge controller, and a central inverter to convert the direct current (DC) provided by PV and batteries into alternating current (AC), as in the national electricity grid. This has the advantage that conventional appliances can be used, maintenance and operation of the power plant can be organised centrally, and later connection to a public electricity grid is technically relatively easy. A weak point is the inverter, through which all the energy has to pass. With the development of a power supply as a common resource in a village, the consumption of electricity must be distributed fairly among the users. This requires village-wide co-operation.

A single house system, often called Solar Home System (SHS, see also Box 8.4), consists of a small PV generator, usually only one module of about 50 Wp, a storage battery with a charge controller, and directly connected DC appliances. Sometimes an inverter is added to operate appliances with larger AC consumption. As it is a family-centred resource, distribution and consumption are organised by the individual users themselves.

---

[2]Wp is peak power attained under standard tests conditions: irradiation of 1000 W/m², cell temperature 25 °C.

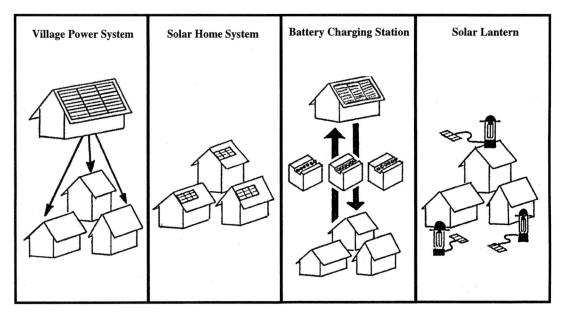

Figure 8.4 Different PV system designs to supply houses with electricity (modified from Martin, 1994)

Box 8.4 Solar Home Systems in the Bolivian rural highlands

Providing a conventional energy supply for the rural population in the Andean highlands has proved to be very difficult and, in many cases, impossible. The lack of basic services is an important factor for rural out-migration with the corresponding social, economic, and environmental costs. In this context, PV Solar Home Systems can help to alleviate this situation by improving living standards and providing access to technical development.

The Bolivian highlands comprise the western high plateau, called the Altiplano, and the neighbouring highland valleys situated between the Andean Cordillera (14 to 23°S). Characteristics of the region include:

- the mean altitude of the inhabited area is in the order of 3800 m (from 2500 up to 4200 m);
- high solar radiation, a typical value is 6.0 kWh/m^2 for the daily average (2200 kWh/m^2/year) with less than 30 percent seasonal variation;
- low ambient temperatures, around 9 °C annual mean with very low night temperatures (minima about −15 °C) and approximately 18–24 °C maximum day temperatures;
- very low precipitation (less than 400 mm/year).

Less than 20 percent of the rural population in the highlands is served by the national grid system, and this situation will not change significantly in the next 20 years. A major part of the Altiplano has been deforested over the last centuries. Previously the plateau was covered with Thola, the most common tree species at this altitude.

On average, 96 percent of the energy is used for cooking, and 3.5 percent for illumination. Dung is of major importance (more than 60 percent of the total energy use) as a cooking fuel, given the state of deforestation.

Energy supply	Energy use per year and per head (KWh)	Percent of total use	Application
dung, wood, charcoal	940	82	100% cooking
city gas, kerosene	180	16	80% cooking, 20% illumination
candles	20	2	100% illumination
batteries	0.2	0.02	100% communication

It is evident that the most important barrier to the extension of the use of PV Solar Home Systems in the Andean rural highlands is the high initial capital cost. To make the application of

Plate 8.4 A solar home system in Orinoca, in the department of Oruro in Bolivia, approx. 3850 m asl, installed by the regional development organisation COREDOR in 1993 (Photograph: K. Preiser)

this kind of energy possible to final users, the joint action of international co-operation, state and development organisations in the area, as well as the energy users themselves, is absolutely necessary.

To date, there are some 2000 SHS installed in the Bolivian highlands, with capacities that range between 22 Wp and 50 Wp. The technical reliability and cultural adaptability have been proved and programmes for the extension of this technology are foreseen.

The Government has an electrification plan, according to which by the year 2006, 77 percent of the households will be electrified compared to 55 percent in 1995. In 1995, 1 percent of the households were supplied with electricity from renewable energy such as PV systems and micro-hydroelectric plants. According to the plan, 30 percent of all households in Bolivia should be supplied by SHS by the year 2006.

Source: Javier Gil
Energetica Cochabamba, Bolivia

Mismanagement affects only the family responsible. Each system can be extended according to the needs of the individual users. This system design ensures that operation and maintenance must be arranged individually.

The concept of PV-powered battery charging stations does not include a PV generator for the individual houses. Portable batteries are used to supply electric appliances, and – after fully discharging – are transported to the central station for recharging. Maintenance and operation of the PV system is centrally organised, so investment costs

for the users are lower than for those of individual systems. However, transporting batteries with liquid acid represents a constant danger and the often non-regulated discharging process reduces the battery lifetime considerably.

PV lanterns, the smallest systems to supply electric light, are intended to be a substitute for the commonly used kerosene lamps or candles. They are portable, easy to handle and, compared to the other systems described above, relatively low in investment costs. Small radios can also be connected to some models.

Solar electricity can also be used directly without energy storage but the appliance operates only if the insolation exceeds a certain value. Water pumping and grain milling are applications that are well suited to this condition, as they do not have to run continually. Their storage is intrinsic to the application itself (water tank and grain sack).

Wind energy

Strong winds are characteristic of all mountain regions. Wind speeds with an average annual value of more than 4 m/s, in principle, are very favourable for wind turbines. However, this enormous energy potential is very difficult to utilise for decentralised systems. Due to the uneven surface of the landscape the wind patterns are very site-specific and complex. Regularly occurring valley wind or upcurrent systems are very different from highly changeable, turbulent and gusty winds. With the extreme climatic conditions (low temperature, ice), the requirements on the mechanical load capability are unusually stringent. Experience shows that most of the turbines in mountain areas did indeed fail very soon after installation, due to wind-induced overloads. Possible direct applications for wind power are water pumping, milling, and operation of other mechanical equipment. Relatively small units of a few kW could be used for a single application or scaled up to community-sized systems.

Wind-power production of electricity could be used either in larger wind generation parks connected to the national grid, which acts as a very large energy store, or decentrally in smaller units to power stand-alone systems. Because wind energy often fluctuates considerably during the day and seasonally, large and expensive energy storage units are necessary in stand-alone systems to ensure a continuous energy supply. This, of course, constitutes a serious obstacle for poor mountain communities.

Geothermal energy

Warm springs are found in many mountain regions. They are often used for ritual bathing, but in principle they can also be used for energy purposes. It is evident that transportation of warm water over long distances is very difficult, but space and water heating in nearby buildings is possible.

Because of the low energy density of warm water, the conversion into electricity requires relatively complex technical systems. To make use of this energy potential, large systems are necessary, such as central power stations, that produce electricity mainly for consumers other than those in the immediate mountain surroundings.

Hybrid electricity systems

For buildings with a large energy demand (hospitals, hotels, and workshops), the installation of hybrid systems is the most secure option. A hybrid system consists of different power producers, such as PV, wind, hydroelectric, and/or diesel generators (Plate 8.5). This combination increases the reliability of the power generated and provided. Especially where diesel generators are already in use, the addition of renewable power sources saves fossil fuels and reduces pollution and noise. In these miniature electricity grids, operation and maintenance is centrally organised. Besides the provision of electricity, the waste heat of thermal electricity converters, such as diesel generators, can contribute to heating purposes, thus increasing the yield of the fossil energy used.

DIFFUSION OF ENERGY INNOVATIONS

The diffusion of any new technology is an evolutionary process. It involves the adaptation of the technology to the pre-existing context, but it is also important that people adopt the new technology. Sustainable integration of a new technology into a society can be achieved only by such an interactive

Plate 8.5 Rotwandhaus, an Alpine hut in the Bavarian Alps situated at 1765 m asl and equipped with a hybrid system comprising wind, photovoltaic, and diesel generator

Box 8.5 Energy supply of German Alpine huts

In 1986, the German Mountaineering Club (DAV) adopted a 10-year programme to protect the environment of the Alps. One of the practical measures proposed was an environmentally benign energy supply for mountain lodges. Due to the booming tourist industry in the Alps, the tolerance limits of this ecologically sensitive region have been reached and already overstepped in some areas. For instance, it was recognised that the tendency of the past decades to increase the comfort in alpine huts has developed to undesirable levels, as besides other environmentally harmful effects (for example, increase in disposable waste), the demands have also been constantly increasing. The high energy demand of most alpine lodges could be met only by using diesel generators; these degrade the environment considerably (noise, stench, pollution).

The reduction of the energy consumption of an alpine hut, which is a pre-condition for wise use of regenerative energy sources requires, in most cases, not only technical measures (energy-saving appliances) but also a restriction in the use of electrical appliances. A conscious reduction in comfort and management of the lodges to supply only the most simple needs of trekkers and mountain climbers is considered by the DAV to be a prerequisite for successful application of the agreed technical measures.

With the current status of technology and prices for conventional fuels, the use of regenerative energy sources does not seem to be economically viable at first sight. Detailed considerations, however, show that the prices of energy in many cases (for instance, solar water heating) are now already comparable with conventional energy. Even the very high costs of about 3 to 7 US$/kWh for a solar electricity supply is very often competitive with electricity produced by a diesel generator, due to the high transport costs. However, financial aspects alone are not the central criteria for decisions about energy supply for alpine huts.

Different system concepts are applied to supply alpine buildings with electricity. With increasing requirements on the demand side, the complexity in the system design also increases. For small huts, low voltage (12 V or 24 V) photovoltaic systems without additional energy sources are practicable. For larger lodges, already equipped with AC appliances, the combination of a photovoltaic system with the motor generator that is usually already installed is the best option. The photovoltaic generator provides electricity during the day and surplus energy is stored in the battery from which the appliances are operated during the night or in overcast weather. Continuous supply of the 230 V AC appliances is guaranteed by an inverter. To power larger loads, such as a cable-car for resupply, the motor generator takes over the supply of the whole system including battery charging.

Source: Peter Weber,
Deutscher Alpenverein (DAV)

process, and when continuous further development and adoption is assured. This is especially true for decentralised renewable energy systems that comprise not only purely technical components but also people who install, operate, and maintain them.

The modification of an energy use system implies changes at different levels of society. On the individual level, this development can be understood as a learning process. On the society level, it is a cultural change that arises from the new technological options. The diffusion of a new energy technology is a process in which individual users or user groups decide whether an innovation suits their special needs and fulfils the existing demands in an acceptable way. Whether a technology is applied successfully and diffused broadly within a society depends on different aspects on the product and the user side.

RECOMMENDATIONS FOR FUTURE ACTIONS

As the development of mountain regions has increased with the growing needs and aspirations of mountain communities, energy has become a fundamental factor in development strategies. For three reasons power supplies need to be designed with careful consideration of mountain ecosystems and communities.

First, many energy sources commonly used in mountain regions are becoming scarce. For instance, firewood is being harvested in many areas at faster rates than regeneration of the supply. Also, many mountain inhabitants use dry cell batteries, which pose serious environmental threats and are cumulatively very expensive. Second, while today energy demands in mountain communities are almost limited to cooking and lighting, the demands are rapidly increasing as households seek modern appliances and conveniences and as local industries, tourism and other services expand. Finally, conventional energy production units and distribution systems are designed to serve, densely populated communities; this centralisation is often inappropriate and costly for remote mountain regions. Therefore, mountains require decentralised and locally managed energy supplies.

For the sustainable fulfilment of present and future energy demands, environmentally and culturally sound energy programmes, developed with the participation of the users, are required. They should be based on a sound evaluation of the local needs and potentials; the local populations must be provided with sufficient amounts of renewable energy to support their livelihoods and legitimate aspirations. Programmes should make use of all existing and emerging technologies for harnessing renewable energy resources.

The decision to create a private or a communal energy system should depend on existing social structures. If there is a strong local social structure for use of common resources, it can be assumed that one energy system for the whole community will function well. Prerequisites are: small and stable user groups; dense networks of social interactions; common standards and rules; and sufficient knowledge of the dynamics of the resource. Where communities are already well integrated community-based energy supply systems are to be preferred. In contrast, where individual households are independent, or where the social structure is individually oriented, the resource-use system should be privately organised.

Achievement of a sustainable energy supply, using any of the technologies mentioned above, will depend on the necessary infrastructure being created to overcome the barriers of long distances and often poorly accessible locations. Therefore, integrated strategies are needed on local, regional, and national levels. This will require:

(1) adaptation of existing technology to the needs, experience, and capabilities of local users;
(2) creation of an infrastructure for installation, maintenance, and repair;
(3) financing schemes to overcome the specific characteristics of renewable energy (relatively high investment costs with low operation costs);
(4) training and information on different levels: end-users, decision-makers, industries;
(5) local centres of expertise to accompany rural energy programmes.

Participation of the local people in this process is essential. They are traditionally accustomed to living under harsh environmental conditions and to overcoming the inherent difficulties. Only with their experience, co-operation, and assumption of responsibility will any newly established, or modified, energy system operate in a dependable and long-term manner.

On the national and international level, co-ordination of the different actions is imperative. An interdisciplinary approach, with exchange of experience, co-operation in organisational, technical, and financial aspects, and joint case studies, is important for development of new schemes for sustainable energy supplies for mountain societies.

References

Adhikari, D. (1993). Rural Electrification in Muktinath, Nepal: Micro-Hydro an Environment Friendly Energy Option? *HYDRONET*, **1**: 4

Aitken, J.-M., Cromwell, G. and Wishart, G. (1991). *Mini- and micro-hydropower in Nepal (ICIMOD Occasional paper No. 16)*. ICIMOD: Kathmandu

Inversin, A. R. (1986). *Micro-Hydropower Sourcebook*. NRECA: Washington DC

Martin, R. (1994). Montage von kleinen Solarleuchten SOLUX in den Werkstätten der Dritten Welt.

In Goetzberger, A. (ed.), *Neuntes Symposium Photovoltaische Solarenergie* (S. 239–250). Ostbayerisches Technologie Transfer Institut: Regensburg

Norberg-Hodge, H. (1991). *Ancient Futures: Learning from Ladkah*. Sierra Club Books: San Francisco

Ramsey, W. J. H. (1986). Erosion problems in the Nepal Himalaya – an overview. In Joshi, S. C. (ed.) *Nepal Himalaya: Geoecological Perspectives*. Himalayan Research Group: Nainital, UP, India

Mining in mountains 9

David J. Fox

INTRODUCTION: MOUNTAINS, MINERAL DEPOSITS AND MINING REGIONS

The association between mining and mountains is generic; the same natural forces which have raised mountains have also helped concentrate assemblages of minerals useful to human society. The clash of tectonic plates which over geological time has buckled the margins of continents and created mountains was accompanied by the subduction of crustal materials at depth; extreme pressures and temperatures transformed them into molten magmas which, on migrating surfaceward, progressively solidified into differentiated chemical fractions forming discrete deposits of metallic minerals in the host rock. The world's great metallogenic zones are all related to past mountain-building activity. And just as many mountain ranges have been eroded to become mere topographic shadows of their former selves so nature has removed many of their commercial minerals and redistributed them elsewhere. Thus the current disposition of many valuable minerals, whether of primary or secondary origin, can only be correctly understood through an appreciation of their original montane provenance.

The mines in today's mountains are the major source of many of the world's strategic non-ferrous and precious metals. For example, those in the Chilean Andes meet 30 percent of the Western world's demand for copper (and Chile houses over a fifth of the world's copper reserves); the Peruvian and Bolivian Andes, the Sierra Maestra of Mexico, and the western ranges of the United States are the source of a fifth of the world's lead and zinc; Russia's most important gold region is in the Magadan ranges inland from the Sea of Okhotsk; the mountains of Papua New Guinea and adjoining Indonesia are of growing importance as sources of gold; the western cordillera of the Americas is responsible for almost half the world's silver production; primary molybdenum, used to toughen and harden steel and largely a by-product of copper

mining, comes mainly from the same mountainous spine; half the world's tungsten, also used as an ingredient in specialised steels, is mined in the mountains of Southern China; and highland Bolivia still produces half the vein tin mined in the world today.

Gold, copper, and zinc are the most valuable non-ferrous minerals entering the world economy; gold mining companies have been particularly profitable in the mid-1990s, copper partly so and lead/zinc mines less so. One should be careful not to exaggerate the overall importance of mountains in the world's mining economy, however, for if one reviews the whole gamut of bulky minerals that contribute to world mine production – including the enormous volumes of iron ores and bauxite, the energy minerals such as coal, petroleum, and natural gas, industrial minerals such as phosphate rock, potash, and salt, and some secondary deposits of precious and semi-precious metals, – it is apparent that most come from the lowlands. But as the world economy expands and the commercial demand for minerals grows miners are targeting ever more closely mineral deposits in both the mountains and the lowlands. Copper production has been growing at an annual rate of about 1 percent during the last twenty years; the use of aluminium has been increasing more rapidly; that of lead, zinc, and tin at lesser rates. More efficient use of metals, the reclamation of scrap, and some substitution by other materials has slightly dampened demand for primary metals in the industrialised world but the growth in consumption, most notably in less developed countries, has strengthened. And since advances in technology have made lower grade deposits worth working the actual volume of material coming out of the ground has grown more rapidly than that of their metal content. These circumstances, and others, have given added impetus to the recent activities of miners – in the mountains as elsewhere.

Few parts of the world are immune to the attentions of miners: Antarctica and the deep sea bed are notable exceptions for the moment. The ability of miners to overcome physical difficulties in even the most hostile environments hardly needs exemplification. The prospectors who scoured South America in the footsteps of the conquistadors, the pioneers who put the Yukon on the map, the involuntary miners of the gulags of Siberia, the volunteer coal miners of Spitsbergen, the peasant miners of the Andes and the commercial miners of tropical New Guinea – all demonstrate the tenacity of miners in striving to achieve their objective. A mountainous or non-mountainous location *per se* is neither a significant deterrent nor a positive incentive when it comes to choosing a potential mine site. Recent analyses of the criteria employed by the modern multi-national mining corporations in deciding where to search for and develop new mines, are widely agreed to be the perceived geological potential of the region, the political stability of the host country, and the mineral policies pursued by that country; it is these which are setting the future pattern of world mining (Johnson, 1990). In practice this means that South America, especially Chile, is viewed favourably by the global mining houses; the United States suffers from 'the high costs of environmental controls and ambiguities of current regulations and laws'; and there is a marked reluctance at the present time to commit major new investments to mining projects in the former Soviet Union. About 34 percent of planned capital expenditure in the second half of the 1990s is earmarked for Latin America (mainly for copper projects), 21 percent for North America and 18 per cent for Africa. Other considerations, such as existing infrastructure, geographical location, and past experience may seemingly count for very little. However, scientific and other discoveries are providing a more accurate appreciation of geological structures, international pressures are promoting greater political stability, more open economies encourage the widespread adoption of pro-mining policies, and individual mining projects increase in size and the role of the multinationals grows, with their access to up-to-date technology, capital, and expertise garnered world-wide[1].

Within this decision-making context secondary considerations, including local topography and the wider geographical setting, can become important discriminates in choosing between competing sites for new mining development. The rate of exploitation of the world's mineral wealth is increasing so the need to encourage a responsible attitude by miners to their working environment is correspondingly of enhanced importance. It is increasingly important that the fragile nature of mountains be given proper recognition by miners and this imposes a special duty of care on those who work in them. This is a duty both miners and society are taking much more seriously than heretofore.

THE IMPACT OF MOUNTAIN SITUATIONS ON MINING VENTURES

Prospecting

A mountain situation has an impact on mining at all stages in the industrial process. At the exploration and development stages it may aid or exacerbate physical and legal access to the site and ease or hinder the establishment of reliable reserves. At the extraction stage high altitude and steep slopes may make working conditions both easier and harder. They may reduce the availability of skilled technicians and unskilled labourers and the ability to attract immigrant workers. At the milling and smelting stages, as during extraction, the costs of imported supplies and of shipping production to markets may be higher than for competitors in the lowlands. And the rehabilitation of a mountain site, once the mine is exhausted, brings both opportunities and costs which do not occur elsewhere.

Mountains can offer a number of attractions to mineral prospectors. In many mountainous regions the top soil has been stripped by glaciation and

---

[1] In August 1995 the top three mining companies – BHP, RTZ, and Anglo-American – had a market capitalisation of US$50 billion, and the top 20 US$150 billion (or 52 percent of the total of that of the 212 companies with capitalisation of over US$50 million). Corporate concentration is increasing: in 1993 the top three companies increased their control to 17.2 percent (1992: 16.1 percent) and the top ten to 28.6 percent (1992: 27.9 percent). In 1994 the top 151 companies spent US$2130 million on exploration, or 80 percent of total world-wide spending; the 12 largest spenders accounted for US$821 million or 31 percent of the world total. In 1993 half the western world's copper production was produced by the ten leading copper corporations (as during most of the last 20 years). Most of the large state mining corporations, so important until the mid-1980s, have withered on the bough as governments have adopted neo-liberal, structural adjustment policies.

removed downslope exposing much of the bare rock surface facilitating mineralogical and structural interpretations. Large diurnal changes in temperature, a characteristic of many tropical and temperate mountains, particularly involving alternative freezing and thawing of water, help weather the surface rock to the point of disintegration exposing what lies underneath for geological inspection. In places the dislodged surface material has been sorted by the elements and useful minerals concentrated and milled by nature: such alluvial or morainic deposits may be the first clue as to what lies in the mountains above and so provide the incentive to explore. For the explorer on the ground mountains offer vantage points from which wide areas may be scanned; for him and the airborne investigator the enhanced brightness of the sun and the clarity of thin mountain air – freer from dust, moisture, and pollutants than that at lower altitudes – and the more variable play of light and shadow on the variously inclined slopes all help make accurate visual interpretations easier. The higher reaches of many tropical mountains elevate them into the drier air of the troposphere with concomitant benefits to the surveyor. In contrast, the lower windward slopes of many ranges are drenched with rain, squeezed out of moist air as it is forced upslope, swathed in dense vegetation, frequently dank and cloud covered, and dissected by a close network of torrential streams hindering work; bedrock is obscured and prospecting is more difficult and costly than elsewhere.

Until recently mountains have been relatively inaccessible and this, and other considerations, have delayed their exploration; they remain relatively underexplored and their mining potential continues to be high. Innovations in engineering, the development of cross-country vehicles, the extension of transport networks, and other advances have made mountains more accessible from the ground; the helicopter and airplane have transformed accessibility from the air in spite of the practical difficulties both experience in the less-supportive air at high altitude. Further, the application of the already large array of remote sensing data, increasing in range and definition with time as aerial and satellite observations expand, has obviated much initial disturbance on the ground. The harnessing of exploratory geophysical and geochemical techniques to the interpretation of airborne measurements of magnetic and gravitational gradients and anomalies, the use of multi-spectral images providing observations independent of weather conditions and particularly sensitive to specific minerals, and the integration of complementary techniques at different geographical scales, have all been greatly helped by the enormous analytical power of computers (Watson and Knepper, 1994). Such advances have reduced the need for land-based access and the use of, for example, ground seismic surveys and exploratory excavations when assessing the potential of many mountainous mining prospects. Where fieldwork in mountains proves necessary (and there remains a place for the geological hammer and the seeing eye in mineral prospecting) modern portable communication systems and improved clothing have taken some of the danger and discomfort out of primary exploration.

The search for mineable minerals in mountains may be helped or hindered by the human and legal context which history has given mountains. Many mountain watersheds have been chosen to form seemingly convenient political boundaries between adjoining states; such boundaries have proved inconvenient to miners. Where resistant mineral deposits core the summits, as, for example, does the gold porphyry deposit recently discovered in northwestern Argentina, identical in style and dimensions to the Refugio deposit in adjoining Chile, mining difficulties may arise. It is true that greater international competition in the acquisition of the wherewithal of mine development may make it possible for companies to play one country off against the other in bidding for mining rights. On the other hand, some border areas have been more difficult to obtain prospecting permits because of local sensitivity over sovereignty: indeed until recently concessions to explore for minerals in some frontier zones (for example, in Latin America) have been reserved for nationals or, more narrowly, for particular categories of nationals (typically, the military or political favourites), and in others prohibited altogether on the grounds of national security.

Politically and socially many mountains may have been marginalised and their interests subsumed to others. Many mountain chains are refuge areas for peoples otherwise threatened by outside events and suspicious of the intentions of outsiders. The prospect of change is often unwelcome; however, where, as in the United States, Canada, and Australia native peoples have been able to use the law to protect their interests, mining companies have had to accommodate themselves to what may seem an alien culture. For example, in Alaska the

12 Native Corporations hold the best mineral potential and cover 12 percent (60 Mha) of the entire state: their Red Dog mine on the northern slopes of the Brooks Range, operated by Cominco Alaska and in production since 1989, is the largest zinc producer in North America.

Mountains may be affected by non-ethnic conflicts between provincial and national claims to authority over mining rights as, for example, in Argentina. The higher reaches of mountains have often been deemed by tradition common land at the disposal of the community; the separation of mineral rights from those of the land, and the prospect of mining disturbance for the benefit of a distant market and an alien state, are concepts difficult to accept. The summits, or other prominent features, may have deep religious significance which would preclude access, let alone disturbance, by those not of the same faith (see Chapter 3). On the other hand, some mountains are sufficiently inhospitable to be uninhabited, or to be largely deserted as former residents have migrated to more hospitable lands, affording the geologist carte blanche in his search for exploitable deposits.

Extracting the mineral

The two intrinsic characteristics of mountains – high altitude and comparatively steep slopes – have consequences for the extraction of minerals at both the local and the regional geographical scales.

Mining at high altitudes

High altitude means reduced air pressure and this has a direct impact on both the generation of mechanical power and the productivity of manual labour. Early mechanisation of mining (and of bulk transport) relied on the steam engine, which was much less efficient at high altitude than at sea level. Technology improved in the nineteenth century with the invention of high pressure closed systems and engine efficiency was more closely related to the calorific value of the fuel used. Where coal was not locally available and expensive to ship in, as in many landlocked mountain situations, and where even firewood was in short supply, as in the arid reaches of many mountain chains, other organic material was used; in Bolivia this included resinous shrubs (*yareta*) and dried llama dung (*taquiá*).

Today the internal combustion engine has replaced the steam engine, modified to take account of the reduced oxygen content of air at high altitudes. The depletion is, perhaps surprisingly, large. At 2000 m there is already a fifth reduction in oxygen compared with sea level, at 3000 m a third, and at 5000 m a half: some of the world's highest mines are worked above 5000 m. The physiological effect on people is slightly less drastic. Hypoxia (lack of oxygen) is partially compensated for (in a matter of days or weeks) by stimulating the concentration of haemoglobin in the blood to the extent that at about 4000 m it contains 90 percent of the oxygen it would have had at sea level, and at 5000 m about 83 percent. The physiology of many mountain peoples has seemingly adjusted over the generations by an increase in the volume of lungs allowing the intake of oxygen depleted air to be increased. Incomers from lower altitudes, be they manual labourers or mining engineers, may have to rely on portable oxygen cylinders to make good immediate deficiencies. Although miners have proved their ability to adapt it would be difficult to conclude other than that the stress of working at high altitude is a hinderance to human productivity, and perhaps a longer-term health hazard and a future financial liability to employers.

High altitudes induce lower temperatures and many mines in the mountains, whatever their latitude, have to overcome cold or very cold working conditions; these conditions may differ radically from those in the valleys below which are a few kilometres away. Air temperatures decline at roughly 1 °C per 200 m (the precise rate is dependent on local topography) and sensible temperatures can fall even more rapidly as high winds and loss of warmth through radiation exaggerate the effect of altitude. In tropical America the permanent snowline and the 0 °C isotherm is at about 5000 m and passes through the highest mine worked by Centromin in Peru (Yauricocha); the snowline varies little in position with the season. It rises to about 6000 m at 30°S, the latitude of some of Chile's highest mines, and then declines rapidly southwards to about 600 m in Tierra del Fuego. Once away from the tropics the seasonal difference between winter and summer increases as does the importance of winter temperatures in fixing the snowline. In the tropics the major thermal contrasts affecting work lie between daytime warmth and nocturnal cold, between cold-shaded and warm-sunlit surfaces, and between temperatures above ground and those underground. Freezing temperatures, whether overnight or lasting the winter, mean that machinery must be equipped to deal with these conditions at start up (and perhaps with near-tropical heat at midday or below ground). The precautions taken at, for example the Usibelli Coal Mine at Healy, near Fairbanks, Alaska, as the long winter with temperatures of −50 °C and below approaches – replacing fuel, lubricants, oils and hydraulic fluids, preheating equipment prior to start up, running excavating machinery some time before starting work, and so on – are replicated in the northern ranges of Alaska and the North West Territories. In the mountainous tropics the daily thermal stresses on machinery, which include greater wear on internal and external parts, can be equally severe and demands additional skills from the engineers in charge. Suppliers of machinery from the industrialised world have been known to ship machinery 10 000 km to the Pacific ports of South America with warm weather grade oils and lubricants appropriate to the coastal port but completely useless under the working conditions at their destination a few hundred kilometres inland in the High Andes. But lessons have been learnt. For example, the hydraulic excavators and haulage trucks specified for the Refugio gold project in the Chilean Andes were custom-built to operate at 4400 m (and their productivity derated accordingly in the feasibility study). Chile has a significant number of mines at altitudes around 4000 m and considerable expertise in coping with extreme winter weather.

The impact of thermal stresses varies with the type of mining techniques employed. Traditional underground mining is partly insulated from external temperature fluctuations and the deeper reaches of mines may be virtually isothermal. Machinery that stays in place may need no more than routine maintenance. Miners, however, may be under extreme stress thanks to strong seasonal and diurnal temperature contrasts at the surface. The geothermal lapse rate is generally higher in mountainous regions than elsewhere and working temperatures underground may be, as in Bolivia, 40–45 °C higher than those at the mine head at the freezing cold start of a morning shift. Daily exposure to such huge contrasts increases the miners vulnerability to respiratory and other infections and reduces labour productivity. Where the mine interior is damp and humidity high the contrast with the exterior is even more stressful; where dry silicosis is an ever-present hazard. Bolivian miners relieve stress (and hunger) by chewing coca.

Most modern mines are open-pit workings and their relative importance is growing. The degree to which the ground is frozen clearly is a very important variable in mining costs: it is not only a question of getting the ore out of the ground but of ensuring that it does not freeze solid before it is treated. Mining operations in frozen ground stand or fall on the strength of the equipment used: today, modern excavators exist which can cope with the most inclement of circumstances, as at Usbelli, and it is only the availability of capital to invest in such equipment which may stand in the way of mining. It is when ground becomes almost plastic as freezing alternates with thawing that machinery may be least effective.

Extraction and steep slopes

Steep slopes at the site of the mine can work to the advantage of the miner but normally add costs away from the mine. They may enable the miner to harness gravity to work for him: in particular, bulky material excavated at the surface or mined underground may be moved downhill under its own weight to a central point for further treatment or discarded as waste. In underground mines steep slopes may allow the evacuation level of the mine to be an adit below the main ore body: mined material from higher stopes may be directed to this level reducing underground haulage and perhaps saving the expense of boring shafts. They may also allow any accumulations of water in the mine to be drained in a sough without incurring the heavy expense of pumping. Furthermore, the often strong contrasts in temperature between the surface and the interior can be carefully exploited to engineer natural ventilation systems: these reduce substantially the costly installation (and more costly operational and maintenance charges) of a mechanical system. Open-cast workings may achieve some savings by eating into the mountain sides rather than excavating great holes in the ground. On the other hand, gravity can work against the interests of miners when, for example, loose material – ore or dross – slips uncontrollably, engulfing areas below the mine. In 1992 some 500 gold miners and their families were killed when heavy rains loosened tailings from workings upslope and a landslide swept over an entire peasant mining community near Tipuani in the Andean foothills: Bolivia's worst mining tragedy for many years.

The major negative impact of steep slopes and high relative relief upon the economic viability of mines in mountains is the imposition of relatively high transport costs. Such costs give a strong economic incentive to dispose of waste material –

by far the greatest bulk of material mined – as close to the mine as possible: this is particularly the case with large open-pit workings which, in the United States, produce eight times as much waste per ton of ore as underground mines (Warhurst, 1994: p. 20). Most mined material of value travels by road or rail or, in the case of precious metals, for security and speed, by air. Mining ventures often have to bear not only the high direct costs of transport that circuitous routes and both ascents and descents bring but also those of installing and maintaining transport infrastructure. The absence of additional traffic generated by an otherwise unproductive mountain environment, the need to bring in most consumables and to ship out most production and the imbalance between imports of mining consumables – explosives, timber, food supplies, machinery, and so on – and the return cargoes of mineral ores, concentrates, and finished metals gives added emphasis to transport arrangements. Transport costs may decline or be partially recouped over time. History offers many examples of mining infrastructure becoming the precursor of other economic developments: pioneer railways, and more recently, new access roads built as a precondition of mine development, have opened up virgin territory to others and have been a catalyst of wider regional change.

In some circumstances a high altitude situation can be turned to advantage. In the vicinity of mines aerial ropeways have been used for over a century to transport ore cheaply by gravity to processing plants; at the modern mine Grasberg (4300 m) in Indonesia employees and materials move between the mill and the mine by aerial tramways, while copper concentrates are piped as a slurry 100 km to the coast. Copper concentrates from Escondida (3050 m) are piped mainly under gravity 166 km to the Pacific port of Coloso; this transport system was prefered to building rail or road links.

Steep slopes, particularly where accompanied by a favourable precipitation regime and valleys deepened by glacial activity, do provide the opportunity to generate hydroelectricity and harness it to mine activities. Locally available power is a positive asset and any surplus to mine requirements can be used to establish forward linkages into the local economy.

Processing the ore

A preliminary to processing the metal-bearing mined material is the discarding of barren over-burden or waste country rock either above ground or below. After that, the transformation of mined material into metal normally involves a sequence of reductions and an accumulation of waste. For the common non-ferrous ores the mined rock is first dressed, that is, metal-containing material is separated from the gangue; second, the metal-containing material is concentrated and separated from the tailings; and third, the concentrate is smelted and refined to release the pure metal leaving behind a slag.

Ore-dressing involves the initial grinding and milling of the mineral and the mechanical sorting of ore from barren gangue; the earliest application of mechanical power in mining was the use of water from the hills in medieval stamp and crushing mills. Traditionally, the next stage in processing non-ferrous ores, such as copper and tin, is using gravity by passing the milled material suspended in moving water over shaking tables and jigs where the suspended elements in the slurry settle according to their different densities and can then be collected. Such mills are most conveniently built using the slope of the mountainside: enriched metal concentrates are collected at various points as the material drains downslope and, characteristically, the waste tailings form an apron of debris beyond the lowest part of the mill. For most ores the recovery of metal-bearing material has been improved by installing flotation cells. In this process the milled material is directed through a series of chambers containing flocculated liquids of different densities and acidities, chosen so that the concentrates attach themselves to the rising bubbles and are skimmed off from the resulting froth. Studies show that recovery rates are higher the finer the ore fed into the chambers and careful experimentation can increase the efficiency of the process without extending plant capacity. This is more selective than the traditional method and has allowed many poly-metallic deposits to become worth mining. The reagents are expensive and are conserved. The differential magnetic and electrostatic qualities of minerals are also employed to separate useful concentrates from useless dross. Many mills use combinations of these technologies in their circuits and nowadays waste material from a first processing may be recirculated several times until the residual metal content is too small to be worth extracting.

Another relatively new approach to concentrating minerals from their ores is by using bacterial leaching; this allows the later smelting process to be bypassed. The approach relies upon the ability of iron- and sulfur-oxidising bacteria of the *Thiobacillus* group in acid mine water to oxidize the sulfide minerals of copper, zinc, nickel, and gold (that is, the great majority of their ores) into acid-soluble form: once the effluent from such leaching has been captured it is a relatively simple matter to extract the metals. This process can take place *in situ* but is mainly applied to dumped material, including overburden. It requires less capital and fewer metallurgical skills than the traditional methods and so could make smaller deposits, away from sources of skilled labour, worth mining; it allows byproduct metals to be collected and produces no atmospheric pollution. Unfortunately it takes time for the bacteria to oxidize the ground ores. The bacteria work best in fairly specific environmental conditions – of acidity, temperature, the mineralogy and dimensions of the ore-feed, and nutrients – and particular strains are needed (or may be engineered) to optimise recovery levels from individual mines. The process appears to have been in use in the late nineteenth century at the Hermosa copper mine (1030 m) in the Andacollo porphyry deposit in Chile (Cocha *et al.*, 1991). A small-scale bacterial leach operation at the nearby huge open pit at Chuquicamata produces copper at costs (44 cents/lb) a third less than that elsewhere in Codelco (68 cents/lb) and from much leaner ore (as low as 0.2 percent copper content) than the conventional (of about 1 percent). Other pilot projects use bacteria to extract copper in the Andes and gold in the Rand and Nevada and similar projects in North America use yeasts and algae as collecting agents (Warhurst, 1994: p. 17); they may be pointing the way to reducing the environmental impact of metal extraction in mountainous regions, particularly where the deposits are small in scale.

The average grade of non-ferrous ores entering a mill is low, say about 1 percent or less for tin and copper and about 2.5 percent for lead; the average Bolivian tin concentrate will contain only 25–40 percent metal and half the mined tin is lost in the milling process. The volume of waste is almost the same as the entire volume of ore emerging from the underground mine. In open-pit exploitation an often considerable amount of overburden contributes to a very substantial waste disposal problem. For example, the ore to waste ratio at the new Chilean gold mine at Tambo (4200 m) is approximately 1:10. Sixty-five thousand tons of material is mined daily to provide 6500 tons of feed for the mill from which about 400 oz of gold is eventually recovered.

The third stage in recovering metal from most ore is at the smelter. Most non-ferrous concentrates are sulfides and these are first roasted to reduce the sulfur content: this may be recovered as sulfuric acid or lost to the atmosphere. Careful selection of the site of a smelter may allow mountain winds to disperse and dilute emissions. After roasting, the material is smelted or oxidized to form, in the case of copper, a matte and then an impure blister of, say 98 percent purity; this is then refined in furnaces or through electrolysis to marketable metal. In some more isolated mountain sites neighbouring forests have provided convenient sources of charcoal and fuel until demand outstripped declining local supplies; in some mountain regions hydroelectric power offers a cheap source of energy.

Gold is processed differently. The key to a viable deposit has been the ability to separate gold from any dross on the spot, so obviating otherwise unsustainable transport costs. Alluvial deposits panned by individual miners to recover the specks of paydirt still give employment to hundreds of thousands of peasant miners around the world: such gold is separated from its host by amalgamating it with mercury and then vapourising the mercury leaving the near-pure gold behind. Normally not all the (valuable) mercury is captured and reused – a recent estimate is that half the 300–500 tons of mercury used annually by peasant gold miners is lost to the environment (Dahlberg, 1995). Some may be lost to the atmosphere, some may be abandoned in the tailings slime; liquid mercury is of only low toxicity whereas mercury vapour can be lethal (Breward and Williams, 1994). The same amalgamation process is used on an industrial scale where gold dredges have superseded the panners.

Alluvial deposits downstream have been the clues that miners have used to trace the motherlodes upstream. Vein gold is normally present as auriferous sulfide in arsenopyrites. Gold is leached from the mined ore by crushing the ore and allowing a highly basic solution of sodium cyanide to percolate through it. This dissolves the gold (the same process is used for silver) which is then recovered using activated carbon and electrolysis. This process has become very popular in the last 15 years (although potassium cyanide has been used for this purpose since the late nineteenth century) and is being used increasingly in conjunction with open-cast mining to bring very low grade disseminated ores into production. Some of the cyanide may be discharged as waste and is highly toxic – international safety standards for water deemed suitable for human consumption is a maximum cyanide content of 0.02 ppm. It is a process which can be more effective at higher altitudes where temperatures are lower and evaporation less (Brown and Rayment, 1991). The average gold content of most ore is minuscule: in 1995 reserves held at the mines in the Battle Mountain mining group – Battle Mountain (Nevada), San Luis (Colorado), Kori Kollo (Bolivia), Lihir (Papua New Guinea), and San Cristobal (Chile) – measured between 0.128 and 0.029 troy ounces (that is from 4.0–0.9 gm) of gold per ton of ore and the miners recovered between 70 percent and 90 percent of the gold mined (Battle Mountain, 1996). In 1995 Kori Kollo (3800 m) produced 339 000 oz (10.5 tons) of gold (making it the most productive gold mine in South America); to do so about 8 million tons of ore were treated (and, of course, at least 8 million tons of waste created). The problems posed by the need to dispose of large quantities of waste have been less controversially solved in more remote and less populated mountainous situations than elsewhere. The opportunity for dumping waste out of sight is becoming rarer as the influx of visitors that are drawn to the highlands grows.

Mountains and mining economics

A further impact of a mountain situation on mining economics has often been the need to import a labour force: mountains, even if populated, are rarely reservoirs of would-be miners. In the glory-days of the Spanish Conquest the mines of the Andes used a system inherited from Inca times of extracting a levy of labour (the mita) from a wide area – for example, up to 500 miles distant from Potosí – to supply mine labour (Cobb, 1977). More recently the practice in the Soviet Union was to transplant miners from elsewhere to work newly discovered deposits, often creating enclaves of Russians in an ethnically and culturally very different population. More usually the free market has operated and both drawn immigrant miners into the mountains in search of work and revealed qualities in local peoples they did not know they possessed. Where mining employs relatively primitive technology and relies on human muscle power local people often find little difficulty in mastering the requirements of the job. Elsewhere the acquisition of more sophisticated mining skills may make demands upon those with a traditional lifestyle which are difficult to meet (although it rarely takes more than a generation to acquire those skills). In the higher echelons of modern mine engineering and management the labour market is international and in the mountains, as elsewhere, it is the graduates of the universities of the developed world who direct affairs; their careers often span the globe. Where mining personnel are not available locally, as in many mountain mines, and are brought in from outside, it has been a common practice for the mine to provide many of the social facilities which elsewhere would be the responsibility of the state. Housing, health facilities,

Box 9.1   Turning old silver into new gold – the Cerro Rico de Potosí, Bolivia

The huge expansion in world tourism and an increased curiosity in industrial archaeology is opening up opportunities to turn exhausted mines into cultural attractions and generate new wealth for the mining locality. The salt mines in the Austrian Alps have been tourist attractions for over a century; nowadays even the most out-of-the-way of yesterday's mines are becoming accessible. Few are more remote or more interesting than those of the Cerro Rico de Potosí (4779 m) overlooking the arid high plateau of the Eastern Andes of Bolivia. The discovery of silver in 1544 in the 'Rich Mountain' (or in Quechua Orcko Potocchi, the hill where silver flows) transformed this isolated peak into the treasure house of the Spanish Empire: silver from distant Potosí made Spain a great power while for a century the Villa Real de Potosí at its foot became the most populous city in the Americas. Adits and surface trenches were worked by up to 12 000 Indians under a forced labour (mita) system, the mineral was ground in any of 150 water mills – the water to drive them impounded behind a series of 32 dams in the surrounding mountains, smelted in any of the 15 000 furnaces (huayras) that burnt day and night, and authenticated in the royal mint. The most up-to-date mining technology of sixteenth century Europe was employed at Potosí and the ruins of this remarkable epoch abound. The rich architecture of the dilapidated Villa Real has won it the status of a World Heritage Site. Mining ebbed and flowed during the next five centuries; in the nineteenth century dynamite and mechanical pumps allowed deeper veins to be worked and British and French mining companies drove shafts, excavated new galleries, and continued to ship silver overseas; company housing was built into the urban web. In the

Plate 9.1 Taken from the foot of the mountain with a group of unemployed miners (made redundant following the collapse in the price of tin and the near bankrupcy of the state mining corporation, Comibol) waiting in the sunshine in the early morning to see if they can pick up some casual work in one of the small private mines or mining co-operatives. Spoil heaps from underground workings in the middle ground; glory holes at the summit (centre of photograph) (Photograph: Jane Benton)

twentieth century tin became the staple until the price slumped in the mid-1980s and zinc took over; the main commercial mines became state-owned, development reached 1150 m below the summit (now flooded), but reserves suffered from growing exhaustion; a new smelter was built with Soviet assistance at nearby Karachipampa, but engendered unacceptable levels of pollution and has been mothballed. Today, as in the past, the majority of miners work the upper reaches of the mountain from adits, either with partners or as members of mining co-operatives, using methods little removed from those of their colonial forebearers and under archaic mining laws; others are scavengers reworking the surface waste of yesteryear for what little can be retrieved. Their lives are hard, bereft of comfort or security; hammers, hand drills, dynamite and coca are their working tools. In the last ten years old tailings and waste have been bulldozed and taken to leaching pods where precious metals are reclaimed using the most modern of cyanide processing technology. The Cerro offers the visitor a live panorama of the whole gamut of mining techniques and mineral processing over the centuries in a spectacular and historical setting. Potosí has an airstrip, a railway station (but only infrequent trains), and no metaled road connections with the rest of the country: it attracts a growing number of tourists seeking a modest adventure and an appreciation of one of the most remarkable mining regions of the world. Some enterprising peasant miners have begun taking visitors into the mines; the local tourist board has persuaded the World Tourism Organization to fund a study for an in-mine museum (García-Guinea, 1995). The opportunity exists of turning the legacy of centuries of mining into a new source of income; although mining continues to be the life-blood of Potosí that blood is increasingly anaemic and needs a far-sighted injection of help from outside to restore its vitality.

Source: David J. Fox,
University of Manchester, UK

schools, commissaries, sporting facilities, air strips, roads – all these, and others, may form part of an expensive infrastructure for the mine to provide and maintain. Two things have reduced this responsibility in recent years. The first is the move away from underground mining and towards open-cast workings: the latter often have a shorter life and finite reserves; provision does not have to be made for a permanent labour supply. The second is improved transport which allows miners to commute to the mine on a weekly, fortnightly or, perhaps, seasonal basis leaving their families and the responsibilities they bring elsewhere[2]. Accommodation and services at the mine sites are limited to those for the miners and none is provided for their dependants. The overall savings are considerable and reduce the additional mining costs imposed by a mountain location.

Reclamation, rehabilitation, and restoration

It is only relatively recently that miners have taken seriously a responsibility for the mines which have reached the end of their economic lives. It is also an unfortunate fact that many mines die because metal prices fall and company finances become depleted: even where the will exists to make good a redundant mine site there may not be the resources to do so. It may have been easier to abandon mines in the mountains where people are scarcer and political interests weaker than in more densely populated lowlands. Today the growing recognition of the aesthetic value of mountains and their growing popularity as holiday destinations means that they cannot be treated in such a cavalier fashion.

IMPACT OF MINING ON MOUNTAINOUS ENVIRONMENTS

The environmental impact of mining increases and extends as the phases of mine development advance. Exploration may have only a limited effect and adverse changes can be fairly easily controlled and reversed. Extraction brings with it the environmental challenges associated with the disposal of large quantities of mine waste – overbur-

den, tailings, and minewater. Smelting, refining, and leaching may generate noxious fumes polluting the atmosphere and toxic solutions may poison the groundwaters. The direct impact of mining may be exaggerated in mountainous environments by the greater fragility of montane ecosystems consequent upon their more limited biotic composition and their longer recovery times after disturbance. Part of that impact may be localised and of limited concern, but part may transcend regional and national borders and may even be of global import.

The exploratory activity of the world's major mining houses is much more widespread than is actual mining. In the early 1990s the countries attracting the most active interest were the United States, Canada, Australia, Chile, Indonesia, Papua New Guinea, Mexico, and Brazil; in the mid 1990s the mountainous Andean countries and Indonesia were the favourites in the developing world: China, Russia, and India were among the least attractive. World-wide the search for suitable ore deposits costs US$2500 million annually, half devoted to exploration for gold and a fifth for copper; in 1995 BHP Minerals and associated companies spent US$400 million on exploration in 48 different countries and employed 600 people in this task. Yet exploration accounts for only a small part of the activities of the mining industry: for example, in 1994 only about 2.6 percent of the operating costs of 195 separate mineral-producing companies world-wide was spent on exploration, research, and development. The activity is mainly directed to discovering large-scale ('world class') deposits capable of sustaining mining over a longer period; this means that only about one in a hundred of the prospects examined will be developed as working mines, and only after the exploratory phase has been extended over a considerable time, perhaps five years. Exploration may be in virgin mining territory or of existing minor workings; some of it will be from the air, leaving no environmental shadow, while exploration at ground level will cause only very limited disturbance: digging surface trenches and test drilling at depth, running drainage lines, carving out access roads and building temporary accommodation is probably all that most sites demand. If a site within the concession

---

[2]The 2350 employees and contractors who work at Barrick's El Indio underground mine (3960 m) and Tambo open pit (4200 m) in the Chilean Andes 500 km north of Santiago work 12-h shifts but have their homes and families at La Serena on the coast 177 km to the west; those at Rodomiro Tomic mine (3000 m) 25 km north of Calama, are expected to work similar shifts, four days on and four days off.

appears promising then the company will retain a presence in the area and the responsibility for restoring damaged land will remain clear; where development does not follow exploration the company ceases to have an interest in the area and reclamation may be difficult to enforce.

Most money and effort in mining is spent on extracting and processing ore with the life of most mines measured in decades or even centuries. The visual impact of underground mining can be of a mountain pockmarked by unsightly and dangerous glory holes, surface dumps, shafts and adit entrances, and liable to sudden collapse and slow subsidence. The volume of underground disturbance may not be appreciated from the surface: the opening up of a third ore panel at Andina (3200 m) will require the cutting of the equivalent of 57 km of underground tunnels of 4 m bore even before full exploitation can begin. Open-cast workings can invert the natural relief and turn peaks into

flooded pits and plateaus into stepped sinks. Fortunately, with the recent development of ammonium nitrate blasting agents, more precise knowledge of ore distributions and geological structures, and the proper placement of drilling holes, waste and environmental disturbance, is better controlled. Nevertheless the pits can be huge: that at La Caridad (1822 m high in the Sierra Madre Occidental, 125 km south of the United States border) is expected to measure 2×3 km when exhausted and be 600 m deep. Two of the world's earliest pits, Chuquicamata (Chile) and Bingham Canyon (Utah), are already larger than this after 90 years of excavation and vie with each other for the world record.

Open-cast strip-mining of coal is unlike metal mining since the only source of waste is overburden. Once this is removed, the underlying coal is little reduced in volume in processing. The depth of overburden, or of intercalated waste,

Plate 9.2 Taken from about 4700 m up the mountain looking down over the pock-marked upper surface slopes of the Cerro from which oxidised silver and tin ores have been worked by peasant miners for five centries to the buildings (right) of the nineteenth and twentieth century underground workings exploiting the lower sulfide ores. The colonial Villa Imperial and World Heritage Site of Postosi lie at the foot of the mountain. Any original vegetation was long since consumed as fuel (Photograph: Jane Benton)

181

which it is worth removing has increased as technology and capital investment have allowed the benefits of economies of scale to be realised: a ratio of five overburden to one of coal is not exceptional. Considerable work has been done in the last decade to reduce gas (and noise) emissions from the large stripping machines with excavating wheels and huge haulage vehicles which make mining possible. The near-horizontal seams which favour this type of mining tend to lie on the margins of mountains; as the topography becomes more rugged extraction methods are modified (and operating machinery scaled down). Under so-called contour strip-mining – as widely practised in the United States, notably in the central and southern Appalachians – removal of the overburden starts at the outcrop and follows the hillside creating a step with a steep backwall; the coal is removed and overburden from the next step downhill dumped on the higher one. Such mines are often very extensive and their potential for landscape abuse is large. The industry has been notorious in the past for its lack of concern for the environmental damage it created – for example, the removal of forests and top soils and the associated increases in run-off, in increasing the propensity to flooding and landslides, the silting of rivers and the pollution of aquifers and surface waters; the mere disturbance of soils can cause potentially toxic ingredients in stable clays to be mobilised and enter the local drainage systems. Nowadays, however, in most industrialised countries the scope for damage is curbed by planning controls and legal constraints and the restoration of mined areas is mandatory.

In metalliferous regions the waste gangue from the mine is frequently accumulated as conical tips, extended ridges, terraced hillslopes, or raised plateaus, depending upon the method of removal from the mill and the lie of the local land. The mass of the material may make it uneconomic to transport it far; it is rarely of any commercial value. The amounts may be large: in the copper mining regions of the Urals, for example, there is believed to be 700 million cubic metres of dumped overburden, 200 Mtonnes of mill tailings and 95 Mtonnes of slag from the refineries. The Ok Tedi copper–gold mine in the Star Mountains of western Papua New Guinea produces 80 000 tonnes of ore and 110 000 tonnes of waste of various categories each day. The milling rate at nearby Grasberg in Indonesia is about 118 000 tonnes per day. The initial gangue is coarser and more heterogeneous in size and composition than other waste material; if left to form steep slopes, close to the angle of rest of the unconsolidated debris, it can be conditionally unstable. Careful siting of such dumps taking due account of the inclination and aspect of the natural slopes can make their outline blend into the local topography in a way impossible in the lowlands. An even less intrusive way of disguising such waste is by using it as backfill for mined-out areas underground. This disposal method has the added advantage of reducing the possibility of subsidence at the surface.

Waste from the later stages in the concentrating process offers different problems of disposal since the tailings are partly liquid and may contain toxic elements introduced or concentrated during processing. In the past such tailings have often been discharged directly into local streams and rivers with little or no thought for environmental consequences. This practice is still followed by many peasant miners and at some old or remote mines. Until recently the Panguna copper mine on Bougainville, Papua New Guinea, was dumping 130 000 tons of tailings a day into the Kawerong River, which has grown a huge delta at its mouth in consequence; commercial mines in the Peruvian Andes are responsible for adding almost a million cubic metres of tailings a week to the Pacific near Ilo (Warhurst, 1994: p. 24). But most modern mines today are obliged to take more care over their effluent and most discharge their tailings into specially constructed reservoirs where the solid material settles out and the liquid may be separately treated.

Two types of ecological damage may arise from inadequate management of tailings in reservoirs. The first is a rupturing of the tailings dam and the possibility of catastrophic damage downhill; mines situated in mountains have to be particularly careful because of the larger destructive power of materials released from great heights. The second is the more insidious damage done by the percolation of toxic reservoir waters into the local groundwater or drainage system; this may occur unnoticed until remedial action becomes either impossible or very costly.

The potential danger arising from the failure of tailings dams becomes greater as the capacity of underground mines and open-pits becomes larger and the volume of tailings grows. Thirty years ago most tailing dams were constructed using the so-called upstream method. This required the initial construction of a starter dam at the lowest point of the proposed reservoir followed by the progressive raising of the dam wall as the capacity of the reservoir was increased to accommodate new tailings. The disadvantage of this procedure is that the dam wall grows on unconsolidated tailing slimes thus limiting the height of the dam and increasing the possibility of failure. Much safer is the so-called downstream method which raises the dam on the downhill side using waste rock and sand grade tailings, if possible, on a solid foundation. A variant is to raise the level of the dam by adding sand grade waste to both sides, being careful to keep the new slopes stable. Stability can be enhanced by growing vegetation on the dam banks thus protecting the dam from erosion by rain and wind, and by using an appropriate mixture of impermeable, permeable, and graded building materials. One of the technical problems which tailings reservoirs face is the need to decant excess water once the sands and slimes in the tailings have settled. Some mines still discharge that excess through pipelines at the base of the dam but the hydrostatic pressure on the pipes can lead to their rupture. Exceptional local circumstances may make added demands upon dam construction. Occasional very heavy rainfall and associated surges in storm water run off – such as happens, for example, in the Andes during El Niño – or earthquakes – as in the Pacific rim – should be taken into account in setting engineering standards; unfortunately data to calculate such risks are often non-existent for mountainous areas. Where several hazards come together the problems mount, as the developers of Ok Tedi, near the summit of Mount Fubilan in Papua New Guinea, have found to their cost. In the first stages of development in the early

1980s attempts were made to build a tailings dam but were abandoned when initial foundation work collapsed. Subsequent reports suggested that a static conventional tailings dam was too dangerous in an area prone to high rainfall and seismic activity. From 1984, when production started, the mine has dumped its waste (currently about 100 000 tons per day) into the upper tributaries of the Fly River; some is carried the 1000 km to the coast in the Gulf of Paria. Monitoring shows increased levels of copper in some reaches of the river but, according to government consultants, no evidence of any other adverse ecological impact; sediment build up has caused the forest to die back along some reaches of the river. This did not prevent an Australian law firm, on behalf of local claimants, filing a A$4 billion lawsuit in 1994 seeking compensation for the damage to land and fish stocks and the lives of the 70 000 river people downstream from the mine, claiming negligence on the part of the mine operators (BHP). In response, the mining company commissioned consultants to review previous waste management work and suggest changes; the consultants reported that a waste storage system which would retain most of the tailings and waste rock could be built for between A$430 million and A$1 billion but it would not be permanent after the mine closed. Consultants for the lawyers countered that a tailings dam could be safely constructed for about A$500 million. In 1996 the PNG government (which owns 30 percent of the company) introduced legislation invalidating any determination for compensation in foreign courts and rendering any such judgements unenforceable; it also signed an agreement with Ok Tedi Mining Limited establishing a system of compensation for landowners in the Fly River basin affected by the environmental impact of the mine. In contrast to Ok Tedi's difficulties some mountainous situations allow the potential for such damage to be much reduced: for example, Escondida uses a natural enclosed basin nearby to deposit its tailings from which neither tailings nor effluent can escape, obviating the need for a dam.

A more widespread waste product of mining is polluted water. It is probably the most serious environmental hazard associated with mining and can be ecologically harmful even when significantly diluted (Warhurst, 1994: p. 23). It may not only create long-term damage, but it can also be a potent source of bad publicity for an industry with a public image problem. Water which is drained or pumped from mines and water which leaves the mills is

Plate 9.3 The Andina mine (3500–4200 m) in the Andes of Chile. The photograph shows ore from the Sur Sur open pit where it is excavated from below a glacier 40 m thick; climatic conditions at this altitude allow working for only 9 months in the year (Photograph: Cadelco-Chile)

normally highly acidic (pH 2–4) and may be rich in harmful dissolved heavy metals and reagent chemicals. The common sulfide-rich ores are exposed to natural (chemical) oxidation, which may be accelerated a hundredfold by biological action, to form sulfates and sulfuric acid. Arsenical contamination of a thousand times established water quality standards has been recorded in mine drainage in Southern Africa. The degree to which arsenic and mercury pass into the human food chain is a matter for concern and research: it is a less important problem in the cooler waters of the mountains but in tropical latitudes it can contaminate the lowland rivers downstream with ease. Acidity can be neutralised by treatment, for example, with lime: this can result in some of the heavy metals being precipitated as hydroxides and trapped. Other approaches are more expensive and include promoting ion exchange, reverse osmosis, ultrafiltration, electrodialysis, electro-oxidation and electro-reduction, chemical oxidation and reduction, and activated carbon sorbtion systems (Wills, 1979: p. 409). Inorganic reagents which may produce toxic side effects in the tailings include zinc and copper sulfates, sodium dichromate and sodium cyanate; chemical flotation agents are customised according to the nature of the individual ores in question and are very varied. Particular care must be taken with the seepage of cyanide-rich waste waters from the leach pads used to extract gold: the layers of very heavy inert plastic under the dumps are not always safe from failure (as at Summitville, Colorado – see Box 9.2). Regular monitoring and an emergency routine are increasingly standard practice to reduce the risk of accidental spillages and of environmental damage. Controls over the discharge of toxic waters should be imposed from the beginning of mining otherwise sometimes irreversible, or certainly very expensive to reverse, damage is done. Unfortunately, capturing minewaters is made more difficult in mountainous areas since the opportunities for dispersal under gravity are greater than in terrain with low local relief, and where local lithological differences and geological structures may be more complex.

Atmospheric pollution is the major agent of adverse environmental change in the later stages of metal recovery. Sulfide concentrates of, say, copper, mercury, cadmium, lead, or zinc, may be roasted to drive off the more volatile associated metals such as arsenic, bismuth, lead, and antimony, and some of the sulfur as sulfur dioxide (in 1975 a ton of sulfur in gaseous form was produced for every 2 tons of copper (Prain, 1975: p. 201). Where, as is a frequent occurrence in many mountain situations, cold air settles in the valley floors, atmospheric inversions can trap the poisonous fumes from the smelter between particular contour zones with dire local consequences. Subsequent treatment produces a waste slag, rich in iron silicates and very difficult to break down, and soils inimical to vegetation growth. The recent widespread appreciation of and concern about the ecological damage done by acid rain, no matter where its source, has helped keep the pressure on smelters to reduce toxic emissions. The final stage in the metallurgical treatment – refining – is normally an electrolytic process: some mountain sites provide the possibility of cheap hydroelectricity giving them an advantage over others.

Box 9.2 Summitville, Colorado

A salutary example of the operation of the Superfund is the attempt to reclaim the Summitville mining region of south-central Colorado (Summitville, 1993). The region is about 3500 m high in the San Juan Mountains where mining has taken place intermittently since the 1870s; at one time there were 50 underground gold operations, also yielding small amounts of copper, silver, and lead. In 1986 open-pit working began using cyanide heap leaching to recover gold but ceased six years later. The modern excavations intercepted old underground workings and their acid water was added to the waste disposal problems of the new mine; these problems were aggravated by the absence of a lining to the dumps where reactive sulfide waste rock was placed; they were amplified when, within a month of start up, cyanide leaked into the groundwater. The financial surety provided by the company to cover remedial work and final decommissioning proved inadequate and in December 1992 Summitville Consolidated Mining filed for bankruptcy. The State of Colorado requested the US Environmental Protection Agency (EPA), through the provisions of the Comprehensive Environmental Response, Compensation and Liability Act (CERCLA), to take over management of the mine. Subsequent study shows that the level of discharge into the local (Alamosa) river system of water contaminated with acids from the old underground workings and toxic metals from the open-pit was much higher than previously thought and cyanide-contaminated solutions had leaked through punctured leach pads and liners. Inadequate treatment

Plate 9.4 The paralysed open-pit workings at Summitville in the San Juan Mountains (3500 m) of south-central Colorado. (Photograph: IntraSearch-Denver)

and failures in the pumping system had led to an accumulation of waste solutions with concentrations above the permitted discharge standards; when the company collapsed their reservoirs were almost overflowing (in spite of several recent unpermitted discharges). Emergency dewatering, including release into the river system, was one of the first actions of the EPA. The EPA found itself in the unenviable and unusual position of managing a mine site, both attempting to mitigate the on-going damage of contaminated drainage and initiating longer-term remedial actions. The Governor of Colorado blamed the failure of government policy at Summitville on 'over-confidence in an unproved mining technology, failure of existing state statutes to address this technology, inadequate state support for the Mined Land Reclamation Board, state regulatory oversight, and the empty promises of a company

that problems would be fixed'. The cost to the public of that failure – US$40 000 per day in managing the site – has made it a *cause célèbre* and many of the major mining companies, and others with an interest in the impact of gold mining (for example, Gold Fields Mining, Amax Gold, Newmont Mining, the Sierra Club), have set up a Study Group to learn lessons from this disaster. Amongst recommendations emerging are the need for more robust reclamation bonds, more rigorous state monitoring (the Colorado Division of Minerals and Geology had only six Inspectors to oversee 2500 mining and exploration operations in 1993), and the introduction of an environmental risk grading of individual mines based upon site location, type of operation, and material being disturbed. These ideas sit uneasily with plans to cut government spending.

Source: David J. Fox

All mines have a finite life. Life may be prolonged by changes in mining technology, by enlargement of the market for its products, by changes in ownership, by the need to meet political or strategic requirements, and by a variety of predictable and unanticipated circumstances. Conversely, life may be shortened by the development of new mineral sources elsewhere, by the invention or adoption of substitutes, by the perception of undue political risk, by a scarcity of technical expertise, of capital, and by other adverse factors. The geological uncertainties tend to be more pronounced in underground mines, for obvious reasons; the future of open-pit workings is more predictable. When mines die they normally leave behind a landscape of worn-out buildings and spoiled surroundings: deep pits from which the oxidized ore was first won, and gashes marking where mineral veins once outcropped, adit entrances and minehead haulage gear, spoil heaps and tailing dumps, derelict plant and abandoned mills, tracks leading nowhere and railways without wagons, faded signs leading to ghost mining camps. There will be less obvious reminders of past mining below ground. And equally invisible will be the loss of community, typically tight-knit, and the redundancy of the skills which traditional mining demanded. Happily, the skills needed by open-pit miners are not so specialised as those needed underground and are more easily transferable to other sectors of the economy when an open pit closes: they are equipped to work in engineering and the haulage and construction industries. It is only relatively recently in the long history of mining that any serious attempt has been made to make good the environmental damage done and confound its not unjustified sobriquet as a robber economy. Mass mining, increasingly mechanised and automated, clearly has the potential to create major disturbances to the environment. But the same tools and intelligence which has allowed it to dominate current mining also gives it the power to make good damage done and even, with goodwill and encouragement, to improve on nature and so be of double benefit to mankind.

GREEN MINING AND SUSTAINABLE DEVELOPMENT

The environmental ravages wrought by mining have long been evident and largely accepted as an unavoidable corollary of satisfying the ever-growing demand for minerals. But in the last half century, and particularly during the last decade, the link between mining and an inevitable and unavoidable despoilation of the landscape has been questioned. Romantic concern for mountain environments began to flourish two centuries ago and began to have practical impact with the spread of the idea of national parks: Yellowstone in the United States was the first in 1872; the idea gained its first Alpine recruit in Switzerland in 1914, and spread to the Andes (1934), volcanic Japan (1934) and the Caucasus (1935) twenty years later. Inclusion in a National Park has given many mountain areas a legal status which has demanded environmental compromises by miners hoping to tap mineral resources within their bounds. The modern dilemma of many parks – created to preserve the natural environment but increasingly disturbed by the demands of tourism – can be mirrored in the ambivalent relationships between miners and park authorities. The growth of land-use planning during the Depression and following World War II nurtured an increased acceptance of the need to balance public and private interests and heightened awareness of the detrimental aspects of mining in mountains. Public concern expressed through such organisations as the Sierra Club – founded in California in 1892 and one of the oldest and most effective – was strongly reinforced when specific national legislation to curb environmental damage began to be enacted in the industrialised countries in the middle of the twentieth century.

International weight was thrown behind such national initiatives in the 1980s with the publication of the Brundtland Report (WCED, 1987). The report popularised the concept of sustainable development – development which satisfies the needs of the present without compromising the ability of future generations to meet their own. At first glance this may seem a concept difficult to apply to mining since ore bodies are finite in size and mining depletes or exhausts the original resource. Nevertheless, it is possible to view the world's mineral reserves as conditionally infinite in all but the longer term since changing demands and technologies, and conservative methods of exploitation, can turn economically non-viable deposits into potential and actual reserves. In principle, any local environmental damage should be temporary and limited and any adverse global impact on the wider social, cultural, and economic interests should be minimised. Whereas until recently the decision to mine involved two parties – the mining industry and the consumer – nowadays the interests of a third – the environment – must be built into any development equation. This approach was affirmed in Agenda 21 of the 1992 United Nations Earth Summit, which called upon national governments and multi-national mining companies to 'play a major role in reducing impacts on resource use and the environment'.

Limiting environmental damage

There are several approaches to limiting further environmental damage due to mining. The first is through the application of new or improved technology. Such technology normally emerges from scientific research prompted by a fairly clear view of the economic benefits to be won from it; it tends to favour the larger-scale operations and be first adopted by the more innovative of the multinational enterprises. Most of it is directed to reducing production costs (for example, by adopting column flotation and autogenous grinding mills, and improved computer programmes). Improved productivity gives greater scope to pay for the improved environmental standards now demanded. Water management is a major cost in mine operations, as well as a major environmental concern. A second is through legislation: international and national environmental legislation has mushroomed in the last decade. The third is through improved managerial structures giving greater weight to environmental concerns: international financial agencies and mining houses increasingly incorporate an environmental code into their working practices and public bodies are more aware of their responsibilities as trustees of the environment. And the fourth is as a response to public and corporate pressures. In general, it is far easier to reduce the environmental damage done by newer mining ventures than by those further into their life cycle.

Improved technology

Moves towards a greener and cleaner mining technology are mainly directed to two ends: reducing chemical pollution and dealing with waste. The most important source of long-term pollution resulting from mining is acid drainage. The problem often becomes most acute once a mine is closed and 'groundwater rebound' takes place. Technical solutions to this problem include reducing the rate of oxidation of sulfide ores and wastes by excluding their contact with air, say by flooding old workings, and restraining flow; this is rarely a foolproof process. The use of bactericides to inhibit bacterial activity and curtail acid production may prove more profitable. Another approach is to direct the waters into a reducing environment, such as a marshland, and neutralising the acidity: this clearly has an ecological impact on the host. Stabilising the quality of acid waters artificially in water treatment plants is expensive and has been avoided on cost grounds in the past but is now more widely used and required; scavenging (for example, with hydrous ferric oxides) can concentrate base and precious metals to give a subsidiary economic gain (Breward and Williams, 1994)[3].

Probably the most useful technical advance has been improved monitoring of acid water flows so that a much more soundly-based understanding of the hydrologic regimes of mines and mining areas is becoming available. This means that the scale and location of problem areas can now be more exactly known and remedial action more focused; it also means that responsibility for environmental degradation can be more precisely allocated. First results of studies of underground flows suggest that there is little mixing or dilution of acid waters with groundwater and *in situ* remedies are difficult to devise. The practical problem often resolves itself into dealing with contaminated water from two sources – mine dewatering effluent and the run-off from normal precipitation: a solution is simplified if the two are channeled by contouring the mine site such that both accumulate in an underground sump and are treated as a single stream controlled by an appropriate pumping system (Fraser, 1995). In open-pit operations the pit acts as a reservoir from which water must be drained and the problem of disposal is more clearly confronted as part of the mining process. The extraction of gold and silver using cyanide solutions demands particularly high standards of care: the impermeable lining of the pods has benefited from technological advances in

---

[3]Warhurst (1994: p. 75) describes an interesting response to the problem of potential acid water discharge at the new Exxon mine at Los Bronces (Chile) where recycling acid drainage through low grade copper material in a bacterial leaching plant not only reduced the discharge of acid waters into the local drainage system (of the Mantaro, Santiago's source of drinking water) but recovered sufficient copper to yield a profit.

plastics research. In general, the hydrology of mines in mountain areas, and of underground mines in particular, is more difficult to determine than those elsewhere. Such monitoring to establish patterns of contamination is a standard practice in new mining projects and is built into most new developments but there is often a strong financial disincentive for owners (where they can be identified) to install such systems into older or abandoned mines.

The problem of acid drainage has led some mining companies to pay closer attention to the chemical monitoring of waste rock from mining. In the past there has been little differentiation of such waste but in recent years miners have been conscious of the importance of separating chemically stable rock from acid-generating rock waste: one incentive has been the possibility that such benign waste might be marketable in civil engineering works. There are dozens of types of tests available to determine the potential for pollution of rock waste, some technically elegant, others expensive, most providing a measure of the amount of sulfide present; it is perfectly practical to categorise the degree of hazard present in waste. The Mine Environment Neutral Drainage (MEND) programme in Canada has been researching ways in which the damage done by acid-generating materials could be mitigated and concluded that the most effective way is to submerge the material underwater, out of contact with air, and to slow the process of oxidation. A less satisfactory partial solution is to bury the waste under an engineered, multi-layered soil cover, perhaps topping it with an impervious cap to reduce the flow of water through the dump. Mixing acid-generating and alkaline-base waste in calculated proportions may be an option in some geological circumstances. The degree to which neutralisation of the waste is possible depends upon such physical factors as the coarseness and homogeneity of the waste rock as well as chemical composition. A key consideration is the local climatic regime: in general, aridity and low temperatures reduce risk. Where there is sufficient precipitation and the tailings are reasonably impermeable it has proved possible to encourage a build up of organic material over the tails and so the gradual reduction of the waste. Dealing with the accumulation of highly acid mine waters is expensive but costs are reduced if the issue is faced at the planning stage: prevention is cheaper than cure.

Although access to water for mine operations is often made easy by tapping abundant mountain sources this is not the case everywhere. The careful scientific work of Messerli et al. (1993) in the arid Andes has revealed that the mines of the Atacama, including Chuquicamata, are drawing on fossil groundwater reserves which are not being replenished: a less casual approach to sustaining a natural resource, critical not only for the mining industry but for burgeoning irrigated agriculture, is clearly needed.

The physical management of drainage, rock waste, the tailings from mills and the slag from smelters is best done by the application of well-known engineering principles governing the stability of slopes and the application of probability calculations on the occurrence of unusual events. Where the underlying geological structures and tectonic instability make landslides and the fracturing of dams a possibility, where the waste material is heterogeneous and has different degrees of permeability, and where effluent flow is erratic, then clearly these concerns should be recognised and allowed for. For example, where roads cross dumped scree material they should do so at the lowest part of the toe area where there is a higher proportion of coarser material and the slope angles are lower; where they cross much finer material comparable to mudflows they should cross at the highest place possible where the flow is thinnest (but best avoided entirely). Unfortunately, ignorance, inadequate data, and poor construction methods have created disasters in the past – for example, in the Peruvian Andes and Chinese Yunnan – which modern practices could have avoided.

Atmospheric pollution is more amenable to control. At the mining stage the control of emissions from vehicles presents no unsurmountable technical difficulties. For example, Caterpillar have recently announced the development of a new front-loader system which reduces exhaust emissions of carbon dioxide from 4.42 to 3.51 g/kWh, hydrocarbons from 1.74 to 1.5 g/kWh, and nitrous oxide from 11.79 to 7.51 g/kWh: the figures are significant because they are 20 percent below the levels imposed by European Union legislation (Dunn and Hoddinott, 1995). Persistent concentrations, for example of diesel fumes underground, can put miners lives in jeopardy just as can exposure to radioactive radiation in uranium mines or dust from drilling. The causes of silicosis are well known and dampening the working atmosphere and wearing masks to filter out dust have long been standard responses to this occupational hazard in most underground mines; however, peasant miners and those working under lax control frequently ignore to their peril these obvious safety precautions. Dust from drilling, blasting, and hauling ore is a concomitant of open-cast mining, particularly in arid areas, and mine roads are rarely tarmacadamed or sprayed with water to lay dust, although good practice requires this. Although dust is a more localised hazard than noxious gaseous emissions it is no less serious for those who live and work nearby.

Innovations in smelting processes – such as continuous smelting, flash smelting combining roasting, and smelting in closed systems – and alternative extraction processes – such as the leaching of copper from low-grade ores using dilute sulfuric acid and subsequent precipitation of the metal – are part of the modern armoury of metallurgists and reduce pollution. The notorious examples of widespread corrosion, the destruction of vegetation, and a high incidence of chronic respiratory diseases in the neighbourhoods of the smelters at Trail and Sudbury in Canada earlier this century have been heeded but, unfortunately, smelters have a relatively long life and innovations may be expensive or impossible to adopt. The ability to cleanse smelter fumes of most of their injurious sulfurous gases has been dramatically and recently demonstrated by Kennecott at Bingham where sulfur dioxide emissions have been reduced one-hundred-fold, although at high cost (George, 1995). Academic research has convinced a recent President of the Institute of Mining and Metallurgy that 'zero total emission (of gas, liquid, and waste solid) is close at hand for both zinc and copper smelting' (Warner, 1992) and for lead. For him 'the way forward is to smelt high grade zinc concentrate in association with copper concentrates using tonnage oxygen'. The adoption of such a process implies a reduction in the total number of smelting sites and hence fewer locales suffering the possibility of local environmental degradation: the new Kennecott smelter not only replaced three reactors and four converters by only two furnaces but doubled production and increased productivity three-fold. Tracing problems posed by conventional smelters is made easier by the point source of the contaminants; both increasingly precise monitoring devices and more sophisticated computer models of dispersion patterns contribute to solving such problems.

Legislation

A second approach to reducing environmental damage has been through legislation. Legal interest in the environmental impact of mining grew as mining impinged upon protective legislation drawn up for other purposes and was often focused on specific mines or areas. In the last hundred years, for example, town and country regulations in Britain and national park legislation in the United States have placed brakes on some of the detrimental consequences of mining. The experience of state directives during World War II led to the acceptance of regional planning in post-war Western Europe and a growing concern and interest in restraining adverse landscape changes through legislation. In contrast, unfortunately, in the communist bloc where central planning reigned supreme, economic considerations heavily outweighed any concern for the environment and soil, water and air pollution became and remain malign features of most mining areas. Within Western Europe the need to pay more attention to the environmental consequences of new and existing activities was encouraged by the European Community which correctly recognised it as an appropriate and popular realm for international and trans-border co-operation (Redman, 1994). The Single European Act of 1986 strengthened the Community's power to introduce measures of environmental protection and in 1987 – the European Year of the Environment – the Treaty of Rome was amended to incorporate a new article (number 130) to encourage the preservation, protection, and improvement of the environment, and to ensure a prudent and rational utilisation of national resources. To these ends, since 1989 all new mineral projects must be subjected to an environmental assessment. In October 1992 the Heads of Government demanded an environmental action programme and the Maastricht Treaty gave the Community further powers in this respect. The Community's objectives have been underpinned by a large number of environmental regulations (which have the status of law in the growing number of member states) and of directives which have to be incorporated into national codes but may still have legal force of their own; under the principle of subsidiarity recognised at Maastricht, the Community can only legislate on environmental matters when Community objectives cannot be achieved by national legislation alone. Policy statements since the Maastricht Treaty give considerable emphasis to the need for harmonious and balanced economic development to promote sustainable and non-inflationary growth whilst respecting the environment; these are being given teeth under the Community's

Fifth Environmental Programme. The principle of sustainability has become one of the criteria used in assessing planning applications for new mining proposals. The integration of environmental assessment into the planning process is designed not only to optimise resource management and enhance the protection of the environment but to reduce disparities between member states.

In the United States early mining legislation, such as the primary General Mining Law of 1872, was little concerned with the impact of mining on the environment, and although New Deal legislation sixty years later gave a temporary focus to the desirability of reclaiming worked-over mineralised land, particularly in areas of steep slope, it was not until the need to integrate state and national approaches to environmental problems led to the passing of the National Environmental Policy Act (NEPA) of 1969 and the subsequent setting up of the Environmental Protection Agency (EPA) and the Council on Environmental Quality, that the matter came to the fore. One of the first requirements under the Act was the preparation of Environmental Impact Statements (EIS) for any proposed new mining project. Major federal legislation quickly followed: the Clean Air Act in 1970 (followed by a second Clean Air Act in 1990), the Clean Water Act (1972), and the Comprehensive Environmental Response, Compensation and Liability Act 1980 (CERCLA or the 'Superfund'). CERCLA is directed to reclaiming old mining sites and reducing the levels of residual hazards. In 1995 a bill was introduced into the US Senate to charge a 3 percent royalty on the net proceeds of mines, part of which would contribute towards clean-up funds administered by the states; this is part of a general intention to overhaul the 1872 Mining Act. Federal legislation is supplemented, and frequently made more stringent, by state regulations and laws, just as European legislation is often reinforced by more demanding national laws. In Canada the responsibility of enforcing similar legislation rests mainly with the provinces.

The legal restrictions placed upon environmental damage by mining in Western Europe and North America have an impact well beyond their frontiers. First, the majority of the world-class mining companies are incorporated in Western Europe or North America, or have subsidiaries there, and are answerable to the requirements demanded by First World legislation[4]. Standards are set in the First World and even if mining subsidiaries are permitted to reflect 'the local situation' (RTZ–CRA, 1996) this is rarely interpreted as an opportunity for environmental undercutting. Codelco is currently investing US$680 million in cleaning up smelters and introducing new systems of environmental management to comply with what it describes as international standards.

Second, much recent mining development outside the industrialised world is partially financed through such multi-lateral agencies as The World Bank, the various regional development banks, and other official guarantors which demand legal compliance with standards derived from those required in the West; they advise on new environmental legislation. The enactment of an unambiguous environmental legislation is now as important an element in an institutional framework designed to attract overseas mining investment as an appropriate fiscal regime and a clear mining code (Morgan, 1994). This has been a preliminary to the privatisation of many state mining corporations in developing countries which have found it both tempting and sensible early on in the process to borrow heavily from existing legislation elsewhere and establish standards and requirements familiar to potential investors. It can take even longer to recast legislation in formerly socialist countries now adopting a free market economic policy. New legislation must be clear on who is liable for the environmental legacy of past mining activities. There is likely to be a seller's market in mining investment in the immediate future and the position of most developing countries has weakened. They are unwilling to put themselves at a disadvantage by insisting on more costly environmental demands than those made by their competitors: on the other hand, overseas political pressures increasingly require certain minimum standards of environmental care to be met. The apparent opportunity of attracting investment by lowering environmental standards – 'environmental dumping' – is largely false for commodities like minerals which trade in the world market: retaliatory action in other areas of trade by more strictly

[4]In 1993 19 of the largest 25 mining companies in the world were in the industrialised countries (Anglo American Corporation of South Africa and BHP of Australia were included in this category). The rest were state-owned or state-controlled companies in the developing world, including Codelco of Chile and CVRD of Brazil at third and fifth places – rare examples of economic success in this category.

regulated competitors and trading partners is fairly easy to envisage. These and other considerations encourage the standardisation of environmental laws, a concept underpinned by international protocols.

Third, the flowering of democracy in the last decade has given a stronger voice to communities adversely affected by mining contamination and their realisation that they have the political strength to counter casual treatment of their environment. This has been helped by the post-1985 privatisation of most state mining corporations which has removed conflicts of interest within governments: in the past they could more easily ignore local and provincial concerns and sacrifice long-term environmental problems for short-term savings on operating costs and a surplus at the end of the year. Today governments find themselves freer to enact environmental legislation and move towards meeting the wishes of their electorates.

Concern in the industrialised world for environmental matters and the adoption by most developing countries of more open and competitive economies has generated a recent plethora of environmental legislation. Between 1985 and 1994 more than 75 nations either adopted a new mining law, made a major revision of the existing law, or were currently working on a new draft (Otto, 1995), and few do not incorporate regulations relating to environmental protection. Clauses such as that in the 1991 Resources Management Act of New Zealand to 'promote the sustainable management of natural and physical resources' so that 'people and communities can provide for their own well-being within a sustainable bio-physical environment' are normal. General statements of good intent have begun to be followed by precise legal instruments restricting activities which would damage the environment. For example, the Chilean General Law of the Environment (of 1994) consolidates the entire body of decrees, supreme decrees, orders, standards, laws, and resolutions, relative to mining (and other activities) and makes clear what is tolerated and what is not (Chile, 1996).

There are four main requirements of most legislation designed to control and reduce the impact of mining on the environment. The first is the insistence on a rigorous environmental impact assessment in the early stages of a mining project; the second is the containment of emissions and discharges to fall below specific maximum levels; a third, the employment of effective operating practices; and, finally and less universally, that adequate provision be made for the decommissioning of redundant mines and smelters.

The assessment of the environmental impact of a project normally requires mining companies to prepare a detailed technical description of any proposed development addressing the ways in which environmental damage will be avoided, contained, and minimised; usually the plan will be required to be available for public discussion and be binding on the company if accepted. The wider ecological impact is coming to the fore in these assessments: for example, the recent European Community conservation directive to create 'a coherent European ecological network' (called Natura 2000) may be a particularly important constraint on future mining projects in Europe's mountains. If an assessment indicates some inevitable damage to the landscape, this does not mean that the project is abandoned: normally there would be a consideration of alternatives so that the Best Practical Environmental Option (BPEO) could be identified (Ricks, 1995). Where there are no alternative solutions there may be reasons of overriding public interest, such as economic or social conditions, which allow such projects to go forward. The Environmental Impact Statement (EIS) arising from an assessment for a major development may take up to 5 years in preparation, particularly where there is the likelihood of well-informed opposition to development. This requirement has created a new specialisation within the range of mining consultancies (Johnson and Sides, 1991): their products are environmental baseline studies often of considerable scientific and academic value.

Legislation covering the maximum permitted levels of noxious emissions and discharges are usually simply stated as concentrations in air or water: typical upper values are less than 1 mg m^{-3} of toxic metals and 50 mg m^{-3} of particulates in air and over 95 percent of sulfur dioxide removed from smelter fumes; typical upper limits of concentrations in water are 0.05 mg l^{-1} of toxic metals, 5 mg l^{-1} of lead, and 25 mg l^{-1} of copper. In some countries, for example Spain and Germany, maximum emission levels are legally binding, in others, for example the United Kingdom and Italy, they exist for guidance (Warhurst, 1994: p. 64). There appears to be a move away from the command and control model (which set targets and methods of achieving them), as in the US, towards a more consensual approach using economic incentives, such as emission charges, in countries such as Canada and Chile. Fugitive acid mine drainage and the escape of toxic metals following oxidation of tailings (as has happened at Ok Tedi in Papua New Guinea, Gunung Bijih in highland Indonesia, and Andina in Chile) are difficult to predict and simple ceiling measures seem inappropriate. It is easier to control gaseous discharges since modern gas cleansing technology exists which can meet stricter standards than in the past – although at considerable expense. The question of expense is becoming more appreciated in setting the legal framework within which mines and smelters operate, and is now sufficiently familiar to have its own acronyms. Legislation has long incorporated the requirement that the the 'Best Available Technology or Techniques' (BAT) be employed; nowadays the rider 'Not Entailing Excessive Cost' (BATNEEC) is frequently added. This concept means that acceptable emission and discharge levels are likely to fall as technical advances are made.

The need for efficient environmental management is beginning to be recognised in environmental law. Efficient management includes the formulation of an environmental policy, the setting of environmental objectives and targets, and the periodic monitoring of achievements. Compliance requires trained personnel, records of progress, reviews of policy, and an external vetting procedure. It may require assessments of the environmental probity of suppliers and overall company investment policies. This has introduced the concept of eco-audits into the language of mining. The United Nations Conference on Trade and Development (UNCTAD) is developing full-cost pricing and 'green' national accounting protocols, which, as structured in 1995, show mining on the debit side of a profits-and-loss account and as a liability in balance sheet terms.

The fourth element in environmental legislation is that proper provision be made for reclaiming and restoring sites after mining and smelting cease. Unlike many applicants for planning permission, mines can be expected to have a finite life and it is reasonable to plan for the afterlife of their sites and have the costs built in to any financial proposals. New projects are in a position to adopt sound and acceptable practices from the start which will mitigate damage: a new maxim for miners could be that prevention is not only better but cheaper than cure. The initial Environmental Impact Assessment and statement and employment of the BATNEEC approach should allow the cost of the final withdrawal to be incorporated into the business plan. For an underground mine the immediate post-

closure requirements – such as the demolition of buildings, removal of infrastructure, sealing of shafts and openings – can be achieved easily and quickly; the revegetation of spoil tips, the management of water, and the monitoring of change may take up to five or six years. Reclamation of opencast workings may be quicker since much can be done whilst mining is still occurring. New projects can underpin their closure plans by incorporating an insurance bond or banking guarantee in their financial schedule (Ricks, 1995). On the other hand, the generally poor environmental record of many older working mines means that the application of the 'polluter pays' principle to making good sites once mining stops could bankrupt companies prematurely with consequent loss of employment and income. Many current mines, as they approach the end of their useful life, do not have the financial resources needed to reclaim their sites and, as in the past, the local community and the national exchequer have to bear those costs or suffer the effects of dereliction.

The most obvious consequence of making mines responsible for the environmental losses perpetrated by their actions has been in the negotiations transferring ownership of nationalised mining companies to private ownership. In highland Bolivia, for example, the government has been obliged to undertake environmental audits of specific mining sites (where mining may have taken place for almost five hundred years) and separate existing pollution stocks, for which the government has assumed responsibility, from subsequent pollution flows for which any future private operator would be responsible; the privatisation process is unpopular with many communities and it is important for potential operators that the legal definition of their responsibilities is watertight. The Peruvian government has recently announced that it will assume liability for past environmental damage of mines belonging to the state-owned Centromin (Peru's largest mining company, including mines in the Andes and La Oroya smelting and refining complex); the company failed to attract any bids when first offered for sale in 1994. It is unlikely that either Bolivia or Peru have the resources to make good past despoilation.

The problem of organising the clean-up of closed or abandoned mining operations has been tackled head on in the United States under the 'Superfund' (CERCLA) legislation (which applies not only to mining but to a wide range of economic activities). The fund has the responsibility of identifying anticipated or existing pollution hazards, of designing a budgeted programme to remove or reduce actual and potential problems, and to determine who is liable for the costs of the clean-up – individuals or companies, previous or present owners, or the public. The Superfund legislation has spawned numerous environmental consultancies fed partly by the American propensity for litigation. It is a very ambitious programme but suffers from inadequate funding; it does not lend itself as a universal model for other countries to adopt. The experience of the Summitville mining region 3500 m high in the San Juan Mountains of Colorado has not been a happy one (see Box 9.2) and illustrates the high cost of rehabilitating old mining areas.

Regulation through legislation to promote good environmental management is increasingly being supplemented as a government tool by market-based and other incentives. A refinement of the policy to charge the full cost of environmental damage done, a figure almost certainly contentious, is to incorporate pollution taxes, product charges, emission charges, good behaviour deposits, and such like, into mining levies. A second approach is to incorporate the perceived cost of the external effects of mining into the legal liabilities falling on the company and give credits (in the form of tradeable 'pollution rights') if that cost is reduced below the one previously determined. These approaches are gaining some ground since they provide tangible targets for profit maximisation and, when appropriately calculated, should stimulate good practice and research into ways of reducing environmental damage (Wålde, 1993). Similar financial incentives are increasingly incorporated into conditions governing the provision of financial credits, notably from the multi-national agencies and in bilateral agreements between industrial and developing nations, and act as incentives to apply best practices in both the planning and operation of new mining ventures. (Warhurst, 1994: p. 58). The United States Office of Surface Mining annual awards for excellence in surface mining reclamation – increasingly sought after as public opinion takes a greater interest in environmental matters – have proved another way in which good practice is promoted by government.

The response of mining companies

The positive response of mining companies during the last decade is the third force which is promoting a more sustainable approach to mining. Agenda 21 directly appealed to transnational corporations to play their part in reducing environmental damage from mining and 'to operate responsibly and efficiently and to implement longer term policies' to this end. The mining transnationals are responding to this request and they are setting benchmarks for smaller mining companies[5].

Their co-operation is essential if the new technologies are to be adopted and new legislation complied with. Paradoxically, perhaps, it has been some of the publically-owned state corporations and the older nationally-based mining companies which have the poorer record of response to the new demands of environmental care. The ability to conduct or sponsor research and development in the industry and to learn from experience elsewhere clearly helps the larger companies cope better with environmental challenges. The environment is now a standard agenda item at most board meetings in most companies: they are much more conscious of, and better informed about, the environmental impact of mining today than even in the more recent past. Most transnationals have an environmental policy which is explicit and applies to all member companies and sometimes their suppliers (see Box 9.3: Mining and good environmental practice). RTZ–CRA, the largest mining company in the world, has an explicit list of the minimum requirements subsidiaries must meet as part of its overall health, safety and environment (HSE) policy (RTZ–CRA, 1996). This is designed 'to minimise any adverse impacts its activities may have on the

---

[5]The big corporations, either state corporations or multi-national companies, dominate commercial mining. In 1993 the production (in millions of tons) of the leading copper corporations and their share of the western world's total were as follows: Codelco 1.139 (15.1 percent), RTZ 0.604 (7.9 percent), Phelps Dodge 0.593 (7.7 percent), Asarco 0.470 (6.1 percent), Broken Hill Pty 0.383 (5.1 percent), Anglo American 0.360 (4.7 percent), Freeport 0.299 (3.9 percent) and Zimco 0.297 (3.9 percent): together these eight controlled half the western world's copper. The corporations often have only a part share in the ownership of some of the other bigger mines but can exercise a strong influence over policies and so over considerably more than half the world's output. Perhaps two-thirds of their output is won from mines above the 2000 m contour line.

environment . . .' by complying with legal 'regulations and voluntary commitments', by 'using best available practices appropriate to the local situation, by contributing to the development of sound legislation and regulations' and by 'fostering a better understanding of . . . environmental issues pertinent to its activities'. The practical requirements of subsidiaries include incorporating HSE matters into long-term strategy, establishing appropriate programmes and procedures, assessing in advance the potential HSE implications of exploration, development, expansion, acquisition, divestment and closure activities, and implementation of actions to minimise adverse social and environmental impacts of these activities, making proper financial provision for the costs, ensuring the efficient use of inputs and the implementation of pollution prevention programmes, preparing plans to deal with emergencies, conducting regular audits to evaluate compliance with company

Box 9.3   Mining and good environmental practice

Mining produces materials essential to modern life and inevitably involves some disturbance to the environment. Mining practice should be conducted in such a way as to meet reasonable demand at minimum cost in environmental damage. The existing practice of some companies is indicative of the way forward for all (RTZ, 1996; RTZ–CRA, 1996; Heath *et al.* 1994). Mining in mountains demands particular care in view of the fragile nature of such habitats. Recommended practices are:

(1) Environmental impact assessment procedures should be established for mining projects as part of an integrated approach to environmentally sustainable mineral development.
(2) Environmental management should form a central part of national development strategies.
(3) Planning should be long-term, aimed at environmental sustainability and not simply at short-term profitability.
(4) Environmental management procedures addressing the technical, economic, and social impacts of mining should be formalised; manuals of established Best Practice, including that in mountain environments, should be prepared.
(5) The costs of environmental management and post-mining care should be built into financial structures.
(6) Baseline environmental data should be collected as an integral part of mineral exploration.
(7) Cost–benefit analysis of environmental expenditures should be made to assist future planning.
(8) Environmental protection measures should be incorporated into plant design and all levels of management.
(9) Environmental awareness should be raised by training staff and by educating the local population.
(10) There should be continuous monitoring of environmental quality and rapid response to

Plate 9.5   Modern goldmine at 4000 m in the Western Corillera of Chile (27°S)

failures of protective measures. Regular and, if possible, independent environmental audits should be carried out and defects acted upon quickly.
(11) The mining industry should participate positively in the drawing up of new environmental legislation.
(12) In setting international standards, local economic conditions, the level of technological development, and local and national development priorities must be recognised and an acceptable balance struck between the needs of environmental protection and mineral production.
(13) The local communities should fully participate in the preparation and execution of local environmental policies, including those of a socio-economic nature.
(14) Reclamation should be carried out simultaneously with mining wherever possible.
(15) Information on environmental protection matters should be made freely available and further research encouraged.

Source: David J. Fox

requirements, promoting an awareness of the importance of HSE amongst employees and the local communities, and engaging in research to improve HSE performance. In particular, subsidiaries are 'to prepare and maintain a plan for the eventual closure of each operation including: management of social and environmental impacts, estimates of closure costs and financial provisions, and consultation and co-operation with local communities'. The requirements extend to contractors who are also required to implement practices consistent with the company's HSE policy. This has not prevented controversy over some of the company's actions and the company's annual general meetings have afforded a public platform to critics of, for example, the 'irretrievable damage to the environment' at the Grasberg open-pit mine in the highlands of Indonesia, in which RTZ–CRA holds a 12 percent stake (RTZ, AGM, 1996).

There is a growing acceptance by companies that the demographic and economic impacts of mineral development on the human environment, as well as on the biophysical environment, is their responsibility: this requirement was incorporated in the NEPA (1973) legislation in the United States where it has been cleverly used, for example by representatives from Indian Reservations, to promote and protect their interests. Companies are employing a more comprehensive definition of environment nowadays and Social Impact Assessment (SIA) is becoming an established feature of many proposed mining developments; it has been employed in water resource projects for some time. SIA investigations extend well beyond the direct impacts of mining on health and demography (Bisset, 1996). Just as assessment of changes in the physical environment are probably best done by field scientists so social impact assessment is more efficiently done with the strong participation of the local community; a good example of careful preparatory work is the SIA surveys carried out between 1982 and 1995 at Lihir (PNG) prior to the granting of a mining lease. The process of 'modernisation' inherent in introducing state-of-the-art mining into isolated communities, as in the mountains of Papua New Guinea, will probably bring improved material prosperity and better health care and educational opportunities, but can also bring social disintegration and permanent loss of traditional values. It is now recognised that both companies and governments have a responsibility to reduce this type of environmental damage to a minimum.

CONCLUSIONS

The major players in the world's mining industry have recognised the need to integrate the concept of sustainable development into their industrial policies: the minor players must be encouraged to follow suit. The need to be specially sensitive in mountain regions must be part of that realisation. The move towards a more environmentally conscious exploitation of the mineral resources of mountains depends upon a balance being struck between the economic objectives of the miner, the community and the state, on the one hand, and the less easily quantifiable and more widely shared responsibility for the well-being of the natural and human environment. It is easier for the big mining houses and the multi-national political and financial organisations to appreciate the need for economic compromise in the long-term cause of good global stewardship than it may be for the peasant miner, the indebted nation, or the threatened community, driven by the more immediate needs of survival in a hard world. The cost to the corporations of this more enlightened attitude is far from negligible.

Abuse of the environment has been one of the prices the industry and its world consumers have been prepared to tolerate in the past in return for apparently cheap minerals. Historically, mining has produced a poor rate of return to countries dependent upon minerals and this is reflected in the world's 'green' accounts. The cost of this deficit has fallen more heavily on the shoulders of the communities in the producing countries than on the shareholders in the mining corporations: in effect, it has been a subsidy for the industrialised world at the expense of some of the poorer countries, many of whose mineral assets are largely stored in their mountains.

With more of the world's new mineral production coming from a relatively small number of very large mines controlled by the major mining corporations with responsible and acceptable environmental policies, and from countries whose governments have ever easier access to the expertise needed to impose contractual obligations on those mining corporations, the outlook for the environment may seem to be improving. This is fortunate for the world's mountains which yield an increasing volume of the minerals entering the global market. It is, of course, one thing for a company to have good intentions and a comprehensive environmental policy and quite another putting those into practice. And a country may have sound environmental laws but may not have the will or the ability to enforce them. But the revolution in communications and wider availability of information are working to counter laxity or corruption in mistreating the environment. The grass-root voices of host mining regions speaking up for the interest of their natural partner in life is being ever more widely heard.

The more difficult environmental issues faced by the mining industry lie elsewhere than with new projects promoted by the multi-nationals. It is still not difficult in many parts of the world for smaller companies with good political connections to flout legal requirements, to subvert or buy off local communities, to ignore environmental hazards, and to make insufficient allowances for the actual and potential costs of rehabilitation[6]. This situation is most clearly seen amongst the peasant miners, historic anachronisms by the standards of the developed world but important breadwinners in many impoverished parts of the world: in sub-Saharan Africa there are an estimated 1.5 million mining 'artisans', in lowland Amazonia and adjoining highlands there are half a million, in the Bolivian Andes at least three out of four miners are peasants. Those who work the alluvial deposits in the lowlands often lead a nomadic life with no long-term attachment to the land which provides their sustenance; on the other hand, those who work the veins in the highlands have to give a little more thought to the legacy they are bequeathing to their families. In general, however, all eake out a hand-to-mouth existence and have only a marginal concern for the concomitant environmental damage they cause.

Another major concern is how to reach an equitable solution to the problem of repairing the damage done by past mining. In practice, in both poorer and richer countries, the state is usually

Box 9.4 The impact of mining and mineral processing on biochemical cycles – a case-study in the Sikhachi Alin Mountains, Siberia

---

The 1000 km² Rudnaya watershed in the Sikhachi Alin Biosphere Region fronting the Sea of Japan has been the locus of a long-term exercise to evaluate the movements of elements released through mining and smelting (Badenkov, 1976, 1981; Yelpatyevskij, 1993). The river headwaters originate between 800 and 1000 m above sea level and flow through heavily dissected terrain, 80 percent of which is under coniferous forest. Mining of polymetallic ores for lead, zinc and silver began in 1910 and local smelting was introduced in the 1930s. The original gangue contains many subsidiary metals – copper, cadmium, bismuth, antimony, arsenic, tin, mercury, and germanium – and when dumped on the valley sides becomes actively involved in the local biochemical flows. The gangue metals are present mainly as sulfides of various complexity with sulfur as the 'dictator element'. The highly acidic emissions of the mining plants are partially neutralised by the soil and vegetation and are geochemically changed into the less aggressive organic forms of the metals. The monsoon rains transport these heavy metal salts in the surface run-off and into the fresh mountain torrents. These evacuate the metals to the coast where most accumulate in bottom sediments and in marine hydrobionts; others are carried hundreds of kilometres away from the river estuaries. Tracing the movements of these fortuitous metallic byproducts has allowed a keen understanding to be gained of the processes of biochemical migration set in motion by mining and an appreciation gained of the wider region – 'the impact space' – affected by mining and smelting.

Source: Yuri P. Badenkov,
Institute of Geography,
Academy of Sciences, Moscow

---

[6]A recent Canadian survey (Metals Economics Group, 1994) of the environmental (and other) issues affecting 54 larger gold mining companies covering 105 gold projects worldwide found that the average proportion of costs due to environmental requirements at the exploration stage was 3.3 percent, and 11.7 percent at the feasibility stage (with higher proportions where acid mine drainage was a threat); once the decision to mine was taken, environmental considerations added about 10 percent to initial capital costs and 2.7 percent to total operating costs; restoration after mining finishes accounted for 4.2 percent of the total life-of-mine capital costs (with open-pit operations costing considerably more than underground workings). Costs varied widely. A generally agreed cost-minimising strategy is to build environmental technology into production processes rather than employ a pragmatic add-on approach.

saddled with that responsibility and inevitably rehabilitation has to take its place amongst other more-and-less urgent calls upon often scarce public funds. It seems inevitable that landscapes degraded by historic mining and countries suffering from widespread poverty will continue to coincide. There may be room for some optimism, however: the recent move to establish an International Society for Geochemical Reclamation to promote the development and application of techniques for the reclamation of contaminated land, including mine sites, should make the considerable expertise dispersed around the world more accessible and more readily exploitable (Heath *et al.*, 1994).

As mines grow larger and mining reaches into the highest summits and the most remote parts of the world, so the potential for environmental dam-age grows – but so does the ability to avoid it. One key to preventing catastrophe is the acquisition of much more detailed knowledge of the natural processes operating on mountains, the biological relationships of montane habitats, and the human ecology of mountain regions. The recognition of the intricacies of nature lends credence to the industry view that there is a need for flexibility and an adjustment to site-specific circumstances in applying conservative environmental measures in their work: blanket prescriptions are too unselective to staunch most wounds mining inflicts on mountains. The Best Practice approaches to sustainable mining will vary with local circumstances and over time. Future miners will have to extend their mental horizons to include the multifarious skills that sustainable mining in the mountains requires.

References

Much of the factual material on individual mines in this chapter is taken from recent issues of the *Mining Journal* (weekly, London), the *Mining Magazine* (monthly, London), the *Mining Annual Review* (London) and other mining periodicals.

Badenkov, Y. (ed.) (1976). *Geochemistry of the Zone of Hypergenesis and Technological Activities of Man*. Vladivostok, p. 131 [in Russian]

Badenkov, Y. (ed.) (1981). *Sikhachi Alin Biosphere Region*. Vladivostok, p. 148 [in Russian]

Battle Mountain Gold Company (1996). *1995 Annual Report*. Battle Mountain Gold Co.: Houston, TX, p. 72

Beal, C. M. (1994). Human Biology at High Altitudes. In Ives, J. D. (ed.), *Mountains*. Rodale Press: Emmaus, PA, pp. 62–3

Bisset, R. (1996). Social impact assessment and its future. *Mining Environmental Management*, March 1996, pp. 4–11

Breward, N. and Williams, M. (1994). Arsenic and mercury pollution in gold mining. *Mining Environmental Management*, December 1994, pp. 26–7

Brown, A. J. and Rayment, B. (1991). Refugio Gold Project, Chile. *Mining Magazine*, November 1991, pp. 306–12

Chile (1996). Registro de legislacion ambiental vigente applicable a la mineria. Dames and Moore: unpublished report, p. 58

Cobb, G. B. (1977). *Potosí y Huancavelica*. Academia Boliviana de la Historia: La Paz, p. 56

Concha, A., Oyarzum, R., Lunar, R. and Sierra, J. (1991). Over a century of Bioleaching Copper Sulphides at Andacollo. *Mining Magazine*, November 1991, pp. 324–7

Dahlberg, E. H. (1995). SMI attend International Conference on Development, Environment and Mining. *Small Mining International Bulletin*, 8, p. 1

Dunn, A. R. and Hoddinott, P. J. (1995). Surface Mining. *Mining Annual Review*, Mining Journal Ltd. London, pp. 17–27

Fraser, W. (1995). Mine decommissioning: an HBM&S case study. *Minerals Industry International*, Jan 1995, pp. 24–6

García-Guinea, J. (1995). Cerro Rico in-mine museum project, Potosí, Bolivia. *Mining Industry International*, September 1995, pp. 13–18

George, D. B. (1995). Kennecott's smelter and refinery modernization. *Mining Environmental Management*, Dec 1995, pp. 9–13

Heath, M. J., Merefield, J. R. and Stone, I. M. (1994). Environmental protection strategies for mineral development. *Mining Environmental Management*, June 1994, pp. 24–5

IDB Interamerican Development Bank (1996). *Annual Report 1995*. Interamerican Development Bank: Washington DC, p. 176

IDB Interamerican Development Bank (1996a). *1995 Annual Report on the Environment and Natural Resources*. Environment Committee: Washington DC, p. 40

Ives, J. D. (ed.) (1994). *Mountains*. Rodale Press: Emmaus, PA, p. 160

Johnson, J. (1990). Ranking Countries for Mineral Exploration. *Natural Resources Forum*, **13** (2): 166–8 (quoted in Warhurst, 1994)

Johnson, M. S. and Sides, A. (1991). Environmental assessment of new mining projects – code of practice for appointment and management of environmental consultants. *Minerals Industry International*, September 1991, pp. 13–17

Messerli, B., Grosjean, M., Bonani, G., Bürgi, A., Geyh, M. A., Graf, K., Ramseyer, K., Romero, H., Schotterer, U., Schreier, H. and Vuille, M. (1993). Climate change and natural resource dynamics of the Atacama Altiplano during the last 18 000 years: a preliminary synthesis. *Mountain Research and Development*, **13** (2): 117–27

Metals Economics Group (1994). Counting the cost of environmental management. *Mining Journal*, **322**: 827–87

MII (1995). Regional Report USA. *Mining Industry International*, **1027**: 24–7

Morgan, C. A. (1994). Privatization of state mining companies. *Mining Industry International*, July 1994, pp. 13–17

Otto, J. M. (1995). Legal risk analysis for mining projects. *Minerals Industry International*, July 1995, pp. 18–22

Prain, R. (1975). *Copper – the Anatomy of an Industry*. Mining Journal Books: London, p. 298

Redman, M. (1994). European Union environmental law and the mining industry. *Mining Environmental Management*, June 1994, pp. 18–21

Ricks, G. (1995). Closure considerations in environmental impact statements. *Minerals Industry International*, January 1995, pp. 5–10

RTZ, Rio Tinto Zinc (1996). *RTZ and the Environment*. p. 21

RTZ–CRA (1996). *Health, Safety and Environment Policy Statement*. p. 2

Summitville (1993). Summitville: a disaster for the industry. *Mining Environmental Management*, June 1993, pp. 9–11

United Nations Council for Economic Development (1992). *Earth Summit '92*. Regency Press: London

Wålde, T. (1993). The development of environmental policies. *Mining Environmental Management*, March 1993, pp. 12–13

Warhurst, A. (1994). *Environmental Degradation from Mining and Mineral Processing in Developing Countries: Corporate Responses and National Policies*. Organisation for Economic Co-operation and Development: Paris, p. 89

Warner, N. A. (1992). Towards cleaner and greener technology – some personal initiatives in advanced smelting. *Mining Industry International*, **1007**: 5–11

Watson, K. and Knepper D. H. (eds.) (1994). Airborne Remote Sensing for Geology and the Environment – Present and Future. *USGS Bulletin* **1926**, Washington DC

Wills, B. A. (1979). *Mineral Processing Technology*. Pergamon Press: Oxford, p. 418

World Commission on Environment and Development, WCED (1987). *Our Common Future*. Oxford University Press: Oxford

Yelpatyevskij, P. (1993). *Geochemistry of the Migrational Flows in the Natural and Technogenic Geosystems*. Doctoral Thesis, Moscow, p. 271 [in Russian]

The diversity of mountain life 10

Jan Jeník

INTRODUCTION

Ridges, deep valleys, and snow-capped summits have long been important landmarks for people viewing the mountains from the lowlands. Many topographic names refer to these conspicuous features, their shapes and colours, and even the howl of the wind, all of which signify a different world. From a closer viewpoint, the first landmarks recognised by mountain people and visitors often include the prominent sculpture of the rocks, streams, waterfalls, and lakes. Superimposed on this physical structure, the vegetation of the mountains – its colours, textures, scents and even sounds – also helps to define mountains. In many mid-latitude countries, the presence or absence of an upper treeline has led to a distinction between 'black' mountains, covered by closed canopy coniferous forest, and 'white' or 'snow' mountains with snow- or rime-covered areas protruding above the treeline (Plate 10.1). The Alps (Latin: *Alpes*, from *albus*, meaning white) are a classic example. In both humid and arid mountains, the green colour of the grassy summits attracts the shepherds and

Plate 10.1 The High Tatras of Slovakia, showing the characteristic pattern of 'white' European mountains marked in winter by snow-capped summits above the 'black' montane coniferous taiga. Natural avalanche track to the right, ski run to the left. In the foreground is a typical piedmont rangeland. (Photograph: J. Havel)

herdsmen at the beginning of the grazing season. In the Near East, the timing of the nomads' way of life very much depends on the colours of vegetation and indirect signals from migratory birds or insects (Křikavová, 1995). In mountain areas around the world, the distribution of prominent plant species and even individual growth forms of trees have often served as property boundaries, or as signposts for people collecting edible and medicinal plants or hunting game.

All of these features and variations in the visible nature of mountain areas represent elements of their biodiversity. Biodiversity can be characterised at the scale of entire mountain systems, their component ranges, and individual peaks and valleys; each of these can be regarded as a more or less isolated 'ecological island' amidst the surrounding lowlands or on the inland margins of seascapes. Mountain systems are separated from each other; individual peaks and valleys are also partly isolated; and these again are divided into local ecosystems defined by their natural fragmentation or human land uses. For those developing priorities for nature conservation, land-use policies, and management strategies, this leads to problems of defining the comparability and uniqueness of ecosystems at different scales. Because of the large dimensions and variety of mountain 'islands', national and international biodiversity projects tend to concentrate on selected areas. At the global scale, the WorldWide Fund for Nature (WWF) and the World Conservation Union (IUCN) have given priority to the Centres of Plant Diversity, selected according to preliminary inventories and after intensive consultation with experts around the world (WWF/IUCN, 1994). A significant proportion of these centres are located in mountain areas.

Three types of biodiversity can be distinguished: genetic, species, and ecosystem diversity (Groombridge, 1992). Most of this chapter discusses various aspects of species and ecosystem diversity. However, 'species' of plants and animals are taxonomic units whose genotype may include great internal variation. Accordingly, the diversity of individual mountain species has to be defined through studies of chromosome behaviour during mitosis and meiosis. While basic chromosome numbers have been counted for many mountain plants (Löve and Löve, 1966), detailed molecular analyses of their deoxyribonucleic acid (DNA) are available for only a few key species or commonly cultivated crops. There are many opportunities for deeper taxonomic analyses of micro-species, sub-species, and varieties in mountain areas. Even in the well explored European mountains, taxonomic revision still leads to the description of new taxa; and new taxa are still evolving, through microevolutionary processes in individual glacial cirques (Jeník, 1983a).

Following this introduction, this chapter will consider species and ecosystem diversity and their assessment; the distribution of species and ecosystems; ecological interactions; endangered ecosystems; and management and conservation.

SPECIES DIVERSITY

An understanding of the biological diversity of a mountain area, from an individual valley to an entire range, begins with identification of the species living there and the preparation of lists of these species and, often, the locations where they are found. In mountain areas around the world, such information has been passed down as traditional knowledge which also includes the various uses of each species of plant, animal, bird, or other organisms. The 'scientific' approach began in Europe, when travellers and local people began to compile 'herbaria' which described plant 'species' and their habitats. Such books were published from the mid-sixteenth century for the Alps and the Central European middle mountains. They allow us to follow the stability or changes in distribution of certain mountain plants and animals over four centuries. This historical evidence can be useful in characterising the status of a species. For instance, the cloudberry (*Rubus chamaemorus*), a species which is mainly found in boreal and subarctic areas, has been described as indigenous to the Giant Mountains along the Czech–Polish border since 1600 (Schwenckfeld, 1600). Historical data are available for shorter periods in other mountains that could be reached from centres of scholarship, such as the White Mountains of the northern Appalachians, with a history of botanical exploration over two centuries (Löve and Löve, 1966).

The occurrence and sometimes even the last specimens of easily identified large mammals are recorded in many historical documents. However, scientific evaluations of the distribution of vertebrates (mammals, birds, amphibians, reptiles, fishes) are available only from the early nineteenth century in Europe (e.g. Gloger, 1827) and even more recently in other continents. The history of the enumeration and taxonomic evaluation of invertebrates, bryophytes, algae, fungi, and protists is much shorter.

Reasonably complete lists of species belonging to various groups of life-forms have only become available in recent decades. These enable us to compare mountain ranges in biogeographical terms; to trace the origins and long-term evolution of species; and to speculate about the conservation values of endemic and relict species. Neighbouring mountain systems can be analysed on the basis of both present and absent families, genera, and species (Box 10.1). Many flora contain drawings and maps showing the details of newly described spe-

cies and the locations of rare species. More detailed treatments include chromosome data and detailed maps of distribution, for instance for the flora of Switzerland (Hess *et al.*, 1967–1972).

Tables with total numbers of families and genera are the first step towards the evaluation of species diversity for conservation purposes. However, direct comparison of the species lists and numbers of species in different mountain ranges or valleys is difficult because of their great variety of topography and size. For example, the Italian Dolomites host about 2500 vascular plant species (Brandmayr, 1988) not only because of their ruggedness,

Box 10.1 The two faces of the Pamir

The core of Central Asia is formed by the Pamiro-Alai mountain system, whose southeastern part is dominated by the Pamir mountains. Situated mostly in Tajikistan, between the borders of Afghanistan and China, the Pamir consists of several ranges with individual peaks reaching above 7000 m. The highest peak is Kungur Tagh (7719 m) in the eastern Pamir, in China.

The biotic diversity of the Pamir is determined by both its geographical position at the junction of great mountain ranges – the Himalaya to the east, Iranian Highlands to the west, and Tian Shan to the northeast – and its climatic diversity. The resulting landscapes provide great contrasts. They range from dry cold continental semi-deserts (< 150 mm annual precipitation), on the eastern plateau, to relatively small areas on the western slopes which are covered with dense coniferous forests, and well supplied by precipitation (> 2500 mm/yr) from westerly air masses.

Like the Alps and the Himalaya, the Pamir originated in the Tertiary period. Their uplift continues, at up to 20 mm/yr in some areas. Uplift has been followed by intensive denudation, especially in the less-glaciated eastern and southern Pamir. The northern Pamir has been and remains heavily glaciated, with nearly 7000 glaciers which cover a fifth of its area. The snowline, limiting life by permanent low temperature, is at 3600 m in the west and rises considerably to the east, reaching more than 5200 m.

Species diversity is high across Central Asia (5500 species of flowering plants in the mountains), especially in Tajikistan, where more than 4200 species have been recorded. About 2000 of these occur in the western Pamir, but only 800 in the dry eastern Pamir. A similar trend holds for animals. However, while the east is species-poor, some valuable and

endangered species, such as argali *(Ovis ammonpolii)*, still survive in remote areas.

Altitudinal zonation is markedly different in the western and eastern Pamir. A continuous forest belt is hardly developed except for the wettest parts of the western ranges. Instead, scattered juniper bushes *(Juniperus)* with steppe vegetation, or mountain semi-deserts with numerous *Artemisia* species, dominate at elevations up to 3000–3500 m. In the general absence of a forest belt, species migration along altitudinal gradients and along large rivers has been quite extensive. Altitudinal zonation is relatively poor, as many species occur from the lowlands to mid- or high altitudes. The percentage of endemic flowering plants is low at lower elevations, increasing in the subalpine belt (5–10 percent of species) and decreasing again in the alpine belt (2–5 percent). At higher elevations, steppe and mountain semi-desert vegetation with xerophytic plants dominates. Cryophilous dwarf and cushion plants prevail on the eastern cold plateau.

Most of the land at higher altitudes, especially the wide eastern valleys and plains, is utilised for grazing by the herds of sparse populations of nomads. However, large areas of the Pamir, including those with dense forests in the west, are virtually untouched due to their remoteness and extremely difficult access. Maintenance of these characteristics appears to be the most efficient way of protection. Where water and minerals are easily accessible, human influences may quickly disrupt the fragile mountain ecosystems which, except for moderate grazing managed in a traditional way, have very limited capacity to buffer any human activities.

Source: Leoš Klimeš, Division of Ecology,
Academy of Sciences,
Třeboň, Czech Republic

but due to their size which is much greater than that of the Giant Mountains, which have 1150 species (Šourek, 1969). The richness of the vascular flora of the Kashmir Himalaya must be remarkable, considering that 1610 species have been identified in the 'alpine/subalpine' region alone (Dhar and Kachroo, 1983). A similar restriction of the area of detailed floristic enumeration is characteristic of tropical mountains whose flanks are covered by a forest flora which is species-rich and little explored. For example, Vareschi (1970) compiled a flora of the Andean páramos, but the species richness of rain and cloud forests on the flanks of the Andes is far from complete enumeration. Overall, in spite of the lack of experienced taxonomists, specialists in mammals, birds, and reptiles have made reasonable recordings of the diversity of fauna of even remote mountains; and systematic collections of invertebrates have provided fascinating biogeographical syntheses (e.g. Franz, 1979; Kingdon, 1989; Müller, 1981).

Assessment of the significance and value of mountain biodiversity requires comparison within general biogeographical evaluations of whole continents (e.g. *Flora Europaea*), adjacent sub-continental regions (e.g. *Flora of the USSR, Flora Malaysiana*), or a single sub-continental region (e.g. *Flora of Tropical West Africa, Mammals of East Africa*). Detailed biogeographical data are available also for many groups of animals, such as Neotropical butterflies (Brown, in Prance, 1982). Such large-scale integrated treatment permits the identification of areas of high biodiversity, and suggests possible centres of speciation and probable routes of migration of different groups of plants and animals.

These approaches show that mountains include many centres of endemism: areas with a high concentration of species with highly restricted distribution. Endemic species may be described in taxonomic terms either as products of 'recent' local speciation (new endemics), or relics of past wider distribution (old endemics). Summits and sheltered refuges in cirques or valleys also include many discrete forerunner populations which are spreading beyond the boundaries of earlier distribution. Endemic, relict, and forerunner populations are the most valuable treasure of mountain biodiversity. They provide important arguments for the assessment of the past and enable the prediction of possible future developments. Most have been recorded as endangered or threatened in the 'Red Lists' and are strictly protected by nature conservation agencies.

Measuring species diversity

To assess species diversity, calculation of the total number of species (i.e. species richness) remains the most workable measure (Solbrig, 1991). Consequently, a census of taxonomically 'accepted species' is the first approximation towards the assessment of species diversity. However, in the complex landscapes of mountain areas, further quantification of species diversity and straightforward application of commonly used statistical approaches can be very difficult. Problems may arise with the identification of particular species, the discrimination of individuals in the population, the selection of an appropriate and comparable size for sample plots, and the season and frequency of sampling. Some of the problematic questions to be resolved include the following: (a) how to compare the species diversity of plants in an alpine tundra with a subalpine forest crowded with epiphytes; (b) how to count individuals in clonal plants, such as páramo grasses or sedges in mires; (c) in temperate and subarctic mountains, in which season and how often should a seasonally changing biotope be sampled; and (d) how large is a satisfactory sample area on the flanks of an 'evergreen' tropical mountain?

Specialised methods in various fields of biology and ecology have facilitated the sampling and complementary laboratory analyses of mountain organisms. Methods for sampling lichens, for example, differ greatly from those used for flying birds or burrowing rodents. Distinctions relating to 'within habitat', 'between habitat', and 'regional' diversity have proved appropriate for comparing animal species (Colinvaux, 1993). However, there are no widely-accepted standardised approaches which would be easily applicable for conservation management and decision making. The value of the species diversity relates not only to absolute numbers, but also to the existence and fitness of small populations, such as the relics from earlier eras and endemic taxa which occur nowhere else (Dhar and Kachroo, 1983). Conversely, an area with a high number of alien species which invaded through human action may be regarded as having devalued biodiversity.

ECOSYSTEM DIVERSITY

Mountains are distinguished from the largely horizontal lowlands by three-dimensional variation in all environmental characteristics. The geophysical factors of latitude, longitude and altitude are unique for each location, though certain geological and climatic situations may create similar physical frameworks. As discussed in Chapter 1, mountains are topographic eminences rising *considerably* above the surrounding country. A sizable landform is always assumed, yet its actual dimensions are seldom defined in linear, square, or cubic measures. In mid-latitude countries, a distinction is frequently made between *high mountains* with many vegetation belts, usually including woodlands, treeless tundra, and snow-rich summits; and *middle mountains* which are covered by a few altitudinal belts of the same physiognomy, usually by woodlands, or have the potential, if deforested, to be overgrown by continuous forest. In addition, *hills* have lower altitudes and lack any altitudinal zonation. This three-member classification suggests that ecosystem diversity is one of the few criteria readily available for classifying elevated landforms.

In polar, tropical, and arid regions, the physiognomy of vegetation is often less differentiated. Near the Arctic Circle, for example in the Scandinavian and Ural Mountains, treeless arctic–alpine tundra appears even on low hills with altitudes from 100 to 200 m (Gorchakovsky, 1966). Many tropical humid ranges, such as the western ranges of the northern Andes or the northern ranges of New Guinea, are continuously covered by closed-canopy forests stretching from sea level to the summits. The parched and treeless mountains of the western Hindu Kush also appear, from a distance, very uniform over an altitudinal range of several thousand metres.

On stabilised relief lacking cliffs and in ranges undisturbed by natural catastrophes or human interference, a gradual change of the vegetation with altitude is typical. Except where affected by fire, clearcutting, or cultivation, the eastern flanks of the Andes exhibit a continuous change from lowland rainforest to the páramo or puna. In the grassy alpine steppe of the rolling uplands of Central Asia, the species composition of ecosystems alters with elevation and aspect, but no clear boundaries or patches can be discerned from the air. In contrast, some highlands display clearly outlined vegetation types caused by differences in soil depth, wind action, landslides, avalanches, flooding, fire, or differential erosion of rock strata. A particular example is provided by the *tepuis* (table mountains) of the Guiana Highlands, which have a stepped pattern according to the different resistance of the ancient rocks to tropical weathering.

The evolution of scientific knowledge of ecosystem diversity in the mountains has proceeded in parallel with evaluations of species diversity and investigations into other environmental variables. Such work began in the second half of the eighteenth century in Europe, and spread to tropical America at the end of the century, when von Humboldt and Bonpland (1807) produced a colour cross-section of the Andes between 10°N and 10°S. This model of altitudinal zonation, linking physical parameters and both botanical and zoological aspects, laid the foundations of the new science of biogeography. A century and a half later, Troll (1959) described a three-dimensional regionalisation of South American and East African mountains, based mainly on a parallel evaluation of climatic factors and the distribution of lifeforms and plant species. This approach formed the basis of the new paradigm of geoecology (Troll, 1968).

These holistic approaches have shown the importance of plants as indicators of similar settings of environmental factors, providing a framework for synthesis and generalisation because meteorological observations and soil analyses can only be conducted on a limited number of sites. For example, after parallel microclimatic measurements and analysis of plant communities in the Langtang Himal of the Central Himalaya, Miehe (1990) distinguished 250 vegetation units which provided useful indicators of environmental status over the entire region. However, while vascular plants are the most conspicuous and 'immobile' organisms, many animals and microorganisms can also be used as indicators of environmental changes (e.g. Wielgolaski, 1975). For instance, an altitudinal zonation of the Italian Dolomites has been described by reference to the distribution of carabid beetles (Brandmayr, 1988).

Patterns of ecosystems

In most mountains, detailed investigation shows that the broad-scale zonation of ecosystems described above is interrupted, to a greater or lesser extent, by patches deriving from the distribution of springs, streams, natural clearings after tree-fall, clumps of clonal plants (e.g. grasses, bamboo, or krummholz), grazing herbivores, burrowing animals, and so on. In addition, human land uses, such as clearcutting, cultivation, fencing, and plantations are always clearly imprinted on the surface of mountains.

One way to interpret this ecological pattern is the patch–corridor–matrix model (Forman, 1995), which states that most ecosystem mosaics consist of:

(1) patches of relatively homogeneous non-linear ecosystems that differ from the surroundings;
(2) corridors (strips) of a particular ecosystem that differ from the adjacent land on both sides; and
(3) a matrix representing a more extensive uniform 'background' ecosystem.

Patches and corridors differ in size, shape, number, and distribution within the matrix. In different locations, their configuration reflects the local ecological processes and trends in the development of diversity. This model can be usefully used in conjunction with the theory of island biogeography, which states that large islands have more species than small islands, and that islands near a mainland have more species than those which are more distant (MacArthur and Wilson, 1967). For mountain areas, the dominant 'mainland' can be far from a small 'island' habitat; for example, subarctic tundra will be far from an alpine summit. A specific example comes from the Appalachians, where the species richness of the treeless summits of the Great Smoky Mountains, small solitary balds of the Spruce Knob type (Western Virginia), and the alpine tundra of the White Mountains (New Hampshire), is proportionate to the size and predicted history of these patches, which are situated in a matrix of broad-leaved montane forests.

Occasional random disturbances, such as tree-falls or rock-falls, change only the local pattern. In contrast, catastrophic events, such as volcanic eruptions, wind storms, major fires, and earthquakes, change the pattern over large regions. Major alterations of the vegetation mosaic can also be caused by population explosions of parasites (insects, fungi) or by the invasion of an alien competitor. For example, invasive *Rhododendron ponticum* shrubs have changed the appearance of Snowdonia National Park, North Wales. This species was first planted as an ornamental shrub in large estates, but is now mixing into the woodlands and covers the open flanks of the mountains (Gritten in Pyšek *et al.*, 1995).

In densely inhabited countries, most major changes in the patch–corridor–matrix configuration are due to changes in land use and tenure. One striking example is the Bohemian Forest, a massif lying astride the boundary of Austria, the Czech Republic, and Germany. For 60 years, due to political confrontation prior to World War II and throughout the period of the Cold War, very different patterns of land use and resulting ecosystems have developed in the parts of the massif in each of the three countries, with contrasting political and economic systems and demographic evolution (German MAB National Committee, 1994).

The size and number of patches play an essential role in the conservation of biodiversity. Before a new nature reserve is established, the question, 'one large area or several small areas?' often arises. The benefits of a single larger reserve are that it can protect the flora and fauna in the core against environmental edge effects, and can accommodate populations large enough to protect the gene pool. In contrast, small but more numerous reserves tend to protect rare species against accidental disturbance, stress, and disease. An integrated area can secure stabilisation of the genotypes; fragmented ecosystems enhance the genetic drift, a process which favours micro-evolution. In the past, the numbers and distances of ecologically distinct mountain 'islands' have positively affected the evolution of unique mountain biota, such as the afro-alpine flora and fauna on the summits of African mountains (Kingdon, 1989; Hedberg, 1995).

Corridors are another important element in mountain mosaics. Elongated ridges and valleys which follow the direction of essential physical vectors positively affect the migration of species from centres of distribution towards marginal habitats. Air channels, streams, and avalanche tracks serve as ways of communication between summits, valleys, and plains. Running waters not only wash down and deposit soil, but also clear treeless corridors for migrating animals and for seeds, fruits, and plant tussocks to be transported from the upper belts through the forested montane and submontane belts. On both sides of the High Tatras, many heliophilous alpine species are found far into the Polish and Slovak lowlands, having travelled along river valleys which traverse closed-canopy montane forests (Walas, 1938; Jeník, 1955).

An important distinction must be made between corridors of the high mountains in the mid-latitude/boreal zone and those of the tropical zone. In the former, cold air flows and snow avalanches tend to clear the bottom of gullies and valleys while, in the latter, the forests, scrub, and krummholz climb high into the afro-alpine/páramo belt, along the bottom of the gullies which offer multiple shelter and are less endangered by snow drifts and avalanches (Troll, 1968). However, the drainage of cold air weakens the vitality of forests in larger valleys, so that they support natural grasslands (Hope, in van Royen *et al.*, 1980).

Measuring ecosystem diversity

As for species diversity, the identification of individual life-forms and preparation of inventories of their populations are the first steps towards assessing ecosystem diversity. However, ecosystems are defined not by individual species, but in terms of a community of organisms interacting with their physical environment (Odum, 1971). One simple characterisation is in terms of formations: aggregations of key plants, named according to their physiognomy or by a common local term, such as fjell, páramo, krummholz, or elfin forest. Biomes are formations including plants, animals and microorganisms (Clements, 1916). This general approach is made more precise when accompanied by quantitative estimates of variables such as biomass and species abundance, and taxonomic approaches.

An important approach in this regard is plant sociology: the hierarchical classification of plant communities in a similar manner to idiotaxonomy for individual organisms. This discipline, based on standardised field procedures, was developed in the Swiss and French Alps (Braun-Blanquet, 1951). It has since been used in most European mountains, as well as the Japanese Alps (Miyawaki and Okuda, 1979), Himalaya (Miehe, 1990), and other ranges. In tropical mountains, this approach is appropriate in the afro-alpine regions (e.g. Schnell, 1952) and Andean páramos, characterised by simple layering and relatively low species diversity. However, in tropical forests with complicated layering and much greater tree diversity, such semi-quantitative estimates are more difficult to apply, especially when, as is often the case, idiotaxonomic knowledge is lacking. The vegetation of multi-layered and species-rich rain or cloud forests must be analysed by detailed forest mensuration methods and profile diagrams (Richards *et al.*, 1940). More recent computer-based approaches, using sophisticated numerical classification and ordination techniques, have also been used to classify mountain plant associations in North America and Great Britain (e.g. Rodwell, 1992), as well as West African middle-mountain rain and montane forests (Hall and Swaine, 1981).

The identification of plant communities facilitates fine-scale mapping and provides a tool for better monitoring of biodiversity for management and conservation. The outline of plant communities shows the patchiness of environmental situations and allows sites to be selected for parallel chemical analyses of rocks and soils, and measurements of microclimatic factors. Results of these environmental analyses and measurements at a few selected sites can then be extrapolated (Hadač, 1969; Miehe, 1990). Sampling for other biological disciplines, such as entomology and soil biology, is also facilitated.

Even though many detailed methods exist, there are still many challenges in estimating and evaluating ecosystem diversity in the mountains. A single spring area can be subdivided into smaller zones according to the thermostatic effects of the rising water; a mire complex according to hummock-and-hollow structure; a subalpine krummholz and forest according to the mosaic of shaded patches and clearings. However, the entire spring, mire, or forest may function as a natural complex – which in turn can be described as part of a larger mountain landscape or macroecosystem. There is no unique definition and clear delimitation of individual biomes which might be calculated in all locations and with regard to all life-forms. The territory of a bird and large mammal may extend over large ridges and valleys, while the habitat of an endemic moss may be restricted to a single cliff. However, using vegetation, biomes may be compared at local and regional levels. Unfortunately, activists in mountain conservation often support their proposals with rather local inventories of biotic diversity, defending their interests without comparative analysis.

At larger geographic scales, remote sensing from aircraft and satellites has become essential for inventories and evaluations of ecosystem diversity (Forman, 1995). The resulting air photographs and satellite images permit the identification of typical vegetation mosaics even on remote ridges and in inaccessible valleys, and the monitoring of ecological changes over a period of years. For example, in the Giant Mountains, air photographs taken over five decades facilitated the surveying and rehabilitation of the primary upper forest line (Jeník and Lokvenc, 1962) and assessment of the string-and-flark pattern in mires related to subarctic peatlands (Jeník and Soukupová, 1992). Satellite images, incorporated into a geographic information system, have been of great value in understanding the form and pattern of ecosystems in Glacier National Park, Montana, USA (Walsh *et al.*, 1994).

Such approaches are also important in the many mountain ranges which are traversed by political frontiers which not only divide them along ridge lines and streams but also across natural landscape units. In such cases, transboundary co-operation of mountain biologists and ecologists is a necessity, and air photographs or satellite images can provide a common tool for assessing ecosystem diversity, developing appropriate policies and management strategies, and monitoring their results. For example, the 'Iron Curtain' along the Czech frontier was overcome by a trilateral project sponsored by UNESCO's Man and the Biosphere (MAB) Programme (German MAB National Committee, 1994). Similarly, under the terms of the 1991 Alpine Convention, the countries of the Alps have established the Alpine Information and Observation System. In North America, the Long-Term Ecological Research Programme also assists in overcoming the distances between isolated mountain systems.

THE DISTRIBUTION OF MOUNTAIN SPECIES AND ECOSYSTEMS

Recent research has led to increasing knowledge of the geographical distribution of all taxonomic groups, and especially to the comparison of the occurrence of similar plant and animal species in mountain ranges located far apart. Many boreoalpine and arctic-alpine species are found on temperate latitude mountains around the northern hemisphere; for example, mountain sorrel (*Oxyria digyna*) is found on the summits of both North American and Eurasian mountains. Many mosses and lichens are present in mountains on most continents (except Antarctica): *Racomitrium lanuginosum* and *Polytrichum juniperum*, for example, are frequent in the European as well as the New Zealand Alps.

Understanding the distribution of species in open treeless areas and areas with closed-canopy woodlands presents another set of challenges. Many alpine or páramo areas appear to function like isolated patches amidst the matrix of forested slopes and valleys. Species lists from smaller ecological 'islands' have raised the issue of the past history of the surrounding landscapes and the origin of non-forest habitats in the matrix of woodlands. As recently confirmed by palaeoecological research, a major cause was the fluctuation of

altitudinal belts and, particularly, upward and downward shifting of the upper treeline.

Montane forests also harbour species with restricted distributions, whose affinity to other montane regions is marked at species and intraspecies taxonomic level, such as subspecies, race, or variety. In Africa, the mountains of the Upper Guinea and the Lower Guinea, each with very different sets of endemics, are separated by the dry Dahomey Interval (Hall and Swaine, 1981). Differences between European middle mountains can often be recognised only at intraspecies levels. These details are not only important for the general biogeography and evolution theories, but also play an important role in nature conservation and the exploitation of biotic resources.

Thus, an explanation of current patterns of biodiversity requires an understanding of the interaction of a vast series of factors at many time scales, from the billions of years of geological history to the years, or decades, of recent micro-evolution and the introduction of species by people. The northern mid- and high-latitude mountains have been profoundly affected by the climatic changes of the postglacial era and the subsequent succession of key organisms and biomes. Another source of explanation of current taxonomic composition is provided by observations of ecological and biotic interactions in areas disturbed by the action of harsh physical forces.

This section discusses some of the geological and ecological factors and processes which have influenced the distribution of species and ecosystems, and also some of the adaptations of plants and animals to harsh mountain environments. The following section focuses on various ways in which ecological and human processes have interacted to influence the current distribution of species and ecosystems.

Geological evolution and climatic changes

The affinities and origin of mountain biotas have been greatly clarified through geophysical, geological, and geomorphological research. The formation of today's land masses and major orogenic events can now be explained in terms of plate tectonics. The drifting and colliding continents set and reset the physical stages for the mountain biodiversity of today's continents, including the changing distribution of cool and warm climates in different regions and altitudes. Current mountain biotas have evolved from the Tertiary stock of species inhabiting the lowlands of Laurasia and Gondwana, but these two ancient super-continents split and locally collided, stimulating orogenesis. Tectonic uplift and regional vulcanism created alpine landforms which have developed into the world's fragmented mountain systems.

These systems subsequently created corridors and barriers affecting the speciation, migration, and extinction of biotas. The adaptive evolutionary process of 'orophilization' (Agakhanjanz and Breckle, in Chapin and Körner, 1995) is due to variations in humidity and aridity, the effects of periglacial environments, the onset and decline of glaciation, and the contrasting environmental impacts of closed canopy forest and open treeless communities. Plants and animals slowly adapted to the changing ecological conditions and, in the isolated mountain ranges, centres of speciation developed and refuges for surviving species were maintained. However, occasional exchanges of genetic material between distant centres of biodiversity took place along more or less distinct 'corridors'. The location and density of the 'stepping stones' affected the rate and taxonomic range of these exchanges. For many groups of biota, there was a marked difference between the evolutionary events in the east–west oriented European mountains and the north–south American ranges. Fossils of highland organisms are scarce, but past and current centres of speciation for various mountain organisms are readily estimated from biogeographical analyses of areas of distribution (Müller, 1981). The most reliable insights into the past biodiversity of mountains can be obtained from Tertiary and Quaternary geological deposits, which contain macro-fossils of plants, some mammals and a few insects, and abundant micro-fossils of pollen and spores of flowering plants and ferns.

At the beginning of the Quaternary, profound climatic changes initiated a new era in many parts of the world. Shield and valley glaciation caused frontal migration of woodlands and alpine ecosystems in the northern mountains. Biogeographical zones and altitudinal belts shifted up and down several times, both impoverishing and enriching mountain floras. On the treeless summits of many European and American mountains, a mixed arctic–alpine biota developed whose primary origin is sometimes difficult to trace. The existence of specialised arcto-alpine vascular plants (amphi-atlantic flora) on the western mountains of Scandinavia led to the hypothesis that present day species distributions could best be explained by assuming that

they had survived the Ice Ages, or at least the last glacial period, on mountain tops or in specially protected ice-free areas, while most of northwest Europe was mantled by a continental ice sheet. This 'Nunatak Hypothesis' was extended to Iceland, Greenland, Svalbard, and the east coast mountains of Labrador and Baffin Island, among other areas (Löve and Löve, 1963). Controversy continues over this interpretation, but the amphi-atlantic plants in the mountains of Norway are endangered and protected by law.

After the last glacial period, the floristic and faunistic elements of the montane woodlands returned from various southern refuges. These routes of immigration are recorded in the gene pools of different varieties of plants and races of animals. This is particularly important with regard to dominant trees and game species which often belong to different genotypes and could only be identified after many years' provenance trials or breeding, e.g. certain European and American conifers or the numerous races of red deer. Even minute genetic differences play an important role in mountain silviculture and game management (Oldeman, 1990).

Chromosome studies have proved that, within the few millennia of the postglacial period, new endemic races and micro-species have developed in the isolated subalpine and alpine areas of many northern mountains. In the Central European middle mountains, numerous micro-species of the apomictic genera (*Hieracium, Sorbus*), subspecies, and hybrids have developed on the treeless summits and, particularly, in former glacial cirques (Soukupová *et al.*, 1995). One major threat to the survival and undisturbed micro-evolution of arcto-alpine flora and fauna in these mountains was the competition from woodland species, which ascended the summits during the warm phases of the postglacial period (Plate 10.2).

Even in mountains far from the ice sheets which covered much of the northern hemisphere, profound ecological changes took place during the Quaternary, with consequent changes of biodiversity. Expanding valley glaciers affected the development of the flora and fauna in continental Central Asia, as well as the temperate mountains of Australasia, Africa, and South America. The changes of temperature and humidity in the Pleistocene and Holocene affected even today's humid equatorial highlands, where periods of semi-arid climate markedly changed the proportion of rain forest, savanna, and páramo (Van der Hammen, in

Plate 10.2 *Sorbus sudetica*, an endemic shrub of the Giant Mountains. Chromosomal and cytochemical studies have shown it is a descendant of two parent species which lived in the area in the mid-Holocene period. (Photograph: J. Štursa)

Prance, 1982). These shifts strongly affected micro-evolution in plants and animals. For example, the South American butterflies have undergone a remarkable evolutionary radiation under the influence of climatic deterioration, and due to parallel differential extinction of other flora and fauna. Many geographical races have been identified in the widely distributed and conspicuous *Heliconii* forest butterflies, which have their endemic centres in mountains and lowlands on both sides of the equator (Brown, in Prance, 1982).

Layers of peaty sediments with preserved pollen and various macro-fossils of plants and animals offer insights into development during the past thousand years, the naturalness of the mountain ecosystems, and behaviour in response to environmental changes (Ammann, in Chapin and Körner, 1995). These data sources provide information on the development of individual bogs and lakes, and also with regard to vegetational history of larger landscapes surrounding the mountains. Improved taxonomic resolution in herbaceous pollen identification and radionuclide dating have permitted the study of trends in vegetation in relation to climatic changes and human interference not only in the Alps, Vosges, and Scandinavian and Scottish highlands of Europe (Kral, 1979; Valk, 1981; Birks, in McConnell and Conroy, 1996) but also in more remote highlands of the world, such as those of Tasmania and Madagascar (Macphall, in Barlow, 1986; Straka, 1996).

Speciation, dispersal, survival

As discussed above, the measurement of numbers of taxonomic units, especially species, is central to defining biodiversity. 'Each species is a biological experiment' (Mayr, 1988); and mountains are remarkable experimental sites. The presence of a particular species is a result of past and/or ongoing speciation, migration, and the existence of positive factors for survival. While scientific theory is interested in the entire time-scale of these processes, practical management and conservation in mountains are concerned with potential events which might affect biodiversity.

Speciation is the process of the evolutionary formation of a natural group of organisms which share a common gene pool, actually or potentially interbreed, and are reproductively isolated from other such groups. Two main modes are important: sympatric speciation, proceeding without geographic isolation; and allopatric speciation, following geographic isolation or due to strong ecological contrast. Because of the environmental and regional fragmentation of mountains, allopatric speciation has been especially important, although sympatric and other types of speciation have also occurred, especially on homogeneous mountain plateaux and forested areas.

Detailed taxonomic and chromosome studies show that new species come into existence just once, at a single location. Successful species reproduce and, often, migrate. In short-distance migration, species use similar habitats as stepping stones to expand their area of distribution. Long-distance migration overcomes valleys and the often great distances between ranges, requiring use of resources across large areas. Some species invade a mountain region from the same direction, and thus belong to one 'migro-element'. Species which arrived centuries or millennia ago are considered as 'indigenous', and are the most natural component of mountain biodiversity. However, conservation is also concerned with recent migrants, which can create undesirable competition or genetic erosion in mountains influenced by humans.

Once established in the area of origin and/or reproduced elsewhere, the population of a species undergoes further screening by harsh physical factors and biotic competition – a kind of global and local 'sieve' (Chapin and Körner, 1995). Species have to survive and retain appropriate fitness in order to maintain occupied habitats and expand. As this is only possible if the environmental conditions do not change beyond certain limits tolerable to the gene pool, some species lose territory and may become extinct. This can improve the fitness of other species or affect natural selection within the available habitat (Turner, in Prance, 1982).

In the alpine and subnival belts, landslides, rockfalls, and avalanches can dramatically damage local habitats and their populations. As discussed below, variations in humidity and temperature, accompanied by physical stress, lead to shifts of the upper treeline and, in continental or tropical areas, of the boundary between woodland and, respectively, steppes or savannas. For heliophilous alpine/páramo organisms which cannot survive under the shade of closed canopy forest, the continuity of treelessness is as important as the stability of temperature or humidity. For instance, the summits of the *tepuis* in the Guiana Highlands have remained treeless and waterlogged because of a special cool perhumid mesoclimate, which has preserved old non-forest endemics.

In contrast, the biodiversity of mountain forests depends on the persistence of the forested area. The impressive species richness of some of the flanks and lower ranges of the Andes can be explained by the 'refuge theory' which holds that tropical South American forests underwent profound reduction during the late Pleistocene and early Holocene (Haffer, in Prance, 1982). Spectacular forest refuges have developed on the western and eastern flanks of the Andes, and the lower slopes of the Guiana Highlands.

Primary and secondary succession

The harsh physical factors and enhanced exogeodynamic processes of mountains offer remarkable opportunities to follow the invasion, competition, and extinction of populations which partly control their environmental limits and tend to culminate in stabilised ecosystems. The ecosystems of exposed slopes and sheltered valleys have very different rates of change and sequence (sere). On precipitous rocks and scree, the sere seldom culminates in a stabilised ecosystem with a stratified mineral organic soil. In mountains, the progress towards the terminal ecosystem (climax) is frequently interrupted by catastrophic events.

Primary ecological succession has been studied in areas characterised by: retreating glaciers which leave extensive rocky surfaces and moraines; volcanic action (fresh lava, new volcanic ash fall, mudflow); rockfalls, congelifraction, landslides, and avalanches; and the deposition of alluvial sediments (Box 10.2). Primary succession in the severe alpine and subnival belt is very slow or blocked by limited ion and energy sources or the absence of suitable organisms. On moraines in the New Zealand Alps, the sequence from mosses and pioneer grasses to the tall tussock alpine grassland and herbfield is estimated to last several thousand years (Wardle, in Barlow, 1986).

At about 5000 m in the Langtang Himal, three phases of progressive colonisation on previously glaciated surfaces can be identified, using the diameter of a lichen (*Rhizocarpon geographicum*) as an indicator (Miehe, 1990; Beschel, 1961). In

Box 10.2   Paektu-san, three centuries since the last eruption

Paektu-san (2744 m), the highest mountain of the Korean Peninsula, is an old volcano in the Changbai Mountains, on the border between North Korea and China. It is a remarkable example of biotic relationships between the species of the alpine and arctic tundra of eastern Asia and western North America. Of the 131 species of vascular plants in its upper zone, 51 percent are shared with the arctic tundra of Alaska, 41 percent are common in the Northwest Territories of Canada, and 33 percent are also found in the alpine tundra of the Rocky Mountains. Unlike the surrounding lowlands of North Korea and China, Paektu-san hosts a significant number of endemic organisms: 13 percent of the summit tundra flora species are confined only to this one mountain.

The mountain lies in the eastern part of a basalt plateau, 190 km by 140 km. Since the fifteenth century, the volcano has erupted four times. The last eruption, in 1702, caused large-scale damage to the forests, depositing a layer of basalt 50 to 200 m thick, and covering the highest parts of the plateau with mud, lava, and pumice layers 1.5–7 m deep. Remnants of buried pines, spruces, larches, and broad-leaved trees have been found in the pumice layers and ignimbrite sheets at several sites high above the present upper treeline.

The Changbai Mountains are in the northern temperate zone, and their vegetation creates an ecotone between the cool-temperate broad-leaved woodlands and subarctic coniferous forests. Up to 2000 m, Paektu-san is covered by subarctic coniferous forest, while the uppermost belt is alpine tundra. The transition between the forest and tundra is rather gradual, and the structure of the timberline varies according to the slope, aspect, and gullied relief.

Elevations below 1100 m are covered by species-rich mixed woodlands with *Pinus koraiensis, Populus davidiana,* and *Betula platyphyllos*. Moist and north-facing slopes have an astounding variety of other broad-leaved trees. The most important forest tree in the upper montane belt is *Larix olgensis*, an endemic larch restricted to a very small region of the Far East. This lofty tree creates single dominant stands or a mixed taiga with two species of spruce, a fir, and a birch.

Almost three centuries since the last eruption, the alpine zone, on nearly bare pumice, is dominated by a species-rich community consisting of small shrubs and herbs displaying spectacular flowers in summer. The subalpine krummholz belt contains, beside *Larix olgensis* and the two spruce species, *Juniperus sibirica* and, on the northern Chinese side, *Betula ermanii, Alnus mandshurica,* and *Sorbus decora*. *Pinus pumila* is no longer found on the Korean side because of destruction during the last eruption. Today, human activities are having a greater influence: unorganised tourism and occasional military activities markedly affect the status of the naturally restored biodiversity on top of Paektu-san.

Source: M. Šrůtek, Academy of Sciences, Třeboň, Czech Republic

recently deglaciated areas, primary succession starts with widespread sand-fixing mosses, crustose lichens, and pioneer *Saxifraga, Draba,* and *Saussurea* species. A second phase correlates with *Rhizocarpon* diameters of 1 cm and sparse creeping rosette-cushions of *Sibbaldia, Rhodiola, Saxifraga,* and *Anaphalis* species. In the third phase, indicated by *Rhizocarpon* diameters of 2 to 3 cm, mats of sedges, and dwarf shrubs cover about 30 percent of the ground.

Determination of the absolute chronology of primary succession has been very much assisted by tree ring measurements and pollen analysis of samples from bogs and by measurements near the upper treeline. At Glacier Bay, Alaska, succession has been monitored on moraines and glacial deposits left by valley glaciers which retreated about 200 years ago (Crocker and Major, in Colinvaux, 1993). The succession goes from pioneer moss polsters and arctic dwarf shrubs, through almost pure stands of alder, to pure stands of Sitka spruce and, after 200 years, to spruce–hemlock forest – the local climax community. During this sequence, soil acidity increases as a result of plant growth and plant detritus. This example shows the bilateral linkage of biotic and physical factors, and the relatively rapid progress of primary succession under stable conditions.

Identification of the climax – the final stable phase of primary succession, in equilibrium with the physical factors of the environment – has been frequently examined yet seldom clarified in mountain regions. Instead of the initial 'monoclimax' theory, ecologists now prefer several final stages of biomes: a single *climatic climax* on mature stratified soil, with an average supply of nutrients and exposure to atmospheric factors; and several *edaphic climaxes* on limited or particularly mineral-rich soils, and/or with a deficit or surplus of moisture, and/or with extreme exposure or shelter with regard to climatic factors. Under the given climatic and lithologic conditions in a particular altitudinal belt, the climax 'model' is often used as a measure of 'naturalness' and provides a goal for nature conservation or management strategies.

In the humid montane and subalpine belts of temperate zone mountains, both climatic and edaphic climax are characterised by stands dominated by broad-leaved or coniferous trees of a single species. In disturbed or human-transformed montane belts, maps of surviving climaxes or potential climaxes, reconstructed from the relicts of natural ecosystems, are often prepared and used for land-use planning and management. However, in the highest belts, the concept of climax is of little use, as their physical and biotic diversity does not allow a uniform developmental trend to be satisfactorily conceived. Predictions can only be made from the detailed classification and mapping of existing ecosystem types.

In mountains with a long and continuing history of human use, secondary regrowth takes place on exploited peatlands, abandoned arable land, overgrazed pastures, ski runs and in clear-cut forests. A spontaneous 'secondary succession' can return the ecosystem to the predicted climax, but this requires that the soil has not been degraded or eroded, the water regime is undamaged and air is unpolluted, and the necessary genetic material is still available in the surrounding ecosystems. However, goal-oriented plantations and restoration measures are often required.

Remarkable growth forms

Mountain plants have evolved many physical structures and life strategies to survive in harsh conditions. In the temperate mountains of the northern hemisphere, these characteristics include the dominance of perennial species, stunted tree growth, narrow annual rings in woody plants, prostrate growth, reduced length of internodes, multiple protection of leaves against radiation, and elongation of roots (Schroeter, 1926; Rougerie, 1990). Many of these characteristics have been adopted by krummholz, dwarf shrubs, cushion plants, tussock grasses, and rosette plants in the alpine belt. Marginal situations in cool, windswept, rocky sites are inhabited by mosses, algae, and lichens; but hardy consumers and decomposers are also numerous due to their mobility and seasonal adaptability.

Tropical mountains are characterised by marked diurnal fluctuations of radiation, tempera-ture, and moisture (Hedberg, 1995; Rundel et al., 1994). In the upper 'afro-alpine' or 'páramo' belt, hard frosts occur on many or most nights. Shortly after daybreak, the intense insolation causes rapid warming of the surface of the soil and vegetation. One response to this harsh alternation of temperature is shown by the dominance of large tussock grasses, such as Festuca pilgeri ssp. pilgeri in East Africa, and various Festuca, Calamagrostis, and Stipa species in the páramo and puna grasslands of the Andes. Their precursory shoots are formed inside a dense tussock, where they are insulated against diurnal changes of temperature and moisture.

Another spectacular growth form providing insulation against diurnal climatic fluctuations has evolved in Senecio and Lobelia species in Africa, and Espeletia species in South America. These plants are characterised by upright or partly prostrate, sparsely branched stems, each bearing a huge leaf rosette, occasionally terminated by a large inflorescence. The rosettes may be up to a metre wide, and function as a huge bud which opens during the day and closes at night. In some Andean rosette plants, this and other anatomic and physiological adaptations for thermo-regulation are combined with the growth of tank roots in humus within the rosette which supply water and nutrients (Rundel et al., 1994).

A contrasting adaptation of vascular plants above treeline in tropical mountains is a shortened stem, particularly as shown by rosette plants in the Andean puna and páramo (Troll, 1958) and on the East African summits (Hedberg, 1995). These have leaf rosettes at or slightly below the soil surface; their stout subterranean stem and root can contract downwards. The much shortened overground stem bears flowers or a very flat inflorescence. The subterranean stem can branch, creating a group of rosettes or even a cushion.

Cushion plants are a growth form with a much wider geographical distribution. They have densely crowded branches with very short internodes, terminated by dense leaf rosettes, and sometimes by flowers or an inflorescence. The space between the branches is filled with decaying leaf remains and soil which reinforce the solidity of the cushion. Cushion plants occur in the tropics, but are most frequent in the north temperate and subarctic regions, sometimes with a circumpolar distribution, such as Silene acaulis. At 5000 m in the Langtang Himal, several plant communities are dominated by cushion plants (Miehe, 1990). In New Zealand, remarkably large cushions of *Haastia pulvinaris* are called 'vegetable sheep'. As shown by experiments, one reason for the shortened internodes in this and other species is the high levels of UV radiation at high altitudes. High levels of solar radiation are also counteracted by leaves which are pubescent and at suitable protective angles.

The responses of animals

Most animals are mobile, and less limited to the patchiness of mountain rocks, soils, and topoclimate than plants. The responses of animals are often described in terms of life strategies, such as horizontal or vertical diurnal migration, seasonal migration, or hibernation. The distances moved by migratory birds, bats, wild goats, snow leopards, or small insects, for example, vary greatly. Sophisticated monitoring techniques have shown astonishing migration routes of animals across mountain systems and, in the case of birds and bats, across continents. Such spatial transitions make the concept of ecosystems rather complicated.

Solar radiation plays an important role in the behaviour of mountain animals. Excessive doses of ultraviolet (UV) radiation are harmful to most animals. This is one reason why many insects live in the shelter of soil, rock crevices, tree bark, or plant tussocks. While the body of epigeal beetles is protected by a dark, thick cuticle, most cave dwellers lack natural colouring, and die if they come into the sunlight. In the evening, bright light attracts mountain insects towards peaks, and shiny snowbeds similarly function as traps, whose icy surface kills unadapted victims, or renders them lethargic and easy prey for predators.

No animals can live in permanent frost, but their metabolism and life strategies enable them to utilise shelter and favourable microclimates which are only briefly available, and to survive temporary exposure to extreme conditions. Adapted species tolerate low temperatures and can exploit even a brief warming of the soil surface. This is particularly important for cold-blooded animals which cannot regulate their body temperature. Most mountain animals are active only during favourable hours of the day and in favourable seasons. The toughest insects are springtails (*Collembola*) which can survive on snow and ice even well above 6000 m (Mani, 1974; Franz, 1979). Other very hardy insects are flies (*Diptera*), butterflies (*Lepidoptera*) and beetles (*Coleoptera*). Spiders have also been found on the summits of icy

peaks, isolated rocks, and nunataks surrounded by ice sheets and glaciers.

While relatively few cold-blooded reptiles and amphibians live in mountain areas due to the adverse thermal conditions, warm-blooded vertebrates are often found high in the mountains. Some survive by seasonal migration or hibernation, but many mountain mammals and birds are active year-round. For instance, capriform antelopes and some birds can temporarily tolerate frosts below −50 °C. Watching the chamois in the frosty air, inspecting fox tracks on fresh snow, and listening to the cries of the ptarmigan in high valleys are among the best experiences of the European mountains. The high Himalaya is inhabited by seven ungulates: two sheep (argali, urial), three goats (markhor, ibex, and bezoar goat) and the closely related Himalayan tahr and bharal. Perhaps the enigmatic yeti also lives in the high mountains, but its existence remains unconfirmed; identification of a new species requires more than a few hairs.

Other environmental factors, such as the partial pressure of oxygen, very likely limit the high elevation occurrences of mammals. At 5800 m, air pressure is only about half that at sea level. The lack of oxygen appears to be counter-balanced by increases in lung capacity, heart size and activity, blood volume, and number of smaller red blood corpuscles (Franz, 1979). Thus, some mammals live at very high altitudes; for instance, the hare *Ochotona wollastoni* at 6125 m on Mount Everest, and its relative *Ochotona ladacensis* at 6000 m in Ladakh. On Kilimanjaro, wild dogs are found above 5500 m. Mountain lion and wild dogs visit the top of volcanoes in the Andes.

In summary, although physical factors such as radiation, temperature, pressure, or wind are frequently used to explain the form and behaviour of mountain animals, there are also always complex physical and biotic interactions deeply rooted in evolution.

ECOLOGICAL INTERACTIONS

Visitors, managers, conservationists, and scientists in mountain areas frequently ask: why do certain species live in rather severe or extreme habitats; which environmental factors affect their patchy distribution; and what roles do these organisms play in the creation of mountain scenery? These questions are tied up with explaining past events, predicting future events, and thus assisting in developing goal-oriented projects. Many direct ecological relationships are encountered at the level of species, populations, and ecosystems, but the response to 'why' is more complicated.

One approach to answering these questions is to consider the interactions between the genetic disposition of organisms (genotypes) and environmental factors. The observable structural and functional properties of mountain organisms (phenotypes) are the ultimate effects of inherited features and physical–chemical factors. The genotype sets the 'norm of reaction' which can be studied by experimental cultivation and breeding; botanical gardens are maintained in many mountain areas. Area of origin, family relationships, co-adaptations, past migration, and survival in refuges can explain the micro-evolution of a particular genotype.

The ecological aspects of these questions have been addressed in a number of integrated studies during the International Biological Programme (succeeded by the current International Geosphere Biosphere Programme) and UNESCO's MAB Programme (Price, 1995). The Hubbard Brook Ecosystem Study in the White Mountains, New Hampshire, USA was vital in providing a new interpretation of physical and biotic processes in areas covered by different vegetation (Bormann and Likens, 1979). The hydrological and ecological consequences of reforestation have been studied in many watersheds, such as the Coweeta basin, North Carolina, where interdisciplinary efforts have taken place for over 60 years (Stickney *et al.*, 1994). Drawing on these and other experience, this section discusses some of the ways in which geological, biophysical, and human processes interact to influence the distribution of species and ecosystems.

Climate

Even a casual visitor to the mountains notes that certain organisms are found in locations characterised by wind on summits, low temperature in a frost pocket, reduced light in caves, thick snowpack on leeward slopes, or compression and shear on avalanche tracks. Many mountain plants and animals can be characterised with regard to their 'autecological' preference for different aspects of the microclimate that we can measure near to the ground – temperature, moisture, snow, and air currents – and according to comparative observations over the seasons. However, a proper understanding of the reciprocal relationships between

mountain organisms and atmospheric factors requires long-term meteorological measurements, synthesised into climatological statistics. Similarly, momentaneous breathlessness at around 3000 m may suggest oxygen deficiency, but the reduced partial pressure of this life-supporting gas must be measured over a longer period. And, while the effects of decreasing air density or increasing global solar radiation at high altitude can be anticipated, sophisticated equipment is needed to detect their impacts.

The inadequacies of data and our understanding of mountain climates are considered in Chapter 17, and will be not discussed here. Altitude, latitude, and orography are the main controlling meteorological factors (Figure 10.1), but other modifications of the microclimate are plentiful, and some of these are due to vegetation. The mid-altitude mountains, such as the Alps, are the most studied in this respect (e.g. Enders, 1979). In the Central European middle mountains, detailed measurements have been conducted in connection with the ecological impacts of air pollution, especially the large-scale decline of spruce forests (Plate 10.3) (Fischer, 1995). Other studies of note have considered the ecological impacts of human-induced snowpack augmentation (Steinhoff and Ives, 1976)

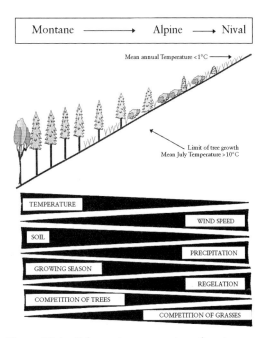

Figure 10.1 Schematic presentation of environmental gradients and the response of life-forms in mid-latitude mountains

and the topoclimate of the East African mountains (Hedberg, 1995).

Various averages and indices, calculated from temperature and precipitation measurements, have been used to characterise mountain ecosystems and to support altitudinal zonation models. Much research has shown that maximum and minimum values play decisive roles in the distribution and the behaviour of mountain organisms. Nevertheless, a single annual or monthly average isotherm rarely accounts for the macro-scale occurrence of particular mountain species or biomes. The only common exception appears to be the coincidence of the alpine (and polar) treeline with the 10 °C July isotherm in the northern hemisphere, as discussed below. Even this indicator isotherm, however, should be regarded only as an approximation to represent the total available energy derived from the length and warmth of the growing season. At the local scale, slope aspect and angle are decisive factors in thermal and moisture budgets and their impact on the patchiness of mountain ecosystems. There are also strong relationships between the structure of vegetation and prevailing winds. Near the upper limit of closed subalpine forests, stands become more open, and exposed trees frequently have a characteristic 'flag' structure, with the crown aligned in the direction of prevailing winds. At higher elevations, wind-trained trees become progressively stunted, and the branches of the krummholz tend to proliferate on the leeward side. Such deformed trees have been used to survey near-ground air currents, particularly in the Alps, Rocky Mountains, and Japanese Alps (Miyawaki and Okuda, 1979; Holtmeier, 1981).

Growth rings in woody plants provide another means for evaluating the long-term impact of habitat factors on life. Old conifers are particularly suitable for such dendrochronological studies (Fritts, 1976). Old stems of bristlecone pine have been used to elucidate climate fluctuations over centuries and even millennia. Current investigations use ring width and wood density to estimate even small environmental changes, including those perpetrated by human intervention, such as air pollution (Sander et al., 1995).

Rocks and soils

On mountain summits and steep slopes, the outcropping rocks are an important component of the topography. Hidden under the mantle of weathered materials and soil, the bedrock acts as the foundation of the landscape, as well as the pool of mineral nutrients for plants, the producers of biomass which support other members of the ecosystem. Relief and soil-forming processes are strongly affected by rock stratification and texture, but the physical and chemical metabolism of ecosystems is controlled by mineralogical composition of the geological substrate.

Between-habitat and regional biodiversity are strongly affected by the diversity of rock types. Some mountains, such as the Hercynian mountains of Europe, appear rather monotonous due to the dominance of silica-rich rocks underlying a smoothly sculptured topography. To their south, the calcareous Alps largely consist of limestones which produce rugged relief and induce a high between-habitat biodiversity. The slopes of individual volcanoes and even large massifs, such as the central Caucasus or the mountains surrounding the Vatnajökull ice cap of Iceland, are underlain by basalts of different age and varying mineralogy, and thus soil fertility; their influence on the between-habitat biodiversity may be considerable.

The distribution of individual species is often explained by chemical and physical effects of the calcium ions in the substrate. Plants and plant communities on calcium-rich substrates are defined as 'calcicolous'; and many 'petrophilous' earthworms, molluscs, grasshoppers, butterflies, beetles, and also certain amphibians and reptiles, prefer calcium-rich rocks (Franz, 1979). Snails are particularly good indicators of calcium-rich substrates, and their presence can indicate even small calcium-rich veins amidst siliceous schists or granites. However, over time, the carbonate minerals are leached from the surface soil horizons, acidity increases, and acidophilous organisms invade the site. Comparably, detailed geological surveys have assisted in the interpretation of regional floristic and faunistic rarities. For example, small outcrops of marble and limestone in glacial cirques enrich the otherwise uniform flora and fauna of the prevailingly silica-rich Lesser Tauern in the Eastern Alps. Encouraged by geological discoveries, botanists and zoologists are able to describe new taxa and even classify 'endemic' units of ecosystems.

Gravity, erosion, and deflation keep the rocks on the mountain tops and slopes devoid of weathered layers. On the bare ground and boulders live an astonishing variety of life-forms: bacteria, algae, fungi, lichens, and small invertebrates. While cliffs in temperate zone mountains are typically colonised by mats of lichens and mosses, those in tropical ranges may host vascular plants, such as ferns, orchids, and bromeliads, which are often identical with the epiphytes in the surrounding forests (Longman and Jeník, 1987). Crevices are rapidly inhabited by small ferns and cryptogamic plants and offer suitable refuges for animal life. Large hollows in the scree may host a variety of invertebrates, including predatory carabids living on detritus and other insects swept from the surrounding alpine tundra. Finally, solid rocks form the dark habitats of mountain caves which create 'amputated' ecosystems lacking primary producers, but with a species-rich consumer and decomposer community supplied by energy sources and nutrients imported by bats.

Even temporary snow fields can offer an ephemeral substrate for life in the spring and summer. For example, at higher altitudes in the Alps, Tatras, and many other mountain ranges, patches of red snow are created by the seasonal reproduction of *Chlamydomonas nivalis*, an alga containing a red pigment which is supposed to protect the otherwise green micro-organism against UV radiation damage. Other species of algae, micro-organisms, and fungi produce a spectrum of colours in a landscape which is otherwise largely snow-white (Fott, 1967). In contrast to this seasonal phenomenon, very old soils are sometimes preserved on upland plateaux, and traces of past millennia can be found in the subsoil. For example, minerals from the Tertiary and cryogenic structures from the last glacial

period, identifiable by patterns of vegetation, occur in the unconsolidated layers on the plateaux of the Giant Mountains. In the same region, air pollution has led to high levels of lead, cadmium, and zinc in the soils which, combined with the leaching of aluminium, is a suspected cause of the decline of forests (Fischer, 1995) and impoverishment of the alpine ecosystems (Box 10.3).

At the regional scale, the greatest biodiversity occurs in mountain systems which are fractured and/or composed of a mixture of sedimentary, igneous, and metamorphic strata: such as the Western Carpathians. In tropical mountains, the old and deeply weathered soils tend to obscure the variety of parent rocks, but the barren rocks in the higher alpine, páramo, or subnival belts differ in their pioneer inhabitants. Alpine humus soils, peats, and snowpatch soils have convergent development on different rocks in mountains around the world. Even in temperate regions, on moderately inclined slopes with stabilised and deeper soil, differences between base-rich and silica-rich substrates tend to disappear under the impact of leaching, podzolisation, and other soil-forming processes. Podzolic soils tend to develop under coniferous trees, and brown soils under broad-leaved trees.

Altitude

In spite of the considerable species diversity locally observed on summits, particularly during the peak

Box 10.3 The impact of air pollution on soil fauna in the Tatras

Soil animals are sensitive to ecological stress and disturbance, including global climatic change and air pollution. Since 1959, long-term studies of soil fauna, chemistry, and micro-structure have been underway in 25 plant communities of the High Tatras of Slovakia.

Between 1959 and 1977, the chemical and biological characteristics of the ecosystems did not change greatly. However, by 1990 and subsequently, significant changes of soil mesofauna (animals living in the air spaces and in the soil water film) have been identified in certain ecosystems. Signalling an ecological catastrophe, these changes have various origins. Some seem to result from acid deposition and subsequent increases in soil acidity, others from parallel increases in nitrogen loading; some could not be explained by observed changes in climate or pollutant input.

Environmental alteration is more marked on limestones than on granite. The greatest increase of soil acidity was in an endemic grassy community (*Festucetum versicoloris*), where the composition of the springtails (order Collembola), a typical group of soil mesofauna, changed dramatically. The calciphilic *Folsomia alpina*, a species bound in the past to this endemic ecosystem, died out in the study area. In contrast, *Tetracanthella fjellbergi*, formerly restricted to soils on silicate bedrock, invaded acidified soils on the limestone.

Profound ecosystem changes also took place on northern slopes underlain by granite. The mossy and peaty *Sphagno-Empetretum hermaphroditi* community has completely changed, very likely due to the high nitrogen inputs from the atmosphere. *Sphagnum* mosses disappeared, and the decomposition of the 60 cm thick peat layer by soil micro-organisms was enhanced. A new grassy community (*Oreochloetum distichae*) has developed, with quite different plant and animal components. The former low-density and species-poor springtail fauna,

Plate 10.3   Forest decline at the alpine treeline of the Giant Mountains, caused by air pollution in Central Europe. (Photograph: J. Štursa)

typical for wet bog habitats, has been replaced by dense, diverse populations of springtails associated with more nutrient-rich soil. Similar dramatic changes have been observed in other ecosystems on granite, such as gullies occupied by the grassy *Calamagrostietum villosae tatricum* community, where the density of springtails increased almost four times, reaching maximum values of 301 600 individuals per m².

Certain springtails, such as *Folsomia penicula*, which used to be frequent only in lowland and lower montane forests, are now also dominant in alpine ecosystems. Many other lowland species have been recorded at high altitudes. However, the arcto-alpine *Folsomia sensibilis* has disappeared from the lower alpine ecosystems.

Overall, these changes are most pronounced in the runoff gullies and karst sinkholes filled by snow drifts and thus receiving higher air pollution loads. The endemic ecosystems with the rarest species are most fragile.

Source: Josef Rusek, Academy of Sciences,
České Budějovice,
Czech Republic

of growing/breeding/nesting seasons, one general trend is clear: species richness generally declines with increasing altitude, particularly towards the stressful environment of cold, windswept, and/or arid summits. This trend is in agreement with the poleward decrease in numbers of terrestrial species (Collins and Morris, 1985).

Reasons for the reduction in biodiversity with altitude in mountains are similar to those in other rigorous habitats, such as hot springs, salt flats, or caves (Colinvaux, 1993). Comparable severe habitats are rare, sparsely scattered, and sometimes ephemeral. Their relatively small size, low biomass production, and lack of ecosystem stratification constrain reproduction and population density, resulting in high extinction rates. Migration between these very specific places is difficult or impossible. Slow arrival by immigration and high local extinction cannot be surmounted by speciation, leading to the limited number of species present in alpine and subnival habitats at any time.

However, the general trend does not apply everywhere. Due to irregularities in rainfall or the distribution of competitive species, the upwards decline may not be gradual. For example, in the high mountains of Arizona, a transect from 700 to 2700 m showed increasing plant diversity from 1000 to 1500 m, where the high diversity Sonoran desert biome is inserted between the species-poorer oak and pine dominated forests (Whittaker, 1977). In Peru, the floristic diversity on the Pacific flanks of the Andes is low in the foothills of the Atacama desert; high in the montane, oreal, and subalpine woodlands; and relatively low in the subpáramo and puna.

Possibly the most marked evidence of the influence of altitude is the boundary or transition between forested and treeless ecosystems, defined variously as alpine timberline, upper treeline, or upper forest line. Its nature and structure have been examined and surveyed in mountains around the world (e.g. Daubenmire, 1954; Plesník, 1971; Wardle, in Ives and Barry, 1974; Morriset and Payette, 1983; Tranquillini, 1979). In principle, there are three possible causes for the upper survival limits of trees in temperate zone mountains: reduced vegetation period, negative carbon balance, and inadequate resistance to deleterious factors. The needles of coniferous trees overcome some of these causes, and can become photosynthetically active even during short seasonal or diurnal periods of positive temperature.

In the American tropics, the dicotyledonous *Polylepis sericea* reaches the highest altitudinal limits of arborescent growth. Its special features include an exceptionally high photosynthetic capacity; the production of cryoprotective substances; osmotic adjustments to enhance the cooling capacity of the leaves, and to lower the freezing point of the cell sap at night; and seasonal changes of osmotic potential that maintain water uptake and turgor, and thus a positive carbon balance during the dry season (Goldstein *et al.*, in Rundel *et al.*, 1994). However, many types of upper treeline reflect the inherent role of the available species. For example, *Eucalyptus globulus*, a tree introduced from Australia, is more successful at the treeline in the equatorial Andes than indigenous woody species.

In mountains whose topmost forests are dominated by one widely-adapted woody species, such as European beech (*Fagus sylvatica*) or southern beeches (e.g. *Nothofagus solandri* in New Zealand), the upper treeline is a linear boundary between trees rapidly declining in height and non-arborescent scrub. A similar boundary marks patches and corridors of closed canopy woods occasionally scattered within the matrix of grassy tundra. Fire, grazing, and clear cutting often emphasise the linear boundary between subalpine and alpine belts or between the elements of the patch–corridor–matrix mosaic.

In contrast, when many species with broad ecological requirements are available, the upper 'treeline' tends to be a wide transition zone, leading to difficulties in detailed mapping of altitudinal belts (Jeník and Lokvenc, 1962). This is a common case in tropical mountains where cloud forest grades into elfin scrub, bamboo stands, and for páramo grasslands. In mid-latitude mountains such as Mount Paektu, the highest summit of the Korean Peninsula, the transition from forest to the alpine tundra is also gradual (Šrůtek and Kolbek, 1994). Another problem with defining 'treeline' concerns the definition of a 'tree'. Especially in the tropics, where various branched woody life-forms create the treeline (e.g. members of the *Ericaceae* and *Podocarpaceae*), the simple morphological criterion of the single-trunked woody growth habit may not be applicable. Other criteria, such as stem density and cover, and stand size, are necessary to define a 'forest', and to identify other woody formations, such as scrub, tall scrub, shrubland, and krummholz (Jeník and Lokvenc, 1962; Plesník, 1971; Hope, in van Royen, 1980).

As discussed above, models of altitudinal zonation are of great value for summarising knowledge in ecologically complicated mountains, especially those which have experienced much human disturbance, such as the Dolomites (Brandmayr, 1988); as well as less accessible ranges, such as the Langtang Himal (Miehe, 1990). These models are generally based on the distribution of vegetation – especially trees – but other life-forms, such as sedges of the alpine tundra, scree plants, epilithic lichens, and invertebrate animals, also reach defined altitudinal limits and can be used to generalise the altitudinal/vertical zonation of mountains for comparative and educational purposes. A common scheme applied in European mountains is the following series of belts: *colline–montane–alpine–nival*. However, other intermediate classes, such as *submontane, subalpine, supra-alpine*, and *subnival*, are occasionally used. In the alpine tundra of Fennoscandia, *subalpine, low alpine, mid-alpine*, and *high alpine* belts can be distinguished (Dahl, in

Wielgolaski, 1975). Though rainfall totals and average temperature may be valuable in defining such belts, the species composition of the different ecosystems is the most practical criterion for delimiting them. A vertical zonation for the Swiss Alps, defined by the upper limits of characteristic widespread species, is shown in Figure 10.2 and Table 10.1 (Landolt, 1983).

In larger mountain systems, a simple vertical zonation seldom satisfies the requirements for a plausible model, as the limits of belts vary with a variety of factors, including latitudinal position; orientation (southern *versus* northern situation); and the position of massifs within a group of ranges and with regard to exposure to prevailing oceanic winds (windward/oceanic *versus* leeward/continental). The 'continental mountain belt' inserted into the model of the Swiss Alps (Figure 10.2) is a

clear example; this belt is well marked in all aspects of biodiversity and even in human colonisation. The three-dimensional pattern of the Alps varies greatly, as illustrated on numerous north–south and west–east cross-sections and maps (Ozenda, 1988). The Urals show a north–south variation between 51 and 69°N, as well as a considerable difference between western and eastern slopes (Gorčakovskij, 1966). Under the influence of large water bodies, such as Lake Baikal, a reverse altitudinal zonation can develop (Box 10.4). In western North America, both north–south changes along the Pacific coast and west–east diversity across the Coast Range and Rocky Mountains are strongly superimposed on the altitudinal variation of biomes.

In the tropical mountains, geographers and ecologists use different terminologies to describe the altitudinal belts, e.g. *afro-alpine, subpáramo, páramo, suprapáramo,* or *puna* (Müller, 1981). In the High Andes of Peru and northern Bolivia, *tierra templada, tierra fria,* and *tierra helada* have been described, from bottom to top (Troll, 1968). However, European scientists have also used the typical European scheme for New Guinea (van Royen, 1980) and the Andes of Ecuador and Peru (Ellenberg, 1979). At an increased level of detail, the most realistic approach to describing surface patterns is based on the distribution of dominant life forms and biotic communities (Richards, 1996; Hope, in van Royen, 1980).

Though wild plants remain the skeleton of most zonation models, cultivated plants and even domesticated animals are also used as indicators of particular altitudinal belts. For instance, in the Andes, the upper limit of corn cultivation, the belt of temperate tuber plants and cereals, and pastures for llamas and sheep (up to the upper limit of vegetation cover in the subnival belt) can be

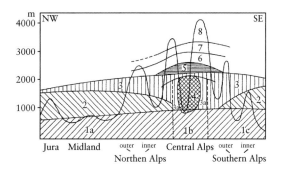

Figure 10.2 Altitudinal and latitudinal zonation in the Swiss Alps: 1 – colline belt (a: Northern Alps, b: Central Alps, c: Southern Alps), 2 – montane belt, 3- subalpine belt (3a: Central Alps), 4 – inner valleys with continental influence, 5 – supra-subalpine belt, 6 – alpine belt, 7 – subnival belt, 8 – nival belt (after Landolt, 1983)

Table 10.1 Altitudinal zonation in the Swiss Alps (after Landolt, 1983)

Belt	Plants	Upper limit (m asl)
Colline	Oak (*Quercus* sp.)	500–1000
Montane	Beech (*Fagus sylvatica*)	900–1500
Subalpine	Spruce (*Picea abies*)	1600–2300
Continetal mountain	Scots pine (*Pinus sylvestris*)	800–2100
Supra-subalpine	Cembra pine (*Pinus cembra*)	2100–2600
Alpine	Patches of alpine meadows, e.g. *Carex sempervivens, Carex curvula, Elyna myosuroides*	2600–2900
Subnival	Plants of alpine scree e.g. *Saxifraga oppositifolia, Androsace alpina)*	2900–3300
Nival	Cryptogams	no upper limit

Box 10.4 Contrasts and challenges in the mountains around Lake Baikal

Lake Baikal is the deepest lake in the world (1620 m), with the water level at an altitude of 455 m. It contains a tenth of global lake water reserves. Originating at the beginning of the Tertiary, numerous ancient life-forms are found in its depths, signifying a stable aquatic environment where they have survived as relicts; others have evolved into numerous endemic lineages. In contrast, the region around the lake has experienced severe environmental changes, caused by both tectonic activities and climatic fluctuations, including two glacial periods when a large proportion of the surrounding mountains were glaciated.

This vast natural reservoir has modified the local mesoclimate. While Mongolian steppes occupy the large valleys parallel to the lake, at distances up to 200 m – and sometimes even 600 m – from the lake shore, the high continental summer temperatures are 5 °C lower and winter temperatures 10 °C higher on average. Around the northern part of the lake, this modified mesoclimate is responsible for the phenomenon of 'vegetation belt reversal'. Close to the lake is a subalpine belt of coniferous krummholz. Above this are continental forests with deciduous larch, followed by evergreen coniferous taiga, another krummholz belt, and the alpine zone.

The lake also enhances cloudiness and precipitation in the Chamar–Daban range to the south, as dry air masses coming from the north and northwest gain moisture over the lake and deposit this as they cool at higher elevations. As a result, precipitation levels are very high (> 1500 mm/yr). In contrast, the large islands in the central part of the lake are mostly covered by steppe vegetation, as they receive less than 200 mm of annual precipitation.

Along the shore, several endemic plant species are restricted to specific dune habitats. However, as endemism in the surrounding mountains is not higher than in other central and south Siberian mountain ranges, it does not appear that the lake has induced evolutionary processes resulting in plant speciation. Nevertheless, the great ecosystem diversity – ranging from dry continental steppes and large wetlands along the shore and in river deltas, to deciduous and evergreen forests, and large alpine areas – is mirrored by the high species diversity (2359 species and subspecies of flowering plants). The high level of endemism of alpine plants in Central Siberia (13.5 percent of species) contrasts with much lower values calculated for individual mountain ranges, indicating that speciation was rather intensive. However, the endemic species were able to migrate along the mountain ranges extending east–west for hundreds of kilometres, and only relatively few local endemics remained restricted to small geographical areas.

Nature in Central Siberia is still largely undisturbed and wild. Human activities have been concentrated along the Trans-Siberian and Baikal–Amurian Railways. Access to the mineral and biotic resources in the mountains remains very difficult. However, this may change soon, as economic progress brings efficient technology to the region and international companies come to play a dominant role in exploiting vast forests and rich mineral resources. As shown by environmental problems in the populated lowland areas, the natural balance in Central Siberia can easily be disturbed in a way which is difficult to restore. While some of the mountains have been declared as national protected areas, co-operation between industrial companies and nature conservation agencies has been limited; if more efficient means of exploitation become available, few institutional forms of protection will prove effective.

Source: Leoš Klimeš, Division of Ecology,
Academy of Sciences,
Třeboň, Czech Republic

distinguished (Troll, 1968). Uhlig (1978) applied these concepts to the cultivation of subsistence crops, especially hill rice, in the inner valleys of the Himalaya and other areas of Southeast Asia (see also Chapter 2).

Human activities

Chronicles, old maps, and management plans and other archival documents record events which can help to explain the present day status of mountain forests, or the distribution of alpine plant and animal species (e.g. Jeník and Hampel, 1992).

On all continents, major changes in mountain areas followed extensive deforestation, hunting, the introduction of large domestic herbivores, and the cultivation of new crops. Also, the distribution of wild plants and animals changed in the historical period due to unintentional introduction of alien species. For example, the mountain meadows of the Appalachians contain numerous European herbaceous species brought by colonisers in recent centuries. These changes can be explained by palynologists and historians and clearly analysed by using the results of earlier botanical and zoological research. More recent impacts

Box 10.5   Human influence on alpine pasture

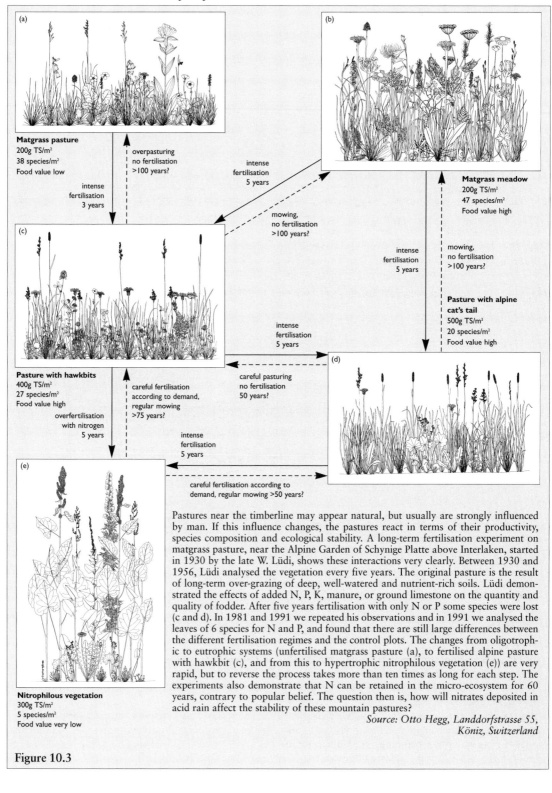

Matgrass pasture
200g TS/m²
38 species/m²
Food value low

overpasturing
no fertilisation
>100 years?

intense
fertilisation
3 years

intense
fertilisation
5 years

mowing,
no fertilisation
>100 years?

Matgrass meadow
200g TS/m²
47 species/m²
Food value high

intense
fertilisation
5 years

mowing,
no fertilisation
>100 years?

Pasture with hawkbits
400g TS/m²
27 species/m²
Food value high

intense
fertilisation
5 years

**Pasture with alpine
cat's tail**
500g TS/m²
20 species/m²
Food value high

careful fertilisation
according to demand,
regular mowing
>75 years?

careful pasturing
no fertilisation
50 years?

overfertilisation
with nitrogen
5 years

intense
fertilisation
5 years

careful fertilisation according to
demand, regular mowing >50 years?

Nitrophilous vegetation
300g TS/m²
5 species/m²
Food value very low

Pastures near the timberline may appear natural, but usually are strongly influenced by man. If this influence changes, the pastures react in terms of their productivity, species composition and ecological stability. A long-term fertilisation experiment on matgrass pasture, near the Alpine Garden of Schynige Platte above Interlaken, started in 1930 by the late W. Lüdi, shows these interactions very clearly. Between 1930 and 1956, Lüdi analysed the vegetation every five years. The original pasture is the result of long-term over-grazing of deep, well-watered and nutrient-rich soils. Lüdi demonstrated the effects of added N, P, K, manure, or ground limestone on the quantity and quality of fodder. After five years fertilisation with only N or P some species were lost (c and d). In 1981 and 1991 we repeated his observations and in 1991 we analysed the leaves of 6 species for N and P, and found that there are still large differences between the different fertilisation regimes and the control plots. The changes from oligotrophic to eutrophic systems (unfertilised matgrass pasture (a), to fertilised alpine pasture with hawkbit (c), and from this to hypertrophic nitrophilous vegetation (e)) are very rapid, but to reverse the process takes more than ten times as long for each step. The experiments also demonstrate that N can be retained in the micro-ecosystem for 60 years, contrary to popular belief. The question then is, how will nitrates deposited in acid rain affect the stability of these mountain pastures?

*Source: Otto Hegg, Landdorfstrasse 55,
Köniz, Switzerland*

Figure 10.3

of human activities are discussed in the following section and many of the other chapters of this book.

ENDANGERED ECOSYSTEMS

The current status of mountain biodiversity is a product of both continual changes and sudden events. The former are measured in centuries, millennia, or millions of years – compared to a human life span, the slow rates of tectonic processes and generally slow rates of speciation escape our attention. However, sudden changes – measured in seconds, years, or decades – are readily recorded and remembered, particularly if human life, property, and associated biota were threatened or lost. Although landslides, avalanches, and floods are regarded as dangerous events with disastrous effects, the same recurrent factors are responsible for mountain scenery and ecosystem patterns. However, while these physical processes may be repeated, biological evolutionary events are irreversible; whenever the existence of a biotic population or ecosystem is at risk, we are justified in saying that it is endangered.

In the global context, mountain ecosystems are among the most fragile and endangered. In recent years, significant changes have been recorded and monitored in the mountain ecosystems of every continent (Stone, 1992; Denniston, 1995; IUCN, 1995). High rates of change are becoming apparent because of the frequency of observations and the availability of comparative data. Dramatic fluctuations in meteorological parameters have been recorded, together with a higher frequency of extreme events which stress biotic and human communities and disturb their habitats. Internal biotic factors also occasionally destabilise and disrupt mountain ecosystems, for instance through population invasions, explosions, or extinctions (Box 10.6).

Humankind plays a central role in the alteration of contemporary mountain ecosystems. Increases in the density of mountain inhabitants and visitors, as well as distant industry, threaten mountain species and ecosystems both directly and indirectly,

Box 10.6 The isolated mountains of southern Western Australia

A biological survey of mountain protected areas in southern Western Australia has been in progress since 1994, with the aims of assessing the nature conservation values of selected peaks and quantifying threats to these mountains. The study is being conducted by the Western Australia Department of Conservation and Land Management, with the financial assistance of the Australian Nature Conservation Agency. Although the mountains are small by world standards (the maximum elevation is only 1770 m), they are very isolated, being the highest peaks for several thousand kilometres. The region is renowned for its species richness; for instance, 1517 plant species have been recorded from the Stirling Range National Park alone.

The survey has identified a high number of endemics with narrow ranges on mountain peaks. Of some 750 species surveyed, 100 were restricted to a particular mountain range or group of mountains. High levels of speciation may be attributed to geographical isolation and fluctuating climatic conditions in the past. The mountains are refugia, providing a more mesic environment than the surrounding lowlands. Fifteen mammal species, including four threatened species, have been recorded, using a range of techniques including hair sampling and scat analysis, as standard trapping techniques are of limited use in this mountain environment.

An invertebrate survey identified a significant number of endemics and Gondwanan relict species, particularly spiders and snails. Pockets of habitat in sheltered gullies and slopes with a more mesic climate provide refuges for invertebrates that can no longer exist in drier sites. Many species have closer relationship to groups in mountainous areas of eastern Australia, Tasmania, New Zealand, and other Gondwanan habitats than to species in the surrounding lowlands.

The major threat to the mountain plant communities is the introduced fungal dieback disease (*Phytophthora cinnamoni*). Where active, it is dramatically altering these communities and threatening rare and endemic species with extinction. Post-fire regeneration is extremely slow on the higher peaks and may also influence the impact of the disease. The management of recreational access is paramount to control. Strategies include the permanent or seasonal closure of uninfected areas to walkers, and the construction of boot cleaning stations in order to prevent the introduction of the fungus in soil on footwear.

Source: Sarah Barret, Department of Conservation and Land Management, South Coast Region, Australia

intentionally and unintentionally (see also Chapters 12 and 17). Civilisation has indirectly affected mountain ecosystems by pollution, the dispersal of viable seeds, and the spread of cultural information. However, anthropogenic factors should not be indiscriminately identified as having negative impacts (Figure 10.3). The status of biodiversity must be flexibly assessed in relation to global nature conservation objectives and with regard to the cultural needs of resident people and their lowland neighbours.

To point out some typical examples of acute threats to biodiversity, this section concentrates on a few characteristic types of ecosystem, mainly in the alpine/páramo belt and at the upper treeline. The forested montane belt and valleys with their complex issues of deforestation, agriculture, and urbanisation are tackled only marginally, as they are considered in more detail in Chapters 12, 13, 14, and 15.

Cliffs, rock faces, and caves

The colonisation of rocky peaks, cliffs, ravines, and caves by hardy organisms is strongly controlled by the mineral composition and surface features of the rocks. Mountain peaks play a unique role in the life strategy of many flying insects, such as flies and beetles, during the mating period. Treeless peaks or solitary cliffs usually offer an important refuge for rare birds of prey, such as golden eagle or eagle owl, especially in countries where these magnificent predators have otherwise been eradicated (Jeník and Price, 1994).

Most cliffs consist of a bare surface, crevices, and ledges. These microhabitats are colonised not only by specialised species exclusively adapted to them, but also by species which use them as a refuge against predators; for example, nesting birds and monkeys staying overnight on a cliff. The small organisms in the crevices and on the cliff surface play important roles in the weathering of solid rocks and the formation of unconsolidated humus-rich soil. Nowhere is this link between inanimate rock and life more fragile.

One major threat to these organisms is the removal of the rocky substrate itself. Quarrying, open-pit mining, and road construction remove whole cliffs or destroy deep ravines. However, in the wooded montane belt, the first phases of succession, including the varied flora of tropical epiphytes which readily colonise moist rock faces, can be seen on newly exposed rock faces and roadcuts, where many plants and animals have been discovered. Quarrying and mining often start on natural rock faces, and are probably the most destructive economic activities in mountain areas (see Chapter 9). An open-pit copper and gold mine in the Star Mountains of western Papua New Guinea will virtually level a 2330 m mountain which is sacred for the Wopkaimin people and also the cradle and home of indigenous biota (Denniston, 1995).

While tiny alpine saxicolous herbs seldom destroy their own habitat, the crevices penetrated by the expanding roots of woody plants actively contribute to rock falls and abrupt changes in the subalpine belt. Similarly, cliffs can be damaged by pegs hammered in by rock climbers. This has created serious problems on heavily used routes in the Alps and Himalaya, and is also a sensitive issue in lower mountains, where small refuges of alpine plants have been damaged or destroyed. Concentrated tourism also affects look-out points where visitors pollute the soil, trample vegetation, and disturb animals. Yet plant collectors have also devastated many rare species.

The rugged karst relief in calcareous mountains is very rich in overhanging cliffs, caves, and ravines, which provide shelter for highly specialised plants and animals. The lairs of hibernating bears or pandas are easily plundered by hunters; and bats, birds, reptiles, and amphibians living in caves have often been disturbed or collected for commercial purposes. Cave floors are often important palaeontological sites because of their long-term animal or human residents. Climbers' bivouacs and the excavations of amateur archaeologists have frequently destroyed these archives of the past.

Glacial cirques and avalanche tracks

Glacial cirques offer a great variety of habitats and serve as refuges of outstanding biodiversity. In the Central European middle mountains – such as the Black Forest, Bohemian Forest, and Sudetes – the flora and fauna of these sites include species with a close affinity both to alpine and arctic areas, and to ecosystems to the south. This is explained by the long-term impact of avalanches on treelessness and the exposure of a wide variety of outcropping rocks during the much warmer mid-Holocene (Jeník, 1961). Some cirques on Scottish mountains, such as Ben Lawers, have played a similar role, with rare species on steep headwalls protected from grazing by red deer and sheep (Poore, in Mardon, 1986).

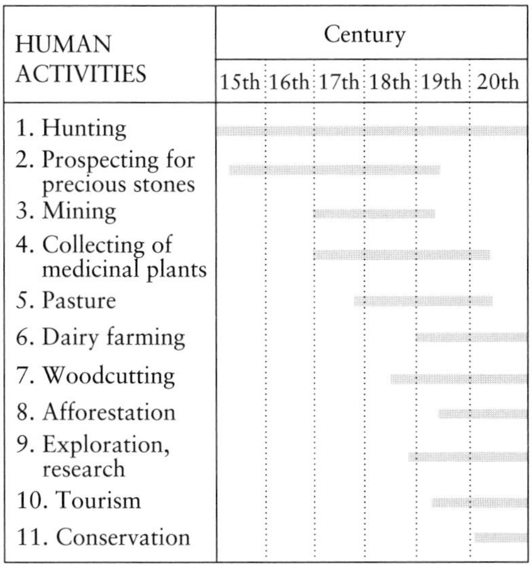

HUMAN ACTIVITIES	Century					
	15th	16th	17th	18th	19th	20th
1. Hunting						
2. Prospecting for precious stones						
3. Mining						
4. Collecting of medicinal plants						
5. Pasture						
6. Dairy farming						
7. Woodcutting						
8. Afforestation						
9. Exploration, research						
10. Tourism						
11. Conservation						

Figure 10.4 Human activities influencing biological diversity in the upper zone of the Sudetes, Central Europe (after Jeník and Hampel, 1992)

In recent decades, the biodiversity of the middle mountain cirques has been threatened by avalanche defence measures indiscriminately introduced from the Alps, with the aim of improving the safety of winter tourists and skiers. Foresters undertook large-scale afforestation, often using trees of alien species, in the subalpine belt in order to influence snow deposition, keep avalanches from starting, and 'protect' woody plants on avalanche tracks from mechanical stress. After this major misunderstanding between nature conservation and forest management was recognised in the Sudetes, afforestation of natural avalanche tracks ceased, and the plantations are being cut down.

In many mountains, mining, tourism, and skiing facilities threaten the biodiversity of the cirques, which are naturally snow-rich and thus attractive for skiing, particularly in middle mountains and marginal ranges with less snow and a short winter season. In the appropriately named Gross Kar (Great Cirque) in the Eastern Alps, cable cars, ski-lifts, and down-hill ski runs have occupied a snow-rich area which was famous for its botanical wealth. While most cirques in the Scottish Cairngorms are still inaccessible to tourists, the cirques and summits on the western margin and the northeastern ski area of Aviemore experience an immense tourist pressure, and suffer from trampling and erosion along impromptu trails.

Renowned botanical localities are threatened also by plant collectors and gardeners. Relict populations of rare plants, such as *Rhodiola rosea* and *Gentiana* species, have disappeared into medical laboratories and/or factories producing liqueurs. Many ornamental species of *Delphinium, Campanula, Doronicum*, and *Aconitum* have first been collected in the cirques' spectacular tall herb communities, which are continually depleted unless declared as strict nature reserves. Hunters also use the sheltered cirques of the Alps and many of the Central European mountains as bases for chasing wild animals and game. Many of the 'last' specimens of large mammals, game birds, and even birds of prey have been shot in these refuges of biodiversity. Animals introduced for hunting, such as chamois in certain central and eastern European ranges, can also have serious impacts on the indigenous biota of species-rich cirques.

Mountain tundra

Mountain tundra, including a variety of ecosystems on flat areas and moderate slopes, is the most widespread type of vegetation on the world's mountains and high plateaus. These natural and human-induced ecosystems range from the discontinuous vegetation of stony 'fjell-fields' to more-or-less closed communities of mosses, grasses, sedges, and dwarf shrubs, both above and below the natural treeline. Latitude, altitude, and orientation modify the severity of the climate and the variety of soils. Species composition is influenced by the speciation, immigration, and survival of species similar to the biota of the Arctic tundra. While species diversity in alpine tundra is lower than in the ecosystems below the treeline, the image of monotonous grasslands or heaths is misleading. In the Austrian Central Alps, alpine sedge mats harbour representatives of most of the orders of invertebrates present at lower altitudes in the same massif (Meyer and Thaler, in Chapin and Körner, 1995).

Mountain tundras have a long history of human exploitation. For people living in fertile lowlands and valleys, the land above the timberline was valuable for hunting, medicinal plants, and minerals. Large mammalian herbivores – llamas, goats, sheep, antelopes, wild yaks, and horses – were hunted, and later replaced by domesticated and imported animals. The large grazers are important in maintaining and even improving grassland ecosystems; a moderate degree of grazing promotes a series of positive feedbacks that favour grasses over mosses, stimulate the nutrient cycle, and improve primary productivity (Zimov *et al.*, in Chapin and Körner, 1995).

Recognition of the various roles of ungulates is central to understanding the natural appearance of the alpine belt. Even in virgin mountains, herds of large herbivores maintain grasslands much below the potential growth of subalpine forests. For example, large open areas along the Yellowstone River, in the Rocky Mountains, are maintained by buffalo, wapiti, and pronghorn antelope. Wherever feasible, people have extended similar pastures at the treeline by burning and cutting subalpine or montane woody ecosystems, thus creating large semi-natural grasslands on the Andean altiplanos, the plateaus of the Rocky Mountains, the Himalayan and Tibetan Plateaus, the Carpathians, and the Alps.

The grasslands and heaths of the Scottish Highlands were largely created in historic times, partly in the ninth to eleventh centuries, and again between the fifteenth and sixteenth centuries until the end of the eighteenth century (Fraser Darling and Morton Boyd, 1969). Large-scale burning and cutting of pine and oak forests abruptly decreased the biodiversity to levels which are maintained by excessive populations of red deer which browse all naturally regenerated or planted trees. However, relationships between species diversity and sheep and cattle grazing are always complex. In the Polish Eastern Carpathians, the flower-rich polonina grasslands, which developed through centuries of sheep grazing, rapidly lost their species diversity after grazing was terminated and competitive clonal grasses invaded in a newly declared national park (Jeník, 1983b).

Human settlement has ascended high into the mountain tundra, and a particular economy combining agriculture and pasture has developed in the mountains of every continent (see Chapter 14). In the Alps, this *Almwirtschaft* (or *Alpwirtschaft*, *Berglandwirtschaft*) maintains natural alpine tundras and man-induced meadows, guaranteeing modest yet sustainable yields. Special geographic, economic, and sociological projects and plans have been adopted for the optimum development of this economy (Naegeli-Oertle, 1986). The conservation of wildlife requires consensus between private proprietors and nature conservation agencies.

Around the world, indigenous agriculture has developed on moderate montane and alpine slopes, and careful management of the limited fertile soil and reasonable manuring have provided sustainable yields for centuries (see Chapter 14). However, in many developing countries, under the pressure of population growth and a fast-growing global economy, traditional methods of cultivating staple crops and raising sheep and goats are often being abandoned and more profitable cash crops and cattle introduced. The adoption of intensive farming requires high inputs of pesticides, herbicides, and fertilisers, resulting in soil degradation, driving the peasants to migrate and clear more marginal land. However, traditional practices can also decrease biodiversity. For example, to regenerate pastures, the Indian inhabitants of the Andean altiplanos regularly set fire to mature stands of grasses upon entering the páramo. This kills many indigenous herbs and invertebrates living between the huge tussocks of páramo grasses. With continued burning and increasing numbers of cattle, the biodiversity of the mountain tundra becomes impoverished. There are fewer problems with the conservation of indigenous biota in the loosely vegetated and less inflammable supra-páramo.

Wetlands and peatlands

Even dry tropical and temperate highlands possess scattered springs and temporary streams. All European mountains and the humid ranges of other continents have wet areas resulting from high precipitation and upwelling groundwater, as well as the melting of glaciers and late snowbeds. Alpine and subalpine springs and flushes feed permanent or seasonal streams which eventually merge to form rivers or fill lakes. All these wetlands and waterbodies are inhabited by specialised organisms which markedly extend the terrestrial biodiversity of mountains and deserve protection.

The size and duration of glaciers and late snowbeds fluctuate periodically or annually according to temperature and snowfall patterns. Their margins are sooner or later colonised by organisms which tolerate enormous ranges of temperature, radiation, and physical stress. Due to the recurrent distribution of late snowbeds in the lee of ridges, stabilised communities of relict mosses and liverworts are accompanied by soil organisms which survive the long-lasting snowpack (e.g. Mucina *et al.*, 1993; Rodwell, 1992). In mountain ranges affected by industrial emissions, these ecosystems are endangered by the cumulative effects of pollutants deposited together with the snow (Fischer, 1995).

Springs, flushes, and pools on mountain plateaus and flanks greatly enrich mountain biodiversity. Flowing through and over various rock types, the water is charged with different nutrients, creating patches of aquatic and semi-terrestrial biota within the matrix of the rather monotonous mountain tundra and montane forests. Due to the long-term stability of temperature and chemistry of water, some of these species are relicts of earlier periods (Franz, 1979). Springs and flushes require protection against both human-induced drainage and herds of thirsty livestock which trample the plants and import alien seeds.

The threat of introduced species is a major issue in mountain lakes and streams, as many high mountain lakes were never colonised by indigenous fish, which could not ascend small streams and waterfalls. Keen anglers and fishermen soon discovered this 'deficiency' and introduced one or several species of fish. As well as the indigenous trout, American trout species have been successfully introduced in some glacial lakes in the Alps and Carpathians. These new predatory inhabitants have totally changed the limno-biological diversity in all groups of aquatic plants and animals. Some of these biologically foreign ecosystems exist amidst terrestrial nature reserves considered as unique, such as the Tatra National Park in Poland (Szafer, 1962).

The aquatic bio-corridors of the mountain streams have frequently been disturbed or interrupted by human activities. Mountains attract most of the world's hydroelectric projects. Reservoirs inundate long stretches of agricultural and grazing land, as well as riparian and alluvial forests. In some tropical mountain valleys, large areas of species-rich rain forests and/or savanna ecosystems have been inundated, with further deforestation for the power lines.

Although mountain streams have a great capacity for self-purification, aquatic life suffers from domestic sewage and industrial waste. Many mountain streams of the Giant Mountains, affected by acid rain and heavy metals emitted by Central European industry, have lost all aquatic life. Remarkably, the streams flowing across limestone areas have preserved their original biodiversity.

Peatlands (mires) are wetland ecosystems found in humid mountains, both above the treeline and in the montane belt. They are dominated by *Sphagnum* mosses, sedges, and dwarf shrubs. They also occur in several tropical mountain ranges. In the Andes, the bogs of the páramo are created by compact cushions of plantain (*Plantago*), tussocks of sedges (*Oreobolus*), and *Eriocaulon* species. The same genera are represented in the subalpine mires of New Guinea, where robust cushions of *Astelia alpina* are locally dominant (Hope, in van Royen, 1980). Around the world, the biodiversity of mountain peatlands is threatened by exploitation of peat for fuel, agriculture, or gardening, and by artificial drainage intended to eliminate the inhibiting effects of waterlogging on tree growth.

Woody ecosystems

Woody ecosystems of all kinds – subalpine scrub, krummholz, bamboo stands, elfin forest, cloud forests, montane matorral, and subalpine parklands – mark a significant transition between two worlds of non-forest and forest diversity. Even ecosystems dominated by a single woody species are structurally and taxonomically richer than the neighbouring alpine tundra, if the cryptogamic flora (algae, mosses, lichens, fungi), invertebrate soil fauna, and insects are considered (Meyer and Thaler, in Chapin and Körner, 1995). Because of their greater underground and above-ground stratification, the variety of food and shelter, and the patchiness caused by irregular disturbance, these marginal woody ecosystems, especially the tropical cloud forests (Hamilton *et al.*, 1996) exhibit high biodiversity.

These marginal ecosystems face two kinds of threat: destruction by cutting, burning, and lopping, mainly to extend pastures and agriculture; and deforestation for plantations, mostly of alien species. Marginal subalpine forests seldom provide timber, but in densely inhabited areas of the Andes, Himalaya, and Central Asia, they are an important source of fuel (see Chapter 13). In the tropical mountains of Southeast Asia and Latin America, colonists from crowded lowlands have pushed upland farmers, whose land tenure is often insecure, ever higher into the montane and subalpine belts. The uplanders, in turn, have often encroached on the homelands of indigenous people and the ecosystems on which they depend.

The biodiversity of the marginal woody ecosystems is also severely threatened by the introduction of alien species, often from distant continents and different latitudes and altitudes. For the narrow goals of forestry, some of these plantations may appear successful; for example, *Eucalyptus globulus* has been planted in Ecuador at over 3000 m for more than a century, producing straight timber, in contrast to the indigenous flora which provides only crooked growth forms of shruby *Chusquea* bamboo vegetation. Similarly, recently planted *Pinus radiata* seems to be successful in protection against deflation and erosion on the altiplano. Yet such alien woody species could eliminate indigenous flora and fauna of the páramo and puna in the national parks of Ecuador and Peru.

In the European mountains, the conversion of the biodiversity of the subalpine woody ecosystems was generally less extreme. Cut or burned marginal forests and krummholz were artificially regenerated with the original species, but with plants grown from seeds whose provenance was unsuitable. Plantations of Norway spruce (*Picea abies*), Austrian pine (*Pinus nigra*), and dwarf pine (*Pinus mugo*) have a similar physiognomy to the original stands, but their growth, health, and ecological stability proved to be a failure, as shown by frequent population explosions of both insect and fungal pests.

MANAGEMENT AND CONSERVATION

Life in the mountains is in a constant state of change. The majority of these variations are intrinsic features of the physical environment and the life strategies of organisms. Much of the 'fragility' of mountain ecosystems derives from the sensitivity of their biota to stress and disturbance, not only from physical factors, but also from the activities of other organisms, especially human beings. The human factor is increasing in importance, with strong influences on biodiversity.

Individual organisms, exposed to competition for life-supporting resources and situated in the foodchain of producers and consumers, necessarily face damage or death: herbivores graze and browse; carnivores bite and swallow; aggressive insects, fungi, and bacteria infest host populations. 'Victims', 'pests', and 'vermin', the usual derogatory descriptors applied in agro-ecosystems, are not appropriate in mountain ecosystems which are not managed. Sustainably disturbed mountain ecosystems have various compensating processes which overcome disruptions, enabling sustainable development and the evolution of new biota. However, people who enter and disrupt virgin ecosystems must be regarded as external offenders.

While individuals die but are usually succeeded by their offspring, the extinction of a whole population is an irreparable event, and an extinct species disappears forever from the global community.

Such events have always happened and are important in evolution. The vacant niche may or may not be filled by another species, yet speciation generally takes millennia. It appears probable that rates of extinction have increased globally due to human actions; the human-induced loss of native species is a frequent reason for the collapse of fragile mountain ecosystems.

Human responsibility and awareness of the endangerment of mountain species and ecosystems varies according to the land uses in different types of mountain areas (Table 10.2). In densely populated and heavily exploited mountain ecosystems, resident people have an intimate knowledge of plants and animals, which are central to their economies and cultural heritage. They will have learned to cope with soil losses on steep slopes, or surpluses or deficits of water, but have no substitute for eradicated animals or devastated forests.

Management priorities for mountain biodiversity vary according to environmental conditions and land-use practices. In cool uninhabited tundra areas, only the adequate control of hunting and the limitation of greenhouse gases produced by distant industrial centres are generally required; there are few constraints to the establishment of large national parks. In cool forested areas in industrialised countries, the priorities should be to stop the fragmentation of ecosystems caused by excessive logging, develop sound forestry, and maintain sound farming with complementary tourism and recreation; protected areas require adequate zonation, with core zones reserved for wildlife. In arid and semi-arid highlands, there is a need to stop desertification by ending overgrazing and creating new nature reserves, whose biodiversity might be restored by the reintroduction of locally eradicated species. In the high mountains of the humid tropics, species diversity has mainly been affected in the forested zones, where logging and badly-practised shifting cultivation cause much extinction; in the páramo, fires and overgrazing have significant impacts near villages and roads, but remote areas and inaccessible cliffs and summits still have much space for wildlife, even though many species experience severe pressure from hunting.

Recently, awareness of species extinction and responsibility for maintaining biodiversity has greatly increased, influencing many national and international organisations, both governmental and non-governmental. In many programmes, projects, and documents, mountains are mentioned as seriously endangered landscapes where increases in

Table 10.2 Types of mountain areas and priorities for the management and conservation of biodiversity

Ecosystems and land-use level	Major centres	Biodiversity status	Human interference	Management priorities	Conservation priorities
Cool tundra inaccessible, unexploited, uninhabited	High latitudes and/or altitudes in Europe, North America, Central Asia	Explored at species level, undisturbed	Hunting, decline of carnivores, global warming	Tourism, adventure trekking, free access	National parks, protection of migratory birds and mammals
Cool, snow-rich tundra and forest accessible, exploited, uninhabited	High altitudes in mid-latitudes of Europe, North and South America	Monitored at species and ecosystem level, regionally disturbed, multiple production species	Deforestation, human-caused fire, fragmentation of habitats, exotic species, air pollution, acidification, soil erosion	Sustainable logging, reforestation, small-scale farming, ecotourism, recreation, watershed management	Zonation of protected areas, reintroduction of species, stabilisation of upper treeline
Arid, semi-arid steppe and scrub accessible, semi-sedentary agriculture, nomadic transhumance	West Asia, Asia Minor, Ethiopia, Australia, mid-latitudes of the Andes	Less known, transformed ecosystems, genetic resource of useful plants and animals	Overgrazing, human-caused fire, desertification, hybridisation between wild and domestic ungulates	Farming, traditional crops irrigation, soil conservation, animal husbandry, afforestation	Rehabilitation of desert, afforestation, reintroduction of eradicated animals, preservation of germplasm
Páramo, hot and humid forests less accessible, sedentary agriculture, nomadic hill tribes	SE Asia, NE Australia, Central America, tropical South America	Little known, species-rich, increasingly disturbed, traditional crop varieties, medicinal species	Fire, grazing, logging, soil erosion, nutrient depletion, landslides, rice cultivation	Shifting cultivation, on-farm management, agroforestry, hedgerows, grass strips, traditional craftsmanship	Large protected areas, wildlife corridors, *in situ* maintenance of biodiversity, enhanced research

the diversity and populations of species could ensure economic benefits and uninterrupted environmental processes. Through fruitful interaction between theory and practice, the two concepts of biodiversity and sustainable development have become important underpinnings of global policy.

In 1992, the World Resources Institute, IUCN – The World Conservation Union – and the UN Environment Programme, in consultation with the Food and Agriculture Organization (FAO) of the UN and UNESCO, published the *Global Biodiversity Strategy: Guidelines to Save, Study, and Use the Earth's Biotic Wealth Sustainably and Equitably* (1992). In the same year, at the UN Conference on Environment and Development, over 160 nations agreed to take steps to conserve biological diversity by signing the Convention on Biological Diversity. The same conference produced 'Agenda 21', whose Chapter 13 refers to fragile mountain ecosystems and led, among other actions, to the preparation of this book. The documents deriving

from all subsequent events and initiatives dealing with mountain issues, at both regional and global scales (e.g. Banskota and Karki, 1994; The Mountain Institute, 1995; Council of Europe/UNEP/ European Centre for Nature Conservation, 1995), have given particular attention to the legal, scientific, administrative, and financial aspects of the biological assets of the world's mountains.

All of these documents agree that appropriate management is needed in all mountain areas affected by civilisation, and that on a planet influenced at the global scale by human activities, more-or-less strict conservation can be applied very selectively. As discussed in Chapter 11, while mountain protected areas cover about 250 million hectares worldwide, their conservation regime varies. An increasingly valuable and accepted model is that of 'biosphere reserves', consisting of core, buffer, and transition zones. This recognises the need for different levels of protection and, conversely, human activity in each zone of the protected ecosystems, from a strongly protected core to the surrounding landscape where a wide range of economic activities can take place, but with due regard to conservation objectives (Price, 1996).

The values of minute rock plants, flower-rich herbs, lofty trees, colourful insects, singing birds, wild ungulates, and dangerous beasts of prey have long received widespread recognition. However, comprehensive methods for reasonable assessment of the value of wildlife in protected areas have only recently been proposed (Munasinghe and McNeely, 1994). Such benefit–cost analysis and non-market valuation techniques are being applied to quantify the value of the biodiversity of protected areas. The positive results of such assessments can be clearly demonstrated for well-known protected areas, for example in Nepal and Africa. Yet, even in the absence of quantitative estimates, it is clear that the sheer variety of even inconspicuous organisms is of immense value in safeguarding the productivity and multiple services provided by all mountain ecosystems.

References

Banskota, M. and Karki, A. S. (1994). *Sustainable Development of Fragile Mountain Areas of Asia: Regional Conference Report.* ICIMOD: Kathmandu

Barlow, B. A. (ed.) (1986). *Flora and Fauna of Alpine Australasia: Ages and Origins.* CSIRO: Melbourne

Barthlott, W., Lauer, W. and Placke, A. (1996). Global distribution of species diversity in vascular plants: towards a world map of phytodiversity. *Erdkunde*, 50:317–27

Beschel, R. (1961). Dating rock surfaces by lichen growth and its application to glaciology and physiography (lichenometry). In Raasch, G. O. (ed.), *Geology of the Arctic*, Vol. 2. University of Toronto Press: Toronto, pp. 1044–62

Bormann, F. H. and Likens, G. E. (1979). *Pattern and Process in a Forested Ecosystem.* Springer Verlag: New York

Brandmayr, P. (ed.) (1988). *Zoocenosi e paesaggio I: Le Dolomiti, Val di Fiemme – Pale di S.Martino.* Museo Tridentino di Sc. nat.: Trento

Braun-Blanquet, J. (1951). *Pflanzenökologie.* Springer Verlag: Wien

Chapin, F. S. and Körner, C. (eds.) (1995). *Arctic and Alpine Biodiversity.* Ecological Studies 113. Springer Verlag: Berlin/Heidelberg

Clements, F. E. (1916). *Plant Succession: An Analysis of the Development of Vegetation.* Carnegie Institution: Washington DC

Colinvaux, P. (1993). *Ecology 2.* John Wiley and Sons: New York/Chichester

Collins, N. M. and Morris, M. G. (1985). *Threatened Swallowtail Butterflies of the World.* The IUCN Red Data Book: Gland

Council of Europe/UNEP/European Centre for Nature Conservation (1995). *The Pan-European Biological and Landscape Diversity Strategy.* Council of Europe: Strasbourg

Daubenmire, R. (1954). Alpine timberlines in the Americas and their interpretation. *Butler University Botanical Studies*, 2: 119–36

Denniston, D. (1995). *High priorities: Conserving Mountain Ecosystems and Cultures.* WorldWatch Paper 123. WorldWatch Institute: Washington DC

Dhar, U. and Kachroo, P. (1983). *Alpine Flora of Kashmir Himalaya.* Scientific Publishers: Jodhpur

Enders, G. (1979). *Theoretische Topoklimatologie.* Nationalparkverwaltung: Berchtegaden

Ellenberg, H. (1975). Vegetationsstufen in perhumiden bis perariden Bereichen der tropischen Anden. *Phytocoenosis*, 2: 368–87

Ellenberg, H. (1979). Man's influence on tropical mountain ecosystems in South America. *Journal of Ecology*, 67: 401–16

Fischer, Z. (ed.) (1995). *Problemy ekologiczne wysokogórskiej cześci Karkonoszy.* Institut Ekologii, Polish Academy of Sciences: Warsaw

Forman, R. T. T. (1995). *Land Mosaics: the Ecology of Landscapes and Regions.* Cambridge University Press: New York/Cambridge

Franz, H. (1979). *Ökologie der Hochgebirge.* Verlag Eugen Ulmer: Stuttgart

Fraser Darling, F. and Morton Boyd, J. (1969). *The Highlands and Islands.* Collins: London

Fritts, H. C. (1976). *Tree Rings and Climate.* Academic Press: London/New York

Fott, B. (1967). *Algae and Cyanophytes*. Academia: Praha [In Czech]

German MAB National Committee (1994). *Development concept: Bavarian Forest, Bohemian Forest, Mühlviertel*. German MAB National Committee: Bonn

Gloger, C. (1827). Ueber die auf dem Hochgebirge der Sudeten lebenden Säugethiere und die während des Sommers daselbst vorkommenden Vögel mit Angabe ihres Vorkommens nach Höhenbestimmungen, nebst einigen Bemerkungen über manche der neuen Arten von Brehm und das Erscheinen einiger seltenen Species in Schlesien. *Isis von Oken*, **20**: 566–609

Gorčakovskij, P. L. (1966). *Flora i rastitel 'nost' vysokogoriy Urala*. Akad. Nauk SSSR: Sverdlovsk

Groombridge, B. (ed.) (1992). *Global Biodiversity: Status of the Earth's Living Resources*. Chapman and Hall: London/Glasgow

Hadač, E. (1969). *Die Pflanzengesellschaften des Tales 'Dolina Siedmich prameòov' in der Belauer Tatra*. Vyd. Slov. Akad. Vied: Bratislava

Hall, J. B. and Swaine, M. D. (1981). *Distribution and Ecology of Vascular Plants in a Tropical Rain Forest*. Dr W. Junk: The Hague/Boston

Hamilton, L. S., Juvik, J. O. and Scatena, F. N. (eds.) (1996). *Tropical Montane Cloud Forests*. Springer Verlag: Berlin

Hedberg, O. (1995). *Features of Afroalpine Plant Ecology. Acta Phytogeographia Sulcica*, **49** p. 144. [Facsimile edition of original, 1964, published with new forward]. Ekblad and Co.: Vastervik, Sweden

Hess, H. E., Landolt, E. and Hirzel, R. (1967–1972). *Flora der Schweiz*, vol. 1–3, Basel/Stuttgart

Holtmeier, F.-K. (1981). What does the term 'krummholz' really mean? Observations with special reference to the Alps and the Colorado Front Range. *Mountain Research and Development*, **1**: 253–60

IUCN (1995). *Mountains of Central and Eastern Europe*. IUCN: Gland

Ives, J. D. and Barry, R. G. (eds.) (1974). *Arctic and Alpine Environments*. Methuen: London, p. 999

Jeník, J. (1955). Succession of Plants in the Alluvium of Belá river in the Tatras. *Acta Univ. Carol. Biol.*, **4**: 1–55 [In Czech]

Jeník, J. (1961). *Alpine Vegetation of the High Sudetes: Theory of Anemo-orographic Systems*. Nakladatelství Czech. Academy of Science: Praha [In Czech with German summary]

Jeník, J. (1983a). Microevolutionary scene of the Sudetic corries. *Biol. Listy*, Praha, **48**: 241–8 [In Czech]

Jeník, J. (1983b). Succession on the polonina balds in the Western Bieszczady, the Eastern Carpathians. *Tuexenia*, **3**: 207–16

Jeník, J. and Hampel, R. (1992). *Die waldfreien Kammlagen des Altvatergebirges*. Mähr.-Schles. Sudetengebirgsverein: Kirchheim/Teck

Jeník, J. and Lokvenc, T. (1962). Die alpine Waldgrenze im Krkonoše Genbirge. *Rozpravy Cs. Akad.Sci.*, **72**/1: 1–65

Jeník, J. and Price, M. F. (1994). *Biosphere Reserves on the Crossroads of Central Europe*. Empora: Praha

Jeník, J. and Soukupová, L. (1992). Microtopography of subalpine mires in the Krkonoše Mountains, the Sudetes. *Preslia*, **64**: 313–26

Kalin Arroyo, M. T., Squeo, F. A., Armesto, J. J. and Villagran, C. (1988). Effects of aridity on plant diversity in the northern Chilean Andes: Results of a natural experiment. *Annals of the Missouri Botanical Garden*, **75**: 55–78

Kingdon, J. (1989). *Island Africa: the Evolution of Africa's Rare Animals and Plants*. Princeton University Press: Princeton, NJ

Klötzli, F. (1976). Vielseitige Tropenvegetation. Vom Regenwald bis zum Dornbusch – vom Nebelwald bis zur Puna. In Dolder, W. (ed.), *Tropenwelt. Fauna und Flora zwischen den Wendekreisen*. Kümmerly and Frey: Bern, pp. 33–72

Klötzli, F. (1991). African Mountain Grasslands in their Global Context with an Overview on Puna as an Orobiome. In Ojany, F. F. *et al.* (eds.), *Proceedings of the International Workshop on Ecology and Socio-Economy of Mount Kenya Area*. UNESCO/IUBS: Paris, pp. 75–80

Kral, F. (1979). *Spät- und postglaziale Waldgeschichte der Alpen auf Grund der bisherigen Pollenanalyse*. Veröff. Institut für Waldbau, Universität für Bodenkultur: Wien

Křikavová, A. (1995). Some aspects of the Man–Nature relationships in the Islamic world. *Archív orientální*, **63**: 251–85

Landolt, E. (1983). Probleme der Höhenstufen in den Alpen. *Botanica Helvetica*, **93**: 255–68

Lauer, W. (1981). Ecoclimatological conditions of the Paramo belt in the tropical high mountains. *Mountain Research and Development*, **1**: 209–21

Longman, K. A. and Jeník, J. (1987). *Tropical Forest and its Environment*. (2nd ed.) Longman Scientific and Technical: London

Löve, A. and Löve, D. (1963). *North Atlantic Biota and their History*. Pergamon: Oxford

Löve, A. and Löve, D. (1966). *Cytotaxonomy of the Alpine Vascular Plants of Mount Washington*. University of Colorado Studies 24: Boulder

MacArthur, R. H. and Wilson, E. O. (1967). *The Theory of Island Biogeography*. Princeton University Press: Princeton

McConnell, J. and Conroy, J. W. H. (eds.) (1996). Environmental history of the Cairngorms. *Botanical Journal of Scotland*, **48**: 1–154

Mani, M. S. (1974). *Fundamentals of High Altitude Biology*. Oxford annd IBH Publishing Co.: New Delhi/Bombay

Mardon, D. (ed.) (1986). *Ben Lawers*. (3rd ed.). National Trust for Scotland: Edinburgh

Mayr, E. (1988). *Toward a New Philosophy of Biology*. Harvard University Press: Cambridge, MA

Miehe, G. (1990). *Langtang Himal: Flora und Vegetation als Klimazeiger und -zeugen im Himalaya*. J. Cramer, Dissertationes botanicae 158, Berlin/Stuttgart

Miyawaki, A. and Okuda, S. (1979) *Vegetation und Landschaft Japans*. The Yokohama Phytosociological Society: Yokohama

Morriset, P. and Payette, S. (1983). *Tree-line Ecology*. Collection Nordicana 47: Québec

Mucina, L., Grabherr, G., *et al.* (eds.) (1993). *Die Pflanzengesellschaften Österreichs*. 3 vols. G. Fischer: Jena/Stuttgart

Müller, P. (1981). *Arealsysteme und Biogeographie*. Verlag Eugen Ulmer: Stuttgart

Munasinghe, M. and McNeely, J. (eds.) (1994). *Protected Area Economics and Policy*. The World Bank: Washington DC

Naegeli-Oertle, R. (1986). *Die Berglandwirtschaft und Alpwirtschaft*. Schlussbericht zum Schweizerischen MAB-Programm 21: Bern

Odum, E. P. (1971). *Fundamentals of Ecology* (3rd ed.). W. B. Saunders: Philadelphia/London

Oldeman, R. A. A. (1990). *Forests: Elements of Silvology*. Springer Verlag: Berlin/Heidelberg

Ozenda, P. (1988). *Die Vegetation der Alpen im europäischen Gebirgsraum*. Gustav Fischer Verlag: Stuttgart

Plesník, P. (1971). *Horná hranica lesa*. Vydav. Sl.Akad.Vied: Bratislava

Prance, G. T. (ed.) (1982). *Biological Diversification in the Tropics*. Columbia University Press: New York

Price, M. F. (ed.) (1995). *Mountain Research in Europe*. UNESCO/Parthenon: Paris/Carnforth

Price, M. F. (1996). People in biosphere reserves: an evolving concept. *Society and Natural Resources*, **9**: 645–54

Pyšek, P., Prach, K., Rejmánek, M. and Wade, M. (1995). *Plant Invasions*. SPB Academic Publishing: Amsterdam

Richards, P. W. (1996). *The Tropical Rain Forest*, (2nd ed.). Cambridge University Press: Cambridge/New York

Richards, P. W., Tansley, A. G. and Watt, A. S. (1940). The recording of structure, life form and flora of tropical communities as a basis for their classification. *Journal of Ecology*, **28**: 224–39

Rodwell, J. S. (ed.) (1992). *British Plant Communities, vol.2: Grasslands and Montane Communities*. Cambridge University Press: Cambridge, New York

Rougerie, G. (1990). *Les montagnes dans la biosphere*. Armand Colin Éditeur: Paris

Rundel, P. W., Smith, A. P. and Meinzer, F. C. (eds.) (1994). *Tropical Alpine Environments: Plant Form and Function*. Cambridge University Press: New York

Sander, C., Eckstein, D., Kyncl, J. and Dobrý, J. (1995). The growth of spruce (*Picea abies*) in the Krkonoše Mountains as indicated by ring width and wood density. *Annales de Sciences Forestieres*, **52**: 401–10

Schnell, R. (1952). *Végétation et flore de la région montagneuse du Nimba*. Mémoires de l'Institut Français d'Afrique Noire, 22: Dakar

Schroeter, C. (1926). *Das Pflanzenleben der Alpen: eine Schilderung der Hochgebirgsflora.* Verlag A. Raustein: Zürich

Schwenckfeld, K. (1600). *Stirpium et fossilium Silesiae catalogus.* Lipsiae

Solbrig, O. T. (1991). *From Genes to Ecosystems: a Research Agenda for Biodiversity.* IUBS: Paris

Soukupová, L., Kociánová, M., Jeník, J. and Sekyra, J. (eds.) (1995). Arctic–alpine tundra in the Krkonoše, the Sudetes. *Opera corcontica*, **32**: 5–88

Steinhoff, H. W. and Ives, J. D. (1976). *Ecological Impact of Snowfall Augmentation in the San Juan Mountains of Colorado.* Colorado State University: Boulder

Stickney, P. L., Lloyd, W. S. and Wayne, T. S. (1994). *Annotated Bibliography of Publications on Watershed Management and Ecological Studies at Coweeta Hydrologic Laboratory, 1934–1994.* USA Department of Agriculture, Forest Service, SE Forest Experiment Station, Asheville, NC

Stone, P. B. (ed.) (1992). *The State of the World's Mountains. A Global Report.* Zed Books: London/NJ

Straka, H. (1996). Histoire de la végétation de Madagascar oriental dans les dernier millenia. In Laureno, W. R. (ed.), *Biogéographie de Madagascar.* ORS-TOM Éditions: Paris, pp. 37–47

Szafer, W. (ed.) (1962). *Tatrzanski Park Narodowy.* Polska Akademia Nauk: Kraków

Šourek, J. (1969). *Flora of the Krkonose Mountains.* Academia: Praha [In Czech]

Šrůtek, M. and Kolbek, J. (1994). Vegetation structure along the altitudinal gradient at the treeline on Mount Paektu, North Korea. *Ecological Research*, **9**: 303–10

The Mountain Institute (1995). *International NGO Consultation on the Mountain Agenda: Summary Report and Recommendations to the United Nations Commission on Sustainable Development.* The Mountain Institute: Franklin, WV

Tranquillini, W. (1979). *Physiological Ecology of the Alpine Timberline.* Ecological Studies 31. Springer Verlag: Berlin/Heidelberg

Troll, C. (1958). *Die Physiognomik der Tropengewächse.* Jahresber. Gesellschaft Freunde u. Förd. d. Univ. Bonn, 1958, pp. 1–75

Troll, C. (1959). Die tropischen Gebirge: Ihre dreidimensionale klimatische und pflanzengeographische Zonierung. *Bonner Geographischer Abhandlunge*, **25**: 169–204

Troll, C. (1968). The cordilleras of the tropical Americas: aspects of climatic, phytogeographical and agrarian ecology, In Troll, C. (ed.), *Geo-ecology of the Mountainous Regions of the Tropical Americas.* Ferd. Dümmlers Verlag: Bonn, pp. 15–56

Uhlig, H. (1978). Geoecological controls on high-altitude rice cultivation in the Himalayas and mountain regions of Southeast Asia. *Arctic and Alpine Research*, **10**: 519–29

Valk, E. J. (1981). *Late Holocene and Present Vegetation of the Kastelberg (Vosges, France).* State University of Utrecht: Utrecht

van Royen, P. (ed.) (1980). *The Alpine Flora of New Guinea*, vol. 1: General. J. Cramer: Vaduz

Vareschi, V. (1970). *Flora de los páramos de Venezuela.* University de los Andes: Merida

von Humboldt, A. and Bonpland, A. (1807). *Ideen zu einer Geographie der Pflanzen nebst einem Naturgemälde der Tropenländer.* F. G. Cotta: Tübingen

Walas, J. (1938). *Migration of Mountain Plants along the Tatra's Rivers.* Kom. Fizjogr. PAU: Kraków [In Polish]

Walsh, S.J., *et al.* (1994). Form and pattern in the alpine environment: an integrated approach to spatial analysis and modelling in Glacier National Park, USA. In Price, M. F. and Heywood, D. I. (eds.), *Mountain Environments and Geographic Information Systems.* Taylor and Francis: London, pp. 189–216

Walter, H. and Breckle, S.-W. (1983–91). *Ökologie der Erde*, vols. 1–4. G. Fischer: Stuttgart

Whittaker, R. H. (1977). Evolution of species diversity in land communities. *Evolution Biology*, **10**: 1–67

Wielgolaski, F. E. (ed.) (1975). *Fennoscandian Tundra Ecosystems, vols. 1 and 2.* Ecological Studies 16 and 17. Springer Verlag: Berlin/Heidelberg

WRI, IUCN and UNEP (1992). *Global Biodiversity Strategy.* World Resources Institute: USA

WWF and IUCN (1994). *Centres of Plant Diversity. A Guide and Strategy for their Conservation, vol. 1.* IUCN Publication Unit: Cambridge

Annex

Frank Klötzli

Biodiversity and vegetation belts in tropical and subtropical mountains

INTRODUCTION

With the highest number of vascular plants per unit area (Box 10.7 and Global Biodiversity map), tropical and subtropical mountains are globally unique with respect to biodiversity at the genetic, species, community, and ecosystem levels. Among the most important areas are the Northern and Central Andes, the Himalaya, and the mountain areas of Central America, Southeast Asia (Malaysia, Papua New Guinea), East Africa, and Cameroon.

The high biodiversity of these areas is due to extremely steep geoecological gradients. Within a few tens of kilometres, the geoecological belts may range from low elevation warm humid conditions to high elevation cold environments. On the highest tropical mountains, which reach altitudes of 6000–7000 m, almost all combinations of temperature and precipitation, as a function of altitude, are found over a short horizontal distance. Compared with the lowlands, such a broad variety of ecotopes would span from the warm tropical rainforest to the arctic tundra, with a horizontal distance of several thousands of kilometers. Superimposed on this large-scale pattern is a small-scale mosaic deriving from local topography, geology, and microclimates, further increasing the biodiversity.

Vegetation on tropical mountains may range from near sea level to 7000 m, thus encompassing a far larger variety of potential habitats than most mid- or high-latitude mountains. Many tropical mountains, particularly the volcanoes (e.g. Mount Cameroon, Kilimanjaro, Mount Kenya) and table mountains (e.g. Guiana, Ethiopia) are isolated. Thus, the upper vegetation belts, in particular, are sensitive centres of endemism. The higher elevation geoecological belts also include many harsh environments with extreme climatic conditions, e.g. large diurnal temperature changes, high number of frost cycles, high insolation rates. In response

various special physical structures and remarkable survival strategies have developed.

Changes in precipitation and temperature as a function of altitude explain much of the variance of plant communities and are, therefore, suitable for the classification of the vegetation of these mountains (Troll, 1959; Ellenberg, 1975; Klötzli, 1976, 1991; Lauer, 1981; Walter and Breckle, 1983–91). The following sections present the main types (Figure 10.5).

VEGETATION BELTS ON THE MOUNTAINS OF THE HUMID TROPICS

In the mountains of the humid tropics, precipitation and the moisture regime do not limit plant growth. The vegetation belts are exclusively controlled by decreasing temperatures with increasing altitude. These mountain areas are often related to relatively young landscapes. Especially in the upper vegetation belts, soils are often more fertile than the deeply weathered soils of the lowlands.

The montane tropical rain forest in the lowest zone (300–2000 m) is the most luxuriant form of tropical forest, with one or two tree strata. Ferns, palms, vines, and epiphytic orchids (bromeliads in the Neotropics) are abundant. Typically, 40–60 species plus 10–20 epiphytes occur on a relevé area of 500 m^2.

The upper montane rain forest (cloud forest, 2000–2800 m) is characterised by frequent mist and drizzle from clouds. Average annual temperatures are lower (around 16 °C), and frost may occur sporadically. Mixtures of broad-leaved genera from the temperate zone (*Prunus, Juglans, Ilex, Quercus*) and conifers (e.g. *Podocarpus*) are frequent. This forest is rich in vines and epiphytes, particularly mossy epiphytes in the crowns of the

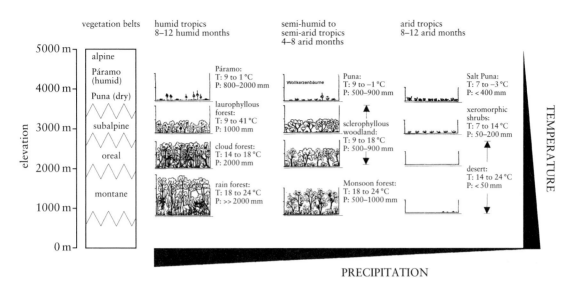

Figure 10.5 Vegetation belts in humid, semi-humid to semi-arid and arid tropical mountains. Note: the given ranges for annual mean temperatures (T) and annual precipitation (P) may vary significantly at local scales (after Ellenberg, 1975; Klötzli, 1976, 1991)

trees. Around 30 to 50 species (plus *c.* 10 epiphytes) per 500 m^2 relevé are typical. 'Elfin woodlands' are a special form of cloud forest found on exposed ridges and crests, where rainfall is very low and transpiration reduced. This leads to the typical life-forms of dwarfed trees and shrubs.

Evergreen, laurophyllous (or laurineous) subalpine forests are found at 2800–3500 m, but much lower in isolated situations (e.g. 1700 m on Marsabit, Kenya). Annual precipitation is about 1000 mm, mean annual temperatures are around 12 °C, and frost is not unusual. Ericaceous, laurineous, and sclerophyllous trees and ferns are abundant, and the shrub layer is usually well developed. About 10 tree species, 3–5 epiphytes, and 30–40 plant species are typically found on relevés.

The tropical alpine (páramo in the northern Andes, puna in the arid central Andes) ranges from 3800 to 5000 m, where annual mean temperatures are still around 1 °C. Cushion plants may be found even higher. This belt is dominated by large numbers of frost cycles and high diurnal temperature amplitudes. Precipitation may be more than 1200 mm/yr. On cloud-free days, insolation may reach values close to the solar constant. Specific plant life-forms, physical structures, and modes of metabolism are the result of such harsh environmental conditions. The vegetation is dominated by tussock grasses (*Andropogon, Calamagrostis,*

Stipa, Danthonia), giant groundsels (e.g. *Senecio* in Africa), giant rosette plants, candle-like *Lobelia* and *Puya*, cushion plants (e.g. *Azorella* in the Andes) and many shrubs. Wetlands and bogs with *Sphagnum, Carex,* and *Juncaceae* are often found in depressions and on gentle slopes. The diversity of the páramo grassland typically reaches 30–40 species on relevés.

VEGETATION BELTS ON THE MOUNTAINS OF THE SEMI-HUMID TO SEMI-ARID TROPICS

The vegetation belts on the mountains of the semi-humid to semi-arid tropics are largely controlled by both temperature and precipitation. These areas are characterised by one or two rainy seasons and four to eight humid months. Many of the areas are within the monsoon circulation.

The montane semi-arid forest on the foothills of the mountains (800–2000 m) is less diverse than the montane rain forest; quite frequently, there is one tree stratum with a single prevailing tree species. Both evergreen and deciduous species are found. Typical life-forms, which are partly adaptations to forest fires, are bottle trees (*Bombacaceae*), umbrellas (*Acacia*), columnar and succulent *Cactaceae* and *Euphorbiaceae*, caulescent rosettes (e.g. Agarvaceae, *Yucca, Aloë*-types), fruit trees, and some bamboos. While epiphytes are less frequent, a grassy herb layer with mixtures of *Cyperaceae, Liliaceae, Bromeliaceae,* and ferns are characteristic.

Sclerophyllous mountain forests (2000–3500 m) are mostly found in the transition zone between the semi-arid montane forest and alpine grasslands. Annual mean temperatures are between 16 °C and 10 °C, frost may occur. Shrubs and coniferous trees are typical. Woody sclerophyllous types (e.g. *Ericaceae, Myrtaceae*) are dominant and well-adapted to the harsh environment with fires, frost, and droughts. Other adaptive, but less competitive strategies include malacophyllous species (e.g. *Thymus, Satureja*) which may shed their leaves during long drought periods.

Semi-arid alpine grasslands (puna in the Central Andes) are found at 3500–5000 m. Thorny, partly or evergreen shrubs, succulents and, particularly, sclerophyllous grasses are dominant. Many plant species have hairy or reflective leaves which protect them from the intense solar radiation. The *Polylepis* trees in the Central Andes are a unique feature of global importance, they can be found at altitudes of 4000 m in the humid tropics, and 5000 m on the Bolivian volcano, Sajama.

VEGETATION BELTS ON THE MOUNTAINS OF THE ARID TROPICS

On the mountains of the arid tropics, with 10 to 12 arid months, plant growth and vegetation belts are largely controlled by very limited precipitation and moisture availability. In these areas, mountains often provide favourable conditions compared with the adjacent lowlands, because precipitation increases with elevation. For instance, on the western slope of the Central Andes at 18°S, in the Atacama desert, vegetation cover and the number of plant species are highest at 3500–4000 m (45 percent cover, 140 species), while almost no plants are found at altitudes below 1000 m (Kalin Arroyo *et al.*, 1988). This pattern is similar to arid mid-latitude mountains, but fundamentally different from mountains in the humid tropics and mid- and high latitudes.

Xeromorphic shrubs, tussock grasses, and cushion plants are the dominant life-forms. Many plants are confined to specific sites (e.g. rocks, protected microsites), showing that microclimate plays a key role in this extremely harsh and, at high elevations, cold and dry environment.

Box 10.7 Mountains – The focal point of global biodiversity (see loose leaf map)

The Biodiversity map shows, for each 10 000 km² area of the continents, the number of vascular plant species, derived from regional vegetation data. Despite this relatively rough simplification, the pattern shows important spatial differences, including both centres of extremely high species diversity, and areas with very few plant species. The regions with the highest diversity are all in the inner and outer tropical climatic zones. Secondary maxima appear in the narrow belt of the warm humid mid-latitudes (mediterranean regions). The minima are in the subtropical and polar deserts.

The distribution of species number is not simply a function of the latitudinal geo-ecological gradients, but is strongly differentiated within individual zones; coastal regions around the warm tropical oceans with landward-directed airflows are among the areas of highest species diversity. World-wide, there are six different centres with maximum species diversity: 1) Chocó in Costa Rica; 2) the tropical eastern Andes; 3) the Atlantic coast of Brazil; 4) the east-Himalayan–Yunnan region; 5) North Borneo; 6) Papua New Guinea.

The outstanding influence of mountains stems from their differences from the surrounding lowlands not only in topographic and climatic, but also biological aspects. Geodiversity *and* biodiversity are highly susceptible to zonal modification: for instance, the main regions with highest species numbers per area are the windward slope of the Andes, the Central American cordillera, the mountains of Guyana, the east and southeast African mountains, the central and eastern Himalaya, and the highlands of the Malayan archipelago. However, subtropical arid mountains are also important centres of biodiversity, though with lower species number: the Rocky Mountains, the mountains of the central Sahara, and especially the mountains of middle and central Asia, such as the Pamir. The south-central Andes show the interesting feature of the 'dry diagonal', an arid corridor with extremely low species diversity which separates the relative biodiversity maxima of the northern (tropical) and the southern (extratropical) Andes.

In contrast to the low latitude mountains, those of the central and northern mid-latitudes exhibit low biodiversity relative to the surrounding areas. Whereas this effect is observed to some extent in the Alps, the reduction of species numbers is pronounced in the northern Urals, the central and east

Siberian mountains, and Alaska. While the map does not support this supposition in some of these areas, this derives from the poor database and the coarse resolution.

The zonal and vertical arrangements of plant species in mountains are controlled primarily by climatic conditions, and secondarily by soils and geomorphology. In the sub-humid and semi-arid tropics and subtropics, mountains have significantly higher precipitation than the surrounding lowlands. The favourable hygric-thermal conditions, combined with a mosaic of ecological settings, lead to a large number of plant species, particularly in the forest zone, up to about 3500 m. Higher up, biodiversity decreases again. This suggests that massive deforestation of the semi-arid and arid tropics and sub-tropics may be seriously detrimental to their ecological diversity. Northern mid-latitude and sub-arctic mountains, on the other hand, are climatically rather unfavourable in comparison to the neighbouring lowlands, as thermal limitations significantly limit plant growth, leading to reduced biodiversity.

Although the map – the first differentiated map at this scale – considers only numbers of vascular plant species, it has great ecological significance. In an ecosystem with consumers and decomposers, the number of species is positively correlated with total terrestrial biodiversity. Thus, as the global number of vascular plant species is largely known, even if uncertainty remains concerning total species diversity, this map may be regarded as analogous to a map of overall terrestrial biodiversity. However, as it is based on strictly quantitative criteria, one important aspect is not evident. This is, that mountains are refuges, with high numbers of endemic plant species.

The importance of mountains with regard to species numbers and quality (endemism) is clear. Mountains are core areas of ecological diversity in many parts of the world; yet they belong to the most threatened and vulnerable ecological 'islands'. The protection of these areas with appropriate strategies, including interlinked 'corridors' and connections with the forelands, must be given the highest priority.

Source: Matthias Winiger, Institute of Geography, University of Bonn, Germany

Protection of nature in mountain regions

11

Jim Thorsell

INTRODUCTION

In times past, before humankind had exploited the planet beyond a certain threshold, nature could defend and replenish itself. This is no longer possible, even in the remote mountain regions of the world. It is now an ethical responsibility to protect nature and over the past century, virtually every country in the world has attempted to preserve selected portions of their landscapes and species.

For reasons clearly outlined in Mountain Agenda's 1992 *The State of the World's Mountains* (Stone, 1992), there are various reasons why alpine areas have received special attention by governments when national parks and nature reserves are established. Some of these are listed in Box 11.1. In fact, of the 785 million ha of protected areas in the world today, 260 million ha, almost a third, are in mountains. This is an area equivalent to the State of Alaska and the Province of British Columbia combined.

In conducting this overview of the state of the world's mountain protected area system, five key questions are addressed: (1) How many mountain protected areas exist? (2) Where are they located?

Box 11.1 Why mountains need protection

Conservation of mountain regions requires a number of approaches, one of which is the establishment of protected areas. The rationale for parks and reserves in mountains is especially convincing and includes one or more of the following:

(1) Mountains often harbour many endemic and threatened species, genetic resources, and are nature's last strongholds for those species that have been extirpated in adjacent lowlands (e.g. Sangay National Park in Ecuador and Virunga National Park in Zaire).

(2) Mountains have immense downstream values in terms of soil erosion control and watershed protection. Nature reserves are a useful measure in stabilising upland resource use (e.g. Lore Lindu National Park in Indonesia and Kasungu National Park in Malawi).

(3) Mountains are fragile high energy environments where regulatory controls over potentially disturbing human activities are often needed (e.g. Simien and Bale Mountains National Parks in Ethiopia and Nanda Devi National Park in India).

(4) Mountain regions often harbour a wealth of human tradition and protected areas can pro-

vide a mechanism whereby the alliance between conservation and local cultures can be strengthened (e.g. Bwindi Impenetrable National Park in Uganda and Manu National Park in Peru).

(5) Mountains act as focal points for those seeking aesthetic and recreational benefits and many cultures have a reverence for certain peaks considered 'sacred' (e.g. Huangshan Scenic Natural Area in China and Yosemite National Park, USA).

(6) Mountain ranges in many places act to form the frontiers between countries and there is often value in maintaining them as transboundary protected areas or lightly inhabited buffer zones (e.g. Sajama and Lauca National Park of Bolivia and Chile and Mont Nimba Nature Reserves in Guinea and Côte d'Ivoire).

(7) Mountains are particularly sensitive indicators of global climate change and are ideal settings for research on the impact of global change on species, ecosystems, and hydrology (e.g. Garibaldi Provincial Park in Canada and the Pechoro-Ilychsky Reserve in Russia).

Source: Jim Thorsell, Natural Heritage Programme, IUCN, Gland, Switzerland

(3) How well do they represent the features of each range? (4) Where are the gaps in coverage? and (5) What are the current challenges facing mountain park managers and what actions are needed?

INVENTORY OF MOUNTAIN PARKS

Working with data bases requires that certain arbitrary definitions need to be taken. For mountain parks we used three parameters in compiling a global list. First, the protected area had to have a minimum size of 10 000 ha. Second, it had to display a minimal relative relief of 1500 m within its boundaries. Third, it had to be classified in the 1993 United Nations List of National Parks and Protected Areas as serving primarily a nature conservation function (IUCN Categories i–iv).

When these filters were applied, the data base at the World Conservation Monitoring Centre in Cambridge, UK produced a list of 473 sites totalling 264 million ha (Table 11.1). This list precludes many hundreds of other parks and reserves that may locally be considered mountains and have steep slopes but do not meet one of the above three criteria.

All of the Earth's biogeographic realms (classified using Udvardy's (1975) scheme) contain mountain protected areas. A total of 65 countries are involved.

The Nearctic biogeographic realm has by far the greatest area in mountain parks and this is due to just one site – the 97 million ha Greenland National Park. Even when this one park is subtracted, however, the Nearctic still contains the larger area of mountain parks than any other realm.

Further analysis of the data provides more detailed information on which biogeographic provinces in each of the realms contain mountain

Table 11.1 Mountain Parks

Biogeographical realm (Udvardy classification)	Number	Total area (ha)
Afrotropical	42	20 427 439
Antarctic	15	3 232 582
Australian	3	2 649 148
Indomalayan	42	7 204 043
Nearctic	96	153 804 175
Neotropical	103	34 454 473
Oceanian	8	3 643 048
Palaearctic	164	39 090 448
Total	473	264 505 356

protected areas. An earlier version of the list has been published by Thorsell and Harrison (1992).

INTERNATIONAL DESIGNATIONS

Selected mountain parks are given added prominence through their participation in UNESCO's Biosphere Reserve Programme, inscription on the World Heritage List, or because they are in a trans-frontier location.

Under the framework of the World Heritage Convention (which 145 countries have signed), five individual mountains and thirty-one mountain ranges have been inscribed on the prestigious World Heritage List (see Table 11.2). A further

Table 11.2 Mountain World Heritage sites

Name	Country
Los Glaciares National Park	Argentina
Tasmanian Wilderness	Australia
Pirin National Park	Bulgaria
Rocky Mountain Parks (7)	Canada
St Elias Parks (5)	Canada/USA
Waterton/Glacier International Peace Park	Canada/USA
Huangshan	China
Huanglong	China
Jiuzhaigou	China
Talamanca/Amistad	Panama/Costa Rica
Galapagos National Park	Ecuador
Sangay National Park	Ecuador
Simien National Park	Ethiopia
Nanda Devi National Park	India
Shirakami	Japan
Sagarmatha National Park	Nepal
Te Wahi Pounamu/ SW New Zealand	New Zealand
Tongariro National Park	New Zealand
Aïr Ténéré	Niger
Rio Abiseo National Park	Peru
Huascaran National Park	Peru
Manu National Park	Peru
Virgin Komi Forests	Russia
Kilimanjaro National Park	Tanzania
Rwenzori National Park	Uganda
Yosemite National Park	USA
Hawai'i Volcanoes National Park	USA
Great Smoky Mountains	USA
Olympic National Park	USA
Yellowstone National Park	USA
Grand Canyon National Park	USA
Canaima National Park	Venezuela
Virunga National Park	Zaire
Kahuzi-Biega National Park	Zaire

eight mountain sites have been proposed for consideration in 1996. Many of the mountain parks on the World Heritage List have received substantial benefits such as increased funding and strengthened protection.

The Man and the Biosphere (MAB) Programme of UNESCO has recognised 67 mountain biosphere reserves in 27 countries (see Box 11.2) according to the definition provided on page 237 of this chapter. Biosphere reserves have a particular applicability to mountain environments that have resident human populations and that would be suitable sites for training and research. Even if an area is not officially a Biosphere Reserve, IUCN advocates the biosphere reserve approach to management for all protected areas. This is based on the belief that greater protection for a core zone is usually achieved if an outer buffer zone is established where certain controlled uses are also considered by the park manager.

A third distinction of some mountain parks is their location as a trans-frontier or border park. As the perimeters of many nations often follow the crest line of mountain ranges, adjoining parks can serve to lower border tensions and promote international co-operation in many ways. Examples include joint fire and pest control operations, research co-ordination, and shared education programmes. Management arrangements between neighbouring countries or states, as in the case of Australia, USA and Canada, can also be beneficial to effective conservation. Table 11.3 lists many of the mountain border park 'pairs' between neighbouring countries; there are many others in lower mountain areas, such as Krkonose (Czech Republic) and Karkonosze (Poland), Sumava (Czech

Box 11.2 Coupling conservation with development and science – the biosphere reserve approach

Conservation is often at stake when local populations surrounding protected areas feel that they do not benefit from conservation; if they want to exploit rather than conserve biological and mineral resources. Biosphere reserves try to combine conservation with sustainable use of natural resources. Assisted by research on human–environment interactions and ecosystem functioning, they wish to demonstrate that conservation is best practised if people are involved in area management and decision-making processes so that they protect 'their own' environment. In this, biosphere reserves go beyond the classical protection concept and promote a wider spatial and conceptual approach.

Many biosphere reserves are in mountain areas where they can act as 'hubs' for conservation, research and development. The Southern Appalachian Biosphere Reserve in the USA covers part of six federal states. Under the auspices of the Southern Appalachian Man and the Biosphere (SAMAB) Programme, research projects include the reintroduction of the red wolf into the Great Smoky Mountains National Park, and a habitat assessment for neo-tropical migratory birds. The biosphere reserve co-operative is successfully promoting public awareness through development of educational material for schools and public education programmes.

In the Manu Biosphere Reserve (reaching 4000 m) in Peru, rural development in the buffer zone around the protected areas is considered essential for the viability of conservation measures. Known as the Manu Project and led by the Peruvian

Foundation for the Conservation of Nature since 1989, emphasis has been on the development of sustainable agricultural systems, and health and education services for settlers (mainly Quechuas from the highlands) and indigenous people (Amazonian) living in the buffers zone. The development of ecologically sustainable agriculture was introduced to use the available resources more intensively in a small area and to reduce the pressure on the national park which is the biosphere reserve's core zone.

In southwestern China, the Xishuangbanna Biosphere Reserve with its highest peak at 2429 m is often called the 'kingdom of biodiversity'. With more than 200 species of butterflies throughout the forests of Xishuangbanna, butterfly farming has become an important economic asset and an option for sustainable development for, and species conservation by, local people. Ethnobotany as a discipline in China originated in Xishuangbanna: plant ceremonies, flower-eating, use of plants for communication, and knowledge of herbs by the reserve's minorities (especially the Dai people) are currently studied.

In Eastern and Central Europe transborder collaboration in mountain biosphere reserves on species inventorying and management issues has become one of the main foci in recent years. Biosphere reserves were initiated under UNESCO's intergovernmental Man and the Biosphere (MAB) Programme in the early 1970s and form a network for international collaboration. Presently, there are over 330 sites in 85 countries.

Source: Thomas Schaaf, UNESCO–MAB

Table 11.3 Mountain transfrontier reserves (border parks)

North America	
Wrangel–St Elias/Glacier Bay (USA)	Kluane/Tatshenshini (Canada)
Glacier (USA)	Waterton Lakes (Canada)
Cathedral/Manning/Skagit/Cascade (Canada)	Pasayten, N Cascade (USA)
Arctic Wildlife Refuge (USA)	N Yukon (Canada)

Europe	
Tatrzanski (Poland)	High Tatra (Slovakia)
Pyrenees Occidentales (France)	Ordessa (Spain)
Vanoise (France)	Gran Paradiso (Italy)
Swiss (Switzerland)	Stelvio (Italy)
Sarek, Padjelanta, Stora, Sjöfallet (Sweden)	Rago (Norway)
Berchtesgaden (Germany)	Various sites in Austria

Asia	
Manas (India)	Manas (Bhutan)
Khunjerab (Pakistan)	Taxkorgan (China)
Sagarmatha (Nepal)	Quomolangma (China)

Africa	
Volcanoes (Rwanda)	Gorilla (Uganda)
Virunga (Zaire)	Queen Elizabeth (Uganda)
Nyika (Malawi)	Nyika (Zambia)
Gebel Elba (Egypt)	Proposed Gebel Elba (Sudan)

Latin America	
La Amistad (Costa Rica)	La Amistad (Panama)
La Neblina (Venezuela)	Pico da Neblina (Brazil)
Puyehue and Vincente, Perez Rosales (Chile)	Lanin and Nahuel Huapi (Argentina)
Bernardo O'Higgins and Torres del Paine (Chile)	Los Glaciares (Argentina)
Sajama (Bolivia)	Lauca (Chile)
Los Katios (Colombia)	Darien (Panama)

Australia
Australian Alps National Parks (NSW, SA, Victoria, ACT)

Source: Thorsell, 1990

Republic) and the Bavarian Forest (Germany), Pieninsky (Slovakia) and Pieniny (Poland), and Mercantour (France) and the Maritime Alps (Italy), often with long-established programmes of co-operation. A publication on transboundary co-operation in mountain protected areas has been produced from a recent meeting on the issue held in the Australian Alps in 1995 (Hamilton *et al.*, 1996).

ADEQUACY OF THE MOUNTAIN PROTECTED AREA SYSTEM

Even through there is already an extensive network of mountain parks throughout the world, the coverage in some areas is minimal and there are some 'missing links' in the system. It is not easy to provide a simple percentage figure that

would indicate the proportion of the earth's high mountains that have been given legal protected status. This would require an estimate of the total area of the different high mountain ranges with allowances for ice caps, dissected plateaux and certain oceanic islands. Nevertheless, estimates have been made for the Andes (Thorsell and Paine, 1995) where a substantial 11.4 percent of the total range is under protected status. In Africa a similar estimate of 4 percent has been made (Thorsell, 1993).

The figures in Table 11.1 on the extent of coverage by biogeographic realm do indicate that some mountain areas are much better protected than others. Certainly, a disproportionately high number of parks and reserves are found in the mountain biome compared to other biomes, such as temperate grasslands or lake systems.

The question of how adequate and/or representative is the existing mountain protected area system is thus open-ended. IUCN, UNEP, and WWF have often promoted that 10–12 percent of each habitat or biome within each nation should be under some form of protected status. Within this crude target it must be recognised, that for reasons of biodiversity, any system should be biased in favour of areas with high species richness. This has been particularly well demonstrated by Jonathan Kingdon (1990) in his book, *Island Africa*. But biodiversity is only one measure of 'conservation value' and we must take account of vulnerability, attractiveness, and fragility, all of which are major factors in mountain areas.

Another problem with percentage targets is that they do not necessarily translate into an effective conservation system. Parks may be in a sub-optimal location, receive inadequate funding, or be poorly integrated into the regional landscape. Many were established for reasons of spectacular scenery and exclude areas of biological importance below the treeline. An understanding of all these factors is thus necessary when making assessments of adequacy.

A project initiated by the IUCN Mountain Protected Area Task Force is to identify specific locations in mountain regions that merit investigation as potential new protected areas and where there are design flaws in the boundaries of existing mountain parks. Many mountain parks suffer from 'hole in the doughnut' design where only the higher slopes of the mountain are contained within the boundaries. Classic examples are the national parks of Vanoise (France), Cotopaxi (Ecuador), and Mount Kenya (Kenya). Powell and Bjork (1990) and Stuart *et al.* (1993), moreover, have found seasonal altitudinal movements of mountain forest bird communities in Central America and Africa that depend on existence of often unprotected lower elevation forests. Critical habitat outside park boundaries for migratory species is, of course, a problem common to all biomes, not just mountains.

A preliminary list of prospective new mountain parks and extensions to existing ones is given in Table 11.4. This is certainly not an exhaustive list

Table 11.4 Potential new (or extensions) mountain protected areas

Area	Country	Main natural values
Kokoxili	Qinghai, China	rare species
Arksai	China	snow leopard
Liqiara	China	snow leopard
Arba	China	snow leopard
Gunzi	China	snow leopard
Baoxin	China	snow leopard
Trus Madi	Sabah, Malaysia	biodiversity
W Pamirs	Tajikistan	scenery, glacial processes
K2 – North face	China	scenery, glacial processes
Rinjani	Lombok, Indonesia	flora
Imatong	Sudan	flora, watershed
Atlas	Morocco	flora, watershed
Aconguija	Argentina	endemic species
Sangay extension	Ecuador	wildlife, flora
Isibora Sécure	Bolivia	biodiversity
Blue Mountains	Jamaica	flora
Ojos del Salado	Argentina	endemism, scenery
Mont Blanc	France/Italy/Switzerland	scenery
Tysfjord/Hellemobotn	Norway	scenery, geology
Klamath/Sikiyou	Oregon, USA	biodiversity
S Chilcotin	BC, Canada	wildlife, landscape
Waddington	BC, Canada	scenery, flora
Cummins	BC, Canada	wildlife, scenery
N Rockies	BC, Canada	wildlife, scenery
N Baffin Island	NWT, Canada	landscape, wildlife
Moehau	New Zealand	kauri forest, cultural values
Kaikoura	New Zealand	geology, flora

but is indicative of the work that still needs to be done to 'complete' the system.

MOVING AHEAD: ISSUES AND ACTIONS

What then can be done to promote more and better managed mountain protected areas? Mountain park managers everywhere are facing difficult challenges in protecting what has become an archipelago of highlands surrounded by an expanding sea of human activity. Although the problems have been well studied, solutions are harder to come by. One source of constructive ideas is a recent publication by IUCN of *Guidelines for*

Mountain Protected Areas (Poore, 1992). From this publication five key areas for attention over the next decade stand out:

(1) Completing the system

Despite the large area already protected, there are a number of additional sites and extensions to existing mountain parks that have been proposed. Existing reserves do not cover sufficient area to guarantee biodiversity conservation and many important sites have no form of conservation regime in place. More mountain areas need to be nominated for inscription on the World Heritage List. Innovative approaches, such as the Andes Corridor Initiative being undertaken by the South American

Box 11.3   A mine or a park? Economics and ethics on the Tatshenshini

In the furthest northwestern corner of British Colombia on the border with the Yukon and Alaska is a remote 10 000 km² mountain wilderness known as the Tatshenshini. In the early 1990s the area became the scene of a major land-use controversy: either develop a major open-pit copper mine or protect the area as part of the adjoining St Elias

Mountains World Heritage site. There was no room for compromise; it was a black and white decision as both a mine and a park could not be accommodated.

Developing the proposed mine in such a remote, mountainous, and earthquake-prone valley, such as the Tatshenshini, would have been a major

Plate 11.1   Campsite at the confluence of the Tatshenshini-Alsek Rivers, St Elias Mountains World Heritage site, British Columbia, Canada (Photograph: J. Thorsell/IUCN)

engineering challenge. Storage of 225 millions tons of acid-generating waste rock would be required. So would a new 100 km road up the Tatshenshini River and a 250 km pipeline to ship copper and slurry effluent to Haines, Alaska.

Many construction jobs would be created and an estimated US$ 6.5 billion worth of minerals could be extracted over 20 years. The environmental risks, however, were enormous and a downstream fishery worth an estimated US$ 82 million annually would be endangered.

Other factors in the equation were the interests of the local First Nations people who seasonally use part of the area for hunting and fishing. Recreational benefits were also studied and estimated to amount to Can$ 2.2 million in annual direct revenues from river rafting as well as Can$ 11.4 million in non-use preservation values.

Complicating the decision even further were four trans-boundary international legal agreements between Canada and the USA that teams of lawyers from all sides were lining up to investigate.

In the end the Government of British Colombia weighted the various factors and in June, 1993 cancelled the mine proposal, created a provincial wilderness park, and nominated the area for World Heritage status. The mining company was eventually provided with Can$ 26 million in compensation. The policy 'fallout', however, is still being debated by the various factions. The Tatshenshini thus served as both a natural and social watershed in demonstrating the North American public's desire for protecting large, wild, natural areas.

Source: Kevin McNamee (1994), Canadian Nature Federation, Canadian Parks and Wilderness Society

members of IUCN and the Yellowstone to Yukon Biodiversity Corridor should be encouraged. Such bioregional approaches are being increasingly promoted by conservation biologists in order to tackle the larger issues of climatic change and biodiversity loss.

(2) Co-operation

Fostering regional co-operation in mountain protected area management can be done by promoting border parks, twinning programmes, staff exchanges, and regional workshops. The Alpine Convention in Europe with its provision for a Protocol on Protected Areas is another mechanism. The existing regional mountain centres, such as ICIMOD, should ensure that the role of protected areas is recognised in their programmes. The coalition of groups coming together under 'The Mountain Forum', along with its electronic networking initiative will also serve a useful function in stimulating mountain conservation (The Mountain Institute, 1995). All of these initiatives are assisting in placing mountain protected areas into the broader framework of the regional landscape where they are not seen as 'islands' but as essential components of a healthy environment.

(3) Science

The application of science to mountain protected areas management needs to be strengthened. As UNEP (1980), IMS founded in 1981, UNU, IGU Commission on Mountains, and others have emphasised, mountains are not nearly as well-researched as they deserve to be. Many do not even

have basic species inventories or vegetation maps. The tropical Andes, as one of the world's twelve major crop gene centres, merit special attention. Alliances between mountain protected area managers and scientists need to be improved with the objectives of making managers more science-oriented and scientists more management-oriented (see Harmon, 1994) as is attempted in biosphere reserves.

(4) Social issues

Social issues in mountain protected area management need to be better addressed. Many mountain protected areas have resident human populations and some are used on a seasonal basis by pastoralists and tourists. Conflict resolution to deal with illegal activities is a new required skill of mountain protected area managers. The illegal growing of narcotics is one such example of a social problem within many mountain protected areas. People, as well as peaks and parks are usually inseparable and the participatory approach to planning has become fundamental. Conservation of cultural diversity also goes hand in hand with conservation of biodiversity as McNeely (1995) noted at the recent Mountain Agenda meeting in Lima.

(5) Nature tourism

World-wide there has been a strong growth in the tourism/recreation function of high mountains. This is reflected in the intensive use being made by overseas and local visitors of such parks as Nahuel Huapi and Machu Picchu in the Andes, Yosemite and Banff in North America, Hohe Tauern in the Alps, Huangshan in China, and many others. Mountain protected area managers need to recognise both the potential benefits and adverse impacts

of this functional shift in the human use of mountains and to channel it in a direction that will lead to strengthened protection of the world's mountain heritage.

Box 11.4   Planning a future for Kilimanjaro

Very few mountains are so widely known as Kilimanjaro, the highest mountain in Africa (5895 m). Over 10 000 climbers as well as 20 000 porters and guides spend a week attempting to reach the summit every year. Pressures on the park, not only from the growing numbers of visitors, but from the impacts of surrounding land-use activities, have been accelerating. The Tanzanian Parks Authorities thus initiated a planning process on the adage that 'a park that plans no future has no future'.

As one of Tanzania's first park management plans, the approach used for Kilimanjaro has set an example for others. The preparation of the plan was also seen as a training exercise, with IUCN supplying a team leader funded by the Swedish International Development Agency.

The first step in the process was to assemble an interdisciplinary team of park planners, managers, scientists, and experts. Several interactive workshops were then held to conduct the following:

(1)   identify the issues facing the park;
(2)   define the management objectives;
(3)   design a zoning scheme;
(4)   determine limits of acceptable use;
(5)   outline specific activities required in each zone; and
(6)   analyse the environmental impacts of future developments in the park.

The Plan also underlined the importance of needed boundary adjustments to take in a larger area of the lower montane forest belt and to protect wildlife corridors. The park agency's Community Conservation Service had a special role to play in the planning process in terms of developing a dialogue with the communities surrounding the park and promoting a benefit sharing scheme.

With a formally approved plan published in late-1993, implementation has now begun. Through the

Plate 11.2   Kilimanjaro National Park, Tanzania (Photograph: B. Messerli)

World Heritage Fund, for example, a new entrance gate and information facility has been provided. A model for mountain park management in Africa could be the end product.

Source: Tanzania National Park Authority (1995)
Kilimanjaro National Park General
Management Plan, p. 188

Box 11.5   Peaks, parks, people: participating to protect K2

One of the most exciting new national parks in the world was created by the Government of Pakistan in 1993. The new 3000 km² park will protect an outstanding natural landscape in the Central

Karakoram in Baltistan. The area is heavily glaciated with some of the longest glaciers outside the polar regions. Although K2, the world's second highest mountain, is the centrepiece of the new park, there

Plate 11.3 Hiking near Masherbrum La, Central Karakoram National Park, Pakistan (Photograph: J. Thorsell/IUCN)

are three other peaks over 8000 m and 20 over 7000 m. Remnant wildlife such as urial, ibex, markhor, are still found.

The high mountains and narrow valleys of the Central Karakoram kept the area physically isolated and it remained culturally separate until quite recently. Improved communication services, particularly the Karakoram Highway, are bringing rapid socio-economic changes. An annual influx of 14 000 tourists is rapidly transforming the local economy as well as causing undesirable impact on the fragile upper valleys.

Although there will be no permanent human settlements within the Central Karakoram National Park, there are some 15 000 residents in villages surrounding it. Local people will retain summer grazing and property rights in portions of the park.

The approach being followed in the planning of the Central Karakoram National Park is to consider the local Baltis as an integral part of the ecosystem, along with the flora, fauna, and the peaks. Developing the national park will thus mean developing a regime that respects traditional culture and rights. The initial planning team held a round of meetings in the local villages to obtain their views on boundaries and management of the area, and a two-day workshop in Skardu also incorporated their concerns.

Since then a steering committee has been formed, with local government providing the lead, but also involving conservation NGOs and community leaders. The guiding philosophy is that ensuring ecosystem integrity requires that government, visitors, villagers, and NGOs must all work together to recognise the intrinsic values of the central Karakoram and live in harmony with it.

Source: Fuller S. and Gemin, M. (1995),
IUCN, Pakistan

Box 11.6 The corridor of the Americas concept

Conserving nature in mountain regions is not a simple task. This is particularly the case in the Andes. Here, poverty, civil disturbances, population growth, land tenure issues, and often, the lack of political will have been obstacles to the establishment of a viable network of nature reserves through this

Plate 11.4 Domestic llamas and alpacas grazing in the Sajama National Park, Bolivia, a transfrontier park with Lauca in Chile and a link in the 'Corridor of the Americas' (Photograph: J. Thorsell/IUCN)

biodiverse region. Nevertheless, the framework of a protected area system does exist. Moreover, there are three ambitious regional initiatives that could combine to create a continuous 15 000 km corridor from Cape Horn through the Andes and on through North America to the Bering Strait.

The most advanced portion of what could ultimately become a 'Corridor of the Americas' is the 'Paseo Pantera' (Path of the Panther) in Central America. Here, a 1000 km long biological corridor has been proposed, based on mapping of existing and proposed reserves through the rugged Darien region on the Colombian border, north through the Talamanca ranges and Central cordillera, to the Mexican border. Using GIS techniques and gap analysis, a range of variables, such as native people interests, have been analysed and a data bank has been prepared. Peace dividends, conservation benefits, cultural heritage interest, and indigenous rights are blended and all agendas are combining to promote the concept.

For the Andes, the concept of a continuous chain of protected areas extending throughout their length has been the subject of several regional meetings of IUCN members since 1990. The thrust of the 'Cordillera de los Andes' project is to incorporate the interest of mountain protection within the wider context of sustainable development of the region. In the Canadian and American Rockies, the Yellowstone to Yukon Biodiversity Strategy is being actively promoted by various NGOs to protect the wild heart of North America.

These three bioregional approaches have not yet been linked but a map of a green corridor through the mountains of the Western Hemisphere is emerging. Of course there will be gaps, particularly in Mexico, but such a corridor is already partly in place. To make the Corridor of the Americas a reality what is needed is a vision as bold as the mountains it is designed to protect.

Source: Jim Thorsell (1996), National Heritage Programme, IUCN, Gland, Switzerland

Box 11.7   Protected areas around Mount Everest

As the highest, and one of the most remote points on earth, Mt Everest has long been revered by local inhabitants as a sacred protector of the life which it nurtures on its slopes and in its valleys. Fascination with this mountain massif and the cultures and wildlife it harboured became world-wide with the advent of mountaineering expeditions at the beginning of the twentieth century. However, while local Tibetan, Sherpa and other Nepalese groups had long-standing traditional methods for forest and wildlife protection, the new mountain pilgrims from the West brought in new demands, new weapons, and new forms of waste, which accelerated human pressures on the environment.

Nepal was the first to act in protecting this global heritage by establishing Sagarmatha (the Nepalese name for Mt Everest) National Park in 1976. With assistance from the Himalayan Trust and the New Zealand Government, a 1113 km² area was put under the protection of the Nepalese government and local health, education and park management facilities were developed to serve the local people and the growing number of visitors. As recent research has shown (Byers, 1996), this effort has been remarkably successful. Except for the fragile alpine areas on main mountaineering and tourist routes, there is now more forest cover and wildlife than there was 40 years ago and the economic status of villages on the tourist route has increased manyfold.

In the mid-1980s, joint expeditions of The Mountain Institute with local scientists and ecologists in the surrounding areas in Nepal and the Tibet Autonomous Region of China, revealed that the larger ecosystem surrounding Mt Everest was an extraordinary preserve of biological and cultural diversity intimately connected through low river valleys dissecting the Himalaya and high mountain passes over them. With a vertical rise of 8000 m and rainfall patterns that vary from 250 mm annually to 4500 mm – all within horizontal distances of 10 to 20 km – the area was found to contain over 3200 species of flowering plants, 30 forest types, 440 bird species, and rare and endangered fauna such as snow leopard, red panda, and musk deer, along with a rich variety of amphibians, reptiles, and butterflies. Seven languages and many local dialects are spoken by the numerous ethnic groups that have long inhabited these remote regions, and who have subsisted through agriculture, animal husbandry, and trading.

With assistance from The Mountain Institute, both Nepal and the Tibet Autonomous Region embarked on remarkable endeavours to design and develop new protected areas in partnership with the local people. The result was the establishment of the Qomolangma (pronounced Chomolangma, the Tibetan word for Mt Everest) Nature Preserve in 1989 in Tibet, and the Makalu-Barun National Park and Conservation Area in 1991 in Nepal. Together with the contiguous Langtang National Park in Nepal, these adjacent trans-boundary protected areas aim to conserve a vast mountainous area of approximately 39 000 km², almost the size of Switzerland.

With five of the world's fourteen 8000 m peaks, and some of the most lush and pristine valleys of the Himalaya, these protected areas of the Mt Everest region provide corridors for wildlife and new opportunities for co-operation in conservation and sustainable development. Support from a variety of international donors, including the Global Environment Facility, the governments of The Netherlands, USA, and Canada and The Mountain Institute, is being provided to encourage participatory conservation and development on both sides of the border. These programmes are: training local people as national park guardians, supporting local community management of forests and ecosystem resources, developing sustainable forms of economic livelihood, supporting local conservation of cultural heritage, increasing local benefits from tourism and visitor revenues, building enduring local and government institutions, and working to increase and share knowledge. In addition, there have been innovative efforts to develop new forms of co-operation across the border to address opportunities for transboundary management of conservation and economic development. These transboundary activities have included meetings and exchange visits of adjoining park wardens and officials as well as scientists and local experts.

While much remains to be done to make this vast and varied Himalayan region into the model for community-based biodiversity and cultural conservation which it seeks to become, the last decade has seen an extraordinary transformation of outside perceptions of Mt Everest. From being seen as the ultimate peak to conquer, Mt Everest is increasingly perceived as the unique ecosystem to be protected and supported by local people and the global community. This new harmony of perceptions, which brings the outside world into consonance with traditional values, provides the basis for new partnerships for conserving our global mountain heritage – partnerships for action.

Source: J. Gabriel Campbell, Director of
Asian Environmental Programmes,
The Mountain Institute

References

Byers, A. C. (1996). Final Report: Repeat Photography of Mt Everest National Park. National Geographic Society grant 5309–94

Fuller, S. and Gemin, M. (1995). *Proceedings of the Karakoram Workshop*. Skardu, IUCN: Pakistan, p. 49

Hamilton, L.S., Mackay, J.C., Worboys, G.L., Jones, R.A. and Manson, G.B. (1996). Transborder Protected Area Co-operation. Australian Alps National Parks, and World Conservation Union (IUCN): Cooma, NSW and Gland, p. 64

Harmon, D. (ed.) (1994). *Coordinating Research and Management to Enhance Protected Areas*. IUCN: Gland, p. 116

IUCN (1994). *1993 United Nations List of National Parks and Protected Areas*, p. 313

Kingdon, J. (1990). *Island Africa*. Collins: London, p. 287

McNamee, K. (1994). A Retrospective on the Tatshenshini Campaign: World Class Wilderness versus World Class Ore in Northern Protected Areas and Wilderness. In Peepre, J. and Jickling, B. (eds.). *Canadian Parks and Wilderness Society*, p. 379

Mountain Institute (1995). *Report of the Initial Organising Committee of the Mountain Forum*, p. 27

McNeely, J. (1995). *Conserving Diversity in Mountain Environments: Biological and Cultural Approaches*. Presentation to Mountain Agenda Consultation. Lima, Peru. Feb. 1995, p. 11

Poore, D. (ed.) (1992). *Guidelines for Mountain Protected Areas*. IUCN: Gland, p. 47 (also available in Spanish, Russian and Japanese)

Powell, G. V. N. and Bjork, R. (1990). *A Study for the Design of Viable Montane Reserves in Middle America*. Unpublished Report to RARE

Stone, P. (ed.) (1992). *The State of the World's Mountains*. Zed Books: London, p. 391

Stuart, S. N., Jensen, F. P., Brogger-Jensen, S. and Miller, R. I. (1993). The zoogeography of the montane forest avifauna of eastern Tanzania. In Lovett, J. C. and Wasser, S. K. (eds.) *Biogeography and Ecology of the Rainforests of Tanzania*. Cambridge University Press: Cambridge, pp. 203–28

Thorsell, J. (1990). *Parks on the Borderline: Experience in Transfrontier Conservation*. IUCN: Gland, p. 98

Thorsell, J. (1997). Africa's Mountain Parks and Reserves. In Ojany, F. (ed.), *African Mountains and Highlands*. Workshop Proceedings, Kenya (in press)

Thorsell, J. (1996). North, South, East and West – Toward an Inter-Hemisphere Conservation Corridor. *Wild Earth Journal*, **6** (2)

Thorsell, J. and Harrison, J. (1992). National Parks and Nature Reserves in Mountain Environments. *Geojournal* 27 (1): 113–26

Thorsell, J. and Paine, J. (1995). *Mountain Parks and Reserves in the Andes*. Presented at II International Symposium on Sustainable Mountain Development. Bolivia, April, 1995

Udvardy, M. D. (1975). *A Classification of Biogeographical Provinces of the World*. IUCN Occasional Paper: Gland, p. 18

UNEP (1980). *The Major Problems of Man and Environment Interactions in Mountain Systems*. Unpublished Report. Nairobi, p. 38

Tourism and amenity migration 12

Martin F. Price, Laurence A.G. Moss, and Peter W. Williams

Throughout human history, people have moved in and out of mountain areas. While mountain regions are the centres of society in many tropical and sub-tropical countries, such as Colombia, Ecuador, Ethiopia, and Tibet, in many other parts of the world mountains have been places of refuge: the final homes of peoples with previously widely spread populations, such as the Celts in Europe and the Kirats and Kinnars in the Himalaya (see Chapters 2 and 6). Individuals and groups of people from both lowland and mountain settlements have also temporarily visited higher areas in the mountains for resources such as minerals, timber, or summer pastures (see Chapters 9, 13, and 14), or to visit holy places (see Chapter 3). Yet most have subsequently returned home, especially when the snows arrive or the resources are exhausted.

Most of the chapters in this book present a predominantly one-way flow of people and resources: from the mountains to the lowlands. Yet, since the last century, and increasingly from the 1950s, significant flows of people into the mountains have taken place. The majority come for leisure, contributing to the world's largest industry, which employs 212 million people – a tenth of the world's workforce – with an output of US$ 3.4 trillion. By 2005, it is estimated that the tourism industry will employ 338 million people and have an output of US$ 7.2 trillion (World Travel and Tourism Council, 1995). The mountain tourism industry includes a very wide range of activities. Although the season for tourism is rather short in many mountain areas, this very diverse industry has had, and continues to have, major long-term impacts on mountain peoples and their environments.

While much attention has focused on tourism as an agent of change in mountain regions, a second phenomenon is increasingly recognised as having strong influences on mountain communities and environments in many parts of the world. This is amenity migration: the movement of people into the mountains, because of a perceived high incidence of attractive environmental and/or cultural resources, to live there either seasonally or year-round (Moss, 1994a). Amenity migration is very different from the migrations of previous centuries, driven by needs to find refuge or to extract natural resources. While amenity migrants may be fleeing unpleasant or inhospitable conditions in the plains – typically the stressful life of polluted urban areas – they are attracted not by physical resources which can be extracted, such as minerals, trees, or grass for their livestock, but by the vision of a quiet life, clean air, and a more pleasant environment in which to work and play. Yet, like many tourists, their expectations are often those of urban people; and their lifestyles may change little. Transplanted into an environment where the supply of certain resources may be very limited and seasonal, their attitudes are often very different to those of longer-term residents who, from extensive personal or family experience, recognise and largely accept the constraints of living in the mountains. Furthermore, they are frequently much more affluent than the local people amongst whom they come to live.

Five issues complicate the assessment of both tourism and amenity migration and the development of policies for these complex phenomena. First, differences between the two phenomena are not clear-cut. Tourists and amenity migrants often visit mountain destinations for similar reasons, and use the same resources. Often, amenity migrants move to places which they have previously visited as tourists; and a person who is frequently at the same place in the mountains where he or she has a residence, may be viewed as a tourist according to some criteria and as an amenity migrant by others.

Second, both phenomena are unevenly distributed over space and time, with effects that are more widely distributed than the phenomena themselves; for example, with respect to demands for resources (labour, food, energy, water) and economic costs and benefits at local to national scales. One valley may be heavily influenced by one or both of these phenomena, while they are not present in others nearby; and changes in demand for different types of tourism over time also affect relative impacts. Accessibility is a primary issue, though it is also often a paradoxical one. Neither tourism nor amenity migration is possible unless people are able to gain access to their desired destination or new domicile. Yet relative inaccessibility is often a key attractant for both tourists and amenity migrants; and once accessibility increases beyond a certain point, a destination may lose its clientele – or attract a different type – and the early amenity migrants may fight to limit access in order to try to preserve what they originally came for.

Third, and very broadly, populations of tourists and amenity migrants may be divided into two contrasting types. The first type comprises people who have a similar culture and/or economic status to that of the indigenous people. The second type, including most amenity migrants, comprises people who have a markedly different culture and/or economic status to that of the indigenous people. This categorisation is highly simplified; it may be better to speak of a continuum between these two polar types. Yet the differentiation is crucial with regard both to the cultural, economic, and environmental demands and impacts deriving from these incoming people and therefore to the policies required to minimise negative impacts on, and maximise benefits to, mountain people and their environments. Whether the incomers are similar to, or markedly different from, the indigenous population, the significant increases in population – whether short- or long-term – place great stress on the workforce, infrastructure, and resources needed to service the incomers, as well as on family and community life.

Fourth, statistics on all aspects of both tourism and, especially, amenity migration in mountain areas, are generally very limited and, even when available, not always very useful. Both phenomena are highly dynamic and difficult to characterise. Government agencies rarely collect or provide statistics which are adequately disaggregated to allow the identification of trends for either mountain regions (because reporting or administrative areas often include both lowland and mountain areas) or activities (few of which are only practised in mountain areas). In addition, tourism is a competitive industry, and operators are not always willing to divulge information that is important for maintaining their market advantage. Amenity migration poses a different set of statistical problems, both when people are only seasonally-resident, or

when they work in a series of seasonal jobs in tourism and other economic sectors, becoming long-term residents with a sequence of different roles and, often, residences.

Fifth, both tourism and amenity migration need to be considered in the broader context of the restructuring of mountain economies. As mentioned above, accessibility is a major driving force in the development of both phenomena, allowing greater freedom of movement and increased flows of commodities in and out of the mountains. Thus, many of the impacts on mountain peoples and their environments ascribed to these phenomena might well have happened even without their development. As discussed below, both tourism and amenity migration may be forces for positive change in the mountains; but policies for these phenomena must be integrated with policies not only for other economic sectors but also for transport and other infrastructure, education and training, community and regional planning, socio-cultural change, and many other topics.

The remainder of this chapter is divided into two main sections, the first on tourism and the second on amenity migration. The concluding section presents recommendations for research and policy development for these two complex phenomena.

TOURISM

Sites of pilgrimage: religious or secular tourism?

The longest-established form of tourism in the mountains is religious pilgrimage. As described in Chapter 3, mountains are holy places for all of the major world religions, and also for many minor religions, and many sites have histories of pilgrimage that extend back centuries, if not millennia. At the end of the twentieth century, the line between pilgrimage and other forms of tourism is often blurred. Major sites of pilgrimage now attract large numbers of tourists whose motives for visiting may be only slightly religious at most. These include T'ai Shan in China, now with a cable car to its summit; Mount Fuji in Japan, visited by over a million people a year; Kilauea in Hawai'i, within a national park; and Jebel Musa or Mount Sinai in Egypt, part of the region's tourist circuit. Mount Kailas, the most sacred peak for Hindus – but also of importance to Buddhists – is still visited by thousands of pilgrims a year who make a circuit around the mountain. Yet it is also regarded as a tourist attraction by the Chinese government, and European and North American tourist companies offer it as part of their itineraries. However, other sacred mountains, with no particular aesthetic or cultural characteristics which can be marketed by the tourist industry, predominantly remain sites of pilgrimage: for example, Croagh Patrick in Ireland, Ausangate in Peru, and Mount Athos in Greece (Bernbaum, 1990).

Probably the greatest concentration of pilgrims in any region in the world occurs in India's mountainous states of Himachal Pradesh and Uttar Pradesh, where 9.3 million pilgrims each year arrive at the major entry point into the Garhwal region, Hardwar-Rishikesh. Major sites of pilgrimage (*dhams*) have existed in this region since the beginning of Hindu civilisation, if not before. Today, tourism accounts for nearly half of the domestic product of Uttar Pradesh, and 60 percent of the tourists are pilgrims, almost entirely from other parts of India (Sreedhar *et al.*, 1995b). Badrinath, one of the major sites, is visited by about 450 000 people a year, a three-fold increase over two decades (see Chapter 3).

One of the major reasons for this great increase in numbers is improved accessibility. The road to Badrinath was constructed in 1968, following the 1962 war between India and China. Until 1968, pilgrims generally walked for 20 days from the nearest roadhead, or were carried by *kandiwalas* (porters). With the advent of the road and motorised transport, people living in the villages along the pilgrimage route who provided food and accommodation in *chattis* (lodges), as well as the *kandiwalas*, lost these additional sources of income. These were limited because of the pilgrims' austere needs, but still aided the local economy.

In Badrinath itself, when the major temples are open from May to October, there are now increasing problems of sanitation and waste management. Their resolution is constrained by the limited budget of the local authorities, especially since entry taxes were abolished, and by low investment of the State government. Less than 40 percent of the businesses are run by local people; an increasing number are being constructed and run by immigrants from other parts of India. Local traders and politicians are putting pressure on the government to lengthen the period that the temples are open, noting that the climate is suitable, and that the technology for clearing snow from the roads is available. However, local people and religious in-

terests express strong feelings against this (Sreedhar *et al.*, 1995a).

Tourism: a complex industry

The trends described above for Badrinath – changes in accessibility linked to opportunities for income for most immigrants and many indigenous people, inadequate infrastructure and investments in maintenance, ownership of many facilities by outsiders, pressures to lengthen the tourist season, and tensions between different stakeholders – can be found at many tourist destinations. However, pilgrimage differs from other types of tourism in that the destinations cannot be created by changes in access or demands for new activities. Nevertheless, sites of pilgrimage vary in importance over time for both religious and political reasons and, as noted above, may experience massive increases in visitation when accessibility improves. And both sacred sites and religious festivals often become one element of governments' regional development plans, with the sacred aspects often becoming overwhelmed by secular prerogatives such as increasing foreign exchange earnings, government revenues, employment, and incomes.

Many of the world's governments and mountain communities now regard tourism as an important and integral aspect of their strategies for economic survival and growth. Yet, though they typically extol the economic and employment benefits of tourism, it is more than an economic activity. At its core is a massive and complex interaction of people, who demand a wide range of services, facilities, and inputs. These demands generate opportunities and challenges to host countries and regions, and vary greatly with the types of tourism which develop, or are promoted. These vary from mass tourism to cultural tourism, health tourism, ecotourism, and an immense range of types of sports tourism. With such a vast diversity of operations and products, the social, cultural, economic, and environmental inputs to tourism must be well co-ordinated if this multi-faceted industry is to bring benefits to mountain people and the environment. Without careful management of the processes that govern flows between lowlands and highlands, these benefits almost inevitably slide downhill, leaving mountain people with depleted resources, inflated prices, and cultural and biophysical disruptions (Box 12.1).

At the global scale, a number of factors have led to the rapid and continuing growth of tourism,

Box 12.1   Greening the Winter Olympics

Over the past two decades, winter sports events have been increasingly criticised by community and environmental groups, and in the media. Nature has increasingly become a venue for competition, with a dramatic increase in demand for technical installations. Winter sports belong naturally to mountain areas, with their snow and glaciers, and the steep slopes necessary for gravity sports such as alpine skiing, ski-jumping, luge, and bobsleigh. The specialised sports federations, driven by sponsors and the media, have continued to seek new challenges and criteria for their events. These criteria impose new demands on the landscape, sports venues, and equipment. In particular, technical requirements which aim for uniformity of gradient, curvature, and contour at different venues in order to achieve comparable conditions, have led to major changes in mountain landscapes through bulldozing and clear-cutting.

Competitive events may give economic opportunities to mountain communities, but may also cause many environmental, social, and economic problems. These problems have been particularly highlighted with regard to recent international competitions and Winter Olympic Games. There was intense criticism regarding environmental damage and massive infrastructural development associated with the Games in Grenoble in 1968 and Albertville in 1992. Environmental concerns were predominant in the votes of the citizens of Denver, USA, and Lausanne, Switzerland, against holding the Games in their regions. Today, such concerns are increasingly being recognised as valid and important by international sports organisations. The International Bobsleigh and Luge Federation now urges competition organisers not to construct any new

artificially-frozen bobsleigh runs, and there is a revitalisation of the use of naturally-frozen snow and ice for bobsleigh and luge courses. Similarly, the International Ski Federation has made a declaration calling for the preservation of nature and landscape.

The 1994 Lillehammer Games firmly placed environment issues on the agenda of the sporting world. These Games were important because, for the first time, an ambitious series of environmental actions was planned and undertaken in connection with a major sports event. This initiative in environmental planning and practice represented the start of an important process. The International Olympic Committee (IOC) has signed an agreement of co-operation on sports and the environment with the United Nations Environment Programme (UNEP), and is developing an environmental policy based on the experiences at Lillehammer, with a set of stringent measures concerning the selection and pre- and post-Games activities of Olympic host cities. The bid process now requires candidate cities to define a comprehensive environmental policy and action plan, including carrying out impact analyses and avoiding nature protection areas and important habitats. The re-use of facilities and the restoration of habitats are encouraged.

Success in achieving the goals of the new environmental policy will have to be measured scientifically, and local events organisers, athletes, and the IOC will have to accept the scrutiny of the media, the interested public, and the scientific community. A lack of negative headlines should not be the aim; there should be real environmental, as well as social and economic, benefits from competitive sports events.

Source: Olav Myrholt, Environment Lillehammer

which has reached most, if not all, of the world's mountain ranges. These factors include increases in urban populations, discretionary time and income, mobility, telecommunication technologies, and awareness of the abundant natural and cultural resources of the mountains (Williams, 1992). For specific destinations, critical factors that either encourage or constrain tourism include attractiveness, accessibility, and image.

Attractiveness

A wide variety of consumer and pleasure travel surveys emphasise the aesthetic and recreational benefits of mountain and other natural areas as vacation destinations (Thorsell and Harrison,

1993). Historically, mountain tourists were a mixture of mountaineers and others seeking physical challenges, and those visiting the mountains for their health and to appreciate the beauty of mountain landscapes. Today, up to 40 percent of European long-haul travellers seek mountain destinations. These same factors have helped to create flows of domestic visitors to mountain regions; for instance, 30 percent of domestic French travellers (COFREMCA, 1993), and 95 percent of visitors to Himachal Pradesh and Uttar Pradesh in India (Sreedhar *et al.*, 1995b). Except in small countries, domestic visitors generally outnumber foreign visitors to mountain areas; an exception is Austria, where foreigners account for 77 percent of the 124 million overnight stays (Zimmermann, 1995).

In contrast, about one third of the 300 000 tourists who visit Nepal each year come from India alone, and another third from Western Europe (Banskota and Sharma, 1995).

The same travel surveys also show that 'living cultural resources', such as indigenous inhabitants and their cultures, are an important part of the travel experience for many travellers. For instance, about 65 percent of Asian travellers seek travel experiences which involve cultural components (Williams, 1995). The rich and diverse cultural resources of mountain areas are quite different from those of the lowlands. Urbanised travellers perceive mountain peoples as being intimately connected with nature, and increasingly value them for their perceived innate wisdom with respect to biodiversity, production systems, health, and spiritual sustenance (The Mountain Institute, 1995). When the interactions between mountain cultures and visitors appear hospitable and authentic, visitor flows tend to expand, especially as more services and facilities catering to visitors' needs are installed. The development of lodging, potable water, food and beverage, sewage disposal, recreation and entertainment facilities all attract more visitors to mountain regions, but at the same time alter their innate character.

Accessibility

If a mountain community or region wishes to encourage flows of visitors, not only attractiveness, but also relatively easy access is usually necessary. Improved accessibility by road, air, and/or rail may include either new technologies or the improvement of existing routes. These may contribute either intentionally or unintentionally to tourism.

In Western Europe, the traffic capacities and efficiencies of the major mountain highways have greatly increased in recent decades. Through improved design and construction standards, and features such as gradient tunnels, snowsheds, and avalanche control systems, they have opened up the Alps and the Pyrenees to almost any road vehicle. In Japan, the 'bullet train' (Shinkansen) system links the country's major urban centres to the highest mountains. Similar patterns may be seen in western North America, where aesthetic considerations have been a prime criterion in the location of highways, such as Going-To-The-Sun Highway in Glacier National Park, Montana, USA, and the Icefields Parkway through Canada's Banff and Jasper National Parks.

In Asia, several major highways initially developed for 'strategic' military purposes have subsequently provided an impetus to tourism. These include the Karakorum Highway linking Islamabad, Pakistan and Kashgar, China (Kreutzmann, 1991, 1993); the road from Kashmir to Leh, Ladakh, in India; and the highway from Lhasa, China, to Kathmandu, Nepal (Allan, 1988; Kreutzmann, 1991, 1995). However, as long as roads remain strategic, they may not be open to tourists; and continued or renewed military action is a great impediment to tourism, as seen in recent years in Afghanistan and Kashmir. The expansion of tourism may also be a primary reason for the construction of major highways in developing countries. For example, the construction of the 1000 km-long 'Carretera Austral' has opened a vast area of Chilean Patagonia for tourism, as well as forestry.

While visitor flows to mountain regions are dependent on major land and air transportation networks to facilitate interaction with more urbanised regions, it is often the availability and character of local transportation systems that determine the extent to which local communities can tap these markets. In Western Europe, exits from the major mountain highways lead to a wide range of local cableways, chairlifts, community roads, hiking and biking trails, and even helicopter services, which connect visitors to mountain communities, forests and meadows, and peaks. In western North America, well-co-ordinated bus, air shuttle, car rental, and rail transportation networks play an important role in determining the choice of mountain destination. Once in the mountains, tourists may use former logging and mining roads for new types of recreation, such as back-country and off-road driving, cross-country skiing, and snowmobiling.

In developing countries, road and trail networks have been major factors in stimulating tourism. Some governments have constructed roads and hiking trails specifically to encourage the development of tourism, for instance in Sulawesi, Indonesia (Crystal, 1989), and in Peru, where improved highway and public bus services have provided better access for international skiers and mountain climbers from Lima to Huaraz and the gateway to the Cordillera Blanca and Huascaran National Park. Also, increasingly, the availability of air transport is of great importance. For example, Nepal now has a number of private airline companies which use fixed-wing aircraft and helicopters to deliver tourists to locations that were previously many days' walk from the nearest road. This allows people, as well as food grains and construction materials, to reach their destinations much more quickly, but it also means that villages on former trekking trails have lost important sources of income (Dixit, 1995; see also Swope, *et al.*, 1997).

Tourism may also be identified as a secondary reason for building roads into the mountains for large projects; a recent example is the Lesotho Highlands Development Project, whose primary purpose is to create a reservoir to provide water to South Africa. In all cases, however, the availability of access routes is only one prerequisite; marketing or information is essential. For instance, the production of guidebooks, with or without input from local people or government agencies, is crucial in determining where tourists go.

Image

Visitors' images of mountain regions also influence their interactions with these areas and their inhabitants. Image is a multi-dimensional concept including the 'sum of beliefs, ideas, and impressions a person has of a destination' (Crompton, 1979). Images can be more powerful psychological forces in determining the choice of destination than the realities which exist there (Hunt, 1975). To varying degrees, all tourism managers and promoters are involved in strategically creating positive images that can influence the extent and types of tourists' interactions with potential markets. In North America, preferred tourist destinations tend to have positive images of accessibility, climate, environment, personal safety, hygiene, resident receptiveness, lodging, recreational amenities, and price/value, so that these dimensions of the destination product image are stressed in marketing (McGinnis, 1992). In other areas, initiatives aiming to attract tourists stress images of the mystique of mountains and their societies.

Conversely, less formalised images of mountain destinations play significant roles in affecting flows of travellers to mountain areas. Once a negative image becomes established, it is hard to overcome. For instance, when a ski area experiences poor snow conditions early in the season, it is often difficult to overcome this negative image later in the year when skiing conditions improve. At longer time-scales, perceptions of environmental damage and excessive expansion have resulted in reductions in the number of visitors to some mountain resorts in France (Cockerell, 1994). In Austria and

Switzerland, the perceived high cost of mountain vacations has affected travel for some markets (Weiermair, 1983). In Central and Eastern Europe, images of poor economic conditions, political instability, and visa problems have led to significant reductions in national and international tourism to many mountain regions in the 1990s (Price, 1995). In recent years, high risks to personal safety, whether real or imagined, have limited or even stopped tourism in many countries, such as Afghanistan, Bosnia, Colombia, Croatia, Ethiopia, Peru, and Rwanda. The perceived risks may not even be near the destination, as shown by significant decreases in travel to mountain and other destinations throughout Europe and Asia by North Americans in 1991, during the Gulf War.

Interactive processes and impacts

Most tourist areas go through a life cycle that begins with a period of growth. Growth, however, inevitably slows or stops, to be followed by later phases which may involve consolidation, rejuvenation, stagnation, and/or decline. Studies in mountain tourism destinations around the world have examined various dimensions of each of these phases and their economic, social, and cultural consequences. These studies suggest that a wide range of factors influence the type of effects on the visited populations and environments, as well as the direction and rate of these effects. These factors include: the characteristics and activities of the tourists; the relative strength of local cultural and social institutions compared to those of the visitors; and the resiliency of the affected ecosystems.

Tourist characteristics

Even relatively small groups of tourists can affect local communities if their motivations and activities conflict with community norms and goals (Whinney, 1996). However, as mountain destinations become more recognised and popular, entrepreneurs tend to expand the range of facilities and services in an attempt to maintain competitive advantage. The combination of an expanded range of attractions, greater accessibility, and heightened image (with associated market expectations) often leads to losses in cultural authenticity and environmental quality and, linked to these, changes in the types of tourists. At the same time, interactions between hosts and guests tend to become more commercialised, which can engender conflicts between and among different types of users, and between visitors and local residents.

Visitor activities

The number and types of mountain activities pursued by tourists have increased dramatically in recent years. Historically, these activities primarily included sightseeing, hiking, hunting, fishing, climbing, skiing, horseback riding, camping, canoeing, and kayaking. Now, visitors may take part in both 'passive' activities – including bird- and animal watching, flower identification, painting, meditation, and concert-going – and new 'active' activities, including mountain biking, hang-gliding, golfing, para-sailing, off-road driving, heli- and summer glacier skiing, snow-boarding, snow-shoeing, orienteering, hydrospeeding, and rafting. This range of options continues to increase as new technologies emerge and destinations compete in global and regional markets.

As the technological adaptations of the equipment associated with the different 'active' activities increases, so do the number of adherents and their demands for new areas in which to play – and each activity creates its own range of biophysical and social effects. For instance, while the impact of cross-country skiing on natural areas in North America is relatively limited, it is socially incompatible with other winter activities, such as snowmobiling (Ives, 1974; Jackson and Wong, 1986). In many parts of Europe, heli-skiing is prohibited because of its environmental effects – especially on over-wintering animals – and perceived incompatibility with most other winter activities. Conversely, snow-boarders and skiers participate in activities with divergent equipment and styles of participation, but their impact on the others' on-slope experience and the physical environment can be limited if wisely managed (Williams *et al.*, 1995). To orchestrate the optimum blend of products and services, managers need to understand the relative social and environmental compatibilities and susceptibility for conflict of different activities.

Economic, social, and cultural impacts

Tourism is part of an overall process of development that brings external influences, serving as a catalyst for changes to traditional lifestyles and often leading to various forms of urbanisation. Both local and national institutions encourage tourism for economic reasons. A common example is the search for new and supplementary sources of income and employment, which are seen as a partial solution to problems of out-migration, especially of young people, and of real or perceived inequities facing mountain societies in comparison to lowland populations. This attitude often persists in potential or actual tourism destination communities despite the prospects and realities of economic leakages, escalations in the prices of goods and services, and increased taxation to pay for infrastructure improvements. Yet many of those living in these communities believe that these disadvantages can be offset by improvements in living conditions, public health, food and nutrition, and employment opportunities.

Whether or not these improvements appear depends to a large extent on the policies of national and regional governments, which may be more concerned with improving their balance of payments through tourism, rather than recycling the resulting revenues to mountain communities. A particular example is the imposition of fees for climbing particular mountains, which may be quite substantial; for instance, the fees collected by the Government of Pakistan in 1995 amounted to US$ 389 751 (Government of Pakistan, 1995). Such fees are collected by agencies in capital cities, and it is difficult to assess whether they provide any benefit to mountain communities; though the expeditions provide employment: about 4700 porters, guides, and others in Pakistan in 1995. In contrast, fees collected at mountain destinations may be used directly for local conservation and development projects, as in Nepal's Annapurna Conservation Area (Gurung and De Coursey, 1994).

While economic changes which result from the introduction and development of tourism are not easy to direct or gauge, the inevitable cultural changes are even more difficult to predict and assess. Over time, the functional and symbolic tools of the tourists tend to be adopted by many of their hosts. One particularly obvious set of changes occurs as 'western' or lowland clothing and footwear become available. These may be worn as status symbols, because the demands of work in tourism make the production of traditional clothing impossible, or simply because they are more suitable for mountain environments. An alternative reason for the disappearance of traditional clothing may be its sale to tourists in search of authentic souvenirs (Dearden and Harron, 1994). However, souvenirs may not be locally made; those for sale in the mountains of India, Nepal, and Thailand, for example, may be imported from considerable distances and even from completely different cultures (Dearden, 1996; Fisher, 1990; Singh and Kaur, 1986). Yet, while this may bring no benefit to local artisans, it is preferable to the theft or sale of cultural, and especially religious, artifacts reported from communities in many mountain regions.

The commoditisation of culture also occurs when cultural activities, especially festivals, become mere performances, losing cultural meaning. This is a widespread phenomenon, whose effects are intensified when priority is given to paying tourists, rather than local people or pilgrims. To avoid this loss of cultural value, some communities have decided that the disadvantages of tourists outweigh any potential benefits, and have excluded them from festivals, as in Zuni, USA (Mallari and Enote, 1996). For similar reasons, most of Bhutan's monasteries are closed to tourists.

In most mountain communities influenced by tourism, there are also distinct changes in building styles and settlement patterns. In traditional communities, the latter generally represent the result of many generations' appreciation of the rigours of the mountain climate, and both the design and ornamentation of structures are characteristic of a community or region – and part of the image that tourists expect. In general, the influx of tourism is reflected in the homogenisation of housing design and ornament, the import of foreign (especially lowland) building styles, and the deterioration of old, but repairable buildings. The building and layout of new resorts may have little, if anything, in common with traditional communities.

Nevertheless, tourism does not necessarily lead to a loss of traditional forms of expression, and those involved in tourism around the world are recognising that diversity, rather than homogeneity, is part of the attraction of mountain destinations. Mountain people may discover that adhering to their traditional activities and costumes can help them maintain their cultural identity (Gamper, 1981; Moser, 1987; Stevens, 1993), as well as contributing to the image that tourists expect. Local production, rather than import, of handicrafts and food for both local and tourists' use can decrease economic leakages, and have other benefits. For instance, the rejuvenation of artisanal trades can provide new employment opportunities, especially outside the tourist season, and can be encouraged by local regulations – as in Santa Fe, and many of the pueblos of the southwestern states

of the USA – which only allow the sale of authentic local products. In Ladakh, India, tourists' demands for vegetables led to local production, which has been beneficial to the health of local people (Singh and Kaur, 1985). Finally, new buildings may be constructed in traditional styles, but taking advantage of new technologies to provide benefits to health and the efficient use of energy (Norberg-Hodge, 1991) (see Chapter 8).

Environmental effects

Until recently, most of the environmental effects of mountain tourism have occurred on the lower portions and floors of mountain valleys, where the construction and on-going operation of most transportation, lodging, and other infrastructure are concentrated. There have been numerous studies of impacts on wildlife, vegetation, soils, and air and water quality deriving from the construction and maintenance of transportation corridors, accommodation, trails, and golf and other recreation facilities (Bayfield and Barrow, 1985). With the advent of downhill skiing, such impacts have taken place across whole slopes, up to the summits of mountains, through the construction of chairlifts and cableways; the logging and bulldozing of ski runs; the introduction of equipment to create and spread artificial snow; and the disturbance of wildlife and vegetation by skiers and machines (Felber et al., 1991; Mosimann, 1991).

Thus, as technology has increased the ability of the tourism industry to penetrate more remote, rugged, and higher mountain regions, its environmental impacts are spreading. This is true not only in industrialised countries – where the unprecedented access of mountain bikers, snow-mobilers, off-road drivers, and backcountry skiers to formerly remote areas results in a range of impacts – but also in the highest and most remote mountain ranges. The advent of large numbers of tourists, each with specialised equipment and needs for lodging, personal hygiene, cooked food, and warmth, has led to significant levels of pollution and other environmental damage in the Himalaya, the Andes, and elsewhere. One impact that is often cited is the removal of trees and shrubs for cooking and heating in the Himalaya (Banskota and Sharma, 1995). However, tourism is only one of many factors in deforestation, and its relative importance has sometimes been overstated (Byers, 1987; Ives and Messerli, 1989; Shrestha, 1995). The Government of Nepal now requires expeditions and trekking groups to use kerosene or bottled gas, and to remove their waste. However, while such regulations succeed to some extent, they are not rigidly enforced. One paradoxical result is a new form of tourism: special treks, and even expeditions to the world's highest mountains, to remove the litter left by tourists and mountaineers (McConnell, 1991; Bishop and Naumann, 1996).

Planning for the future

While changes are inevitable when tourism is introduced to mountain regions, their direction, rate, and extent are often unpredictable. Despite many of the dramatic changes that have occurred in some tourist destinations, mountain people, their cultures, and the mountains themselves survive and, in some cases, even prosper in spite of – and sometimes because of – the presence of tourism. When policies for tourism are developed and implemented, it is crucial that they consider existing and likely cultural, economic, and environmental conditions, recognising that tourism is an 'industry of fashion' in which demands for different activities

and types of accommodation not only emerge and grow, but may later change or decline, often over a period of a few years (Harrison and Price, 1996). As discussed below, such policies, which place tourism in the context of sustainable mountain development, may focus on visitor management, regional and tourism development, or social and cultural issues; and, preferably, these concerns should be integrated at local, regional, and even national scales.

Visitor management

Each type of mountain tourist has a different set of needs and is likely to have specific impacts on mountain communities and environments. The mountaineering community is giving increased consideration to such impacts (Box 12.2), and some trekking agencies are also showing interest in developing 'codes of conduct', sometimes in collaboration with NGOs in mountain regions, such as the Kathmandu Environmental Education Project. However, mountaineers comprise only a small proportion of mountain tourists. A far larger

Box 12.2 The mountaineers' role

Mountaineering organisations have an important role to play in promoting awareness of the values of mountains and encouraging their better use. Many people enjoy their first contact with the mountain environment by joining a local club or section of a national club or federation. Others may use the services of mountain guides, outdoor centres, trekking organisations, and other commercial ventures to facilitate their mountain experience. At the international level, the International Association of Alpinist Associations (UIAA) provides a focus for collective action by the mountaineering federations of different nations.

With the increasing development of mountaineering has come a concern about the state of mountain environments and their people, both in terms of the mountaineers' own impacts and the effects of other demands on the resources of mountain regions. One example is the 'Clean Mountain Days' co-ordinated by the UIAA, in which local groups remove litter from the mountains, a response to poor standards of behaviour by visitors to the most popular areas. Even in the more remote parts of the Himalaya and Andes, the debris of old camp sites has become a significant intrusion, and special expeditions have been organised to clean them up, a sad reflection of how earlier mountaineers and

trekkers have treated some of the wildest, most unspoilt scenery on earth (McConnell, 1991; Bishop and Naumann, 1996).

The UIAA and individual mountaineering organisations are making increased efforts to promote codes of best practice. For example, rock climbers have been working with conservation organisations to establish and promote seasonal restrictions on climbing where nesting birds could be threatened. Efforts to reduce environmental impacts have concentrated on the development of minimal impact techniques, using the slogan 'leave only footprints'. For expeditions, this means better planning, including the minimisation of packaging of food and equipment, reduced dependence on local fuelwood, and the removal of all waste to acceptable disposal locations at the end of the expedition. Improving technologies – especially the use of solar energy, recycling materials, and the installation of dry toilets – can help to mimimise impacts. The promotion of good conduct codes must also be accompanied by regular monitoring, especially of expedition campsites and trekking routes. However, ensuring compliance with agreed procedures can be difficult, especially in more remote locations. Grant-awarding and sponsoring organisations can play useful roles in helping the regulatory organisations meet this need.

As environmental awareness increases among mountaineers, wider concerns have emerged with regard to the future of mountain land and its people. For example, mountaineering organisations have sometimes been at the forefront of campaigns to prevent the construction of water reservoirs, hydro-electric schemes, and downhill skiing facilities. Similarly, mountaineers have been involved in efforts to persuade policy makers to give greater attention to the local economy of mountain areas and the need to sustain local populations and their characteristic land-use patterns. A particular concern is to ensure that money collected as peak or trekking fees is directed towards the specific needs of mountain areas. Such arrangements will help to encourage mountaineers to adopt the highest standards of environmental care, thus maximising the benefits of mountaineering to mountain people, and securing the freedom to enjoy mountaineering for the future.

The environmental ethic of mountaineers was particularly well expressed by the British mountaineer, Peter Boardman (1978: p. 173), as he described his final activities following his ascent of Changabang, in India's Gahrwal Himalaya, in 1976:

> At Advance Camp I packed the remains of the equipment into my sack and collected all our rubbish together. There was a bottle of paraffin left and I poured the liquid all over the rubbish and then set it alight. Flames leapt up and licked the sky. I tended the fire for a while, until I was sure that everything would burn into disintegration. Then I knocked over the cairns that had marked the site of our camp and guided us towards it across the glacier. I was determined, if at all possible, to leave no sign of our passing.

Source: Dave Morris, UIAA,
Mountain Protection Commission

segment of the market comprises the 65–70 million downhill skiers world-wide, including 20 million in North America, 14 million in Japan, and perhaps 25 million in Europe (Cockerell, 1994). Throughout its history, the downhill skiing industry has been perceived by many to expand with little consideration to local people or environments. Yet Germans, who dominate the European skiing market, are now showing increasing concern about environmental damage. Driven partly by such concerns, some ski companies now provide information in their brochures and at destinations which aim to minimise pollution, impacts to animals and vegetation, and water use – recognising that, in winter especially, water is a relatively scarce commodity in the high mountains, and that sewage treatment in cold conditions presents significant problems (Todd and Williams, 1996).

As discussed above, the image of a destination is a vital influence on an individual's decision whether or not to go there, and which company to go with. Once visitors arrive at the destination, their actions may be guided by a range of types of information, provided in the hotel, at cultural tourist sites, at the beginning of trails, or in interpretive centres (e.g. Ham, 1992; Sharpe, 1982). These means of communication are not specific to particular mountain destinations, though the message may often be; for instance, that alpine vegetation is easily damaged by skis or hikers' boots; that it is culturally inappropriate to wear scanty clothing or take photographs of people; or that water supplies are limited. Such messages can often be included

within other information tailored to the specific market; part of the package that many tourists expect.

Informed and integrated planning

Tourism is never a free-standing economic sector, and should never be regarded as such. Yet there are innumerable policies for the development of tourism, at local to national scales, that have not recognised that the activities of tourism are introduced into communities with pre-existing sets of cultural norms and economic activities, and that the infrastructure for tourism is constructed in a landscape that is used for many other purposes, especially outside the main tourist season.

The strong seasonality of most mountain tourist activities is a crucial reason for ensuring that tourism is placed in a wider economic and societal context. This seasonality derives from two sets of factors. First, the vacation periods of many tourists are constrained by the timing of national and/or religious holidays and school terms. Second, most of the resources required for tourism are climate-dependent: their availability may be affected in the short- and long-term by variability, extremes, and long-term changes in climatic parameters (see Chapter 17). These resources include the landscape of natural and anthropogenic ecosystems, water, weather conditions that are suitable for specific activities and aid the dispersion of pollutants, and snow for skiing. These resources are important not only for tourism, but for many other aspects of the life of mountain communities.

Climate is a central factor in the annual cycle of economic activities in mountain areas. However, the times of the year that are best-suited to traditional economic activities, such as agriculture and forestry, are often those that are most attractive to tourists – and therefore when the new opportunities for employment become available, often at wages that far surpass those available from traditional activities. In the Swiss Alps, the availability of jobs in the skiing industry has been an important factor in the change of the main forest harvesting season from winter to summer, when far greater damage is caused to standing trees (Price, 1987). In the Himalaya and Hindu Kush, many men now prefer to work as porters or guides during the trekking and climbing season, leaving their fields to be cultivated by their wives, or by hired workers who have far less interest in the careful maintenance of terraces and irrigation channels. This may also lead to changes in the types and amount of crops cultivated, and often to decreased yields; as well as increasing demands on women, especially when they also go into the business of providing food or accommodation (Box 12.3 and Chapter 4; Farida Hewitt: Box 4.1).

The recognition that tourism is a seasonal activity that interacts with many others is a major justification for comprehensive national, regional, and local policies and plans for mountain areas. Regional policies in the Alps include the Bavarian

Box 12.3 Developing women's entrepreneurship in tourism (DWET): An experience in the Annapurna region of Nepal

Mountain tourism was first recognised as an important potential for income for Nepal in 1959, and is now Nepal's major source of foreign exchange. It has provided job opportunities for thousands of Nepalese, particularly through offering scope for diversifying the participation and involvement of women in non-traditional activities.

Large numbers of women are involved in the tourism industry. But their employment rate in the direct and formal sectors of the industry is very low, and their involvement is often invisible and unaccounted for. Officially, only 18.8 percent of the tourism workforce is female. Regular government and non-governmental tourism initiatives neither focus directly on women nor assist them to benefit from tourism. Yet women's involvement in tourism has been encouraged in a few special cases, such as the DWET programme of the Annapurna Conservation Area Project (ACAP).

The DWET programme was launched in 1991 by His Majesty's Government of Nepal, with financial and technical assistance from the UN Development Programme and the International Labour Organisation. The programme is specially geared towards developing the entrepreneurial skills of women and training them to use available opportunities. It encourages and assists the trainees to initiate new businesses or improve the profitability of existing female-owned ventures. It also provides the trained entrepreneurs with ongoing business assistance for new or existing enterprises, and links them to credit via banks or revolving loan funds.

The DWET programme, the first of its kind in Nepal to specifically and deliberately target women, was carefully designed to allow women to fully participate in tourism-related enterprises. For example, credit facilities (without collateral) have allowed women to start enterprises of their choice; the participatory approach used to select trainees has helped the project personnel to avoid the intervention of village leaders, ensuring the selection of genuine trainees; and the recruitment of four full-time field-based trainers ensures regular monitoring and supervision of the trainees and that they are provided with guidance. More than 200 women have directly benefited from the programme which, together with ACAP's gender-related activities, has enabled women to own and manage enterprises, enhanced their decision-making capabilities, and empowered them both economically and socially.

Although the DWET programme has succeeded in making tourism enterprises more accessible to the women of the ACAP region, it has some critical shortcomings. As most adult Nepali women are illiterate, the basic education requirement for the training course was a major constraint. For women from poor families, especially from occupational castes who contribute to their family incomes, the 21-day training period was very long. Consequently, women from relatively well-off households were favoured. In addition, though the training course was very aptly designed for developing women's entrepreneurial skills, there was no provision for them to fully use these skills. For women with no experience or technical skills, this lack of linkages with other aspects of tourism development has hindered them from starting enterprises in new areas.

Plate 12.1 Lodge management training given to DWET trainees in Dhampus village (Photograph: Dibya Gurung)

Plate 12.2 One of the participants who opened a consumer shop after training. She received a loan from the Brahmin Caste where women do not have high status (Photograph: Dibya Gurung)

We have learned that entrepreneurship development programmes cannot be implemented in isolation; they must be linked to other activities such as skill development, business exposure, and regular guidance. Each training course must be area-specific, designed according to the needs of the local people, particularly so that both well-off and poor women can participate; or with separate programmes focusing on the needs of different groups. Two or three years are not sufficient for such programmes; they should be at least four to five years in length, backed up with regular and high-quality monitoring and supervision.

Source: Dibya Gurung,
UNICEF, Nepal

Plate 12.3 Traditional display room-cum-restaurant in Ghandruk. The owner Ms Priya Gurung is shown in the picture. She was abandoned by her husband but now has a high status in her village (Photograph: Dibya Gurung)

Plate 12.4 Participants of the DWET training in Ghodepani (Photograph: Dibya Gurung)

Alpine Plan (1972), the Vorarlberg Tourism Concept (1978), and the Ski Development Principles of Tyrol (1992) (Barker, 1994). Such documents focus on tourism but, over time, have increasingly come to consider it in the broader context of sustainable development, as have plans for smaller areas in the USA (Culbertson *et al.*, 1993; 1996) and Canada (Resort Municipality of Whistler, 1995). They should also be developed in consultation with local communities – not imposed arbitrarily by regional or national governments. Local communities may not want tourism, or certain types of tourism; and their wishes should be heeded (Fitton, 1996). Nevertheless, it must be recognised that mountain communities are not internally homogeneous, and that not all of those concerned with the future of mountain regions are residents; they may also include government agencies, NGOs, and other frequent users (Culbertson *et al.*, 1996).

One element of involving the various stakeholders in developing plans for tourism and other aspects of development is to ensure that a realistic assessment is made of the potential of a mountain region or community for tourism. Such assessments should take both socio-economic and cultural, and environmental aspects into account. At a national scale, one example of the former is the decision of the Royal Government of Bhutan to minimise the cultural effects of tourism by imposing a daily fee of US$ 200 on international tourists, who must take a package holiday organised by one of 33 locally-registered tour operators. This policy contrasts significantly with that of neighbouring Nepal, whose government, while limiting access to many parts of the country, charges a low fee for entry visas and has a large and growing tourism industry. At a smaller scale, the Zuni nation of New Mexico have consciously chosen to limit the number and activities of tourists, in order to retain control over their own destiny (Mallari and Enote, 1996), and various Alpine villages have decided to limit the expansion of tourism – or even to close facilities – for similar reasons, as well as minimising environmental impacts (Barker, 1994; Moser and Moser, 1986).

Environmental considerations may be of various types. One relates to the suitability of a region for a particular activity. For instance, some ski resorts have been developed in locations where a cursory evaluation of historical snowfall records would have indicated that the likelihood of reliable snowfall – and therefore profits – was marginal in many years. At this point, a second consideration may be whether there is adequate water for snow-making. Other considerations concern the potential for severe impacts to significant species or ecosystems, water supplies, air quality, or slope stability. However, such impacts may not be caused directly by activities associated with tourism; secondary impacts may result, for instance, from changes in land-use practices that derive inadvertently from its introduction.

Strategies and plans for tourism development have often been, and continue to be, based on limited information. However, more holistic approaches are available, and have often been used for developing such strategies and plans, including 'carrying capacity' approaches (Sharma, 1995), growth management (Gill and Williams, 1994), and computer-based geographic information systems and community debate (Culbertson *et al.*, 1996). All such approaches begin with an initial classification and mapping of a region's biophysical, socio-economic/cultural, and infrastructural characteristics. They then evaluate tolerance levels and critical areas, preferably in consultation with stakeholders, before establishing policies and plans, including frameworks for implementation and monitoring. These must be not only on paper, but based on community institutions which can identify and deal effectively with problems and conflicts associated with the growth and development of tourism. Increasingly, such institutions may include non-governmental organisations, as well as more traditional structures.

Sustaining communities and cultures

There is a common perception that mountain communities were generally isolated from external economies before the arrival of tourism. However, this is not always accurate; in the Alps, Tibet, and elsewhere, many communities had centuries-old traditions of trading and seasonal work along mountain trading routes and in adjacent lowlands (van Spengen, 1992; Viazzo, 1989). Thus, a number of ways of life necessarily co-existed in mountain communities, each allowing individuals to benefit from diverse opportunities for income and subsistence, depending on available resources within the framework of community and external institutions (Price and Thompson, 1997), especially extensive networks of community and mutual co-operation (Beaver and Purrington, 1984; Guillet, 1983; Viazzo, 1989).

This dynamic framework was certainly not egalitarian; strong hierarchies exist in many mountain societies. However, tourism typically creates a new range of forces which lead to tensions within communities. These may occur between the newly wealthy and those who live a more traditional lifestyle; those who earn enough income to live the desired lifestyle in a new, higher-priced system and those who cannot or do not wish to; those who are self-employed and the wage-earners; and long-term residents and immigrants (Moser and Moser, 1986). In addition, in mountain areas around the world, tourism is a major factor in the redistribution of economic activities between the sexes (Price, 1992). Further inequities are often introduced as prices for food and other commodities increase. Similarly, increased prices for accommodation and land can lead to great inequities, when those working either full- or part-time in tourism, or who are not involved in tourism at all, can no longer afford the available housing; or members of long-established families in mountain communities cannot invest in the new sector because they have no access to capital.

One aim of sustainable development is to minimise inequities both in the present and between generations. Consequently, approaches to the development of tourism should attempt to minimise the types of tensions and inequities mentioned above. The need to involve the diverse elements of the community in the various stages of planning and policy implementation has been mentioned previously, though this must be done carefully, recognising the constraints of cultural norms (see Box 12.3). However, many of these inequities may originate outside communities; for instance, national policies – and subsidy and incentive schemes – for agriculture may not take into consideration the importance of agriculture for maintaining the landscape that tourists come to see, bringing large amounts of foreign exchange with them. Such relationships have been considered at the regional scale, for instance in the Vorarlberg of Austria, where communities which benefit directly from tourism make equalisation payments to other communities which maintain the landscape, but derive no direct income. These payments are divided between individual farmers, and matched by the provincial government. Throughout the mountains of Europe, there are suggestions that similar payments should be made by regional and central governments.

The furtherance of close links between agriculture and tourism offers many other opportunities for contributing to sustainable communities. Farmers can benefit from tourism by marketing their produce locally; with added value if local 'brands' from guaranteed sources can be sold for higher prices. The provision of accommodation also offers a chance for supplementary income, and can be expanded so that tourists actually take part in agricultural activities: variously known as agri-tourism in many parts of Europe, ranch tourism in the western USA, or as part of 'ecotourism' (Oberacher, 1995; Bryan, 1991; Cater and Lowman, 1994). Such initiatives may also arise from within

the tourism industry, for instance, companies which include the construction of water supplies, schools, and health facilities as part of the trekking vacation in Nepal offered to their clients. These initiatives are very promising but, like any innovative idea which creates a new infrastructure, must also ensure that adequate operating and maintenance budgets, as well as human resources – for instance, teachers and health workers – are available.

Obviously, institutions alone are not sufficient to ensure sustainable communities; resources are also essential. A major issue is whether these are available within mountain communities, or can be provided preferentially to them through government programmes, especially financial subsidies and incentives; or whether the 'free market' is allowed to dictate the sources of investments, which usually means that these are from outside mountain areas. One of the greatest challenges to mountain communities experiencing the arrival and development of tourism is the possibility of loss of control. A valuable contrast is provided by Kalam and Hunza in northern Pakistan (Al-Jalaly et al., 1995). In Kalam, most of the facilities for tourism are being built by outsiders, on land leased from local people. As land prices are increasing, and probably no more than 20 percent of the income generated by tourism remains in the area, local people are deriving few benefits. In Hunza, local rules do not allow the sale of land, and most facilities for tourism have been built by Hunzakuts. It is estimated that most of the income remains in the area. However, the tourist season is short, and most of the available jobs are in the service sector, so that emigration, especially of educated young men, continues (Kreutzmann, 1993). Thus, tourism may be one element in ensuring a sustainable future, but it must also include other economic activities.

In the Alps, there are also examples of communities which have maintained some degree of control over the development of tourism through strong local plans and by limiting the extent of external investment. To some extent, this may be linked to regional mountain cultures and government policies (Barker, 1994). For example, rooms in private houses are a major component of the accommodation supply in Austria, where federal subsidies foster local participation and control. In contrast, lodging in private houses is insignificant in France, where the regional development policy of the central government encouraged the construction of large ski resorts with external capital, overwhelming small communities, or outside them. Today, the real estate market seems saturated, and government subsidies are necessary to ensure profitability (Chaspoul, 1993).

A major issue in these resorts, as in others around the world which were built primarily for skiing, is that investments must be paid off, and their infrastructure maintained, year-round. However, until recently, they were not designed with any great regard to their attractiveness in the snow-free season. Many are now trying to move towards year-round tourism. One example is Whistler, British Columbia, Canada, which has actively pursued a strategy of creating a broad portfolio of cultural and recreational events and activities, designed to meet the changing demands of existing and emerging tourism markets. Its competitive advantage is based more on attention to consumer research than on the region's inherent natural qualities; Whistler has few innate advantages over many other mountain destinations except for its awareness of changing travel trends and a willingness to adapt to these. Thus, in some cases, tourism in the mountains can come to be driven primarily by the demands of the tourism marketing profession, with minor dependence on 'natural resources' such as landscape and climate, and little, if any, integration with other economic sectors. This may allow communities to survive, and even expand, but their reliance on a single industry – even one with many diverse segments which have been identified – may not offer long-term prospects for a sustainable future.

AMENITY MIGRATION: AN EMERGING FORCE FOR CHANGE IN THE WORLD'S MOUNTAINS

The mountains of the world are the destination of a new kind of migrant – the amenity migrant. The perceived pristine beauty and wild remoteness of the higher reaches of our planet have historically attracted lowlanders, but they have been comparatively few in number, and those who came to settle were even fewer. The hunter, miner, and logger, along with the poet and meditator, have long taken their aspirations to the mountains. More recently, these seekers have been joined by the alpinist and tourist. Typically however, after a period of time they left the heights: some engrandised, others not. Their physical and cultural impacts remain in the fragile highland environments, usually for the long term. Recently a new influx has become apparent, people attracted by a higher quality of natural environment and greater cultural distinctiveness. Their numbers are growing and they intend to remain.

Around the world, places that until recently were veiled in comparative remoteness are now the focus of migrants seeking to enjoy our planet's dwindling natural beauty and cultural diversity. Ironically, in their process of settlement they typically threaten the continued existence of the very attributes that attract them. This condition is the result of a complex configuration of factors, which *per se*, do not dictate positive or negative outcomes. Unfortunately, the public and private stakeholders who are most influential in responding to and promoting this new phenomenon characteristically are focused on marketing destinations and accommodating the resulting migrants, without understanding the larger context of their actions explaining the demand for these special places and likely outcomes. With such a paradigm, more sensitive and informed analysis, planning, and management should be possible, and could assist in reducing the threats of amenity migration while realising its inherent opportunities.

Amenity migration is a significant contemporary societal phenomenon deriving principally from the attractiveness of the cultural and environmental resources of a place or a bioregion: a geographical area whose cultural and biotic characteristics are symbiotically linked within a regional framework. Individuals may consider a bioregion as their 'place' of residence and spiritual affinity (Devall, 1988). *Environmental resources* are the perceived attractive natural, physical attributes of a place, including terrestrial and aquatic landscape, climate, air and water quality and quantity, and biodiversity. *Cultural resources* are tangible and intangible manifestations of human groups or communities, considered culturally significant by either the originators or others who value the manifestations. Tangible resources are art objects and the constructed or significantly altered physical environment. The intangible resources are more difficult to define. At the more visually perceivable end of a continuum are the performing arts or cultural spectacles, such as rites of passage. Towards the other end are audible language, the silent language of gestures, and other constructs shared by groups, such as aesthetics or organisational paradigms. The criteria for valuing the attractiveness of both environmental and cultural resources may be intrinsic or utilitarian (Moss, 1994a).

Amenity migration has been identified as a globally emerging phenomenon, beginning with a study of the Santa Fe, New Mexico, USA, bioregion, whose contemporary economic success flows from the attractiveness of its cultures and natural environment (Moss, 1987). This and subsequent research suggests that the construct of amenity migration offers considerable understanding of the phenomenon, and insights for formulating strategies that can aid in sustaining the culture and natural environment of the places attracting these new residents. However, there are still few case studies of amenity migration. Along with Santa Fe (Moss, 1987; 1994a), the others are of bioregions centred on Baguio, northern Luzon, Philippines (Dimaculangan, 1993); Chiang Mai, northern Thailand (Moss, 1993); the Sumava Mountains in the southwestern Czech Republic (Glorioso and Moss, 1995, Moss 1994a); and the Okanagan Valley of south-central British Columbia, Canada (Beck, 1995).

Several other analyses of specific places also shed light on amenity migration: the Bow River corridor between Banff and Calgary, Alberta, Canada (Webster, 1986; 1992); Paradise Valley, Montana, USA (Parham, 1994; Wright, 1993); the Costa del Sol, Spain (Oliver-Smith *et al.*, 1989); and the Puncak highland of western Java, Indonesia (Babor *et al.*, 1988). This material focuses on regional development and tourism. These two subject areas were also the source of two notions that reinforced Moss in his formulation of the amenity migration paradigm. In the regional economics literature, Ulmann (1954) and Perloff and Wingo (1964) identified amenity as a factor in regional economic growth, coining the term 'amenity resources'. In the 1970s, there were several discussions of the impact of foreign retirees in 'less developed countries', where they are referred to as 'permanent tourists' (Ball, 1971), and 'settled tourists' (de Kadt, 1979), principally because they neither earn their livings at their destinations, nor share the cultural characteristics of the local people.

The above material, together with observations of other amenity migration destinations by Moss, and additional secondary source information, strongly indicates these migrants have a considerable impact on the cultural and environmental attributes of their destinations. Moreover, in this respect, these new settlers appear to be as important, or more important, than tourists, the other main amenity movement into the mountains. These migrants are a formidable force for change; one that appears to be growing in size and intensity as they target a diminishing host. Furthermore, research indicates amenity migrants are especially attracted to mountainous and coastal locales. While they are not yet actually dwelling in great numbers in the highest and most remote mountain reaches, this could soon change.

Given the limited awareness of this new societal driving force, and its likely considerable impact on the world's mountains and their indigenous communities, the reasons for its emergence and growth and its primary characteristics are outlined below.

Characteristics of amenity migrants

The tourists travelling principally for amenity may be described as migrating for a short period of time, without the intention to reside or to earn a living in their destinations. On the other hand, the longer-term amenity migrants can be characterised as settled in their destination, where they reside for various periods of time, either periodically or permanently. They typically perceive themselves as residents of the high-amenity place they have chosen. The residence may be the migrant's sole dwelling, or one of two or more. Three types of amenity migrant are suggested: permanent, seasonal, and intermittent. The first type reside most of their time in the high-amenity place. Seasonal ones reside there for one or more periods each year, such as the summer, the ski season, or the Opera season. The intermittent type move among their residences more frequently.

Wealth and its expenditure differ among the amenity migrants, but they are typically from the middle and elite economic strata, and characteristically originate in metropolitan regions in both the more and less economically developed countries of the world. While the proportion of the middle income class of the poorer nations participating appears to be smaller, it is increasing. This population sector seems to be growing, especially in the countries of East and Southeast Asia, whose economies are sustaining high growth and the middle class is expanding, with consumer value trends shifting to greater commonality with Western nations. While these amenity migrants principally seek higher amenity destinations in Asia, they also appear to be of increasing importance in other regions.

These migrants may earn an income in the high amenity places full-time, part-time, or not at all. A high percentage are not employed locally, but live from income earned elsewhere. Many are still economically active elsewhere, typically in large cities; the wealthier among them have been called 'flexecutives' (Leinberger, 1994).

Why amenity migration?

Migration for greater amenity probably has persisted for centuries. This is evidenced by the leisure settlements of the elites in early China and Greece and it was a primary reason for the popularity of the vacation-cum-retirement cottage in industrialising Europe and North America. The important difference today is a considerable increase in numbers, precipitated by changing motivation and increasing ability to migrate, with resulting greater impacts on the host communities. In this context, mountain areas in particular are recipients of significant increases for, in the past, migration for amenity typically targeted more accessible coastal and pastoral environs.

Six factors have been identified, which appear to coalesce into two dominant societal forces driving amenity migration:

Increasing motivation for amenity migration:
(1) higher valuing of the natural environment;
(2) higher valuing of cultural differentiation; and
(3) higher valuing of leisure, learning, and spirituality.
Greater facilitation of mobility:
(4) increasing discretionary time;
(5) increasing discretionary wealth; and
(6) increasing access through improving and less expensive communications and transportation technology.

In the societal force of increasing motivation the value of the natural environment and culture seem common as paramount driving factors, with some additional contribution from others, especially the existence of economic opportunity. These factors almost always exist together, although their relative weights vary among amenity migrants in different case studies. This condition seems readily manifest in the predominant attributes of particular destinations. For example, the Bow River Corridor and Paradise Valley attract those primarily seeking a spectacular and relatively pristine natural environment. Cultural distinction is secondary, although 'the Western lifestyle' is also important. The Santa Fe bioregion has three such cultures: Native American, Hispanic, and Western American. Chiang Mai is cited as having both the Lana culture of the northern Lao people and the cultures of the hill tribes. The Sumava bioregion attracts Czechs principally seeking beautiful and unpolluted mountain landscapes, while foreigners are also interested in their cultural attributes (Box 12.4). Generally, in the case studies, and from less comprehensive analyses, informants more easily express or explain their attraction to and concern for environmental attributes, finding it difficult to articulate clearly what they value in the cultural realm. Nevertheless, this does not seem to hamper those who wish to express concern for the state of cultural attributes.

The climate of mountains is of special interest to amenity migrants. While its variability and unpredictability have historically tended to inhibit lowlanders from settling, as distinct from sojourning, improved communications and energy technologies are reducing the significance of this variable, and will probably continue to do so. The comparative coolness of the highlands in the tropics has always made them especially attractive, leading to them being major areas of settlement in many parts of the world, and to the establishment of the hill stations of the colonial elites. Dimaculangan (1993) clearly identifies climate as an important variable for contemporary amenity migration into the Cordillera of Luzon in the Philippines (Box 12.5). However, she also establishes that the six factors identified above are all causally involved. The same colonial impetus for development is apparent in other elevated tropical locations, such as Dalat in Vietnam, Bangalore and Darjeeling in India, and the Cameron Highlands in Malaysia. In this latter case, the national government has gone further, using it as a model for developing new amenity-based settlements in the highlands of both the isthmus and the island of Borneo. The comparatively cooler summer weather of the Santa Fe bioregion has also been significant for its development as an amenity destination.

The motivational roles of leisure, learning, and spiritualism are as yet less understood. While definitely identified, their systemic relationships with each other, and with the natural environment and cultural differentiation factors, are complex and remain unclear. The importance and interrelationships of these three motivational factors are also seen as characteristic of the post-industrial or

Box 12.4   Amenity migration in the Sumava bioregion, Czech Republic

Amenity migration in the Sumava bioregion is centred on the region's mountains. With a mean altitude of 1400 m, they include a Protected Landscape Area and the National Park (1630 km²): an area 90 km long and 10–20 km wide. Often referred to as the 'Green Roof of Europe', this area is one of the largest forests, with some of the best-preserved ecosystems, in Central Europe. Human settlements dependent on its amenity resources are also a significant part of this bioregion, especially the culturally rich towns in the surrounding foothills.

For most of the past four decades, much of the bioregion experienced little human habitation. This permitted spontaneous secondary succession towards forest to take place, resulting in the proliferation of highly diverse forest ecosystems of unusual landscape value. From 1946 to 1989, the core of this beautiful landscape was thinly populated by the Czech military, while its periphery was the site of numerous small Czech public corporate resorts and private second homes. Although discretionary income was limited, leisure time was significant and the government encouraged Czechs to spend it in non-urban vacation homes. Use of the bioregion was also encouraged by good transportation facilities into the lower mountains and by inexpensive petrol. Thus, many Czechs spent their weekends and whole seasons in Sumava: skiing in winter, and otherwise tending their gardens, hiking, or fishing.

Since the 'Velvet Revolution' in 1989, there has been rapid change, with greater pressure on the unusually rich amenity resources of the bioregion. A shift to more utilitarian and consumptive socioeconomic behaviour appears to be reducing concern for the environment and the effective management of areas such as Sumava. There is less control over the use of amenity resources in the bioregion outside the National Park. For example, any number of people can boat or sail on Lake Lipno or the Vltava River, or camp out. In the towns and villages, buildings designed as single-family dwellings are rapidly and randomly being turned into restaurants,

Plate 12.5   Sumava bioregion, Czech Republic: a traditional farmhouse converted for intermittent and seasonal amenity migration (Photograph: L. A. G. Moss, 1996)

pensions, and souvenir shops. The general goal appears to be to increase personal wealth as swiftly as possible, with little concern for present or future environmental or cultural impacts. Even the Czech proto-amenity migrants are decreasing in number due to considerably reduced leisure time for sojourning in second homes, and a desire to experience foreign locales now that they are again free to do so.

At the same time, increasing numbers of people – especially from neighbouring countries – who seek amenity resources and were excluded from this large comparatively pristine environment for over four decades, are visiting the bioregion for recreation. While individual foreigners are not allowed to own property in the Czech Republic, they can have private companies. This has increased possibilities for some local inhabitants to earn income, but it appears that the capital needs, to respond to this growing external demand, are being met principally by non-residents of the mountain range. A change in the land ownership law could further disadvantage the indigenous inhabitants, mainly due to their considerably lower wealth and access to information. In the meantime, local commodity prices increase, and facilities and services in the settlements of the bioregion typically face increased strain without adequate fiscal and human resources.

Yet, while potential opportunities are being ardently pursued, there is evidence of awareness of negative impacts, and an intention to manage them. The town of Cesky Krumlov, Plate 12.6, the second most popular tourist destination in the Czech Republic and a leading location of nascent amenity migration, is crafting a long-range sustainable development strategy in which the growing global demand for amenity resources is clearly recognised as a major driving force. Similarly, the national Ministry of the Environment and Parks Service are taking amenity migration into consideration as they develop a strategy for sustaining the biodiversity of Sumava National Park.

Plate 12.6 Cesky Krumlov, Czech Republic: a premier tourist and amenity migrant destination in the Sumava Bioregion (Photograph: L. A. G. Moss, 1996)

This is a time of considerable change and uncertainty in this Central European nation, and the Sumava bioregion reflects this condition.

Source: Romella S. Glorioso, International Cultural Resources Institute, USA

information society (e.g. Capra, 1983; Clifford, 1988; Devall, 1988; Harman, 1988; MacCannell, 1976; Moss, 1994b).

Regarding spiritualism in particular, the significance of mountains in numerous cultures, as a historical and re-emerging post-industrial value, is manifest by many amenity migrants. It appears to be part of a larger desire, or concern, for lifestyles having greater attachment to the land or natural landscape. Typically neither the amenity migrants nor other informants are able to express themselves well on this subject, which is now receiving attention from social and environmental researchers

(e.g. Atwood, 1991; Bernbaum, 1990; Capra, 1983; Tobias, 1985; Tuan, 1993). The case study interviews also indicate that those amenity migrants concerned with protecting or otherwise sustaining the amenity resources may also be characterised as being motivated by spiritual and learning objectives in their decisions to move. Equally, such new residents frequently indicate a willingness to trade-off income for time to undertake these activities in their new homes. Others, not motivated by greater attention to these two objectives, appear less, or not, involved in activities to protect amenity resources.

Box 12.5   Amenity migrants in the Baguio bioregion, the Philippines

'Baguio . . . a wonderland of eternal spring and beautiful fall, which knows no heat of summer nor chill of winter, and where you breathe in misty mountain air'. (Nakahara, Z., Early impressions of a Japanese visitor, *c.* 1920)

Situated at an altitude of 1600 m, with an average annual temperature of 18 °C, Baguio is about 12 °C cooler than the rest of the Philippines. This unusual temperate climate in a tropical country, at a site within reach of the capital, Manila, caused the American colonial administration to choose Baguio as its summer location. Considerable resources were spent in developing and maintaining its natural beauty, especially the preservation of the pine forests. By the 1920s, Baguio was acclaimed one of the foremost hill stations in tropical Asia, attracting foreign visitors from Hong Kong, Indochina, and other Southeast Asian colonies. Many American and Filipino elites and pioneer amenity migrants, came regularly for a season, while others came for frequent but shorter intervals. In short, Baguio became a centre for elite amenity migrants. And with this distinction began its rise and fall.

Since World War II, telecommunications and transportation facilitiies have considerably improved in Baguio, giving rise to more amenity migrants staying both permanently and intermittently. The local tribal cultures also became more pronounced as a motivating factor in this period, though they have become less and less differentiated. This factor is especially influential in the expansion of the amenity migrants' bioregion into central Benguet, northeast Ifugao, and Mountain provinces. However, because of the lower level of access and comfort in these parts of the bioregion, most amenity migrants, particularly the foreigners, continue to reside in Baguio. These migrants play a major role in Baguio's 'economic success', which in turn attracts a multitude of poor economic migrants fleeing national poverty.

Although the phenomenon and its impacts are apparent in the whole bioregion, it is centred on Baguio, with an area of 49 km². Due to the high demand of amenity migrants for a 'prime view', mountain top and hillside sites, including watershed protection areas, are being settled (Plate 12.7). And because of high development cost and the high demand of middle-income amenity migrants for housing, land is being divided into smaller and smaller units, with forests being clear cut, steep slopes graded, and dense multiple-storey buildings erected. However, land prices are far above the reach of the poorer economic migrants. For political expediency, in the early 1980s, the city's administration opened up Baguio's main watershed to squatters and promoted cash crop agriculture to provide income to these less fortunate migrants (Plate 12.8). This has overtaxed the water supply. Although about 70 percent of Baguio is watershed, only about 20 percent is still covered with forest and is relatively unpolluted. Furthermore, local administrations have failed to upgrade the pre-World War II physical facilities, constructed to support a population of 20 000. As a result, Baguio's 300 000 residents experience severe water shortage, especially in summer; long inundations during the wet season; and ground subsidence, due to excessive extraction of groundwater. In 1988, the Philippine Department

Plate 12.7   Baguio, Philippines: new sub-division for amenity migrants (Photograph: R. S. Glorioso, 1993)

Plate 12.8 Baguio, Philippines: poor amenity migrants squatting on a watershed (Photograph: R. S. Glorioso, 1993)

of Environmental Resources declared Baguio an 'environmentally critical area'.

In 1990, a strong earthquake hit Luzon, with its epicentre near Baguio. Massive landslides occurred, multiple-storey buildings collapsed, major roads fissured, telecommunications were severed, and thousands were left homeless. Baguio was isolated for several months. To many, this was the end of Baguio. Numerous foreign amenity migrants left. But to some, this series of events offered an opportunity to pause and ask 'what direction should we take, and how do we get there'? The Baguio economic and political elites, especially environmentally-aware Filipino amenity migrants, seem in agreement to seriously curtail tourism and redirect Baguio towards knowledge-intensive industries, capitalising on her seven institutions of higher learning and the town's superior climate and historical image. The central issue now is how to manage the amenity migrants and the poor people who follow them.

Source: Romella S. Glorioso, International Cultural Resources Institute, USA

The increases in value of the natural environment and well differentiated culture – both living and material – seem to correspond with a general growing realisation of their increasing scarcity. Those who are able to do so are improving their proximity to these resources before they either disappear, or are priced too high in the marketplace. This desire for propinquity typically is manifest by individuals wanting to own their piece of an increasingly valued commodity: the environment. This suggests some value ambiguity or conflict among migrants arising from the acquisitive propensities of commercialised and commoditised value systems.

These two prime motivating factors of amenity migration are joined by four other factors that make up the societal driving force of greater facilitation of this migration. Increases in discretionary time and wealth, and access to places due to improved communications and transportation technologies are clearly identified. However, the communications factor is not as evident in less economically-developed countries. This is due principally to the comparative lack of access to telecommunications technology, especially telephone and telefacsimile, and transportation. In the Chiang Mai region, the rapid improvement of these access facilitators has been important in the recent, considerable growth of amenity migration. In addition, discretionary wealth, especially among the upper-income strata, may still result from inheritance as much as from post-industrial increases in discretionary time and money.

Increasing levels of comfort are also considered by some amenity migrants to be an important facilitating factor – especially older, or retired people. They appear to be attracted in part by improvements in comfort level, including personal safety, infrastructure, and services – such as electricity, supermarkets, medical facilities, fire stations – of some amenity-rich places. For others, improved comfort is irrelevant, and for some, an irritant. In the research by Dimaculangan (1993) and Moss (1987, 1993) on mountain amenity migrants this factor did not rank as significant. However, the travel and resort promotional literature usually draws attention to these comfort amenities, especially when treating the fashionable watering holes of the European socio-economic elite, such as San Moritz and Cortina d'Ampezzo, and historical mountain health spas.

In general, all six factors seem more characteristic in economically highly-developed countries, especially for their middle and elite economic classes. The Czech case is an exception as second home ownership has been high for some fifty years. But these factors also seem to be increasingly apparent among the same population groups in the poorer nations. This is well demonstrated in studies of the Philippines and Thailand, with a greater number from the economic elite involved in the latter case.

Major impacts

The impacts of amenity migration are increasingly global, appearing virtually wherever places exhibit rich caches of amenity resources. One result is that the new residents are also becoming more apparent in high-amenity places of poorer countries, such as the mountain and coastal regions of Costa Rica, Ecuador, Indonesia, Thailand, The Philippines, and Sri Lanka. The latter two countries have specific governmental programmes to promote the long-term residence of foreigners, although their promotional literature to date focuses on seaside rather than highland locations. Foreigners constitute a sizeable part of the amenity migration population in the poorer countries. In addition to the attractions of more benign climates and exotic flora and fauna, increasing ease of access plays a critical role. There is some evidence, however, that these foreign migrants form weaker attachments to their new homes. Very few involve themselves in local community affairs, including protection of the special attributes that attracted them. Moreover, they appear to leave swiftly when they perceive conditions turning adverse, especially a degradation of the qualities that attracted them to the locale. This characteristic appears even more pronounced than in the more general case, which is discussed below.

Amenity migrants significantly increase the use of local amenities and basic resources. As residents they are *in situ* more than tourists or other visitors, and they often have considerable leisure time. Much of it is spent using the natural environment, and their imprint is most pronounced when they engage in active forms of outdoor leisure, such as skiing, snow-mobiling, hunting, or boating.

The amenity migrants' role in land-use change is quite significant in both amount and pattern. Their land use, especially that of the wealthier among them, is considerable, and may be characterised as extensive when compared to the usual intensive land use of indigenous inhabitants. Particularly in the mountain condition of comparative scarcity, this behaviour increases the cost of real estate and converts land use at the periphery of traditional settlements in a pronounced manner. Their ownership pattern usually takes land out of other uses; most importantly, agriculture, watershed protection, and residential use by the indigenous people (Moss, 1987; Oliver-Smith *et al.*, 1989). The displacement of Hispanic people from their historic core and peripheral neighbourhoods of Santa Fe, due principally to increased real estate prices and resulting higher property tax, is an example of the latter. It is also interesting to note that, in the wealthiest amenity places, this change in the ownership of land also appears to remove it from tourism use.

In addition to land, the new residents typically consume more of everything, both indigenous and imported; especially local foods, water, fuel, and labour. Moreover, this consumption pattern is also often one of excess, especially in the scarcity context of most mountainous locales. In this process, local people are forced to pay higher prices for basic commodities and services, to do without, or to move away.

At the same time, there is a tendency for new jobs to be generated where amenity migrants locate; employment for both the amenity migrants and the indigenous populations. The economic activities of the amenity migrants who work locally tend to be knowledge intensive, and are typically environmentally-friendly. Formal and alternative learning institutions, undertaking both research and training activities, are common. A 1988 survey indicated that Santa Fe was the location of some 50 such knowledge-intensive entities. Examples of the formal type are The College of Santa Fe, St John's College, and the Museum of New Mexico. The alternative learning type included Blue Cloud Meditation Center, Pecos River Learning Center, and the International Institute of Chinese Medicine. Many focus on spiritual and philosophical training, drawing on both Eastern and Western traditions. Informant interviews undertaken in 1995 indicate that these numbers have increased, especially the alternative type. The amenity migrant, however, may also be engaged in the more mundane consumer services sector, such as the food and beverage, construction, and travel and tourism industries.

Santa Fe is also characterised by a large number of art galleries, clothing boutiques, and restaurants. In 1988, when its population was only 64 000, among its 575 retail establishments, 200 were restaurants and 150 were art galleries. These figures only include establishments located in the town itself. Many more are located elsewhere in the bioregion, where most of the increase in such establishments has subsequently taken place, as indicated by the 1995 interviews. These magnitudes far exceeded that of the other case studies, but appear closer to the norm of other places where amenity migration is well developed, such as Banff, Canada, and Aspen, USA.

This new economic development, however, seems to offer employment requiring few skills and providing low incomes to indigenous people. The higher paying jobs in the knowledge-intensive sector are rarely occupied by the earlier inhabitants, as they usually do not have the necessary level of education to compete. Likewise, the managerial positions associated with the more mundane economic activities are usually occupied by the better-educated or wealthier new migrants.

Another important impact of the new migration to date is seen at its worst in the Asian hill stations.

Increasing wealth due to their amenity also attracts the poor from the less fortunate, larger surroundings, frequently the entire nation. Their search for shelter – typically resulting in squatting – and income considerably increases deforestation, serious hillside erosion, and water shortage (Box 12.5). More globally and less dramatically, the wealth possessed or generated by amenity migrants in their new locales attracts many others, who can be characterised primarily as 'economic migrants'. They may or may not also be interested in the amenities of these special places, but seem to create a further burden on these attributes due to both their numbers and their behaviour.

Amenity migrants may be characterised as usually urban and cosmopolitan, typically aware that their values and behaviour differ from those of the local people, but still expecting to have their related needs and wants gratified locally. This puts considerable pressure for change on the more traditional, agrarian values and folkways of a mountain community, both by example and also by direct demands for local socio-cultural patterns to change to the more dominant ones migrants bring with them. Work, leisure, and family customs are three specific areas of considerable stress. It is difficult from the available information to distinguish the extent to which the source of these changes is amenity migration or, more generally, modernisation and post-industrialisation. Nevertheless, the case studies indicate a causal relationship between amenity migration and such changes, and that it is characterised by rapid change. Aside from the question of the desirability of such changes *per se* to local people, the high rate of change reduces the time in which those being pressured have the possibility to consider alternatives, and can also create additional psychological stress.

With respect to impacts on indigenous communities, the amenity resources of small and poorer societies, typically ethnically or racially distinct, in particular seem to suffer, and at times are victimised. These groups usually have little political or economic power, particularly in relation to the newcomers. Where considerable and rapid development occurs, their water and land rights are often manipulated and their intellectual property exploited for commercial profit of the outsiders (Bodley, 1990; Greaves, 1994; Moss, 1987; Oliver-Smith *et al.*, 1989). In northern New Mexico, the Native American and rural Hispanic communities are at greatest risk; likewise, the highland tribal communities in the Chiang Mai and Baguio bioregions. While observations by Moss of the impacts of amenity migrants on other mountain peoples suggest a global pattern, the few investigations of such impacts have typically considered the short-term amenity and adventure seeking tourist (e.g. Kempf, 1993; Michaud, 1993; Shackley, 1994; Van den Berghe, 1992).

Cultural change, particularly change in living culture, as distinct from cultural artefacts, is a basic issue in coping with the impacts of amenity migration. The kind of changes outlined in the two paragraphs above is now recognised by, and reasonably well described in, the literature on socio-economic change in development, especially that of applied anthropology. As noted above, some of this literature describes changes brought about or accentuated by tourism, and this body of information includes specific reports on change in traditional mountain communities. Yet, there is very little that moves beyond description to the prescription of means to cope with such change from the perspective of the local people (see especially Hufford, 1994; but also Clifford, 1988; Moss 1994b). This is an important weakness for proceeding beyond the descriptive stage with amenity migration.

Some new residents seem the ones in the community most inclined toward activity to protect or otherwise sustain local amenities. And they tend to have relevant knowledge and skills. In some bioregions, such as Baguio and Santa Fe, resident amenity migrants play a principal role in initiating and maintaining environmental protection and rehabilitation activities. They are also involved in activities to sustain local cultures. However, their focus is usually on inanimate manifestations, such as art objects and historical structures, which tends to change both the material objects and deeper cultural traits so they have greater function for the appropriating, dominant cultural group than the originators (Clifford, 1988; Moss, 1994b). In this process, although some income is generated for the indigenous people, attention to the problems of the living is typically limited. For example, few amenity migrants seem to be assisting indigenous people in their struggle for economic and political empowerment.

In their activities to sustain amenity resources, amenity migrants use both formal political means and NGOs, especially local community-based organisations (CBOs). As noted earlier, they are also responsible for consciously fostering environmental resource-friendly income-generating activity in these bioregions, especially formal and alternative

learning institutions. However, it seems that only a minority of amenity migrants actively involve themselves in protecting the attributes that have attracted them to these special places. More typically, they do not undertake community or individual acts of protection or conservation. While the phenomenon is complex and still little studied, it appears there are two general types of longer-term amenity migrant: the resource conserver and the resource consumer. Both types are frequently at odds with the indigenous residents over differences in values and behaviour. Similarly, they are also often in conflict with each other. For the earlier inhabitants of these special mountain locations, the arrival of amenity migrants, in addition to tourists, has brought about a marked decrease in their quietude.

While the number, values, and behaviour of these migrants are important in creating these issues, two other conditions appear crucial at the societal and policy level (especially for the mountains and their indigenous peoples). First, while headway is being made in society generally and among planners and decision-makers, there is still typically a lack of perception that the natural environment and living culture are resources; both utilitarian and intrinsic. This is especially true for culture, particularly the living culture of ethnic or native communities. Second, even where the environment and culture are perceived as resources, a resource-conserving ethic and legal sanctions to control careless and abusive attitudes are absent or not sufficiently developed. Cultural resources in particular are still typically viewed as 'free goods' or 'common property'. By comparison, a protection ethic, typically one of conservation, and related sanction and reward systems are being established for sustaining environmental resources. In poor nations, both environmental and cultural resources still seem to obtain little protection from contemporary societal perception and sanction and reward systems (de Sousa, 1993; Moss, 1994b). Whether or not this is the product of economic necessity is the subject of considerable debate.

Although the construct of amenity migration appears to identify and explain a significant societal phenomenon, and should better facilitate systemic understanding of opportunities and threats being created by these migrants, local, regional, and national planners and decision-makers responsible for the special places studied are not utilising this construct to understand, plan, and act.

The Sumava and Baguio situations may be exceptions characterised by considerable awareness, but even in these places action has been quite limited.

Too many of the key stakeholders still use simplistic economic development constructs, and in this context are too intent on tourism as the basis for the economic and, sometimes, the socio-economic well-being of these special places. However, a few realise that tourism is a double-edged sword, socio-culturally, economically, and environmentally; one to be used with care. Whether or not planners and decision-makers show this realisation, their excessive focus on tourism means they are using a very partial and static paradigm of reality and, therefore, are at a considerable disadvantage for developing strategic solutions. Even better funded and more highly skilled development bodies, still typically view tourism as a primary 'solution' for more remote regions and human settlements (e.g. European Commission, 1994).

The future

More information is needed about amenity migration. However, while this is being obtained, not afterwards, strategies need to be formulated and implemented in a small number of specific places where amenity migrants are beginning to congregate. Such a proactive project will need to be both experimental in nature and strongly supported by a sponsor willing to commit to an intensive and difficult task for probably a five- to six-year period, followed by a less expensive commitment over the subsequent decade. This undertaking would also shed considerable light on the phenomenon, but this should not be its prime mission. In following this recommendation, it will be crucial to move beyond the all too frequent prescription for integrated, comprehensive planning. It has not worked, principally due to the considerable human and financial resources required, coupled with inherent weaknesses as an intellectual construct. Strategic analysis and planning has a superior, though still limited, success record, primarily in the private sector. But it will most likely also fail without a commitment to the difficult and time-consuming task of bringing the indigenous inhabitants into the core of the planning and management process.

Although amenity migration merits urgent response and greater study, many of the main factors for its growth and development also suggest it is probably vulnerable to reversal or termination. Among the formulations of relevant alternative future scenarios, a very plausible one is the considerable reduction, or virtual elimination, of such migration. Principal factors in such a future would be a significant decline in discretionary income due to rapid and considerable inflation; the inability of global economic restructuring to significantly distribute the wealth earned by our intelligent machines; a wide acceptance of virtual travel and related experience; or, more generally, a marked shift away from the phenomenon's causal societal values. Of course, the mere degradation of the amenity attributes to a level below the attraction threshold of existing and potential amenity migrants alone – or this combined with any or all of the above – would terminate this driving force. Nevertheless, while amenity migration may not remain as a societal driving force in a century, against a closer horizon, it appears to be an important force for change in our mountains.

GENERAL CONCLUSIONS AND RECOMMENDATIONS

Tourism and amenity migration are complex phenomena, with a number of aspects in common, yet many profound differences. The extent of these differences requires careful consideration, based on improved knowledge. For tourism, a thorough evaluation of existing literature and experience in mountain areas is necessary, to draw together the very disparate and dispersed knowledge which is still largely based on case studies, few of which have been longitudinal, allowing analysis of changes over time. Existing models are often useful for descriptive purposes, but not for prescriptive action. Knowledge of almost all aspects of amenity migration remains limited and, as proposed above, a multi-year project aimed at providing greater understanding of the phenomenon is needed, to be undertaken concurrently with the formulation and implementation of strategic plans for a select number of locations where amenity migration is taking place. Overall, in order to develop strategies to manipulate tourism and amenity migration to their greatest potential for mountain communities and environments, it is vital to be able to differentiate between and compare tourists, economic migrants, and amenity migrants – and the different groups in these general categories – in terms of the nature and magnitude of their diverse impacts, both positive and negative. Equally, better understanding of the interactions between the two phenomena are needed, as discussed below.

If mountain people and others who are responsible for the future of mountain areas are to address these phenomena in ways that maximise their benefits and minimise their negative impacts, adequate, consistent, and transparent information is an essential prerequisite. Communities, regions, and nations need to clearly define the diverse types of tourists and amenity migrants, and their characteristics (place of origin, economic status, activities), in order to begin to compile data on trends which can be used in planning for the future. Such data need to be collected frequently and be widely accessible in order to keep track of, and respond to, these dynamic phenomena.

Mountain communities and regions which wish to attract tourists also need to better understand visitor expectations both before and after arrival at their destinations. Such information is essential to define potential impacts and, therefore, the desired types of tourism, along with forming and maintaining the appropriate image to attract such visitors. In doing so, careful attention must be given to the wants and needs of the indigenous people. Similarly, to respond proactively to amenity migration, considerable research is needed to understand the ways in which different environmental and cultural resources attract and hold these newcomers. The other key factors of increasing discretionary time and wealth, and increasing access through communications and transport, also need to be examined in greater detail. Improved knowledge of the interactions and relative importance of all of these factors in different populations is also essential.

If tourism and amenity migration are to be forces for positive change, research into their many societal effects must be conducted and utilised. Both phenomena can bring benefits to mountain communities, such as improved infrastructure and services and additional opportunities for employment and income generation, often in new fields. Such benefits may be multiple, especially when new activities can be linked directly to traditional knowledge and activities, and the complementarity of activities in different seasons can be enhanced. Yet tourism and amenity migration can also lead to serious negative impacts, such as competition for employment, loss of affordable housing, significant pressures on infrastructure, increased prices for vital commodities, and the loss of cultural resources, and restricted access of locals to environmental resources. Amenity migrants and economic migrants can be important actors in influencing the

rates at which benefits and impacts accrue; their impacts on local, and even regional and national, political economies may often be greater overall than those of tourists simply because they remain, and often invest, in their destinations for longer periods. For example, the migration of highly-educated people to mountain areas to establish or work in knowledge-based industries – as well as economic migrants who move in the hope of finding employment in the associated service sector may have significant impacts on the economies and societies not only of their destination communities, but also the urban communities they leave behind.

Detailed longitudinal studies on these societal issues – preferably with local people being directly involved, not just acting as informants – are essential in communities experiencing tourism, amenity migration, or both simultaneously, not merely to identify benefits and impacts, but also to evaluate social, institutional, and technical mechanisms in order to respond to them. A particular objective of such studies is a clearer understanding of the relationships between tourism and amenity migration. For instance, does longer-term amenity migration tend to develop from amenity-based tourism and short-term amenity migration? If this is so, how does this reflect growth in the societal driving forces described above; is amenity migration likely to come to dominate or replace tourism; and, if so, what are the central causal factors? The results of such studies, both in specific communities and regions, and in comparable locations, should then inform the actions and policies of local authorities, private entrepreneurs, NGOs, and governments active in these communities and regions. The critical issues they must confront are why, how, and how much tourism or amenity migration, or what mixture of the two, are desirable and possible? If decision-makers wish to control or limit their growth, clear understanding of means – physical, fiscal, or political – to limit access or to focus on qualititative rather than quantitative growth is essential. Information concerning such means should be obtained from analysis of past and current experiences and careful experimentation.

The potential environmental impacts must be recognised. Yet these also can be seen as opportunities to test and develop new technologies and strategies in diverse fields, including waste and water storage and treatment; energy production and utilisation; building construction; traffic management; and trail and road design. There are many challenges to living in a mountain environment, and increased population levels, whether transitory or permanent, accentuate these. Nevertheless, investments in the construction and maintenance of physical infrastructure have to be made; and these can be designed to derive new benefits, rather than repeating old approaches and, often, mistakes. This refers not only to the investments themselves, but also to their origins; for instance, how financial instruments, such as transfer payments or preferential loans or taxation, may be used to give mountain people economic opportunities or to compensate them for providing 'free goods' to other mountain communities and those living in lowland areas.

As underlined in the preceding sections, policies which address tourism, amenity migration, or both together have to be linked, and preferably integrated, with policies for other economic sectors, transport and other infrastructure development, training and education, community and regional planning, and so on. In other words, the futures of mountain communities and regions must be addressed holistically and strategically. This will often require a move beyond current and well-known approaches which have tended to be rather myopic, sectoral, and geographically-limited. It also implies the involvement of not only professional 'planners' and policy-makers, but also the tourism industry and the diverse stakeholders involved in mountain communities, from the earliest stages of issue definition through to the implementation and monitoring of policies. NGOs and CBOs have important roles to play in such processes, particularly in identifying existing and potential conflicts between stakeholders, and resolving these.

Finally, there is a need for better transfer of information, policy approaches, and models for 'development' both within and between mountain regions. Most of the research proposed above will have to be conducted as case studies. However, there is considerable scope for comparative and collaborative approaches and, where possible, studies should be designed with consideration of their wider relevance. Each place has its own specific characteristics, yet in a post-industrial world in which it is possible to travel to, or communicate with, any mountain community using modern technologies, there are many lessons that can be learned from experiences elsewhere. In mountains around the world, tourism and amenity migration are powerful forces for change. Yet, even if we can grow to understand and direct them better, they will always be unpredictable, and those who participate in them should be prepared for surprises.

ACKNOWLEDGEMENTS

The authors of the sections of this chapter are as follows: tourism – Martin Price and Peter Williams; amenity migration – Laurence Moss. Laurence Moss wishes to thank the following institutions for their financial support of his research on amenity migration between 1992 and 1995: the Canadian International Development Agency, the Canadian Universities Consortium, and the International Cultural Resources Institute. Special acknowledgement is also due to Romella S. Glorioso (Dimaculangan), whose research is heavily drawn on in the section on amenity migration.

References

Al-Jalaly, S. Z., Nazeer, M. M. and Qutub, S. A. (1995). *Tourism for Local Community Development in the Mountain Areas of NWFP and the Northern Areas of Pakistan. Phase Two – Case Studies of Kalam and Hunza.* Discussion Paper MEI 95/12. ICIMOD: Kathmandu

Allan, N. J. R. (1988). Highways to the sky: The impact of tourism on South Asian mountain culture. *Tourism Recreation Research*, **13**: 11–16

Atwood, R. (1991). *The Futures of Environmental Concern.* 2nd edition. University of Georgia Press: Athens, GA

Babor, D. C., Llanto, R. B. N. and Paudyal, D. R. (1988). *The Upper Cliwung Watershed of West Java: Agenda for Development.* Asian Institute of Technology: Bangkok

Ball, D. A. (1971). Permanent tourism: A new export diversification for the less developed countries. *International Development Review*, **4**: 20–3

Banskota, K. and Sharma, B. (1995). *Mountain Tourism in Nepal: An Overview*. Discussion Paper MEI 95/7. ICIMOD: Kathmandu

Barker, M. L. (1994). Strategic tourism planning and limits to growth in the Alps. *Tourism Recreation Research*, **19**: 43–9.

Bayfield, N. G. and Barrow, G. C. (eds.) (1985). *The Ecological Impacts of Outdoor Recreation on Mountain Areas in Europe and North America*. RERG Report 9: Wye College, UK

Beaver, P. D. and Purrington, B. L. (eds.) (1984). *Cultural Adaptation to Mountain Environments*. University of Georgia Press: Athens, GA

Beck, G. (1995). *Amenity Migration in the Okanagan Valley, BC, and the Implications for Strategic Planning*. Unpublished MSc thesis, University of Calgary

Bernbaum, E. (1990). *Sacred Mountains of the World*. Sierra Club Books: San Francisco

Bishop, B. and Naumann, C. (1996). Mount Everest: Reclamation of the world's highest junk yard. *Mountain Research and Development*, **16**(3):323–7

Boardman, P. (1978). *The Shining Mountain: Two Men on Changabang's West Wall*. Hodder and Stoughton: London

Bodley, J. (1990). *Victims of Progress*. Mountain View: Mayfield

Bryan, B. (1991). Ecotourism on family farms and ranches in the American West. In Whelan, T. (ed.), *Nature tourism: managing for the environment*. Island Press: Washington DC, pp. 75–85

Byers, A. (1987). An assessment of landscape change in the Khumbu region of Nepal using repeat photography. *Mountain Research and Development*, **7**(1):77–81

Capra, F. (1983). *The Turning Point: Science, Society and the Rising Culture*. Bantam: New York

Cater, E. and Lowman, G. (eds.) (1994). *Ecotourism: A Sustainable Option?* John Wiley and Sons: Chichester

Chaspoul, C. (1993). Immobilier de loisirs. *Cahiers d'Espaces*, **32**

Clifford, J. (1988). *The Predicament of Culture: Twentieth-Century Ethnography, Literature and Art*. Harvard University Press: Cambridge, MA

Cockerell, N. (1994). The international ski market in Europe. *Travel and Tourism Analyst*, **3**: 34–55

COFREMCA (1993). *Pour un Repositionnement de l'Offre Tourisme-Loisirs des Alpes Françaises*. COFREMCA: Paris

Crompton, J. L. (1979). An assessment of Mexico as a vacation destination, and the influence of geographic location on that image. *Journal of Travel Research*, **17**(4): 18–23

Crystal, E. (1989). Tourism in Toraja (Sulawesi, Indonesia). In Smith, V. (ed.), *Hosts and Guests: The Anthropology of Tourism*. University of Pennsylvania Press: Philadelphia, pp. 139–68

Culbertson, K., Turner, D. and Kolberg, J. (1993). Toward a definition of sustainable development in the Yampa Valley of Colorado. *Mountain Research and Development*, **13**: 359–69

Culbertson, K., Snyder, D., Mullen, S., Kane, B., Zeller, M. and Richman, S. (1996). Finding common ground in the last best place: The Flathead County, Montana, Master Plan. In Price, M. F. (ed.), *People and Tourism in Fragile Environments*. John Wiley and Sons: Chichester, pp. 139–58

Dearden, P. (1996). Trekking in northern Thailand: Impact distribution and evolution over time. In Parnwell, M. J. G. (ed.), *Uneven development in Thailand*. Avebury: Aldershot, pp. 204–25

Dearden, P. and Harron, S. (1994). Alternative tourism and adaptive change. *Annals of Tourism Research*, **21**: 81–102

de Kadt, E. (1979). Introduction. In de Kadt, E. (ed.), *Tourism: Passport to Development?* UNESCO and IBRD: Washington DC, pp. 1–10

de Sousa, D. (1993). Modern mass and luxury tourism: Blessing or curse, a third world perspective. *Contours*, **5**(8): 4–7

Devall, W. (1988). *Simple in Means, Rich in Ends*. Peregrine: Salt Lake City

Dimaculangan, R. S. (1993). *Key Policy Implications for Strategic Use of Amenity Resources: A Study of Longer-term Amenity Migration, Baguio Bioregion, The Philippines*. Unpublished MSc thesis, Asian Institute of Technology, Bangkok

Dixit, K. M. (1995). The porter's burden. *Himal*, **8**(6): 32–8

European Commission (1994). *Europe 2000+: Cooperation for European Territorial Development.* European Commission: Strasbourg

Felber, H. U., Hirsch, M. and Walther, P. (1991). *Modifications du Paysage en Faveur de la Pratique du Ski: Directives pour la Protection de la Nature et du Paysage.* Département fédéral de l'intérieur: Bern

Fisher, J. F. (1990). *Sherpas: Reflections on Change in Himalayan Nepal.* University of California Press: Los Angeles

Fitton, M. (1996). Does our community want tourism? Examples from South Wales. In Price, M. F. (ed.), *People and Tourism in Fragile Environments.* John Wiley and Sons: Chichester, pp. 159–74

Gamper, J. A. (1981). Tourism in Austria – A case study of the influence of tourism on ethnic relations. *Annals of Tourism Research,* 8: 432–46

Gill, A. and Williams, P. W. (1994). Managing growth in mountain tourism communities. *Tourism Management,* 15: 212–20

Glorioso, R. S. and Moss, L. A. G. (1995). *Notes on Amenity Migration in the Sumava.* International Cultural Resources Institute: Santa Fe

Government of Pakistan (1995). *Mountaineering and Trekking in Pakistan during 1995.* Press release, 30 October 1995, Sports and Tourism Division, Government of Pakistan: Islamabad

Greaves, T. (ed.) (1994). *Intellectual Property Rights for Indigenous Peoples: A Source Book.* Society for Applied Anthropology: Oklahoma City

Guillet, D. (1983). Toward a cultural ecology of mountains: The Central Andes and the Himalaya compared. *Current Anthropology,* 24: 561–74

Gurung, C. P. and De Coursey, M. (1994) The Annapurna Conservation Area Project: A pioneering example of sustainable tourism? In Cater, E. and Lowman, G. (eds.), *Ecotourism: A Sustainable Option?* John Wiley and Sons: Chichester, pp. 177–94

Ham, S. (1992). *Environmental Interpretation.* North American Press: Golden

Harman, W. W. (1982). Transforming the lifestyles of industrial society. In Harman, W. W. (ed.), *Optimistic Outlooks.* Global Futures Networks: Toronto

Harman, W. W. (1988). *Global mind change.* Knowledge Systems: Indianapolis

Harrison, D. and Price, M. F. (1996). Fragile environments, fragile commnunities? An introduction. In Price, M. F. (ed.), *People and Tourism in Fragile Environments.* John Wiley and Sons: Chichester, pp. 1–18

Hufford, M. (ed.) (1994). *Conserving Culture: A New Discourse on Heritage.* University of Illinois Press: Urbana and Chicago

Hunt, J. D. (1975). Image as a factor in tourism development. *Journal of Travel Research,* 18(3): 18–23

Ives, J. D. (1974). The impact of motor vehicles on the tundra environment. In Ives, J. D. and Barry, R. G. (eds.), *Arctic and Alpine Environments.* Methuen: London and New York, pp. 907–10

Ives, J. D. and Messerli, B. (1989). *The Himalayan Dilemma: Reconciling development and conservation.* Routledge and United Nations University Press: London and New York, p. 296

Jackson, E. L. and Wong, R. A. C. (1986). Perceived conflict between urban cross country skiers and snowmobilers in Alberta. *Journal of Leisure Research,* 14: 47–62

Kempf, E. (ed.) (1993). *The Law of The Mother: Protecting Indigenous Peoples in Protected Areas.* Sierra Club Books: San Francisco

Kreutzmann, H. (1991). The Karakoram Highway: The impact of road construction on mountain societies. *Modern Asian Studies,* 25(4):711–36

Kreutzmann, H. (1993). Challenge and response in the Karakoram: Socioeconomic transformation in Hunza, Northern Areas, Pakistan. *Mountain Research and Development,* 13(1):19–39

Kreutzmann, H. (1995). Globalization, spatial integration, and sustainable development in Northern Pakistan. *Mountain Research and Development,* 15(3):213–27

Leinberger, C. B. (1994). Flexecutives: Redefining the American dream. *Urban Land,* (August 1994): 51–4

MacCannell, D. (1976). *The Tourist: A New Theory of the Leisure Class.* MacMillan: London

Mallari, A. A. and Enote, J. E. (1996). Maintaining control: Culture and tourism in the Pueblo of Zuni, New Mexico. In Price, M. F. (ed.), *People and Tourism in Fragile Environments.* John Wiley and Sons: Chichester, pp. 19–32

McConnell, R. M. (1991). Solving environmental problems caused by adventure travel in developing countries: the Everest Environmental Expedition. *Mountain Research and Development*, **11**(4):359–66

McGinnis, D. (1992). The changing image of Jackson Hole, Wyoming. In Gill, A. and Hartmann, E. (eds.), *Mountain Resort Development*. Centre for Tourism Policy and Research, Simon Fraser University: Burnaby, pp. 126–36

Michaud, J. (1993). The Social Anchoring of the Trekking Tourist Business in a Hmong Community of Northern Thailand. Unpublished manuscript, GERAC, Laval University, Quebec City

Moser, P. and Moser, W. (1986). Reflections on the MAB-6 Obergurgl project and tourism in an Alpine environment. *Mountain Research and Development*, **6**: 101–18

Moser, W. (1987). Chronik von MAB-6 Obergurgl. In Patzelt, G. (ed.), *MAB-Projekt Obergurgl*. Veröffentlichungen des Oesterreichischen MAB-Programms 10. Universitätsverlag Wagner: Innsbruck, pp. 7–24

Mosimann, T. (1991). *Beschneiungsanlagen in der Schweiz*. Geographisches Institut: Hannover

Moss, L. A. G. (1987). *Santa Fe, New Mexico, Post-industrial Culture Based Town: Myth or Model?* Department of Economic Development and Trade, Government of Alberta, Edmonton, and International Cultural Resources Institute: Santa Fe

Moss, L. A. G. (1993). *Notes on Amenity Migration in the Chiang Mai Bioregion*. Asian Institute of Technology: Bangkok

Moss, L. A. G. (1994a). Beyond tourism: The amenity migrants. In Mannermaa, M., Inayatullah, S. and Slaugther, R. (eds.), *Chaos in Our Uncommon Futures*. University of Economics: Turku, pp. 121–8

Moss, L. A. G. (1994b). International art collecting, tourism and a tribal region in Indonesia. In Taylor, P. (ed.), *Fragile Traditions*. University of Hawaii Press: Honolulu, pp. 91–121

Norberg-Hodge, H. (1991). *Ancient Futures: Learning from Ladakh*. Sierra Club Books: San Francisco

Oberacher, A. (1995). Landwirtschaft und Tourismus im Alpenraum – Widerspruch oder Symbiose. *Forderungsdienst*, **43**: 109–15

Oliver-Smith, A., Arroñez, F. J. and Araal, J. L. (1989). Tourist development and the struggle for local resource control. *Human Organization*, **48**: 345–51

Parham, D. W. (1994). *Growth Issues in the Rocky Mountains*. Educational Policy Forum Series 631, Urban Land Institute: Washington DC

Perloff, H. and Wingo, L. (1964). Natural resource endowment and regional growth. In Friedmann, J. and Alonso, W. (eds.), *Regional Development and Planning: A Reader*. MIT Press: Cambridge, MA, pp. 215–39

Price, M. F. (1987). Tourism and forestry in the Swiss Alps: Parasitism or symbiosis? *Mountain Research and Development*, **7**: 1–12

Price, M. F. (1992). Patterns of the development of tourism in mountain environments. *GeoJournal*, **27**: 87–96

Price, M. F. (1995). *The Mountains of Central and Eastern Europe*. Environmental Research Series 9, IUCN European Programme: Gland

Price, M. F. and Thompson, M. (1997). The complex life: human land uses in mountain ecosystems. *Global Ecology and Biogeography Letters*: in press

Resort Municipality of Whistler (1995). *1995 Community and Resort Monitoring Program*. Planning Department: Resort Municipality of Whistler

Shackley, M. (1994). The land of Lo, Nepal/Tibet: The first eight months of tourism. *Tourism Management*, **15**: 17–26

Sharma, P. (1995). *A Framework for Tourism Carrying Capacity Analysis*. Discussion Paper MEI 95/1. ICIMOD: Kathmandu

Sharpe, G. W. (1982). *Interpreting the Environment*. John Wiley and Sons: New York

Shrestha, T. (1995). *Mountain Tourism and Environment in Nepal*. Discussion Paper MEI 95/4. ICIMOD: Kathmandu

Singh, T. V. and Kaur, J. (1985). In search of holistic tourism for the Himalaya. In Singh, T. V. and Kaur, J. (eds.), *Integrated Mountain Development*. Himalayan Books: New Delhi, pp. 365–401

Sreedhar, R. *et al.* (1995a). *Mountain Tourism in Himachal Pradesh and the Hill Districts of Uttar Pradesh*. Discussion Paper MEI 95/6. ICIMOD: Kathmandu

Sreedhar, R. *et al.* (1995b). *Mountain Tourism for Local Community Development: A Report on Case Studies in Kinnaur District H.P. and the Badrinath Tourist Zone.* Discussion Paper MEI 95/10. ICIMOD: Kathmandu

Stevens, S. F. (1993). *Claiming the High Ground: Sherpas, Subsistence, and Environmental Change in the Highest Himalaya.* University of California Press: Berkeley

Swope, L. and Swain, M. B., Yang, F. and Ives, J. D. (1997). Uncommon property rights in Lijiang, SW China: Tourists and Trees. In Johnston, B. R. (ed.), *Life and Death Matters: Human Rights and the Environment at the end of the Millennium.* Altamira Press: One Thousand Oaks, CA, pp. 46–69

The Mountain Institute (1995). *International Consultation on the Mountain Agenda: Summary Report and Recommendations to the United Nations Commission on Sustainable Development.* The Mountain Institute: Franklin, WV

Thorsell, J. and Harrison, J. (1993). National parks and nature reserves in mountain regions of the world. In Hamilton, L. S., Bauer, D. P. and Takeuchi, H. F. (eds.), *Parks, Peaks and People.* East–West Center: Honolulu, pp. 3–23

Tobias, M. (ed.) (1985). *Mountain People.* University of Oklahoma Press: Norman

Todd, S. and Williams, P. W. (1996). From white to green: a proposed environmental management system framework for ski areas. *Journal of Sustainable Tourism,* 4(2): 1–27

Tuan, Y.-F. (1993). *Passing Strange and Wonderful.* Island Press: New York

Ulmann, E. (1954). Amenities as a factor in regional growth. *Geographical Review,* 44: 119–32

Van den Berghe, P. L. (1992). Tourism and the ethnic division of labour. *Annals of Tourism Research,* 19: 234–49

van Spengen, W. (1992). *Tibetan Border Worlds: A Geo-historical Analysis of Trade and Traders.* Academisch Proefschrift, Universiteit van Amsterdam: Amsterdam

Viazzo, P. P. (1989). *Upland Communities: Environment, Population and Social Structure in the Alps since the Sixteenth Century.* Cambridge University Press: Cambridge

Wallace, M. B. (1983). Managing resources that are common property: From Kathmandu to Capitol Hill. *Journal of Policy Analysis and Management,* 2

Webster, D. (1986). *Calgary: Can Amenity Replace Energy as the Propulsive Sector?* Faculty of Environmental Design, University of Calgary: Calgary

Webster, D. (1992). The role of amenity in Canadian regional development. *Plan Canada* (July 1992): 6–15

Weiermair, K. (1983). Some reflections on measures of competitiveness for wintersports resorts in overseas markets. *The Tourist Review,* 4: 35–41

Whinney, C. (1996). Good intentions in a competitive market: Training for people and tourism in fragile environments. In Price, M. F. (ed.), *People and Tourism in Fragile Environments.* John Wiley and Sons: Chichester, pp. 221–9

Williams, P. W. (1992). Emerging North American tourism market: Demassifying travel. In Gill, A. and Hartmann, E. (eds.), *Mountain Resort Development.* Centre for Tourism Policy and Research, Simon Fraser University: Burnaby, pp. 1–6

Williams, P. W. (1995). International native-based tourism markets. In *Proceedings, Conference on First Nations Tourism and Resort Development.* Native Investment and Trade Association, Vancouver, pp. 12–35

Williams P. W., Dossa, K. B. and Fulton, A. (1995). Tension on the slopes: managing conflict between skiers and snowboarders. *Journal of Applied Recreation Research,* 19: 191–213

World Travel and Tourism Council (1995). *Travel and Tourism's Economic Perspective.* World Travel and Tourism Council: Brussels

Wright, J. B. (1993). *Rocky Mountain Divide: Selling and Saving the West.* University of Texas Press: Austin

Zimmermann, F. M. (1995). The Alpine region: Regional restructuring opportunities and constraints in a fragile environment. In Montanari, A. and Williams, A. M. (eds.), *European Tourism: Regions, Spaces, and Restructuring.* John Wiley and Sons: Chichester, pp. 19–40

Montane forests and forestry   13

Lawrence S. Hamilton, Donald A. Gilmour, and David S. Cassells

INTRODUCTION

Montane forests are capable of providing society with a wide range of goods and services. The values of these different outputs frequently are not fully expressed in the market place and, as a result, policy interventions are often required to protect indirect and non-tangible values. Table 13.1 summarises the range of outputs provided by montane forests in relation to the ability of markets to provide values for particular products and the size of the community that uses them. A recurring theme of this chapter will be the changing community perceptions of these various forest values.

The popularly perceived picture of a mountain shorn of its forest in many places is a valid one. But the situation in montane forests world-wide is far more complex; simple categorisation and naive broad-brush solutions are not warranted. Nonetheless, a generalised assessment of conditions today is presented below.

The extent of forest cover in mountains of *temperate zone* industrialised countries, in general, has stabilised or is increasing slightly. Lands abandoned by highland agriculture and grazing, amounting to thousands of hectares annually, have been reforested gradually by natural regeneration. Artificial reforestation has also added to the forest extent in many countries. In Switzerland there is even concern that the increase of forest cover is degrading the rural landscape which, with its mosaic of fields and forests, makes the Swiss middle-mountain scenery so appealing to most eyes (tourists and natives alike) (Plate 13.1). This process has been going on for decades in highland areas of North America, Japan, Western Europe, and New Zealand. Conditions are more difficult to ascertain in Australia, temperate South America, Eastern Europe, and northern Asia where on-going montane forest clearing is at least partially off-set

Table 13.1 Classification of joint products of mountain forests

OUTPUT	TYPE OF GOOD		
	Private (market)	*Impure public*	*Pure public*
Ecosystem diversity			Option/existence
Fish	As input to economy (sold)	Recreational use	
Forage	Grazing permits sold on open market	Community use (local public good)	
Game	As input to economy (sold)	Recreational use	
Genetic diversity			Option/existence
Hazard protection		Individuals' life, property, safety	Public land, facilities
Landscape		Limited access viewpoints	Public access viewpoints
Recreation	Developed: ski areas, private campgrounds, etc.	Undeveloped: trails, campsites, picnic areas	
Water quality	Industrial, municipal, domestic use	Recreational use	Perception
Water quantity	Industrial, irrigation, municipal use	Recreational use (type of craft)	Perception
Wilderness		Perceived environment for recreation	Existence value
Wood	Sold on market stumpage fees, market products	Community use (local public good)	Long-term security of supply

Source: Price, M. 1990

Plate 13.1   Mosaic of forests and farms in Swiss mountain landscape. (Photograph: L. S. Hamilton)

by extensive reforestation. Even if the area of montane forests is relatively stable the quality has deteriorated in many areas. Natural forest ecosystems have become less diverse due to slow recolonisation with the full complement of late-succession species of both plants and animals, and the use of monocultures in the artificially reforested areas. Moreover, air pollution has damaged many montane forests of the industrialised world.

In the *tropical mountains*, loss of forest cover is still a serious factor. The most recent loss assessment by FAO indicates that upland forests are disappearing at an annual rate of 1.1 percent, a rate greater than for all other forest biomes in the tropics, including the lowland tropical rainforest (FAO, 1993). But the picture here also is complex and varies from region to region and country to country. The FAO assessment indicated that the greatest losses in montane and upland forest area between 1980 and 1990 occurred in Latin America and the Caribbean, with the most extensive depletion in the Central American highlands, Jamaica, Haiti, Bolivia, and Venezuela. Wunder (1996), however, has shown that detailed site survey in Ecuador, for instance, demonstrates the problem of

nation-wide, or regional, aggregation of data. In Africa, montane forests continued to be cleared for agriculture at significant rates in Uganda, Ethiopia, Tanzania, Malawi, and Zambia; and in Asia, the highest rates of loss appear to be in Pakistan, Philippines, Vietnam, Malaysia, and Thailand. However, in contrast, Bhutan has recently announced a new policy of maintaining the country's forest cover (largely mountain forest) at 60 percent of the land area, including 26 percent in formal protected areas (see Box 13.1 by Tobgay Namgyal). In Nepal, where mountain forests at one time achieved world-wide notoriety over the alleged rate of deforestation (Eckholm, 1976), there has been a remarkable re-assessment of what was happening, and a marked shift in landowner practices. In the Middle Mountains, for instance, farmers now augment government reforestation efforts by planting trees on their own lands (Gilmour, 1995). In certain areas of the tropical uplands, the effects of agricultural intensification in the valleys and lower slopes, stall-feeding of livestock, extension of agroforestry practices, tree planting for farm needs, and development of community-based management of 'social forests', are reducing, and even preventing,

Box 13.1   Policy of forest conservation in Bhutan

Roughly the size of Switzerland, the Himalayan kingdom of Bhutan has a population of less than 1 million people, 90 percent of whom are primarily engaged in subsistence agriculture. This small kingdom's unique economic development philosophy is clear in the words of HM King Jigme Singye Wangchuk: 'Gross National Happiness is more important than Gross National Product'. Recently opened to the rest of the world, Bhutan regards material consumerism as a major threat to the nation's cultural identity.

Known for its cultural and scenic splendour, Bhutan could generate more of the foreign currency it requires for its development programmes by increased exports of valuable timber and forest products (more than 60 percent of the country is covered by natural forests), or by ushering in free-market tourism. Having done neither makes Bhutan one of the few countries in the world where conservation of nature, combined with the socio-economic development of its people, is the highest priority on the national development agenda.

Nevertheless, increasing exposure to the world and new development opportunities have introduced consumerism into Bhutanese society. Unsustainable land-use practices are exerting increased

pressure on limited arable land and now pose a greater threat to conservation. This is in part the result of a change from manual farm labour to mechanised and chemical inputs; less crop rotation and loss of valuable topsoil through cultivation on steep slopes; growing livestock and grazing pressure on wildlands; and increasing fuelwood consumption. A high annual population growth rate of 3.1 percent, increasing urbanisation, and an alarming dependence on imported industrial products also contribute significantly to longer-term threats to conservation.

Bhutan has vigorously implemented sustainable development as a practical strategy to counter threats to the country's fragile mountain environment. Government policies promote gradual economic growth for the longer term and discourage steps that could be environmentally and culturally destructive. A decision to maintain the country's rich forest cover at 60 percent of the land area, including 26 percent in formal protection areas, exemplifies Bhutan's genuine commitment to conservation.

Source: Tobgay S. Namgyal
East and Southeast Asia Program
World Wildlife Fund/USA

serious degradation and forest loss. Elsewhere, much remains to be improved in terms of negative impact of mountain farming.

The logging of forests in mountains has a greater potential for damage both on-site and off-site, than in gentler topography. On steep slopes, where soils are unstable, the roads and skidding tracks used to remove the wood often initiate large mass movement of soil which is not a factor in the lowlands. Reasonably erosion-free logging techniques for mountains have been developed, but are not generally applied, even in industrialised countries, so that much mountain logging and road construction is still having serious adverse impacts.

Logging, fuelwood, and leaf fodder collection; slash-and-burn subsistence agriculture; land clearing for commercial agriculture; clearing or drowning land for reservoirs; bulldozing for new mountain roads; mining; felling for transmission lines or pipelines; harvesting forest game; grazing by livestock; and burning – these are all processes that augment forest clearance or degradation, and are still occurring in almost all countries of the world.

In addition, there is evidence of deterioration of montane forests due to atmospheric pollution, mainly from increased acid precipitation, but also ozone, photo-oxidants, and heavy metal deposition. Global warming will have more pronounced effects on montane forests than on any other kind of forest due to the altitudinal zonation and stressed upper-elevation situations.

In mountain areas of many parts of the world there seems to be a shift from economic production functions to social/ecological functions where environmental services (such as carbon fixation and watershed protection), scenic resources, and cultural heritage values are increasing in importance. Some of these develop economic functions through tourism, improved water quality, and carbon sequestering offsets, and these values often outweigh the direct product values. And, in most of the world, mountains are often the last bastion of undisturbed nature, islands in a sea of transformed landscape, and are thus of great importance in biodiversity conservation of wild species. Many of the aforementioned facets of forest functions are described in the following sections.

THE HETEROGENEITY OF MOUNTAIN FORESTS

Mountains are often differentiated from hills not only by range of relief, but by virtue of having different altitudinal zones, or belts, of vegetation. These different altitudinal belts are primarily due to temperature differences, since the altitudinal lapse rate ensures an average temperature decrease of 0.6 °C for every 100 m increase in elevation. In addition, as elevation increases there may be major differences in soils and in soil depths due to slopes, and in the steepest portions there may be talus, or no soil at all. Where there has been glaciation, one may move from coarse outwash soils on lower slopes and valleys to glacial tills at higher elevation and these give rise to quite different types of vegetation. Toward the summits, at or near treeline, poor growing conditions and exposure may produce a 'krummholz' of short-statured, twisted-stem, foliage-flagged trees. And above treeline a belt of alpine herbs and shrubs of great variety occurs, depending on rock type and latitude. Forests at treeline, and this alpine tundra belt, are possibly the most fragile of the mountain's vegetation because they recover very slowly following any disturbance. Any lower treeline, influenced mainly by moisture deficiency, is also an especially sensitive ecotone.

Heterogeneity is also increased by aspect. This results in receipt of different amounts of solar radiation and moisture (often a wet side and a dry side if there are prevailing winds) especially for large mountains or mountain ranges (see Chapter 17). These factors also give rise to contrasting types of forests, or other vegetation, on the different sides of a mountain. For instance, where there are wind-driven clouds at a certain altitude, montane cloud forest may occur, especially in the tropics. These have an unusual hydrological and biological importance and are picturesquely termed 'eyebrows of the mountain' (bosque de ceja montaña).

An example of the variation in montane forests and vegetation that occurs on only one side of a mountain is that of Mauna Kea in Hawai'i; it is shown diagrammatically in Figure 13.1. The west slope (dry side) has a quite different profile. The variability in major vegetation types in the tropical Andes is demonstrated in Figure 13.2 which shows the western slope along a north–south transect from Colombia to Chile (4°N to 30°S latitude). In tropical mountains the vegetation can vary from tropical rainforests at the base to summit vegetation which has its greatest affinity to high-latitude arctic tundra. Mount Kenya and Mount Kilimanjaro in East Africa are dramatic examples.

This heterogeneity of forests and other mountain vegetation implies great biological diversity, especially when one considers the variety of micro-habitats that exist particularly in mountain areas (for example, in caves, gorges and waterfalls, behind boulders on exposed slopes, and so on). The importance of mountains as a treasure-house of biodiversity at both the species and ecosystem levels, is discussed in Chapter 10.

One of the major types of forests that is unique to mountains is the cloud forest, also known by the terms, mossy forest, elfin forest, and many others in different languages, such as, nebelwald, fôret néphéliphile, or bosque de ceja montaña. These usually are short-statured, twisted, epiphyte-laden, dense-canopied forests of hard-leafed species and occur where persistent wind-driven cloud strikes the mountains. They are usually found in a relatively narrow altitudinal range, but their actual position varies widely from tropics to temperate zones, and from island mountains in oceans to large inland mountain systems. Cloud forests have developed as low as 500 m above sea level on small islands in humid equatorial locations (e.g. Fiji). More often they are found at elevations of 2000–3500 m in mountain ranges such as the Andes or the Ruwenzoris (see Andes schematic in Figure 13.2). From the movement of saturated air masses these trees, with their epiphytes, lichens, and mosses, are able to condense (or 'strip') water which drips or runs to the ground and is added to the water budget of the area. The hydrologic role of these forests is described in Chapter 15. In areas of low rainfall where closed canopy forests do not develop, even scattered individual trees in the cloud belt can capture this horizontal, or occult, precipitation and provide water for wildlife, domestic stock, or even human consumption. Where these forests have been removed, this water-capture function can be restored by reforestation. In their natural state these cloud forests, because of their unusual environment, structure, and species, have a great importance for biodiversity. Unusual wildlife species occur here, many of them endemic to a single isolated mountain or to a mountain range. The mountain gorilla and the resplendent quetzal are examples. A recent biodiversity project by BirdLife International shows the importance of tropical montane cloud forest to restricted range and threatened bird species. Long (1995) reported

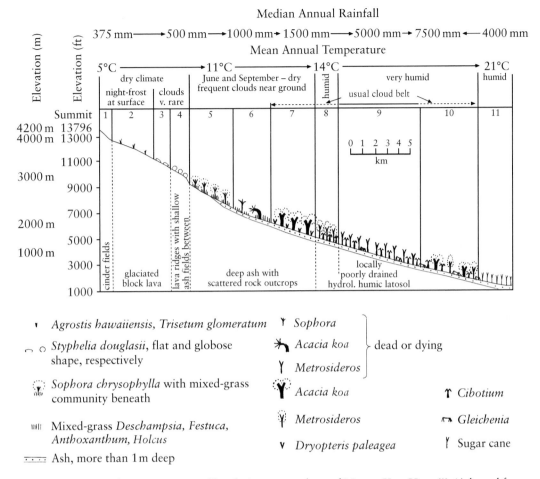

Figure 13.1 Topographic ecosystem profile relating to east slope of Mauna Kea, Hawai'i. (Adapted from Müller-Dombois and Krajina, 1968)

that ten percent of all restricted range species in the world have cloud forest as their sole, or main, habitat. Unfortunately for both hydrologic and biologic functions, these forests are rapidly disappearing and, combined with losses of other upland forests of the tropical world, the total rate of loss exceeds that of other classes of forest, including the much more publicised tropical lowland rainforest (FAO, 1993). A compilation of the state of knowledge of tropical montane cloud forests and a synthesis have been prepared by Hamilton *et al.* (1995). Well-known mountains with cloud forests include: Mount Elgon (Africa), Mount Kinabalu (Asia), and Pico de Neblina (South America).

The upper treeline belt represents another interesting and very fragile environment. Often the transition from forests to alpine meadow, or tundra, is quite abrupt; this is typical of the southern beech forests (*Nothofagus*) of New Zealand's Southern Alps. In some areas abrupt artificial treelines have been created by humans who have developed upper pastures for livestock. In others, the transition is more gradual, with the closed forest deteriorating to deformed island groups (often a single tree that has rooted its lower branches) or wind-blasted and dwarfed individuals. The name 'krummholz' has generally been applied to this form of vegetation. And elsewhere, in this harsh climate, dwarf varieties of tree species that are almost prostrate have developed. Relatively small trees may be hundreds, or even thousands, of years old, and thus many are useful in dendroclimatological work where the variation in annual ring width of the stem gives indications of past climatic conditions. It is in this upper elevational belt at treeline that the

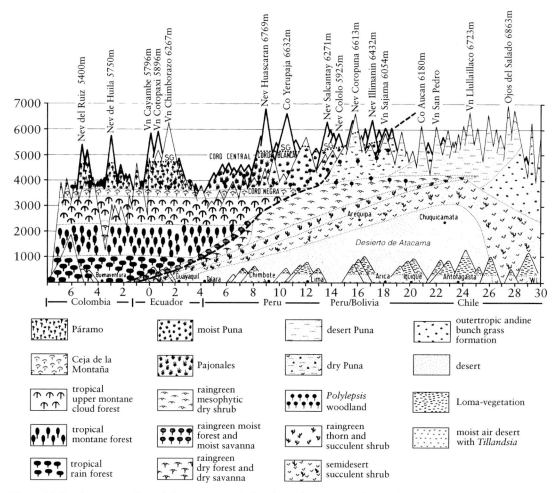

Figure 13.2 A cross-section of the vegetation belts along the western slope of the tropical Andes. (W. Lauer, 1993)

bristlecone pine (*Pinus aristata* var. *longaeva*) is found in the mountains of Utah, Nevada, and eastern California, USA. (Plate 13.2). There are living specimens here with ages of over 4000 years which have provided a wealth of climatic data. Techniques for correlating living specimens with dead trees have pushed the record back to about 10 000 years. These dendrochronological techniques have been used to demonstrate that the carbon-14 method of dating was flawed and thus facilitated the incorporation of corrections. Until 1995 it was thought that the bristlecone pine was the oldest living organism, but this distinction is now given to a recently discovered huon pine (*Lagarostrobus franklinii*), one of the Podocarps, found on Mount Read in Tasmania (Dr J. Kirkpatrick, 1996, personal communication). This species is a vegetative reproducer (as well as by seed) and it is estimated that although the original stem has not persisted, a single individual has existed for about 10 000 years and occupies more than a hectare. And now, in another remote corner of Tasmania a shrub, up to 8 m in height, and as a clone covering over a square kilometre in two gullies, was found in October, 1996. *Lomatia tasmania,* the King's holly, is reputed to be about 40 000 years old (Reuters, October 18, 1996). As natural archives of climate, these old growth forests near treeline need protection, and fortunately many are being given national park or other protected area status. It is especially important to protect such forests on tropical mountains for climate research and monitoring, since trees at lower elevations do not exhibit reliable annual rings.

Plate 13.2   Bristlecone pine at treeline in White Mountains, California, USA. (Photograph: L. S. Hamilton)

The severe climatic conditions of middle and high mountains in the temperate zones and high mountains in the tropics result in forests in which coniferous species are dominant, due to their adaptations to cold, snow-loads, and dryness due to frozen ground. Thus much of the mountain forests of the world at the higher elevations are evergreen forests of needle-leaved species, the most common genera being *Pinus* (pines), *Picea* (spruces), and *Abies* (firs). *Larix* (larches), another fairly common genus is a deciduous conifer, and *Tsuga* (hemlocks) also occur in the mountains of some parts of the western United States and Canada, Japan, Bhutan, Nepal, and China. Evergreen forests of a broad-leaved genus, *Nothofagus*, are found at high elevations in New Zealand's Southern Alps and in parts of the Southern Andes. *Podocarpus*, a conifer, also characterises many mountain forests in the Andes and Africa. Southeast Asia has its own interesting high-elevation evergreen, *Dacrycarpus*, while in Australia several species of the ubiquitous genus *Eucalyptus* occur at the treeline. The numerous species of the genus *Rhododendron* occur both as ecotonal shrubs and in arboreal form throughout the treeline belt of the Middle East, the Himalaya, and the Hengduan Mountains of southwestern China.

The generally more favourable climate and deeper soils at lower elevations and in mountain valleys produce yet other kinds of forests. In areas of high rainfall in the tropics where there is no serious moisture stress, lower montane rainforests develop with the species complexity/rarity conditions that come close to their relatives, the lowland tropical rainforests, which have been the focus of well-merited concern for the scientific community and environmentalists. However, tropical seasonal forests world-wide have received historically much heavier impact, and it is now the upland and montane forests which are disappearing at the most rapid rate and require urgent attention.

The threat of global warming has rather serious implications for mountain forests. Warming would cause an upward migration and, because mountains taper from a broader base to a small summit, the present forest belts, if they were able to successfully migrate (that is, if the changes in climate are not too rapid), would have increasingly reduced space available under many mountain topographic situations. This upward progression would reduce the total populations of the associated species, thus adding to the danger of extinction due to both genetic and environmental pressures. In mountain ranges running north–south there would be opportunity for poleward migration. Studies of migration rates (particularly of treeline) following the last glacial maximum are shedding light on this topic, but the prognosis is not good even if the conservative scenarios of the Intergovernmental Panel on Climate Change are used. Moreover, warming is only one postulated effect of climate change, for also there would be changes in precipitation amounts and distribution. See Chapter 17 on Climate Change for further insights into the likelihood of climate change and its impact on mountains.

The major exception to this general description of the montane, and especially of the treeline forest communities, is the high mountains of the inner tropics. Here, and particularly in the equatorial Andes and the mountains of East Africa, are found the giant rosette forms of the Andean páramos and the Afro-Alpine vegetation, the latter so well described by Hedberg (1995).

MONTANE AGROFORESTS: TREES IN MOUNTAIN AGRICULTURE AND GRAZING

The land-use pattern in many of the heavily populated parts of the world's mountain regions is often a mixture of agriculture, grazing land, and forest. Trees generally play a critical role in such situations, providing essential elements for the sustenance of the farming system (see Chapter 14). Fuelwood, construction timber, fodder, fruit, and a multiplicity of other tree and forest products provide subsistence goods as well as products for cash sale. In most situations the land-use pattern is very dynamic with changes constantly taking place between the various components of the system. A common evolution of this pattern is that, as human populations expand, the natural forests are steadily reduced in both quality and areal extent to make way for agriculture and grazing. This process often continues until tree and forest products are in critically short supply. At this point communities may perceive that a problem exists and choose to change their land-use practices.

Two of the common responses are to plant trees, or encourage their natural regeneration, on the crop land and to institute protection or controlled utilisation of the remnant forests (see the section on Community Forests for more details on this latter response). There are numerous examples in the highlands of East Africa (Warner, 1993), Ethiopia (Poschen-Eiche, 1987), and the Himalaya (Gilmour, 1995), where this dynamic agroforestry pattern has evolved. However, farmers are able to invest in such long-term enterprises as tree planting and forest protection only if certain social and institutional conditions are met. For tree planting on private land they need to have secure tenure to ensure that they can harvest the results of their labours, given the substantial time lag from planting to harvest. This is the situation in much of the mountain areas of East Africa and the Himalaya, with the result that quite sophisticated agroforestry systems have developed with little input from outside agents. In some of the mountain areas of the Philippines, by contrast, the farming land is almost totally devoid of trees. The inhabitants are largely tenant farmers and the land is owned by absentee landowners, so that the farmers are not prepared to invest in such a long-term crop as trees, although they recognise the severe shortage of tree products.

There are important lessons to be learned from agroforestry experiences in mountain areas. First, farmers invariably have the ability to assess their own needs in terms of tree cover (that is, they rarely need to be 'educated' on this topic). Second, technical inputs can often help the process of agroforestry development by providing previously unavailable species which are desirable additions to the indigenous plants. Third, there needs to be an enabling political environment to ensure that the farmers have sufficient security over their lands and so benefit from their labours in tree planting. Several of these facets are illustrated in Box 13.2 by Jeannette Gurung for some agroforestry systems in the Himalaya.

MONTANE PRODUCTION FORESTS

In many countries, agricultural and urban expansion have occurred at the expense of forest ecosystems and there has been a gradual loss of forests in the more accessible locations. There is often an inverse relationship between steepness of slope and degree of forest disturbance and clearing (see Sader and Joyce, 1983, for an example from Costa Rica). Because of these patterns, mountain forests may be able to provide a disproportionate level of the forest

Box 13.2 Mountain agroforestry in the Himalaya

Agroforestry systems in the hills and mountains of the Himalaya have been in practice for longer than is often recognised by foresters and development workers who are now promoting widespread agroforestry interventions throughout the region. Most commonly encountered is the agro–silvo–pastoral system that combines elements of perennial and annual crops with animals and pasture. The crop/tree/livestock mix typically found around homesteads, presents an example of this. The practice of grazing animals on fallow croplands, the establishment and retention of fodder trees around fields, and the use of woody shrubs and hedges for mulch, browse, and green manure, point to the multi-purpose uses of this diverse system. Mountain agriculture in the past has been sustained through the effective linkages amongst components of crop production, horticulture, forestry, and animal husbandry. The organic integrity of mountain agriculture has suffered in the past due to increased human and livestock population growth, market forces, and public interventions. The result has been reduced quantity and quality of forests, reduced diversity of production systems, and decreased soil fertility. More recently new trends are in evidence. In some areas livestock numbers are declining, and stall-feeding has removed some animal pressure on the land. Agroforestry is now promoted as a means to contribute to both productivity and conservation in the mountain regions.

Farmers' strategies focus on maintenance of soil fertility, largely through the transfer of nutrients from forests and grazing lands to agricultural lands via livestock. The fodder element is the linchpin in the farming system. Fodder from grasses and tree leaves constitute well over 50 percent of the diet of livestock on an annual basis, which is then recycled to produce the valuable manure on which farmers depend. The majority of farmers do not rely on commercial fertilisers as they are unavailable (except where there are roads), expensive, and disliked due to negative effects on soil structure and pH.

Trees are often the sole suppliers of timber and fuel as well as providing fodder. Other uses of agro-forest trees are for fruit, income generation, medicines, food, fencing, and construction materials.

As farmers' land-extensive and subsistence-oriented strategies are now becoming less feasible, agroforestry emerges as an option to ensure the biomass-generating functions of forests and pastures through the intensified use of smaller land units. Agroforestry is particularly well suited to the special circumstances characterising mountain habitats, such as inaccessibility, diverse agro-ecological conditions, and environmental hazards. As such, it is now being promoted by government and non-government agencies in most countries of the Himalaya.

In eastern Bhutan and northeastern India, the common sequential agroforestry practice of *jhum*, or shifting cultivation, is being discouraged by government agencies, who offer subsidies for conversion of land into terraced, irrigated fields. Improved agroforestry practices, horticulture development, and integrated livestock farming are promoted.

State governments of India have actively promoted development of horticulture in the hill states since 1970. The farmers of Himachal Pradesh have greatly benefited from the recently introduced apple industry, combining silviculture and horticulture with animal husbandry. Fodder trees and grasses are being planted on marginal lands throughout the Uttar Pradesh hills, with assistance from the multitude of local NGOs. A programme launched by a matchbox manufacturer encourages farmers to follow a model of alley cropping wheat with poplar seedlings on a buy-back guarantee with loans, technical advice, and quality seedlings.

Farmers in Nepal are improving their agroforestry systems with no prodding from government agencies. Faced with biomass shortages and the need for cash, they have increased significantly the quantity of trees and grasses growing on terraces and bunds as well as the borders of their agricultural land. A very lucrative system has been the production of cardamom under the cover of *Alnus nepalensis*, a multi-purpose tree which provides shade, nitrogen and humus to this profitable cash crop while producing timber, fuelwood, and small poles for subsistence needs. Sericulture practices in eastern Nepal have brought income to women farmers there who raise mulberry trees for silk worm production. These two agroforestry systems have been replicated in communities across the Nepal and Indian hills, as farmers learn of the cash opportunities available while improving the productivity and conservation of farm lands.

Gender issues in agroforestry, which affect women's participation and control over resources and benefits, are beginning to be addressed by institutions responsible for the promotion of agriculture and forestry programmes. Because women are the primary farmers in this region, their exclusion from decision-making processes and from the benefits of agroforestry programmes has affected the efficacy of these programmes. In fact, agroforestry has the potential to reduce substantially the excessive workloads of mountain women, by providing biomass resources close to their homes so as to minimise time and energy spent carrying fuelwood and fodder from distant sources.

The observed reasons for farmers' non-adoption of improved agroforestry practices are numerous, and represent a broad spectrum of constraints that differ from country to country. At the farm level, a basic problem is that of limited landholdings as well as a high preference for cereal crops. Increasingly, male out-migration is creating critical labour shortages. Planned agroforestry interventions face problems of co-ordination among concerned agencies. Inappropriate extension systems, inadequately equipped to reach farmers in remote areas, and female farmers, in particular, are a common malady in this region.

In general, agroforestry programmes in the Himalayan region do not receive nearly the degree of donor support and national attention given to the community forestry programmes. Yet, if agroforestry projects start with an adequate understanding of the strategies and knowledge already in use by mountain farmers; if they employ participatory methods to ensure the maximum participation by farmers in research and extension; and if they offer a choice of feasible technologies, species, and market schemes, there is a great potential for conservation, agricultural production, and income generation on private lands in ways that programmes on community-managed lands cannot replicate.

Source: Jeannette Denholm Gurung,
Mountain Farming Systems,
International Centre for Integrated
Mountain Development

goods and services demanded by society, despite environmental fragility.

Mountain forests are important sources of timber in many countries. In Austria, for instance, the majority of forests are located in mountainous areas, with just over 50 percent of the productive forests on slopes in excess of 36 percent (Tuchy, 1982). In Albania, most of the timber harvesting occurs in relatively difficult mountainous country with some 80 percent of forests occurring on slopes between 20 and 50 percent (World Bank, 1995). In Slovakia, more than 60 percent of the nation's forests occurs on steeplands with slopes in excess of 60 percent (World Bank, 1994).

Extension of commercial forestry activities into montane forests has also occurred in many tropical areas. In Malaysia, the uniform silvicultural system, successful for lowland forests, was inappropriate for hill forests, which became targeted for forest products as agriculture and tree crops took over in the lowlands (Lamprecht, 1989).

In many countries, forests have been modified for centuries and few truly natural forests are left in Europe and parts of North America (see Dudley, 1992, for estimates of the proportion of old growth forests remaining in selected temperate forest countries). In addition, many modified forests have been biologically simplified and degraded due to unsustainable logging, air pollution, and fragmentation caused by urban growth. As a result, many authors have suggested that there is a biodiversity crisis in forests of both tropical (Wilson, 1992; Myers, 1988) and temperate (Dudley, 1992) regions.

Reductions in forest area and in the availability of forest goods and services have historically led to the development of institutional arrangements to facilitate forest management to meet the needs of various social interests. Westoby (1989) notes how, as early as the eleventh century, William the Conqueror and later monarchs restricted the customary use of British forests by the peasantry by setting aside large tracts for royal use and creating forest laws with stiff penalties for trespassers. These patterns were repeated elsewhere and there were many peasant revolts throughout Europe between the twelfth and the fourteenth centuries in protest over the loss of customary rights. Both the Church and large landholders appropriated forest lands for agriculture and obtained title through enclosure. However, peasant resistance and forest clearance continued until the 1800s in Europe as forest laws multiplied to punish those who trespassed in the forest for food, fuel, or forage. Westoby describes riots in France in which peasants cut thousands of trees and notes that Karl Marx reported that five-sixths of all crimes in Prussia were for forest theft.

Similar social alienation due to restricted access to forest resources occurred as the European colonial powers turned their attention to the developing world for plantation agriculture and timber resources such as teak, mahogany, and sandalwood. Until relatively recent times, forests in many developing countries were communal or tribal domains in which members of the community or the tribe had defined customary rights of access and use (Panayotou and Ashton, 1992). However, over the last two centuries or so, colonial and national administrations have brought the majority of the world's forests under government ownership. Forest legislation has given special legal status to particular forest areas in the various forms of reserves and national parks. In some countries, areas officially designated as government forests may cover more than half of the national land area.

This expropriation to government ownership has frequently caused significant resentment amongst local people living in or near the forest because of failure to recognise or accommodate customary rights. Nowhere has this been more in evidence than in the mountains traditionally remote from government demands and enforcement. One of the most widely-known protests, initiated by the 'tree-huggers' of the Chipko movement, arose in the Indian Himalaya (Shiva and Bandyopadhyay, 1986). In addition, governments in many countries did not have the capacity to exercise their ownership rights and enforce appropriate forest laws and regulations. Indeed, in many areas, the only direct government involvement with forest lands is the granting of inadequately supervised concessions to private logging companies. Thus, many nominally government forest areas have become *de facto* open access resources with an undefined but large number of non-exclusive claimants. As a result, forest destruction and degradation through inadequate management, log poaching, encroachment, and squatting have become pervasive in many countries, especially in the highlands.

Sustainability has long been a focal concern of the forest sector and the forestry profession. However, what individual foresters mean by sustained yield can vary quite significantly. Cassells *et al.* (1988) provide a review of the evolution of sustained yield concepts and how they relate to Old World forests, New World forests, and tropical forests. In its earliest and simplest forms, sustained yield forestry aimed to balance harvesting with growth. While this approach was effective in many of the European forests, it was not effective in many forests in Russia, in North America, Australia, and the tropics. This was because old growth forests in these areas had accumulated substantial volumes of timber, with large trees that had developed over the centuries without having been subjected to industrial scale utilisation pressures. In these forests, the initial timber management challenge was to oversee the orderly utilisation of the accumulated volume, in an extended transition period,

that would ultimately lead to timber yield allocations balanced with managed growth.

The fundamental weakness of sustained yield forestry is that it gave undue emphasis to industrial wood production and largely ignored the other non-wood forest values and uses. Additional problems occurred because forest managers lacked a detailed understanding of both the ecological and social processes that govern resource sustainability.

In this situation of inadequate but rapidly evolving understanding of ecosystem processes, growing demands for all forest goods and services are leading to rapid changes in both forest ecosystems and the policy environment that governs their management. New insights from both the ecological and social sciences have undermined the authority of forestry professionals and the institutional arrangements for forest management in many countries. Concerns about the interests of non-industrial stakeholders and the rights of peoples dependent on forests have become increasingly enmeshed with concerns about forest biodiversity and fragile habitats. Nowhere is this more in evidence than in the montane forests of the world.

As a result, forest policy issues, which only a few decades ago were viewed as technical and remote from the concerns of community life, have aroused considerable public interest in both the developed and the developing world. In many countries, the public is increasingly demanding participation in decisions that have long been the purview of professional foresters and government officials. In countries like the United States, these social changes have led forest policy makers, forest managers, and the general public to seek new alternatives for management, indeed to calls for a 'New Forestry' (Brooks and Grant, 1992a and b). Globally, they have led to heightened concern about the ecological, social, and economic sustainability of forest management systems.

Recent statements on sustainable forest management (Poore *et al.*, 1989; ITTO, 1991, 1992; IUCN, 1992) go well beyond concern with the mere maintenance of timber volumes and site productivity; they focus on the sustainability of forests as ecosystems. They suggest that for timber production from natural forests to be considered sustainable, it should only be permitted after land planning has allocated forest lands for the protection of biodiversity and fragile environments, such as steep slopes and critical watersheds. This ecosystem focus is perhaps best summarised by Botkin and Talbot (1992) when they note that sustainability of the forest ecosystem refers to 'maintaining the integrity of the natural forest in terms of its structure, composition (i.e. its species composition and biological diversity), ecological processes, along with the environmental services it provides'.

In mountain areas, these general concerns about the impact of industrial utilisation on forest ecosystems need to be amplified to match concerns about the specific characteristics of mountain environments, such as their relative inaccessibility, fragility, marginality, and heterogeneity (Jodha, 1992). The more active hydrologic and erosional dynamics of mountain environments require special attention to engineering design when planning forest utilisation. In addition, the ecological and social diversity of montane forests suggests the need for a specific mountain forest perspective to govern management (Table 13.2).

Fuelwood

Globally, fuelwood and charcoal production consume nearly 3000 million m^3 of timber each year and account for some two-thirds of estimated global wood consumption (Sharma *et al.*, 1992). Some three billion people depend on wood as their primary source of energy, and with rapid population growth in developing countries, a deepening fuelwood availability crisis has been predicted for many years (de Montalembert and Clement, 1983).

Fuelwood gathering has unquestionably degraded forests in many mountain environments, including East Africa, the Himalaya, and the Andes. From the late-1970s onward, these concerns have provided the conceptual motivation for many aid-sponsored interventions to redress problems of tree scarcity and land degradation associated with 'deforestation'.

Table 13.2 Dominant features of conventional and mountain perspective based approaches to forest management

CONVENTIONAL APPROACH	MOUNTAIN PERSPECTIVE BASED APPROACH
Primary focus and concern	
Forest treated as an isolated, revenue- generating sector of a region; yield of selected key products (e.g. timber) focused	Forest as an integral component of ecosystem; inseparability of sustainability of the two; emphasis on both 'service' and 'product' functions of forests
Dominant products and usage system	
Timber and other high value products; market-directed over-extractions; insensitivity to negative side effects; isolated sectoral activity run through legal and administrative superstructures	Diversified biomass-based, interlinked activity patterns (e.g. farm–forest linkages); compatibility with ecosystem needs, people sustenance strategies, and user perspectives
Valuation norms/yardsticks	
Market-based narrow yardsticks for pricing products, compensating for extractions, and determining investment and subsidies; unequal terms of exchange (compensations); insensitivity to local concerns	Focus on health and stability of total system and interlinked activities with concern for multiple externalities; compensation mechanisms also involve biophysical components
Research and development approach	
'Extraction'-oriented approach with focus on monocultures of selected attributes (e.g. timber with high value); with little concern for folk knowledge, local needs	Focus on sustained biodiversity and linkages, regeneration and conservation, and people-centred possibilities; effective use of folk knowledge (folk agronomy, ethnoecology etc.)
Sustainability prospects	
Emergence of indicators of unsustainability (i.e. persistent negative changes in the health, productivity, usage pattern of forest)	Possibility of restoring sustainability by sensitising forest interventions to mountain perspective (mountain specificities)

Source: Jodha, N. S. 1992

Many of these programmes have not been successful because of their overly simplistic assumptions about the interactions between rural people and forest resources. Gilmour and Fisher (1991) note that in the late-1970s, the solutions to these problems were seen as simply to plant more trees and they suggested that, in Nepal at least, the nature and condition of the remaining natural forest was largely ignored. The 'technical fix' of reforestation fitted neatly into the programme requirements of aid agencies but ignored the broader farming system and social context in which forestry needed to be practised (Chapter 14). In many areas, labour availability, preferences for alternative fuels, and assignment of higher priorities to other problems, such as the need for cash income, or for more animal grazing, or for shelter, have all undermined the success of tree planting efforts designed to address apparent fuelwood availability problems.

From this experience, it is clear that foresters and development planners require a much more detailed understanding of the interdependencies between trees and people in mountain environments. For management programmes to be successful, they must be based on a thorough understanding of people–environment relationships and be able to respond to the broad range of household needs for both wood and non-wood forest products.

Fuelwood is a major output from mountain forests for local use whether in the tropics or the temperate zone. And, fuelwood and charcoal (which is more easily transported per unit of heat supplied, or for speciality uses) continue to be exported from mountain forests to lower elevations depending on the price of alternative fuels or the romantic desires of lowlanders for a wood fire. The Green Mountains of Vermont, USA, are now supplying wood as chips for a modern wood-fired electrical generating plant that produces 50 mega-watts (MW) and consumes about 75 tonnes per hour. Fuelwood removal does not usually entail such potentially damaging heavy equipment as does commercial timber production and, if undertaken carefully, does not represent as great an erosional hazard. It does require restraint in terms of having a sustainable yield from a given area. If done wisely and well, as a stand improvement measure, it can enhance the value of the forest for subsequent commercial logging, and can have little adverse impact on other forest environmental services.

Non-wood forest products

Mountain forests provide rural people with a wide range of non-wood forest products including fruit, nuts, mushrooms, roots, sap, cane, resins, gums, spices, medicinal plants, and fodder. Dr Pei Shengji (1994, personal communication) found that the mountain minorities of Xishuanbanna (China) brought to the markets for sale or barter some 170 different plant species from their forests and forest farms. These were classified: 106 species as food related; 20 species as fibre and wood items; 14 species as spices or dyes; 20 species as herbal medicines; and 10 species as associated with cultural activities. These products constituted 70–80 percent of the total goods in the mountain/valley markets. Although individual non-wood products are usually less economically significant than timber, as a group they provide more regular income and can sometimes contribute more to domestic and international economies on a per hectare basis than timber. Lampietti and Dixon (1995) have produced a valuable review of this topic. See also Box 13.3 for an example from Uganda's Mount Eglon.

Most non-wood products are consumed locally, but international markets for products such as rattan, mushrooms, maple syrup, nuts, ornamental

---

Box 13.3 Importance of non-wood montane forest products

A study was conducted in 1994 in a number of communities collecting forest products from Mount Elgon National Park, Uganda. The study focused on the importance of the forest for local community socio-economic and cultural benefit. It was found that the following products were the most commonly collected and traded: bamboo shoots, bamboo stems, poles, firewood, medicinal plants, mushrooms, honey, green vegetables, ropes, timber.

Grazing and hunting were also important. The net value of the forest products collected by local communities averaged US$ 60 per household per year (gross value of US$ 100 less labour of US$ 40). This amount aggregated to US$ 1.5 million annually for the local collectors (gross value of US$ 2.7 million less labour). This is seven times the estimated value which would be derived from the sustainable offtake of timber.

A distinction was made between the 'near' villages (those villages adjacent to the park boundary), and the 'far' villages (those villages distant from the park boundary, but still within parishes adjacent to the park). In the 'near' villages, 86 percent of the households sampled were users (either collectors or buyers) of eight main products. It was determined that 54 percent of the households were collectors of these products and 32 percent were buyers. In-forest activities consumed 40 percent of household productive labour time for the 'near' villages. Most of the collection was for subsistence use (only 15 percent of the labour time spent in the forest was for collecting resources to sell). By comparison, in the 'far' villages, purchase of forest products was more frequently observed: for the eight most important products, 70 percent of the villagers were users of these products, with 53 percent being buyers and only 17 percent being collectors. In-forest activities consumed 32 percent of household productive labour time for the 'far' villages, if grazing is included, but only 12 percent of labour time without grazing.

In a separate study on bamboo, it was estimated that the minimum annual gross value of bamboo to neighbouring communities was approximately US$ 750 000 for a population of 45 533 households. This figure is more than twice that of the gross value of the potential sustainable timber harvest from the Mount Elgon National Park.

Source: Scott, P. (1997)

---

plants, resin, and medicinal species already exist, and new markets for fruits, medicinal plants, and chemical extracts are developing. India, China, Indonesia, Malaysia, Thailand, Sudan, Brazil, and Guatemala already have significant exports of non-wood forest products from their highlands. Maple syrup, produced on the lower slopes of mountains of the northeastern USA and eastern Canada, can be purchased in Austria and Australia.

The harvesting of non-wood forest products from montane forests in the tropics faces many of the same problems that are associated with timber production in these areas. The basic characteristics of tropical plant populations limit the process and intensity of resource exploitation. The major challenges include the diversity and low density of any one tree species, the complexity of flowering and fruiting, the specificity of micro-sites for successful regeneration and growth, and the response of population structure to changes in the level of recruitment (Peters, 1996). To date, in the temperate forests much of the commercial exploitation of non-wood resources is plagued by destructive harvesting, over-exploitation, and a basic disregard for the functional ecology of plant populations. Nevertheless, like timber extraction from both temperate and tropical forests, the limiting factors are often economic and institutional rather than technological. Marketing arrangements can be especially disadvantageous for indigenous mountain collectors of non-wood products, as indicated in a Nepalese study by Edwards (1996).

A considerable amount of indigenous knowledge of potentially sustainable management systems is frequently available (Peters, 1996). However, in many cases, the production of non-wood forest products by people living in or near the forest tends to be marginalised, and products with significant market appeal tend to be drawn into the external economy through the development of plantations and other forms of estate production (Dove, 1993). Even small-scale family mushroom collection in the coastal mountain forests of Oregon/Washington, USA, and British Columbia, Canada, is becoming commercialised due to lucrative overseas markets, with large companies sending hundreds of harvesters into the woods. There is inadequate knowledge of sustainable levels of mushroom harvesting and of the effects of trampling. Perhaps the licensing system in effect in some areas of the Appenines in Italy, or the secretive truffle-collecting individualism of the French foragers, offer paths to conservation. The sustainability of mountain forest production systems is thus intimately related to the institutional arrangements, tenure, and political marginalisation of people living in and near these forests, and to the external forces seeking to profit from their exploitation. There are many examples where outside concessionaires are given the legal authority to harvest non-wood products and usurp the indigenous, traditional people who have often harvested in a sustainable fashion. These local people are then often hired, and knowingly become involved in levels of exploitation that are unsustainable. Because of the political and institutional factors, they are powerless to do otherwise, and subsequently they receive a large share of the blame for the degradation of the resource (for example, orchid harvesting in the tropical Andean cloud forests).

THE PROTECTION FUNCTION OF MONTANE FORESTS

In addition to the tangible products from forest harvesting just discussed, montane forests provide a host of intangible, or indirect goods and services, or that are tangible but difficult to quantify. These range from game hunted for sport or sustenance to animal watching, from recreational hiking or riding to scenic landscapes and sacred groves. It is common to set aside forests under some kind of protected status to assure that these valued benefits are sustained despite forest exploitation that would impair or destroy them.

While several other chapters in this book deal with associated aspects of montane forests, here the emphasis is on their protective role with respect to snow, water, and soil that persistently move downhill due to gravity. The three-dimensional form of mountains is one of the basic reasons why they need special attention as they undergo human transformations in land use and infrastructure.

Those who live or work in mountains know well the hazards of avalanches, torrents, landslides, rockfall, and the like, when snow, water, or soil move at great speed down steep slopes. Less dramatic, but also of great significance, is the process of surface soil erosion that is responsible for shifting productivity downslope, leaving the area above degraded and, in extreme cases, unusable (Chapter 15). The role of montane forests in providing the most stable kind of cover to reduce the impact of these natural downslope processes has been long recognised. There are allusions to their positive role in the face of natural hazards in ancient Greek, Chinese, Hebrew, and Roman literature. For instance, the removal of the Lebanon cedars in Biblical and Roman times, caused extensive land degradation, augmented by the spread of overgrazing by goats (Lowdermilk, 1953). More recently, in the 1850s, forests were protected in the European Alps to reduce the hazard of avalanches and mountain torrents; local regulations were codified into law in Switzerland in 1872 (Zwerman and Richard, 1959). At about this time, across the Atlantic, a major portion (all the state-owned land representing some 1 021 500 ha) of the Adirondack Mountains in the State of New York was recommended for protection as a Forest Reserve to safeguard the headwaters of the navigable rivers (see Chapter 15). This action recognised the role of mountain forests as 'regulators' of the quality (sediment) and timing of water flow in streams and rivers to benefit the area downstream. Today, at the end of the twentieth century, most countries, states, or local governmental units have established protection forests in hills and mountains to alleviate problems related to negative effects of snow, water, or soil movement. Major portions of the annual budget (up to 11 percent) of mountain prefectures in Japan are allocated to restoration or maintenance of protection forests on the landslip-prone soils (Dr Y. Tsukomoto, 1987, personal communication).

It is important to recognise that forests will not prevent avalanches, floods, rockfall, landslides (and other kinds of mass movement, such as debris flows, soil creep) or even surface erosion. Nevertheless, under forest cover, the incidence and magnitude of these events is less than under any other land cover or land use: hence the frequent designation of 'protection forest' where there is a high probability of such hazards.

Construction of housing or ski resorts below the forest in avalanche-prone areas remains unwise even where there is protected forest upslope. Avalanche hazard mapping and avoidance of these areas is usually the best policy, although rapid growth in winter tourism has prompted heavy investment in artificial structures in many areas. Nonetheless, forests do reduce the likelihood of fast-flowing snow and, to a limited extent, divert or mitigate movement from upslope. Elimination, or partial clearing of the forests, increases the risk. In one study in the European Alps, 60 percent of avalanches and mudflows were released within the formerly forested area (Professor F-K. Holtmeier, 1995, personal communication). However, in certain areas tree planting has been successful in reducing the risk by restoration of a 'protection' forest.

Similarly, where there are shallow landslips, tree roots can produce a shear strength safety margin to many slip-prone soils. Loss of this shear strength following tree cutting, and the subsequent greater incidence of slope failure, has been studied in the mountains of New Zealand, Alaska, Japan, and western North America (O'Loughlin and Ziemer, 1982). Such hazard-prone areas can be identified in surveys and need to be red-flagged against forest removal. The catastrophic damage from thousands of landslips that occurred in Thailand in 1988 and aggravated flooding there, led to a complete ban on logging public lands; but this was caused not by logging, but by the replacement of forests with agriculture and young rubber plantations on these slip-prone soils (Hamilton, 1992). However, in major storm events, large landslides and other mass soil movements will still occur, even in forests.

Local flooding is no different. Where soil storage capacity is exceeded by total rainfall amount and slopes are steep, there will be rapid surface and soil movement of water to streams. If the precipitation is prolonged (or of high intensity), stream bank capacity may be exceeded from the runoff and there will be local flooding, even under undisturbed forest cover. But, under forest, the soil storage capacity is in the best condition to receive water on the greatest number of occasions (especially if there is deep soil) due to the high use of water by trees. There are, therefore, usually fewer small flash floods under these conditions (see Chapter 15 for further elaboration). The importance of forest cover in reducing soil erosion is also described in Chapter 15 and will not be detailed here. The undisturbed montane forest is an excellent cover to deter watershed erosion, but it is more appropriate to regard forests as having low erosion rates rather than no erosion; and when contemplating reforestation of mountain land, to plan for erosion reduction, rather than prevention.

While undisturbed forests on steep slopes have the greatest protective role for reducing downslope movement of soil, water, and snow, it must be recognised that they are also important sources of livelihood for mountain dwellers; they often are vital for providing timber for a larger economy that may include lowland consumers and fabricators. Although not all mountain forests can be preserved as protection forests, in spite of their fine hydrological and erosional functions, it is very important that, in the most sensitive areas, many should be designated as watershed preserves. Many of these might be harvested on a sustainable basis for their medicinal plants, their wildlife, or their non-wood forest products. Others might be lightly logged, but harvested with great care so that their protective function is not unduly reduced. Ecologically sound harvesting techniques are available to keep risk of damage low, although the timber industry has been unacceptably slow to adopt them.

Mountain forest survey, assessment, and classification into most appropriate use is necessary for establishing protected forest areas and for determining what level of protection or control is needed. In such a process, those forests with spiritual value, high levels of biodiversity, great scenic attraction, and critical soil/water/snow protection values warrant special consideration.

Many mountain areas are prime sites for establishing protected forests for several reasons. The altitudinal variation associated with mountains produces a wide range of ecological conditions with a resulting rich assemblage of plants and animals. Mountain areas, by their very nature, are often remote and have been subjected to less development pressures than lowland ecosystems and, thus, are frequently more nearly 'intact'. However, this same isolation limits the influence of government agencies responsible for the management of protected areas.

AMENITY FORESTS

Montane forests are now used for outdoor recreation by an increasingly urban and affluent society. It is certainly true that traditional rural landowners derived amenities from their woodlands and forests, but they would abjure the word 'recreation'. This is probably particularly true for owners of mountain forests who, by force of circumstances, were obliged to regard them as sources of exploitable products whether wood or non-wood. In contrast, in publicly owned forests of the developed countries, there has been a major shift from a former almost exclusive orientation to wood production to management policies that increasingly emphasise their 'service functions', such as water quality, scenic amenity, erosion and avalanche protection, wildlife habitat and, to a rapidly growing extent, the provision of outdoor recreational opportunity. This trend has begun also in many developing countries. It is well illustrated in the anecdote that in two generations the axe once used to cut down trees by a grandfather is now only used by his grandchildren to pound in tent pegs in forest campgrounds.

Chapter 12, dealing with Tourism and Amenity Migration, discusses the role of forests as sites for recreational pursuits. Since scenic mountain vistas are so important to recreational users, the totally closed-canopy forest is not a preferred locale (see Plate 13.1). Open woodlands, or periodically created openings, to provide vistas are preferred by recreationalists. Ski runs and lifts require wide tracts cut through the forest. Much activity is concentrated in alpine meadows and in the forest–alpine meadow ecotone. These are all fragile environments and require control of visitor activities in order to maintain the natural or semi-natural character that determines their recreational and scientific biological interests. Here, trail location and rehabilitation, control of trail damage by restriction of number of users, or by regulation of certain uses (e.g. motor- and mountain bikes) in sensitive areas, are often required. On mountain trail expeditions in several areas of the United States, llamas from the Andes have been substituted for the conventional mule or horse as pack animal, because of the lower hoof impact of these animals. In 1993 there were roughly 30 million visits to 10 mountain national parks in the USA, and most were in what is, or what was, a forest ecosystem (Hornback, 1994). This is a heavy impact, though it was mainly confined to limited visitation areas.

For those activities where cooking or warmth are required, controls on collection or cutting of fuel have become necessary in most popular amenity forest areas. Severe impacts on forest vegetation, for instance, have occurred along the popular trekking routes in the Nepal Himalaya; for instance, for the Annapurna Sanctuary, Gurung and de Coursey (1994) have pointed out how serious these effects are, especially on the showcase rhododendron forests. The same is true for the juniper shrublands along the Sagarmatha (Mount Everest) trekking route. Mountain users increasingly are being required (and always urged where controls are not enforced) to bring in their own fuel in order to reduce their adverse impact on mountain forests. Similarly, solar energy, wind, and small 'hydel' installations, are being increasingly used by the tourism service industry to reduce reliance and impact on mountain forests where hydrocarbon fuels are expensive, not available, or because they represent a softer energy source (see Chapter 8 on energy resources for remote highland areas).

Even recreational pursuits not normally associated with mountain forests are having an impact. Forest clearing for golf courses or resort hotels in the cloud forest areas of the Genting and Cameron Highlands of Malaysia and Mount Kinabalu in Sabah (Malaysia) represent a new threat in the tropics (Hamilton, *et al.*, 1995). Golf courses proliferate in the European Alps, mainly in the forest/farm zone, where they increased from 100 in 1990 to 250 by 1992, and 500 are expected by the end of 1996 (International Environment Report, 1994).

In some cultures there is no serious incompatibility between many amenity uses and controlled forest harvesting for timber or other forest products. Examples include the Black Forest of Germany and most parts of Switzerland. In other cases, such as in the USA, much of the amenity-oriented community, through its environmental advocacy groups, is asking that mountain forests not be exploited, but be reserved as 'unmanaged' wildland for wilderness recreation and biodiversity conservation.

Many species of forest animals also are involved in the amenity use of mountain forests, both for viewing (non-consumptive) or for hunting or trapping.

FOREST WILDLIFE HABITAT

The different slopes, exposures, and vegetation belts of mountains produce a variety of habitats and, therefore, a high diversity in fauna. Among the most interesting and rare species are the mountain gorillas of the Virungas in Zaire, Rwanda, and Uganda; the resplendent quetzal of Central America's mountain archipelago; the red panda of Eastern Himalayan forests; the Andean spectacled bear; and the European lynx found in scattered locations in mountains of Central Europe. Mountain forests are often the last wild refuge of hitherto wide-ranging species, especially top carnivores. Many forest protected areas are given this conservation status in order to maintain populations of some of these rare animals.

In the tropics, where mountain forests are still being cleared for agriculture or grazing, the resulting shrinkage and fragmentation of habitat is taking serious toll of wild animal populations. In many temperate zone mountains of the industrialised countries, the forested area is increasing as agricultural lands are abandoned, and this provides additional habitat for the natural recolonisation of formerly extirpated wildlife, or for its reintroduction. A good example of the former is the spread of the wolf in the Appenines of Italy which now has moved into the Maritime Alps of both Italy and France. A further example is the recent release of wolves from Alberta, Canada, into the mountain wildlands of Idaho and Wyoming, USA. This, however, has caused nervous reactions from stock graziers, and there have been some confrontations.

Subsistence or sport hunting for game in mountain forests is an ancient and continuing activity. Game management techniques to permit sustainable use are known and practised in many areas where there is a hunting tradition, for example, in the Austrian Alps or the Western Highlands of Scotland. In mountain subsistence economies, food needs, coupled with modern weapons and lack of game law enforcement, has put severe pressure on many species. This has become especially critical where large numbers of military personnel are stationed in mountain frontier zones. In many parts of Africa, where bushmeat provides the major protein source for most of the population, commercial hunting has replaced subsistence harvesting. This has resulted in a substantial reduction of wildlife in mountain areas which have become accessible to commercial hunters. In such situations governments often have little capacity to enforce wildlife laws. Innovative approaches, involving collaboration between governments and local communities, seem to hold some of the few hopes for improvement. Nevertheless, perhaps one of the worst features is the systematic poaching, aided by official corruption, for endangered trophy species such as Virunga gorillas and snow leopards.

World-wide, mountain wildlife is increasingly valued as a resource for viewing or for shooting only with cameras. Wildlife tours offer a way in which some monetary value can accrue to mountain people and communities. Before the recent warfare in Rwanda, the viewing of the mountain gorilla in Parc National des Volcans earned about US$1 million in entrance fees and US$ 2–3 million in other expenditures – a major part of the country's foreign exchange earnings (WTO/UNEP, 1992).

Where wildlife values are high, for any of the above reasons, forest harvesting has often been modified, restricted severely, or even banned. In some cases, somewhat modified commercial harvesting can enhance certain kinds of wildlife and both wood and animal production in the same area is compatible. The case of the spotted owl in the forests of the northwestern United States has become a *cause celebre*. Where the wildlife species involved are sensitive to disturbance and cannot easily shift location to an undisturbed area, harvesting of wood and even non-wood products may have to be curtailed. This is usually done by establishing national parks or wildlife sanctuaries. The Andean spectacled bear is an example, and a linked system is being proposed in Venezuela involving five national parks with three corridors and then a possible extension across the border into Colombia (Yerena, 1994). Development of such approaches, however, will require full involvement of the local communities, a point clearly demonstrated by the recent conflicts surrounding efforts to establish the Khunjerab and Central Karakorum parks in Northern Pakistan.

SACRED GROVES AND FORESTS

Trees and forests are important to humans for deep spiritual and psychological reasons. Religious beliefs or fears affect the way many forests are perceived and managed. Most of the world's major religions and early cultures held certain trees, groves, or forests in special reverence (or fear) and accorded them special protection. Very often these were in mountain locations due to the greater visibility and their connection with the associated spiritual power (see Chapter 3).

The Hindus revere the groves and forests derived from clumps of the mountain dropped by the monkey-general Hanuman, as recounted in the *Ramayama*. Devout Buddhists have a great reverence of the life-force trees, god-trees, and Naga trees, and for groves and trees that are homes of the guardian spirits. Very many temples have an associated or surrounding protected forest. For instance, almost every temple in Northeast Thailand has a wooded area ranging from 0.5 to 10 ha.

In the classical world of the Greeks and Romans, areas of forest set aside as sacred to the various gods dotted the landscape, especially on hills and mountains. The Dionysiac cult was famous for its 'mountain dancing' in the forests high on the mountain slopes. Usually not only were all trees protected, but all other living organisms in the grove or forest. In many, statues of the god or goddess were erected; then shelters to protect these; and finally temples, and thus these became temple forests.

Shinto shrine groves are numerous in Japan and probably can be traced back to animistic beliefs of souls in nature and beliefs that the souls of the dead resided in special natural places. Shrine groves are used as centres for community gatherings and festivals, and may also be community parks.

While Jewish and Christian spiritual leaders went to wilderness areas (on mountains) for meditation and renewal, the concept of the sacred forest, grove, or tree did not strongly develop; nor were temples or churches particularly associated with them. Rather, in modern times environmental philosophers have fostered the emergence of the idea of 'cathedral' groves, the purity of wild areas, and the setting aside and protection of inspirational wildland sites in government-designated reserves. But these too are 'sanctified' by lovers of nature who draw renewal of the spirit from such places. The giant sequoia groves of the Sierra Nevada or the ancient bristlecone pine forests of the White Mountains in the western USA are examples of protected areas where modern-day 'Druid tree lovers' find joy. And many a pilgrimage is made to special groves in Europe where ancient trees, or trees of unusual shape, are found.

Many sacred forests are protected by taboos that threaten misfortune for any who cuts trees, hunts, or removes any living thing. In Venezuela the mountain forest sacred to the cult goddess Maria Lionza is a *de facto* total preserve since campesinos, who might practice slash-and-burn on other public land, do not do so here out of fear. Consequently, this area, now the Maria Lionza National Monument, is possibly the best protected unit in all of Venezuela's park system. In other places, the sacred forests provide a wide range of products used in religious rituals (sandalwood or juniper for incense, or plants for healing). If the temple needs repair or money, the grove or forest may be harvested for timber. Wood for funereal cremation or fuelwood for festivals may also be obtained from these forests. In Karnataka (India) and many parts of Nepal, religious forests are managed to provide trust funds to support religious institutions, or ceremonies.

As traditional religions have lost some of their strict controls through modernisation trends, protected forests have deteriorated from exploitation. In Nepal, there has been significant modification and degradation of understorey vegetation and soils, particularly where the forests are small in size This is due to increased grazing pressure, even though tree cutting may have been restricted by sanctions to serve only religious purposes. Forest conservation campaigns might well seize the opportunities to build on past protection, rebuild any degraded forests, and support reinstitution of community controls. In this process the mountain culture also may be maintained or even strengthened.

Chapter 3, on the sacredness of mountains, presents information relevant to the topic of sacred forests.

RESTORED FORESTS

Many mountain areas have been subject to heavy population pressures which have resulted in inappropriate clearing and associated degradation. Among the best known examples of large-scale deforestation are the Himalaya, the African highlands, and the Andes. In these regions, but particularly in the Himalaya, efforts have been made to restore the forest cover, driven by two forces. One

has been the desire to restore on-site productivity for the benefit of local communities; the other has been a perception that restored forests will provide downstream benefits of improved streamflow regimes and reduced sedimentation. There is no doubt about the local benefit derived from restoration through an increase in the availability of forest products and improved on-site productivity. This is achieved by decreasing erosion and restoring nutrient cycles through the re-establishment of a fully functioning forest ecosystem. However, large-scale reduction of downstream flooding and river sedimentation as a result of reforestation of small mountain watersheds is less likely. This is discussed in more detail in Chapter 15.

There are numerous examples where degraded mountain watersheds have been successfully restored and some of these will be mentioned before commenting on what is necessary to achieve a successful outcome.

The Himalaya is a region which has suffered quite spectacular loss of tree cover during the past several hundred years, but there have been many attempts at restoration. Early attempts in the 1960s and 1970s were essentially 'top down' approaches where those making the decisions (government officials and foreign advisers) failed to appreciate the complexities of land-use practices in the mountain regions; they chose largely exotic species (often eucalypts); they presumed that because land was bare of trees and uncultivated it was unused and un-owned; they did not consult with local communities on their perception of the problems or the solutions. Many thousands of hectares of seedlings were planted during these reforestation efforts but most of the plants died, or were killed by grazing or fire, often deliberately.

A shift by the government towards a community based focus in the 1980s facilitated a radically different approach to forest restoration and management of degraded forest landscapes which focused on recognition of the need for dialogue between agency officials and community groups. A process of social learning evolved so that each of the groups learned about the interests and perceptions of the other. The process was not perfect but, as experience was acquired, the restoration efforts became more sustainable and better adapted to the overall requirements of local communities. Many thousands of hectares of new forests have been not only established but also protected, and many more thousands of hectares of severely degraded forest lands have been placed under village protection.

Some of the most successful efforts derive from local and entirely independent initiative.

The mountains of Switzerland were severely degraded by the middle of the nineteenth century. Photographs of that time show a scene not much different to that in modern day Nepal, and contemporary records indicate the growing concern of both the first Swiss foresters and civic authorities. The forests had been heavily cut for house construction and fuelwood; leaf litter was collected for animal bedding for stall-fed animals during winter, and sustained summer grazing prevented natural regeneration. Heavy rains were causing substantial on-site erosion through surface wash and increased torrent flows. It is noteworthy that the first efforts to restore the ecological balance by reforesting the bare mountain slopes ended in failure for reasons similar to the first efforts in Nepal 100 years later. Foresters tried to impose a technical solution on local communities without adequate understanding of the local land-use context, and in particular, without an adequate understanding of the growing schism developing between urban and rural interests (see Chapter 2 for a detailed discussion of these points).

The Swiss foresters learned from their early failures (as did the Nepali foresters and their expatriate advisors a century later) and now Switzerland presents a well-forested mountain landscape, which is a tribute to the collaboration between its foresters, its law makers, and the local communities. Probably few people would realise that there is a greater forest cover in Switzerland now than there has been during the past three or four centuries. Virtually all of the forests are a result of human intervention and few could be categorised as 'natural' in the strict sense of the word. However, they provide an integrated mix of production, amenity, and catchment protection benefits which are managed jointly by communes, cantons, and the federal government, as well as by private individuals.

There are a number of lessons that can be derived from the example of restored montane forests around the world:

(1) the importance of fitting restoration attempts into the overall land-use context;
(2) the importance of matching the species selected for restoration with the site and the local needs;

Plate 13.3   Trees and groves in Nepalese farming system (Photograph: D. Gilmour)

(3) the need to engage in an effective dialogue with local communities and to involve them in both planning and implementation;

(4) the need to have a policy and legislative framework which helps the restoration efforts, and particularly to assign authority and responsibility to affected stakeholders;

(5) the importance of a clear definition of tenure (legal and/or usufruct) so that access and use rights over trees on private land, as well as on communal or government land, are defined to ensure that those who are asked to commit resources to restoration activities have a guaranteed right to the benefits.

Most of these lessons are equally valid even when government agencies are undertaking the restoration on public (government) land.

POLLUTION-IMPACTED FORESTS

Mountains generally receive more precipitation than other landforms, and thus acquire a greater load of any chemicals and particulates from atmospheric moisture. Mountain vegetation, including forests, suffer the effects of both direct deposition on foliage, branches and stems, and the uptake of soil moisture which contains some of these air-borne materials. In many areas, precipitation has become highly acid and contains sulfur, nitrogen oxides, ammonia, and heavy metals. In addition, where clouds enshroud mountains and are intercepted by trees (cloud forests), this additional 'soup' of pollutants compounds the effect. Ozone formed as a secondary reaction is also a major component. The sources of these pollutants may include motor vehicle exhausts, coal and oil burning power plants, industrial boilers, stack emissions from industry, domestic heating with coal, wood, or oil, and burning of agricultural wastes. The source may be local, especially in mountain valleys where there are temperature inversions that reduce dispersal, or may be thousands of kilometres away, even involving transfer across international boundaries. Many of these pollutants are long-lived in the atmosphere. The adverse impact on mountain forests from long-range transborder transmission of air pollutants has become an important international issue in Europe and North America, and an inter-regional one within the United States and China (Plate 13.4).

The long-term impacts of air pollutants on montane forests are not known with any degree of certainty. Forest dieback, disappearance of lichens, and reduction or elimination of soil organisms or aquatic organisms (including important fish and invertebrates) have been well documented and attributed to air-borne pollutants.

A 1994 survey of almost 30 000 sample plots in 32 countries in Europe indicated that 26.4 percent of trees had more than 25 percent defoliation (an increase of 3.8 percent since the previous year), and 12.1 percent of trees had more than 10 percent discoloration (UNECE and European Commission, 1995). Most of this degradation was in montane forests, and the most badly affected countries were the Czech Republic, Poland, and Slovakia. Box 13.4 presents a more detailed account of the damage in the Krkonoše or Giant Mountains of the Czech/Polish border.

The state of knowledge of negative impact on living organisms other than the trees is very scanty, but indications are that it is extremely serious (for instance, a study by Flousek (1994), reports a decline in both breeding bird density and bird

Plate 13.4 Mountain spruce forests killed by air pollution in the Giant (Krkonoše) Mountains in the Czech Republic. (Photograph: J. Flousek)

Box 13.4   Impact of air pollution on montane forests

Forest decline is currently regarded as a serious environmental problem in Europe and North America. A typical example of this phenomenon can be found in Central Europe, in the so-called 'Black Triangle', – a chain of mountain ranges along the border of the Czech Republic, southeast Germany and southwest Poland (including the Krkonoše Mountains) where 100 percent of the forests are adversely affected.

The natural composition of montane forests in the region (especially Norway spruce *Picea abies*, European beech *Fagus sylvatica*, sycamore maple *Acer pseudoplatanus*, silver fir *Abies alba*, European mountain ash *Sorbus aucuparia*) has been simplified and replaced by nearly mono-cultural spruce stands during the last century. Mixed beech–spruce forests remained in the lower elevations (up to 800 m), and remnants of native Norway spruce stands are present in remote valleys, glacial cirques, and along the upper timberline (at about 1200 m).

The region is located in the vicinity of important industrial centres in all three countries. Air pollution, brought to the mountains by prevailing air currents, adversely affects the quality of forest stands and soil and water chemistry, not only on windward slopes and mountain ridges but also in leeward valleys. The first visible forest damage was registered in the early 1970s.

What is the main reason for the ecological impacts? It is related to the burning of fossil fuels and air-borne toxins from industrial plants. There are tens of theories explaining the forest decline but the majority of them are based on synergistic ozone, air dust (heavy metals), and other pollutants in mountain regions, combined with extreme climatic conditions.

Mechanisms of tree damage can be roughly divided into two main groups: synergistic impact of the gases, dust particles, and acid rain ($pH < 5$) directly on assimilatory organs and bark; and indirectly on abiotic and biotic systems including soil, root system, and nutrition regime. As a result, photosynthetic processes and water balance are impaired, tree growth is retarded and defoliation increases (shortening, deformation and loss of needles/leaves). Viability of seeds decreases and natural regeneration is reduced or totally absent. Mountain forests, weakened by the impact of air pollution, are more sensitive to a complex of other factors, especially weather (wind, snow, frost) and insect pests (bud moths, spruce webworms, bark beetles). The whole process results in a rapid and extensive forest decline on a large scale.

The degree of forest damage increases with increasing altitude, especially at elevations where pollutants are concentrated in condensed precipitation, for example, fog or hoar frost. Coniferous species are much more impaired (and silver fir especially) than broad-leaved trees. Pure Norway spruce stands have been the most heavily affected; the situation is better in mixed and deciduous forests. Decline is more pronounced in artificially planted non-native stands than in native stands.

As individual trees die out, the forest structure becomes markedly simpler, microclimatic conditions become more extreme, and plant species composition becomes simpler (with development of a very dense and nearly mono-cultural herb layer, especially of a grass *Calamagrostis villosa*). In such conditions, declining forest fauna are replaced by simpler open-habitat dwelling species (e.g. abundance and species diversity of breeding bird communities decrease significantly with increasing forest damage). Decaying forests and newly formed clearings offer conditions for penetration of some species from subalpine to montane vegetation belts.

What are the possibilities for restoring air-polluted and damaged mountain forests in the region? The most important priority, gradual reduction of air pollution, is regarded as a logical and necessary beginning. Long-term goals indicate two main approaches: (1) gradual reconstruction of native stands and restoration of destroyed forests in close-to-nature species composition, all-aged structure, and with respect to ecological conditions; (2) conservation of original/natural tree populations.

There are several useful and successful methods for forest restoration. The majority of them are combined with a support of natural regeneration (if possible). First, under-planting of young trees under the protection of heavily damaged or dead 'mother' stands (without any clearcutting) is highly appropriate. Another approach supports establishment of new forest groups (bio-centres) on clearcuts. These 'starting points' of restoration are gradually enlarged by additional planting, and later interconnected into a network of bio-centres using planted bio-corridors. Layering at the timberline, stabilisation of forest edges on debris soils, or additional planting of target species into stands of pioneer (ameliorating) species have been tried as well. On the contrary, mechanical preparation of soil for plantations, widely used in other areas, is not recommended in mountains, due to erosion hazard.

However, there are many problems facing the realisation of forest management activities mentioned above. Soil erosion has reduced productivity. Animal damage can be serious, as in the case of deer browsing. Extensive clearings offer optimal conditions for voles; their abundance increases, and in peak years they are able to damage up to 80 percent of young broad-leaved trees, especially beech and mountain ash. (Application of rodenticides is restricted or forbidden due to water supply constraints or the presence of national parks in the mountains.)

Forest decline is not confined to the 'Black Triangle'. Based on data from 1990, 53 percent and 66 percent of forests are affected in western and eastern Germany respectively, 49 percent in Austria, 50 percent in Hungary, 23 percent in Slovakia and 86 percent in Poland have been subject to different degrees of damage. Taking the Czech Republic as an example, the process over time has been: 1970 – 4 percent, 1975 – 11 percent, 1980 – 25 percent, 1985 – 46 percent, 1990 – 59 percent, 1994 – 63 percent damaged stands. The situation has been improving in the last few years, but to restore the mountain forests in the region is a task for the next one hundred years at least. Increased control of the emission sources is imperative to any attempt.

Source: Jiri Flousek, Krkonoše National Park,
Czech Republic

diversity with increasing forest damage). Moreover, where much of the precipitation occurs in the form of snow that accumulates on the ground in the forest, the surge of pollutants that accompanies snowmelt (e.g. high levels of aluminium, lead, or cadmium) can exceed thresholds of toxicity for many organisms, reach groundwater or streams, and accumulate in mountain lakes. While air pollution is most serious in industrialised countries it may affect trees in the mountains of developing countries where they are close to traffic-congested cities or where pollution control for new industries is not firmly enforced. Suggestions for amelioration and management of affected montane forests are included in Box 13.4.

FOREST CLEARANCE IN THE UPLANDS

The sustainability of all forest values is at risk when the forest itself ceases to exist. The apparent acceleration in forest clearing in tropical mountain areas is a matter for considerable concern. The results of the most recent FAO forest assessment have already been mentioned in the Introduction to this Chapter.

The agents for mountain deforestation vary enormously from clearing for commercial grazing and cropping to small-scale clearing by shifting cultivators and landless refugees (Plates 13.5 and 13.6). Clearing for illegal crops is also part of the mountain scene because of the lower level of law enforcement and the relative inaccessibility. Poverty and population growth are frequently cited as underlying causal factors of deforestation. However, it is clear that the relationships between these factors and forest clearance is complex and significantly influenced by external factors such as government pricing and incentive policies, land tenure, and the demands of the market economy

Plate 13.5 Slash-and-burn agriculture in the mountains of NW Thailand. (Photograph: B. Messerli)

(Rowe *et al.*, 1992). Indeed, it has been suggested that much deforestation is caused by the political marginalisation of forest dwelling people and that the degradation and destruction of forest resources is the cause rather than result of their poverty (Dove, 1993).

Plate 13.6   Slash-and-burn agriculture in Venezuela (Photograph: L. S. Hamilton)

Multiple scenarios of deforestation in different mountain environments require multiple solutions. Policy reforms are required to ensure that land tenure, resource pricing, and taxation patterns provide undistorted incentive structures to forest user groups. In the long term, montane forests will be preserved only if their benefits and sustainable uses outweigh the costs to the people who live within or nearby them. Where significant benefits from maintaining forest cover accrue to people living away from the mountains, creative institutional linkages between these stakeholders and the forest communities need to be developed to ensure that the local people receive adequate payments for the environmental services they provide.

Distorted perceptions of the nature and extent of deforestation in the Middle Mountains of Nepal in the 1970s and 1980s led to considerable investment in interventions that did not address the most pressing needs. In some areas where deforestation was thought to be rampant, forest recovery through both natural regeneration and tree planting by local farmers was in progress. Regular monitoring of the extent and condition of the forest resource is a necessary prerequisite for developing appropriate management policies. Participatory monitoring involving the local communities can provide much information; many of the inaccurate perceptions in Nepal, for instance, could have been avoided if such approaches had been used.

Mountains have traditionally been regions of physical, infrastructural, political, and economic marginalisation. However, although the broader environmental benefits have now been recognised, these values will continue to be eroded unless creative institutional arrangements are developed to overcome the pressures of marginalisation. The conservation and sustainable management of mountain forests will depend, therefore, on the assurance of an equitable distribution of benefits.

COMMUNITY INVOLVEMENT IN MANAGEMENT OF MONTANE FORESTS

As many governments around the world are embracing various forms of economic rationalism, they are seeking new, innovative, and cost effective ways of policy implementation, and are moving to decentralised forms of government that engage society (including community groups, NGOs, and the corporate sector) more actively in management.

There is now a much wider range of interested parties than previously.

The development and application of policy needs to take into account the increasing number of interested parties; a review of the mandate of government in terms of re-defining the authority and responsibilities of the various stakeholders will be required. This can be very challenging for government officials previously accustomed to centralised forms of authority. The change is very noticeable in many parts of the former USSR, but also applies in other countries. In many mountain areas indigenous and local communities have had a long-standing association with, and dependence on, the forest for providing some part of their livelihood, and they have developed their own responses to forest management problems. A brief account of how forests can be degraded following government control taking over local management in Uganda is given in Box 13.5. A similar scenario has been played out in many other countries (e.g. Nepal and India), and shows the importance of community participation in management of the forest resource. While economic benefits derived from the forests may be a major driving force for creating changes in indigenous forest management approaches, religious and cultural values are often very important. It is quite common to find that local communities have developed their own institutional arrangements which define access and usage rights of the forests.

Such responses give rise to what are called indigenous management systems (see Fisher, 1989 for a discussion of this topic for Nepal; Plate 13.3). However, there are certain social and institutional preconditions for these responses to emerge and

Box 13.5 Degradation of indigenous management systems

The Forest Department appropriated control of Mount Elgon, Uganda, without consultation or discussion with the local residents. Community members were not involved in decisions regarding the boundary or the activities that could continue beyond it. Many of the former residents migrated back down to their main base in the lower areas, but no compensation was offered for the foregone land, over which people believed they had traditional rights. Land was still abundant, and according to local elders, the Forest Department had maintained that the area under their control would revert to the community if future generations required land, and there was therefore minimal resistance.

Once control had been completely removed from local hands, the indigenous systems of forest resource management gradually disintegrated. With no authority over the forest, the elders were not in a position to enforce regulations that had previously been in place. A new generation of local residents emerged without instruction on the importance of maintaining sustainable use systems.

The bulk of the damaging activities which have left scars on Mount Elgon have occurred during the period that central government has officially administered the forest as a protected area. This is due to a combination of factors, including escalating population pressure, local civil disruptions, and outsiders taking advantage of limited Forest Department capacity and gradually increasing their activities within the forest. Over the past few decades, government capacity for controlling activities has degenerated even further, and increased illegal activities have resulted in the degradation of many resources, animals in particular. Pit-sawing, primarily by external interests, reached its peak in the late 1980s. Agricultural encroachment has become a major issue in the area, due to population expansion, tribal upheavals between the residents of Kapchorwa and Mbale, and increasing insecurity within the lower plains. It was beyond the capacity of the Forest Department, crippled for several decades by civil unrest, to prevent these developments.

In one of the parishes surveyed during the resource use assessment (Mutushet), the community elders, the former 'managers' of the forest, took a stand against mounting pressure for expansion of agriculture into the forest. The power vacuum during the period of unrest left space for local management initiatives to flourish, and by default, increased the authority that the community could exercise over the forest. This is a clear example of the capacity of the community to police their own, and others' activities, – that they can be motivated to act rationally if they are sufficiently empowered, either by default or decree.

One general conclusion can be drawn: for a combination of reasons, the management strategy which has been followed for the past couple of decades, has not been effective in conserving the forest, and with growing population pressure and little hope of a substantial and sustainable increase in financial resource flowing to the official management body, it is very unlikely to work in the future.

Source: Scott, P. (1997)

thrive. There needs to be a strong element of social stability for communally agreed arrangements to function, and a cultural ability for different members of the community to work together for the common good. There also needs to be sufficient internal 'power' to maintain effective local control, free from outside powerful groups, such as merchants, the military, or government which may wish to expropriate the resources for the benefit of outsiders. These conditions are most often met in isolated mountain communities remote from outside influences. Indigenous approaches to forest management provide ideal building blocks for community forestry, that is, where government authorities, such as forest departments, develop some form of negotiated agreement with local communities. In most cases modifications to government policy and legislation are needed to provide an enabling environment for community forestry to flourish. Fisher (1995) provides a thorough discussion of the various forms of collaboration between government and local communities to achieve both conservation and development objectives.

The situation described above applies primarily to developing countries where local communities have a dependence on the forests for various socio-economic, spiritual, or cultural benefits. However, similar principles can be applied in developed countries where a wider range of stakeholders (community groups, environmentalists, corporate sector, etc.) is demanding a broader consensus in mountain forest management. It is no longer sufficient, and in many countries no longer possible, for authorities to be the sole arbiters of management decisions. A new view of management is needed and new skills among those entrusted with the prime management mandate. Thus, abilities in stakeholder identification and analysis, negotiation, and conflict management need to be added to the previous technical skills which most resource managers bring to their job.

Sharing of power (authority and responsibility) for forest management between government and other interested stakeholders often requires a review and revision of policy, and possibly new legislation. In some countries this is already well advanced (see Box 13.6, with an example from Nepal).

SUMMARY OF MANAGEMENT AND POLICY STRATEGIES

Perspectives on the role of mountain forests have changed over time and will continue to evolve. In mountain areas in many parts of the world there is a shift from economic production functions to social/ecological functions, where environmental

---

Box 13.6   Nepal's evolving policy framework for community forestry

The first attempt to define a coherent forest policy in Nepal occurred in 1952–53 following the country's emergence from a feudal state. This recognised three categories of forest, including 'community forests' which were to be set aside to satisfy community needs, and their protection and management was to be entrusted to village panchayats. This policy was never formalised nor implemented, and it predated by 25 years subsequent attempts to implement a community forestry policy. During the intervening years Nepal's forest policy essentially followed models from the West, where forest ownership was vested in the State and management authority was placed in the hands of the Forest Department. The forests were nationalised in 1957, although the prime motive for this was to take back into State control the large area (almost one-third) of Nepal's forest and agricultural lands which were held under feudal tenure arrangements, – 75 percent of it belonging to one family.

The Forest Department was charged with performing a policy and licensing role, but at the time with only about five or six professionally trained forest officers in the country it was clearly an impossible task. The Forest Act, 1961, introduced into legislation the idea of transferring government forest land to village panchayats for their use. However, no steps were taken to implement these provisions and the legal status of the forests was not addressed for a further 16 years. Nonetheless, the attempt to recognise the legitimacy of local communities in having a role in forest management was important, and it was built upon in later policy and legislative changes.

The Forest Preservation Act, 1967, was introduced to define forest offences and prescribe penalties, thus strengthening the role of the Forest Department as a policy and law enforcement agency. The first formal national Forestry Plan was promulgated in 1976 and this proposed the establishment of 'Panchayat Forests' for the benefit of local communities. This was followed in 1978 by a set of rules and regulations which would govern the handing over of limited areas of government forest land to the control of panchayats. Thus for the first time, formal recognition was given to the rights of villagers to manage their own forest resources with technical assistance (where necessary) being provided by the Forest Department. A large number of field projects, supported by both bi-lateral and multi-lateral donors, began implementation during the late 1970s and 1980s. These accumulated considerable experience which was influential in shaping future policy and legislative changes. A major master planning exercise for the forestry sector was completed in 1990 and this indicated the strong emphasis which the government wished to give to community forestry. A number of very clear statements signalled the principles the government intended to follow in implementing community forestry:

(1) Phased handing over of all the accessible hill forests to the communities to the extent that they are able and willing to manage them;
(2) To entrust the users with the task of protecting and managing forests. The users to receive all of the income;
(3) Retraining the entire staff of the Ministry of Forests and Soil Conservation for their new role as advisers and extensionists.

The handing over of forests to communities for their management became formalised in legislation passed in 1993 (Forest Act, 1993) and this was followed by the promulgation of a set of Forest Rules

in 1995. These rules outline the procedures by which forests are handed over for community management. It should not be presumed that because there were very few forest officers in the hill regions in the decade immediately following 1950, there was no interest in protection and management of forests. While this presumption was (and is) widespread, the reality is somewhat different. Investigations during the past 15 years have revealed that a great many village communities took the situation into their own hands and put into place local institutional arrangements to ensure that many hill forests were given basic protection and that access and use rights were prescribed. It was, after all, in their own best interests to do so. Many of these indigenous management systems have survived, often with modifications, for more than 35 years, – not perfectly, but with outcomes better than could be achieved by a fledgling Forest Department.

Current approaches to community forestry attempt to bring these two strands of interest (the official and the community) together, with each recognising the interests and ability of the other. During the past decade about 4000 agreements covering almost 200 000 ha of forest have been negotiated between local village user groups and the Forest Department, and formalised as Operational Plans. These legitimised the authority of user groups to manage specified areas of forests.

Adapted from: Gilmour, D. A. and Fisher, R. J. (1991)

services (such as carbon fixation and watershed protection), scenic resources, and cultural heritage values are increasing in importance. Economic functions are introduced through tourism, improvement in water quality, and carbon sequestering offsets, and these values often outweigh the direct product values. And, in most of the world, mountains are often the last bastion of nature, islands in a sea of transformed landscape, and are thus of great importance in biodiversity conservation of wild species. A summary of these changed perspectives is presented in Table 13.2.

Many mountain areas are prime sites for locating protected areas for several reasons. The altitudinal variation produces a wide range of ecological conditions with a resulting rich assemblage of plants and animals. By their very nature they are often remote and have not been subjected to the strong development pressures exerted on lowland ecosystems. Consequently, mountain areas are frequently more 'intact' than lowlands. However, this same remoteness also restricts the

influence of government agencies responsible for protected area management. Watersheds that are important for water supplies are in the 'safest' condition under forest cover which has some kind of protected or conservation regime (see Chapter 15).

The more-or-less 'intact' mountain forests, as well as being important sources of biodiversity for conservation and hydrological regulatory functions, may also provide a range of subsistence and other goods to surrounding communities. Thus, management of these forests increasingly requires collaboration between governments and communities. This is one of the real challenges for the future and there are encouraging signs that a number of governments are attempting to meet the challenge by exploring different forms of collaboration and by reviewing their policies. A further elaboration of this is given in Box 13.5 on government control in Uganda.

There are concerns about qualitative degradation in forests even though the areal extent may remain unchanged. This stems from the pauperisation of forest ecosystems as they are subject to increasingly intensive management regimes that may involve such things as:

- removal of last vestiges of old-growth trees;
- harvest of complex forests and replacement with one, or at least fewer, species for commercial exploitation;
- use of herbicides to eliminate unwanted regrowth of competing plants;
- inadvertent introduction of alien species (plant or animal) in disturbed areas;
- deliberate introduction of alien species for some commercial purpose (e.g. non-native game animal for sport hunting);
- nutrient depletion from intensive short-rotation, total biomass logging;
- increased use of pesticides that may adversely affect non-target organisms.

On the positive side are:

- an increasing amount of mountain forest area being given greater protection in reserves and national parks;
- improved forest harvesting technology that does less damage to the forest floor and residual stands through removing products by aerial systems designed for steepland logging;
- increasing change in management policies away from a timber-production dominance to accommodate other forest values;
- progressive forest ecosystem science that is altering forest management policies to take into account 'forest ecosystem health' and 'sustainable ecosystem management' in some parts of the world;
- increased restoration, or reforestation, with native species rather than aliens and with deliberate mixtures rather than monocultures.

Most urgently needed are better methods and wider application of monitoring, in order to detect signs of unsustainability over several cycles of harvesting or treatment. The changed perspectives on the role of forests as a major constituent of the mountain environment is rapidly becoming reality in many of the industrialised countries. In this process there are often increased demands for wood products from the forests of less developed countries, thereby placing a heavy impediment to a changing perspective in those countries. Malaysia and Thailand are also beginning the same process. But, changes are also occurring in the tropical developing countries, as indicated by the policy of Bhutan (Box 13.1).

The World Wide Fund for Nature (WWF) and the World Conservation Union (IUCN) have recently completed a joint forest policy study (Dudley *et al.*, 1996). While it is not specific to mountain forests, it is most applicable to these forests and is offered below as a vision for the future:

Forest management

It is vital that forest management systems are based on the principle of sustainability. This means that management must be appropriate to the environment, benefit society and be economically viable. Management systems must seek to conserve biodiversity at genetic, species and ecosystem levels. Such systems can only come into being if foresters alter the way they approach forests. Current management practices focus principally on producing timber. Other forest products and services are almost always regarded as secondary and are only utilised when this does not interfere with timber production. Future forest management plans and systems should concentrate first on conserving the rich natural diversity of forests and on the environmental functions they perform. All human use of forests depends on these two crucial elements.

Forest quality and extent

In the future the world's forests should be both more extensive and of a higher quality than now. There should be a greater proportion of natural and semi-natural forest. This should largely be made up of existing natural and old-growth forests, but may be supplemented by restored secondary forest, which has been permanently set aside to develop a natural dynamic. It will be necessary to carry out ecologically and socially appropriate reforestation activities (preferably using mixes of native species), to encourage regeneration, and to promote widespread use of sustainable forest management.

Timber production

A proportion of the world's forests is likely to remain under fairly intensive timber production. This should be based around more rigorous social and environmental safeguards than is usually the case at present. Only in exceptional circumstances or where native species are unsuitable should non-native (exotic) species be used. Management areas should be defined by natural boundaries, such as watersheds, and management should follow the principles of landscape ecology.

Management policies

A range of management policies will be needed. These must involve the full participation of local communities. Greater priority should be given to non-timber values. The impacts of pollution should be reduced. At the policy level, there should be enhanced regional and cross-sectoral collaboration.

Not all the elements identified above can be catered for in every forest stand. They should, however, be reflected in national and regional forest policies around the world.

References

Botkin, D. B. and Talbot, L. M. (1992). Biological diversity and forests. In Sharma, N. P. (ed.), *Managing the World's Forests: Looking for Balance Between Conservation and Development*. Kendal/Hunt Publishing Company: Dubuque, IA, pp. 47–74

Brooks, D. J. and Grant, G. E. (1992a). New approaches to forest management – part one. *Journal of Forestry*, 90(1):25–58

Brooks, D. J. and Grant, G. E. (1992b). New approaches to forest management – part two. *Journal of Forestry*, 90(2):21–9

Cassells, D. S., Bonell, M., Gilmour, D. A. and Valentine, P. S. (1988). Conservation of Australia's tropical rainforests – local realities and global responsibilities. *Proceedings, Ecological Society of Australia*, 15: 313–26

Dove, M. R. (1993). A revisionist view of tropical deforestation and development. *Environmental Conservation*, 20(1):17–24

Dudley, N. (1992). *Forests in Trouble: A Review of the Status of Temperate Forests Worldwide*. WWF International: Gland

Dudley, N., Gilmour, D. and Jeanrenaud, J.-P. (1996). *Forests for Life, the WWF/IUCN Forest Policy Book*. WWF/IUCN, WWF-UK. Panda House: Godalming, Surrey

Eckholm, E. (1976). *Losing Ground*. WorldWatch Institute, W. W. Norton: New York

Edwards, (1996). The trade in non-timber forest products from Nepal. *Mountain Research and Development*, 16(4): 383–94

FAO (1993). *Forest Resources Assessment 1990 – Tropical Countries*. Forestry Paper No. 112. Food and Agriculture Organization of the United Nations: Rome

Fisher, R. J. (1989). Indigenous systems of common property forest management in Nepal. *Working Paper No. 18*, Environment and Policy Institute, East–West Center: Honolulu

Fisher, R. J. (1995). *Collaborative Management of Forests for Conservation and Development, Issues in Forest Conservation*. IUCN and WWF: Gland and Cambridge

Flousek, J. (1994). Breeding bird communities and air pollution in the Krkonoše Mountains (Czech Republic) in 1983–1992. In Hagenmeijer, E. J. M. and Verstrael, T. J. (eds.), *Bird Numbers 1992, Distribution, Monitoring and Ecological Aspects*. Statistics Netherlands: Voorburg/Heerlen, pp. 233–9

Gilmour, D. A. (1995). Rearranging trees in the landscape of the middle hills of Nepal. In Arnold, J. E. M. and Dewees, P. A. (eds.), *Tree Management in Farmer Strategies: Responses to Agricultural Intensification*. Oxford University Press: London, pp. 21–42

Gilmour, D. A. and Fisher, R. J. (1991). *Villages, Forests and Foresters – The Philosophy, Process and Practice of Community Forestry in Nepal*. Sahayogi Press: Kathmandu

Gurung, C. P. and de Coursey, M. (1994). The Annapurna Conservation Area Project: a pioneering example of sustainable tourism? In Cater, E. and Lowman, G. (eds.), *Ecotourism: A Sustainable Option?* John Wiley and Sons: New York, pp. 177–94

Hamilton, L. S. (1992). The wrong villain? *Journal of Forestry*, 90(2):7

Hamilton, L. S., Juvik, J. O. and Scatena, F. N. (eds.) (1995). *Tropical Montane Cloud Forests.* Springer-Verlag: New York

Hedberg, O. (1995). *Features of the Afroalpine Plant Ecology*, Acta Phytogeographia Suecica, 49, p. 144. [Facsimile edition of original 1964 publ. with new forward] Ekblad and Co.: Vastervik, Sweden

Hornback, K. E. (1994). U.S. National Park Service Statistics. Unpublished. *Cited in* Denniston, D. (1995). *High Priorities: Conserving Mountain Ecosystems and Cultures.* WorldWatch Paper 123. WorldWatch Institute: Washington DC

International Environment Report (1994). Seven-nation convention to protect Alps expected to come into force by end of year. September 21, 1994 newsletter

ITTO (1991). *Guidelines for the Sustainable Management of Natural Tropical Forests.* ITTO Policy Development Series No. 1. International Tropical Timber Organization: Yokohama, Japan

ITTO (1992). *Criteria for the Measurement of Sustainable Tropical Forest Management.* ITTO Policy Development Series No. 3. International Tropical Timber Organization: Yokohama, Japan

IUCN (1992). *Guidelines for the Ecological Sustainability of Non-Consumptive and Consumptive Uses of Wild Species.* Draft Guidelines Presented to IUCN General Assembly, Buenos Aires, Argentina

Jodha, N. S. (1992). Sustainability of Himalayan forests: some perspectives. In Agarwal, A. (ed.), *The Price of Forests.* Centre for Science and Environment: New Delhi, pp. 285–90

Lamphrecht, H. (1989). *Silviculture in the Tropics – Tropical Forest Ecosystems and Their Tree Species – Possibilities and Methods for Their Long-Term Utilization.* Deutsche Gesellschaft für Techniche Zusammenarbeit (GTZ) GmbH, Eschborn, Germany

Lampietti, J. A. and Dixon, J. A. (1995). *To See the Forest for the Trees: a Guide to Non-Timber Forest Benefits.* Environment Department Paper No. 013. The World Bank: Washington DC

Lauer, W. (1993). Human development and environment in the Andes: a geoecological overview. *Mountain Research and Development*, 13(2):157–66

Long, A. (1995). The importance of tropical montane cloud forests for endemic and threatened birds. In Hamilton, L. S., Juvik, J. O. and Scatena, F. N. (eds.), *Tropical Montane Cloud Forests.* Springer Verlag: New York, pp. 79–106

Lowdermilk, W. C. (1953). *Conquest of the Land Through Seven Thousand Years.* Agriculture Information Bulletin 99. US Department of Agriculture: Washington DC

de Montalembert, M. R. and Clement, J. (1983). *Fuelwood Supplies in Developing Countries.* Forestry Paper 42. FAO: Rome

Mueller-Dombois, D. and Krajina, V. J. (1968). Comparison of east-flank vegetations on Mauna Loa and Mauna Kea, Hawai'i. *Recent Advances in Tropical Ecology*, 2:508–20

Myers, N. (1988). Tropical forests: much more than stocks of wood. *Journal of Tropical Ecology*, 4:1–13

O'Loughlin, C. and Ziemer, R. R. (1982). The importance of root strength and deterioration rates upon edaphic stability in steepland forests. In Waring, R. H. (ed.), *Carbon Uptake and Allocation: a Key to Management of Subalpine Ecosystems.* Oregon State University: Corvallis, pp. 70–8

Panayotou, T. and Ashton, P. S. (1992). *Not by Timber Alone – Economics and Ecology for Sustaining Tropical Forests.* Island Press: Washington DC

Peters, C. M. (1996). *The Ecology and Management of Non-Timber Forest Resources.* Technical Paper No. 322. The World Bank: Washington, DC

Poore, D., Burgess, P., Palmer, J., Rietbergen, S. and Synnott, T. (1989). *No Timber Without Trees – Sustainability in the Tropical Forest.* Earthscan Publications: London

Poschen-Eiche, P. (1987). *The Application of Farming Systems Research to Community Forestry.* Tropical Agriculture (1). Triops Verlag: Langen, Germany

Price, M. F. (1990). Temperate mountain forests: common-pool resources with changing multiple outputs for changing communities. *Natural Resources Journal*, 30:685–707

Rowe, R., Sharma, N. P. and Browder, J. (1992). Deforestation: problems, causes and concerns. In Sharma, N. P. (ed.), *Managing the World's Forests – Looking for Balance Between Conservation and Development*. Kendall/Hunt Publishing: Dubuque, IA, pp. 33–45

Sader, S. A. and Joyce, P. B. (1983). Deforestation rates and trends in Costa Rica, 1940 to 1983. *Biotropica*, **20**:11–19

Scott, P. (1997). *People–Forest Interactions on Mount Elgon, Uganda: Moving Towards a Collaborative Approach to Management*. IUCN: Nairobi

Sharma, N. P., Rowe, R., Openshaw, K. and Jacobson, M. (1992). World forests in perspective. In Sharma, N. P. (ed.), *Managing the World's Forests – Looking for Balance Between Conservation and Development*. Kendall/Hunt Publishing: Dubuque. IA, pp. 17–31

Shiva, V. and Bandyopadhyay, J. (1986). The Evolution, Structure, and Impact of the Chipko Movement. *Mountain Research and Development*, **6**(2): 133–142

Tuchy, E. (1982). Forestry and ecology in mountainous areas. In Heinrich, R. (ed.), *Logging of Mountain Forests*. FAO Forestry Paper 33, Rome, pp. 9–14

UNECE and European Commission (1995). *Forest Condition in Europe: Results of the 1994 Survey: Executive Report*. Convention on Long-Range Transboundary Air Pollution, UN Economic Commission for Europe and European Commission

Warner, K. (1993). *Patterns of Farmer Tree Growing in Eastern Africa: A Socio-economic Analysis*. OFI and ICRAF Tropical Forestry Papers No. 27, Nairobi

Westoby, J. (1989). *Introduction to World Forestry: People and their Trees*. Blackwell Press: Oxford

Wilson, E. O. (1992). *The Diversity of Life*. Harvard University Press: Cambridge

The World Bank (1994). *Slovak Republic: Environmental Forestry Development Project Staff Appraisal Report*. The World Bank: Washington DC

The World Bank (1995). *Albania Forests Project Staff Appraisal Report*. The World Bank: Washington DC

WTO/UNEP (1992). *Guidelines: Development of National Parks and Protected Areas for Tourism*. World Tourism Organization, Madrid and United Nations Environment Programme, Paris

Wunder, S. (1996). Deforestation and the Uses of Wood in the Ecuadorian Andes. *Mountain Research and Development*, **16**(4): 367–82

Yerena, E. (1994). *Corredores Ecológicos en los Andes de Venezuela*. Instituto Nacional de Parques: Caracas

Zwerman, P. J. and Richard, F. (1959). Protective forestry in Switzerland. *Soil and Water Conservation*, **14**:220–3

Mountain agriculture

14

Narpat S. Jodha

INTRODUCTION AND THEMATIC FOCUS

This chapter describes and analyses the dominant scenarios characterising mountain/hill agriculture. The latter is broadly defined as covering all organically or functionally interlinked land-based activities, (ranging from crop farming to forestry, to animal husbandry). The central purpose is to identify approaches and strategies conducive to sustainable agricultural and natural resource use in mountain areas. Sustainability, in turn, is understood as the ability of mountain agriculture to maintain or enhance its performance without depleting its natural resource base. The discussion is focused on shifting patterns of resource use and their compatibility/incompatibility with imperatives of mountain circumstances; the factors and processes contributing to these changes as well as their consequences; and possible ways to mitigate the negative consequences of the process of change. This is attempted through a synthesis of observed and documented situations in different mountain areas rather than by presenting statistical profiles of the variables related to specific administrative units (countries, districts). For reasons of ready availability of information, the discussion draws more on experiences of countries in the Hindu Kush–Himalaya, supplemented by context-specific evidence from other areas to reflect the situation of mountain regions in developing countries in general. The somewhat different circumstances of mountain agriculture in the industrialised world are not addressed; rather, they receive attention, at least in part, in Chapters 2, 4, and 5.

The thematic framework of the discussion is built around the specific features of mountain areas, such as difficulty of access, fragility, diversity, and so on, and the societal responses to these mountain imperatives. These characteristics not only differentiate mountains from plains, but influence the patterns of resource use, as well as perspectives or approaches taken by the internal and external users of mountain resources. Sustain-ability of mountain resources and dependent activities (including agriculture) are linked to the latter's compatibility with the imperatives of the aforementioned mountain conditions. Accordingly, the current status and likely future of mountain agriculture (as a dominant system of natural resource use) can be seen in terms of the shifting nature of a two-way adaptation process where (a) human needs and mechanisms to meet them are adapted to the mountain conditions, and (b) the latter are adapted (or rather manipulated) to suit the changing human needs. Under (a) biophysical factors influence the pace and pattern of resource use, as with the traditional farming systems. Under (b), as reflected through the present day situation, the man-made factors determine the intensity and extraction rates of resource use in mountain areas. The latter, due to their insensitivity to the imperatives of mountain conditions, tend to promote rapid resource depletion and seem to be inherently unsustainable. An understanding of the factors and processes contributing to unsustainability scenarios in mountain areas can help to identify the approaches to arrest and reverse the negative trends. The central focus of such approaches, in this chapter, must be on reducing the degree of incompatibility between imperatives of mountain conditions and attributes of current high intensity resource-use systems, through searching new areas and levels of two-way adaptation. Without minimising the importance of the internal diversities, we attempt this search by presenting a generalised picture of the mountain situation in different areas.

The discussion is divided into five sections. The first briefly comments on conditions specific to mountains and the imperatives set by these conditions. The second projects the implications of mountain conditions for agriculture, both in terms of constraints and opportunities. Here we also comment on societal responses in terms of measures and resource-use practices to manage the constraints and harness the opportunities. The traditional practices directed to low intensity, diversified resource-use systems are contrasted with the current tendency toward high intensity resource-use systems. The third section describes the changing socio-economic circumstances that encourage indiscriminate resource-use intensification and their consequences in terms of emerging prospects of unsustainability for mountain agriculture. In the fourth section we examine the post-UNCED developments that potentially will affect mountain areas. Based on the inferences from the discussion under the preceding sections, the final section elaborates on possible future approaches to resource use and agriculture in mountain areas and their operational and policy implications. Here we also allude to the search areas for potential options to be established by on-going and new policies and programmes.

THE MOUNTAIN CONDITIONS AND THEIR IMPERATIVES

Mountain agriculture, as mentioned earlier, is broadly defined as covering all land-based activities such as cropping, animal husbandry, horticulture, and forestry. Owing to their organic and functional linkages, created and reinforced through biophysical features of mountain areas and harnessed by resource users, the above activities cannot be meaningfully segregated and sustainably managed in a sectoral mode. Hence, we perceive mountain agriculture as an integrated system of natural resource use, along with its man-made support facilities. Viewed this way, and recognising the overall expanse of the above activities, mountain agriculture not only constitutes a major occupation and a source of sustenance for the bulk of the mountain communities, but also represents a primary form of natural resource use in the mountain areas of developing countries. If externally designed and implemented activities, such as hydropower production, mining, and commercial plantation are excluded, the pace and pattern of overall resource use in mountain areas would not be very different from those of agricultural land use. Hence, the terms such as mountain resource use and agricultural resource use are used interchangeably in the following discussion.

The pace and pattern of agricultural transformation, as well as its performance, in mountain areas is conditioned by the biophysical features of these areas and human ability to adapt to them. These interrelated key features include restricted accessibility, fragility, marginality, diversity, specific

niche, and a combination of steep slopes and altitudinally reduced temperature (and also patterns of human adaptation to these features). These features may not be confined to mountains alone, but their high degree and significant impacts on resource-use patterns as well as on the nature of production and exchange activities, including external linkages, differentiate mountains from other areas. These features create objective circumstances, which, in turn, present a range of opportunities and constraints and influence human responses directed to use of mountain resources. A detailed description of these features called 'mountain specificities' and their operational implications are elaborated elsewhere (Jodha, 1990; Jodha et al., 1992) and are summarised in Table 14.1. This is adapted from Jodha and Shrestha (1994) and considers: (a) the biophysical foundations of mountain specificities; (b) their manifestation and implications seen as objective circumstances; and (c) the latter's imperatives in terms of appropriate responses to manage the above features in terms of choice and methods of resource use, including formal development interventions.

AGRICULTURE IN THE MOUNTAIN CONTEXT

It is instructive to juxtapose the imperatives of mountain specificities with the conditions historically associated with the high performance of agriculture (e.g. farming in prime lands), in order to more sharply project the implications of the mountain condition for agricultural performance and development. Accordingly, as indicated by Table 14.2, mountain conditions tend to limit: (a) the capacity of agricultural systems to absorb inputs; (b) the scope for resource-use intensification and upgrading (transformation) through infrastructure; (c) the production opportunities and gains associated with the scale of production systems similar to the green revolution in the plains; (d) the exposure to and replicability of development strategies from the plains; (e) generation of surplus and its exchange at favourable terms of trade. Some of these conditions can be modified through a flow of productive resources and relevant technologies from outside or through internal stock of investible resources which again depends on the initial levels of output, and its tradability as well as institutional arrangements to facilitate the same. However, these possibilities also are very limited in mountain areas. The circumstances created by restricted

Table 14.1 Mountain specificities and their imperatives

1A. Limited accessibility

a) Product of: slope, altitude, terrain conditions, seasonal hazards, etc. (and lack of prior investment to overcome them)

b) Manifestations and implications
isolation, semi-closedness, poor mobility
high cost of mobility, infrastructural logistics, support systems, and production/exchange activities
limited access to, and dependability on, external support (products, inputs, resources)
detrimental to harnessing niche and gains from trade
invisibility of problems/potentials to outsiders

c) Imperatives (appropriate responses)
local resource centred, diversified production/consumption activities
local resource regeneration, protection, regulated use; recycling
focus on low-weight/volume and high value products for trade
nature and scale of operations as permitted by the degree of mobility and local resource availability
development interventions with a focus on:
decentralisation and local participation: inaccessibility reduction with sensitivity to other mountain conditions (e.g. fragility), and changed development norms and investment yardsticks

1B. Fragility and marginality

a) Product of: combined operations of slope/altitude, and geologic, edaphic, and biotic factors; biophysical constraints create socio-economic marginality

b) Manifestations and implications
resources highly vulnerable to rapid degradation, unsuited to high intensity/productivity uses: low carrying capacity, low input absorption
limited, low productivity high risk production options: little surplus generation or reinvestment
high overhead cost of resource use: obstacles to infrastructural development; under-investment, subsistence orientation of economy
people's low resource capacity preventing use of high cost, high productivity options; disregard by 'mainstream' societies

c) Imperatives (appropriate responses)
resource upgrading and usage regulation (e.g. by terracing)
focus on low intensity, high stability uses (e.g. forestry, pasture, perennials)
diversification involving a mix of high and low intensity land uses, a mix of production and conservation measures, low cost, local resource use
local resource regeneration, recycling, regulated use, dependence on nature's regenerative processes, and collective measures
different norms for investment to take care of high overhead costs
focus on vulnerable areas, and people, and their demarginalisation

1C. Diversity

a) Product of: interactions between different factors ranging from elevation and altitude to geologic and edaphic conditions, as well as biological and human adaptations to them

b) Manifestations and implications
a basis for spatially and temporally diversified and interlinked activities
heterogeneity-induced strong location specificity of production and consumption activities
limited applicability of activities meant for wider application, and limited benefits associated with scale

c) Imperatives (appropriate responses)
territorial diagnosis followed by diversified interventions and decentralised arrangements (technologies, infrastructure, and institutions)
small scale, interlinked diversified production/consumption activities: temporally and/or spatially differentiated activities for fuller use of environment
location-specific integrated multiple activities with a focus on performance of total production system

(Continued)

Table 14.1 *Continued*

1D. 'Niche' (activities/products with comparative advantage)

a) Product of:	unique environment and resource characteristics of biophysical conditions (people's traditional practices for adaptation to specific mountain conditions also part of 'niche')
b) Manifestations and implications	potential for unique products/activities (ranging from hydropower production to tourism and horticulture, to medicinal herbs, and also indigenous knowledge systems), which if properly harnessed can offer significant comparative advantages to mountain areas
	the bulk of the potential remains under-utilised for want of resources and infrastructure (or selective over-extraction by external agencies).
c) Imperatives (appropriate responses)	harnessing of 'niche' as an integral part of diversified resource use, using the rationale of traditional systems and modern science and technology
	infrastructural support for harnessing the 'niche'; gains to local people through participatory approach

Note: Based on evidence and inferences from Whiteman (1988); Sanwal (1989); Pant (1935); Jochim (1981); Guillet (1983); Bjønness (1983); Jodha (1990); Jodha and Partap (1993); Allan, *et al.* (1988); Mayer (1979); Dollfus (1982); Camino (1992); Rhoades (1992); Tapia (1992); Brush (1977); Thompson and Warburton (1985); Mateo and Tapia (1989); Getahun (1991); and many others referred to in Jodha and Shrestha (1994).

Plate 14.1 Intense land use in the Middle Mountains of Nepal (Photograph: J. D. Ives)

Plate 14.2 Slash-and-burn agriculture in the mountains of north-west Thailand (Photograph: J. D. Ives)

accessibility, fragility, marginality, and to an extent, diversity, are a primary source of the above limitations. Nevertheless, as Table 14.2 also indicates, diversity, niche, and people's capacity to adapt to objective situations also have potential which, if properly harnessed through diversified, location specific strategies, can satisfy some of the basic conditions associated with high performance agriculture.

The whole history of agricultural evolution and transformation, or rather that of the dynamics of natural resource use in mountain areas in general, can be viewed in terms of the ways in which communities have managed the constraints and opportunities created by mountain conditions (Brush, 1977; Guillet, 1983; Jochim, 1981; Hewitt, 1988; Allan *et al.*, 1988; Ohtsuka, 1995; Jodha, 1990; Camino, 1992). To understand community

Table 14.2 Mountain specificities and the conditions of high agricultural performance

Mountain specificities – generated constraints/ opportunities	Conditions associated with high performance agriculture					
	Production enhancing factors				Abilities to link with wider systems	
	Resource-use intensity	Input absorption capacity	Infra-structure	Scale economies	Surplus generation/ trade	Replicating external experiences (tech)
Limited Accessibility Distance, semi-closedness, high cost of mobility and operational logistics, low dependability of external support, or supplies	$(-)^a$	(−)	(−)	(−)	(−)	(−)
Fragility Vulnerable to degradation with intensity of use, limited low productivity/ pay-off options	(−)	(−)	(−)		(−)	(−)
Marginality Limited, low pay-off options; resource scarcities and uncertainties, cut off from the 'mainstream'		(−)	(−)	(−)	(−)	(−)
Diversity High location specificity, potential for temporally and spatially inter-linked diversified products/activities	$(+)^a$	(+)		(−)	(+)	(−)
Niche Potential for numerous, unique products/activities requiring capacities to harness them	(+)	(+)		(+)	(+)	(−)
Human adaptation mechanisms Traditional resource management practices – folk agronomy, diversification, recycling, demand rationing, etc.	(+)	(+)		(−)		(+)

Source: adapted from Jodha (1990).
[a](−) and (+) respectively indicate extremely limited and relatively increased degree of convergence between imperatives of mountain conditions and the conditions associated with high performance of agriculture. The constraints indicated for the primary production sector also applies to the secondary and tertiary sector activities, such as product processing and marketing

responses to them, these constraints and opportunities can be presented in terms of the following focused questions:

(1) In view of fragility, marginality and, to a certain extent, accessibility problems, how are the use-intensity and (physical and economic) input absorption capacity of land to be enhanced without negative side effects that lead to resource degradation?

(2) How are the diversification strategies and options to be evolved in order to: (a) harness the potential of diversity and niche offered by mountain resources without over-extracting them; (b) ensure high productivity despite low land-use intensity and low input regimes

(especially external inputs); and (c) cope with periodic shocks (scarcities) and rising pressure on fragile resources?

(3) What are the forms and patterns of external linkages and how are they to be developed, to ensure accomplishment of potential options under (1) and (2) above?

THE SOCIETAL RESPONSES

Even when not formulated in the above manner, these issues have been central to the traditional farming systems of the past. In recent decades agricultural development strategies in mountain areas have also formally addressed these issues. However, relative understanding of the problems and mechanisms of responses differ considerably under the two cases presented below. Table 14.3 summarises the relevant components of these two approaches directed to resource management and productivity growth in mountain agriculture and reveals both their strengths and weaknesses. A synthesis of the strengths can help in identifying the directions and possible first order options to enhance the performance and sustainability of mountain agriculture.

Traditional measures and practices indicated here in Table 14.3 have been evolved by people through informal experimentation over generations. Hence, they are better adapted to limitations and potentialities of the mountain resources. Broadly speaking, they are location-specific and small in scale; diversified and interlinked in their structure and operations; often land-extensive and locally renewable resource-centred; mainly supported by folk knowledge and informal social sanctions; and generally have lower input use and lower but stable productivity (Jochim, 1981; Guillet, l983; Camino, 1992; Whiteman, 1988; Price, 1981; Thompson and Warburton, 1985; Yanhua, 1992). For the above reasons, these measures are conducive to sustainable resource use under low pressure of demand in relatively isolated or inaccessible situations. But they are becoming increasingly unfeasible and ineffective in the context of rising pressure on mountain land resources. Faced with resource degradation and rising resource scarcity, the people have started undertaking more intensive and exploitative practices (Jodha, 1995a and b; Carson, 1992; Zimmerer, 1992; Rerkasem and Rerkasem, 1995). It is ironic that public intervention measures for mountain agriculture also, in their own way, have followed the people's resource

Plate 14.3 Traditional and ceremonial commencement of potato planting in the Altiplano of Bolivia at 3800 m (Photograph: B. Messerli)

intensification or exploitative approach. The measures promoted through conventional public interventions in the mountains, therefore, generally represent the extension of land-intensive production system characteristics of relatively better agricultural areas (Sanwal, 1989; Tapia, 1992; Denniston, 1995; Jodha, 1995c). They are not well adapted to the fragile and diverse mountain resources.

The broad differences between traditional and present day farming systems, their resource-use patterns, their underlying driving forces as well as consequences are easy to infer from Table 14.3. If the orientation and impacts of practices followed under these two systems are related to the imperatives of mountain conditions, the traditional systems appear more sensitive to present day systems. These, including the formal public interventions encouraging them lack, however, the mountain perspective. This could be simply stated as a lack of understanding, and incorporation of

Table 14.3 Measures directed to manage constraints and opportunities under traditional and present day agriculture in mountain areas

Measures adopted	
Traditional farming systems	*Present day farming systems including development interventions*

A. Enhancement of use intensity/input absorption capacity of land

Small scale, location-specific, community oriented/supported resource amendments using ethno-engineering measures: terracing/ridging/ drainage management, community irrigation, agroforestry, etc.; reduced feasibility with rising pressure on land and weakening of local level collective arrangements	Weakened traditional measures supplemented/substituted by selective, larger scale resource upgrading (e.g. irrigation, infrastructure, watershed development) designed and implemented by external support; use of modern science and technology, and public subsidy; a number of social and environmental side effects of change

B. Usage and management of low use capability lands

Diversified, interlinked, land-based activities; folk agronomy involving measures with low land intensity and low (local and affordable) input regimes; integration of low intensity–high intensity land uses (based on annual–perennial plants, crop– fallow rotations, slash and burn, indigenous agroforestry, common property resources; social sanctions for resource-use regulation; conservation; migration/ transhumance	Compelled by increased population and land shortage, rapid increase in indiscriminate intensification of land use; sectorally separated production programmes; high intensity uses promoted through new technology inputs/incentives or subsidies; limited conservation-oriented initiatives (forests/pastures/ watersheds), largely in project mode

C. Options to harness diversity and niche

Folk agronomy – diversified cropping, focus on multiple use species; complementarity of cropping – livestock/forestry/horticulture; emphasis on biomass in choice of land-use and cropping patterns; complementarity of spatially/temporally differentiated land-based activities; stability-oriented, location-specific choices, harnessing niches for tradable surplus	Reduced diversification and narrowed focus on cropping driven by: (i) subsistence needs (e.g. in the case of food crops); (ii) commercialisation (as in horticulture); and (iii) public interventions. Sectorally segregated programmes and their support systems (R&D, input supplies, crop marketing), focus on selected species and selected attributes (e.g. monoculture, high grain:stalk ratio); extension of generalised development experience of other habitats with high subsidy

D. Managing isolation, external links and demand pressure

Living with general state of relative inaccessibility and isolation from mainstream market; limited linkages through tradable surplus from harnessing niche; scarcity period – external dependence through transhumance, migration, and remittance economy; insignificant surpluses, but petty trade in niche-based products	Reduced risks of isolation due to improved physical and market linkages: integration of mountain economy with other systems; highly uneven, but improved opportunities for relaxing internal constraints through technology, resource transfer, interactions with other systems; inducement for fuller use of niche through external demand; closer integration with mainstream
Subsistence strategies focused on diversification and linkages of land-based activities; flexibility in scale, operations, or input use; local renewable resources, recycling or inputs/products, self-provisioning; crisis period collective sharing arrangements, social regulations for rationed use and protection of fragile resources; release of periodic/seasonal pressure by migration, transhumance, remittance economy; emphasis on managing 'demand'	Reduced sole dependence on local resources, due to public relief and support during crisis/scarcities; public interventions replacing traditional self-help strategies and informal regulatory measures; decline of resource regenerative recycling practices
	Increased dependency for subsistence on external resources; encouragement for perpetual growth of pressure on fragile resources: indifference to local self-help initiative

Source: based on evidence and inferences cited in Table 14.1 and others in Jodha and Shrestha (1994)

imperatives of mountain specificities into resource-use planning and practices (Jodha *et al.* 1992). However, the whole context in which traditional practices worked well (e.g. low population pressure, multiple risk-reducing strategies, greater migration possibilities, low extraction capacities) has changed today, reducing their feasibility and efficacy. The changed man-made factors, which have influenced the whole process of resource use, production systems and demand patterns, as well as exchange relations between mountain areas and the rest of the world, can be grouped into three categories, namely: demographic changes; market forces and processes; and public interventions, or enhanced role of the state. An indication of how these socio-economic factors (changes) interact with relatively unchanged biophysical factors is given by Table 14.4.

THE CHANGING SOCIO-ECONOMIC CONTEXT AND DRIVING FORCES

Demographic changes, such as an increase in the number of people, or a change in their expectation profiles and attitudes that are reflected by growing individualistic tendencies, are products of several factors. These include: the health revolution (in spite of inadequate medical facilities) that reduced death rates without reducing birth rates; and increased contacts with the external world via market integration and State interventions (Sharma and Banskota, 1992). Besides generating direct pressure on mountain resources, they have eroded the effectiveness of social sanctions and institutional measures which traditionally helped to protect and regulate the use of resources. The mismatch of these changes with the imperatives of specific mountain features, such as fragility and limited accessibility and their impacts, are indicated in Table 14.4.

The enhanced role of the State in people's lives is manifested by the increased institutional and technological interventions by the State in the name of public administration, development, and welfare. The main elements of State intervention, in the present context, are extension or imposition of generalised development approaches (as mentioned earlier); the physical and market integration of mountain areas with mainstream/urban areas; the institutional, administrative, and fiscal measures influencing people's decisions and actions *vis-à-vis* their natural resources and production systems; the technologies and resource management/extraction systems introduced and sustained

with support and patronage of the State; and the relief and welfare activities in mountain areas, as part of the mandate of a modern nation state (Banskota and Jodha, 1992; Denniston, 1995). Their side effects, or conflicts with the imperatives of mountain circumstances (biophysical conditions), are presented in Table 14.4 under public interventions a, b, and c.

Market forces which increasingly play a decisive role in mountain areas (particularly in more accessible ones), have two relevant elements, namely inter-regional linkages (integration) and promotion of extractive patterns of harnessing mountain resources, as dictated by distant market signals (Bjønness, 1983; Collier, 1990; Banskota, 1989). Market penetration has introduced a range of incentives that induce people not only to over-extract resources but also to disregard the traditional, informal institutional measures directed to resource conservation and usage. Their impacts on the mountain situation through violation of imperatives of mountain resource features are presented in Table 14.4.

The results of the combined operation of the man-made factors listed under Table 14.4 are manifested by persistent negative trends relating to the resource base, production flows, and resource management options in agriculture and other sectors. Even a quick glance at Table 14.5 will reveal the nature and type of what are described as indicators of unsustainability. While some of them represent negative consequences, others indicate the processes leading to such consequences.

WHAT IS NEW ABOUT THE CHANGING SOCIO-ECONOMIC CONTEXTS?

Associating the present crisis of mountain areas to the post-World War II developments, i.e. the 'population explosion', enhanced market intrusion, and public interventions imposed from above, does not mean that in the absence of these changes mountain agriculturalists (transhumants, pastoral nomads, or mixed farming peasants) had comfortable, stable, crisis-free, subsistence existence in the past. Due to their very nature, the biophysical circumstances of mountain areas constrained the survival strategies of the people whenever their number exceeded the carrying capacity of the agricultural resource base at given levels of technological and organisational capabilities. Mountain communities responded to these imbalances by resorting to

Table 14.4 Interactions and implications of rapidly changing socio-economic circumstances in mountain areas

Socio-economic changes interacting with biophysical factors, i.e. increased human interventions in mountain areas	Biophysical factors			
	Limited accessibility Semi-closedness; limited dependability of external support; local resource focus of activities	*Fragility and marginality* Incompatibility with high intensity uses, focus on diversified, low cost, low risk activities	*Diversity* High potential for diversified, interlinked activities; location specificity	*'Niche'* Products; activities with comparative advantage, including human adaptation measures
Population growth changed expectation levels/attitudes; per capita increased activities guided by profit motive or forced by poverty	Excess pressure on local resources with limited outlets; resource-use intensification, over-extraction, degradation (croplands, pasture, forest)	Indiscriminate resource-use intensification; disregard of resource-extensive, diversified cropping practices; dependence on external subsidy; discard of usage regulations; group action	Pressure of food needs; reduced diversification of cropping; reduced resource regeneration; diversity of food systems replaced by limited grain types	Pressure of food needs; disregard or misuse of natural potential for diversified and better suited activities and their complementarities
Market forces trade links; pressure of external demand; changes in people's attitudes, expectations	Integration with mainstream market situation despite low physical accessibility; additional pressure on resources; market driven corridors of change; intra-regional disparities	Distant demand induced over use of resources; backlash of selective commercialisation of agriculture; decline of environment-sensitive agronomic practices; poverty of ethnic minorities and women with decline of common lands	Market-driven, narrow crop specialisation, reduced diversification; marginalisation of traditional knowledge, and practices supporting interlinked diversified land uses, favouring accessible areas	External demand-induced over-exploitation of major niche, e.g. hydropower, horticulture, marginalisation of petty 'niche', local concerns, traditional small scale activities, biodiversity
Public Interventions a) Imposition of generalised development interventions, including investment priorities, technology choices, macro-economic policies–price, tax, trade, resource extraction	Reduced isolation, increased integration and level of activities, leading to unmanaged increase in pressure on land resources – crop land, forest, pasture	Promotion of increased use intensity; degradation of fragile/marginal resources; public relief, subsidy encouraging pressure on land without upgrading resource base	Subsidies, incentives for intensification, reduced diversification of crops, access-determined narrow specialisation (e.g. horticulture) with backlash on food supplies, food systems; widening gaps between areas with different accessibility	Over-exploitation of areas with high potential (e.g. horticulture, high value crops); disregard of side effects of local concerns; emergence of a dual sector economy (accessible–inaccessible areas)
b) Infrastructure for accessibility, integration, market-driven harnessing of 'niche'	Application for improved mobility, integration; priority to areas with high potential; regional inequities, emergence of dual systems	Priority to production over conservation; indifference to resource limitations and local community concerns; increased intensity of fragile resource use	High cost, external input-based, narrow specialisation; focus on limited crops and attributes; disregard of traditional know-how, and institutional arrangements for diversification of land use, cropping	Market-driven over-extraction of 'niche', disregard of side effects on environment and people's survival strategies; traditional know-how; limited local participation/benefits
c) Technology and institutional support: narrow focus, directed to short-term needs, sectoral orientation, external origin/orientation, non-participatory	Inaccessibility-induced invisibility of problems/opportunities making development measures as inappropriate impositions	Focus on current production, high use-intensity; disregard of resource limitations and long-term consequences; sustained through subsidies, e.g. cropping on steep slopes; disregard of traditional know-how	Narrow specialisation, through incentives and support systems, technologies/R&D disregarding organic linkages and performance of total cropping systems; marginalisation of traditional farming systems; increased dependency on subsidisation	Focus on revenue generation; meeting external demand; extraction levels disregarding the side effects; locally useful area-specific potential given low priorities

Source: adapted from Jodha and Shrestha (1994), based on synthesis of evidence and inferences from more than 30 studies and documents (some of them cited in Table 14.1) covering mountain areas in Asia, Latin America, and Africa

Table 14.5 Negative changes as indicators of emerging unsustainability of agriculture/current resource-use systems in mountain areas

	Change related to[a]		
Visibility of change	Resource base	Production flows	Resource-use management practices/options
Directly visible changes	Increased landslides and other forms of land degradation; abandoned terraces; per capita reduced availability and fragmentation of land; changed botanical composition of forest/pasture, reduced biodiversity. Reduced water flows for irrigation, domestic uses, and grinding mills	Prolonged negative trend in crop yields, livestock, etc.; increased input need per unit of production; increased time and distance involved in food, fodder, fuel gathering; reduced capacity and period of grinding/saw mills operated on water flow; lower per capita availability of biomass, and range of agricultural products	Reduced extent of: fallowing, crop rotation, intercropping, diversified resource-management practices; extension of plough to submarginal lands; replacement of social sanctions for resource use by legal measures; unbalanced and high intensity of input use, subsidisation
Changes concealed by responses to change	Substitution of: cattle by sheep/goat; deep-rooted crops by shallow-rooted crops; shift to non-local inputs; choice for inferior options. Substitution of water flow by fossil fuel in grinding mills; or manure by chemical fertilisers	Increased seasonal migration; introduction of externally supported public distribution systems (food, inputs); intensive cash cropping on limited areas; additional production by using marginal areas	Shifts in cropping pattern and composition of livestock; reduced diversity, increased specialisation in mono-cropping; promotion of policies/programmes with successful record outside, without required adaptation
Development initiatives, i.e. processes with potentially negative consequences[b]	New systems without linkages to other diversified activities and regenerative processes; generating excessive dependence on outside resource (fertiliser/pesticide-based technologies, subsidies); ignoring traditional adaptation experiences (new irrigation structure); programmes focused mainly on resource extraction	Agricultural measures directed to short-term quick results; primarily production (as against resource)-centred approaches to development; service-centred activities (e.g. tourism) with negative side effects; focus on food self-sufficiency ignoring environmental stability	Indifference of programme and policies to mountain specificities; focus on short-term gains; high centralisation; excessive and crucial dependence on external resources and advice ignoring self-help and traditional wisdom; generating permanent dependence on subsidies

[a]Most of the changes are interrelated and could fit more than one column.
[b]Changes under this category differ from the previous two categories, in the sense that they are yet to take place, and their potential emergence could be understood by examining the involved resources-use practices, in relation to specific mountain characteristics. Thus they represent the 'process' dimension rather than the consequence dimension of unsustainability.
Source: adapted from Jodha (1990), Jodha and Shrestha (1994), based on data or description in over 45 studies from countries covered by ICIMOD, namely: Nepal (18), China (15), India (7), Pakistan (3), Bhutan, Bangladesh, and Myanmar (1 each), as synthesised by Jodha and Shrestha (1994). As a part of the preparation for the present chapter, most of the above aspects were broadly reconfirmed in relation to nearly a dozen studies/documents of the Andes and African mountains (Collier, 1990; Dollfus, 1982; Mateo and Tapia, 1989; Tapia, 1992; Zimmerer, 1992; Davis, 1991; Salas, 1993; Camino, 1992; Getahun, 1984, 1991; Messerli and Hurni, 1988)

age-old two-way adaptation strategies. They rationed demand through various social sanctions and through migration. They upgraded the production resource base through introduction of terracing, community irrigation, and added new crops such as maize and potato to their cropping systems.

However, the situation in the post-World War II period is very different from the earlier situation, both qualitatively and quantitatively (Jodha, 1995a, b, c, and d). First, the pace of change is too rapid to leave enough lead time to evolve effective adaptation measures. Second, both the complexity of the change and its magnitude are far beyond the technical and institutional capacities of mountain people to fully understand and effectively respond. Third, the key driving forces (including indirect inducement for population growth) fall far outside the mountain people's sphere of influence and control. Fourth, responses to the crisis (mostly originating from outside) focus on one-way adaptation, namely adapting the resources (by extractive manipulation) to rising demands, rather than the other way round. The powerless mountain people simply join the supply-focused process adaptation rather than addressing the demand side of the equation.

An important inference suggested by the preceding description and also revealed by Table 14.4 is that, unlike the former situation, decisions and actions of present-day farmers or communities in mountain areas are very strongly influenced by the decisions and actions conceived and implemented at levels far away from them (Sanwal, 1989; Stone, 1992; Mountain Agenda, 1992; Denniston, 1995). In other words, the policy environment and programme decisions in which they have little involvement, influence their resource-use decisions more than the immediate biophysical circumstances. For instance, subsidy to 'grow more food' induces cropping on steep slopes, despite their unsuitability for this purpose.

This has two important consequences. First, the measures to amend the resource users' (e.g. the farmers') approach to mountain resources will have to focus first on the enabling policy environment to influence their decisions and actions (Jodha, 1995c). Second, since the conventional mountain development policy–programme framework lacks mountain perspective and, therefore has little sensitivity to imperatives of mountain conditions (Jodha *et al.*, 1992; Denniston, 1995; Thompson and Warburton, 1985), the filling of this major gap is a crucial requirement of changing the policy environment, to make resource users' decisions conducive to sustainable use of mountain resources.

These aspects are synthesised by Denniston (1995) although without the above terminology (e.g. mountain perspective, mountain specificities), and addressed by several studies and policy advocacy work, though without any particular success in achieving change. The essence of the above message was also behind the *Mountain Agenda* presented at the Earth Summit in Rio in 1992. The Earth Summit, being a high profile and concern-generating event with a variety of follow-up actions, will be a highly appropriate object for examination of its possible impact on the world's mountains. Hence, before we comment on the future strategies for sustainable agriculture and resource use in mountain areas, it will be appropriate to briefly review the post-UNCED scenario in mountain areas.

THE POST-UNCED SCENARIO

Five years (after the Earth Summit in 1992) is too short a period to assess changes in the state of the world's mountains, particularly in view of the invisibility of the mountain situation for statistical systems in different developing countries. However, even without concrete statistical proof, one can state that as a part of 'business as usual', the processes (manifested by unabated population growth, public interventions, selective over-extraction of the mountain niche) seen prior to 1992 are still in progress in mountain areas. UNCED has so far made little difference. Accordingly, the post-Rio situation can be seen more in terms of a few formal decisions and (yet to be implemented) activities with potential impacts for mountain areas. Table 14.5 lists the important developments in this regard. Though one can easily see the direct links between mountain agriculture and the main elements of the following discussion, its focus, due to the very nature of the issues involved and their relevance in the wider context of overall resource use and sustainable development of mountain areas, will be on the broader aspects of the new developments and the trends represented by them.

The post-Earth Summit development relating to mountain areas (Table 14.6) can be discussed in different contexts, namely: changes directly related to the Earth Summit (as follow-up) and those unrelated to it; developments directly or exclusively focused on mountains and the general global

Table 14.6 Some post-UNCED developments and the mountain areas*

Developments/changes	Implications for mountain areas
A. Developments as follow-up to the Earth Summit	
consolidation of mountain advocacy – networks, forums to study, project, and plead for the mountain cause	enhanced understanding, concerns and visibility of mountain problems; emerging lobby for mountains
global treaties/conventions on environmental resources, such as biodiversity, climate, desertification also covering mountains	top-down approach; universalisation of problems and aggregation of solutions, but likely to miss mountain perspectives; positive impact: resource-transfer and attention to mountain areas
global debates and discourses on indigenous people, women, population, urban habitats	attention to marginal areas and people, also conducive to projection of mountain problems
B. New developments unrelated to the Earth Summit	
globalisation, liberalisation of world economies, World Trade Organisation (WTO) – according primacy to market forces; decline of public sector initiatives, support systems	primacy to market forces likely to promote inequities, accentuate selective over-extraction of mountain resources ignoring imperatives of diversity, or customary rights to nature in mountain areas shrinkage of public sector resources affecting marginal entities – mountains and their people positive impacts: promotion of high pay-off, value-adding options, trade not charity
visible growth of civil society – NGO activities, user-groups for natural resource management; agricultural R&D	opening of opportunities for diversified, location-specific, participatory development
C. Unabated/accentuated 'business as usual'	
population growth, public interventions without mountain perspective	over-extraction, resource degradation and accentuation of unsustainability prospects
rapid /commercialisation of accessible areas, narrow specialisation	widening gap between accessible and inaccessible areas

*excludes bilateral initiatives supported by individual countries, foundations, and a number of projects funded by multilateral sources, such as The World Bank and GEF

initiatives extending to mountains; and changes as a part of the 'the business as usual', that is, continuance of processes of the past.

'Business as usual' trends

Let us first comment on item (c) in Table 14.6, dealing with the 'continuation of business as usual'. Accordingly, a large part of the post-UNCED situation in mountain areas represents the continuation of pre-existing trends in terms of unabated population growth; imposition of externally conceived and designed interventions; rapid commercialisation of the more accessible areas and their backlash effects on other areas; and increasingly unsustainable use of resources due to over-extraction. In these contexts, the Rio meeting and its *Mountain Agenda* do not seem to have made any difference other than sounding an alarm: thus, the 'missing mountain perspective' of development policies and programmes, i.e. poverty and profit-

driven extraction of resources, continue to be the dominant features of mountain areas five years after the Earth Summit (Jodha, 1995a; Jodha and Shrestha, 1994).

Enhanced advocacy and visibility of mountains

If one ignores the global treaties and conventions on environmental resources (commented on later), the most important development concerning mountain areas which could be directly attributed to the Earth Summit, relates to the increased debate and enhanced visibility received by the mountains through national and global NGO activities – supported by the Commission on Sustainable Development (CSD), and other agencies. If the number of mountain advocacy forums and national and international participation therein is any indicator, the post-Rio period has created more 'space' for the mountain agenda (The Mountain Institute, 1995).

The related global initiatives on indigenous people, population growth, and women and gender equity, also have elements related to the mountain situation. This may ultimately help in moving mountains from the margins of public consciousness to a more central place on national and international agendas (Denniston, 1995).

Local initiatives

Closely related to the above is the faster spread of NGO activities and resource-user group organisations in many mountain areas. They partly represent the local people's movements toward sustainable resource management and sustenance security. They are visible through a number of success stories scattered in different mountain areas. This represents a positive change with great potential to facilitate participatory development (The Mountain Institute, 1996). Since these initiatives are more locally oriented, they could prove a vehicle for local resource-centred, diversified activities (including participatory agricultural research and extension), and mobilisation of resources for conservation and development. The Mountain Institute (1996) recently organised an electronic conference that assembled and exchanged experiences of a large number of such initiatives from mountain areas in different parts of the world. Similarly, the motivation underlying the production of this book and its accompanying policy document has vital potential.

Conventions and treaties on environmental resources

Apart from the previously discussed post-Rio initiatives directly focused on mountain areas, the follow-up to the Earth Summit also includes international treaties and conventions formally binding the countries to participate in the international arrangements dealing with environmental issues. Since mountains constitute an important ecosystem with huge off-site impacts on environmental stability, their problems are also addressed as a part of the overall strategies conceived through the global treaties (including the Global Environment Facility – GEF). The more important elements among them relate to biodiversity conservation, desertification and climate change (see Chapters 10, 15, and 17 respectively). Some of them have provisions which may affect mountain areas, mountain communities, and their main occupation (agriculture), both positively and negatively. The negative possibilities

stem from the fact that most of these global initiatives share the same limitation that characterises the conventional development approaches, namely, they are missing a specific mountain perspective. Hence despite good intentions, these provisions may by-pass the concerns and perspectives of mountain people and the imperatives of mountain conditions. The potentially positive impacts of the global initiatives are closely associated with the better recognition of mountain problems, focused resource transfer to these areas, and involvement of local communities in designing and implementing different provisions of the treaties at the grassroots level (The Mountain Institute, 1996).

Globalisation–privatisation

The most important set of post-UNCED changes (though not directly emerging from the Earth Summit) which may affect the mountain areas includes the post-Cold War trends toward globalisation and liberalisation/privatisation of the developing country economies, including formal arrangements, such as the World Trade Organisation (WTO). Irrespective of the capability of a country or a region to participate in the process and the initial discriminatory possibilities it may face, every country on its own, or through pressures from international arrangements, is engaging in the process. Mountains cannot escape their impact. The primary mechanisms which could influence mountain areas in terms of: over-extraction of niche (e.g. medicinal herbs, fine timber) and exchange on terms unfavourable to mountain areas; enhancement of internal inequities (i.e. between accessible and inaccessible areas); promotion of selective, narrow specialisation (e.g. cash crops, horticulture) ignoring imperatives of diversity and marginality, and so on, are associated with the unrestricted role of market forces promoted by the above developments. Markets generally are neither diversity-friendly nor sensitive to negative externalities or side effects.

Second, with the gradual decline of the public sector in the liberalisation process, the resource allocation to mountain areas may decline substantially, especially to those areas with poor accessibility and that are not producing items demanded by the wider market. Due to application of standard investment norms and performance evaluation yardsticks (unsuited to marginal entities) which characterise the competitive framework to be promoted by the globalisation process, mountain areas may lose in the process. On the positive side, one can think of incentives and durable opportunities for better management and harnessing of niche; scope for value-adding products and activities in mountain areas which, if accompanied by fair terms of trade with the rest of the world, can help stop the unequal downward flow of mountain resources. However, realisation of such gains is closely linked to the sensitivity of interventions to mountain conditions and enhanced capabilities of mountain communities to adapt to new circumstances.

The debates and discourses

Chapter 13 of UNCED's Agenda 21 addressed the basic issues of poverty, discrimination, marginalisation of communities, and degrading quality of life for the bulk of the world's mountain population. Satisfactory solutions to these problems at present are beyond the realm of the existing political will and material resources available to developing countries. Hence, for practical purposes, the post-Rio activities relating to these subjects were more for information and awareness generation, sensitisation of policy makers and mobilisation of support through international meetings and summits. They related to social issues (Copenhagen summit), women (Beijing summit), population (Cairo summit), urban habitat (Istanbul), indigenous peoples, amongst others. Mountain areas have a disproportionately higher share of the problems debated and covered by the above subjects. But due to their already mentioned hitherto 'invisibility' and lack of sufficient research, they found inadequate representation and space in these discourses. Yet these meetings have the potential to help mobilise support for the major social issues and to assist NGO networking arrangements to facilitate exchange of experiences, resource mobilisation, planning, and action, even in mountain areas.

The brief account of post-UNCED developments presented above has some important features relevant to the thematic focus of this chapter. First, the developments relating to globalisation and liberalisation are guided by the same fundamental driving forces (e.g. market forces and the state's desire to integrate and extract revenue from mountain resources: see Chapter 5), which significantly contributed to indiscriminate intensification and degradation of mountain resources in the past. Second, the formal global arrangements – treaties

and conventions (notwithstanding their scientific and environmental concerns) share with the conventional development strategies their top-down approach and lack of a mountain perspective. Hence, their consequences may not be very different from those of past developments (Table 14.4). The most significant implication of the above assessment relates to the future strategies for mountain areas. Accordingly, the measures designed to arrest the negative side-effects of conventional development approaches (i.e. indicators of unsustainability) can also help to address the potential backlash effects of the post-UNCED developments that are influencing mountain areas. Furthermore, the positive developments of the post-UNCED period, such as the increased awareness of mountain issues and the enlarged space available for them through advocacy and dialogue prompted by NGOs and international discourses, can facilitate the promotion of measures directed against the negative trends.

FUTURE APPROACH TO AGRICULTURE AND RESOURCE USE IN MOUNTAIN AREAS

The above discussion has shown that, due to basic similarities in their orientation and approach, the conventional development interventions, as well as the new initiatives (globalisation, and liberalisation of world economies, global treaties on environmental resources to the extent they involve mountain areas), may have broadly similar consequences. Hence, to reverse the trend toward unsustainability generated by past interventions and to guard against the potentially negative side effects of some of the post-UNCED developments, a common strategy must evolve. The central thrust of such a strategy (with location-specific variations) should be to adapt the resource-use systems, as well as development policies and programmes to the imperatives of mountain conditions. Such a thrust would entail creation of environment and support systems for final resource users in mountain areas to facilitate adaptation (or rather re-adaptation) of their activities to their local conditions. This also would help to reconcile development needs and conservation concerns in mountain areas (Ives and Messerli, 1989).

The basic premises behind advocacy of 'adaptations' in the changed socio-economic context are as follows:

(1) There is an apparent incompatibility between the imperatives of biophysical conditions in mountains (e.g. focus on diversification) and the attributes of the present-day socio-economic circumstances represented by the conventional development paradigm and the resource-use patterns in agriculture (e.g. focus on indiscriminate intensification). See Table 14.7.

(2) If the terms and concepts used to describe and understand the imperatives, implications, and contents of the two sides (i.e. mountain conditions and current usage systems) are redefined and reoriented in today's context, much of the incompatibility between the two will disappear.

As indicated by Table 14.1, the key imperatives of mountain conditions (as manifested by traditional farming systems) are: (i) diversification of resource-use systems; (ii) focus of activities on local resources; (iii) adaptation of technological and institutional options to mountain conditions; and (iv) management of demand or pressure on resources. The development interventions and present day resource-use patterns, on the other hand, are characterised by: (i) a generalised (and not mountain-specific) approach of development interventions; (ii) indiscriminate intensification of resource use; (iii) physical and market integration of mountain areas with the mainstream economy without concern for uncontrolled side effects, especially unregulated demand pressure; and (iv) focus on supply aspects disregarding demand management as a component of development strategy. These imperatives and attributes have their own rationale as well as limitations. Building on the features outlined above, an approach can be evolved to reconcile and effectively use them for sustainable development of mountain areas. To facilitate thinking in this direction, it may be useful to juxtapose the imperatives of mountain conditions and the dominant features of both the current resource-use systems and development interventions; and identify match or mismatch between the two. This is attempted through Table 14.7.

A closer look at the rationale behind the situations summarised under Table 14.7, suggests that the imperatives of the mountain condition can be accommodated with the requirements of the present-day situation, namely a focus on resource-use intensification and economic integration of mountains with other systems. Similarly, development

Plate 14.4 Tea plantation on Mt Kenya. The 'tea belt' separates the intense subsistence agriculture below and the protected mountain forest above. This works only as long as the local population is interested and involved in the production of this cash crop (Photograph: B. Messerli)

interventions and current resource-use systems can be reoriented to focus more on diversification and be made more sensitive to the local resource situation as well as to issues of demand management.

Reconciliation of imperatives/implications

Based on details elaborated and illustrated elsewhere (Jodha and Shrestha, 1994), we summarise the discussion on reconciliation of the imperatives and implications of the current biophysical and socio-economic circumstances obtaining in mountain areas. We indicate different focal areas which should be linchpins of the strategies for sustainable (and more productive) mountain agriculture. Accordingly, to be relevant and effective, future development strategies should address specifically the issues of diversification, intensification, local resource focus, integration and intersystemic linkages, and management of demand pressure. Table 14.8 provides an inventory of issues arising from both mountain specificities, and compulsions of present socio-economic circumstances that can help to identify concrete components for policy and programme activities.

The important point from Table 14.8 is that all the focal areas of concern, irrespective of their conventional nomenclatures, need to be redefined in the present context. The unifying and underlying approach behind all the suggestions that follow is to ensure an appropriate match between the resource-use systems and the imperatives of mountain resource features.

Table 14.7   Apparent conflicts between the imperatives of mountain conditions and attributes/implications of present approaches to resource use and development in mountain areas

Imperatives of mountain conditions, i.e. appropriate responses to objective circumstances created by limited accessibility, fragility, and diversity	Attributes of resource use/development approaches			
	Generalised development interventions (technology institutions) without mountain perspective	Indiscriminate resource use intensification to meet increased demands	Integration with the mainstream economy with uncontrolled side effects (over-extraction)	Focus on supply enhancement ignoring demand management
Diversification of resource use, production systems; demand structure to harness resource diversity, adapt to low carrying capacity/resources, and withstand relative isolation and poor mobility conditions	X	X		
Strong local resource focus to facilitate conservation, regeneration, and recycling of resources to evolve interlinked, diversified activities; greater stakes in local resources due to low external dependability	X	X	X	
Evolution and adaptation of technical and institutional measures to biophysical constraints and opportunities, especially due to fragility and diversity and low applicability of external experiences	X			X
Rationing/regulating pressure on resources by institutional and other means in view of low carrying capacity, resource degradation risks, and low external dependability			X	X

Accordingly, the term **diversification** need not be confined to diversified production activities on the land but should include the combination of primary, secondary, and tertiary sector activities with high value-adding potential, and linkages with other regions (production systems). Furthermore, functional linkages between diversification of resource use, production systems, and demand structure must be recognised. Diversification in the development context will also mean greater sensitivity to specific local circumstances and greater decentralisation in designing and implementing development activities. Decentralisation also implies greater local participation. This will need different priorities for areas with different degrees of accessibility.

Similarly, the term **intensification** should not be confined to land-intensive, high input use, individual crop activities. Instead, it should focus on intensification of the total system (e.g. interlinked, diversified activities) rather than on individual components of the farming system (e.g. raising grain yield). It will also involve a concerted focus on high value, value-adding activities. Finally, even during the increased intensification, the balancing of land intensification and land extensification will have to be addressed, both to harness their complementarity and avoid their undesirable impacts.

A **specific resource focus**, as one of the key elements, implies three things: (i) maintenance and increase in resource productivity, as well as local resource regeneration and balancing resource protection and productivity needs; (ii) rehabilitation and upgrading of fragile and marginal resources to raise their input absorption capacity for higher productivity without resource degradation; and (iii) special attention to resource-based niche opportunities. This would mean identification and enhancement of recognised niche, appropriate technological, institutional, and infrastructural

Table 14.8 Strategies for sustainable mountain agriculture: focal areas and their operational implications

Focal areas and components	Operational implications
Diversification/Decentralisation	
a) Overall development strategies and priorities to match the variability and area-specific circumstances b) Resource use and production patterns focused on: interlinked activities with high productivity and high payoff; integration of primary, secondary, and tertiary sector activities; and inter-systemic (regional) linkages	Discard the imposition of generalised development approaches and focus on decentralisation and planning from below Reorientation of agricultural R&D, support systems, infrastructure, and macro-economic policies to suit the requirements of multiple forms of diversification; through production/processing/marketing; inter-regional links Use rationale of traditional systems to make interventions sensitive to local circumstances
Intensification	
High resource-use intensity for high productivity focused on: intensification of total system (interlinked activities) rather than individual components; intensification involving value-adding activities, high payoff activities; and balancing land intensive–land extensive areas	Crucial role for R&D to cover totality of production – processing activities differentiated according to area specificities Infrastructure and support systems to suit intensification, appropriate activities (representing levels of intensification) Focus on 'high value' products/activities Upgrade fragile/marginal resources
Resource focus	
a) Local resource regeneration, balancing resource protection and productivity needs b) Rehabilitation and upgrading of natural resource base c) Harnessing of niche and local access to gains d) Resource features to guide infrastructural design/development	Agricultural R&D and support systems with combined focus on resource-cum-product centred options; focus on biomass productivity and stability, harnessing of local niche Infrastructural support and strong inter-systemic links for fair terms of exchange; regulation of market forces against over-extraction Usage regulation, local participation; adequate fiscal resources and modern technologies to help upgrade resources
Integration/Infrastructure	
Widely spread and dependable infrastructure to ensure: effective, equitable inter-systemic linkages; higher levels of diversification and off-farm employment gainful harnessing of comparative advantages; and participatory development and an even spread of development activities	Multiple means or forms of communication/transportation with minimal negative environmental and socio-economic side effects Provisions to protect areas/people against over-exploitation during the transitional phase Pricing and compensation mechanisms to ensure protection against exploitation Building of local capabilities and skill levels to benefit from integration
Management of demand pressure	
Increased attention to demand side issues for sustainable development to keep the pressure on resources according to their carrying capacity through: diversification of local demand (including food chain); greater focus on off-farm activities to take pressure off the land; control of human and livestock numbers; and regulation of external demands following improved accessibility	Identification and promotion of diversified high payoff activities with comparative advantage Population control measures, including education and economic independence of women External demand management through fiscal and physical measures; control over-extraction levels of niche Improved skill levels and off-farm employment opportunities Local participation in project planning/implementation

Box 14.1 Himachal Pradesh – A case of modern diversification

Himachal Pradesh, a wholly hill state, became one of the better developed and rich states in India within a period of 20 years or so. The strategy involved replacement of traditional biomass-centred, subsistence-oriented diversification by modern, multiple and interlinked activities which have: high pay-off, market orientation, and a strong focus on mountain niche. They make effective use of physical and market linkages with other regions. The new, local resource focused diversification includes activities such as horticulture, off-season vegetable farming; honeybee and angora rabbit-based non-land consuming activities; high yielding grass and stall feeding-based dairying; processing and marketing of above enterprises as value-adding activities; and tourism. Ecological zonation and area-specific focus of activities; R&D and infrastructural support; and policy commitment were the key steps in the process.

Source: Verma and Partap (1992)

Box 14.2 Ningnan County (West Sichuan) – From stagnation to progress

In Ningnan, one of the poorest counties of China, within a period of 15 years, people's income and product availability increased manyfold. The vital emphasis of the development approach was on selecting agricultural activities and overall land-use patterns according to natural suitability, i.e. harnessing the niche and rehabilitation/upgrading of marginal land resources. Decentralisation, people's involvement, use of new technologies and market links were the key instruments. High value crops such as cereals, vegetables, oilseeds, fruit, and other food crops, besides agroforestry, were promoted according to location suitability. Post-harvest processing, marketing and agro-industries further enhanced the overall income and resource generation for reinvestment in a traditionally poor country.

Source: Yanhua et al. (1992)

arrangements to harness niche, with explicit provisions for sharing the gains with mountain communities, and avoiding their negative side effects on the environment. This calls for site-specific integrated resource planning for mountain areas.

Integration (and **intersystemic linkages**) of mountain areas with other areas is an important factor in mountain development. The central role of integration, through an effective and dependable infrastructural network, is to ensure intersystemic (inter-regional) and inter-mountain linkages and mobility. On this hinges the success of most of the steps covered by diversification, intensification, resource harnessing, and equitable gains. However, conventionally, integration involved a strong focus on roads and bridges to facilitate exploitation of niche opportunities offered by mountains, or easy imposition of external interventions. This calls for a complete reorientation of overall policies and programmes directed to integration and infrastructural development in mountain areas. The focus has to be on: (a) diversification and decentralisation of infrastructural facilities to suit diverse situations; and (b) building local capabilities and ensuring local participation in the process of change. This can help arrest and reverse the widening gaps between different areas within a mountain region; and ensure that integration does not act as an instrument for exploitation and inequitable downward flow of mountain resources.

Management of demand pressure is an extremely crucial component of future development strategies in mountain areas, though this has not been a strong point of conventional development strategies. In this respect, one can learn a lot from traditional systems. Management of demand pressure will involve regulating the unprecedented growth in external demands for mountain resources through fiscal means (e.g. realistic pricing of traded mountain resources) and physical means (e.g. biophysical modes of compensation such as planting a new tree for the logged one); reducing/regulating the pressure of local demands through diversification of demand structure as well as through already recognised steps, such as control of human and animal populations; and creating skills and opportunities for off-farm employment, including high income activities in the processing and marketing sectors. In addition and wherever feasible, population transfers from extremely fragile areas of low productivity should be considered.

The strategic approaches suggested above have already been incorporated partly or fully as

Box 14.3 Ilam District (Nepal) – A transforming area within a stagnant region

Compared to the other neighbouring districts, Ilam district has much higher levels of income and assets, better education, health and housing conditions, and an improved environmental situation, as indicated by the status of the forest. The process contributing to this change was focused on a few lead sector activities based on niche or comparative advantages; effective use of upland–lowland linkages; emphasis on sprinkler irrigation (most suited to mountain conditions and crops); use of private enterprise and focused 'associations' of local people with higher education and skill levels; road and market linkages and demonstration effects of neighbouring areas of India. Local resource-focused and market-oriented diversification is exhibited by emphasis on high value cash crops (cardamon, tea, *amlisho*, potato) on mountain slopes, and high-yielding dairy cows and foodcrops using new technologies on the low-lands. Also, agroforestry and sericulture were extended to both zones. Value-adding secondary and tertiary sector activities further strengthen the extent and level of diversification. The area shows much less dependence on public sector subsidies and support.

Source: Koirala (1992)

components of the successful transformation cases in different pockets of mountain areas (Boxes 14.1–14.3).

Operational steps are shown in Table 14.8 (column 2) which also indicates the operational implications of all the steps suggested above. Most of the implications, indicated for individual focal areas (Table 14.8, column 1), are interrelated and multi-purpose in nature. For instance, reorientation of agricultural R&D as a practical step is necessary for diversification, intensification, and appropriate resource focus. Similarly, changes in macro-economic policies for development designs also relate to most of the other steps suggested by Table 14.8. The operational implications can be categorised as follows:

(1) Operational implications in terms of development policies and programmes will involve de-emphasising the application of generalised, externally-conceived approaches and a greater focus on location/area-specific planning from below. This will require measures for decentralisation and local participation.

(2) Reorientation of agricultural R&D in order to: (i) address the issues covered by strategies focused on diversification, intensification, intersystemic linkages, and enhancement of the comparative advantages of mountain areas; and (ii) incorporation of the rationale for traditional technologies and resource-use systems into new technologies.

(3) Macro-economic policies and their regional focus, which will determine the choice and scale of interventions, including the type and extent of infrastructural development; level and method of harnessing mountain 'niche'; mechanisms to involve and compensate mountain communities in the process; and the level and type of investment, as well as the norms and priorities to guide the same.

(4) Changes in overall development thinking and priorities, which can: (a) make development interventions more relevant to mountain circumstances; (b) encourage recognition and application of traditional knowledge systems in the changed context; and (c) facilitate local capacity building as well as participation.

Search areas for operational options

The preceding discussion, while mentioning the operational implications of different components of strategies for sustainable agriculture, has already alluded to a number of steps to facilitate their implementation. However, in order to identify more significant and concrete steps the discussion is extended below. The potential areas in which the search for options could be focused are indicated. Jodha and Shrestha (1994) record a number of successful initiatives in the Hindu Kush–Himalayan region in this regard.

Modern science and technology

The important potential leads for identification of options indicated above can be provided by modern science and technology. For instance, both in terms of plant genetic material and resource management systems, today's world has more synthesised and documented knowledge than ever before. However, since the perspectives of modern agricultural R&D planners are highly skewed, they have been unable to address the issues covered by diversification, resource regeneration, intensification without resource degradation, and improvement of regional comparative advantages in the mountains (Jodha, 1995c; Jodha and Partap, 1993). This under-utilised potential can be harnessed to identify relevant options for mountain agriculture. But this would need R&D policy and planning with a mountain perspective.

Rationale of traditional farming systems

Elements of both the technologies and institutional support systems of traditional agriculture have shown their effectiveness in sustainable resource use in mountain areas under low population pressure. In the changed context, they may have become irrelevant and ineffective, but their rationale represents an unrecognised and under-utilised store of practical wisdom. The invisibility of traditional know-how to the mainstream decision-makers, and the latter's ignorance, arrogance, and paternalistic attitudes, have contributed significantly to the under-utilisation and loss of this asset (Jodha and Partap, 1993; Hoon, 1996). A focused effort to identify, take inventory, and incorporate the rationale of traditional systems can help generate relevant technologies as well as methods of local resource management, participation, and even 'management' of demand pressure (Jodha, 1995L).

Lessons from the 'paradoxes of progress'

Development interventions generating negative side effects and, thus, in total proving counterproductive, are some of the 'paradoxes of progress'. ICIMOD's reviews of development policies and programmes in the Hindu Kush–Himalayan region revealed several cases in which failure to incorporate the imperatives of mountain specificities and their inter-relationships was the primary cause of negative side effects (Jodha et al.,1992). However, a systematic analysis of such paradoxes can offer useful lessons which, in turn, can help to identify user-friendly elements for future strategies and operations.

Experiences of replicable success stories

In contrast to the 'paradoxes of progress', replicable success stories are also a part of the mixed picture of current development efforts and achievements in mountain regions. A review of such cases (Jodha and Shrestha, 1994; Bebbington et al., 1996) shows that, consciously or subconsciously, most of the successful development initiatives incorporated the imperatives of mountain specificities into their design and implementation. ICIMOD, in collaboration with national agencies, invested considerable efforts in identifying, screening, documenting, disseminating, and applying replicable development experiences. The success stories covered were of different magnitudes and gestation periods; they related to product or resource and involved technological or institutional focus. ICIMOD found promotion and use of replicable successes a cost-effective, already tested, and easy to spread approach to helping development. Some of them are documented by Jodha and Shrestha (1994).

CONCLUSIONS AND RECOMMENDATIONS

We can briefly sum up the key inferences from the issues discussed in this chapter:

(1) The central focus of the policy, programme, and action for mountain development (including agricultural development) has to be on reconciling the imperatives of mountain-specific conditions and the implications of development interventions.

(2) To enhance and harness the opportunities that can help in the above reconciliation, the development strategies for mountain areas should focus on: (a) combination of diversification with selective intensification of resource-use systems; (b) local resource-centred approaches to address the imperatives of diversity, to harness niche, and prevent resource degradation; (c) fair and equitable upland–lowland links with involvement of mountain communities to prevent the historically inequitable downward flow of mountain resources; and (d) regulation of the resource-degrading demand pressure on the mountain resources through fiscal and physical means to control external demand, and through diversification of production, exchange, and consumption activities, to control internal demand.

(3) At the operational level, the implementation of the above suggestions would involve several innovative and unconventional steps directed to: (a) reduced emphasis on extension of generalised and externally conceived/designed interventions to mountain areas; (b) macroeconomic policies and regional planning,

which build on the specific constraints and opportunities in mountain areas; (c) reorientation of agricultural R&D to make it more resource-centred and directly focused on issues of productive diversification and selective intensification of resource use and enhancement of niche opportunities; and (d) effective means to promote local participation, and local capability building using NGOs and others, in the process of transformation.

(4) To design concrete options to implement the innovative/unconventional approaches, one can look forward to the opportunities provided by: (a) modern science and technology adapted to mountain conditions; (b) traditional knowledge systems and practices, evolved by the mountain communities through trial and error over generations; (c) lessons from the failures or negative side effects of development interventions, as well as the replicable success stories in different mountain areas.

(5) Finally, there are some initial preconditions which can aid in mountain development: (a) since mountain area development, due to the biophysical features, is a highly information-intensive activity, generation, synthesis, and dissemination of information through focused research should be an important first step in the process of planned change; (b) one of the basic steps which is necessary for the very initiation of the development process in mountain areas would include commitment of the national governments as well as the global community to a high priority for mountain areas; (c) in the practical context, partnership between governments, private sector, NGOs, and the people involved in the task, will be another most useful step to help the process of mountain development.

Most of the measures listed above are already in use in a small and scattered manner. The next step is to build further upon them, and ensure that they are applied on a much wider scale.

References

Allan, N. J. R., Knapp, G. W. and Stadel, C. (eds.) (1988). *Human Impacts on Mountains*. Rowman and Littlefied: NJ, p. 308

Bebbington, A., Quisbert, J. and Trujillo, G. (1996). Technology and Rural Development Strategies in Small Farmer Organisation: Lessons from Bolivia for Rural Policy and Practice. *Public Administration and Development*, 16: 1–19

Banskota, K. (1989). *Hill Agriculture and the Wider Market Economy: Transformation Processes and Experience of the Bagmati Zone in Nepal*. ICIMOD Occasional Paper No. 10. ICIMOD: Kathmandu

Banskota, M. and Jodha, N. S. (1992). Mountain Agricultural Development Strategies: Comparative Perspectives from the Countries of the Hindu Kush–Himalayan Region. In Jodha, N. S., Banskota, M. and Partap, T. (eds.), *Sustainable Mountain Agriculture*. Oxford and IBH Publishing: New Delhi

Bjønness, I. M. (1983). External Economic Dependency and Changing Human Adjustment to Marginal Environments in High Himalaya, Nepal. *Mountain Research and Development*, 3(3): 263–72

Brush, S. B. (1977). *Mountain Field and Family: The Economy and Human Ecology of an Andean Valley*. University of Pennsylvania Press: Pennsylvania

Camino, A. (1992). Andean Farming Systems: Farmers' Strategies and Responses. In Jodha, N. S., Banskota, M. and Partap, T. (eds), *Sustainable Mountain Agriculture*. Oxford and IBH Publishing: New Delhi

Carson, B. (1992). *The Land, the Farmer, and the Future: A Soil Fertility Management Strategy for Nepal*. ICIMOD Occasional Paper No. 21. ICIMOD: Kathmandu

Collier, G. A. (1990)). *Seeking Food and Seeking Money: Changing Relations in a Highland Mexico Community*. UNRISD Discussion Paper 11. United Nations Research Institute for Social Development: Geneva

Davis, S. (1991). *Indigenous Views of Land and Environment*. The World Bank: Washington DC

Denniston, D. (1995). *High Priorities: Conserving Mountain Ecosystems and Cultures*, WorldWatch Paper 123. WorldWatch Institute: Washington DC

Dollfus, O. (1982). Development of Land Use Patterns in the Central Andes. *Mountain Research and Development*, 2(1): 39–48

Getahun, A. (1984). Stability and Instability of Mountain Ecosystems in Ethiopia. *Mountain Research and Development*, 4(1): 39–44

Getahun, A. (1991). Agricultural Growth and Sustainability: Conditions for Their Compatibility in the Tropical East Africa Highlands. In Vosti, S. A., Reardon, T., von Urf, W. and Witcover, J. (eds.), *Agricultural Sustainability Growth, and Poverty Alleviation: Issues and Policies*. SDE and IFPRI: Feldafing

Guillet, D. G. (1983). Towards a Cultural Ecology of Mountains: The Central Andes and the Himalayas Compared. *Current Anthropology*, 24: 561–74

Hewitt, K. (1988). The Study of Mountain Lands and Peoples: A Critical Overview. In Allan, N. R. J., Knapp, G.W. and Stadel, C. (eds.), *Human Impacts on Mountains*. Rowman and Littlefield: NJ, pp. 6–23

Hoon, V. (1996). *Living on the Move: Bhotiyas of the Kumaon Himalaya*. Sage Publications: New Delhi, Thousand Oaks, CA, p. 253

Ives, J. D. and Messerli, B. (1989). *The Himalayan Dilemma: Reconciling Development and Conservation*. Routledge: London, p. 296

Jochim, M. A. (1981). *Strategies for Survival: Cultural Behavior in an Ecological Context*. Academic Press: New York

Jodha, N. S. (1990). Mountain Agriculture: The Search for Sustainability. *Journal of Farming Systems Research Extension*, 1(1): 55–75

Jodha, N. S. (1995a). Environmental Crisis and Unsustainability in Himalayas: Lessons from the Degradation Process. In Hanna, S. and Munasinghe, M. (eds.), *Property Rights in Social and Ecological Context*. vol 2. Case Studies and Design Applications. The Beijer Institute and The World Bank: Washington DC

Jodha, N. S. (1995b). *Transition to Sustainability in the Next Century: Hopes and Dismay in Mountain Regions*. World Resource Institute: Washington DC

Jodha, N. S. (1995c). Enhancing Food Security in a Warmer and More Crowded World: Factors and Processes in Fragile Zones. In Downing, T. H. (ed.), *Climate Change and World Food Security*. Springer Verlag: London

Jodha, N. S. (1995d). The Nepal Middle Mountains. In Kasperson, J. X., Kasperson, R. E. and Turner, II,

B. L. (eds.), *Regions at Risk: Comparison of Threatened Environments*. United Nations University Press: Tokyo

Jodha, N. S., Banskota, M. and Partap, T. (eds.) (1992). *Sustainable Mountain Agriculture*. Vol. I Perspectives and Issues; Vol. II Farmers' Strategies and Innovative Approaches. Oxford and IBH Publishing: New Delhi

Jodha, N. S. and Partap, T. (1993). Folk Agronomy in the Himalayas: Implications for Agricultural Research and Extension. In *Rural People's Knowledge, Agricultural Research and Extension Practice*, IIED Research Series Vol. 1, No. 3. International Institute for Environment and Development: London

Jodha, N. S. and Shrestha, S. (1994). Sustainable and More Productive Mountain Agriculture: Problems and Prospects. In *Proceedings of the International Symposium on Mountain Environment and Development*. ICIMOD: Kathmandu

Koirala, G. (1992). *Transformation of Ilam District: A Search for Replicable Experiences from Nepal*. ICIMOD: Kathmandu

Mateo, N. and Tapia, M. E. (1989). High Mountain Environment and Farming Systems in the Andean Region of Latin America. In Riley, R. W., Mateo, N., Hawtin, G. C. and Yadav, R. (eds.), *Mountain Agriculture and Crop Genetic Resources*. Oxford and IBH Publishing: New Delhi

Mayer, E. (1979). *Land Use in the Andes: Ecology and Agriculture in Montana Valley of Peru with Special Reference to Potato*. International Potato Centre (CIP): Lima

Messerli, B. and Hurni, H. (eds.) (1990). *African Mountains and Highlands: Problems and Prospects*. African Mountain Association: Addis Ababa

Mountain Agenda (1992). *An Appeal for the Mountains*. Geographical Institute, Bern

Mountain Institute, The (1995). *Report of the Initial Organizing Committee of the Mountain Forum*. The Mountain Institute: Franklin, West Virginia

Mountain Institute, The (1996). *Investing in Mountains: Innovative Mechanisms and Promising Examples for Financing Conservation and Sustainable Development*. The Mountain Institute: Franklin, WV

Ohtsuka, R., Inaoka, T., Umezaki, M., Nakada, N. and Abe, T. (1995). Long-Term Subsistence Adaptations to Diversified Papua New Guinea Environment.

Human Ecological Assessment and Prospects. *Global Environmental Change: Human and Policy Dimensions*, 5(4): 347–54

Pant, S. D. (1935). *The Social Economy of Himalayas: Based on a Survey in the Kumaon Himalayas*. George Allen and Unwin: London

Price, L. W. (1981). *Mountains and Man: A Study of Process and Environment*. University of California: Berkeley

Rerkasem, K. and Rerkasem, B. (1995). Montana Mainland South-East Asia: Agro-ecosystem in Transition. *Global Environmental Change: Human and Policy Dimensions*, 5(4): 313–22

Rhoades, R. E. (1992). Thinking Globally, Acting Locally: Technology for Sustainable Mountain Agriculture. In Jodha, N. S., Banskota, M. and Partap, T. (eds.), *Sustainable Mountain Agriculture*. Oxford and IBH Publishing: New Delhi

Salas, M. A. (1993). The Cultural Dimension of the Knowledge Conflict in the Andes. In *Rural People's Knowledge, Agricultural Research and Extension Practice*. IIED: London

Sanwal, M. (1989). What We Know About Mountain Development: Common Property, Investment Priorities, and Institutional Arrangements. *Mountain Research and Development*, 9(1): 3–14

Sharma, P. and Banskota, M. (1992). Population Dynamics and Sustainable Agricultural Development in Mountain Areas. In Jodha, N. S., Banskota, M. and Partap, T. (eds.), *Sustainable Mountain Agriculture*. Oxford and IBH Publishing: New Delhi

Stone, P. B. (1992). *The State of the World's Mountains: A Global Report*. Zed Books: London and New Jersey, p. 491

Tapia, M. E. (1992). Mountain Agricultural Development Strategies: The Andean Perspective. In Jodha, N. S., Banskota, M. and Partap, T. (eds.), *Sustainable Mountain Agriculture, vol. 1*. Oxford and IBH Publishing: New Delhi

Thompson, M. and Warburton, M. (1985). Knowing where to Hit it: A Conceptual Framework for the Sustainable Development of the Himalaya. *Mountain Research and Development*, 5(3): 203–20

Verma, L. R. and Partap, T. (1992). The Experiences of an Area-Based Development Strategy in Himachal Pradesh, India. In Jodha, N. S., Banskota, M. and Partap, T. (eds.), *Sustainable Mountain Agriculture*. Oxford and IBH Publishing: New Delhi

Whiteman, P. T. S. (1988). Mountain Agronomy in Ethiopia, Nepal, and Pakistan. In Allan, N. J. R., Knapp, G. W. and Stadel, C. (eds.) *Human Impacts on Mountains*. Rowan and Littlefield: NJ, pp. 57–82

Yanhua, L., Yang, T., Taichang, W. and Xiangou, G. (1992). *Transformation of Ningnam County, West Sichuan*. An ICIMOD-sponsored Study Report. ICIMOD: Kathmandu

Yanhua, L. (1992). *Dynamics of Highland Agriculture*. (A Study in Tibet, China). ICIMOD Occasional Paper No. 22. ICIMOD. Kathmandu

Zimmerer, K. (1992). Land Use Modification and Labour Shortage Impacts on the Loss of Native Crop Diversity in the Andean Highlands. In Jodha, N. S., Banskota, M. and Partap, T. (eds.), *Sustainable Mountain Agriculture*. Oxford and IBH Publishing: New Delhi

Mountain watersheds – integrating water, soils, gravity, vegetation, and people

15

Lawrence S. Hamilton and L. A. (Sampurno) Bruijnzeel

THE BATTLE AGAINST GRAVITY

The King of Corinth, Sisyphus, who misused his power and deceived the gods, was condemned to roll a stone to the top of a very steep hill in Tartarus; and just as he thought he had succeeded, the rock would slip, roll to the foot of the hill, and he was compelled to renew his exertions in a never-ending task. Like Sisyphus, individuals who attempt to use mountain land sustainably are engaged in a continual and relentless struggle against gravity. The task in domesticated land is to keep productive components (both organic and inorganic) of soil in place and use soil plus water, CO_2, and solar energy to produce the sustenance for supporting human life. And in natural ecosystems on sloping lands, the type and intensity of human intervention (e.g. logging) must be controlled so that natural processes and wild nature continue their evolutionary pathways protected from damaging disturbance and exploitation. In the terrestrial environment the land must be worked to produce the food, fibre, fuel, and other products needed to nourish and sustain all humankind. Like another figure from Greek mythology, the giant Antaeus, we derive our strength from Mother Earth, and that strength is lost when we lose touch with the land. On mountains and steep uplands, soil, and with it productivity, inexorably moves downhill. The process of movement of soil particles, stones, rocks, and soil masses is called erosion and mass wastage. This is a natural phenomenon, but can be greatly accelerated by human actions. One of the primary goals of mountain management is to reduce this impact of erosion as much as is feasible (physically and financially) and to slow the process of natural erosion in some special cases. We need to conserve the soil, which has been termed the 'living membrane' of the landscape by Messerli and Winiger (1992). It is better to strive by understanding and working with natural processes than to struggle like Sisyphus. Most erosion in steeplands occurs when water moves downslope in sheets, or channels of various sizes. Thus the fight against gravity becomes focused on water.

USING WATER ON ITS WAY TO THE SEA

Water is a precious resource of the mountains, for without adequate moisture there is no productivity, and indeed, without sufficient water there is no life (see Chapter 7 on Highland Water Resources). Under the force of gravity the water reaching the land surface moves downhill, both on the surface and as sub-surface flow. It behoves mountain people to make as much use as possible of this life-giving material as it flows in springs, rills, rivulets, streams, and rivers toward the seas, pausing occasionally in ponds, lakes, and wetlands. Fortunately water is returned to the uplands in the hydrologic cycle by processes of evaporation, transpiration, air movement, and condensation as fog, rain, and snow (see Figure 15.1). Natural vegetation makes use of a large part of the soil-water portion of this hydrologic cycle. Cultivated crops or pastures need water, along with the productive capacity of soil, to produce our plant and animal sustenance. Mountain people have been adept at applying co-operative action to capture water from streams and lead it by canals to irrigate upland crops. This is particularly needed in high, dry valleys and plateaus, such as the Hunza area in the Karakorum, or in areas of higher but unreliable moisture in the Andes. A project of successfully restoring the traditional Inca canals near Cuzco, Peru, is described in Box 15.1. (This restoration process is particularly noteworthy because it

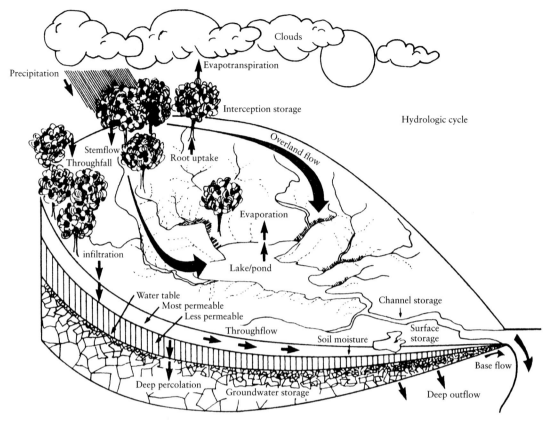

Figure 15.1 The hydrologic cycle in a watershed Source: Warshall, P. (1980)

Box 15.1 Return to traditional Inca watershed management technology

Underlying many of Peru's present economic and political problems is a crisis of poverty and under-development. Agriculture is unable to sustain increasing populations and every year thousands of young people leave the land for the slums of Lima and other cities.

Those familiar with the achievements of Peru's pre-Columbian past will find this a regretable irony, for agricultural innovation and success were the very basis of the remarkable ancient Andean civilisations. By the time of the Incas, vast systems of irrigation canals and agricultural terracing covered the mountain sides of what today is part of Peru; they still occupy perhaps a million hectares of land. Of this, up to 75 percent has been abandoned because of the disruption and depopulation caused by the Spanish conquest. While often buried or broken down, much of this agricultural infrastructure is still in place. Are the canals and terraces to remain as mere relics of past civilisations? Or do they represent a reservoir of agricultural technology that can be re-used today?

In the early 1980s there was just a scattering of families in the Cusichaca Valley (between Cusco and Machu Picchu), little more than subsistence farmers. Dr Ann Kendall, an archaeologist, saw that the productivity of their lands might be improved dramatically if the ancient irrigation canals could run with water once more. Alongside conventional archaeological research she introduced a pilot scheme to restore one of them. The Cusichaca community took on the manual labour of clearing and rebuilding parts of a stone-lined canal, led by a local master mason whose regular job was restoring Inca ruins. The simple techniques they used were very much those of the Incas themselves, employing stone, clay, and sand.

In three years the job was completed. Some 45 ha of land that had been barren for centuries began to flourish under irrigation, producing a wide range of crops, including maize, potatoes, beans, cabbage, and Andean grains, such as quiñoa, kiwicha, and tarwi. The local people are now able to market a sizeable surplus, and the community as a whole has been revitalised. Above all the people have acquired a real sense of what they can achieve, since it was their own efforts that brought about rehabilitation of their lands.

Other farmers along the Urubamba Valley were impressed by the success at Cusichaca. The community in the nearby Patacancha Valley approached Dr Kendall and the fledgling Cusichaca Trust for help in 1989. Now the Patacancha Project is virtually completed. It has taken four years, from 1991, to return 170 ha of terraced land to full production. The end result will be the economic transformation of the valley. In the long term it will also require the training of future master masons, foremen, and engineers to lead restoration projects in other valleys. The project's aim is that in the future such work will be undertaken solely by local people.

From the beginning of the Patacancha Project the community leadership participated in its planning and organisation, and they continued to do so throughout, adapting and making changes to plans when these were required. In working within the project, the community and its leaders have improved their organisational capacity and have the potential to carry out similar projects in the future, including horticultural projects and greenhouses in the high altitude zone. When the results of the agricultural rehabilitation were in sight, the leadership was able to organise an increased number of communal work days to bring the project to completion. From mid-1994, many beneficiaries have participated in this land rehabilitation.

The Cusichaca Trust's approach is to provide technical and organisational expertise throughout a catchment system to promote self-help programmes which augment agricultural production, and improve household incomes while empowering women's role within the community. A major focus has been the restoration of traditional Andean technology and agricultural infrastructure (such as, canals, terraces, and reintroduction of tree species), and an emphasis on an educational and personal approach led by local Quechua speakers. Three principles inform and guide this approach to rural development:

(1) Local environmental sustainability, notably soil conservation through reforestation and restoration of terracing.

(2) The participation and empowerment of women, in their family health and nutrition roles, and in their economic position in the household, leading to increased responsibility in the community.

(3) Reinforcement of the rich local technical and cultural heritage.

Source: Ann Kendall, Director, The Cusichaca Trust, Stourbridge, United Kingdom

---

employed traditional Inca methods and materials and was carried out by local people.) Elaborate systems for hydrologic control of water in wet paddy have been highly developed by upland farmers for crops of rice. These paddy terraces are also excellent erosion control mechanisms, trapping sediments from the rainfed upper parts of the slope.

The need for domestic supplies and clean drinking water for humans and livestock involves tapping into springs or streams, or into underground sources and the creation of storages of various kinds. In the western USA, mountains are increasingly viewed as essential water sources for cities, such as the Rocky Mountains for Denver, Boulder, Albuquerque, and Salt Lake City. Water moving downslope in streams has been a source of mechanical power for small mills that process grain and wood and, in more recent years, for in-stream small-scale generation of electricity. In the Himalaya, traditional water mills are used to rotate prayer wheels as well as to grind grain. On a larger scale, retention dams in mountain valleys are sources for hydropower, flood mitigation, irrigation, and low-flow augmentation for the more populous lowlands. *Use and recycle water as it flows to the sea* should be our watchword.

Surface bodies of water in the mountains are also important for wild animal life. All fauna, from Apennine wolves to Andean bears, need access to water and vegetative cover along banks of streams or shores of lakes and tarns. The importance of these riparian strips of natural vegetation and their extremely beneficial functions will be referenced repeatedly in this chapter. Fish and other aquatic life require water bodies, and permanent water through the dry season is critical to many of them. Mountain wetlands serve as habitats for a rich mix of plant and animal species, and this adds to the ecosystem and species biodiversity in mountains. These wetlands are important storages for water that might otherwise contribute to flooding, and they replenish groundwater. Drainage of wetlands in the uplands for agricultural expansion is an invitation to water problems downstream.

Water quality in the highlands can be impaired by excess amounts of sediment, herbicides, pesticides, mining waste, fertilisers, and human or animal waste. Downstream users of the water incur the harmful effects from such pollution. Sufficiently long storage or adequate stream length, with good aeration, will to some extent reduce contamination. The risks become greater, however, as the upstream polluter and downstream user come closer and closer together in watersheds where human populations are increasing. Appropriate land management practices (including banning the use of chemicals) can reduce or eliminate the problem, but they usually incur associated costs to the upstream land users. Some of these practices are discussed later. Institutional mechanisms for some equalisation of benefits and costs are needed. The watershed offers a model for internalising such accounting.

THE WATERSHED AS A LAND UNIT FOR PLANNING AND MANAGEMENT

The watchword or guide to action, therefore, is to use the water and soil as fully as possible as they move downslope. This is accomplished to a large extent by retarding their downhill progress as much as possible and by capturing them for use on the slopes. Moving soil and water back uphill is an energy-expensive task. These two basic building blocks of our livelihood and our society are interrelated. The term 'soil moisture' indicates this, and it is through this that nutrients become available to plants. The term 'soil erosion' usually infers water transport (soil movement by gravity alone and wind erosion are relatively less important).

A watershed is that area of land from which all the water drains to a point specified by the observer. A watershed therefore can be of any size. It could be the land that drains into a puddle on a slightly concave surface of a few square metres; it could be a small watershed of 75 ha in the Shara Mountains of Serbia draining into a mountain stream. On a large scale, it could be thousands of square kilometres draining into Lake Baikal. 'Catchment' and 'drainage basin' are synonyms, although the term 'catchment' is usually reserved for watersheds of small size, while 'drainage basin' is used for large river systems.

Whatever the size, it is this hydrologic unit called the watershed that best demonstrates the close

relationship between land and water. Moreover, it expresses the physical effects of a given kind of human activity on the land and people who are downslope – the beneficiaries or the harmed. These relationships have been called highland–lowland interactive systems, which was the theme and title of a United Nations University project that has had a significant impact on the prevailing opinions about Himalayan environmental degradation. This concept is discussed in The *Himalayan Dilemma* (Ives and Messerli, 1989) and the *Himalaya–Ganges Problem* (Ives and Ives, 1987). A good review of the literature on the Ganges/Brahmaputra basin has been produced by Bruijnzeel with Bremmer (1989). Studies such as these in the Himalaya have implications for mountains and river basins worldwide.

Why is the watershed a desirable areal unit for planning and implementing sustainable mountain development? Where land and water, and human management of these resources, are a major concern, the watershed hydrological unit is the most appropriate one for study for the following reasons (based on Hamilton and King, 1984):

(1) The physical environment is characterised by variations in climate and features such as soil types in a catena, lithological types in a geologic sequence, or geomorphological forms from deltaic alluvium at the river mouth to craggy outcrops on the ridges. The watershed unit is most successful in linking these gradients. Similarly, climatic variation within the basin is related to elevation and exposure. These physical features provide particular niches for different biota and for diverse human activities, and are integrated within the watershed.

(2) Watersheds best reveal the ecological interrelationships between soil and water, two basic building blocks in all biotic, including human, use of the terrestrial and freshwater environments.

(3) Watersheds can be scaled to different sizes (depending on stream order) to address different magnitudes of development scenarios.

(4) The watershed has utility also as a unit for economic analysis that can internalise the elusive off-site costs or benefits resulting from actions at a higher point on the watershed. Upstream–downstream interactions are not only biophysical but also economic.

(5) Many environmental hazards are manifest in their damage through watershed relationships. Avalanches, rockfalls, landslides, low water yields, and floods are all related to the soil and water processes of a river basin, and such hazards and their consequences are best identified and mapped in a watershed context.

(6) Other hazards such as drought, fire, and earthquakes, are related to climate, vegetation, and geology but can be well accommodated in a watershed framework for control or response, although they are not directly linked to watershed processes.

(7) Watersheds serve as natural movement pathways for many wide-ranging wildlife species that shift from high summer range to lower winter range, or that use riparian habitats. Transhumance (seasonal livestock shifting) patterns are also often confined to a watershed.

(8) Social and cultural considerations are not usually incompatible with the use of watersheds as planning and management units, even though political boundaries bear no relationship to hydrologic units. However, in at least one cultural tradition, that of early Hawai'i, the watershed was the land area that delineated jurisdictional units for resource utilisation and the organisation of social life. The *ahupua'a* (watershed), a pie-shaped jurisdiction district beginning at the coral reef and tapering to the top of the mountain, divided volcanic islands into integrated units, each one having a chief and a land supervisor (Morgan, 1986). In recent times in the United States and elsewhere watershed associations and river basin organisations have been employed effectively, not only for soil and water conservation but also for an array of social and economic purposes (See Easter *et al.*, 1986).

(9) Where specialised production occurs in different altitudinal belts (e.g. grazing lands, forest production belt, horticultural lands, cropland, in roughly descending order) trade exchanges between these units, or economic groups, follow the watershed in a two-way flow.

(10) The use of surface water bodies, whether for potable water, hydroelectric energy, industry, storage for flood control or downstream water delivery, and fisheries management must be integrated for the stream or river, and this is at the watershed scale. Irrigation projects need to be integrated for equitable water allocation, and this requires irrigators to be organised on a watershed basis.

(11) Human-initiated environmental impacts are often best recorded in a watershed context, because many of the underlying processes are expressed through gravity or hydrology. Thus, environmental impact assessment must pay attention to watershed relationships.

The integration of patterns and processes of both natural and social systems can be achieved in planning by using a watershed, without neglecting or minimising the major elements of either system. Unfortunately, institutional processes and patterns already established are normally not organised along watershed boundaries. Thus, implementation, which depends heavily on existing institutions, may be somewhat more difficult. More *planning* should be done on a watershed basis and, where necessary, existing groups, power structures, organisations, and management units should be identified as most likely candidates for *implementation*, and be involved in the action. Wise planning and implementation require great skill and dedication. Integrated planning and management through the establishment of watershed-based institutions has been successful in New Zealand, the United States, and several other countries, particularly where irrigation is involved. Gibbs (1986) provides important institutional and organisational criteria. A generalised watershed management system in terms of physical outputs is presented by Hufschmidt (1986) and given here as Figure 15.2.

Many, if not most, sustainable development planning and implementation strategies should consider the use of the watershed or drainage basin as a suitable operational unit (See Table 15.1). At the very least, the myriad aspects of natural and social processes which are linked in the hydrologic unit, must be recognised and incorporated into plans and programmes. A project funded by the Swiss Development Co-operation in the Bolivian Andes, and presented in Box 15.2, exhibits many elements of an integrated approach.

FACTORS AFFECTING STREAMFLOW

The major determinant of the total yield of water from rivulet, stream, or river and its distribution in time, is the amount, form (snow, mist deposition, or rain), and distribution of precipitation (see Figure 15.1, the hydrologic cycle.) Runoff which becomes streamflow is simply the amount of

Table 15.1 Examples of watershed management tasks required at the planning stage, classified by management activities and management system elements

| Management Activities | Management System Elements | | |
	Resource Management	Implement Tools	Institutional Arrangements
Land-use assignments	• Land capability analysis • Land suitability analysis • Formulation and benefit–cost analysis of alternative land-use plans	Planning for: • Regulation • Economic incentives • Education	Planning for: • Ownership/tenure systems • Public regulation systems • Organisational changes
On-site resource utilisation and management practices	For agroforestry: • Agronomic, forestry, and economic analyses of types, distribution, and rotation of tree and row crops • Planning for methods of tilling, methods of cropping, erosion control practices	Planning for: • Education • Technical help • Economic incentives • Marketing assistance • Regulation	Planning for: • Extension services • Credit/financial aid • Ownership/tenure systems • Soil conservation agency
Off-site management practices	Planning for: • Stream bank vegetation, protection, or revegetation • Channel dredging • Riprapping • Waste water treatment	Planning for: • Education • Technical help • Economic incentives • Public installation and maintenance	Planning for: • Extension services • Credit/financial aid • Soil conservation agency

Source: Easter and Hufschmidt (1985)

Box 15.2 An integrated watershed management programme, Cochabamba (Promic) – a case study from the Bolivian Andes

THE PROBLEM

Cochabamba is Bolivia's third largest city. It is located in an inter-Andean valley at 2500 m asl. The valley, with its 500 000 inhabitants, covers 450 km² including many important agricultural lands, most of them dependent on irrigation. Close to the city a steep mountain ridge rises up to 5000 m, surrounding the northern part of the valley. This ridge is 640 km² in extent and can be divided into 38 small catchments, characterised by very steep slopes (most of them more than 25°). The torrents which descend from these catchments cross the valley, its agricultural areas, and the fast-growing city of Cochabamba.

The annual precipitation is about 400 mm, most falling between December and March. The rest of the year is almost rainless with very low relative humidity. During the dry season severe shortage of water is one of the most serious problems faced by the valley and its city. In contrast, during the rainy season, high rainfall intensities combined with the steep and very degraded catchments of the mountain range frequently cause floods in the valley. Every year more than US$ 1 million is spent in digging and dredging out the river beds and channels in the valley, most of which are inappropriate in size and design. Flood damages appear much more costly, although reliable data are lacking.

The current land-use practices in the upper watershed are largely the result of an increased population of peasant farmers who are recent immigrants following agrarian reform in the 1950s. Many had been mine workers, and others came from other parts of Bolivia. Serious land deterioration has taken place since the 1960s.

THE APPROACH

In 1991 the Regional Development Corporation of Cochabamba, with financial and technical support from the Swiss Development Co-operation (SDC), began an Integrated Watershed Management Programme in order to attack the roots of the problems in the mountain catchments and not, as in the case of earlier attempts, their effects in the valley. Inappropriate land use, a total lack of soil and water conservation practices, lack of torrent control techniques, and loss of most of the natural vegetation cover, combined with very steep slopes and many geologically unstable areas, are the most important biophysical problems of the mountain ridge. The poor peasant farmers who are cultivating the steep slopes must become involved in most watershed management activities so that they develop as 'co-actors', thus guaranteeing long-term success. This presents two basic challenges: first to develop and apply the appropriate techniques, and second to introduce most of them through the local farmers. A third major challenge was to close the gap in basic information; as, for instance: land-use types, deforestation rates, erosion rates and types, hydro-meteorology, and socio-economic factors.

In order to face these challenges and to assure rapid and visible results, a pilot watershed (size 20 km^2) was selected for study and treatment. A 'research-action' method was applied to produce results while, at the same time, important biophysical and socio-economic information had to be gathered, analysed, and documented. In the first three to four years the relevant information was recorded and conservation measures in the pilot watershed were emplaced. On the basis of these experiences and information, a proposal for a regional watershed management programme had to be developed, including identification of those catchments that most urgently needed treatments.

THE RESULTS

A Participatory Rapid Rural Appraisal of the local communities was carried out and the basic biophysical subjects were studied and mapped (principally, land use, vegetation cover, geology and geomorphology, soil loss, and natural hazards). On-farm land treatments were planned together with the farmers, including use of simple maps and airphotographs, and there was an important contribution from local extension workers, and farmer-to-farmer approaches were also applied. Based on this, areas for treatment within the pilot watershed and the river courses were identified, and appropriate measures and techniques were developed and applied. During the same time, a hydro-meteorological research programme was established to produce the missing data within a reasonable time-scale.

After four years the following technical results have been achieved within the pilot watershed:

(1) An extension methodology was developed and implemented with local extension staff which proved to be appropriate to the socio-economic and cultural situation. This fitted the integrated watershed management approach and introduced soil conservation practices and land-use changes by the peasant farmers. It also helped the farmers to appreciate the programme's other activities, thus leading to torrent control measures and new instrumentation, together with hydro-meteorological research, and other studies. Appropriate incentives were used decisively, but carefully; and most of them were progressively reduced over time to discourage any degree of dependency.

(2) The watershed was zoned. Critical areas were designated and treated, using a wide range of appropriate soil conservation techniques which could be adopted by the farmers. At the same time agricultural amelioration techniques were introduced to encourage the poor farmers to adopt soil conservation measures (e.g. crop rotation, including nitrogen-fixing species, mulching, more resistant seed material). Soil conservation was completed by reforestation in different agroforestry systems, mainly with native tree and brush species. Mechanical and biological torrent control measures were carried out taking into account the differing erosional processes in gullies and streams. These control measures, in conjunction with the general conservation practices, were extended into the uppermost parts of the watershed.

(3) All experiences were well documented in a series of more than 70 technical reports; these also included studies of the problems that processes in the pilot watershed created on the valley floor.

On the regional level of the entire mountain ridge and valley the result has been extremely encouraging. A comprehensive study of all 38 catchments and the main valley was completed using satellite images validated by ground truth. The natural hazards were quantified and mapped and a system for establishing priorities among the catchments for future treatment was introduced. Nine catchments were classed as especially hazardous and thus as most urgent for future treatment.

At the institutional level the following results have been obtained:

(1) A capable and well-trained Bolivian technical team has been created, largely of local residents; this includes foresters, agronomists, civil engineers, a geologist, and an economist, among others. The team was trained not only in different technical functions, but also in interdisciplinary thinking and working, in order to obtain

the integrated approach needed for problem solving at the watershed level.

(2) The work has attracted the attention of local, regional, and national level politicians; this has greatly strengthened the institutional importance and position of the programme, despite the current political turbulence that Bolivia is experiencing.

(3) The goals achieved by the programme, and well-documented by its team members, have attracted additional funds from local, national, and international agencies. As a result, it will be possible to treat three additional critical catchments. Furthermore, the Swiss Development Co-operation has agreed to continue its support for four more years.

THE PROSPECTS

The technical and institutional results indicate three basic lessons for the future:

(1) It was possible to gather the necessary information, and to adjust and implement the techniques of integrated watershed management to the local biophysical and socio-economic conditions.

(2) A pilot watershed was treated and extensively controlled, thus establishing a vital representative demonstration project that has ensured strong support from the local farmers.

(3) The integrated watershed management approach is now considered by the local authorities to be the appropriate method to tackle the problem of highland–lowland interactions in the Cochabamba area.

Valuable experience is now available, institutional support has visibly grown, and national and international funding has substantially increased. These results are considered promising, and the well-trained team can be expected to meet the challenge of widening the field of action from the pilot watershed to a regional level. Local and national funds must guarantee maintenance of the investments and the sustainability of the programme; this will require international funding for 10–15 more years in order to facilitate control and management of the nine watersheds that have been identified as most hazardous (see Richards 1997 for further information).

Source: Thomas Stadtmüller,
Inter-cooperation, Switzerland. Formerly:
Integrated Watershed Management
Programme, Cochabamba, Bolivia

precipitation entering the watershed minus the total volume of moisture that is evaporated or transpired back to the atmosphere from vegetated and soil or water surfaces (accounting for the amounts of water that are stored temporarily in the soil or shallow groundwater bodies on their way to the stream). In arid zones, significant amounts of stream water can be lost through transpiration from riparian vegetation. In high mountains with shallow or no soil cover, steep slopes, and little or no vegetation, the runoff or stream yield may be 80–90 percent, or even more, of the precipitation, unless it is locked up as ice and snow. Lower down the watershed, where soils are deeper and where more extensive vegetative cover has developed, the interception, evaporation, and transpiration components increase, and the soil can store more water (which may be used by plants, go to long-term groundwater storage, or even leak out of the watershed). Then stream yields may drop to as low as 10 percent of precipitation (Hibbert, 1967). (See Box 15.3 on a Mount Kenya watershed; and Figure 15.3.)

The form of the precipitation is of special significance in mountain catchments. Where

precipitation arrives as snow, it may accumulate in impressive amounts. Some or all of this may subsequently melt to supply streamflow – especially important to the lower watershed if the warm months are also dry. Much irrigation for agriculture and water for communities and industries comes from mountain snow. Snowpack accumulations may melt rapidly from time to time generating torrents and floods.

The pattern of rainfall governs much of the character of water delivery timing. Fortunate are the mountain watersheds, such as those in the Western Highlands of Scotland, where distribution and intensity has less variability than in the majority of the world's mountains where seasonal variation can be extreme and where violent downpours can be common, especially at lower elevations. Individual storm events of high intensity can initiate mountain torrents as well as trigger massive and numerous landslides. Hurricane rains in the Caribbean mountains and in the Philippines are notorious for their high intensities. Prolonged rains also, as in the monsoons that drench the Himalaya and other uplands in Southeast Asia or South Asia, produce annual flooding in their river basins due

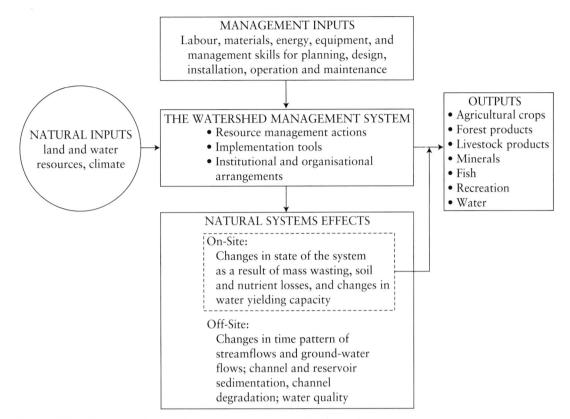

Figure 15.2 Generalised watershed management system in physical output terms.
Note: This schematic can be used to depict a system in the planning, design, installation, or operational stage. Source: Hufschmidt, M. M. (1986)

to saturation of soil storage capacity. This rainfall pattern often entails a dry season when many streams go dry and many others, and major rivers too, have insufficient water for sustaining plant, animal, and human life. Retention storages are built in mountain valleys (often in the foothills) to attempt to alleviate the reduced dry season flow and to lower the flood crests. In higher parts of the watershed small storages for water are often constructed in the form of farm ponds to help reduce these fluctuations in runoff.

Precipitation may also occur in 'horizontal' form. Wind-driven clouds or fog that intersect mountains may have useful water taken from them by barriers of vegetation or even by artificially constructed obstructions. If there are no features to cause condensation, this 'occult' precipitation is not released. In these cloud belts, unusual types of forest and shrubland develop which effectively capture more water from clouds to add to the watershed's budget than they use in evapotranspiration (Hamilton *et al.*, 1995). Even individual trees per-

form the function of 'fountain trees' in the mountains of the Canary Islands. One such tree on El Hierro Island was such an effective harvester of water that it appears on a coat of arms (Gioda *et al.*, 1995). Where these trees or forests have been cut down, this function is lost, and less water is available. Under favourable conditions, with frequent low cloud and persistent trade winds, this form of water capture may produce hundreds of millimetres per year with typical values ranging from 5 to 20 percent of total rainfall (Bruijnzeel and Proctor, 1995). For example, the cloud forest on Sierra San Luis in Venezuela (now protected in a National Park) is a key element in the watershed system of those limestone areas that supply water to the reservoir near its base which in turn supplies the dry peninsula of Paraguana and its important oil industry (Hamilton, 1976). In the coastal ranges of the arid Pacific areas of Chile, Bolivia, and Peru, where a persistent fog, known as the *camanchaca*, strikes the uplands between 500 and 1000 m, artificial nets and screens are erected, or trees planted,

345

Box 15.3 Vegetation, soil, and water in the Mount Kenya–Ewaso Ng'iro watershed

Changes in soil water (i.e. recharge and discharge) are extremely important in the lower forest and the foothill zone. Here, soils have a high storage capacity; there is a water surplus during the wet season and a deficit during the dry season. The recharged water storage capacity provides a source of water for use during the dry season. In the savanna zone, most of the soil water storage is already exhausted by the end of the wet season and little is carried forward into the dry season. Soils of the moorland and alpine zones generally remain moist throughout the year.

Potential evapotranspiration is highest in the lowland zone, averaging between 6 and 7 mm per day; it decreases to less than 0.5 mm per day in the alpine zone. Figure 15.3 depicts an excess of available water (rainfall minus evaporation) above the lower forest line and a deficit in the piedmont zone and the savanna region. During the wet season the excess of water in the upper forest belt is approximately 400 mm, whereas the deficit in the savanna during the dry season is about 300 mm. The greatest excess of water in the upper forest and the moorland zone during the wet season is a result of high rainfall, low evapotranspiration, and medium to low soil water storage capacity. About half the 400 mm of excess water is discharged to the river, either by direct overland flow or via local groundwater flow. The remaining half of the rainfall recharges local and regional groundwater aquifers. In the alpine zone, the water retention function and discharge of the glaciers are calculated as part of the groundwater change. Overall, there is thus a clear period of recharge during the cooler wet season. During the dry season, river flow is maintained by glacial meltwater and discharge from the groundwater in the lower forest zone. Little groundwater accumulates in the moorland, where even during the dry season there is an excess of water and deep percolation. See also Figure 15.3.

An excess of water in the upper basin and a water deficit in the lower basin implies a wide potential range of water resource utilisation within the same basin. The life-giving perennial rivers emanating from the tropical mountains have created rich downstream ecosystems due to the combination of savanna and riverine habitats. This combination of dry and wet habitats provides niches for a wide variety of animals and plants. The whole highland–lowland ecosystem of the Ewaso Ng'iro is thus characterised by its great biodiversity.

Source: Hanspeter Liniger,
University of Bern, Switzerland

in order to capture water for potable domestic use (Schemenauer *et al.*, 1988).

Precipitation, unfortunately, also can bring pollution into the watershed from the industrialised and mechanised world. Concentration of solutes tends to be much higher in fog water than in normal rain. Acid deposition is adversely impacting soil, vegetation, and lacustrine fauna in mountain areas, particularly because clouds bearing the harmful products often intersect these lofty landforms. The mountains of Central and Eastern Europe and of eastern North America have experienced significant air pollution (See Chapter 13 for air pollution impacts on mountain forests).

The underlying rock type and soil depth also affect streamflow. Many types of bedrock are impervious so that any infiltrating water (where there is soil) moves as subsurface flow along the soil–bedrock interface. Limestone bedrock (karst) has solution channels which may conduct any percolating soil water to storage at depth or to underground river channels; under such conditions water moving into sinkholes in one watershed may appear as springs in an adjacent watershed. Pollutants, such as bacteria, are not well filtered out in these areas and hence springs in karst may be easily polluted by land uses such as grazing or septic systems, or where large quantities of chemicals are applied. Shallow soils that quickly become saturated deliver water more rapidly to watercourses than do deep soils. The process is complex and quite specific to a given watershed. Human interference that drains soil or wetlands artificially, or that cuts across subsurface water flows with roads, can alter the process, and such actions need to be carefully considered before being implemented.

Most attention in traditional watershed management, however, is given to the vegetation – which affects evapotranspiration. There has been a strong belief that land-cover manipulations have a great influence on water yield and timing, and therefore on flooding and lack of water in streams and rivers. This is a large and complex subject, and space limitations prevent it being covered exhaustively in this chapter. Those who would pursue this in more detail are directed toward the following reviews: Hamilton with King (1983); Bruijnzeel (1990); and Brooks *et al.* (1991). Nevertheless, some brief observation will be made

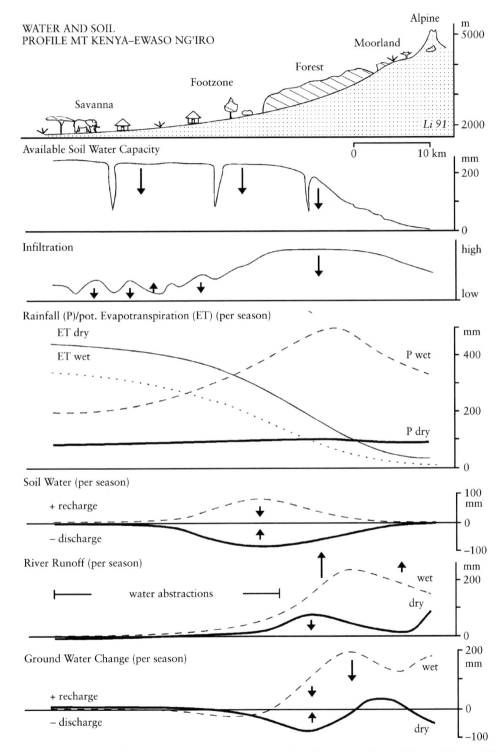

Figure 15.3 Water and soil profile of Mt Kenya–Ewaso Ng'iro (Naro Moru basin) Source: Liniger, H. P. (1995)

herein on the role of vegetation with respect to water yield and timing.

There is no scientific evidence to support the conventional wisdom which states that the presence of forests in mountain watersheds increases the precipitation, except for the aforementioned cloud forests. Removal of trees, or other watershed vegetative cover, increases the total yield of water since it reduces evapotranspiration losses in proportion to the degree of vegetation change. Maximum yield might be achieved by eliminating soil infiltration channels by conversion to a car parking lot. Indeed, where there is a critical shortage of water, metal roofs may conduct all rainfall (less evaporation) into storages, and even small catchments have been concreted. This is a ridiculous objective even for a small area, for it totally removes land from any other kind of useful production. Moreover, removing all, or much, of the vegetative cover exposes the land to serious erosion. Water resources must be shared with plants and animals, but it is human control that determines the type and amount of vegetative cover and the land use, and this has an effect on total yield. Highly productive vegetation, such as lush maize crops, dense fertilised pasture, or fast-growing, site-adapted forests, will consume large amounts of water. In general, large deep-rooted vegetation (forests) will use more water than other kinds of cover. It is difficult, therefore, to understand why the myth persists that if forests are cut down and replaced with other vegetation, water yields decline and droughts ensue or that, conversely, planting trees will increase water yields.

Conventional wisdom maintains that when trees are removed, not only does water yield decline, but flooding increases and dry-season flows diminish. This is supposedly due to the 'sponge effect' of tree roots which soak up excess water during rainfall events, or in the rainy season, and release it slowly during subsequent dry periods (Spears, 1982; Myers, 1983). This myth persists in spite of extensive research evidence to the contrary as summarised by Hamilton with King (1983), Hamilton (1986), and others. Tree roots are a 'pump' rather than a 'sponge'. All carefully conducted research on this topic has shown that replacement of forests by other vegetation gives increases in streamflow, not only during the periods of high rainfall but also during periods of low rainfall. Conversely, reforestation of watersheds results in somewhat lower yields during both wet and dry seasons (Hamilton and Pearce, 1987). However, if forest clearing is followed by a destructive land use that compacts the soil, then precipitation quickly runs off and low season flows are often reduced. If land degradation continues to the deep gully stage, then the gullies can act as 'drains' and again deplete low season and after-rainfall flows.

Stormflow peaks in streams following a rain event are usually somewhat lowered, as well as being slightly delayed, by the presence of undisturbed forest in comparison with grazing, cropland, or other land uses. This beneficial effect in flood reduction is realised when rainfall events are not major, the receiving watershed is small, and the soils are deep and not already saturated. When storms are intense, or when soils are shallow or wetted thoroughly by previous rainfall, or when the area being considered is large (e.g. a river basin) the actual effect of land cover will be small. It is a question of degree and scale (Hamilton, 1990). It is important to remember that it is the soil mantle that has the storage capacity to accommodate precipitation. Steep, high mountain slopes give little opportunity for vegetation or their shallow soils to play a significant role in reducing stormflow peaks and volumes. Also in arid mountain foothills, with sparse vegetation, intermittent streams routinely produce walls of water in the form of flash floods. In deeper soils on gentler slopes of humid regions, vegetation has a greater effect. If forest cover can reduce a minor storm flow peak by 5 cm, this may be significant if the peak is only 30 cm, but when the peak flood is 1 m or more, the reduction of 5 cm is insignificant. Major storm events will produce flooding irrespective of whether the land use is untouched forest, harvested forest, grazing lands, or croplands. This is why it is unfair and incorrect to blame mountain farmers for causing the catastrophic river flooding in distant lowlands or large basins following monsoon rains or unusual rainfall events (Hamilton, 1987; Bruijnzeel with Bremmer, 1989). Mountain land use directly influences erosion and water quality, but its effect in major flood disasters in the lowlands has been vastly overstated. The presence of forests over accumulated snow does slow the rate of snowmelt, and this can ameliorate spring flooding especially where extensive snowpack occurs. The degree of the effect depends on the aspect of the slope and the sequence of melting and runoff from the network of various small catchments that contribute to a larger watershed or drainage basin.

It is largely because of this mosaic network of small catchments with their differing characteristics and orientations in mountain topography that the runoff delay factor is not more important in reducing flood damage. The delayed peak flow caused by forest or other vegetation in one catchment could cause it to coincide more exactly with the peak discharge coming from another one, thus aggravating the problem; with a different storm path, the effect might be reversed.

Land-use practices that severely compact the soil (grazing, agriculture, settlements, or roads, including logging roads) do hasten runoff and reduce soil storage, thus promoting more 'flash' flooding in single storms in small watersheds (Plate 15.1). Good soil conservation practices, especially terracing, maintaining vegetative cover and soil infiltration, and handling accumulated water on compacted surfaces such as roads so that it is diverted appropriately, will all help to prevent flooding. A well-developed forest is hydrologically the safest land cover; nevertheless, inappropriate logging and under-forest grazing reduce the water benefits. Well-managed grazing and cropping under soil and water conservation regimes represent good watershed land use where soils are not slip-prone, and are also benign in their effect on water. Well-managed mosaics of cropping, grazing, and forest husbandry have produced some of the most visually attractive landscapes in the mountain world, and they are hydrologically sound (Plate 15.2).

SOIL EROSION AND SEDIMENT

With the relatively large incoming water supply, frequent high rainfall intensities, and substantial snowpack accumulation due to low temperatures (characteristic of higher altitudes), the steep and long slopes of mountains, often with highly erodible soils, set the stage for what can be rapid soil movement downslope.

Human use of mountain lands entails measures to reduce natural erosion and to minimise loss of productivity and physical impediments, such as gullies, from erosion triggered by the use itself. Natural erosion is commonly in the form of stream channel adjustments, surface wash when there are overland flows, landslides, debris avalanches, and other mass movements (such as creep) that occur in response to earth tremors, high intensity or prolonged storms, and other natural catastrophic events. Natural erosion may occur even under undisturbed forest or alpine grassland, depending on rainfall, soil type, and rate of humus formation, but in mountain regions the natural vegetation pro-

Plate 15.1   Flash flood in small upland watershed in the Philippines causes local damage. Good land use can reduce negative impact in the minor, frequent storms in local areas, but not in major, infrequent storms and in areas well downriver (Photograph: L. S. Hamilton)

vides the most effective control of erosion; mountains, however, are high energy landscapes and are characterised by high rates of downslope movement (see Chapter 1). The natural vegetation cover, nevertheless, is certainly more effective than any alternative cover produced by human use of mountain slopes. For instance, in the Ethiopian Highlands, Hurni (1988) estimates annual soil loss to average 1 tonne per ha under forest, the lowest loss for seven categories of land use.

Surface erosion occurs when individual soil particles or small aggregates of soil are detached from the main body of the soil and moved downslope by gravity. Detachment can occur through frost action, by raindrop impact, or the scouring action of water movement over the soil surface. Splash erosion displaces particles and can begin the transport process on sloping soils. Good vegetation cover or plant litter can reduce all of these processes. The types of surface erosion are sheet, rill, and gully. While sheet erosion involves films of unchannelled water moving downslope, rills and gullies represent increasingly severe forms of erosion by channelled water. Rilling can be initiated by surface irregularities and a thickening of sheet flow. These forms of erosion are frequently initiated by human activity whereby too much of the vegetation cover is removed through grazing, cropping, logging, burning, or clearing for infrastructure (Plate 15.3).

Mass erosion occurs when soil is displaced downslope by gravity, often associated with saturated soil conditions. Mass movements may be rapid (debris flows, mudflows), or almost imperceptible (soil creep). They may be shallow landslips or deep-seated landslides or earthflows. Much mass wasting, particularly the deep movements, is caused by inherent soil and bedrock properties, seismic activity, and rainfall events beyond the influence of humans (Carson, 1985; Chapter 1). Hurricanes, for instance, often induce numerous slope failures, even in forested watersheds. However, human activity can aggravate the situation greatly. Mountain roads, in particular, can initiate numerous and persistent landslides, often mistakenly attributed to peasant agricultural activities. Plate 15.4 shows a large number of slope failures along a mountain road in Venezuela. The removal of forest, or even partial removal of trees as in logging, can also initiate shallow landslips on susceptible soils due to soil water pore pressure and textural and structural characteristics. Tree roots increase the shear strength of the soil (O'Loughlin and Ziemer, 1982) and if they are killed by tree felling, this safety margin is lost. The increased soil moisture content, because of the absence of trees, also aggravates the situation so that a major storm

event can lead to slope failures. There are many examples of such catastrophes, including a dramatic, devastated landscape in high country of New Zealand's east coast following Hurricane

Plate 15.3 Overgrazing has initiated serious erosion in this mountain watershed forest, and no remedial steps have been taken to reduce livestock numbers or to restore land cover (Photograph: L. S. Hamilton)

Plate 15.2 Intensively used yet erosionally and hydrologically benign, and visually attractive, mountain land use (Photograph: L. S. Hamilton)

Bola in 1988 (Trustrum and Page, 1992). Erosion and sediment damage from this one event was estimated at US$ 80 million (NZ$ 120 million). One portion of this large area which had been cleared of trees for grazing purposes is shown in Plate 15.5.

Sediment is eroded material which is carried by water and deposited lower in the watershed. That sediment in turn may be picked up again subsequently and moved further downstream or downslope. An excellent diagrammatic presentation of this process was produced by Megahan (1981) and this is reproduced as Figure 15.4. Deposition of sediment is sometimes beneficial, providing a new medium for growing plants and enriching existing sites (floodplains), but it is often very harmful. It can damage or kill valuable aquatic life (including fisheries); impair water quality for drinking, domestic use, and industrial processes; reduce reservoir capacity for important flood, hydropower, or irrigation storages; clog irrigation canals and channels; shorten the useful life of hydroelectric turbines and pumps; interfere with navigation and recreation; and aggrade river channels, thus aggravating flooding. Most of these negative effects are in the downstream area or lower basin where the greater part of the wealth, population, and political power is found. There is, therefore, a substantial societal or national interest in reducing harmful sediment by minimising accelerated, human-caused erosion in mountain watersheds. A study in the Upper Konto watershed in Indonesia by Rijskijk and Bruijnzeel (1991) indicated that the sediment load of the Konto River had probably doubled as a result of clearing for agricultural and residential purposes although the watershed was still two-thirds forested; the contributions of sediment were: by mass wasting and bank erosion, 15 percent; surface erosion from rainfed fields, 50 percent; and surface erosion from areas with houses and from roads, 35 percent. The extents of built-up areas and roads were only 2 and 3 percent of the watershed respectively, yet they were identified as disproportionately large sources of sediment.

Measures can be taken to control the amounts of sediment produced, even by road construction. Soil and water conservation techniques for forestry, grazing, agricultural cropping, and roading are known and can be used to reduce erosion to low levels. A later section of this chapter briefly reviews the most useful of these. Unfortunately, these techniques, by and large, are not being ade-

Plate 15.4   This picture in the Venezuelan Andes was taken by the author because this site had been used to show the erosional consequences of mountain farming. In actuality all of these slope failures are due to the road construction and maintenance practice (Photograph: L. S. Hamilton)

Plate 15.5   Hundreds of landslips and a large landslide following Hurricane Bola in 1988 in a small portion of affected New Zealand steeplands (Photograph: Noel Trustrum)

quately applied in mountain watersheds worldwide because they incur additional costs over prevailing practices in the short term, and the direct benefits usually accrue only in the long term. One must note, however, that sustainable, low-erosion methods were widely practised in many indigenous, traditional mountain cultures before new and inappropriate technologies were introduced from the lowlands (see, for instance, Photo 15.2). There is substantial justification, therefore, for those living in the lower watershed, who will benefit from reduced sediment damage, to compensate the mountain communities for the costs of improved practices used in their efforts to reduce erosion in the uplands. Internalising the benefit–cost

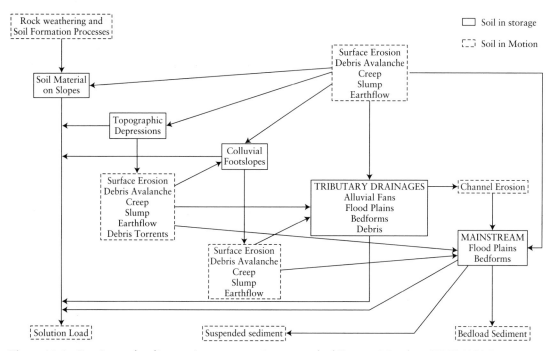

Figure 15.4 Erosion and sedimentation processes in a watershed Source: Megahan, W. F. (1981)

accounting in a watershed supports the viability of this hydrologic unit for planning and management. This is often achieved, for example, in the United States, by subsidies to the landowners who adopt new long-term conservation practices.

Warning! Reduction of erosion in the upper watershed does not immediately manifest itself in less sediment downstream. Looking once more at Megahan's Figure 15.4, one can see a multitude of storages for sediment from past erosion (the rectangular boxes). For years, even decades, these will continue to be sources of sediment when unusual storm events occur. Pearce (1986) suggests, for instance, that on-farm soil conservation practices will have only a very modest effect on sediment yield from watersheds larger than 100 km² for at least decades. In mountain watersheds dominated by mass wastage (for instance, the Brahmaputra) extreme events determine the sediment condition (Bruijnzeel with Bremmer, 1989; Chapter 1, this book). The time lag between erosion and deposition of sediment and the importance of natural processes do not imply that watershed rehabilitation is not worthwhile in large drainage basins. Rather, it is recommended that the process of reducing erosion should begin as soon as possible because reduced sediment is not evident for many years.

MAJOR WATERSHED LAND USES

The major land uses in mountain watersheds are: forestry (protection in preserves or parks, hunting, harvesting of wood and non-wood forest products, outdoor recreation); livestock grazing or wildlife grazing in natural or cultivated grasslands or open forests; crop production of annuals or perennials that involves frequent exposure or turning of the soil; agroforestry or horticulture; mining; reservoirs for urban water, energy production, irrigation or flood control; and roads. Many of these factors are discussed in other chapters of this book. The major land uses to be considered herein will be forest harvesting, grazing, cropping, agroforestry, and roads. Moreover, only the effects of these land uses on soil and water will be considered here, since all of the factors are featured as mountain enterprises or activities in other chapters, where biodiversity, economic, social, and cultural aspects are discussed.

Preservation of a buffer strip of undisturbed vegetation along perennial water bodies (lakes, streams, rivers) is a much needed land management practice. Human and livestock activity should be severely restricted, if not completely prohibited. These riparian buffer strips can reduce streambank erosion and can trap sheetwash and rill sediments (but not that from gullies) from upslope. Moreover, these strips serve as filters for other upslope water contaminants such as pesticides, herbicides, and fertilisers. They also moderate light and temperature regimes for aquatic life, and provide cover for access to water for terrestrial and avian wildlife, as well as corridors for movement. Such buffer strips are guardians of water quality. It is generally suggested that 20–40 m on each side of perennial streams, depending on slope, should be protected (Hamilton and Pearce, 1987). Riparian buffers may be lost in narrow canyons when major flood surges scour out entire stream channels, but they have an important role to play in effective watershed management, and are a high priority for protection.

Forest harvesting

Where conditions are humid, forests of various kinds occupy, or once occupied, large portions of mountain watersheds. Chapter 13, Montane Forests and Forestry, is pertinent at this point. Forests usually represent the most reliable vegetative cover for retention of soil and water because the frequency and duration of disturbance by forest uses is the least of all the major impacts. Even under intensive forest harvesting (whether for non-wood products, fuelwood, or commercial timber) most of the soil cover remains intact most of the time. Because of the relatively slow growth of commercial timber, the impact of logging is widely spaced in time, with a long recovery period, of 20 years or more, for restoration of hydrological and erosional stability before the next commercial harvest. Cropping and grazing disturbance is more or less continuous.

When logging or commercial fuelwood extraction occur, however, the impact can be serious. Such harvesting, with its heavy equipment, roads, skidding tracks, and log-yarding areas, causes hydrological and erosional damage. Exposure of soil, compaction, gullying, and diversion of sediment into streams are a few of the impacts. While many people still believe that the tall tree canopy protects the forest soil from raindrop splash erosion, this theory has been discounted by evidence that it is rather the ground vegetation, the forest litter and debris, and the organic layer that prevent particle detachment by splash erosion (Mosley, 1982; Wiersum, 1984). Poorly located, constructed, and maintained roads are causes of rapid and channelised surface runoff, of cuts into subsurface water

pathways, and of destabilised slopes that lead to serious erosion. Relatively benign logging practices are known and new techniques that completely or partially lift the wood off the ground, rather than dragging it, are in place in many areas. In sensitive areas with high-value forest products, helicopter logging may be employed. Helicopters have been used to remove individual sandalwood logs from the slopes of Mauna Loa in Hawai'i, for instance, and are in use in the Papua New Guinea Highlands. 'Best practice' guidelines in watershed logging have been determined and are gradually being required as a condition of mountain forest harvesting. These are examined in FAO's *Guidelines for Watershed Management* (Kunkle and Thames, 1977), especially chapters by Gilmour (13), Megahan (14), and Rice (16). More recent guidelines have changed relatively little from these, although there has been increased emphasis on maintaining unlogged riparian buffer zones.

Fuelwood harvesting is usually undertaken on a small scale by individuals or families, and extraction and transport are by hand, by animals, or by small tractors without specially constructed roads. The protective soil cover, therefore, is usually less impacted and erosional consequences are much less severe. Degradation increases according to the extent that trees or small logs are dragged on the ground, or the operation becomes mechanised. Fortunately, thus far the intensive, fast-growing, short-rotation, all-tree harvesting, biomass fuel plantations have been established mainly in the lowlands or hill lands. Bruijnzeel (1996) discusses the soil and hydrologic impacts of harvesting plantations. 'Non-wood forest harvesting in general has little impact unless soil is extensively exposed, as for example, in some areas where litter is collected for livestock bedding or fuel, as in the Himalaya, or for making mosquito repellent coils in the Philippines'.

The effects of tree cutting and removal on water yield and timing have been discussed earlier, especially pertaining to cloud forests. These effects tend to be local rather than widespread and are felt more in small watersheds than large ones. It should be noted, however, that in areas where snowpack is important to downstream uses, such as irrigation or water supply, some manipulation of snow accumulation by forest management has been undertaken, as for example in national forests in the Rocky Mountains, USA. Closed forests of conifers can intercept snow on the canopy, where it will re-evaporate back to the atmosphere in larger amounts than from other types of vegetation. On the other hand, the shading effect of the forest canopy will reduce melting and prolong the duration of snow cover into the early summer. Management can reduce forest densities or create openings within the forest, and in the Rockies it has been shown to increase snowpack and resulting streamflow by 7–111 percent (Ffolliott *et al.*, 1989). This wide range of increase reflects different degrees of cutting, and the effects of slope, aspect, and exposure. When too much forest is opened up, such as in clearcutting, high spring streamflow is less prolonged. Whether or not such management is applicable in mountain forests other than in the Rocky Mountains depends on the precipitation regime, the value of water, accessibility, available technical expertise, and competing uses. Thus, whenever water from mountain forests has a particularly high value, some modification of forest management may be instituted. Where potable domestic water supply (good quality and in quantity) is required, downstream 'customers' may acquire and instigate special protection for watersheds. Large areas around reservoirs in the Catskill Mountains, USA, are under special authority, although they are over 200 km from New York City, in order to protect the excellent quality of that large city's water supply. Much of the state-owned land in the Catskills is designated as forever wild, with no forest harvesting permitted. Owners of adjacent private lands are subjected to strong educational campaigns to encourage conservation logging. Ensuring water quality through watershed protection is usually a preferred alternative to the costly systems of purification treatment, filtration, and sedimentation basins.

Grazing

In the upper mountain watersheds that are not in forest or under protected status, where there are steep slopes and generally thin soils, together with a short growing season, a large percentage of the land that is cleared, or being cleared, is devoted to livestock grazing. Indeed, the local need for high energy food and warm clothing (wool, hides) has favoured animal husbandry. Alpine meadows have been extended by intensive forest grazing and by clearing to create mountain pastures. At high altitudes with harsh climate, transhumance and mixed mountain farming (*alpwirtschaft*) have been traditional grazing practices, with livestock using the upper areas during the growing season and being

moved to lower or lowland pastures in winter. In some cases this movement encompassed a hundred or more kilometres and was a way of life for semi-nomadic peoples (see Chapter 2). There are many animals, from yaks to llamas to highland cattle, in the mountain watersheds of the world and grazing is a major land use. Wild herbivores also utilise the mountain grasslands; these include ibex, chamois, Marco Polo sheep, huemul, vicuña, and many others.

Grasslands are excellent watershed cover from the perspective of both hydrology and erosion control, as long as the surface characteristics of the soil remain intact. Indeed, much of the large area of 'degraded' tropical steepland overtaken by *imperata* or other unpalatable and iniquitous grasses is not 'bad' watershed cover. The water yield from grassland is usually greater than that from forest, in both the wet and dry seasons, and the stormflow peaks and total volumes are usually only slightly increased. Grasses usually provide total and compact soil cover, and ensure low erosion rates. However, too many animals are often put on these grasslands, and the weight of animals continually at the same location for long periods of time during the grazing season compacts the soil and reduces infiltration. Grazing, if not managed well, reduces plant cover, exposes the soil, and accelerates erosion. World-wide, overgrazing characterises mountain livestock husbandry and results in land degradation. For instance, an area of the Middle Atlas Mountains, during winter grazing, was subjected to sheep densities of up to 4 head per ha, while the most optimistic estimates of carrying capacity are of 1 animal per ha (Bencherifa and Johnson, 1993). Livestock trails and terracettes add to degradation. When graziers use fire to 'green up the grass', or to extend pasture into forest land, the adverse effects are multiplied. Where cattle or pigs are concentrated in feedlots adjacent to watercourses, not only is there an erosion hazard, but the water is contaminated unless special and expensive measures are taken. Pereira (1989) documents the problems of lack of good animal husbandry and sound watershed management in his book on tropical watersheds, and presents some case studies from several parts of the mountain world. Plates 15.3 and 15.4 illustrate the kind of watershed damage that can occur.

Livestock are a traditional and essential part of mountain land use. (See Chapter 14 on mountain agriculture.) They not only provide meat, milk, butter, curd, cheese, and animal power, but they are an essential source of organic fertiliser, where chemical fertiliser is not available or is not economically feasible. Dung is also used as fuel in some parts of the world. In several mountain societies many animals are owned for cultural reasons and this aggravates the over-stocking problem. There may also be legal tenurial reasons for some high-altitude grazing, and it is only by grazing of animals or hay-making that villagers at lower elevations maintain their 'rights' to the lands; this is characteristic of some areas of the European Alps. Where a remote government has jurisdiction over the lands that are grazed by lessees or traditional graziers, the tenure or usufruct uncertainty that prevails promotes over-stocking and hence watershed damage. This is especially true when many livestock owners have animals on a common property, unless co-operative management regimes are in place and working effectively. Sound watershed management in such areas may consist of strengthening these traditional regimes rather than allowing them to decline as social change breaks down old customs. Privatisation of the open-access rangelands has both an 'up' side and a 'down' side, depending on the values which society encodes into law and acceptable practice. A representative case study of this process in an *agdal* (collective pasture) in the western High Atlas Mountains, where the government is weakening the control of the grazing council, is given by Gilles *et al.* (1986).

In mountain grazing systems, tree and shrub leaves are often used to supplement fodder. Lopping of trees on both private and public land is widely practised in tropical mountains, and trees in agroforestry systems may be selected based on their palatability and nutritive value to livestock. Forest litter and live biomass are also used extensively for animal bedding. Unfortunately, in some areas mountain forests are being degraded and gradually cleared due to the assault of fodder-lopping, in addition to grazing and burning. (Carson, 1992).

An important development that bodes well for improved watershed health is the increase in stall feeding of animals, with the farmers cutting and carrying fodder to the livestock. This allows pastures to recover during the dry season. This technique is increasingly used in Costa Rica (Kunkle, personal communication, 1996) and in many other mountain areas. The manure can be readily collected and applied to cultivated land. This trend is also evident in both Nepal and Pakistan (Dani and Campbell, 1986) and, together with the introduction of improved cattle breeds, is

resulting in increased production. In the Peruvian and Bolivian Andes, there is some shift back to alpaca and llama herding to replace the dominant cattle ranching introduced and fostered by the Spanish. These native livestock are less damaging to the mountain grasslands and soils than are cattle, and are better adapted to the high altitudes.

Watershed damage from overgrazing results from too many livestock on the land or stock being on the land for too long with no recovery period, whether by the sedentary rancher (grazier) or the nomadic pastoralist. How to estimate stocking rates and time lengths that are sustainable is a complex problem in which economic pressure is a major force, but not the only one. The ecologically acceptable carrying capacity of the mountain soils and vegetation must be taken into account, or the grazing enterprise will not be sustainable, and the watershed degraded.

Livestock need access to water but, from the standpoint of public health, their access needs to be controlled. Kunkle (1972) found that 200 cows on a small watershed with a protected riparian buffer strip along the stream produced much less (one-eighth) non-point pollution (nitrates and coliform bacteria) than did 50 cows on an adjacent watershed with no buffer strip which gave the cows direct access to the stream. Creation of specific cattle watering points, or diversion of water to stock tanks, can greatly reduce the need for unlimited stream access, thereby permitting a buffer strip to develop, and play its beneficial role.

Cropping

Possibly no one is more engaged in the mountain watershed struggle against gravity than the mountain farmers who grow annual agricultural crops. They harvest all or most of the protective covering of the soil each year, cultivate the land for planting and weed control, and thus leave the soil susceptible to erosion during these periods. The pre-monsoon rains in Nepal's Middle Mountains occur when the fields are not yet adequately vegetated and therefore soil loss can be greater than during heavier monsoon rains (Carver and Nakarmi, 1995). In some catchments this results in stream sediment concentrations in the pre-monsoon season that are almost an order of magnitude higher than during the monsoons. The steep slopes that characterise mountains and the usually higher rainfall amounts and variability make it difficult to maintain soil stability, and hence productivity.

In the history of high mountain agriculture, traditional subsistence mountain farmers quickly passed through the stage of shifting cultivation and adopted sedentary systems, where keeping the soil in place for use and applying water to it were important aspects of management. Whiteman (1988) discusses the mountain agronomy and farming strategies in Ethiopia, Nepal, and Pakistan. Terracing was one practice that was developed as a very successful soil and water conservation technique that ensured much higher productivity. Erosion protection on the terrace risers is vital to success. These terraced landscapes are often marvels of conservation engineering, considering the materials available and the steepness of many of the slopes, whether Inca terraces in the Andes (see Box 15.1), in the Middle Mountains of Nepal, or the Banaue in the Philippine Central Cordillera; wherever traditional farmers still occupy the land.

In many places terraces have been maintained on slopes up to and exceeding 45° (Ives and Messerli, 1989). Indigenous farming systems employ many other soil and water conservation practices, such as the diverting of sediment-laden waters onto their lands. Carson (1992) presents a good overview of cropping practices in Nepal as they relate to watershed values. In some parts of the world, however, sedentary farmers do not have a long history of land occupancy; they have moved, often with government encouragement, from lower elevations to clear forest and plant crops in steep uplands. This migration has taken place in much of Latin American, the Caribbean, Africa, and Southeast Asia, as populations have burgeoned and lowlands have become intensively farmed. Mountain 'unoccupied' land, or areas occupied only by 'backward' indigenous hill tribes with no recognised tenure, provided an outlet for the land hunger of lowlanders and a new chance for escapees from urban poverty. These new farmers brought with them methods of farming which were not appropriate to steep slopes. Much of the degradation of mountain farming land around the world results from this process over the last 100 years, and especially the last 50 years. The serious degradation in the Cochabamba Watershed (see Box 15.2) is a product of this process. Ryder (1994), in a study in the Dominican Republic's Central Cordillera, found that although farmers were aware of soil erosion, there was little interest in soil conservation, and only 16 percent had adopted conservation practices. He pointed out that the Las Cuevas watershed farmers have tilled that land for less than 60 years and have no accumulated mountain farming tradition.

Shifting cultivation in mountain watersheds is often practised by farmers 'new' to the mountains, where the agricultural frontier is moving from the lowlands into the uplands. The slash-and-burn farmers frequently have no security of tenure and are operating on lands belonging to the government or large land enterprises, as for example in the Andes of Venezuela. Also, in many areas of Southeast Asia the 'new' slash-and-burn activity (with tacit government inaction) is displacing the indigenous mountain and hill farmers who have no legal title to their traditional lands under a centralist government. 'Stable' shifting agriculture has largely given way to unstable 'shiftless' agriculture where new colonists degrade one cleared, eroded, and increasingly infertile area, and then move on to another. These problems are accentuated when power structures in the lowlands find it convenient to blame the indigenous hill tribals for forest degradation as, for example, in Northern Thailand (Chapman and Sabhasri, 1983). One of the best general discussions of shifting cultivation for Latin America, including the Andes, is an FAO publication by Watters (1971). Ramakrishnan (1992) has carried out a comprehensive study in northeastern India. In various regions this practice is known as swiddening, *jhum, conuco, kaingin*, forest fallow, and slash-and-burn. Increased population, access to new technology (e.g. chain saws), institutional factors that keep tenure uncertain, and land speculation by large landowners, together have spelled the end of shifting cultivation as a sustainable and appropriate watershed land use. Fallow periods have been drastically shortened or have been eliminated. Shifting cultivation in mountain watersheds has become damaging, not only to the individuals involved, but also to those downstream; the productivity of the land is progressively reduced. The consequences of shifting cultivation on soil and water have been presented by Hamilton with King (1983).

When new land is cleared for agricultural cropping there is potentially a high risk of erosion. Where this is done by low technology methods, such as in traditional slash-and-burn, the erosional and hydrological impacts are less serious. The greatest problems occur when heavy equipment is used for large-scale clearing associated with commercial farming. Such enterprises have been moving into mountain areas to replace smallholder farms, for example, for cultivation of temperate vegetables on the slopes of Mount Kinabalu in Borneo, tea plantations in Sri Lankan highlands, and commercial coffee in many parts of the mountain world. The effects on soil and water of converting forest land to annual cropping have been reviewed by Hamilton with King (1983), Bruijnzeel (1990), and Critchley and Bruijnzeel (1996). These effects include increase in total water yield, in proportion to the area of any small watershed cleared; greater stormflow volumes and higher peak levels; greater surface erosion and risk of mass erosion events; and increased sediment in streams and downstream sedimentation. It is important to complete the conversion from forest to cropping rapidly because sloping land is vulnerable; new vegetation should be in place as soon as possible, once converted; if cover is maintained, erosion becomes less likely (Critchley and Bruijnzeel, 1996).

The soil and water conservation practices that are needed for watershed 'health' (low erosion/high quality and 'regulated' water) are well known and widely published (El Swaify et al., 1985; Hudson, 1995). They include: contour ploughing, terracing, grassed waterways, cover crops, conservation tillage (crop residue cover) mulching and vegetative filter strips or strip cropping, and introducing trees into the cropping system (see next section on agroforestry).

Occasionally viable practices have broken down under the pressures of increased population, changes in government policy, out-migration of younger males, or introduction of a new technology by well-meaning foreign assistance programmes. An interesting example of the breakdown of conservation practices in Uganda's Kabale Highlands is given in Box 15.4. Haagsma (1995) in the Cape Verde mountain watersheds of Santo Antao, describes the effective indigenous terraces, cropping patterns and water management, and discusses how state intervention with 'improved' practices has had a negative effect.

It is important to note that watershed 'health' embraces the factor of productivity. Soil can be kept in place, but can become less fertile unless there is good management. Maintaining or increasing productivity is, of course, in the best interest of the mountain farmer. Schreier et al. (1995), in their watershed project work in Nepal, have emphasised the productivity–erosion linkage

Box 15.4 Deterioration of soil conservation practices in the Kabale Highlands, Uganda

Kabale Highland in Uganda is underlain mainly by older pre-Cambrian rocks. The non-volcanic part comprises the Karagwe–Ankorean system which was folded and uplifted to 2000 m or more; in places the highland is extremely rugged, with relief of 600–700 m. The Tertiary (2–62 million years old) Muhavura, or Mufrumbiro, mountain (4120 m) rises above this non-volcanic area. It is the most spectacular feature as well as one of the most important watersheds and game reserves in the southwestern part of the country.

As much as 50 percent of the land surface of the highland is too steep for adequate soil development; it is also too unstable to support farming without risk of accelerated erosion. The percentage of arable land in a basically peasant farming society is an important factor. Stocking (1975) has observed that, in such areas, actual population pressure on the land is normally greater than indicated by the official statistics since they aggregate both arable and non-arable land. For instance, in Rukiga (part of Kabale Highland) the official figure for population density in 1980 was 164 km^{-2}, which is extremely high for such a fragile environment. Yet, in some local areas, such as Kisoro near Mt Muhavura, the density is as high as 500 km^{-2}, or even higher.

It is believed that until about 500 years ago the highland was forested, with bamboo, mahogany, and other hardwoods; but centuries of human use has led to serious degradation of the natural vegetation. Today, except for a few surviving patches under protection, most of the area has a very poor cover of various species of human-manipulated vegetation.

Poor soil management is one of the major causes of accelerated erosion in Kabale Highland (Bagoora, 1988). Soil conservation practices, which had reached an advanced stage by 1949, have degenerated since the mid-1960s. The most serious problem is the deliberate destruction of contour bunds and terraces which form the predominant conservation feature in the highland. These were established from vegetation/plant residue and stones, on which grass was allowed to grow or was planted, to trap silt from runoff water (Hudson, 1995). Over time, a combination of soil accumulation from the upper part of peasant farm plots, and undercutting of the bund from below during hoe cultivation, has resulted in a pattern similar to some of the conventional terraces described by Morgan (1986), Hudson (1995), and Schwab et al. (1981). Following deterioration, or outright destruction, of the bunds and terraces, extensive soil erosion has occurred, because of the higher slope gradients and the unchecked runoff from the upper parts of the farm plot.

It appears that the main motive behind this destruction is to use the more productive soils that have accumulated above the bunds or terraces. This has been prompted by the ever-increasing population pressure, and has led to excessive overuse of the land and decline in productivity; this has been especially pronounced on the upper and middle parts of peasant plots. Another cause of damage to the conservation structures is trampling by animals, especially cattle, which are normally grazed on crop residues on the farm plots after harvest.

Another factor that may have influenced this apparently escalating destruction of bunds and terrace walls is the traditional land tenure system, which provides for individual land ownership. Family land is divided into small plots and shared between the farmer's surviving children. The number of parcels increases and the size of individual parcels correspondingly decreases with successive generations. This tenure system inhibits land consolidation and, therefore, prevents the possibility for commercially viable transformations. Also, as the land ownership pattern becomes more and more fragmented, family survival is threatened, prompting land sales. Individual farmers may sell their land to settle outstanding debts or crises; and sales may be made by families who decide to migrate to areas where land shortage is not yet serious. Individuals often buy widely scattered plots sometimes several kilometres from their homesteads. Consequently, one of the world's most fragmented land-holding patterns has developed here. In one study site of about a square kilometre, there were 28 different owners. It is common, therefore, to find many bunds and terrace walls acting as boundaries between farm plots. Such structures often receive little care so that further degeneration is rapid.

While the consequences of soil erosion in Kabala Highland are complex, they clearly include a decline in productivity, food shortage, and a general reduction in the quality of the human environment.

Source: D. K. Festus, Bagoora,
Makerere University Kampala, Uganda

and the serious consequences of productivity decline. Reduced yields due to excessive erosion are also leading to food deficits in the Ethiopian Highlands (Hurni, 1988), and Harden (1988) suggests that the productive future of steepland agriculture under prevailing practices in the Río Ambato watershed of Ecuador may be only 10–75 years. To be successful, watershed rehabilitation and management progammes and projects must assist farmers to maintain and raise productivity through soil and water conservation, rather than to focus on the mechanics of stopping erosion and sedimentation. This is a small distinction of perspective but it is a significant difference in approach.

Agroforestry

Agroforestry is one of the oldest forms of human land use and may be defined simply as combining woody plants (usually trees) with food or forage crops. Although sometimes the practice includes sequential (temporal) combinations, more usually trees and annual or perennial crops are combined in the same field at the same time. Agroforestry has been rediscovered since the 1970s by modern western resource management professionals and overseas development assistance agencies. It is promoted as a land-use system that will solve many problems of sustainability on steep land while im-proving farm livelihood and reducing risk. Many major centres of research and extension promote agroforestry; these include the International Centre for Research in Agroforestry in Nairobi, Kenya, and the Centro Agronómico Tropical de Investigación y Enseñanza in Turrialba, Costa Rica. Agroforestry system management is discussed in Chapter 13; a description of prevailing systems in Asia and the Pacific is given in Mellink *et al.* (1991).

There is little doubt that adding, or retaining, deep-rooted trees in pastures or croplands will provide some margin of protection against shallow landslips (O'Loughlin and Ziemer, 1982) and perhaps also shallow creep or slump. The degree of risk reduction will depend on tree density, soil characteristics, and slope angle, which should be balanced with the needs of the understorey crop. It is speculated that scattered trees, as few as 20 per hectare, might have considerably reduced the dramatic number of shallow landslips in New Zealand's cleared steepland pastures following Hurricane Bola in 1988 (see, for example, Plates 15.5 and also 15.6 from Taiwan).

It is important to note that the mere presence of trees will not reduce surface erosion *per se*. Serious erosion can occur under trees where litter is removed or is thin, or is cleared for cropping. Standard soil and water conservation practices for cropland, and good animal husbandry and forage management for grasslands are essential. It is the *vegetation and litter cover on the soil and close to the soil* that is the critical factor. Tall tree canopies could increase the potential for splash erosion impact if the soils underneath were bare. Scattered trees planted on 80 percent slopes that are overgrazed will not significantly reduce surface erosion. Hydrologically, agroforestry systems if well managed to control erosion, provide an excellent compromise between the heavy water use of forests and their beneficial local effect to reduce stormflows. A mosaic of land uses is encapsulated in an agroforestry system and, from a water management standpoint, this is an excellent arrangement.

Roads and other aspects of access

While the area occupied by roads in a mountain watershed is usually quite small, these features have disproportionate significance for erosional and hydrological processes. Fortunately perhaps, construction and maintenance of roads in mountain areas is expensive and they are not undertaken as readily as in the lowlands. Nonetheless, the demand for access, for moving people and products up and down the mountain, has led to a substantial proliferation of roads. These are replacing older forms of access that had less impact on soil and water, such as footpaths and animal pack routes, cable systems of various degrees of sophistication and, more recently, aircraft landing strips. Some of the older traffic ways that traversed mountains, for example, the Inca Trail in the Andes, and the trade routes from China to South Asia across the Himalaya, or from Russia across the Caucasus to Asia Minor, were extremely well constructed and maintained and caused little landscape impairment. The roads and railroads that replaced them often induce serious slope instability.

With adequate money, engineering skill, and conservation planning, roads and railways can be constructed in rugged mountain terrain with no negative environmental impact. The excellent network of roads and rails in the European Alps, especially Switzerland, is testimony to how well it can be done (Stvan, 1991). In less developed countries, and where seismic activity is frequent, the environmental impact may be severe. India is a good example of this latter situation. Valdiya (1987) estimated that 44 000 km of roads had been constructed in the previous 25 years and that for each kilometre, 40 000–80 000 m³ of debris had to be removed; this was usually dumped downslope,

Plate 15.6 Landslides initiated by the Cross Island Highway in the central mountains of Taiwan (Photograph: Yue-Joe Hsia)

thereby further accelerating slope instability. Valdiya estimated that this led to an average of 550 m³ of erosion material per km of road for the three highways studied, in the West, Central, and East Himalaya. Plate 15.6 shows the extent of land movements associated with the relatively well-engineered Cross Island Highway in the central mountains of Taiwan. Carson (1985) documented the recently-constructed Dharan–Dhankuta road which cost over US$ 1 million per km, that could not be maintained because the Nepalese government did not have adequate resources. This road was closed during parts of the 1983 and 1984 monsoon seasons and, during the latter, slope failure caused over US$ 5 million of damage. Adequate, low-cost, low-maintenance roads can be built for rural uses, including logging roads in humid areas, if properly drained; there are good examples from such mountainous areas as West Virginia and Honduras (Kunkle, personal communication, 1996).

One of the major mistakes that has been made in assessing watershed land use and identifying sources of downstream sediment has been to blame mountain farmers for what is often a problem created by road construction. The research in the Upper Konto watershed in Indonesia has already been referred to (Rijskijk and Bruijnzeel, 1991) where roads which occupied only 3 percent of the total area contributed 15 percent of the sediment. In mountainous Northern Thailand, the prevailing swidden practices of the hill tribes, such as the Yao and Lisu, have been blamed for causing erosion. Forsyth (1994) has argued that much of the damage was from gullies due primarily to roads or the natural weathering of the granite. Plate 15.4 shows a number of slope failures caused by a road in the Andes of Venezuela; this type of erosion in the past has been attributed to the *conuqueros* (shifting cultivators). During a recent World Bank watershed project review in the area around Medellín, Cali, and Pasto in the Colombian Andes, Cassells (1996) remarked that the most serious incidences of erosion and slope instability were associated with roads. He suggested that watershed projects need to give as much or more attention to road engineering and maintenance, and riparian zone management in both upper and lower catchment areas as to land use in the upper catchment. This recommendation applies to most watershed projects in developing countries.

Likewise, logging roads in forests are potentially the most damaging part of forest harvesting (see earlier section). It is possible to identify areas of slope instability and to avoid using them, or to minimise disturbance (Megahan and King, 1985). With public roads, this becomes more difficult because alignments are less flexible. Nonetheless, it is necessary to select an appropriate location, and then exercise care in construction by balancing cuts and fills and by disposing the excess material safely.

Any construction activity which removes the vegetation and does not keep soil in place or maintain it on site by vegetative barriers will result in soil erosion and deposition of sediment. Examples of this process include golf course construction in the Genting Highlands of Malaysia, vegetation clearing prior to flooding to create a mountain reservoir in the Peruvian Andes, road building in the Caucasus, ski development in the Tatras, and establishment of mountain resorts in the northern Japanese Alps. Special precautions are needed during the 'vulnerable' period before the new use has become established and has matured. Prolonged or high-intensity precipitation during this period of transformation can initiate very serious surface disturbance and, in some cases, mass erosion as unvegetated slopes fail. Erosion reduction and prevention codes have been promulgated in some places, for some types of activity, and much more of this regulation is needed in upland watersheds.

Even well-used foot trails, whether frequented by local residents or by visiting hikers, can be badly eroded and provide channels for downhill movement of water and sediment. Continued use kills the vegetation, prevents it from recovery, and compacts the soil. Fragile alpine areas are most easily and severely damaged. In many mountain national parks with high numbers of visitors, trail maintenance and rehabilitation is a necessary and common management practice. Pack animals, whether donkey trade caravans in the Himalaya or mules and horses in the Rocky Mountains, add to this problem. In several areas of the United States, llama from the Andes are now used as pack animals because they are lighter and their hooves create less impact than horses. Ski runs are normally vegetated with grasses or herbs, and would seem not to be very susceptible to erosion, but Ries (1996) has shown that, in the Black Forest, improper management, when snow cover is inadequate, results in damage by both skis and grading equipment. Erosion and needle-ice solifluction occur, and the effect is aggravated by any subsequent grazing during the non-skiing season. Clearing forest for ski runs and lifts increases the risk of both avalanches and mass erosion.

Off-road recreational vehicles that use mountain trails, or make their own, are another hazard. All-terrain vehicles, snowmobiles, and modified 4-wheel drive trucks and wagons can take recreationists from the developed world into sensitive mountain environments unless controls on use are established and enforced. Mountain bikes, as the name implies, have opened up steep and rough trails to a new array of users during the past decade, and erosion on these steep trails is a growing concern in mountain protected areas.

In addition to the factors of erosion and sedimentation resulting from access by recreational users, there is also the impact on water quality from human waste products. The sanitation practices of many back-country recreationists and resort managers often ignore the public health standards established for drinking water supplies for downstream residents. The rapid spread of the parasite *Giardia* throughout the mountain wildlands of the United States and Canada by recreationists and its establishment in the wild animal populations (where it is spread especially by the beaver) has rendered mountain stream water unsafe. *Giardia* is now present in many of the world's wildlands, wherever human access exists.

MANAGING AND REHABILITATING MOUNTAIN WATERSHEDS

Mountain cultivators, graziers, and woodcutters have known for centuries that there is a relationship between the vegetation cover and use of the land in one location, and its erosional and hydrological consequences – especially downslope or downstream. There are allusions to the need for conservation practices in ancient Greek, Chinese, Hebrew, and Roman literature. Watershed conservation, however, did not enter prominently onto the political and public scene until the mid-1800s, at least in the West. At that time, individuals, such as George Perkins Marsh in his book *Man and Nature, or Physical Geography as Modified by Human Action* (1864) began sounding a clarion call to action. Marsh wrote this work from his perspective, living in the Green Mountains of Vermont, USA, which at that time had been largely cleared of forest cover for farming. It is interesting to note that the State of Vermont has recovered from being roughly 75 percent cleared of forest to 82 percent forested in less than 150 years. In the late 1800s concern for the water in New York State's navigable rivers led to the setting aside of virtually the whole mountain area as the Adirondack Park, and within it, all State-owned land, some 1 021 500 ha, was designated to remain forested forever without logging; this was embodied in the State Constitution.

In the United States and Australia, farmland conservation, focused on erosion control and water management practices, entered the political–public arena in a major way during the 1930s. It was institutionalised in the USA by the establishment of a national Soil Erosion Service in 1933, which became the Soil Conservation Service two years later (Dana, 1956), and in 1995 was renamed the Natural Resources Conservation Service. It began an organised research, extension, and technical service programme and has produced much of the material on which agricultural projects of soil and water conservation are based throughout the world. There have been many publications and voluminous reports dealing with upland watersheds which have been used internationally. One of the seminal books, entitled *Soil Conservation* was first published in 1971 and there have been 10 subsequent reprintings, revisions, and new editions (Hudson, 1995). The first FAO Conservation Guide, entitled *Guidelines for Watershed Management* appeared in 1977 (Kunkle and Thames, 1977), and since then a series of these guides and field manuals, on topics ranging from torrent control to people's participation in watershed management and to economic appraisal of watershed projects, has been produced by FAO. Today, almost

every country has its own fund of literature on how to achieve better soil and water conservation in steeplands. If the written word could heal the degraded lands, the problems of degradation would have been solved.

Years of field experience and recording in many disciplines and regions have become increasingly integrated; as a result, conservation in cropping, grazing, and forest harvesting systems have been brought together, as well as the specifications for small engineering structures for environmental protection in stream channels such as retention dams and debris traps. In the last decade problems in urban or suburban watersheds have also been addressed. The list of practices promoted includes: terracing; conservation tillage (leaving crop residues as cover during and after planting); contour cultivation; undisturbed vegetative buffer strips along streams; grassed waterways; mulching; winter cover crops; gully plugging; agroforestry; reforestation of areas unsuited to agriculture; controlled animal stocking rates and rotational grazing; stall feeding; reduced use of chemicals through integrated pest management; reduced but more effective fertiliser use; sediment and debris traps; retention basins and ponds; infiltration trenches, and several others (Hudson, 1992, 1995.) An interesting set of preventative and restorative practices for dealing with typical theoretical watershed problems is given in Box 15.5. The activities to be carried out in a watershed component of a project in the Colombian Andes are presented in Box 15.6.

Reforestation is an important activity in watershed rehabilitation. It is often needed in order to provide wood, particularly fuel, for local inhabitants of mountain watersheds. But even more important for soil and water conservation, it is a process undertaken to stabilise eroding or erosion-prone areas or to restore cover to severely degraded sites. Properly planned, properly planted, and properly maintained watershed plantations can fulfill a number of beneficial purposes, and will continue to play an important role. It is also apparent that in humid climates, the removal of human and animal pressures, such as fire, grazing, and cultivation, that was preventing tree and shrub re-growth, would allow natural revegetation or supplement strategic planting. Only within the past decade or so has it reached professional consciousness that one negative consequence of forestation is the reduction in dry season flow in streams. (Note: it has often been more quickly recognised by affected landowners who cut down 'thirsty'

Box 15.5 Alternative preventative or restorative practices for dealing with problem situations encountered on a watershed

PROBLEM	ALTERNATIVE SOLUTIONS	ASSOCIATED MANAGEMENT OBJECTIVES
Deficient Water Supplies	Reservoir storage and transport	Minimise sediment delivery to reservoir site, – maintain watershed vegetative cover
	Water harvesting	Develop localised collection and storage facilities
	Vegetation manipulation, – ET reduction	Convert from deep-rooted to shallow-rooted species or from conifers to deciduous trees
	Cloud seeding	Maintain vegetative cover to minimise erosion
Flooding	Reservoir storage	Minimise sediment delivery to reservoir site, – maintain watershed vegetative cover
	Construct levees, channel improvement	Minimise sedimentation of down-stream channels
	Flood plain management	Zoning of lands to restrict human activities in flood-prone areas; minimise sedimentation of channels
	Revegetate disturbed areas	Afforestation or reforestation of denuded watersheds; encourage natural revegetation
Energy Shortages	Utilise wood for fuel	Plant perpetual fast-growing tree species that maintain productivity of sites by minimising erosion
	Develop hydroelectric power project	Minimise sediment delivery to reservoir pools and river channels
Food Shortages	Develop agroforestry	Maintain site productivity, – minimise erosion (nutrient losses); develop crops compatible with soils/climate of area
	Increase cultivation	Restructure steep hill slopes and other areas susceptible to erosion; utilise contour ploughing, terraces
	Increase livestock production	Develop herding/grazing systems for sustained yield and productivity
	Import food from outside watershed	Develop forest resources for pulp, wood products, to provide economic base
Erosion/ Sedimentation from Devegetated Landscapes	Erosion control structures	Maintain life of structures by revegetation and management
	Contour terracing	Revegetate, stabilise slopes, terraces, and institute land-use guidelines
	Revegetate	Protect vegetation cover until site recovers; use reseeding, fertilisation, etc.
Poor Quality Drinking Water	Develop alternative supplies from wells, springs	Protect groundwater from contamination
	Treat water supplies	Protect catchments from contamination

Source: Working Group on Watershed Management and Development of the University of Minnesota's Forestry and Sustainable Development Program. Unpublished, October 30, 1987

trees in several reforestation project areas.) Following early pioneer work in South Africa (Banks and Kromhout, 1963), a number of investigations in New York State in the early 1970s, in India in the late 1970s, and in Fiji in the early 1980s, it has gradually become apparent that when well-adapted trees are planted over large areas in a watershed, decreased low-season flows are a consequence, beginning in about the fifth year. Bruijnzeel (1996) has produced a comprehensive review of the hydrology of forest plantations in the tropics.

Hundreds of watershed management and rehabilitation projects have been instituted around the world, with varying degrees of success. Many of them involve technical assistance and funding from donor and aid agencies of the developed world, to be applied in critical areas of the host country. Whether in the Sierra Nevada de Santa Marta of Colombia, high up in the Loukos River watershed

Box 15.6 Range of project-proposed watershed management activities

(1) Participatory planning with landholders and other key catchment stakeholders to produce catchment land-use plans on land capability and land suitability analyses.

(2) Assisting landholders to develop integrated farm development plans consistent with the catchment management requirements defined in the catchment plans for the area and considerations of farm-scale land capability and land suitability characteristics and issues.

(3) Protect or re-establish stable forest cover in key riparian zone areas.

(4) Protect and manage areas of natural forest regeneration; activities to re-establish forests or other appropriate vegetation covers on severely degraded sites not suited to agricultural, horticultural, or pastoral use.

(5) Encourage a variety of tree planting activities (plantations, agroforestry, silvopastoral plantings) on lands suitable for these purposes to enhance on-farm production of fuelwood, fodder, fruits, mining timbers, fenceposts, and building materials for sale or on-farm use.

(6) Improve pasture management and animal husbandry on lands suitable for pastoral use.

(7) Intensify agricultural cropping on land suitable for cropping.

(8) Install appropriate soil conservation measures in all areas including the creation of live vegetation barriers, gully control structures, gabions, hillside terraces, and grass planting.

(9) Improve road location, design, and maintenance, including installation and maintenance of adequate drainage structures and stream crossings.

Source/reference: Cassells (1996)

in Morocco, or the headwater areas of the Irrawaddy of Myanmar, technical and financial 'packages' of the aforementioned conservation practices have been introduced in various combinations, and at tremendous cost. Success has been very mixed. One of the more successful projects (in the authors' opinion) is in the Cochabamba watershed in Bolivia as summarised in Box 15.2. Many results of earlier projects were disappointing because the approaches were paternalistic, 'imposed' on peasants who were assumed to be degrading their watersheds. They were technical fixes installed with outside money, such as, food-for-work programmes, and imposed a structure or major change in practice. Very often the structure was not subsequently maintained and the practice was discontinued because it did not fit the needs of the land user, economically, culturally, or practically. International technocrats, researchers, and politicians were unable to obtain willing collaboration from the local land users.

Recently there has been a substantial reorientation in watershed management and rehabilitation in both developed and developing countries. Past experiences in watershed projects have been documented and used to effect change. FAO (1991) has produced a study of the reasons for success or failure of soil conservation projects, and their most recent guidelines for land-use planning projects indicate a new approach. Hudson (1992), reflecting on a long career in the soil conservation field, has produced a very useful analysis in his book,

Land Husbandry. Box 15.7 lists some reasons for failure of projects.

The 'new' approach has several elements, which have been mentioned briefly in previous sections of this chapter. First, it recognises that where mountain people have farmed the land for many generations, many of their management strategies and practices are sound and include soil and water conservation that ensures a sustainable livelihood.

Second, the existing practice needs to be modified and extended to increase productivity. Often only minor adjustments may achieve additional hydrological benefits and reduce erosion. Research that documents the decreases in productivity and income caused by soil erosion or poor water management can influence land users. A good example is the work in New Zealand's landslip-prone steeplands under treeless grazing regimes; it was shown that, although the eroded areas did revegetate to pasture, on recent scars grass production decreased by 80 percent; and despite eventual recovery, even after 50 years, production was still 23 percent below that of uneroded areas (Trustrum and Lambert, 1982).

Third, in large watersheds, participatory planning is essential. This is now widely recognised as the 'bottom-up' approach, but is honoured more in words than in practice (Liniger, 1995). Rao, working in the Southeast Asian uplands, offers a definition of local participation: the process by which the rural poor are able to organise themselves and, through their own institutions, are able to identify

Box 15.7   Water development programmes: reasons for failure

The lessons learned so far from failed attempts to improve water resource management have been documented by different organisations. Some of the major problems can be summarised as follows:

(1) Too little information available and/or insufficiently applied to reduce direct and indirect degradation.
(2) Too few multi-objective (integrated) analyses and planning approaches which consider all natural, social, and economic factors.
(3) Too much political interference in resource management.
(4) Too much reliance on engineers and technocrats at the planning stage.

(5) Too much fragmentation of institutional responsibilities.
(6) Lack of leadership.
(7) Projects often too large and too rapidly implemented.
(8) Too little consideration of local ecological and socio-economic situations.
(9) Too much emphasis on natural resource conservation rather than on sustainable productivity (especially in developing countries).
(10) Too little attention paid to overall economic and social aspects, including direct and indirect impacts.
(11) Too little involvement on the part of land users.

Source/reference: Hanspeter Liniger (1995)

their own needs, share in the design, implementation, and evaluation of the participatory action (Rao, 1988). This statement is true not just for Southeast Asia, nor only for the rural poor, but has also been recognised in Australia, New Zealand, South Africa, Europe, and North America. In the United States, locally organised and operated small watershed associations have proliferated and are, by and large, extremely successful in marshalling technical and financial assistance from many sources to carry out 'grassroots', locally-determined activities in restoring watershed health. As an important side effect, those involved are enhancing a sense of community and a respect for the land.

In addition to modifying existing practices, if needed, to promote soil and water conservation, it is important that new uses for existing wild lands are viable and sustainable. Where existing forest affords protection to erosion-prone areas, any conversion to an alternative use or any major disturbance, such as logging, must be very carefully considered. Megahan and King (1985) have produced an extremely useful set of criteria for identifying critical areas on forest land for control of non-point sources of pollution; for instance, it includes practical suggestions for identifying areas prone to shallow landslips, based on slope form (concave or convex), degree of slope, likelihood of threshold rain intensity and duration levels, and past evidence of slips.

In implementing soil and watershed conservation practices to benefit downstream areas, it is extremely important to consider the question of spatial scale. This was well stated by Professor Bruno Messerli at the Mohonk (New York State, USA) conference on *The Himalaya – Ganges Problem* in 1986; it was subsequently paraphrased (Hamilton, 1987) and is repeated here:

(1) At the local level (< 50 km^2) sediment load is strongly influenced by human activity, stream discharge characteristics are much less, but that on the whole, for flood and sedimentation problems in the Himalaya, human activity has less impact than natural factors.
(2) At the medium level (50–20 000 km^2) downstream of the catchment receiving the impact, we are still uncertain of the quantitative effects of human activity, but the high variability of natural factors dominates stream discharge and sediment load.
(3) At the macro-level (> 20 000 km^2) in large basins, one postulates that human impacts in the upper watershed are insignificant on lowland floods, low flow, and sediment, but that these effects can be significantly influenced by human activity in the lower reaches of the river.

This concept is supported by data, and elaborated in more detail in *The Himalayan Dilemma: Reconciling Development and Conservation*, by Ives and Messerli (1989). It seems appropriate to other mountain systems as well. Not only size, but geology, relief, and shape of catchment will influence the highland–lowland interaction within a given climatic regime. Because of this variability, it is prudent to refrain from sweeping generalisations about the extension of the protective role of land treatments to areas far downstream.

SUMMARY AND FUTURE DIRECTIONS

Where *environmental impact assessments* are being made for any type of landscape modification, from mining to construction of roads or installation of transmission lines, it is essential to consider the possible effects that are transmitted through soil and water, and to involve people familiar with watershed processes. Improved methods of assessing *cumulative impacts* in watersheds are being developed and should replace the erroneous, discrete, one-factor-at-a-time analysis. Cumulative impact analysis is of critical importance because of the speed of development, population increase, and other aspects of global change, and an increased emphasis on such research is warranted.

The importance of undisturbed, or largely undisturbed, *riparian buffer zones* is increasingly recognised by water quality experts, wildlife and fisheries managers, and landscape architects. Their valuable roles have been listed previously. Further research is needed to determine under what conditions and how best to manage the land to reduce the loads of nutrients (such as nitrogen from agriculture) or heavy metals (such as lead from automobile emissions) that move through overland or subsurface flow.

The important functions of *montane cloud forests* with respect to water capture and conservation of unusual biodiversity is only slowly becoming recognised by governments and the public. This awareness needs to be greatly accelerated because these ecosystems are being rapidly destroyed.

Pathogens, such as *Giardia* and *Cryptosporidium* that may be injurious to healthy persons who drink contaminated water, are often lethal to persons with less robust immune systems. Drinking water supplies in villages and cities can be maintained at acceptable standards by effective watershed management. *Improved human sewage treatment* upstream and *improved livestock manure management* are required. One such programme is that being undertaken on behalf of the 9 million people of New York City on its 4900 km^2 mountain watershed, much of it over 200 km distant from the city.

Projects of watershed management and rehabilitation must, as Dani and Campbell (1986) stated:

> . . . *build on upland residents' existing motivations* for sustaining their upland environments through increasing the value, renewability, security, manageability, and equity of

resources. As additional understanding is gained through on-going projects and future studies, the most effective leverage points, incentives and policy measures can be identified. The need for this ability to be responsive to the motivations and behaviours of the upland resource users thus emerges as the central lesson from this study.

While these remarks were applied to the Hindu Kush–Himalaya Region, they are equally valid for projects anywhere in the world. Attempts to restructure priorities and to introduce new behavioural patterns through heavy subsidies or coercion do not have lasting results.

Watersheds are managed by land users, not by agency technicians, and the planning and implementation of new practices must be undertaken by the local people – with technical help but not technical control. *Grassroots, or bottom-up, planning* is a practical tactic rather than a political slogan.

Participatory planning groups, or watershed associations, should include *representatives, or a cross-section, of all those who have something at stake.* This may include persons or groups not resident in the watershed but who use the watershed and are willing to contribute to its health; for example, those who hunt, fish, use the area for recreation, and downstream water users.

Where land or water rights are uncertain, *greater tenurial security* is required to encourage investments in conservation. Where open-access or common property rights are well managed as, for example, with grazing or water use, measures to strengthen the social controls that foster self-regulation by the community should be supported. Noticeable improvements in resource management have been realised when resource rights were legally transferred to the people, although this was not an easy process. Examples include the recent Panchayat Forestry legislation in Nepal, the introduction of the Responsibility System in China, and the allocation of government land to long-term agroforestry in Bangladesh (Dani and Campbell, 1986).

Of great significance is the *economic question of who receives benefits and who incurs costs in the upstream–downstream interaction* for soil and water conservation. If there can be more precision in accounting and internalising the analysis on a watershed basis, then compensation and transfer of funds can be addressed. *High quality water is not a free resource* and this must be acknowledged. Additional *research in the biophysical realm* will be needed to provide information for economic analysis. The water capture and supply function of *mountain cloud forests* is one specific area where more research is required.

Easily-applied methods of *assessing potential sources of non-point pollution*, particularly sediment, are needed to identify the potential hazard for new proposed land uses, especially on soils prone to shallow landslips where retention of forests can play a role in maintaining stability.

In view of the extensive forest plantation establishment on steeplands for the purpose of watershed rehabilitation or carbon offsetting, further *investigation of the hydrologic effects of new plantations* is required.

A greater understanding needs to be fostered among politicians, bureaucrats, the media, and the public about the *problems of spatial scale*; as an example, the transfer of findings from plots or small watersheds to large river basins is inappropriate and is not justified: this must be more widely appreciated. The scientific community should acknowledge this, and interpret and disseminate the results of watershed research. A good example has been given in the final paragraphs of the last section.

When *watersheds cross international or interstate boundaries*, new mechanisms for co-operation are needed in order to ensure that watershed management is comprehensive. The use of water from international streams or rivers is a particularly sensitive issue.

Sustainable mountain development occurs when the needs and aspirations of mountain people and the capacity of the natural resource base to supply them are in balance over time. Overlaying this are the needs and desires of a larger public, particularly those whose interests lie at lower altitude in the landscape, but who are affected by actions higher up. *Watersheds provide a viable framework* for approaching sustainable development through integrated watershed planning and management. The gamut of actions undertaken in such programmes range from establishing and maintaining totally protected areas, to raising productivity of fruit orchards, or increasing yields of barley, to maintaining grazing productivity. Implementation involves biophysical practices, institutional mechanisms and socio-economic and cultural factors. There is a great need for *political and administrative innovations for watershed 'governance'* at the local level and at the national level. Non-governmental organisations (NGOs) have proved extremely helpful in effecting change and have fostered education and implementation of new programmes.

ACKNOWLEDGEMENT

The assistance of Dr Sam Kunkle, Watershed Consultant and Adjunct Professor, Department of Earth Sciences, Colorado State University, Fort Collins, Colorado, USA is gratefully acknowledged for his conscientious review of this manuscript and suggestions for its improvement.

References

Bagoora, F. D. K. (1988). Soil erosion and mass wasting risk in the highland area of Uganda. *Mountain Research and Development*, 8(2/3): 173–98

Banks, C. H. and Kromhout, C. (1963). The effect of afforestation with *Pinus radiata* on summer baseflow and total annual discharge from Jonkershoek catchments. *Forestry in South Africa*, 3: 43–65

Bencherifa, A. and Johnson, D. L. (1993). Environment, population pressure and resource use strategies in the Middle Atlas Mountains of Morocco. In Bencherifa, A. (ed.), *Montagnes et Hauts-Pays de l'Afrique (2), Utilisation et Conservation des Ressources*. Université Mohammed V, Rabat, pp. 101–21

Brooks, K. N., Ffolliott, P. F., Gregersen, H. M. and Thames, J. L. (1991). *Hydrology and the Management of Watersheds*. Iowa State University Press: Ames

Bruijnzeel, L. A. (1990). *Hydrology of Moist Tropical Forests and Effects of Conversion: A State of Knowledge Review*. UNESCO International Hydrological Programme: Paris

Bruijnzeel, L. A. (1996). Hydrology of forest plantations in the humid tropics. *Better Management of Soil, Water and Nutrients in Tropical Plantation Forests.* ACIAR–CIFOR: Puncak, Indonesia

Bruijnzeel, L. A. with Bremmer, C. N. (1989). *Highland–Lowland Interactions in the Ganges–Brahmaputra River Basin: a Review of Published Literature.* Occasional Paper No. 11. ICIMOD: Kathmandu

Bruijnzeel, L. A. and Proctor, J. (1995). Hydrology and biogeochemistry of tropical montane cloud forests: what do we really know? In Hamilton, L. S., Juvik, J. O. and Scatena, F. N. (eds.), *Tropical Montane Cloud Forests.* Springer Verlag: New York, pp. 38–78

Carson, B. (1985). *Erosion and Sedimentation Processes in the Nepalese Himalaya.* Occasional Paper No. 1. ICIMOD: Kathmandu

Carson, B. (1992). *The Land, the Farmer and the Future. A Soil Fertility Management Strategy for Nepal.* Occasional Paper No. 21. ICIMOD: Kathmandu

Carver, M. and Nakarmi, G. (1995). The effect of surface conditions on soil erosion and stream suspended sediments. In Schreier, H., Shah, P. B. and Brown, S. (eds.), *Challenges in Mountain Resource Management in Nepal.* ICIMOD: Kathmandu, pp. 155–62

Cassells, D. (1996). Review of the watershed management component of the Natural Resources Management Project, Colombia. Internal Document. The World Bank: Washington DC

Chapman, E. C. and Sabhasri, Sanga (1983). Natural Resource Development and Environmental Stability in the Highlands of Northern Thailand. Proceedings of a Workshop organised by Chiang Mai University and the United Nations University. *Mountain Research and Development,* 3(4): 309–431

Critchley, W. R. S. and Bruijnzeel, L. A. (1996). *Environmental Impacts of Converting Moist Tropical Forest to Agriculture and Plantations.* IHP Humid Tropics Programme Series No. 10, UNESCO: Paris

Dana, S. T. (1956). *Forest and Range Policy, Its Development in the United States.* McGraw-Hill: New York

Dani, A. and Campbell, J. G. (1986). *Sustaining Upland Resources, People's Participation in Watershed Management.* ICIMOD Occasional paper no. 3. ICIMOD: Kathmandu

Easter, K. W. and Hufschmidt, M. M. (1985). *Integrated Watershed Management Research for Developing Countries.* East–West Center Workshop Report: Honolulu

Easter, K. W., Dixon, J. A. and Hufschmidt, M. M. (eds.) (1986). *Watershed Resources Management. An Integrated Framework with Studies from Asia and the Pacific.* Westview Press: Boulder

El Swaify, S. A., Moldenhauer, W. C. and Lo, A. (eds.) (1985). *Soil Erosion and Conservation.* Soil Conservation Society of America: Ankeny, Iowa

FAO (1989). *Guidelines for Land Use Planning.* Inter-Departmental Working Group on Land Use Planning. FAO: Rome

FAO (1991). *A study of Reasons for Success or Failure of Soil Conservation Projects.* Soils Bulletin, 64. FAO: Rome

Ffolliott, P. F., Gottfried, G. J. and Baker Jr., M. B. (1989). Water yield from forest snowpack management: research findings in Arizona and New Mexico. *Water Resources Research,* 25(9): 1999–2007

Forsyth, T. J. (1994). The use of cesium-137 measurements of soil erosion and farmers' perceptions to indicate land degradation amongst shifting cultivators in Northern Thailand. *Mountain Research and Development,* 14(3): 229–44

Gibbs, C. J. N. (1986). Institutional and organizational concerns in upper watershed management. In Easter, K. W., Dixon, J. A. and Hufschmidt, M. M. (eds.), *Watershed Resources Management.* Westview Press: Boulder, pp. 91–102

Gilles, J. L., Hammoudi, A. and Mahdi, M. (1986). Oukaimedene, Morocco: a high mountain *agdal.* In *Proceedings of the Conference on Common Property Resource Management.* National Academy Press: Washington, DC, pp. 281–303

Gioda, A., Maley, J., Espejo Guasp, R. and Acosta Baladon, A. (1995). Some low elevation fog forests of dry environment: applications to African paleo-environments. In Hamilton, L. S., Juvik, J. O., and Scatena, F. N. (eds.), *Tropical Montane Cloud Forests.* Springer Verlag: New York, pp. 156–64

Haagsma, B. (1995). Traditional water management and state intervention: the case of Santo Antao, Cape

Verde. *Mountain Research and Development*, **15**(1): 39–56

Hamilton, L. S. (1976). *Tropical Rainforest Use and Preservation: a Study of Problems and Practices in Venezuela.* International Series No. 4. Sierra Club: San Francisco

Hamilton, L. S. (1986). Toward clarifying the appropriate mandate in forestry for watershed rehabilitation and management. In Michaelsen, T. and Brooks, K. N. (eds.), *Strategies, Approaches and Systems in Integrated Watershed Management.* FAO Conservation Guide No. 14. FAO: Rome, pp. 33–51

Hamilton, L. S. (1987). What are the impacts of Himalayan deforestation on the Ganges–Brahmaputra lowlands and delta? Assumptions and facts. *Mountain Research and Development*, **7**(3): 256–63

Hamilton, L. S. (1990). Human activity *versus* natural processes. *Mountain Research and Development*, **10**(4): 359–60

Hamilton, L. S. with King, P. N. (1983). *Tropical Forested Watersheds: Hydrologic and Soils Response to Major Uses or Conversions.* Westview Press: Boulder

Hamilton, L. S. and King, P. N. (1984). Watersheds and rural development planning. *The Environmentalist*, **4**, Supplement No. 7, pp. 80–6

Hamilton, L. S. and Pearce, A. J. (1987). What are the soil and water benefits of planting trees in developing country watersheds? In Southgate, D. D. and Disinger, J. D. (eds.), *Sustainable Development of Natural Resources in the Third World.* Westview Press: Boulder, pp. 39–58

Hamilton, L. S., Juvik, J. O. and Scatena, F. N. (1995). The Puerto Rico Tropical Cloud Forest Symposium: introduction and workshop synthesis. In Hamilton, L. S., Juvik, J. O. and Scatena, F. N. (eds.), *Tropical Montane Cloud Forests.* Springer Verlag: New York, pp. 1–23

Harden, C. (1988). Mesoscale estimation of soil erosion in the Río Ambato drainage, Ecuadorian Sierra. *Mountain Research and Development*, **8**(4): 331–41

Hibbert, A. R. (1967). Forest treatment effects on water yield. In Sopper, W. E. and Lull, H. W. (eds.), *International Symposium on Forest Hydrology.* Pergamon Press: Oxford, pp. 527–543

Hudson, N. (1992). *Land Husbandry.* B. T. Batsford: London

Hudson, N. (1995). *Soil Conservation.* Third Edition. Iowa State University Press: Ames

Hufschmidt, M. M. (1986). A conceptual framework for watershed management. In Easter, K. W., Dixon, J. A. and Hufschmidt, M. M. (eds.), *Watershed Resources Management.* Westview Press: Boulder, pp. 17–31

Hurni, H. (1988). Degradation and conservation of soil resources in the Ethiopian highlands. *Mountain Research and Development*, **8**(2/3): 123–30

Ives, J. D. and Ives, P. (1987). 1987: The Himalaya–Ganges Problem. Proceedings of a conference, Mohonk Mountain House, New Paltz, New York, USA. *Mountain Research and Development*, **7**(3): 181–344

Ives, J. D. and Messerli, B. (1989). *The Himalayan Dilemma.* Routledge: London

Kunkle, S. H. (1972). Sources and transport of bacterial indicators in rural streams. In *Proceedings, Interdisciplinary Aspects of Watershed Management.* American Society of Civil Engineers: New York, p. 31

Kunkle, S. H. and Thames, J. L. (1977). *Guidelines for Watershed Management.* FAO Conservation Guide No. 1. FAO: Rome

Liniger, H. (1995). *Endangered Water.* Development and Environment Reports No. 12, Institute of Geography, Bern

Marsh, G. P. (1864). *Man and Nature, or Physical Geography as Modified by Human Action.* Charles Scribner: New York

Megahan, W. F. (1981). Nonpoint source pollution from forestry activities in the western United States: results of recent research and research needs. In *US Forestry and Water Quality: What Course in the '80s?* Proceedings, Water Pollution Control Federation, Washington DC, pp. 92–151

Megahan, W. F. and King, P. N. (1985). Identification of critical areas on forest lands for control of nonpoint sources of pollution. *Environmental Management*, **9**(1): 7–18

Mellink, W., Rao, Y. S. and MacDicken, K. G. (eds.) (1991). *Agroforestry in Asia and the Pacific.* RAPA Publication 1991/5. FAO Regional Office: Bangkok

Messerli, B. and Winiger, M. (1992). Climate, environmental change, and resources of the African mountains from the Mediterranean to the equator. *Mountain Research and Development*, **12**(4): 315–36

Morgan, J. (1986). Watersheds in Hawai'i: an historical example of integrated management. In Easter, K. W., Dixon, J. A. and Hufschmidt, M. M. (eds.), *Watershed Resources Management*. Westview Press: Boulder, pp. 133–44

Morgan, R. P. C. (1986). *Soil Erosion and Conservation*. Longman: Burnt Mill, USA

Mosley, M. P. (1982). The effect of a New Zealand beech forest canopy on the kinetic energy of water drops and on surface erosion. *Earth Surface Processes and Landforms*, 7(2): 103–7

Myers, N. (1983). Tropical moist forests: over-exploited and under-utilized? *Forest Ecology and Management*, 6(1): 59–79

O'Loughlin, C. L. and Ziemer, R. R. (1982). The importance of root strength and deterioration rates upon edaphic stability in steepland forests. In Waring, R. H. (ed.), *Carbon Uptake and Allocation: A Key to Management of Subalpine Ecosystems*. Oregon State University: Corvallis, Oregon, pp. 70–8

Pearce, A. J. (1986). *Erosion and sedimentation*. Environment and Policy Institute Working Paper. East–West Center: Honolulu

Pereira, H. C. (1989). *Policy and Practice in the Management of Tropical Watersheds*. Belhaven Press: London

Ramakrishnan, P. S. (1992). *Shifting Agriculture and Sustainable Development. An Interdisciplinary Study from North-eastern India*. MAB Series Vol. 10. UNESCO/Parthenon: Paris/Carnforth, UK

Rao, Y. S. (1988). People's participation in upland conservation: regional overview. In *FAO/Finland Workshop on People's Participation in Upland Conservation*. FAO: Bangkok, pp. 14–18

Richards, M. (1997). The potential for economic evaluation of watershed protection in mountainous areas: A case study from Bolivia. *Mountain Research and Development*, 17 (1): 19–30

Ries, J. B. (1996). Landscape damage by skiing at the Schauinsland in the Black Forest, Germany. *Mountain Research and Development*, 16(1): 27–40

Rijskijk, A. and Bruijnzeel, L. A. (1991). Erosion, sediment yield and land-use patterns in the Upper Konto watershed, East Java, Indonesia. Part III, results of the 1989–1990 measuring campaign. *Konto River Project Communication Series No. 18*. DHV Consultants: The Netherlands

Ryder, R. (1994). Farmer perceptions of soils in the mountains of the Dominican Republic. *Mountain Research and Development*, 14(3): 261–6

Schemenauer, R. S., Frienzalida, H. and Cereceda, P. (1988). A neglected water resource: the Camanchaca of South America. *Bulletin of the American Meteorological Society*, 69(2): 138–47

Schreier, H., Shah, P. B. and Brown, S. (eds.) (1995). *Challenges in Mountain Resource Management in Nepal*. ICIMOD: Kathmandu

Schwab, G. O., Frevert, R. K., Edminster, T. W. and Barnes, K. K. (1981). *Soil and Water Conservation Engineering*. John Wiley: New York

Spears, J. (1982). Rehabilitating watersheds. *Finance and Development*, 19(1): 30–3

Stocking, M. A. (1975). Prediction and estimation of erosion in subtropical Africa: problems and prospects. *Proceedings: Symposium on Environmental Geomorphology in Tropical Regions*. Lubumbashi, pp. 161–74

Stvan, J. (1991). *Les Alpes Apprivoisées, Impact des Infrastructures Techniques sur l'Environment Alpin Suisse*. Institut Ecoplan: Geneva

Trustrum, N. A. and Lambert, M. G. (1982). Erosion and productivity. *Streamland*, No. 3. National Water and Soil Conservation Organization: Wellington

Trustrum, N. A. and Page, M. J. (1992). The long term erosion history of Lake Tutira watershed: implications for sustainable land use management. In Henriques, P. (ed.), *Proceedings of International Conference on Sustainable Land Management, November 17–23, 1991*. Napier: New Zealand, pp. 212–5

Valdiya, K. S. (1987). *Environmental Geology: Indian Context*. Tata-McGraw Hill: New Delhi

Warshall, P. (1980). Streaming wisdom: watershed consciousness in the twentieth century. In Brand, S. (ed.), *The Next Whole Earth Catalogue: Access to Tools*. Sausalito: California, pp. 64–7

Watters, R. F. (1971). *Shifting Cultivation in Latin America*. FAO Forestry Development Paper No. 17. FAO: Rome

Whiteman, P. T. S. (1988). Mountain agronomy in Ethiopia, Nepal and Pakistan. In Allan, N. J. R., Knapp, G. W. and Stadel, C. (eds.), *Human Impacts on Mountains*. Rowman and Littlefield: Totowa, New Jersey, pp. 57–82

Wiersum, K. F. (1984). Surface erosion under various tropical agroforestry systems. In O'Loughlin, C. L. and Pearce, A. J. (eds.), *Proceedings, Symposium on Effects of Forest Land Use on Erosion and Slope Stability*. East–West Center: Honolulu, pp. 231–9

Risk and disasters in mountain lands

16

Kenneth Hewitt

INTRODUCTION: DISASTER AND SUSTAINABILITY

The world's mountain lands include regions of exceptional risk for human activities and some unique dangers. We cannot say they are necessarily **more** dangerous than other regions. Well-adapted mountain societies have often benefited from greater security, more diverse resources, and a healthier environment compared to surrounding lowlands. Nevertheless, evidence assembled here suggests that many mountain regions have become increasingly disaster-prone in the twentieth century. Disproportionate numbers of natural disasters occur in mountain lands. In part that reflects the forces at work, or certain very destructive natural processes. They include the largest share of earthquakes, volcanic eruptions, large landslides, natural dams and dam bursts, flash floods, and threats associated with valley glaciers. Avalanches and catastrophic rockslides are almost entirely mountain hazards. Mainly, however, the forms and modern increases in danger relate to human activity, especially societal and ecological changes. These tend to put more people and property at risk. Habitat abuse and other societal processes have magnified or artificially released some of the more destructive forces, as with floods and landslides from deforested watersheds. Often they render mountain peoples more vulnerable to natural extremes.

Some technological risks are also more common in mountains. Notable disasters have been associated with mountain highways and aviation, dams, tunnels, tourist resorts and activities, mines and timber extraction. However, the worst calamities have been brought about by war and other uses of armed violence, whether measured by casualties, displacement of populations, economic, or habitat devastation (Chapter 6). More than 70 percent of almost 8 million war deaths in mountain lands since 1945 have been civilian residents. Indeed, it will be shown that mountain peoples have experienced a quite disproportionate share of these disasters; disproportionate, that is, in relation to land area, population, and their responsibility for the dangers involved.

Disease hazards may represent the greatest of all threats, in themselves or in association with other disasters. Some of the major twentieth century epidemics and pandemics have involved mountain populations. That includes influenza, AIDS, and other sexually transmitted diseases, measles, malaria, hepatitis, cholera, and other dysentery-causing diseases. However, we should not ignore the way various endemic diseases and chronic illnesses take an ever-larger toll in impoverished, crowded settlements, and afflict children especially. They also provide favourable conditions for harbouring and eventual outbreak of epidemic disease. If this contrasts with the historical association of highland regions with healthier environments and fewer diseases, it also relates mainly to humanly induced changes. Of particular concern are resurgent or emerging infectious diseases. Their spread or severity is aggravated by malnutrition, and they threaten epidemic outbreaks in association with other disasters, social upheavals and, especially, wars.

In that regard, mountain lands have had more than their share of 'complex emergencies'. These combine famine and disease outbreaks, armed conflict, economic collapse, huge displacements of populations and refugee crises (UNDHA, 1992). Rather than isolated emergencies, they involve whole countries and populations in a plethora of misfortunes. One can point out multiple, late-twentieth century examples in the highlands of East Africa and Central America. Others include the continuing calamities in former Yugoslavia, the Caucasus, Kurdistan, Afghanistan, Tajikistan, Kashmir, Myanmar, and several of the mountainous islands of Southeast Asia, notably East Timor. These are the largest of calamities in a global perspective. They dominate the demands upon humanitarian assistance and should do so, perhaps, in any agenda to reduce disaster in the mountains (Francois, 1992, 1993; International Federation of Red Cross and Red Crescent Societies, IFRCRCS, 1994). Disasters, especially of such scale and complexity, are of singular concern for sustainability. If sustainable development is an ecological and cultural possibility, recurring disasters are major threats or they record its failure. The indications are that the scope and incidence of most forms of disaster have become more severe through the century, though not equally everywhere or for all mountain peoples That seems to signal a pattern of unsustainable developments.

It is not intended, by such broad generalisations, to deny the actual diversity of mountain environments, nor the specific and variable risks that arise for each place and group of people. Diversity of conditions and cultures, after all, is a defining

Plate 16.1 Mountain hazards were one of the main preoccupations of mountain people and travellers. Often this knowledge and experience has been lost or overwhelmed by 'modern' technological progress and impatient tourist development – compare with Plate 16.2 (Engraving by E. Rittmeyer and W. Gregory in F. Tschudi, *Das Tierleben der Alpenwelt*, 1854)

feature of the highlands, whether viewed locally, regionally or globally. It is associated with very different kinds or degrees of danger and survival strategies in different parts of the world, indeed, within any community. In most cases, the victims of disaster are from among particular, vulnerable persons and groups. Some, if not all, are children and elderly, women, minorities, indigenous peoples, and less influential classes or regions, which prove to be unusually at risk. And that draws attention to the specific, pre-existing conditions of living in each place and society, rather than just to the disasters (Wisner, 1993; Blaikie *et al.*, 1994; Hewitt, 1996).

Equally compelling, however, is a recurrence of similar forms of disaster, of hazardous ecological and societal change, in many different settings. Risks related to forest destruction, toxic contaminants, motor vehicle emissions, armed violence, infectious diseases, or rapid urbanisation are world-wide problems. Growing and emerging threats in the mountains are associated with the penetration of similar technologies, administrative practices, economic exploitation, and militarisation. These create or aggravate similar dangers in widely separated regions. They reveal common forms of vulnerability in otherwise different peoples. Places and groups suffering unusual disadvantage related to age or gender, minority status or economic marginalisation, experience the same adverse shifts. This globalisation of risks has to be a central theme of an overview and assessment of the state of mountain land risks.

However, we can hardly proceed without some idea of the forms, distributions, and relative severity of disasters in mountain lands. With that in mind, the discussion will review and reflect upon records of destructive events in the twentieth century, or its last decades. That seems a necessary, if not always satisfactory, way to establish an empirical basis for identifying questions of interpretation, priorities, and responsibility.

DISASTERS IN MOUNTAIN LANDS

A useful starting point is the history of 'major disasters' in the world from 1900 to 1988 prepared for the United States' Office of Foreign Disaster Assistance Agency (OFDA, 1988). Events are classified by initiating agent or hazard, various damage criteria and country. Predominantly mountainous countries were identified, or those with populations concentrated in highland areas.

The distribution of disasters by initiating agent gives overwhelming importance to earthquakes and floods: about half the events and some 60 percent of natural disasters (Appendix 1). Atmospheric hazards of drought, tropical cyclones and other storms, and heat waves comprised about 15 percent; and geological hazards of volcanic eruption, landslides and snow avalanches about the same. 'Technological accidents' comprise about 7 percent and two types of 'social hazard', civil strife and displaced populations, just 3.5 percent of the events.

Another measure of the importance of different agents concerns the most severe disasters in each country (Table 16.1). Again, earthquakes dominate in every category of impact, followed by floods in all but one. However, droughts, civil strife, tropical cyclones and volcanic eruptions feature more prominently than their number of disasters suggests. That reflects the unusual risks they pose in particular regions.

The inventory probably accords fairly well with common perceptions of, and the bulk of research into, mountain land disasters. Various reasons why the classes of disasters and their importance may be misleading are addressed below. However, we will pursue this, recognising that most work has focused on the roles of the various major forms of damaging agent.

Plate 16.2 Krössbach in the Stubaital, Austria. A tourist development financed by inexperienced foreigners forced the state government to construct an expensive avalanche protection dam (Photograph: J. D. Ives)

Natural disasters

Earthquake and flood are the natural hazards that initiate the largest numbers of major disasters in mountain lands and cause the greatest damages. They have also been areas of huge research, technical, and managerial efforts in wealthy mountain countries. In regions with recurring earthquake disasters, such as Japan or California, there has been some limited success in disaster mitigation (Alexander, 1993). However, most of this work treats the problem as one of the incidence and nature of seismic and hydrological events, and modern counter-measures: much less of the societal and habitat conditions in the places where disasters

Table 16.1 Disasters in mountain lands, 1900–1988: number of events ranked by largest impact in each country, by estimates of death toll, numbers made homeless, total number of persons affected, and cost, from data in Appendix 1 (after OFDA, 1988)

Initiating agent	Largest death toll	Made homeless	Persons affected	Economic losses
Earthquake	17	16	8	13
Volcanic eruption	5	3	3	2
Technological accident	2		2	
Civil strife	2		4	2
Drought + famine	2		5	3
Avalanche	2		1	
Landslide	1		1	
Flood	1	12	7	8
Fire	1			1
Tropical cyclone	1	1	2	2
Displaced persons	1	1		
Heat wave	1			
Storm			1	1

occur and in terms of the global majorities of persons and property at risk. The role of mountain land environments has been considered mainly in terms of higher levels of seismicity, active faulting, or flood-prone hydrology. Yet there are other specific, even decisive, conditions affecting the forms and degrees of risk from earthquake or flood in mountains. They include habitat conditions relating to topography, vegetation cover, ground and slope stability, season and microclimate. In addition, cultural conditions and their transformations are the keys to whom and what will suffer damage in disaster. The case of earthquake is examined in this light (Box 16.1).

While giving pre-eminence to natural disasters, the OFDA data underestimates the role of certain natural processes in the mountains. The global importance of slope failure and destructive mass movements is surely missed (Eisbacher and Clague, 1985; Schuster and Fleming, 1986). In part that is because these have their greatest and more catastrophic incidence in earthquake, flood, and severe storm events. In fact, they substantially decide the destructiveness of those agents in the mountains, and are often the main source of damage in disasters (Keefer, 1984; Hewitt, 1984).

Box 16.1   Earthquake disasters in the mountains – intervening factors and vulnerability in natural hazards

Earthquake is the natural hazard identified with the most lethal, destructive and expensive disasters in the world's mountain regions (see text Tables 16.1 and 16.4). Meanwhile, at least nine out of ten earthquake disasters occur partly or wholly in mountain lands (Table 16.2). The damage zones in 236 inidents, or 97 percent, involved areas where the local relief was at least 1000 m. The greatest earthquake disasters may be those that strike the cities of the plains, or densely settled agricultural lowlands. But the majority of damaging earthquakes affect the people of mountain and mountain fringe areas.

Of primary importance, of course, is the association of seismic activity with mountain zones; most earthquakes being related to the tectonic forces involved in mountain building. No doubt that is a main reason why so many earthquake disasters occur in and around the mountains. Nevertheless, a more careful look at earthquake disasters and damages indicates that there is only a weak relation between the magnitude and frequency of earthquakes within mountain lands, and the scale and incidence of disasters. While the greatest potential danger lies in the largest magnitude events, in most parts of the world and most years, the worst disasters are often associated with more moderate earthquakes. That reflects the fact that, in the mountains especially, other aspects of the environment and society intervene to decide whether an earthquake will cause damage and, if so, how severely, where, and to whom.

Foremost among the conditions that influence earthquake risk are factors affecting the stability of the land surface and the vulnerability of built structures. These intervening factors are both fundamental to the dangers posed by earthquakes and most of them are conditions or constraints in which societal conditions and human activity play a large role. The stability of soils and slopes are crucial in earthquake risk. Much, and often the greater part of, damage in earthquakes is due to the collapse of foundations or of slopes, and destructive mass movements that follow. These, or how human settlement and activity relate to them, are decisive for danger and damage in mountain lands.

Steep slopes, and variable patterns in the types and stability of surface materials, are the rule in steepland areas. Their inherent stability or instability is largely dependent upon natural conditions discussed elsewhere in this volume. But human activity may alter their stability by ill-considered action, for instance removing stabilising vegetation or undermining slopes through road building and other types of construction. More generally, the patterns and degrees of risk depend upon the geography of settlement and land uses. These decide whether people and property are located in places and on sites that are, or are not, relatively stable.

Earthquake is, above all, a threat to the built environment. A majority of those who suffer death or injury in earthquakes do so as a result of the collapse of buildings, most often their own homes. The major public costs are from destruction of the built infrastructure, notably roads. The siting and quality of buildings are, therefore, always a decisive factor in earthquake risk. In the mountains this becomes both a more critical and more difficult problem. We must highlight the combinations of steep slopes, complex and variable bedrock and superficial deposits and associated water tables, and such factors as uneven or degraded vegetation covers, variable and sometimes severe weather, and dangers from earthquake-generated hazards such as landslides or natural damming. These magnify the demands upon sensitive location and appropriate construction or land use. Historic patterns of

Table 16.2 Earthquake disasters: selected examples of large earthquake disasters affecting mountainous areas, especially the most lethal, with estimates of losses and indications of the scale and scope of damage conditions

Place	Date	Disaster features	Deaths
Italy, Sicily and Calabria	1908 Dec 28	Worst twentieth century disaster in Italy. Greatest devastation in Messina with 98 percent homes wrecked, most public and commercial buildings collapsed, and at least 45 200 dead, possibly 58 000. Almost as severe in Reggio di Calabria with about 10 000 deaths. Great fires cause much damage, and tsunamis along coasts drowning 220 persons.	47 000–82 000 (?)
China, Kansu	1920 Dec 16	One of most severe this century, and the second most lethal. Ten ancient cities and thousands of villages razed. Immense landslides buried whole communities and dammed rivers. Many thousands perished from exposure to freezing temperatures.	180 000
Japan, Kanto Plain	1923 Sept 1–2	Earthquake and strong winds fanning mass fires in congested areas of capital city and port city of Yokohama caused most lethal Japanese earthquake, injured 200 000 and destroyed homes of 500 000. 44 percent of Tokyo and 28 percent of Yokohama burnt out. 'Firestorm' engulfed and burnt to death 40 000 sheltering in military clothing depot. Thousands also died seeking safety from fires in canals and waterfront when oil storage depot exploded and spilled flaming oil into bay. 1.7 million had homes destroyed or severely damaged. Inland great devastation from ground failure and landslides, and from tsunamis along shorelines.	143 000
Pakistan Quetta (British India)	1935 May 30	Great devastation in capital of Baluchistan to poorly constructed buildings on weak alluvium of sediment fans; landslide devastation in surrounding mountains. Ancient irrigation systems (Qanats) severely damaged.	30 000
Chile	1939 Jan 25	Second most lethal in Western Hemisphere, and *70 percent of the dead were children*, while 60 000 persons injured, 700 000 lost their homes. Five major centres devastated. Chillan, closest to epicentre, razed except three buildings, and Concepción 70 percent destroyed.	50 000
Turkey, Erzincan	1939 Dec 27	Most lethal known in this country, with hundreds of thousands injured and made homeless. Erzincan city razed, all of its doctors and nurses killed. Large numbers of deaths due to severe winter weather and blizzard immediately following earthquake.	50 000
Turkmenistan and Iran, Ashkhabad Koppet Dagh	1984 Oct 5		19 800
Iran (NE) Dasht-e Bayaz	1968 Aug 31	Some 200 towns and villages damaged or razed, leaving 75 000+ homeless. In Dasht-e Bayaz 3000 of 3500 people killed, and 4000 of 6500 in Kalhk which was totally destroyed. Damages attributed partly to poor adobe construction but also to location on alluvium with high water tables.	10 000

Table 16.2 *Continued*

Place	Date	Disaster features	Deaths
Peru, Pacific coast to Cordillera Blanca	1970 May 31	Most lethal in Western Hemisphere. Also 50 000 injured and 200 000 left homeless. Major losses in port cities, especially Chimbote, but greatest casualties, disruption, and devastation due to countless landslides in mountainous interior. Includes the most lethal example known, a catastrophic avalanche-debris flow which totally annihilated and buried the Andean town of Yungay and killed about 18 000 persons. Cold, destroyed communications, and inadequate relief measures added to plight of survivors.	70 000
Iran, Ghir	1972 Apr 10	Lethal and destructive through Fars Province. Many landslides.	17 000
Guatemala, Guatemala City	1976 Feb 4	More than 1 million people made homeless, 70 000 injured. Majority poor residents and squatter areas on steep bluffs and unstable volcanic ash soils in Guatemala City, but also in poorer mountainous, rural areas. Note that Guatemala City underwent near total devastation on three previous occasions.	22 000
Mexico, Mexico City	1985 Sept 19–20	Widespread damage along Pacific coast but most and worst in Mexico City. Injured 40 000 and left 30 000+ homeless in capital. Collapse of one apartment building, Nuevo Leon, killed nearly everyone in 200 families. The General Hospital collapsed burying 600 staff and patients in rubble. Losses of US$4 billion.	7000+
USSR, Armenia	1988 Dec 12	Major devastation and most casualties in towns of Leninakan, Spitak and Kirovakan. Dozens of villages badly damaged. 130 000 injured, 250 000+ homeless. Poor official emergency response. US$11 billion in losses.	55 000
Iran (NW), Caspian Sea area	1990 June 21	105 000 injured and 500 000 made homeless. Devastated 100+ towns and villages. Major landslide damage. Dam burst near hardest-hit town of Rasht causing deaths and destruction.	40 000
California, Northridge	1994 Jan 17	Thousands injured, 25 000 homeless, six important freeways and highway intersections closed. Tens of thousands without water or electricity. Estimated losses US$17 billion. Federal emergency fund requested US$6.6 billion.	56
Japan, Kobe	1995 Jan 17	Massive devastation to port city including homes, railways, freeways, factories, port buildings. Major fires raged in residential areas. 50 000 injured, 300 000 made homeless. Damages US$100 billion+. Essentially due to ill-advised development on coastal fill. Negligible damage in mountainous hinterland.	6000+

mountain settlement show that many societies have been well-aware of these dangers. They have adopted building methods and settlement sites that will minimise them. The modern world and the worst losses in disasters seem to reflect world-wide and growing developments that ignore the special dangers from mountain land conditions. Rather, earthquake dangers are increased by vulnerable construction on dangerous sites in cities and in the countryside. Moreover, it is society, its members or the constraints imposed upon them by socio-economic conditions, that have the main responsibility for the quality and siting of built structures.

The global inventory of the most destructive earthquakes reported between January 1950 and December 31st 1990, offers a preliminary basis for illustrating the roles of these intervening and societal conditions of earthquake risk. Using such reports as are available in the seismological and disasters literatures, supplemented with contemporary news reports, one can identify conditions repeatedly associated with the disasters, or upon which damage

Plate 16.3 In Armenia in the so-called 'land of stone', an earthquake on 7 December 1988 completely destroyed the town of Spitak, and killed 3600 people, 20 percent of the inhabitants (Photograph: J. D. Ives)

was blamed. Twelve parameters emerged as important in a global sense (Table 16.3). The data are certainly incomplete, but point toward the crucial factors in, and a potential ranking of them for, earthquake risk.

The first attribute, as already described, indicates whether damage occurred partly or wholly within areas of high relief. The second is less obvious, but it is of great significance for identifying the nature of risks and unsustainable developments in mountain lands. This is the frequent association of the worst damages with settlements in mountain fringe, or mountain foot and intermontane basin areas; that includes cities, towns and villages. It is suggested that this concentration of danger and damages in mountain foot or fringe areas reflects the overlap of several important ingredients of earthquake risk. These are locations within or close to the higher seismicity of the mountain belt, but with greater opportunities for denser settlement, often higher value land uses and buildings, compared to more rugged hinterlands. In the modern world, these are also the places with the greatest concentrations of development, urbanisation, and other intensifications of land use associated with mountain lands. Here too, are the towns and institutions with major responsibilities for influences upon (the highland–lowland relations and governance of) the mountain hinterlands. Not only are damages often greatest in these places, but it can critically affect the well-being of the mountainous interiors, or responsiveness to damage in them.

The third ranking attribute identified in these disasters is related to the above, but involves the surprising discovery that a majority of earthquake damage zones occur where climate is semi-arid or sub-humid, usually with a pronounced dry season. For example, an exceptional number, in comparison to seismicity, land area, and human populations, occur in 'Mediterranean' and 'sub-mediterranean' climatic regimes – mid-latitude areas with summer drought and winter rains or snowfall. This is a surprising, or counter-intuitive discovery. That is partly because of the common association of mountains with higher humidity, partly because there is no known causal relation between earthquake and climate. It is suggested that it reflects the special advantages of mountain foot settlement in mountain lands generally, and those with surrounding dryland areas in particular. And these advantages are associated with ancient patterns of human settlement throughout the world.

Of particular note are settlement patterns in the Asian and Mediterranean mountain lands, from western China and northern Pakistan, through Afghanistan and Iran, Turkey and Greece, to Algeria and Morocco. This is the zone associated with the greatest numbers of earthquake disasters. It is also the belt of Mediterranean and sub-mediterranean climates. Semi-arid or arid lowlands often lie below the mountains. Maps of population density, towns, and cities in this belt, show the bulk of the people to be in mountain foot and intermontane basin locations. Similar historic patterns of settlement are found in California, Central America, and the drier parts of the Andes.

Closely related to this is another geographical and habitat correlate of the disasters (Rank 6), damage in settlements in coastal mountains, most often coastline communities living at the junction of mountains and sea. Again, there are many particular attractions for settlement. The Mediterranean lands and related climates elsewhere in the world are strongly represented in the earthquake disasters affecting such areas.

These relationships to the natural environment may seem merely to suggest an expanded 'environmental determinist view' – one in which earthquake risk is partly a matter of seismic conditions, partly of other environmental controls upon human settlement. However, the set of conditions identified as Rank 4 indicate the limitations, if not the fallacy, of such a deterministic view. The fourth factor records the disasters where reports attribute actual damages

377

Table 16.3 Disaster geography: geographical and damage attributes of the 243 most destructive earthquake disasters reported between 1 January, 1950 and 31 December, 1990 (after Hewitt, 1992)[1]

Rank	Damage-associated attribute[2]	n (total 243)	% of cases
1	Mountain land	236	97.12
2	Piedmont/intermontane basin	183	75.31
3	'Dry land'	173	71.79
4	'Development' and change stresses	143	58.85
5	Landsliding/slope instability	106	43.62
6	Coastal mountain settlements	73	30.04
7	'Susceptible' regolith	72	29.63
8	Mountainous interior settlements	68	27.98
9	Severe weather/exposure	42	17.28
10	Tsunami damage	26	10.70
11	Damming/dam bursts	20	8.23
12	Fires	12	4.94

[1]The examples are counted only where reports indicate substantial damage associated with the attribute. In many cases, however, there is only a verbal report, rather than a careful survey. Data must, therefore, be considered incomplete for all attributes except 1, 2 and 3, which are derived from cartographic location.
[2]Attributes of Rank:

(1) *Mountain land*: damage occurs partly or wholly in areas with local relief exceeding 1000 m;

(2) *Piedmont/intermontane basin*: substantial and often most damage in settlements located in mountain foot, foothills or intermontane basin sites;

(3) *'Dry land' relation*: where regional climate of the damage zone has an arid to subhumid climate, with a seasonal (145 events) or perennial (28 events) moisture deficit, and specifically with a Budyko–Lettau Dryness Ratio of 1 or greater (Hewitt, 1982);

(4) *'Development' stresses*: where recent and on-going changes in settlement, economic activity and habitat are prominent features of the damaged area, and cited as causes of vulnerability in reports;

(5 and 7) *'Microseismic' responses*: slope and earth surface materials that prove unstable under seismic shaking, resulting in substantial damage;

(6) *Coastal settlements*: where substantial, or all, damage reported is in the mountainous coastline settlements;

(8) *Mountain interior*: where substantial damage is reported in high-relief and steep slope areas;

(9) *Severe weather*: where substantial damage or human casualties and hardship are reported from weather conditions and exposure following the earthquakes;

(10) *Tsunami damage*: substantial damage to coastal settlements from seismic water waves;

(11) *Damming/dam bursts*: reports of devastation due to flooding induced by earthquake-triggered mass movements and dam-break floods;

(12) *Fires*: fire damage reported following earthquakes. Some of the most devastating urban earthquake disasters have involved mass fires, but they are relatively rare in this set.

to recent, usually rapid and on-going social and humanly-induced environmental change in the area of damage, or where there is independent evidence of that. The damages in at least half of all earthquake disasters are identified with recent construction, land-use changes, habitat abuse, or deteriorating social and economic conditions. It is almost certain that we underestimate the importance of this because of the lack of positive evidence. It is always an important factor where there is detailed evidence of the forms and distribution of damages. This is where natural hazards and disasters involve and strongly overlap with other environmental and societal crises in the mountains. People whose lives, homes, and livelihoods are damaged or destroyed by earthquake are most often those suffering from other forms of harm or stresses that make them vulnerable to earthquake (Jones and Tomazevic, 1982; Wijkman and Timberlake, 1984; Wisner, 1993).

The same arguments are relevant to the other factors of risk. On the one hand, all of them except, possibly Rank 12, involve environmental variables. Their importance for our understanding lies partly in the way they reflect conditions that are special to, or particularly marked in, mountain environments. That is: conditions identified with the steep slopes and complex patterns of weak soils deposited below the slopes or along streams and coastal zones.

They include problems of changeable and severe weather; tsunamis along mountainous coastlines, and the dangers of large, earthquake-triggered landslides and their damming rivers (*cf.* Box 16.2). On the other hand, in any given community or area of settlement, there are always sites, types of building, or forms of land use not at risk, or not as severely endangered, by these conditions. Developments in the socio-economic domain tend to ensure, or fail to ensure, that people and property stay in safer areas, or have the protection and other social benefits where they are required to take greater risks. Invariably, in fact, the people and buildings who survive earthquake disaster, with little or no direct harm, occupied safer sites or less vulnerable structures, or had the benefit of rapid and effective assistance in the emergency. The opposite is generally true of those who are harmed.

The history of earthquake disasters, and especially those in the twentieth century, reveal the consequences of construction, land uses, and activities that do not take sufficient account of the intervening environmental factors in the mountains that may magnify or reduce earthquake dangers. Most mountains appear to be undergoing developments that increase the numbers of vulnerable persons and property on sites that are unstable in relation to earthquake shaking, and poorly served by other measures that might reduce the dangers. That applies especially to many rapidly growing cities, or parts of them, but also to expanding agricultural villages, modern infrastructure, resource, tourist, and military uses of the mountains. To find ways to reduce such extreme risks as earthquake disasters pose is integral to the broader problem of sustainability of habitats and cultures in the mountains and, in many instances, social developments and habitat relations that are sustainable in other respects are likely to improve earthquake resistance, or to be aided by developments specifically intended to do that.

Source: Kenneth Hewitt, Cold Regions Research Centre, Wilfrid Laurier University, Waterloo, Ontario

An attempt to provide a comprehensive survey of the occurrence and economic significance of landslides showed that mountainous terrain, and to a lesser extent hilly areas and steepsided valleys, dominate their role world-wide (Brabb and Harrod, 1989). Densities of some 1–10 landslides per square kilometre apply in the mountainous and hilly areas of the United States, suggesting that perhaps 20 million individual events have affected its present land surface (Brabb and Harrod, 1989: p.31).

However, the damage toll in most individual destructive landslides is usually much smaller than those in a 'major disasters' inventory and even the catastrophic examples rarely affect more than a few square kilometres. If the natural process is more extensive, damage is usually confined to a tiny part of its path, most of which tends to lie in steep, rugged, and uninhabited terrain. Hence, though devastating to everything they touch, landslide damages rarely appear large in global terms. Awareness of the problem can be improved by including events with a lower threshold of loss (Table 16.4). In this inventory of natural disasters, earthquake and flood still remain predominant. But the number and proportion of disasters identified with landslide and avalanche appear much greater.

Damages must also be considered relative to the size and capacities of the social unit at risk and the frequency with which they are impacted. Mountain lands generally have more dispersed and smaller pockets of population and property. Landslide events are relatively more frequent and widespread. Meanwhile, one needs a longer historical perspective to obtain a balanced view of the most catastrophic events. The detailed inventory and case study by Li Tianchi for China illustrates how such events must be recognised as perennial, if relatively infrequent, sources of major disasters throughout mountain ranges (Box 16.2).

In general, it would be a serious error to accept the apparently small importance of slope instability and landslides suggested by the OFDA inventory. Indeed, the prevalence of steep slopes is the single most important environmental hazard in mountains. Directly it creates a pervasive danger of slope failure. It is the main contributor to the higher frequencies, greater power and magnitude of earth surface processes, especially mass movements. Indirectly, it exists as a constraint upon most activities and endangers areas of lower slope where settlement and communication routes are likely to be placed. Finally, the inventory does not recognise specific types of destructive 'landslide', such as debris flows, rock avalanches, or earth slumps (Voight, 1978; Bonnard, 1991; Walling *et al.*, 1992). However, the conditions and risks they involve are distinct, and vary greatly between mountain regions, particular ones assuming major

Table 16.4 Distribution of major natural disasters in Mountain Regions 1953–1988

Region	Natural Agent						
	Earthquake	Volcano	Landslide	Avalanche	Flood	Storm	TOTAL
Mediterranean Basin (excl. Turkey)	21	1	4	–	13	1	40
SW, S Asia	49	–	15	2	44	3	113
E Asia	11	2	19	2	14	10	58
SE Asia, Australia and Oceania	21	8	7	1	9	18	64
Africa	3	2	1	–	5	2	13
Europe (excl. Mediterranean)	2	–	4	6	5	2	19
SC America	33	5	16	7	22	4	87
North America	14	1	4	–	6	1	26
TOTALS	154	19	70	18	118	41	

Note: The events are those reported in the *New York Times*, *Globe* and *Mail* (Toronto), and *The Times* of London. Thresholds of damage were at least 10 deaths and 50 injuries, or more than US$ 1 million damage and emergency assistance from outside the damage zone

Box 16.2 Mountain hazards in China

China has a population of more than one billion and a land area of about 9.6 million km². The country comprises both an extensive land mass adjacent to the Pacific Ocean and a huge continental area within Asia. Moreover, the larger fraction of this land area, or 66.5 percent, is mountainous. The mountains contain 33 percent of the population and 40 percent of the cultivable land.

Geologically, Chinese mountain areas are very complex. Rocks range in age from the Precambrian to the Holocene, and there have been repeated, large deformations in tectonic episodes from the Palaeozoic to the present. Rugged topography, in combination with commonly occurring earthquakes and high intensity rainfall, areas of weak earth materials and complex geological structure, contribute to significant mountain hazards. In the higher, inner Asian mountains, heavy snowfall, meltwater and glacier dam-burst floods, can be important.

Landslides, debris flows, torrents, and floods from the failure of natural dams are the main types of natural hazards in the mountains. However, in the past 50 years, at least, other factors have contributed. Population pressure has forced the expansion of agriculture on to the steeper slopes, often at the expense of forests. At the same time, investment in development projects, such as road and reservoir construction, or minerals and forest exploitation, have accelerated some mountain hazards and increased the people and property at risk. That, in turn, is reflected in increased damages. This brief overview will focus especially on the problem of destructive mass movements.

Landslides are mainly caused by earthquakes and heavy rainfall, although the full impact of the landslides is often masked by other earthquake- and flood-induced damages and lumped together with them in gross statistics of loss (*cf.* Chapter text; Tianchi, 1986, 1989, 1990, 1992). Some of the most lethal landslides have been in certain steepland areas of the well-known loess deposits. The instability of the material, great size and rapid travel of the landslips generated by earthquakes, are important factors. Even more important are relatively high densities of human settlement. In the twentieth century, the greatest number of deaths due to landslides occurred in the 1920 Haiyun, Ningxia earthquake. A series of massive loess landslides triggered by this Richter Magnitude 8.5 earthquake killed at least 100 000 persons, or about half of the total death toll in the disaster (*cf.* Box 16.1).

The Qinghai-Xizang Plateau includes, or is adjacent to, the highest and most rugged mountains, such as the Kun Lun and Karakoram Himalaya, where some of the most extreme mass movement activity takes place. However, large landslides and floods of these mountains rarely feature in major disasters because they occur in sparsely populated high mountain and barren plateau areas. A notable exception was in 1954, when a glacier dam-burst flood occurred at Sewang in Jiangzhi district. It killed at least 450 people.

In many mountainous areas, damages to transportation facilities, especially railways and highways, are a significant cost of landslides. One report, for the years 1974–76, recorded more than 1000 landslides of medium or large scale along China's railways. The estimated cost of stabilising them was US$2 billion. Landslides along a 154 km section of the Baoji to Tanishu railway alone, in the three decades to 1989, interrupted traffic for almost 4700 h and involved repair costs of some US$675 million. Landslide costs for highways may be greater, given their greater extent in the mountains and heavier traffic. Much of the 2413 km Sichuan-Xizang Highway, from Chengdu to Lhasa, passes through mountainous areas. It is seriously affected by landslides, avalanches, and debris flows in every year.

River traffic is also of major importance for China's economy and includes large rivers that pass through mountainous terrain. Landslides which enter and block, or partly block, their flow can be serious hazards for waterborne traffic. Part of a large destructive landslide in the Yunyang district of Sichuan in 1982 slid into the Changjiang River. It disrupted traffic on the river for four years. Costs of stabilisation and dredging to restore navigation were US$32 million. At least five major landslides are known to have blocked navigation on this important river during the past 1700 years. Some, that also created landslide lakes, obstructed movement for several decades. The worst, a great landslide of 1542, obstructed navigation for 82 years and prevented it entirely in the dry season.

Some of the worst destruction and loss of life is associated with landslides that dam rivers, and especially in floods from the sudden failure of the dams. Large landslides have frequently dammed the narrow, steep-walled headwater streams of the Yangtze and Yellow Rivers or their tributaries. The Yangtze is China's largest river and among the largest in the world. It is notorious for historic outburst flood disasters. In 1786, its tribu-

tary the Dadu River, was dammed by an earthquake-induced rockslide/avalanche. Catastrophic breaching after ten days resulted in a flood that killed 7000 persons and caused damage for more than 1000 km downstream. Five major blockages have occurred on the Yangtze's mountainous tributaries in the twentieth century. In August 1933, a series of landslides triggered by an earthquake dammed the Min River gorges at several places. They resulted in a complex pattern of lake formation, dam failure and downstream flooding. The worst disaster occurred in October with the failure of the Deixi Lake dam. That generated a flood wave that was 60 m high 3 km below the dam, and travelling at as much as 30 km an hour. It killed at least 2423 persons. These are also examples of natural hazards in the mountains that may have calamitous results and their greatest costs in lowlands downstream.

On the basis of published and unpublished data, the author has estimated that, between 1951 and 1987, the direct and indirect losses due to landslides exceeded US$0.5 billion annually. Nearly all of these costs were in mountainous areas. The exceptional historical records available in China also show that mountain hazards have been a major source of social and economic loss and death in the mountain areas for at least 4000 years. Some of the better-documented examples of major catastrophes are presented in Table 16.5, below.

Source: Li Tianchi, ICIMOD,
PO Box 3226, Kathmandu, Nepal

Table 16.5 Deaths in China caused by forty-two major mountain disasters

Year	Province (Autonomous region)	Affected areas	Type of mountain hazard	No. of deaths
186 BC	Gansu	Wudu	Rock and debris avalanche	760
100	Hubei	Zigui	Rockslide and avalanche	> 100
689	Shaanxi	Huaxian	Loess and rockslide	> 100
1072	Shaanxi	Huaxian	Rockslide and avalanche	> 900
1310	Hubei	Zigui	Rockslide and avalanche	3466
1558	Hubei	Zigui	Rockslide and avalanche	> 300
1561	Hubei	Zigui	Rockslide and avalanche	> 1000
1718	Gansu	Tongwei	Earthquake-induced landslide	40 000
1786	Sichuan	Luding	Flood resulting from landslide dam failure	100 000
1847	Qinghai	Beichuan	Loess and rockslide	hundreds
1856	Sichuan	Qianjiang	Rockslide induced by earthquake	> 1000
1870	Sichuan	Batang	Rockslide induced by earthquake	> 2000
1888	Beijing	Fangshan	Debris flow	> 1000
1891	Sichuan	Xichang	Debris flow	> 1000
1897	Gansu	Ningyuan	Loess and rockslide	> 100
1917	Yunnan	Daguan	Rockslide	1800
1920	Ningxia	Haiyuan	Loess landslide induced	100 000
1926	Sichuan	Ganlu	Debris flow	230

Table 16.5 *Continued*

Year	Province (Autonomous region)	Affected areas	Type of mountain hazard	No. of deaths
1933	Sichuan	Maowen	Flood resulting from landslide dam failure	2429
1935	Sichuan	Huili	Rock and debris slide	250
1943	Qinghai	Gonghe	Loess and mudstone slide	123
1951	Taiwan	Tsao-Ling	Flood caused by landslide dam failure	154
1954	Xinzang (Tibet)	Jiangzhi	Flood caused by glacier dam failure	450
1964	Gansu	Lanzhou	Landslide and debris flow	137
1965	Yunnan	Laquan	Rockslide	444
1965	Sichuan	Huidong	Rock landslide	400
1966	Gansu	Lanzhou	landslide and debris flow	134
1968	Sichuan	Yuexi	Debris flow	120
1970	Sichuan	Mianning	Debris flow	102
1972	Sichuan	Lugu	Debris flow	123
1974	Sichuan	Nanjiang	Landslide	195
1975	Gansu	Zhuang-lang	Loess slide caused flooding along the shores of the reservoir and downstream	> 500
1971	Sichuan	Ya'an	Debris flow	159
1980	Hubei	Yuan'an	Rockslide and avalanche	284
1981	Sichuan	Ganlu	Debris flow	360
1983	Gansu	Dong-xiang	Loess landslide	277
1984	Yunnan	Yinmin	Debris flow	121
1984	Sichuan	Ganlu	Debris flow	> 300
1987	Sichuan	Wuxi	Rock avalanche	102
1989	Sichuan	Xikou	Rockslide and debris flow	221
1996	Yunnan	Zhaotong	Landslide	216
1996	Yunnan	Yuanyan	Landslide	156

Source: Collated by author from historical and technical records

importance in some regions (Coates, 1977; Brabb and Harrod, 1989).

Some classes of event are much more important in particular regions, for instance, disasters associated with tropical cyclones in the mountainous islands and coasts of Southeast Asia and the Caribbean. The threats from volcanic eruption also warrant more emphasis than is suggested by their share of major disasters. Volcanic hazards are largely mountain land risks. A majority of all dangerous eruptions occur in mountain lands, particularly the cordilleras and mountainous islands rimming the Pacific Ocean, the Mediterranean and Caribbean seas. Meanwhile, the distinctive varieties of hazardous agent need to be brought out. Explosive eruptions, and the high speed processes of *nuées ardentes* ('glowing ash clouds'), hot volcanic mudflows (*lahars*), and pyroclastic (hot ash) flows, threaten catastrophic devastation, if relatively rare in time and space. They pose unique dangers in specific volcanic regions and for settlements on particular volcanoes. Even in a case like Mount St Helen's, 1980, which was predicted well in advance and was the most thoroughly monitored eruption in history, the catastrophic processes surprised observers and overwhelmed people who thought they were prepared. The event showed how the volcanic mountain itself and steepland environment generally, contribute to the exceptional power and reach of these hazards (Sheets and Grayson, 1979; Blong, 1984; McCall *et al.*, 1992). Other natural hazards that pose singular threats in some regions are hardly represented in the global inventory. There is little indication that snow and glacier-related hazards are important, such as ice dams and catastrophic outburst floods, glacier surges and ice avalanches (Tables 16.1 and 16.2). However, these are processes and events that have received a good deal of attention by the mountain research community (Hewitt *et al.*, 1994; Price, 1981; Young, 1993). It may be salutary to acknowledge that they do fall well behind some other dangers in overall threats to mountain lands.

It must also be recognised that the balance of research and preparedness in mountains, as elsewhere, has not been decided by the global scope

Plate 16.4   The eruption of Lascar in Chile in 1993, produced a heavy fallout up to a distance of 200 km on the Argentinian side of the Andes (Photograph: Mina et Laco)

and scale of disasters. For the most part, as has been said of development, disaster studies and mitigations have been deployed '. . . according to how the rich nations feel' and, more generally, the pre-occupations of the richer, more powerful interests and sciences in each country and economy (Sachs, 1990: p.26; Hewitt, 1995). Moreover, if debris flows, avalanches, fog, and glacier hazards rarely bring great disasters, there are places where they do threaten some of the highest profile enterprises and groups. Notable among them are the users of modern transportation systems and activities dependent on them, most obviously strategic and political control, resource extraction, tourism and recreation. This is especially clear in the case of snow avalanches and the 'modernisation' of risks from them.

Furthermore, it is the hazards of the higher and more rugged areas, the treeline and alpine zones, the snow and ice environments, that have attracted a majority of mountain researchers and given a distinctive orientation to our work. Of course, most mountain people and activities are not in those areas, but in the less rugged ones. Their

settlements and land uses are concentrated on the lower slopes; on sediment fans, terraces, valley and intermontane basin floors; in foothills, piedmont and plateau areas. Assessments of risk that fail to reflect this can give a less-than-balanced sense of the overall dangers and exposure of people to hazards. The inventory of major disasters and broad national units at least has the virtue of taking that into account. Indeed, the picture for earthquake disasters suggests that a majority of events cause most destruction to settlements in the valley floor and mountain fringe areas, where the steepland environment is moderated (see Box 16.1).

Another set of problems arises from fluctuations in environmental conditions, especially climate change, and for risk assessment changes in human land uses. These can alter both the occurrence, severity, and forms of hazards, and the forms and extent of human exposure to them. As a result, the experience of recent generations, or inventories of disasters covering even a century or two, may be unrepresentative of what coming generations will face.

For example, the greatest historical floods on the Himalayan Indus streams have been outbursts from glacier and landslide dams (Hewitt, 1982). However, beyond minor, local events, there has been no dam burst flood disaster in almost seventy years. Should we now ignore the danger? The people and wealth exposed to these flood waves has increased enormously since the last occurrences.

On a millennial time scale, there have been dozens of catastrophic rockslides in the habitable areas of the same region. Many of the present day towns and villages lie amid the rubble of those events. But none has affected the habitable zone since an earthquake-triggered rock avalanche in 1841. Most others are prehistoric. Does that mean they are only dangers of distant times and conditions? Indications are that they may have occurred as series of events triggered by more extreme climatic or seismic activity. The potential for destruction by such events is enormous.

Problems like this, of rare but 'worst-case' disasters, are inherent in the exceptional forces that may be released in high mountains. However, the difficulties with, say, climate fluctuations, pale before those of societal changes and habitat abuse; the numbers of vulnerable people and structures newly exposed to greater risks, and the disasters that result.

A final problem with the evidence presented so far is that, above all, it ignores or underestimates mountain land threats other than natural hazards. Of outstanding concern are disasters associated with displacement of populations, armed violence, and epidemic disease.

Displacement of populations

While the OFDA survey includes displacements of population, it surely under-estimates their full significance. In 1983, well within the time frame considered, some 7.5 million acknowledged refugees were located in the 25 mountain countries that gave asylum to at least 1000 (World Refugee Survey, 1987). Nine countries hosted over 100 000 refugees. Countries whence large numbers had fled included 18 mountain lands. Even larger numbers were displaced within their own countries. By 1988 refugee populations involving mountain lands had surpassed 10 million, and have continued to grow through the 1990s.

The conflicts in former Yugoslavia drove 1.5 million people from their homes. In 1993, in the three states of Georgia, Armenia, and Azerbaijan,

there were some 1.3 million displaced persons, while 600 000 fled to other countries. Even greater numbers of Afghan refugees remain in camps, continuing a desperate crisis that is more than a decade old. And these were just a small part of the scope and scale of the problem world-wide (Table 16.6). The share of such tragedies borne by mountain lands continues to be staggering. Moreover, official reports do not include some vast, irregular, self-directed displacements. 'Internal refugees', rarely recognised and usually ignored by the international community, are believed to be at least twice as numerous as those seeking asylum (Clark, 1989). Whole mountain communities, trying to escape ruthless landlords, collapse of traditional economies, war or draconian 'security' measures in the countryside, have taken up a precarious existence in cities from Lima to Bangkok. Sometimes they are at some advantage in remaining in their country of birth, knowing its language, perhaps helped by relatives and members of the same culture previously resettled. Sometimes their wellbeing is a priority of the government in power, the charitable and human rights groups in their own country. Often, however, they are worse off, lacking even the benefits and protections afforded by international agencies, emergency assistance, and resettlement programmes. They may be subject to brutal treatment or quickly sink into a destitute, pariah underclass, their women forced into prostitution, their children into scavenging and petty crime. Such evidence as exists suggests that they are unusually subject to malnutrition, infectious diseases, war injuries and persecution by the authorities (Muecke, 1992; Toole, 1993). Also, they may only speak the language of their minority or ethnic group.

Given the varied cultural mosaic of mountain areas, displaced peoples are more likely to find themselves in an unsympathetic cultural milieu in lowlands and distant cities where their skills are unwanted, their destitution and appearance excuses for hostility and worse. Though we lack any systematic surveys, there are well-known examples from the mountains of Central America, East Africa, the Middle East, Central and Southeast Asia (*ibid*; van den Berghe, 1990; Pizzarello, 1986; Sabo and Kibirge, 1989; IFRCRCS, 1994, Pts. VIII and XII).

Also missing from the figures is the huge exodus of down-country and expatriate workers, and those displaced by mega-projects, land grabs and resource exploitation. The 'development and

Table 16.6 Estimates of major refugee populations from mountain lands, and internally displaced persons within them; recognised and unrecognised refugees hosted by mountain lands; 1991–1992 (after US Committee for Refugees, 1992 and 1993)

Region/mountain country	Refugees from country	Internally displaced	Refugees hosted	Unrecognised hosted
AFRICA (Total refugees 5.7 million)				
Algeria			165 000	
Burundi	208 500		107 350	187 000
Cameroon			1500	35 000
Ethiopia/Eritrea	752 400	1 000 000	416 000	
Kenya			422 900	
Rwanda	203 900	100 000	24 500	
Uganda	14 900	300 000	179 600	
ASIA, EAST and SOUTH EAST (Total Refugees 398 600)				
Indonesia			15 600	
Malaysia			16 700	
Papua New Guinea			3800	
Philippines		1 000 000	5600	
ASIA, SOUTH and SOUTH WEST (Total refugees 7.9 million)				
Afghanistan	6 600 800	2 000 000	52 000	
India*		85 00	375 000	
Iran	50 000		2 781 800	
Lebanon		750 000	322 900	40 000
Nepal			89 000	
Pakistan*			1 577 000	10 000
Turkey			31 700	100 000
Yemen			52 500	
EUROPE (Total refugees 3.3 million)				
Armenia			300 000	
Austria			82 100	8000
Azerbaijan			246 000	
Italy			19 000	
Norway			5200	
Slovenia			68 900	
Switzerland			81 700	
Yugoslavia[1]	1 000 000	557 000		
Bosnia			70 000	
Croatia			420 000	
Macedonia			32 700	
Serbia/Montenegro			621 000	
THE AMERICAS (Total refugees 107 700)				
Belize			6100	28 000
Colombia		150 000		
Costa Rica			34 350	80 000
El Salvador	24 200		250	20 000
Guatemala	46 700	150 000	4900	250 000
Honduras			150	50 000
Mexico			47 300	340 000
Nicaragua	25 400		5850	16 000
Peru		200 000		

[1]Former, *Northern mountainous areas

economic refugees' may also suffer traumatic upheavals and life-threatening risks (Partridge, 1989; Timm, 1991).

In fact, behind all such figures lie some of the worst and most lethal tragedies of recent decades: enforced uprootings of long-settled communities; starvation, illness and high death tolls on route and in camps or squatter settlements; the grief that comes from loss of loved ones and homeland (Sliwinski, 1989). The displacements have brought cultural and ecological collapse in the places where they originate. They may be a source of civil strife and environmental damage in reception countries, problems widely discussed in relation to the huge number of war refugees from Afghanistan in neighbouring countries (Allan, 1987; Azhar, 1990; Weinbaum, 1994). They have placed the largest of all demands upon international humanitarian assistance, and it is rarely adequate (Harrell-Bond, 1986; Francois, 1992, 1993; UNDHA, 1994). Any attempt to assess disasters, the calamitous loss of sustainable communities in mountain lands, and the possibilities of future sustainable living there, must address these problems. Meanwhile, the 'hazards' are primarily human. Problems have been aggravated by drought, exposure to harsh weather, and the incidence of other natural disasters, famine and disease. Political changes and economic downturn, debt or sanctions, have played a larger role. But the main, direct cause of enforced displacements has been armed violence or its threat – as distinct from 'civil strife'. Most involve deliberate, organised uses of military and paramilitary forces.

War on civilians and habitats

In addition to forced uprooting of populations, warfare has been the largest source of sudden, untimely death and injury among mountain peoples and of considerable devastation to their habitats. In any year of the early 1990s there were 30 or more armed conflicts taking place in mountain lands (Hewitt, 1992). Since 1945, available estimates show almost 8 million civilians killed in 105 separate conflicts involving the highlands (Table 16.7). Many of these wars have also been of unusually long duration, averaging almost four years and some continuing for a decade or more (Brogan, 1989; Sivard, 1993).

While broader issues of mountain wars are discussed in Chapter 6, their central role in risk and disaster cannot be ignored here. Armed violence is not solely a problem for military geography and history. It is not just a problem of mountain warfare or national defence and security. It is not only the concern of the fighting forces and struggles between comparable states and alliances. To an ever-increasing extent, civilians, settlements, ways of life, and the habitat suffer the greatest losses in armed conflicts (Elliot, 1972; Ahlström, 1991; Hewitt, 1983, 1996). Warfare undermines every aspect of sustainable living. Military spending and a preoccupation of governments with war risks and options, divert attention and resources from other dangers.

Resident civilians, especially women, children, the elderly and disabled – the definitive 'non-combatants' – are the main victims of modern conflicts. They have comprised over 70 percent of all deaths in mountain land wars since 1945 (Sivard, 1992). Moreover, the harm to them and the habitat tends to occur in sudden, extreme, and unmanageable bouts of destruction. Recent wars have brought countless bombings of settlements, indiscriminate search and destroy missions, deployments of chemical and anti-personnel weapons, and environmental warfare (Westing, 1984). There have been massacres of defenceless civilians and prisoners, rape of women, torture, extrajudicial executions and 'disappearances', all on a huge scale. In scope and experience, these events can only be placed among the worst kinds of disaster (Zwi and Ugalde, 1991).

The role of mountain areas in insurgency or guerrilla wars has received considerable attention. That is not so true of the use of regular state forces against their resident populations, though it is more widespread and disastrous. Systematic 'ethnic cleansing' as a goal of war, in Bosnia and elsewhere, has continued the practice of exterminating 'racial enemies' in the mountain context. Other atrocities involve what may be termed 'ideological' and 'religious cleansing', or ruthless assaults upon political opposition. These have decimated whole social groups, generations of students and intelligentsias of mountain lands, forcing huge numbers into exile. Examples range from Chile to Uganda and from Greece to Tibet (Amnesty International, 1975; 1981; 1985). Of major concern for social risk, and a world-wide calamity in its own right, is the interrelation of violence and illicit drug production in mountain lands, especially of heroin and cocaine (McCoy, 1972; Evans, 1989; Rance, 1991; Morales, 1992).

The disasters of war are not confined to violent actions. Devastated habitats and ruined

Table 16.7 Armed conflicts in mountain lands involving regular forces of the states listed, 1945–1992, and new or on-going wars in 1995. Estimates of civilians killed after Sivard (1993). Wars in 1995 after Project Ploughshares (1996). In predominantly mountain countries all wars are included. In those with substantial lowland areas, only conflicts in or partly in mountainous areas are included

Region/country	No. of conflicts	Years of war	Death toll	
			Civilians	Total
AFRICA				
Algeria	3	10	85 000	105 000
Burundi	2	5	108 000	118 000
Cameroon	1	5	16 000	32 000
Ethiopia	2	18	515 000	614 000
Kenya	2	12	3000+	16 000
Morocco[1]	2	15	6000	19 000
Rwanda	2	10	104 000	107 000
Uganda	4	14	601 000	613 000
ASIA, EAST and SOUTH EAST				
China	8	11	2 020 000	2 610 000
Indonesia	6+	21	604 000+	691 000+
Laos	1	13	18 000	30 000
Malaysia	2	12	7500	15 000
Myanmar (Burma)		3	10 000	20 000
Philippines	3	22	45 000	84 000
Taiwan	3	2	20 000+	26 000
Vietnam	5	26	1 509 000	2 485 000
ASIA, SOUTH and SOUTH WEST				
Afghanistan	2	14	1 000 000+	1 505 000
Bangladesh[2]	1	18	30 000+	30 000+
Cyprus*	2	6	3000+	5000
India*	5	13	28 000+	43 000
Iran	2	21	120 000+	588 000
Iraq*	3	11	109 000	115 000
Lebanon	3	4	155 000	246 000
Pakistan*	3	15	7000+	12 000+
Turkey	2	11	5000+	10 000
Yemen	3	9	16 000	20 000
Tajikistan	1	1	20 000	20 000
EUROPE				
Azerbaijan	1	3	3500	7000
Bosnia	1	1	100 000	120 000
Croatia	1	2	20 000	25 000
Georgia[3]	2	2	?	1000
Greece	1	4	100 000	120 000
USSR/Russia*	1	1		1000
THE AMERICAS				
Bolivia	1	1	1000	2000
Chile	3	3	26 000	28 000
Colombia	3	10	214 000	322 000
Costa Rica	1	1	1000	2000
Ecuador[4]				
El Salvador	1	3	50 000	75 000
Guatemala	2	26	100 500	141 000
Honduras	1	1	3000	5000
Nicaragua	2	10	40 000+	80 000
Peru[4]	1	9	2000	26 000
TOTALS	105	398	7 818 000	11 114 000

[1]Includes Western Sahara, [2]Hill tracts, [3]Chechnya, [4]Ecuador–Peru, *Mountainous areas

infrastructure may limit or prevent recovery, and discourage people from returning to their homes. One of the greatest calamities in recent years is the 'mines plague': a deadly legacy of unexploded munitions, especially anti-personnel mines (Table 16.8). These have killed and maimed tens of thousands of mountain folk to date. They prevent millions of others, displaced by war, from returning home (Epps, 1993; UNDHA, 1994: 21; IFRCRCS, 1994: pp.59–72).

In the industrialised nations, a substantial number of the sites with land and waters contaminated by military, nuclear, and other toxic wastes are in the mountains. Many of the testing and training areas, bases and depots, have been located there and are severely contaminated (Ehrlich and Birks, 1990). Meanwhile, the upheavals of war and its victims feature prominently in disease crises, perhaps the most serious of all hazards threatening mountain people.

Emergent and epidemic diseases

There is a common perception of the mountains as places of clean air and water, and an absence of many diseases or disease carriers of the warmer lowlands. Those able to afford it have gone up to the hills to escape the heat and maladies of summer, for healthy recreation, or to recover from illness. Perhaps this still applies in many places. It is an important factor in the relative healthiness of some mountain populations. Nevertheless, adverse trends and imported diseases are changing the picture drastically for many others.

Disaster studies and emergency assistance are directed mainly to sudden, unmanageable outbreaks and epidemic diseases. Nevertheless, we must also recognise the warning signs in worsening public health conditions and chronic disease; the appearance of new diseases, and pockets of endemic disease that may turn into major outbreaks (Wilson et al., 1994). Of concern in this regard is the ever-growing size of settlements and crowded populations in them, problems widely aggravated by influx of displaced and destitute families. While some mountain lands, such as Nepal or Uganda, still have 70–80 percent of their people in rural, mountain valley settings, in many others over half are in urban centres. Cities that act as gateways to and out of mountainous hinterlands become increasingly parts of their demographic, behavioural, and disease environments.

Table 16.8 The land mines 'plague' in mountain lands. Estimates of the numbers of mines in 1993 for selected mountain countries and mountainous sub-regions (after, IFRCRCS, 1994: pp. 59–72)

Mountain country	Number of mines
Afghanistan	9–10 million
Bosnia and Herzegovina	1.3 million
Croatia	1 million
Serbia and Montenegro	500 000
Ethiopia and Eritrea	500 000
Nicaragua	120 000
Armenia and Azerbaijan	50 000+
Iran	'very extensive'[1]
El Salvador	20 000
Morocco	10 000
Rwanda	'hundreds'
Myanmar (Burma)	'forty-year legacy'

Significant problem in mountainous sub-regions	
Cambodia	4–7 million[2]
Iraq	5–10 million[3]
Laos	'tens of thousands'
Vietnam	'hundreds of thousands'
Russia/Chechen Ingush	'many civilian casualties'

[1]Iran on borders with Iraq and Afghanistan, [2]Cambodia, largely in hilly north and northwest borders, [3]Iraq; major minefields in mountainous Kurdistan, and borders with Turkey and Iran

A calamitous illustration was the cholera pandemic that began in Lima, Peru in 1991. Eventually it spread to affect populations throughout the Latin American cordilleras from Chile to Mexico. The World Health Organisation estimated that, by late 1993, some 901 000 people had contracted the disease and over 8000 had died. These figures are believed to be under-estimates. The crisis continued through 1994; it can now be seen as part of a global pandemic that had not run its course by 1996.

The 'El Tor' strain of cholera was involved. It had appeared in the Celebes Islands, Indonesia, in 1961. The strain is highly resistant to several commonly used antibiotics and to chlorination of water supplies. Epidemiological detective work found that it was brought to Peru in the bilge water of a Chinese freighter, probably pumped aboard in Dhaka harbour, Bangladesh. This was drained into the harbour at Callao, the port of Lima. The disease was first passed to Peruvians in seafood, and on from there in the largely untreated public water supply of Lima and other centres. Distinctive in itself, the history of this disaster points up the sorts of risks and problems from a variety of resistant, infectious diseases (Table 16.9).

The range of conditions that may lead to, or aggravate, the spread of disease in mountain lands involves adverse habitat changes and public health crises in crowded settlements, rural impoverishment, and increasing human mobility. The numbers of outside visitors continues to grow exponentially, and from an ever-wider geographical range. Mountain people themselves are becoming much more mobile or, at least, the young men and, in some cases, young women. Such is the context in which the late twentieth century plague, AIDS – acquired immunodeficiency syndrome – afflicts mountain populations.

AIDS is commonly identified with cities and is predominantly an urban pandemic (Mann, 1992: p.882). However, the disease has spread to most of the mountain lands, their rural hinterlands as well as cities (Table 16.10). By 1992, more than 80 000 AIDS victims had been reported. It is generally believed that the actual numbers were five to ten times greater. For every AIDS victim, there are usually ten or more others who are HIV-positive;

that is, infected with the human immunodeficiency virus implicated as the cause of the disease. It may be present without obvious symptoms for as much as a decade. Under-reportage and unrecognised or hidden patterns of infection apply almost universally. The figures reported for Indonesia, Malaysia, India, and Pakistan, among others, are considered two or more orders of magnitude below reality. Estimates of 5000 cases in Nepal in 1994 are more than ten-times those officially reported (Seddon, 1995). Myanmar (Burma) was officially AIDS-free in 1992, but there is good reason to accept estimates of 150 000 HIV-infected persons by mid-1993.

The mountainous country with the highest incidence, Uganda, was one of the first to be struck by the epidemic, and is unusual in having more HIV–AIDS victims in rural areas than cities. Of course, 88 percent of its people are classed as rural. However, the proportions of infected persons are much higher in cities and they have played the major role in spreading the disease. The initial, rapid spread through the East African highlands was largely a silent partner in the social upheavals, wars,

Table 16.9 Examples of late-twentieth century disease outbreaks, epidemics and pandemics in or affecting mountain lands (after Seaman, 1984; Mann, 1992; Garrett, 1994; Wilson *et al.*, 1994)

Place	Year	Disaster features	Deaths
Turkey	1964–65	Measles epidemic	100 524
Guatemala + 7 other Central American countries	1969	*Shigella* dysentery; 500 000 cases. Deaths especially from a strain resistant to leading antibiotics and sulfa drugs.	8000+
Ecuador	1969	Equine encephalitis contracted by some 80 000 persons	400
East African Highlands	1983–present	AIDS epidemic, with over 100 000 reported, actual estimated x10, and some 6–7 million HIV infected. Associated epidemics of other sexually transmitted diseases, tuberculosis, cancers (e.g. Kaposi's sarcoma), and malaria.	1–2 million
World-wide in tropical highlands (below c.2000 m)	Early 1980s–present	Malaria pandemic. Tens of millions of cases. Huge resurgence of malaria including drug-resistant strains and (new?) cerebral malaria. Cases and mortality rates doubling every two–three years. Historic highest rates in East African Highlands (Kenya, Rwanda) and some other areas by 1993, including Nepal and Burma.	Hundreds of thousands
Bolivia	mid-1980s	Chagas disease widespread in cities, bloodbank infection rate of 63% in Santa Cruz.	?
Ethiopia	1989	Meningitis epidemic.	42 000
Latin America	1991–92	El Tor cholera epidemic infecting 901 000+, mainly in Andean South and Central American countries.	8000+
Thailand, Chiang Mai, etc.	1991–present	HIV/AIDS epidemic with high rates among female sex workers, including 9–12-year-olds, many recruited from impoverished indigenous hill peoples, often by force or deception.	Thousands predicted as victims enter AIDS stage
Ethiopia	1992	Malaria epidemic. High incidence drug-resistant strain. Death toll in six months –	20 000+

Table 16.10 AIDS victims reported for selected mountainous regions to January, 1992. Most figures should be considered a small fraction of the actual numbers. Some idea of the problem can be deduced from World Health Organisation estimates of unreported AIDS cases. The 'percentage actual' is the proportion of that 1992 figure from the WHO estimates for the region concerned. Rates of under-reporting may well be higher in many mountain countries. Estimates of the ratio of best estimates for actual numbers of AIDS victims to total numbers infected with HIV, indicates the full scope of the disaster (after Mann, 1992)

Region/ country	First report	AIDS victims up to 1992	% Actual AIDS cases	HIV+/AIDS ratio
AFRICA		114 522	10%	7 : 1
Algeria		92		
Burundi		3305		
Cameroon	1986	429		
Ethiopia	1987	1818		
Kenya		9139		
Morocco	1988	98		
Rwanda	1986	6578		
Uganda	1983	30 190		
ASIA, EAST and SOUTH EAST		4461	75%	10 : 1
Bhutan				
Indonesia		21		
Laos				
Malaysia		28		
Myanmar (Burma)				
Nepal		5		
New Zealand	1986	274		
Papua New Guinea	1987	57		
Vietnam				
ASIA, SOUTH and SOUTH WEST		813	20%	10 : 1
Afghanistan				
Iran		44		
Lebanon		24		
Pakistan	1988	18		
Turkey		62		
Yemen				
EUROPE		65 895	65%	7–10 : 1
Albania				
Austria		707		
Greece		559		
Italy	1982	11 809		
Norway	1983	252		
Portugal		816		
Switzerland	1980	2228		
Yugoslavia (former)		2236		
LATIN AMERICA		41 603	25%	6 : 1
Bolivia		41		
Chile	1984	500		
Colombia	1985	2189		
Costa Rica		315		
Ecuador		155		
El Salvador		323		
Guatemala		236		
Honduras		1595		
Mexico	1983	9073		
Nicaragua		24		
Peru		541		
TOTALS		227 294		

displacements, and ecological damages of the 1960s, 70s and 80s. HIV–AIDS followed the trail of conflicts, its spread accelerated by widespread rape of women and a rapid growth in prostitution, drug trafficking, destitution, and collapse of the social fabric. Women and children, driven from ravaged traditional economies in mountain areas, have struggled to stay alive or to help their families by becoming sex workers around military bases and for the mobile male workers, drawn along expanding highways and into the cities. Ill-equipped and inadequately staffed clinics and hospitals helped spread HIV before its existence was even known. Immunisation campaigns against other diseases, that routinely used the same, un-sterilised needle for numerous individuals, may have had a still larger role (Fleming *et al.*, 1988; Gould, 1993).

In 1996, it is likely that there are not less than 5 million, and possibly more than 10 million HIV/AIDS victims in mountain lands. After a decade of vigorous research action in some countries, and more or less successful campaigns to slow or stop the spread in a few, the disease is moving rapidly into and through many mountain lands by well-known but unchecked processes. Other priorities or political indifference, stigmatisation, lack of resources, and failure to inform and assist vulnerable groups all contribute.

There do not seem to be any specific mountain conditions associated with the pandemic, but some special vulnerabilities exist there. Spread of the disease can be by visitors, tourists and tour operators, soldiers and officials assigned to mountain areas, miners, forest and construction workers. They include persons most likely to be looking for, and able to afford, casual sexual liaisons, hard drugs, or emergency hospital care. Men and women from mountain communities, working down-country or overseas, especially in desperate or dependent circumstances, are drawn into sexual or drug-taking activities where they will contract HIV. They keep such activities quiet when they return home, and rarely know they have the disease until they have infected a wife and, through her, their children.

Most tropical mountain cultures still depend heavily upon traditional forms of household labour for food production and animal husbandry, or manual labour for cash earnings in the modernising economy. The disease is spread to them mainly by heterosexual activity and in families. Young men and women in their most active and fertile years, are the main victims. AIDS in rural areas of high-land East Africa, for example, infects as many or more women than men. Women are often the principal food producers and their loss can devastate the household and community economies. Families and communities afflicted by the disease become less and less able to maintain their basic economic and household activities. As husband or wife enter the later stages of the disease and die, the family economy sinks into destitution. Often the survivors are ostracised then, if not before (Fleming *et al.*, 1988). The disease is passed to children, or they are left as AIDS orphans. Over one third of AIDS victims in sub-Saharan Africa are children. The implications of HIV/AIDS for the distinctive agricultural and pastoral societies of mountain lands, or prospects for their improvement through modernisation, are all extremely bleak.

AIDS is also accompanied by other serious diseases, usually the actual killers of the victims. It may be a large factor in increasing the spread and the virulence of certain cancers, other types of sexually transmitted diseases and a new pandemic of drug-resistant tuberculosis (Mann, 1992: pp.148–63). In the East African Highlands, these were often the first cause for alarm, before physicians realised they were spreading among people whose immune systems were destroyed by AIDS. However, others without AIDS are also at risk from the same diseases, some of which, like tuberculosis, are spreading even faster (Garrett, 1994).

Finally, ecological changes being wrought by modernisation, urbanisation, forest and other habitat destruction, change, and increase the spread of this and other serious diseases (Usher, 1994). Some believe that to be the main threat from global climatic warming. For example, if there is a danger that malaria- or encephalitis-carrying mosquitoes will spread into higher latitudes, the same may apply to their altitudinal range (Lovejoy, 1993). However, it is essential to emphasise again that socio-economic changes are the main or most threatening sources of epidemic as well as chronic diseases. Social and ecological changes render increasing numbers of mountain dwellers vulnerable to new and old diseases. Few places provide adequate protection against them, or public health measures in keeping with transformed ways of living. Of course, socio-economic improvement is also the only area of hope for reversing these dangerous developments.

Technological hazards in the mountains

All of the ways in which structures, industrial processes, transportation, and other infrastructure may fail or cause harm, can occur in mountains, but certain dangers or risks of failure are more likely. In the OFDA inventory cited earlier, roughly seven percent of major disasters in mountain lands were attributed to technological accident, rather less than one event per year. Most were transportation 'accidents' and almost one third air crashes. That technological risks, the larger preoccupation of richer countries and professional risk assessment, seem to have so small an overall role may restore a certain balance. Once more, however, the evidence is somewhat misleading.

Plate 16.5 Highway reconstruction below Darjeeling after the catastrophic precipitation events of 1968. Such projects can lead to enormous maintenance costs and risk of massive slope instability (Photograph: J. D. Ives)

Much as with landslides, the damage in a majority of technological disasters is concentrated in relatively small areas. The lower death tolls reflect smaller numbers exposed. If we take all disasters reported in the media, it is found that many more events have technological rather than natural initiating agents (Table 16.11). Moreover, roughly one third occur in mountain lands, also out of proportion to their area and population. The kinds of events reported, however, mainly affect high profile persons and modern activities.

Perhaps the most widespread, frequent and, in total, expensive, damages involve mountain roads. Studies of 'natural disasters' are increasingly focused upon damage and closures of arterial mountain roads (Sowers and Carter, 1979; Jones et al., 1983; de Scally and Gardner, 1994). The extent of new construction into highland areas over the past half century has been unprecedented, not least in the use of explosives and heavy equipment. The roads move through previously unserved areas, often the most rugged, and traffic has increased exponentially. Modern administrative, military, commercial, and recreational interests tend to expect that highways be open at all times. The changeable conditions and extreme seasonal and attitudinal variations in mountains add unusual dangers or costs for all-weather highways. The main result is expensive but routine monitoring and warning; rapid reaction to treacherous conditions and restoration of damaged communications. These are put in place because roads have the attention of governments, national security, mining or forestry, and other resource interests, and the tourism business. In most countries the physical hazards are in the hands of army engineers, amply provided with heavy equipment, explosives, and human labour. For highways serving winter resort areas, strategic roads and key mountain passes, professional avalanche monitoring, defences and artificial release with artillery, are

Table 16.11 Natural, technological, and epidemic disasters and wars reported world-wide 1989–1993 (after Encyclopaedia Britannica Yearbooks and Project Ploughshares, 1996)

	1989	1990	1991	1992	1993	Totals	World Average
Natural disasters	46	52	54	32	51	235	47
Technological disasters	128	123	98	109	93	551	110
(New) epidemics	2	–	1	1	1	5	1
(New or continuing) wars*	32	31	30	30	35		32
TOTAL							190

*Those that caused at least 1000 deaths

standard. Even so, there are multiple closures, more rarely serious casualties, in every major mountain area. In general it is difficult to see many of these highways as 'sustainable' in the absence of expensive maintenance and response systems.

The same picture is well-established for mountain railways. One study reports 4000–5000 or more landslides affecting Chinese railways, almost all in mountainous parts (Lasa and Jingfang, 1992: p.414). Casualties are rarely large, but delays and costs to the railroads are substantial.

Major road and rail tunnels and high bridges are concentrated in the mountains, as are most accidents associated with them. One of the worst tunnel disasters in recent years occurred on November 2nd, 1982. It involved the Salang Tunnel (3370 m elevation) in the Hindu Kush, Afghanistan. A vehicle in a Soviet military convoy struck a fuel truck near one entrance. The explosion and burning fuel instantly killed dozens of soldiers and civilians in nearby vehicles. Mistaking the explosion for an attack by the Afghan rebel forces, the Soviet army sealed both ends of the tunnel. Hundreds of private motor vehicles and buses packed with passengers were inside the tunnel. Many trapped civilians were burned by the spreading flames. Meanwhile, the tunnel's ventilation systems were not working and it was extremely cold. Vehicles distant from the explosion sat with their engines running to keep warm, adding to carbon monoxide from the fires. As a result, many more persons died of asphyxiation and carbon monoxide poisoning. Estimates of the total killed range from 2000 to 3000.

The disaster shows the compounding of technological hazards with extreme mountain conditions and wartime perceptions of risk. Another disaster occurred at the Salang Tunnel in March, 1993. Though triggered by an avalanche it was equally about technological risk and social risk-taking.

In fact, natural, mountain environment risks are rarely distinct from technological and societal hazards. Coal mining under Turtle Mountain, Alberta is thought to have aided in the occurrence of the great Frank rockslide and disaster of 1905, though there is dispute about to what extent. Construction projects may trigger, or suddenly accelerate disastrous landslides. Costly examples in the Bolivian Andes, Montana and Colorado, US are described in Voight, 1979, vol.2. The great rockslide and flood disaster at Vaiont Dam, Italian Alps, in 1963, was brought on by the building and filling of the reservoir.

Disasters relating to the failure of dams are also concentrated in mountain regions (Table 16.12). Dam projects are increasingly attracted by their favourable sites or untapped water and power potential. In relation to the importance of mining in mountain lands, some especially lethal disasters have been due to failure of their reservoirs and tailings ponds (Chapter 9). In the Buffalo Creek, West Virginia event of 1972, the role of heavy rains and the steep canyon down which the flood wave moved, compounded technological and environmental risks.

INTERPRETATIONS OF MOUNTAIN CALAMITIES

Though disaster is at the heart of our concerns, it is not necessarily the heart of our work. Equally important is what we understand or believe about the origins of calamities. What are the conditions that influence their nature, their severity, where do they occur and whom do they affect?

The records of disasters, treated cautiously, give us a sense of the range and balance of extreme risks in mountains. However, the limitations of inventories and event descriptions are not confined to the data itself. Even more problematic are these ways of classifying and measuring disasters. The geography of states is not a good reflection of the geography of mountain environments, peoples, and the risks they face. There are the gross differences between states of differing size, wealth, influence, and geographical setting. But larger or more crucial differences may occur between disasters in a given state, than between it and others: indeed, for different people in any one disaster area. Disaster inventories rarely reveal the more pressing humanitarian issues, let alone the concerns of given peoples and places. However, this is also to raise questions of how we approach and interpret the nature of disaster.

Careful reconstruction of disaster events, if not patently obvious facts, show that preexisting social conditions and human actions have a large measure of responsibility, even for natural disasters. In most cases, the victims of disaster come mainly from among particular types and groups of people. They are singled out less by the flood or storm, than by unusual exposure to these hazards; by weakness or lack of protections and, more generally, by a range of social disadvantages. If so, then the conditions and forms of disaster are partly, if not largely, set up by material life and social conditions *before* the

Table 16.12 Examples of the failure of artificial dams in mountainous terrain, with catastrophic outbreak floods (after Jansen, 1980; Serafim, 1984)

Dam/Location	Date	Dam height (m)	Attendant conditions	Deaths
South Fork near Johnstown, PA, USA	1874 May 31	14.2	Heavy rains, prior flooding, outbreak flood steepened in canyon	2200
Gleno near Bergamo, ITALY	1923 Dec. 1	43.6	Heavy rains and floods, sudden break, several villages destroyed	600
St Francis near Los Angeles, USA	1928 March 12	62.5	Reservoir full at time of collapse. No warning. Flood wave 28 m high	450
Alla Sella Zerbino Appenines, near Genoa, ITALY	1935 Aug. 13	42	Heavy rains and floods into dam. Dam overtopped, full height failure	100+
Möhne Ruhr, GERMANY	1943 May 17	40.3	'Dam-busters' bombing raid released 10 m+ flood wave	1200+
Malpasset near Fréjus, FRANCE	1959 Dec. 2	61	Heavy rains. Weak rock layers in foundations. No warning for public	421
Hyokiri Namwon, S. KOREA	1961 July 12	?	Sudden failure. No details	114
Vaiont near Belluno, ITALY	1963 Oct. 9	265	Catastrophic rockslide into reservoir. Water forced over dam. Flood wave below caused deaths, most destruction	2600
Sempor near Mangebong, Java, INDONESIA	1967 Dec. 1	54	Heavy monsoon rains, construction, delays, part-finished dam failed	2000+
Frias near Mendoza, ARGENTINA	1970 Jan. 4	35	Torrential rains, torrents, flood wave into town full of tourists	42+
Buffalo Creek WV, USA	1972 Feb. 26	46	Mine tailings dam, burning embankment burst after torrential rains, 6 m flood wave, 4000 made homeless	125
Stava Dolomites, ITALY	1985 July 19	?	Earth walls of fluorite mine settlement tank failed. Flood wave into resort full of tourists	250+
(Rimac R.) near Lima, PERU	1987 March 9	?	Several earthen dams failed. Torrents carried boulders into narrow valleys devastating small towns	100+
(Shanxi Province) CHINA	1989 June 20	?	Dam burst in mountainous area	38

earthquake, epidemic, or other destructive agent strikes. Otherwise, damage would be socially indiscriminate or it would vary only as the power of the destructive agent varies. That is rarely so. The social order, and relations to the habitat sets up the conditions that are usually decisive for whether a disaster occurs, its severity, whom it will affect, and how well they can respond (Quarantelli, 1978; Turner, 1978; Wijkman and Timberlake, 1984; Hewitt, 1996).

Moreover, while organised, external emergency aid may be desirable, if not essential, the responses of survivors in, and immediately around, damaged places is usually far more critical. They save the lives of a majority of those trapped, injured, or otherwise helpless. If this is not done in the hours – often, in fact, it takes days, if not weeks – before outside emergency aid can be delivered to the victims, few survive to be rescued at all. The unimpaired, knowledgeable, and desperately concerned survivors are the only ones who can do this work, and usually they do. Rebuilding of homes and essential facilities, evacuation of injured and homeless, has often been carried out by surviving community members before outside emergency aid arrived. Much more of the shelter, food, and care for the sick is provided by relatives, friends, and helpful persons in surrounding, undamaged areas.

For these, as well as for ethical reasons, there is a growing call for greater sensitivity in science and administration to the concerns and predicaments of residents and others on the ground in hazardous areas (IFRCRC, 1994, Parts III and IV). Some of the worst consequences of disaster in the mountains have been due to insensitive concepts or policies determined in distant capitals or by visiting 'experts' with little or no knowledge of the cultures and habitats involved. This becomes an acute problem where there are indigenous peoples with a distinct linguistic, religious, and material life. Being resident for centuries, sometimes millennia, they have systems of knowledge and practices uniquely adapted and attached to their mountain habitat. Such observations direct our attention to the contexts, vulnerabilities, and powers of mountain communities, not merely to extreme events.

Context and intervening conditions

Mountain 'specificities', or the highland context, systematically described elsewhere in this volume, are integral to problems of risk and disaster in mountains. They include high relief and steep slopes, strong climatic and other environmental gradients, frequent and extreme earth surface processes. Each region includes a mosaic of microhabitats, a complex adaptive maze of sharp contrasts and fine shadings of risk. The difficulties or higher costs of outside access to, and movement between, mountain areas become acute under modernising conditions. In general, all such regionally specific conditions intervene to moderate or exaggerate destructive processes, to complicate and frustrate human responses to them. These considerations, even more than unique dangers and particular disasters, single out and identify what is distinctive about mountain land risks. Moreover, they have geographical and societal patterns that are more or less independent of dangerous agents (Allan *et al.*, 1988; Ives and Messerli, 1989). For instance, the severity of earthquakes can be strongly influenced by rock and soil type, topography, vegetation cover, and water tables. But, as distinctive mountain patterns, these conditions have no direct relation to seismicity. Similar arguments apply to the way topography and tree cover can intervene to magnify or moderate the impact of landslides, severe snow storms or forest fires.

The prevalence of small settlements and relatively dispersed populations in mountain regions raises special questions. Fewer persons and less property may be exposed in given events, and hence accounted in losses. But the resources present in larger, richer, and more extensive communities may also be absent. Threats to small but culturally and historically special communities and sites are hardly recognised in national or global summaries. That includes the widespread occurrence of sacred places in mountains (see Chapter 3) and the many small, distinct groups of indigenous peoples (see Chapter 2). They may become especially vulnerable to modern dangers, and relatively small events. Meanwhile, the mountain habitat places severe constraints upon survivors and emergency assistance so that smaller events may entail relatively greater loss and hardship for survivors. This point has been demonstrated in relation to an avalanche that killed 'only' 57 persons, but representing 15 percent of the population of a small Alpine village. Death and irreplaceable loss affected every survivor directly. Erikson (1976) showed the 'catastrophic' scope of a West Virginia flood of February 1972, which destroyed the relatively small mountain settlement of Buffalo Creek. Again, the death toll of 125 persons might not seem 'catastrophic', but it affected everyone, wiping out whole families and almost every house. The 4500 persons killed in the catastrophic landslide at Yungay, Peru in 1970, comprised a fifth of all those killed in the event, and about 10 percent of deaths in the earthquake disaster during which it occurred. Yet they represented more than 90 percent of that town's population. All but a handful of its buildings were annihilated and completely buried under the debris. Even those not among its few survivors recognise this annihilation of the community of Yungay as a special tragedy, significant far beyond its share of the whole disaster (Oliver-Smith, 1986; Bode, 1989). Such issues remind us of the faces behind the places. Danger and disaster are occasions of extreme experience that can only be felt by persons, and fully understood by sympathetically and ethically engaged observers.

Vulnerability in the mountains

In mountains as elsewhere, some persons and groups suffer far more in disasters than others. Certain kinds of technology or land use are more often, or more severely, damaged. When some members, or parts of communities, are harder hit, disaster cannot be regarded as socially indiscriminate. When it is found that women and children, minorities, less influential classes, or impoverished regions, are disproportionately harmed, that has to reflect their prior social situation. Then, the ways in which (some) people are exposed to, and poorly protected against, dangers are of primary concern.

Over the past decade or so, vulnerability has become identified with a distinctive view of risk and disaster (Bohle, 1993; Blaikie *et al.*, 1994). A vulnerability perspective focuses upon how communities are put at risk, rather than the character of natural, technological, or violent human agents. When people are exposed, the conditions that influence their protection and coping capacities are important. When disasters occur, vulnerability is revealed, and we look first to the people and places most affected. Emphasis is given to the social constraints and capacities, especially the weaknesses and lack of protection of those harmed (Table 16.13).

Table 16.13 The forms and conditions of vulnerability that may apply to mountain peoples (after Hewitt, 1996)

FORMS

Exposure to hazards – through occupation, lifestyle, location, or siting
Weakness and susceptibility – genetic predisposition, disability, poor nutrition, poorly designed buildings, insecure practices
Disadvantage or 'structural weakness' – poverty, dependency, lack of resources, skills, rights, and access to information
Defencelessness – lack of physical protection or of safety regulations and codes; approval and monitoring of dangerous substances; systems of care for inherently weaker members; lack of warning and emergency preparedness
Lack of response capabilities – limited options, impaired resilience, dysfunctional communities
Powerlessness – inability to influence sources of danger or protection
Enforced vulnerability – exploitation and dispossession; forced resettlement, uprootings and expulsions; people ostracised or denied the usual rights and privileges of citizenship; people exposed to arbitrary violence

SOCIAL SPACE OF VULNERABILITY

Individual – genetic and physiological, age, gender, family situation, and social status, personality, (lack of) experience, training, readiness
Domestic – family well-being, status
Gender – social position, rights, responsibilities and treatment of women as compared to men
Social space – communal, urban, rural, class, hierarchy
Economic – (lack of) skills, employment security, resources owned or entitled; economies marginalised or dependent on volatile markets
Ethnic or Cultural – linguistic, religious, ethical (e.g. pacifist); status and treatment of minorities, indigenous peoples

SOCIAL CONTROL OR CHOICE OF VULNERABILITY

Voluntary – risk-taking or risk-accepting (contractual) activity
Involuntary – as in:
  i) systemic or everyday vulnerability (= 'structural' vulnerability),
  ii) enforced vulnerability and
  iii) subject and captive peoples at risk of enslavement, forced and indentured labour, military occupation, 'reservations'

GEOGRAPHIES OF VULNERABILITY

Exposure in hazardous locations – e.g. flood-prone areas, dangerous facilities, war zone
Locational disadvantage – e.g. with respect to having or obtaining protection
Marginalised areas and peoples
Impoverished and impaired habitats
Defenceless sites, neighbourhoods, areas, regions
Enforced – as in:
  Regions of misrule: corrupt, unjust, violent, and illegitimate governance
  Forced resettlement and expulsions
  Displaced peoples: populations (recently) relocated in an unfamiliar environment, stripped of adaptive knowledge, well-tried practices, and the advantages of belonging
  Ostracised and excluded states, regions, and peoples

A vulnerability perspective sees the safety of homes in earthquake or flood decided by socio-economic constraints upon what was built, where, and who would live there. That may or may not include specific risk-reducing measures such as whether, and for whom, earthquake-resistant construction was available. How people are affected by economic trends or urban congestion may set them up as the more likely victims of disaster. Such conditions of risk also arise from decisions, activities, and powers of (other?) humans.

Likewise, capacities to respond to risks and cope with disaster are embedded in overall relations of the given community to natural, technological, and social environments. This underscores the importance of human defences and resourcefulness in the face of dangers. Equally important, of course, are the ways in which some or all societies deliberately seek to reduce and offset vulnerability, at least for some of their members. Communities at risk may show remarkable adaptive and coping capacities in the face of stress or damage, including societies that might otherwise seem lacking in material supports. Nevertheless, a disaster usually reveals where these capacities are eroded or crushed. The problem of vulnerability is rooted in impaired adaptive capabilities. The people of foremost concern are not just suffering from weaknesses, defencelessness, and structural disadvantage. They lack the means to avoid harm and ability to cope with it. Above all they lack influence upon social policy. As an analysis of hazards in Lima, Peru, concluded:

> The people most vulnerable to the effects of an earthquake in the city are those with limited options in terms of access to housing and employment. The inhabitants of critical areas would not choose to live there if they had any alternative, nor do they deliberately neglect the maintenance of their overcrowded and deteriorated tenements. For them it is the best-of-the-worst of a number of disaster-prone scenarios, such as having nowhere to live, having no way to earn a living, and having nothing to eat . . . Low income families in Lima only have freedom to choose between different kinds of disaster. Within the options available, people seek to minimise vulnerability to one kind of hazard even at the cost of increasing their vulnerability to another [such as earthquake] (Maskrey, 1989: p.12).

Vulnerability to disaster is indicative of severe impairment. The more likely and common victims of disaster tend to suffer from several, if not most, of the forms of vulnerability outlined. Dependant family members or persons of weak constitution, living on dangerous sites, in poorly built homes, without adequate warnings, protections or influence, are the most likely and worst-affected victims of disaster (Wisner, 1993: p.22). Moreover, unusual vulnerability is most often imposed, and is a product of *positive disadvantaging* by social actions that take away options or unfairly allocate risk and resources.

Women and disaster

The sense in which risk is also a *gendered space* has been taken up especially in vulnerability studies. Vulnerabilities are constructed and allocated through gender-based values and actions. That comes out in the plight of women. Few of their vulnerabilities arise because of their sex, even when the dangers involve sexuality. Rather, they result from women's position and treatment in society (Shiva, 1994).

In most mountain societies women have a primary responsibility for domestic space and dependant family members. Even in the many cultures where they have a large responsibility for food production, they spend much of their lives in or near the home. Any dangers that are greater in residential locations tend to affect women most. Moreover, such vulnerabilities as they may have personally and as women, are compounded in a crisis, by the demands of caring for and saving dependants. Although there are no exact numbers, anecdotal material makes it certain that uncounted thousands of women have died while trying to rescue family members in buildings damaged by earthquake, storm, mass fires, and acts of war. The same applies to the millions of women who have headed, or had the main responsibility for, families fleeing disasters and in refugee camps.

Gender-specific data on casualties is available for a very few disasters, but where it exists for earthquakes, for example, it generally records more female victims. Surveys of two disasters in (former) Soviet Central Asia, reported many more women casualties than men. In the 1948 Ashkabad earthquake, 47 percent of the dead were women compared to 18 percent men, and 25 percent more of the casualties in the Tashkent, 1996 disaster were said to be women (Seaman *et al.*, 1985: p.19).

In three villages studied after the Guatemala, 1976 disaster, one third more females died than males, and in all age groups except the over-60s. There were, however, about twice as many more seriously injured women survivors in the latter group (Glass *et al.*, 1977). In the Fruili, Italy, earthquake of the same year, estimates for the mountain community of Venzone show excessive female deaths in the age groups 5–34, 45–54, and over 73 years, and somewhat more in total (Hogg, 1980). The Maharasthra, India, disaster, 1993, was not in mountains but the unusually detailed demographic data can be taken to indicate what probably happens more generally in agricultural villages of South Asia. Females comprised 58.4 percent of 7797 deaths, 63 percent in the age group 25–44. Slightly more male children died (Parasurman, 1995). However, it seems there were no disproportionate casualties by age or gender in mainly mountain villages affected by the 1980 disaster in Southern Italy (Alexander, 1993: p.469). Perhaps that is because, at 19:35 on a Sunday evening in November, nearly everyone in these villages was at home with their families – sharing the risks of poorly built, sited, and maintained domestic buildings.

Unlike the typical picture for AIDS in the West and most cities, in the world's mountain lands, equal or greater numbers of victims are women, and they are at heightened risk. This is associated partly with social and economic changes. These send menfolk out to work and trade in distant areas, where they may contract the disease, then bring it back to their homes, usually infecting the wife, sometimes casual partners. The wife usually ends up caring for the AIDS-stricken husband, perhaps infected children as well, often hiding the fact from relatives and community. Worse, she is often blamed and rejected when they find out. When the husband dies, the extended family may throw her out, especially if she has the disease. She may be left without resources or anyone to care for her when AIDS takes hold. Already, hundreds of thousands of women have lived and died in this terrible predicament in the highlands of Africa. Their stories record the plight of the most vulnerable of the vulnerable.

More than 95 percent of all 'sex workers' are women. They have very high, often the highest, rates of HIV infection. Mountain land examples include Kenya with 88 percent of female sex workers tested HIV-positive, and Rwanda with 80 percent in 1990; Malawi (55 percent in 1984) and Tanzania (38 percent in 1988). Commercial sex and drugs seem to explain the very dramatic rise of HIV-infected persons in Bangkok, seemingly from 2–3 percent to 35 percent in 1988 alone. In the capital a large fraction of female sex workers are recruited from northern mountain communities, to which some return and can carry the infection.

A particular affliction in mountain lands is the spread of sexually transmitted diseases from 'militarised rape'. Assault, many times by different men not only increases the chances of contact with an infected person, violent, resisted, or multiple sex acts, cause worse lesions, greatly increasing the likelihood of infection. For similar reasons women have been and remain at exceptional risk from sexually transmitted diseases of all kinds.

This is also an example of how vulnerability is associated with disempowerment. Few women have the strength or rights to control their sexual relations even within marriage (Raymond, 1993). Those from disadvantaged groups, especially if also subject to racial and religious prejudice, like 'hill tribes' who supply women sex workers, servants and menial labourers in Southeast Asia, are even less able to defend themselves.

Women's vulnerability is often hidden and aggravated by their 'invisibility' in male-dominated societies, at least for outside observers. They can be largely excluded from public life. Disaster studies, most of them by male professionals, have taken that 'public realm' for granted and as representing everyone. They have failed to recognise the special dangers for women. Presumably that helps explain the absence of gender data from most disaster reports and studies. Of course, in traditionally patriarchal societies, only researchers or aid workers who are women can have the opportunity to see 'behind the veil', literally or metaphorically (Hewitt, 1991).

However, their work also stresses the dangers of thinking in terms of uniform or universal stereotypes. Studies of women and famine in India, for example, find that their plight and roles vary enormously from place to place. They may be the most resilient members of society and play a decisive role in coping with, and pulling out of, disasters. The women of mountain peoples include those that have participated fully in efforts to resist oppressive governments, sometimes as soldiers, and to open up commercial opportunities that offset the problems of modernisation. Women figure in a growing number of initiatives to combat economic exploitation and the abuse of mountain environments. They may still remain unusually vulnerable to age-old and modern dangers.

Children in disaster

The scale of uprooting and killing of women and children in recent mountain land conflicts, especially, has been unprecedented. Victims' basic needs far exceed the resources available to humanitarian agencies, even their capacity to identify, let alone adequately assist, the millions of survivors suffering long-term harm from these events (Ahlstrom 1991: p.19; Muecke, 1992).

Death rates in many refugee camps, at least in the early emergency phase, are generally fifteen times greater than in the countries from which refugees come and are predominantly among children. In camps serving Burmese, Ethiopian, and Bhutanese internal refugees, between 15 and 50 percent of children suffered from malnutrition. Deaths among children under 5 years were three to five times higher than overall rates. In a camp for Kurdish refugees in 1991, children accounted for 64 percent of deaths but only 18 percent of the camp population (Toole, 1993: p.148). Hospitals in Kabul, Afghanistan, reported to the Red Cross that a quarter of all victims of anti-personnel mines were children, and another quarter were women (ICRCRCS, 1994: p.62). Maksoud (1992) reported and analysed the almost universal occurrence of 'traumatic war experience' among Lebanese children in recent conflicts.

Here, one last point needs to be stressed. For families and communities involved, hopefully for most observers, the question of *who* dies, or whose lives are blighted, and opportunities cut short, is more than one of mere numbers. Indeed, it is unacceptable that children should be exposed to injury and death in proportion even to their share of population, or *at all*, without every reasonable effort to protect them.

At Aberfan, Wales in 1966, there were 144 deaths in the mudslide from a coal tip on the mountainside above. But of these, 116 were young children in school at the time. Aberfan, like any community, cared for its children in a special way and had surely hoped to protect them before almost anyone else. That they were unable to do so, leaves a huge burden of irreversible loss and unresolved grief.

In war, especially, children should not be exposed deliberately to military operations. That is a crime of war as well as the road to calamity. Indeed,

neither should their mothers be targeted, or the elderly or disabled. In civilised countries, supposedly, all non-combatants are excluded or protected from armed assault. They are, by definition, completely vulnerable. Yet we saw that, in recent mountain wars and displacements of peoples, these have been the majority of victims. They should not be targeted at all, and yet they are dying in the greatest absolute and relative numbers. These are disasters of humanity and civility, as well as of untimely death and unnecessary suffering.

The vulnerability of indigenous peoples

We have already noted the distinctive plight of many indigenous nationalities. There is no intrinsic reason why they should be more vulnerable than others, indeed, good reasons why they can be less vulnerable. They are not only residents of the mountains, but distinguished by cultures attached to the particular mountain land for generations, sometimes millennia, of adaptation to its dangers and benefits. Other things being equal, one might expect them to live more securely in their land. Anthropological and other studies suggest many of them have done so.

However, their vulnerability is generally a function of outside influences, imported diseases, wars, changed economic conditions, and the demands of modern states (Horowitz, 1989; Amnesty International, 1989; 1993). Few have a language and identity that coincides with the state, or administrative unit where they reside. Such differences, added to prejudice and agendas that ignore their needs and values, can leave them with fewer choices and protections, and subject to risks over which they have little or no control.

Historically, some of the worse decimation and exterminations of mountain peoples have been due to diseases brought in by foreign traders and armies. Estimates for the mountain regions of the Americas suggest that, in the century or so after Columbus first reached the Caribbean, the indigenous population declined catastrophically, probably by as much as 80 percent. In this, European and African diseases played a larger role than war, plundering, and displacement. Many groups disappeared altogether, and most that kept their identity have not returned to pre-Columbian numbers to the present day. However, the disasters of conquest, subjugation, exploitation, and genocide continue to the present. In this century, it has mainly been armed violence and arbitrary punishment, that have brought the great disasters for indigenous peoples world-wide, even when intended mainly to ensure their submission and access to their resources and labour. The laws, attitudes, and agendas of most modern states not only fail to provide adequate protection and social justice for indigenous peoples, they serve to permit modernisation that runs roughshod over minorities and fails to protect them against violent exploitation.

The vulnerability of indigenous nationalities, is implicit in the countless atrocities and major humanitarian disasters where they are subjected to violence by regular state forces (Table 16.14). At a minimum, their rights are ignored while their land, labour, and resources are appropriated at gun point. Often that is a prelude to enforced uprootings, brutalisation, deliberate spreading of infectious diseases or failure to check them, and genocide. Unless the international community has the means and will to intervene, the only alternatives for survivors are armed retaliation or abandoning their land and cultural identity (Minority Rights Group, 1984; Nietschmann, 1987, 1987a). These are not only disasters for the peoples targeted. They annihilate unique cultures, destroy irreplaceable knowledge, traditions, skills, and mountain-adapted practices. There are grave implications for sustainable futures. Indigenous cultures are, arguably, 'ground zero' in the destruction of sustainable living. They exist because they have practised sustainability for past centuries and millennia. Their demise seems to be greater cause for concern even than animal extinctions and loss of ecological diversity, although these commonly go hand in hand with destruction of indigenous cultures.

Highland–Lowland interactions and vulnerability

In the larger scheme, it is misleading to treat mountain environments and peoples as a separate problem for disaster or sustainability. The plight of any given mountain area is fully bound up with that of surrounding lands, both lowlands and other highland areas. The importance of 'highland–lowland' interactions, is clearest in the most destructive of human hazards discussed, those of armed violence and enforced displacements. The mountain environment has often conferred some special advantages on mountain warriors and offered protection for their communities. With the scale and forms of modern weaponries, that becomes less and less true. Mountain habitats and people become

increasingly vulnerable to destruction. Likewise, the international trafficking in and addiction plagues of hard drugs, has a unique relation to mountains. The participation of mountain farmers is bringing a range of severe risks to their high valleys, as well as benefits. The existence, and rapid growth of urban centres in and around mountain lands, including large cities, play key roles. They tend to control administrative, strategic, and economic developments in the mountains. Communications, projects, and visitors originate in, or are channelled through them. They are the nodes to which mountain folk go for employment, markets, hospital care and, as we saw, refuge.

Centres like Lima, Mexico City, München, Teheran, Mombasa, or Tokyo; or even more distant ones like New York or Moscow, can be as decisive in a range of mountain land risks, as those in the more rugged, higher altitude settings. That is most obvious in the problems of war and epidemic disease. But even natural and technological hazards confined to mountain areas, become more or less dangerous in relation to the pressures from, and agendas in, centres outside them. Risk evaluation will not suffice if it fails to recognise the place of mountain lands in a larger social and ecological context.

CONCLUDING REMARKS: REDUCING VULNERABILITY

The United Nations, its relevant agencies and member states, have made the 1990s an 'International Decade of Natural Disaster Reduction' (IDNDR). This reflects the belief of prominent political, technical, humanitarian, and other organisations, that much more can be done, and should be done, to reduce the toll of disasters. Seven years into the Decade, most indications are of disasters continuing to increase in numbers and scope. Yet few will say the disasters, or many of their damages, are inevitable. However, rather than any tangible improvements, there is a shift in the balance of opinion as to how they can be achieved. It is a shift towards approaches more in keeping with the requirements for sustainability and towards greater sensitivity to, and involvement of, the places and peoples at risk.

The original IDNDR vision was of 'heroic' applications of the latest, expert understanding and modern organisation, and their transfer to less wealthy countries. It involved a vision of disasters as arising essentially from extreme natural

Table 16.14 Vulnerable peoples: world-wide examples of indigenous mountain peoples subject to armed attacks and systematic human rights violations by regular military and paramilitary forces of the government in power. Many of the same peoples are actively engaged in organised resistance, struggles for greater autonomy or independence, sometimes by peaceful means, sometimes by armed action. However, all cases identified here have involved unarmed, ordinary people of the indigenous group, including women and children, terrorised and attacked by the armed forces; their leaders, students, priests or intellectuals, subject to arbitrary arrest, torture, extrajudicial execution and other violations (after Nietschmann, 1987; Amnesty International, 1993)

Region/country	Indigenous nationality
AFRICA	
Burundi	Hutu
Ethiopia	Afar, Anuak, Eritrean, Oromo, Sidamo, Somali, Tigrayan
Kenya	Somali
Morocco	Saharwis (Western Moroccans)
Rwanda	Tutsi
Uganda	Peoples of northern areas
ASIA, EAST and SOUTH EAST	
Bhutan	Nepali
China	Tibetan, Inner Mongolian (various), Kazakh
Indonesia	East Timorese, Irian Jaya (various), Acehnese
Laos	Hmong
Myanmar (Burma)	Arakan, Chin, Kachun, Karen, Karren, Mon, Pao, Paloung, Shan
Philippines	Cordilleran provinces (various), Kalinga, Moro, Tinggian
Papua New Guinea	Bougainville
Vietnam	Hmong, 'Montagnard'
ASIA, SOUTH and SOUTH WEST	
Afghanistan	Aimaq, Hazaran, Kirghiz, Nuristani, Pathan, Tadjik, Turkmen, Uzbek, Wakhi
Bangladesh	Chakma, Khimi, Khyang, Koch, Mandi, Marma, Murung, Tanchangya, Tripura
India	Gurkha, Kashmiri, Naga
Iraq	Kurd
Iran	Kurd, Turkmen, Baluchi
Lebanon	Druze, Maronite Christian, Palestinian
Pakistan*	Baluchi, Brahui, Khowari, Kohistani, Pathan
Syria	Kurd, Palestinian Kurd
Turkey	Kurd
EUROPE	
Armenia	Azeri
Azerbaijan	Armenian
Bosnia	Bosnian Muslim, Serb, Croat
Croatia	Serb, Bosnian Muslim
Georgia	Abkhazia
Greece	Macedonian
Russia	Chechnyan, Ossetian, Volga German
Serbia/Yugoslavia	Bosnian Muslim, Croat, Macedonian, Montenegran
LATIN AMERICA	
Bolivia	Yuracaré
Chile	Mapuche
Colombia	Cauca (various peoples)
Costa Rica	Guaymi
Ecuador	(various)
El Salvador	Pipil
Guatemala	Mayan, Quiché
Honduras	Maya
Mexico	Ch'ol, Mixe, Triqui, Tzeltal, Tzotzil, Zapotec, Zoque
Nicaragua	Creole, Miskito, Rama, Sumo
Peru	(various)

*Northern and western mountains

agents or technological accident. The latest scientific knowledge, technical, and administrative strategies, were promoted as offering the most promising solutions (Housner, 1989; UNDRO, 1990; McCall *et al.*, 1992).

Now, transfer of modern knowledge and technology surely has a role so long as less wealthy countries continue to build installations like large dams or nuclear power plants, to import industrial products, and to provide their wealthier enclaves with modern infrastructure. However, not only are these prohibitively expensive for the vast majority of disaster victims, their plight involves quite different realities and needs. Indeed, the evidence from mountain lands suggests that *only* improvements in conditions for persons and practices we have identified with unusual disadvantages and vulnerability, offer substantial promise of disaster reduction. A shift in this direction came through more forcefully in the concerns and conclusions of the mid-term IDNDR Conference at Yokohama in 1995 (UNDHA, 1994). To an ever greater extent it is the consensus among local, NGOs, and international humanitarian organisations. It is a plea voiced, above all, by field workers and others dealing with crisis and devastation on the ground (Davis, 1981; Harrell-Bond, 1986; Maskrey, 1989; Francois, 1992, 1993; IFRCRCS, 1994; Blaikie *et al.*, 1994; UNDHA, 1996).

Vulnerability, as we saw, identifies the predicaments that place people in danger and those who lack physical and societal protections. The forms of vulnerability listed earlier, or assessments of them for given persons and places, also prescribe the primary requirements to reduce risk and improve public safety. Mitigation of risks and losses for the vulnerable, turn upon improvements in the most basic of living conditions. Of course, there are natural hazards that we can do little or nothing to control. No technology can be made perfectly safe, however much is spent. There are growing doubts that any or many diseases can be eradicated, and that attempts to do so may have serious or lethal repercussions in other areas of ecological and human health. Rather than the strategies of choice, as they now are, the search for technological control over as much of the environment as possible, and centrally organised solutions to most environmental dangers, appear as crude and unreliable methods of last resort. In terms of disaster reduction, measures that address the predicaments of the most vulnerable persons seem far more promising, if less widely favoured.

Even quite modest uplift and improved security in everyday life for the more vulnerable, could dramatically reduce disaster-proneness. Many of the risks that they face can be reduced or removed by simple avoidance and better land-use choices. Dangers could be diminished or avoided by modest improvements in building quality and siting; by modest but widespread measures to improve domestic and community safety and public health. Much could also be done by providing insurance or social security, education and risk information to those most at risk, and by foregoing costly and dangerous technologies or risky ventures. A major problem, however, is to understand how vulnerable persons and communities in the mountains are situated within larger regional, national, and global contexts. To improve and maintain safety in their own communities, they need protection from, and powers sufficient to counter, unwanted outside pressures. Far from being a prescription for ignoring mountain cultures and institutions, this view argues for taking their relation to modernisation much more seriously. It requires that they be given a voice in what happens in their worlds under pressures of globalisation.

Approaches concerned with reducing vulnerability also require sensitivity to conditions in each place and habitat. They need the kind of knowledge that only long-term residents can provide and the skills they can deploy; serious levels of investment in alternative, culturally and ecological appropriate development options. But these are social rather than technical problems. They recognise that disaster reduction is often more about endangerment (of some) by modernisation, than risks that come from lack of it. Responsible professions and agencies need to implement protocols for paying attention to the concerns, capabilities, and options of those at risk. In most countries, legislation is needed to empower them in matters affecting their safety and habitat. These basic requirements of disaster mitigation are not substantially different from those of sustainable communities and ecosystems. However, it is clear that they are as much matters of social justice and ecological responsibility, as specifically about disaster or sustainability. Hence, many researchers and humanitarian organisations, especially those working on the ground, put social and political improvements ahead of all other requirements for disaster reduction (Cahill, 1993). They see human rights, and civil protections for the more vulnerable members of society, as even more fundamental than scientific knowledge and practices relating to, say, earthquakes and avalanches; toxic chemicals or disease vectors – not least to ensure appropriate development and uses of scientific knowledge. That, again, reflects the evidence that vulnerability is prefigured mainly in the social order, the details and lack of decencies of everyday life (Wisner, 1993; Watts and Bohle, 1993; Hewitt, 1996).

Of course, all of these are utopian notions so long as the more wealthy and influential societies, groups, institutions and professions are preoccupied with other matters. They have no chance in most mountain lands, unless the more powerful economic and political institutions cease to promote developments that increase habitat abuse or impose risks upon inhabitants least able to influence or benefit from changes. Disaster reduction is hardly an option when militarisation and present levels of armed force used against unarmed populations continue. The use of force to impose distant agendas upon mountain cultures is incompatible with safer living. These are also among conditions undermining any chance of sustainable futures for mountain lands.

ACKNOWLEDGEMENTS

Thanks for help with library search and disaster inventories to Ian Gilbart, Tom Hammers and George Yap, Graduate Studies, Wilfrid Laurier University; and to Dr Jodi Decker for help with sources on epidemic diseases.

References

Ahlström, C. (1991). *Casualties of Conflict: Report for the World Campaign for the Protection of Victims of War.* Department of Peace and Conflict Research, Uppsala University: Uppsala

Alexander, D. (1993). *Natural Disasters.* Chapman and Hall: New York

Allan, N. J. R. (1987). Impact of Afghan refugees on vegetation resources of Pakistan's Hindukush–Himalaya. *Mountain Research and Development,* 7: 200–4

Allan, N. J. R., Knapp, G. W. and Stadel, C. (eds.) (1988). *Human Impact on Mountains.* Rowman and Littlefield: Totowa, NJ

Amnesty International (1975). *Report on Torture.* AI: New York

Amnesty International (1981). *Disappearance: A Workbook.* AI: New York

Amnesty International (1985). *East Timor: Violations of Human Rights.* AI: London

Amnesty International (1986). *Unlawful Killings and Torture in the Chittagong Hill Tracts.* AI: London

Amnesty International (1989). Indigenous Peoples: caught in the crossfire. *AI Bulletin,* August/September, 26–8

Amnesty International (1993). *Human Rights Violations Against Indigenous Peoples of the Americas.* AI: New York

Azhar, S. (1990). Afghan refugees in Pakistan: a Pakistani view. Anderson, E. W. and Dupree, N. H. (eds.), *The Cultural Basis of Afghan Nationalism.* Pinter: London, pp. 105–114

Blaikie, P., Cannon, T., Davis, I. and Wisner, B. (1994). *At Risk: natural hazards, people's vulnerability, and disasters.* Routledge: London

Blong, R. J. (1984). *Volcanic Hazards : a source book on the effects of eruptions.* Academic Press: London

Bode, B. (1989). *No Bells to Toll : Destruction and Creation in the Andes.* Charles Scribners: New York

Bohle, H.G. (ed.) (1993). *Worlds of Pain and Hunger: Geographical Perspectives on Disaster Vulnerability and Food Security.* Freiburg Studies in Development Geography 5. Verlag Breitenbach: Saarbrücken

Bonnard, C. (ed.) (1991). *Proceedings, Fifth International Symposium on Landslides.* Lausanne, Switzerland, 10–15 July, 1. A. A. Balkema: Rotterdam, (3 vols.)

Brabb, E. E. and Harrod, B. L. (eds.) (1989). *Landslides: Extent and Economic Significance.* A. A. Balkema: Rotterdam

Brogan, P. (1989). *The Fighting Never Stopped: a comprehensive guide to world conflict since 1945.* Random House: New York

Cahill, K. M. (ed) (1993). *Framework for Survival: health, human rights, and humanitarian assistance in conflicts and disasters.* Basic Books: New York

Clark, I. (1989). Internal refugees – the hidden half. *World Refugee Survey – 1988 in review,* 18–24

Coates, D. R. (ed.) (1977). *Landslides*. Reviews in Engineering Geology, 111. Geological Society of America; Boulder

Davis, I. (ed) (1981). *Disasters and the Small Dwelling*. Pergamon Press: Oxford

de Scally, F. and Gardner, J. S. (1994). Characteristics and mitigation of the snow avalanche hazard in Kaghan Valley, Pakistan Himalaya. *Natural Hazards*, 9: 197–213

Ehrlich, A. H. and Birks, J. W. (eds.) (1990). *Hidden Dangers: environmental consequences of preparing for war*. Sierra Club Books: San Francisco

Eisbacher, G. H. and Clague, J. (1985). *Destructive Mass Movements in High Mountains*. Geological Survey of Canada, Paper 84/16

Elliot, G. (1972). *Twentieth Century Book of the Dead*. Penguin Books: Harmondsworth

Epps, K. (1993). The world's crop of land mines : reaping a deadly harvest. *Ploughshares Monitor* 14(3), 11–14. Project Ploughshares, Conrad Grebel College, Waterloo, Ontario

Erikson, K. T. (1976). *Everything in its Path: destruction of community in the Buffalo Creek Flood*. Simon and Schuster: New York

Evans, R. (1989). The death industry (Narcotics). *Geographical Magazine* May, 10–14

Fleming, A. F., Carballo, M., FitzSimons, D. W., Bailey, M. R. and Mann, J. (eds.) (1988). *The Global Impact of AIDS*. Alan R. Liss: New York

Francois, J. (1992). *Populations in Danger*. J. Libbey: London

Francois, J. (1993). *Life, Death and Aid: The Medecins sans Frontieres Report on World Crises*. Routledge: London

Garrett, L. (1994). *The Coming Plague: Newly Emerging Diseases in a World Out of Balance*. Farrar, Straus and Giroux: New York

Glass, R. I., Urrutia, J., Sibony, S., Smith, H., Garcia, B. and Rizzo, L. (1977). Earthquake injuries related to housing in a Guatemalan village. *Science*, 197: 638–43

Gould, P. (1993). *The Slow Plague; a geography of the AIDS pandemic*. Blackwell: Oxford

Harrell-Bond, B. (1986). *Imposing Aid: Emergency Assistance to Refugees*. Oxford University Press: Oxford

Hewitt, K. (1976). Earthquake Hazards in the Mountains. *Natural History*, 85(5): 30–7

Hewitt, K. (1982). Natural Dams and Outburst Floods of the Karakoram Himalaya. In Glen, J. (ed.), *Hydrological Aspects of High Mountain Areas*. International Association of Scientific Hydrology. IAHS Publ. 138, pp. 259–69

Hewitt, K. (ed.) (1983). *Interpretations of Calamity, from the Viewpoint of Human Ecology*. Allen and Unwin: London

Hewitt, F. (1991). Women in the Landscape: A Karakoram Village Before 'Development'. Unpublished Doctoral Thesis, University of Waterloo, Canada

Hewitt, K. (1992). Mountain hazards. *GeoJournal*, 27: 47–60

Hewitt, K. (1995). Excluded perspective in the social construction of disaster. *International Journal of Mass Emergencies and Disasters*, 13(3): 317–39

Hewitt, K. (1996). *Regions of Risk: a geographical introduction to disasters*. Addison Wesley Longman: London

Hogg, S. J. (1980). Reconstruction following seismic disaster in Venzone, Friuli. *Disasters*, 4: 173–86

Horowitz, M. M. (1989). Victims of development. In *Development Anthropology Network*, 7(2): 1–8

Housner, G. W. (1989). An international decade for natural disaster reduction, 1990–2000. *Natural Hazards*, 2: 45–75

IFRCRCS (International Federation of Red Cross and Red Crescent Societies) (1994). *World Disasters Report, 1994*. Martinus Nijhoff: Dordrecht

Ives, J. D. and Messerli, B. (1989). *The Himalayan Dilemma: reconciling development and conservation*. Routledge: London

Jansen, R. B. (1980). *Dams and Public Safety*. Bureau of Reclamation, US Department of the Interior, Denver, Colorado

Jones, B. G. and Tomazevic, M. (eds.) (1982). *Social and Economic Aspects of Earthquakes. Proceedings of the Third International Conference; Bled, Yugoslavia, 1981*. Cornell University: Ithaca, NY

Jones, D. K. C., Brunsden, D. and Goudie, A. S. (1983). A preliminary geomorphological assessment of part of the Karakoram Highway. *Quarterly Journal of Engineering Geology*, London, 16: 331–55

Keefer, D. K. (1984). Landslides caused by earthquakes. *Geological Society of America Bulletin*, **95**: 406–21

Lasa, J. and Jinfang, N. (1992). A tentative appraisal of the environmental impact of railway construction in mountain areas. In Walling, D. E., Davies, T. R. and Hasholt, B. (eds.), *Erosion, Debris Flows and Environment in Mountain Regions*. International Association of Hydrological Sciences: Wallingford, No. 209. pp. 413–18

Li Tianchi, (1989). Landslides: extent and economic significance in China. In Brabb, E. E. and Harod, B. L., *Landslides: Extent and Economic Significance*. A. A. Balkema: Rotterdam, pp. 271–87

Li Tianchi (1990). *Landslide Management in the Mountain Areas of China*. ICIMOD: Katmandu, p.62

Li Tianchi, Schuster, R. L. and Wu Jushan (1986). Landslide dams in south-central China. In Schuster, R. L. (ed.), *Landslide Dams: Process, Risk, and Mitigation*. Special Publication No. 3, American Society of Civil Engineers: New York, pp. 146–63

Li Tianchi and Wang Schumin (1992). *Landslide Hazards and their Mitigation in China*. Science Press: Beijing, p.84

Lovejoy, T. E. (1993). Global change and epidemiology: nasty synergies. In Morse, S. S. (ed.), *Emerging Viruses*. Oxford University Press: Oxford, pp. 261–8

Macksoud, M. S. (1992). Assessing war trauma in children: a case study of Lebanese children. *Journal of Refugee Studies*, **5**(1): 1–15

Mann, J. M. (ed.) (1992). *Aids in the World*. Harvard University Press: Cambridge

Maskrey, A. (1989). *Disaster Mitigation: A Community Based Approach*. Oxfam: Oxford

McCall, G. J. H., Laming, D. J. C. and Scott, S. C. (eds.) (1992). *Geohazards: natural and manmade*. Chapman Hall: New York

McCoy, A. W. (1972). *The Politics of Heroin in Southeast Asia*. Harper and Row: New York

Morales, W. Q. (1992). Militarising the drug war in Bolivia. *Third World Quarterly*, **13**(2): 353–70

Muecke, M. A. (1992). New Paradigms for refugee health problems. *Social Science Medicine*, **35**(4): 515–23

Nietschmann, B. (1987). Economic Development by Invasion of Indigenous Nations: Cases from Indonesia and Bangladesh. *Cultural Survival Quarterly*, **10**(2): 2–12

OFDA (United States Office of Foreign Disaster Assistance) (1988). *Disaster History: Significant Data on Major Disasters Worldwide, 1900–Present*. Agency for International Development: Washington, DC

Oliver-Smith, A. (1986). *The Martyred City: Death and Rebirth in the Andes*. University of New Mexico Press: Albuquerque, NM

Parasurman, S. (1985). The impact of the 1993 Latur-Osmanabad (Maharashtra) earthquake on lives, livelihoods and property. *Disasters*, **14**(2): 156–69

Partridge, W. L. (1989). Involuntary resettlement in development projects. *Journal of Refugee Studies*, **2**(3): 373–84

Pizzarello, L. D. (1986). Age-specific xerophthalmia rates among displaced Ethiopians. *Archives of Displaced Children*, **61**: 1100–3

Price, L. W. (1981). *Mountains and Man: A Study of Process and Environment*. University of California Press: Berkeley

Project Ploughshare (1996). *Armed Conflict Report, 1996*. Institute of Peace and Conflict Studies, University of Waterloo, Waterloo, Canada

Quarantelli, E. L. (ed.) (1978). *Disasters: theory and research*. Sage: London

Rance, S. (1991). Growing the stuff: coca growers in Bolivia. *New Internationalist* **224**, October, 10–12

Raymond, J. G. (1993). *Women as Wombs: reproductive technologies and the battle over women's freedom*. Harper: San Francisco

Sabo, L. E. and Kibirge, J. S. (1989). Political violence and Eritrean health care. *Social Science Medicine*, **28**: 677–84

Sachs, W. (ed.) (1990). *The Development Dictionary*. Zed Books: London

Schuster, R. L. and Fleming, R. W. (1986). Economic losses and fatalities due to landslides. *Association of Engineering Geologists, Bulletin*, **23**: 11–28

Seaman, J., Leivesey, S. and Hogg, C. (1984). *Epidemiology of Natural Disasters*. Karger: Basle

Seddon, D. (1995). AIDS in Nepal: issues for consideration. *Himalayan Research Bulletin* **15**(2), 2–11. Portland, Oregon

Serafim, J. L. (1984). *Safety of Dams: Proceedings of the International Conference, Coimbra/23–28 April, 1984.* A. A. Balkema: Rotterdam

Sheets, P. D. and Grayson, D. K. (eds.) (1979). *Volcanic Activity and Human Ecology.* Academic Press: New York

Shiva, V. (ed.) (1994). *Close to Home: Women Reconnect Ecology, Health, and Development Worldwide.* New Society Publishers: Philadelphia

Sivard, R. L. (ed.) (1993). *World Military and Social Expenditures 1992.* World Priorities Inc: Washington, DC

Sliwinski, M. (1989). Afghanistan: the decimation of a people. *Orbis*, **33**(1): 39–56

Sowers, G. F. and Carter, B. R. (1979). Paracti Rockslide, Bolivia. In Voight, B. (ed.), *Rockslides and Avalanches*, chapter 10, 401–18. Elsevier: New York

Timm, R. W. (1991). *The Adivasis of Bangladesh.* Minority Rights Group Report, London

Toole, M. J. (1993). The public health consequences of inaction: lessons learned in responding to sudden population displacements. In Cahill, K. M. (ed.), *Framework for Survival: health, human rights, and humanitarian assistance in conflicts and disasters* chapter 9, pp. 144–59. Basic Books: New York

Turner, B. A. (1978). *ManMade Disasters.* Wykeham Publications: London

UNDHA (United Nations Department of Humanitarian Affairs) (1996). Focus: Disasters and Development. *DHA NEWS*, **10** April–May, 3–23

UNDHA (1994). Dignity and Sorrow in Troubled States. *DHA News*, **8**, March–April

UNDHA (1993). Review, Special Edition *DHA News* **7**, January–February

UNDHA (1992). Complex emergencies: a united approach. *DHAUNDRO News* July/August 4–6

UNDRO (United Nations Disaster Relief Organisation) (1990). World Launches International Decade for Natural Disaster Reduction. *UNDRO News Special Issue*, Jan/Feb, Geneva

Usher, A. D. (1994). After the forest: AIDS as ecological collapse in Thailand. In Shiva, V. (ed.), *Close to Home: Women Reconnect Ecology, Health, and Development Worldwide.* New Society Publishers: Philadelphia

van den Berghe, P. L. (1990). *Violence and Ethnicity.* University of Colorado Press: Niwot

Voight. B. (ed.) (1979). Rockslides and Avalanches, vol. 1 *Natural Phenomena.* 2 *Engineering Sites.* Elsevier: New York

Walling, D. E., Davies, T. R. and Hasholt, B. (eds.) (1992). *Erosion, Debris Flows and Environment in Mountain Regions.* International Association of Hydrological Sciences: Wallingford, Oxfordshire, Publication **209**

Watts, M. and Bohle, H. G. (1993). The space of vulnerability: the casual structure of hunger and famine. *Progress in Human Geography*, **17**: 143–67

Wechsberg, J. (1958). *Avalanche.* Weidenfeld and Nicolson: London

Weinbaum, M. G. (1994). *Pakistan and Afghanistan: Resistance and Reconstruction.* Westview: Boulder, CO

Westing, A. H. (ed.) (1984). *Environmental Warfare: a technical, legal and policy appraisal.* Stockholm International Peace Research Institute/Taylor and Francis: London

Wijkman, A. and Timberlake, L. (1984). *Natural Disasters: acts of God or acts of man?* Earthscan: London

Wilson, M. E., Levins, R., Spielman, A. and Eckardt, I. (eds.) (1994). Diseases in Evolution; Global Changes and Emergence of Infectious Diseases. *Annals*, New York Academy of Sciences, **740**, 15 December

Wisner, B. (1993). Disaster vulnerability, geographical scale and existential reality. In Bohle, H.G. (ed.), *Worlds of Pain and Hunger: Geographical Perspectives on Disaster Vulnerability and Food Security.* Verlag Breitenbach: Saarbrücken, pp. 13–52

World Refugee Statistics (1987). *World Refuge Survey.* US Committee for Refugees: Washington DC, pp. 30–4

Young, G. (ed.) (1993). *Snow and Glacier Hydrology.* International Association of Hydrological Sciences, Oxfordshire, Publication **218**

Zwi, A. and Ugalde, A. (1991). Political violence in the Third World a public health issue. *Health Policy and Planning*, **6**(3): 203–17.

APPENDIX 1 Disasters in mountain lands

COUNTRY	FL	DR	EQ	TS	VE	LS	AV	TC	ST	HW	FI	IN	EP	FA	TA	CS	DP	Events	Killed	Killed (largest)	Homeless	Affected	Costs (US$)
Austria	4					1	2											5	225	AV 200		AV 380	
Afghanistan		5	2			1									1			14	3061	(EQ) 2000	FL 80t	DR 600t	FL 52m
Albania			2															2	46	EQ 35			
Algeria	9	1	4									2			3			19	4919	EQ 2633	EQ 443t	EQ 478 948	EQ 5.2m
Bolivia	14	2			2								2		2	2		23	817	FL 250	FL 20t	DR 1583t	DR 500m
Cameroon		1			2							1			3			4	1771	VE 1734	VE 4.6t	VE 4.6	DR 1.5
Canary Islands															3			3	863	TA 863		TA 6.3	
Chile	6	2	20	3	1	1			5					1	1		1	37	40 990	EQ 30 000	EQ 2m	EQ 2.3m	EQ 1.5b
Colombia	14		33		2	9			2		2				4			39	27 508	VE 21 800	FL 105t	FL 5m	VO 1b
Costa Rica	5	1	2		4	1					2							15	168	VE 87	VE 5t	VE 70t	FL 24m
Ecuador	6	1	8		3	5					2	3						30	7830	EQ 6000	EQ 28t	FL 700t	FL 232m
El Salvador	1	4	4					1					1	1	1	1	4	13	22 759	DP 20 000	EQ 250t	TA 990t	EQ 1m
Ethiopia	5	13	1									3	1	1 (5)***	1	2	2	25	403 459	DR/FA 300 000	FL 20t	DR/FA 7.7m	DR 76m
Greece	2	10		2						1 1	2					4	1	16	1747	HW 1000	EQ 16.5t	EQ 600t	EQ 1b
Guatemala	3	1	3						1			1				4	1	20	25 209	EQ 23 000	EQ 1.2m	EQ 3.7m	EQ 1b
Honduras	5	2						3	3			2			1	1	1	15	8470	TC 8000	FL 6t	TC 600t	TC 540m
Iceland			1		1	1					1	1						3	15	LS 12	VE 5.2t	VE 5.2t	VE 24.7m
India**	5	1	6			1	3	2	2		1	12	2		5	1	2	22	375 408	EQ 373 000	EQ 450t	CS 3.5m	EQ 195m
Indonesia	36	8	22	3	16	8	2	2	3		1		2			1	1	110	24 803	DR/FA 8000	FL 150t	DR 3.5m	EQ 3.5m
Iran	9	1	32		1	1		1	3		1				3			53	58 473	EQ 15 000	FL 150t	FL 400t	FL 24b
Italy	8	1	13		1	3		2	5		2				6			38	115 027	EQ 75 000	FL 578t	FL 2.1m	FL 24b
Japan	9	1	16	2	2	3		39	3		5				6			89	182 466	EQ 143 000	TC 1m	FL 3m	EQ 600m
Lebanon	1		4									1			6	5	3	9	61 571	CS 40 000	DP 250t	CS 1.4m	CS 2.5b

*Column groups: INITIATING AGENTS (FL DR EQ TS VE LS AV TC ST HW FI IN EP FA TA CS DP); TOTALS (Events, Killed); LARGEST BY IMPACTS**** (Killed, Homeless, Affected, Costs (US$)).*

APPENDIX 1 (continued)

COUNTRY	FL	DR	EQ	TS	VE	LS	AV	TC	ST	HW	FI	IN	EP	FA	TA	CS	DP	Events	Killed	killed	Homeless	Affected	Costs (US$)
																						(INITIATING AGENTS / TOTALS / LARGEST BY IMPACTS****)		
Mexico	16	1	12	2	2	2		14	6		1	1			6			59	17 057	EQ	EQ 8776	EQ 100t / FL 100t	ST 300t	EQ 4b
Morocco	5	2	2	1									1					1.5	12 459	EQ	FL 12 000	FL 48t	FL 266t	FL 30m
Nepal	9	4	1			5					1	2	2					22	3607	EP	FL 1000	FL 30t	DR 3.5m	EQ 245m
New Zealand	1		2	1	1													4	151	VE	FL 150	FL 2.6t	FL 2.6m	FL 20m
Nicaragua	5		3	1	1			1			1		2	1	3	3		20	43 276	CS	EQ 30 000	EQ 300t	CS 1m	CS 2b
Norway													1		3	3		(633 133)	123	TA	123			
Pakistan**	2	1	4		1	1	2				1					2		10	65 125	EQ	EQ 60 000	EQ 5.2t	CS 1.2m	EQ 3m
Papua N.G.	1	1	5		5	1		2			1						1	15	3554	VE	FL 3000	FL 2t	DR 40t	FL 11.9m
Peru	15	3	20			12	1			1	1		2					55	74 544	EQ/LS	EQ 66 794	FL 75t	EQ 3m	FL 988m
Switzerland	3					3	1		4		1			1	1			10	159	AV	90		LS 48	ST 2m
Taiwan	3		3			1		10			4		5		4			28	7130	EQ	EQ 6000	EQ 23t	TC 40t	TC 405m
Turkey	3		21						2		2				8			43	44 165	EQ	EQ 23 000	EQ 108t	EQ 239t	FI 178m
Venezuela	3	2	2		1	1							1		2			11	1264	FI	FI 500	EQ 80t	EQ 80t	EQ 50m
Yugoslavia	4		8										1		3			16	1638	EQ	EQ 1100	EQ 100t	EQ 300t	EQ 500m
Grand Total	201	48	246	10	46	60	7	62	37	2	31	9	37	7	66	19	13	925	1 598 478				
%	22.5	5.5	28	1	5	8.7	1	6	4	0.2	3.5	1	4	1	7	2	1.5						

* Abbreviations: FL = flood, DR = drought, EQ = earthquake, TS = tsunami, VE = volcanic eruption, LS = landslide, AV = avalanche, TC = tropical cyclone (= hurricane, typhoon), ST = severe storm, HW = heat wave, FI = fire, IN = infestation, EP = epidemic, FA = famine, TA = technological accident, CS = civil strife, DP = displaced persons

** Northern and western mountains only

*** Famines + drought + epidemics

**** t = thousand, m = million, b = billion

Climate change

17

Martin F. Price and Roger G. Barry

INTRODUCTION

A little more than two centuries since the Industrial Revolution began in Western Europe, there is a general consensus among scientists from around the world that human activities have caused, and will continue to cause, irreversible changes in the chemical and physical characteristics of the atmosphere on which humankind depends. In little less than a decade, the issue of anthropogenic climate change has moved from a minor scientific debate to a central place on the global political and scientific agenda.

At the 1992 United Nations Conference on Environment and Development (UNCED, Rio, 1992), the majority of the world's governments signed the United Nations Framework Convention on Climate Change, and are now Parties to the Convention. Although the evidence for global climate change remains uncertain, four undeniable facts suggest that an era of such change is imminent, if it has not already arrived. First, two independent types of research show increasing atmospheric concentrations of certain gases. Air bubbles trapped in polar ice sheets demonstrate that concentrations of carbon dioxide (CO_2) have increased by almost 30 percent from pre-industrial conditions (1750 AD); methane (CH_4) has increased by 245 percent, and nitrous oxide (N_2O) by 11 percent. These data have been complemented over the past 40 years by analyses of air samples collected at isolated locations far from centres of industrialisation or population, including Mauna Loa on Hawai'i. These samples also show a rapid increase in concentrations of chlorofluorocarbons (CFCs), man-made chemicals invented in the 1920s and since used particularly for refrigeration. Second, laboratory experiments indicate that the addition of any of these gases to a volume of air permits it to absorb more infrared radiation; hence, they have become known as greenhouse gases (GHGs). Third, the Earth emits infrared radiation which is absorbed by GHGs (water vapour and those that are increasing due to human activities) in the

atmosphere. The fourth and crucial fact is that everyday human activities – agriculture, cooking, heating, transport – as well as many industrial processes, result in the emission of GHGs. Added together, these facts suggest that, as the Earth's population increases, atmospheric concentrations of GHGs will continue to increase, changing the energy balance of the Earth–atmosphere system in the phenomenon popularly described as global warming.

A major step in increasing knowledge about the possible implications of global climate change was the establishment of the Intergovernmental Panel on Climate Change (IPCC) by the World Meteorological Organization (WMO) and the United Nations Environment Programme (UNEP) in 1988. In its Second Assessment Report, the IPCC records that instrumental records of land surface temperatures show a general warming trend during the past 100 years (Nicholls *et al.*, 1996). The mean annual temperature of land areas in the Northern Hemisphere has increased by 0.7 °C. The last 40 years have also witnessed widespread decreases in the diurnal temperature range, as daily maximum air temperatures have risen less than daily minimum temperatures (Horton, 1995). This may be related to greater cloudiness. Changes in precipitation have varied greatly from region to region. Generally, however, amounts during the century have increased slightly over northern land areas in middle and high latitudes, with a substantial decrease in the northern subtropics since the late 1960s. Snow cover extent in the Northern Hemisphere has decreased since the mid-1980s relative to the 1970s, especially in spring. Nevertheless, snowfall has increased since the 1950s in some northern latitudes, in association with increased precipitation. This observational summary refers to general trends over large regions. Specific changes in mountain areas, where records are available, are discussed below. Nevertheless, there is clear evidence of a global warming tendency, whether this is due to natural or anthropogenic causes, or both.

Based on the observational evidence, the IPCC has concluded that there is 'a discernible human influence on global climate' (Houghton *et al.*, 1996: p. 4). This is likely to continue because of increasing concentrations of most GHGs other than CFCs, whose concentrations began to decrease in 1994 as a result of controls imposed following the 1987 Montreal Protocol (Schimel *et al.*, 1996). From the results of general circulation models (GCMs) run on the world's most powerful computers, the IPCC also concludes that, in spite of many uncertainties, the global climate will continue to change, with a projected increase in global mean temperature of between 1 and 3.5 °C by 2100 AD. Warming is likely to be greater over land areas, with a maximum for high northern latitudes in winter (Kattenberg *et al.*, 1996). This would be accompanied by a decrease in diurnal temperature range, particularly in the extra-tropics. In higher latitudes, winter precipitation is likely to increase, as is annual precipitation in the Asian monsoon region. However, there will be considerable regional variations in changes in all climatic variables, and frequencies of extreme events are especially hard to predict.

As discussed below, assessment of the future climates of mountain regions presents particular difficulties because of their complexity at all spatial scales, of which only the largest can be resolved in existing GCMs. Evaluation of the implications for mountain environments and societies is even more exploratory, partly because relatively little research on this topic has been done in mountain regions. Nevertheless, as recognised by the inclusion of a chapter on mountain regions in the IPCC's Second Assessment Report (Beniston and Fox, 1996), climate change could have highly significant implications not only in mountain regions but also for billions of people downstream.

MOUNTAIN CLIMATES

The climate of a given mountain region depends on its latitude, altitude, and location with respect to the ocean and prevailing wind direction (Barry, 1992b). For example, temperatures decrease by about 1 °C per 150 m altitude and the same per 1.25° of latitude (or 140 km poleward). Summer to winter temperature contrasts in mid-latitudes average between 25 °C and 40 °C, but are lower in coastal ranges, such as those along the west coast

of the Americas; and higher in continental interiors, especially in dry climates like those of Central Asia and the southwestern USA. In equatorial regions, the annual cycle of temperature is much smaller than the diurnal one. This is a result of the small changes in solar altitude and therefore of incoming solar radiation during the year, and the nearly constant day length. The diurnal range of temperature (between day time maximum and nocturnal minimum) on equatorial mountains is commonly at least three times that of the annual range (Sarmiento, 1986). The general characteristics of the vertical gradients of the major climatic elements are described by Richter (1996). In addition, he suggests altitudinal patterns of ecophysiological plant stress and indicators of vegetation diversity. While the relationships are presented in qualitative terms, they provide a useful basis for future application and testing.

The primary controls of cloudiness and precipitation are cyclonic storm systems in mid-latitudes, with generally meso- and small-scale convective systems operative in lower latitudes and in continental interiors in summer. In mid-latitudes, there is a broad tendency for cloudiness and precipitation to decrease inland from the west coasts of the continents as the influence of maritime air masses weakens. In equatorial regions, annual precipitation may be greatest at low elevations, decreasing with altitude, but on mountains such as the Andes, a secondary maximum is observed at higher elevations (Sarmiento, 1986). In the tropics, the wettest zone is commonly between 500 to 1000 m. In mid-latitude mountains, annual precipitation tends to increase with altitude on windward slopes (Barry, 1992c). However, research in the Alps (Blumer, 1994) shows that local topography may give rise to annual totals that increase up to some limiting elevation, remain more or less constant, or even decrease with altitude. The best individual statistical predictors of mean precipitation are the orientation of the watershed and the 'exposure', or openness of the location to moisture transport by the atmosphere (Basist *et al.*, 1994), although algebraic combinations of elevation with exposure, and of slope with watershed orientation give the best overall result. The regression equations for temperate and tropical regions differ only in the respective slope coefficients and constants.

Extensive highland areas, such as the Tibetan Plateau and Bolivian Altiplano, essentially create their own climatic regimes, whereas isolated peaks largely represent the climate of the ambient air.

Plate 17.1 The summit of Snezhka/Snieska (1602 m), the highest peak of the Giant Mountains, on the Czech/Polish border. The climatological observatory on the left is the most recent; the site has a history of data collection since 1824 (Photograph: Martin Price)

High mountain ranges, such as the western cordillera of North America, present a major obstacle to weather systems, and set up areas of precipitation deficit downwind. They also create distinctive localised wind regimes such as the Föhn of the Alps and the Chinook of the high plains bordering the Rocky Mountains. Similar distinctive local winds occur in the lee of many mountain ranges.

In addition to the three large-scale controls of climate, topography plays a strong role in modifying the receipt of solar radiation, diurnal temperature regime, wind velocity, precipitation amounts and snow cover at the local level (terrain elements such as ridgetop, valley slopes and bottom), as well as at individual sites.

Spatio-temporal characteristics of climate

Mountain climates and their vegetation cover typically form a complex mosaic related primarily to slope orientation and angle, but secondarily to altitude and relative relief (Barry, 1992a, b). The mosaic of climates on individual slopes is made up of numerous 'topoclimates', extending over a distance on the order of 100 m, and site-specific microclimates (1–10 m scale) associated with clumps of vegetation, the plant canopy structure, large rocks and irregular topography producing snow patches, for example (Barry and Van Wie, 1974).

Solar radiation received at the surface can be modelled for mountain areas provided that a digital terrain model (DTM) with adequate spatial resolu-

tion is available (Duguay, 1993). Nevertheless, DTMs may under-represent steep slopes and over-represent flat areas, as illustrated by the comparison of a 250×250 m DTM and terrestrial measurements for Switzerland (Brzeziecki *et al.*, 1994). Various computer routines, incorporating the effects of horizon screening, exist to compute daily totals incident over the terrain. Absorbed radiation can be determined similarly, given data on cloud cover and surface albedo (or reflectivity). Similar procedures, incorporating the sky view factor at each grid point, are also available to calculate net outgoing infrared radiation from the surface and thus the net (allwave) radiation budget (Duguay, 1993). Surface temperature regimes can be calculated for idealised conditions from energy budget calculations. The partitioning of the net radiation into turbulent sensible heat transfer or latent heat (evaporative) transfer can also be estimated if the moisture state at the surface is known, or assumed. Alternatively, high resolution GCMs, possibly with a nested mesoscale model, can be used to simulate surface radiative fluxes (Wild *et al.*, 1995) and in future, no doubt, the energy budget and temperature regimes.

Wind regimes in complex terrain are not easily inferred due to the interacting effects of the large-scale pressure gradient, atmospheric stability, the large-scale and local topography, and the relative heating of mountain valley and plain atmospheres and local slopes. In general, wind measurements are essential, even to calibrate mesoscale numerical models.

The climatological features of the distribution of precipitation over mountainous terrain can be estimated with statistical procedures by assuming a series of topographic rules for a given area (Basist *et al.*, 1994; Daly *et al.*, 1994). However, detailed studies in the Alps illustrate the need to exercise great care in using such methods (Blumer, 1994). The incorporation of information on atmospheric circulation characteristics can help to improve statistical procedures (Hay *et al.*, 1991; Barros and Lettenmaier, 1993). Limited-area numerical models, nested within large-scale model analyses (Giorgi, 1990), can give realistic simulations of storm precipitation events provided that an adequate horizontal resolution is used in the mesoscale model and in the terrain data (Katzfey, 1995). Such limited-area models have produced encouraging simulated distributions of precipitation over Europe and North America (Giorgi *et al.*, 1992, 1994; Marinucci *et al.*, 1995).

Data

The general status of mountain climate records has been published by Barry (1992a), who notes that the Alps and parts of the Carpathians have the longest records and most dense networks. The mountains of Britain, the Caucasus, Scandinavia, western North America, and the northern Andes have reasonable station networks, but field research has been more limited in the Andes. In other areas, the station networks are more sparse, and less research has been done. The only recent substantial changes are the readier availability of climate information for mountains in Central Asia (Aizen *et al.*, 1996; Gao and Peng, 1994; Miehe *et al.*, 1996) and the publication of precipitation records for the New Zealand Alps (Wratt *et al.*, 1996). The latter identifies annual totals averaging in excess of 1100 cm/yr in the mountains of South Island. A broader issue concerns the need for topoclimatic information in regions where such observations are lacking. Procedures for extrapolating meteorological data from available stations to adjacent mountainous terrain have been described by Running *et al.* (1987) and tested in Oregon (Glassy and Running, 1994).

EVIDENCE OF CLIMATE CHANGE IN MOUNTAIN AREAS

Observational records

The value of long-term climatic records collected at mountain observatories and other high elevation sites is increasingly being recognised. Valuable syntheses of records of the Sonnblick, Austria (Auer *et al.*, 1990; Böhm, 1986) and the Pic du Midi in the French Pyrenees (Bücher and Dessens, 1991) are now available, although many additional such compilations are required.

For the Pic du Midi (2862 m), Bücher and Dessens (1991) report a mean annual temperature increase of 0.94 °C/100 years (measured between 1882 and 1970), perhaps associated with a cloudiness increase. The warming trend was most marked in spring and autumn and is also shown to be greater at three mountain stations (Pic du Midi, Sonnblick and Säntis [Switzerland]) than at low elevation stations in Western Europe. It is also interesting that mountain stations in the Alps (Säntis, Sonnblick and Zugspitze [Germany]) show an increase of maximum and minimum temperatures, whereas lowland stations in the area – and also the Pic du Midi – show that minimum tempera-

tures increase much more than maximum values (Weber *et al.*, 1994). This asymmetry in response merits further examination. Regrettably, the Pic du Midi ceased to be a manned station in 1985, which will undoubtedly cause a discontinuity in the records from the point of view of climate change monitoring.

Not surprisingly, in view of the enormous economic importance for winter tourism as well as for water supply, records of winter snowfall and snow cover in mid-latitude mountain areas have begun to receive considerable attention. Most records show short-term anomalies of major proportions, in recent years and in the past (Barry, 1992a; Barry *et al.*, 1995), but there are few signs of any persistent long-term trends in the European Alps (Baumgartner *et al.*, 1995; Beniston *et al.*, 1996; Föhn, 1990; Koch and Rüdel, 1990), the Rocky Mountains of the USA (Changnon, 1992), or the Snowy Mountains, Australia (Galloway, 1988; Duus, 1992).

Proxy records

To extend the spatial and temporal coverage provided by climatological records, researchers have utilised various proxy records, particularly evidence of changes in alpine glaciers (Haeberli, 1995; Müller-Lehmans *et al.*, 1996; Serebryanny and Solomina, 1996) and more recently in ice cores from high-altitude ice caps in South America and Central Asia (Thompson *et al.*, 1993). Low latitude, high elevation glaciers in the Sierra Nevada de Merida, Venezuela (Schubert, 1992) and in East Africa (Hastenrath and Rostom, 1990) show dramatic changes in the climate of equatorial latitudes over the last few decades, commensurate with trends observed since the 1920s–30s in many mid-latitude glaciers (Makarevich and Rototaeva, 1986). Only maritime areas, such as the Pacific Northwest of North America and Norway, show recent advances in response to increased winter precipitation (Laumann and Reeh, 1993).

Isotopic records from ice cores from Quelccaya, Peru (14°S, 71°W) and the Dunde (38°N, 96°W) and Guliya (35°N, 81°W) ice caps of western China show that modern temperatures are the highest for at least 1500 years in Peru, and perhaps since the Holocene maximum (6000–7000 BP) in China. The recent, rapid warming in the tropical and subtropical mountains of South America and Central Asia contrasts with records from polar ice sheets which show little change (Thompson *et al.*, 1993). The

Little Ice Age cooling is prominent at Quelccaya from 1550 to 1880 AD, whereas only shorter cold intervals are evident at Dunde.

Recent widespread warming on tropical mountains is confirmed by an analysis of trends in freezing level height, as determined from twice-daily balloon soundings made at 65 radiosonde stations between 30°N and 30°S. Diaz and Graham (1996) find a 4.5 m per year rise in the altitude of the freezing level during 1970 and 1986, with a larger trend between 15°N and 15°S. Additionally, South American records for 1958 to 1990 date the beginning of the increase to the mid-1970s and suggest it was associated with a rise in tropical sea-surface temperatures and possibly an enhancement of the hydrological cycle.

POSSIBLE FUTURE CHANGES IN MOUNTAIN CLIMATES

Model projections

Climate models indicate that, in response to the progressive increase of anthropogenic GHGs in the atmosphere, the Earth's surface and lower atmosphere will warm by between 0.1 and 0.2 °C per decade (Kattenberg et al., 1996). However, considerable uncertainty remains as to the cooling effect of sulfate and dust aerosols and the net effect of changes in cloud cover. Sulfate aerosols originate over the industrial regions of the northern hemisphere, while dust aerosols are concentrated over and downwind of deserts and arid lands in general. Whereas sulfate concentrations may be expected to diminish, through controls on emissions, dust concentrations may slowly rise as a result of desertification.

Clouds exert net radiative cooling on the Earth's climate system; overall, the reflection of incoming solar radiation leads to a greater radiative loss than the trapping effect of the clouds on infrared terrestrial radiation. The net radiative forcing of global cloud cover is about -16 Wm^{-2}, which is four times the calculated effect of CO_2 doubling. Sokolik and Toon (1996) estimate that dust aerosols may cause a direct *positive* forcing on the lower atmosphere of up to 20–40 Wm^{-2} regionally, compared with estimates of about -4 to -10 Wm^{-2} for the local effects of sulfate aerosols. The global effect of sulfate aerosols is about -0.5 to -1.0 Wm^{-2}, and dust aerosols have a positive effect of the same order of magnitude. However, dust aerosols have not yet been adequately incorporated in climate model experiments, although some coupled atmosphere–ocean model simulations have attempted to treat sulfate aerosol forcing. These caveats need to be considered in attempts to project the regional consequences of anthropogenically-induced climate change.

The IPCC model projections all indicate greater warming over land areas, with a maximum effect in high northern latitudes in winter (Kattenberg et al., 1996). This is accompanied by a decrease in diurnal temperature range, particularly in the extra-tropics. There are also increases of winter precipitation in higher latitudes and of annual precipitation in the Asian monsoon region. However, these projections are for the Earth as a whole; the projection of climatic changes from climate models for mountain areas has received relatively little attention, focusing primarily on the Alps and the western cordillera of the USA.

Two approaches can potentially be adopted for projecting climate changes in mountain areas. One procedure involves the use of statistical downscaling to relate local climate variables to large-scale meteorological predictors (Von Storch et al., 1993; Gyalistras et al., 1997). In this approach, large-scale weather situations are defined and their frequencies are determined. Next, a regional model simulation is carried out. Linear-based transfer functions are applied to the model output; for example, the climate characteristics for each weather situation are weighted by their frequency of occurrence in the model output (Frey-Bunes et al., 1995). Alternatively, non-linear methods such as neural networks can be applied, as illustrated by McGinnis (1995) for snowfall over the Colorado Plateau, USA. In future, the accuracy of such down-scaling techniques for precipitation may be improved through consideration of atmospheric circulation characteristics and statistical assessments of orographic processes. Such approaches have been attempted in various mountain areas in order to assess the spatial distribution of precipitation (Barros and Lettenmaier, 1993).

A second procedure is to nest a more detailed regional model within a GCM (Beniston et al., 1996; Giorgi et al., 1992, 1994; Marinucci et al., 1995). Recent research has shown that such models can quite successfully reproduce the orographic effects of topography on precipitation and snow cover for various ranges in the western USA (Leung and Ghan, 1995). This approach has been extended to hydrology and vegetation dynamics (Leung et al., 1996). Nevertheless, all of this work is at a very early stage, and there is a considerable need for collaborative research, involving scientists from diverse disciplines, in different mountain regions, to move towards realistic predictions of future mountain climates.

Impacts on hydrology

In considering the possible impacts of climate change on the hydrology of mountain areas we are faced with the present uncertainty of climate model simulations for regional to local scales (Fitzharris, 1996). This is especially true for precipitation processes because these involve cloud microphysics and mesoscale weather systems modified by complex orography (Kattenberg *et al.*, 1996). Nevertheless, nested model studies and other strategies are beginning to address this issue (Beniston and Fox, 1996). Projecting changes in soil moisture requires adequate treatment of evaporative exchanges as well as the seasonal storage of water in the snowpack. Other questions concern changes in the proportion of precipitation falling as snow and the frequency and intensity of precipitation events.

In the northern hemisphere, with the anticipated increases in GHG concentrations, precipitation totals are expected to increase in high latitudes, and in mid-latitudes during winter. In general, these changes imply increase in snowfalls. In the tropics, there will be increases in the variability of precipitation and, possibly, the frequency of extreme precipitation events. Snow cover duration will decrease, and mid-latitude soil moisture may also decrease in summer (Kattenberg *et al.*, 1996). However, it is premature to attempt quantification of these findings, derived from empirical and model studies, as substantial local variation is to be expected.

Runoff models incorporating changes in climatic parameters indicate that runoff is sensitive to seasonal changes in precipitation, the timing of the accumulation and ablation seasons, and changes in snow-line elevation (Lettenmaier and Gan, 1990). CO_2 doubling is projected to cause increases in spring runoff with lesser summertime flows in Switzerland (Bultot *et al.*, 1992) and the western USA (Rango and van Katwijk, 1990) as a result of reductions in snowfall combined with increased winter rainfall, and earlier, faster snowmelt. Gleick (1989) finds that, for the Sacramento River basin, USA, annual runoff is more sensitive to changes in precipitation than in temperature. Also, that in watersheds with seasonal snowfall, the seasonal timing of runoff is more sensitive to temperature changes since they affect snowmelt.

In mountain drainages of the western USA, flood frequency distributions depend primarily on precipitation amount and intensity rather than physiography (Pitlick, 1994). However, flood frequency curves differ considerably with elevation and the large-scale controls of precipitation. In the western cordilleras of California, characterised by winter precipitation from large-scale frontal cyclones, the 100-year flood is about 3–6 times the mean annual flood. In alpine drainage basins of the Rocky Mountains of Colorado, where snowmelt dominates, the 100-year flood is only 2–3 times the mean annual flood. In contrast, in the dry foothills zone, where intense spring–summer thunderstorms determine the annual precipitation regime, the 100-year flood may be as much as 10 times the mean annual value (Pitlick, 1994). Up to now, there is a very limited capability to model such different regimes and processes adequately in a climate change context. Further progress needs to be made, not only for greater understanding of likely future flow regimes and distributions of extreme events – an essential input to water resources planning (see below) – but also because of the close interactions between hydrological and geomorphic processes in mountain areas (Box 17.1).

Impacts on the cryosphere

Mid-latitude mountains have a seasonal snow cover whose duration increases approximately linearly with altitude. Higher areas may also have partially ice-covered basins. The equilibrium line altitude (ELA) – a climatically-determined altitude on ice caps and glaciers where winter–spring snow accumulation is balanced by summer ablation – is a valuable monitor of the state of a glacier's health. In the Alps (Maisch, 1990), the Caucasus, Altai and on the periphery of the mountains of Central Asia (Kotlyakov *et al.*, 1991; Serebryanny and Solomina, 1996), for example, the ELA has risen 50–80 m since the Little Ice Age (about 1700–1850 AD). The combination of natural climatic variations, and projected future ice recession as a result of greenhouse gas-induced warming, may have important consequences for both the amount and timing of spring runoff, which is dependent upon the snow surface energy balance and air temperature.

In mid-latitudes, especially, snowmelt runoff from the mountains is the primary input to river

Box 17.1 Impacts of climate change on geomorphic processes

In high relief mountain areas with steep energy gradients, many geomorphic processes (e.g. rockfalls, debris flows, snow and ice avalanches) are significant natural hazards. The nature, spatial and temporal distribution of these processes are strongly controlled by climatic factors that involve temperature or precipitation thresholds. Climate change will significantly affect these processes in several different ways.

Plate 17.2 Landslide, Columbia Glacier, Alberta, Canada: slope instability resulting from deglaciation. The initial failure, about 300 m across, occurred in the mid-1950s and transports debris across the lateral moraine onto the glacier below. Moraine crest dates from the mid-nineteenth century and the glacier surface (not visible) has downwasted over 300 m during the twentieth century (Photograph: Brian Luckman, July 1996)

Changes in average conditions (e.g. snow cover, depth, and duration; timing of the snow melt season; freeze–thaw regimes; or glacier extent) will lead to long-term shifts in the spatial distribution, altitudinal zonation and relative importance of specific processes. In addition, changes are anticipated in the magnitude and frequency of meteorological events that trigger specific processes (e.g. summer rainstorms for debris flows; critical rain/snow conditions for snow avalanches; extended dry intervals for forest fires). Every 'extreme' event (e.g. the catastrophic rainfall and landslides in the Darjeeling area of India in 1968 [Starkel, 1972], or the Swiss debris flows in 1987 [Rickenmann and Zimmerman, 1993]) provokes debate as to whether it results from natural variability or is a possible manifestation of human-induced environmental changes (Zimmermann and Haeberli, 1992). Such questions cannot be resolved from short-term records because of the large inter-annual variability of climate. However, they remind us that these high magnitude geomorphic events

have considerable potential for the generation of abrupt, long-term landscape changes: although climate changes may be incremental, the geomorphic response may be non-linear. The continued expansion of human activity into high mountain areas increases the potential for significant interaction with these hazards (Chapter 16).

In addition to the changes that are directly linked to climate events there are collateral effects resulting from landscape changes. The cryosphere is temperature-sensitive and its greatest response to change will be in those environments closest to critical thermal limits, i.e. temperate and tropical glacier/permafrost systems. Apart from obvious direct effects (reduction in glacier area, changes in permafrost thickness and extent, etc.) these rapid changes have other important consequences. Unvegetated, recently deglaciated terrain contains large areas of unstable materials that are reworked by fluvial and mass wasting processes. This results in accelerated sediment yield downstream but the ready availability of large quantities of unconsolidated sediments also has the potential to produce large debris flow events in areas where they have not previously occurred.

Areas adjacent to receding glaciers are extremely dynamic and subject to rapid changes. Lakes can develop on, adjacent to, or under glaciers in relatively short periods of time. Glacial Lake Melbern in the St Elias Mountains of northwest Canada is over

Plate 17.3 Glacial Lake Melbern, NW British Columbia, 1991. The lake drains around Konamoxt Glacier (foreground) and was generated by the rapid calving of debris-covered Melbern Glacier (far end of lake) which receded 7 km between 1979 and 1987. The lake is approximately 1.5 km wide and had an area of 12 km^2 in 1987. It demonstrates the magnitude of local changes associated with glacier recession and the potential for the rapid development of significant new hazards in such environments (Photograph: John Clague, Geological Survey of Canada)

12 km^2 in area, developed in 8 years (1979–1987), and presently terminates in glaciers at both ends. A recently developed supraglacial lake on the Imja Glacier in the Khumbu Himal occupies 0.7 km^2, is up to 100 m deep and may drain catastrophically along the main trekking route through Mt Everest National Park (Watanabe *et al.*, 1995). Elsewhere, in northern Bhutan and western Canada, several formerly stable ice-dammed lakes have drained catastrophically as the impounding glaciers melted back. Large ice avalanches and floods from moraine dammed lakes are also associated with deglaciation. Rapid wastage of glaciers produces changes in the stress fields and thermal conditions of adjacent valley walls that can generate significant slope instability in recently deglaciated areas. This situation has often been inferred for conditions at

the end of the last glaciation but there is increasing evidence of effects associated with recent deglaciation following the Little Ice Age in the Canadian cordillera, Alps, and Himalaya. Similar problems may also be anticipated with the reduction of stability of formerly perennially frozen debris slopes.

Many of the inferred changes in processes are incremental but, especially in glacierised temperate mountains, there is the potential for an accelerated tempo of geomorphic change and associated natural hazards. As present assessments of hazard are often historically or empirically based; such hazard assessments will become progressively less secure as the baseline climate changes (Chapter 16).

Source: Brian H. Luckman,
Department of Geography,
University of Western Ontario, Canada

discharge and is the major source of water for the populations of the adjacent lowlands. Indeed, 80 percent of the water used for agricultural, industrial and domestic purposes in the western USA originates from the high elevation winter–spring snowpacks. In countries like Norway and New Zealand, major hydropower facilities, vital for the national electricity grids, utilise this discharge.

GCM experiments for CO_2 doubling indicate that snowmelt will generally occur earlier because of summer warming over Eurasia and northern North America. Nevertheless, snowfall amounts may increase in west coast mountains as a result of greater atmospheric moisture content and storm activity. Local and regional assessments have also been made of changes in snow cover duration using GCM projections with simple indexing or changes in moisture balance and runoff predicted via snow melt runoff models. For the Victorian Alps, Australia, for example, Whetton *et al.* (1996) find snow cover duration to be highly sensitive to temperature. A 1 °C warming could reduce snow cover duration by 50 percent or more at low to moderate elevations while a 3 °C increase would eliminate the snow cover at sites around 1800–2000 m which have a (simulated) modern duration of 100 days. A precipitation increase of 50 percent is required to offset a 0.5 °C warming at Mt Buller (1800 m), Australia, suggesting that realistic precipitation changes will have a negligible effect.

Comparable studies have been published for Austria (Koch and Rüdel, 1990). They calculate that snow cover duration will decrease by about 25 days/°C warming in winter. For the French Alps, Martin *et al.* (1994) simulate seasonal snow depth

at 37 sites where observations are also available, and evaluate snow cover sensitivity to a doubled CO_2 scenario using the climate responses of a GCM for the region, and incorporating an air temperature increase of 1.8 °C and modified radiation conditions. They find that snow cover duration is reduced by 20 percent in the north (40 percent in the south), or 30–40 days at 1500 m, but only 20–30 days (or 10 percent) at 3000 m. This reduced sensitivity at higher elevations is contrary to the inference that climatic fluctuations on certain time-scales may be more pronounced at higher elevations (Barry, 1990), although temperature trends in the European Alps do show greater increases than in the adjacent lowlands (Weber *et al.*, 1994).

Projections of changes in ice cover in the Alps resulting from global warming provide startling indications of the potential changes in alpine landscapes. Haeberli and Hoelzle (1995) estimate from glacier inventory data that, if the recent warming continues, the ice mass in the Alps could be reduced to only 25 percent of that in 1850 within 30 years (2025 AD). Less obvious, but equally important, consequences may arise as a result of the melting of residual ice within rock glaciers and through the gradual melting of frozen ground, leading to surface subsidence (Haeberli, 1994). For the Upper Engadine, eastern Switzerland, about 29 percent of a 16 km^2 area between 1800 and 3500 m still has permafrost. Only a small decrease is projected by 2025 AD, but by 2100 AD only 10 percent of the area with permafrost might remain (Hoelzle and Haeberli, 1995).

Alpine basins with glaciers that are decreasing in area and volume will undergo reductions in runoff. In the Swiss Alps, Chen and Ohmura (1991) estimate that a 19 percent decrease in ice area from 1916 to 1968 reduced mean summer runoff by 16 percent. However, in the Pamir and Alai mountains of Central Asia, reductions in ice area of 11 to 16 percent, respectively (with corresponding ice volume decreases of 12 and 17 percent), over the last 25–30 years is estimated to have caused runoff reductions of only 5 percent in the Pamirs, with negligible changes in the Zeravshan River basin of the Alai (Konovalov and Shchetinnicov, 1994). An additional factor is summer temperature, which affects snow and ice melt. In glacierised basins of the Upper Rhine, Switzerland, warm/cool summers cause increased/decreased discharge; a 1 °C cooling over a decade can reduce discharge by as much as 25 percent (Collins, 1987).

Recommendations for research and monitoring

Our understanding of the possible impacts of climate change on the climates, hydrology, and cryosphere of mountain areas is limited by a number of factors. The principal ones are: lack of comprehensive data, especially from mountain slopes and ridgetops, as opposed to valleys; inadequate understanding of many of the processes of mountain meteorology and the altitudinal variability of climate change; and the rudimentary treatment of mountain areas in current numerical modelling studies of these complex systems.

Barry (1992a) presented a series of recommendations, noting the essential need for funding to support such activities, and for communicating the importance of sustained measurement and monitoring programmes to appropriate national and international agencies, such as the UN Environment Programme; UNESCO's Man and the Biosphere (MAB) Programme, and Global Environmental Monitoring Programmes; the WMO's World Climate Programme, including the Global Climate Observing System/Global Terrestrial Observing System; and the International Geosphere–Biosphere Programme (IGBP). Existing data collection networks need to be maintained and expanded in critical areas, using automatic stations. To make full use of existing data, a comprehensive inventory of current and completed observation programmes should be prepared. Existing data, often in obscure reports and tabular form, need to be compiled, with quality control, and distributed widely. Finally, national and international institutions should work together to develop the infrastructure of education and training programmes for mountain studies.

POTENTIAL IMPACTS ON MOUNTAIN ECOSYSTEMS

Mountain ecosystems encompass many interacting components, including soil, flora, fauna, air, and water in its different phases. This section considers the first three of these components, with particular emphasis on potential impacts on the Alps and the Fennoscandian Mountains, as most research to date has focused on these mountain ranges (Cebon et al., 1997; Guisan et al., 1995; Holten et al., 1993, Price and Haslett, 1995).

Potential changes in mountain soils

In an era of climate change, changes in mountain soils will be of great significance for both natural and managed mountain ecosystems. Even if new temperature and precipitation regimes are suitable, the growth of natural plant and tree species, and also crops, may be constrained by the availability of appropriate soils. Even if seeds can reach new habitats with suitable climatic conditions, soil conditions may be inappropriate, either because their chemical or physical characteristics are unsuitable; or because they have deteriorated as a result of human activities, such as leaching following the removal of trees above the current anthropogenic timberline, which is as much as 400 m below its climatic optimum in the Alps (Holtmeier, 1994). Timberlines also have been depressed in other mountain ranges. However, even under existing climates, in areas from which trees have been cleared and where soils are appropriate, the growth of trees above the current anthropogenic timberline may be limited by fungal infections as, for instance, in the Alps (Schoenenberger et al., 1990).

The characteristics of soils would be expected to change gradually in response to climate changes, and rates of change will depend not only on the changes in climate but also the physical and chemical characteristics of the soils. For instance, the internal processes of base-rich soils are less sensitive to change than those of acid soils. Consequently, vegetation changes on the former are likely to mainly reflect climate changes, while vegetation changes on the latter will reflect both climate and induced soil changes (Theurillat et al., 1997). Changes in temperature would probably lead to changes in the relative abundance of soil biota, thus affecting trophic relationships in soil food webs. Up to threshold levels, increases in temperature and available moisture would be likely to enhance the decomposition of organic matter, accelerating soil formation, increasing nitrogen mineralisation, raising pH, and thus influencing plant nutrition (Baron *et al.*, 1994; Bowman *et al.*, 1993; Theurillat *et al.*, 1997). Research to test this hypothesis in the sub-arctic mountains of Sweden, using nitrogen fertilisation, showed significant growth of mountain birch (*Betula tortuosa*) at treeline (Sveinbjornsson *et al.*, 1992).

In general, enhanced nutrient supply would likely stimulate any potentially fast-growing plant species, possibly leading to the elimination of others, thus eventually causing substantial changes in vegetation structure (Beniston and Fox, 1996). However, if desiccation increased, mineral recycling could be blocked (Körner, 1989). Consequently, estimates of average precipitation may not be useful in developing assessments of the effects of such changes; estimates of the intensity and length of droughts, or of the intensity and amount of precipitation in extreme events may be of greater importance. Finally, changes in precipitation and temperature regimes could alter the distribution of permanently or seasonally frozen soils, and also the heights of water tables: a critical factor in the distribution of some species, especially those living in wetlands (Pearson, 1994).

Physiological responses of plants to atmospheric change

The natural and cultivated plants of mountain ecosystems may be affected by climate change in two ways: first, by the increased concentrations of CO_2 available for photosynthesis (i.e. physiological responses); and, second, by the actual changes in climate (i.e. physical responses). These two types of responses are very closely interlinked, since physiological processes such as photosynthesis and respiration are directly influenced by temperature, water availability and, in the case of photosynthesis, sunlight.

The potential physiological responses of natural alpine vegetation to atmospheric changes have been summarised by Körner (1994, 1995). Experiments on plants of individual species to investigate effects of CO_2 enrichment have shown that responses are very species-dependent, and that responses of below-ground ecosystem components (below-ground parts of plants, associated my-corrhizal fungi, and other soil organisms) are important. The only study of the effects of elevated levels of CO_2 in an entire alpine ecosystem showed no change in above- or below-ground biomass, leaf area index, or phenology after three years (Körner, 1995). However, the carbon:nitrogen ratio of the plants increased, which would lead to reduced food quality for herbivores and, possibly, to changes in decomposition processes. Overall, while isolated alpine plants in nutrient-rich micro-sites may be stimulated by additional CO_2, possibly leading to more rapid soil stabilisation in disturbed areas, in general only moderate changes in species abundance and in soil processes are likely, as other factors generally provide more severe constraints on growth and development.

For forest ecosystems, no studies of the effects of elevated levels of CO_2 have yet been conducted (Kirschbaum and Fischlin, 1996). Although there is some evidence that the growth rates of trees in high altitude forests have increased over the period of increasing atmospheric CO_2, more likely causal factors appear to include improved temperatures, water relations, successional age, or nitrogen fertilisation from industrial pollution (Innes, 1991; Körner, 1995; Luxmoore *et al.*, 1993).

Responses of plants and animals to changing temperature

Most assessments of the potential impacts of climate change have focused on increasing temperature, for the following reasons. First, this is the most obvious change, which would lead to other changes (e.g. soil moisture, evapotranspiration, fire frequency and severity). Second, it is the change for which there is the greatest agreement between GCMs. Third, it is the change which can be most easily assessed for past (paleo-) and recent climates, from which parallels for future climates can be drawn. Fourth, a large literature exists on the influences of temperature on the distribution of species and on ecological processes. This last point also applies to water; and the distribution of vegetation types can be treated as determined by the interaction of temperature and water regimes (Cramer and Leemans, 1993). Nevertheless, for the distribution and survival of many mountain flora and fauna, temperature may be a less important factor than moisture (Billings, 1974; Woodward, 1992). Yet, while water is crucial in all ecosystem processes and interactions, the outputs of GCMs vary greatly in their projections of changes in precipitation and soil moisture, as noted above.

Overall, the greatest concern is that rates of change in temperature may be greater than the ability of species to adapt or migrate. Adaptation may be a particular problem for long-lived species, such as trees; clonal species; and species which cannot easily migrate, such as those living in waterbodies. For instance, possible negative effects of climatic warming on the forests of British Columbia include increases in frost damage in early spring, incidence of insects and diseases, and damage by wildfire; reduced seed production; and a winter climate too warm to satisfy the chilling requirements of some perennial plants (Kimmins and Lavender, 1992). In the Alps, it is anticipated that an increase in annual mean temperature of 1 °C would extend the growing season by 16–17 days (Theurillat *et al.*, 1997). This could benefit many plant species, but could disadvantage those whose phenology is determined by photoperiod (Körner, 1994). Longer growing seasons may be beneficial to the plants of aquatic ecosystems, allowing greater primary productivity (Byron and Goldman, 1990). Warmer conditions would also lead to increased rates of organic matter processing and nutrient transformation, and could allow fish and other species to expand upstream (Allan, 1995). All of these changes, combined with changes in flow regimes, would be likely to change physical habitats and the species composition of aquatic ecosystems.

Paleo-climatic studies suggest that few forest tree species would be able to disperse as fast as the projected changes in climate, which are likely to be faster than any in recent geological history (Davis and Zabinski, 1992; Roberts, 1989). Migration may be a lesser problem in mountain regions, where species may only have to move a few hundred metres upslope, rather than a few hundred kilometers, as would be the case in flatter areas. Nevertheless, species may become extinct and it is probable that ecotypes and genetic variation will be lost, especially at the edge of their ranges, or on mountain peaks. However, some species will become more abundant, and speciation may occur in response to new conditions.

Studies have been conducted in various mountain regions on changes in the distribution of plant and animal species over the past few decades of increasing temperatures. In the mountains of New Zealand, Switzerland, and the USA, the responses of different tree species to increasing temperatures have varied considerably. In New Zealand, four species have advanced beyond former timberlines (Wardle and Coleman, 1992), as has *Tsuga mertensiana* in California (Taylor, 1995). In Colorado, patches in the upper timberline with adequate moisture have been infilled (Weisberg and Baker, 1995), but the timberline has not moved upwards, as is also the case in Switzerland (Hättenschwiler and Körner, 1995). Evidently, no clear statement concerning advancing timberlines can be made; Holtmeier (1994) suggests that at least 100 years of thermal conditions more favourable than at present are needed. As noted in Box 17.2, however, there appears to have been upwards movement of high alpine species in the Alps in recent decades, although rates of upwards migration are far below those that might be expected from the rate of temperature increase (Grabherr *et al.*, 1995). Animals have also responded to increases in temperature. For instance, in the mountains of northern Finland, bird species have responded to increasing spring temperatures by laying eggs earlier, with larger clutch sizes. Over time, such increased reproductive success, combined with lower winter mortality, may well lead to invasion of new habitats (Järvinen, 1995).

Such long-term research is crucial, not only to evaluate species' responses to changes in climate and other factors, but also to test the results of models, such as those described below. However, future research will need to consider not only the monthly and annual mean temperatures which are typical outputs of GCMs, but also other aspects of temperature, such as ranges and extremes. For many plant and animal species, both actual extremes and the length of the period over which organisms experience them are critical. For instance, timberline is controlled partly by the annual temperature regime, especially the mean temperature of growing season, but also by extreme events (Holtmeier, 1994). More generally, while excessively low temperatures can prevent the germination of seeds, very high soil surface temperatures can prevent seedling survival. Equally, temperature extremes are significant for the dynamics and distribution of many fauna, particularly insects (Mani, 1968; Rubinstein, 1992).

Possible changes in the distribution of plants and animals

Most assessments of changes in the geographical distribution of species due to climate change have

Box 17.2   The upwards movement of alpine plants

Many mountain organisms are strongly influenced by temperature and are therefore good indicators of the impacts of climate change. On the other hand, for exactly the same reason, climate change could threaten the biodiversity and ecological stability of mountain ecosystems. One example refers to the proposition that global warming could induce mountain plant species, and even entire vegetation belts, to move upwards.

In Central Europe, there was a general warming from the end of the nineteenth century to about 1950, followed by a decrease of the annual mean temperature in the 1960s and 1970s, and then a renewed warming until the present. While this warming trend was within natural geoclimatological variability for several decades, the warming trend clearly exceeded the natural range in the 1980s (Houghton *et al.*, 1996).

Over the same period, botanists have undertaken research on the high summits of the Alps, starting in the mid-nineteenth century and intensifying their work in this century. They made exact records of the area they sampled – usually the highest 15 to 20 m of a peak – making exhaustive lists of the vascular plant species, or *florulae* (e.g. Braun, 1913; Reisigl and Pitschmann, 1958) and, on some peaks, noting the exact altitude of the highest specimen of each species (Braun-Blanquet, 1958).

These early botanists had many motivations for studying the plants of the summits, but we decided to use their results and sites as a basis for climate

Plate 17.4   Mount Hohe Wilde, 3480 m: during the last 40 years the number of species increased from 11 to 19 on the uppermost 20 m of the summit

Plate 17.5   *Androsace alpina* restricted to the uppermost vegetation belts in the siliceous high Alps

summit	historical no. of species	present no. of species
Piz dals Lejs	11	34
Piz Tavrü	5	14
Piz Uertsch	2	5
Piz Laschadurella	7	17
Piz Foraz	9	19
Monte Vago	28	51
Piz Nuna	19	34
Hohe Wilde	11	19
Napfspitze	31	53
Piz Kesch	8	13
Stockkogel	12	19
Großer Lenkstein	4	6
Liebenerspitze	2	3
Piz Forun	48	69
Piz Stretta	20	28
Gorihourn	14	18
Munt Pers	15	19
Hint. Spiegelkogel	15	19
Flüela Schwarzhorn	34	43
Wilde Kreuzspitze	8	10
Piz Blaisun	5	6
Piz Plazer	16	19
Hint. Seelenkogel	11	13
Piz Sevenna	17	19
Piz Julier	9	10
Festkogel	58	63
Piz Nair	22	22
Piz Linard	10	10
Piz Trovat	8	7
Radüner Rothorn	8	6

increase in species richness (%)
0   50 100 150 200

Figure 17.1   Change in species richness during the historical record (40 to 100 years) at 30 nival summits in the European Alps

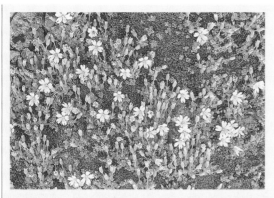

Plate 17.6 *Cerastium uniflorum* is abundant all over the Alps, from the lower alpine zone up to more than 3400 m

change impact research. If the upper limits of alpine and nival plants are determined by temperature, the number of species on summits should increase through invasion from below. In 1992 and 1993, we visited 30 of these summits across a significant proportion of the high Eastern Alps, in the eastern Swiss Alps, the northern Italian Alps, and the western Austrian Alps. Almost all have an altitude over 3000 m, reaching well into the nival zone.

As shown in the Figure 17.1, on 90 percent of the summits, species richness has increased since the first records were made. This change was significant on two-thirds of the peaks (Grabherr *et al.*, 1994, 1995; Pauli *et al.*, 1996). These proportions remain stable even after the new species have been statistically down-weighted, to take account of the possibility that the original research could have overlooked one or more species. Also, the increase in species numbers is not correlated to calcareous or siliceous bedrock, but to geomorphology, being more pronounced on summits with solid slopes and ridges.

These results provide evidence from the high alpine environment that natural species are already reacting to global warming. We can, therefore, really describe mountain ecosystems as 'early warning systems' for climate change.

Sources: Michael Gottfried, Harald Pauli, and Georg Grabherr, Institute of Plant Physiology, University of Vienna, Austria

been based on the concept of ecological niche (Begon *et al.*, 1990). A species' fundamental niche encompasses the environmental conditions (temperature, precipitation, soil conditions, etc.) in which it could grow and reproduce; its realised niche includes those conditions under which it is actually found, and is almost always smaller because of competition with other species. The basic assumption in such assessments is that, as climates change, species – or assemblages of species – will be able to follow their niches, with a lag from years to centuries.

Two general approaches have been taken in research on the future distribution of mountain plants and animals, one focusing on spatial changes, the other on changes in the composition of forest communities. The spatial approach begins with data on the distribution of species or assemblages of species (i.e. communities, ecosystems). First, relationships between these species or assemblages and certain climatic variables (and sometimes other variables, such as soil conditions) are defined. Then, scenarios are used to determine possible future distributions of these variables. Finally, these are used to define the resulting new locations of the species or assemblages, sometimes taking other site factors into account.

For the mountains of southern Norway, Holten (1993) divided 150 plant species into six functional groups, based on correlations between current distribution patterns and three limiting factors: winter temperature, summer temperature, and humidity (annual precipitation). Applying a climate scenario, derived by consensus among Norwegian climatologists, which suggested that the most likely changes were +3–4 °C in mean winter and +2 °C in mean summer temperatures, with a 5–15 percent increase in precipitation depending on the distance from the coast and the time of year, he concluded that three groups would be likely to expand their area: winter thermophilous coastal species, xerophilous species, and summer thermophilous species. Conversely, southwest coast-avoiding species, requiring chilling in winter, would retreat from the coast; humidophilous coastal species would decrease in their vertical range because of decreased snow cover protection; and alpine species would retreat because of competition from less light-demanding boreal species, more frequent high temperatures causing lethal conditions, and longer, drier snow-free seasons. In general, potential vegetation zones would be raised by 300–400 m and, because of the narrowing of mountains towards summits, alpine habitat and species would be particularly endangered; the alpine zone would decrease from 30 to 7 percent of the area of Norway. Similarly, plant communities of the alpine zone, especially on steep slopes, are considered the most endangered in the Alps (Theurillat *et al.*, 1997).

This ecoclimatic approach can be made more realistic and versatile through the use of geographic information systems (GIS). To study potential changes in the distribution of forest communities in Switzerland, Brzeziecki *et al.* (1994) created a GIS including topographic and soil pH data (pixel [grid cell] size 250 m × 250 m), a soil suitability map (500 m × 500 m), and temperature and precipitation maps (1 km²). Together with phytosociological data from about 7500 locations across Switzerland, these data were used to develop a simulated vegetation map at a resolution of 250 m × 250 m. Throughout the process, detailed vegetation maps and the results of field work were compared to the data in the GIS to evaluate and improve its accuracy. To assess the potential impacts of climate change, a 2 °C increase in annual temperature was simulated for communities dominated by beech (*Fagus sylvatica*) and Norway spruce (*Picea abies*). In general, beech forests disappeared from a narrow range of lower altitudes, but appeared in a wide range of higher altitudes which are now generally too cold for growth. Similar patterns were observed for spruce. Yet, in spite of the general upwards movement, the total area of both types of communities changed little.

The results of this study can be compared with the results of research using computer-based ecosystem models of forest succession in the Swiss Alps. However, these models cannot provide detailed information about possible future distributions of species in space. Beginning from the current forest composition, the potential effects of climate change have been simulated by incorporating the down-scaled climate predicted by a GCM at a specified time in the future (2080 AD), and then allowing simulated forests at different locations to adapt to these new conditions (Fischlin, 1995). As with the study by Brzeziecki *et al.* (1994), the results showed only minor changes in species composition at middle altitudes. However, at high altitudes, only two species of the present forest survived in significant numbers, which could lead to slope destabilization and erosion. At low altitudes in the dry central Alps, almost all existing tree species died. A comparable study by Kräuchi and Kienast (1993) also suggests general upwards displacement of tree species in the Swiss Alps, with deciduous trees displacing conifers in the subalpine zone, and conifers moving into the alpine zone if soils were suitable. While such results are generally

consistent across a number of models, they should be regarded principally as initial approaches for assessing the sensitivity of forests to climate change.

Similar modelling work has been done for mountain forests in North America. For a subalpine old growth forest in British Columbia, simulations which mimicked a changing climate showed that tree species vary in their response times to a changing climate, and that rare, long-lived species were favoured by climatic fluctuations at the expense of more common shorter-lived species (Lertzman, 1995). A simulation of a montane forest in the Northern Rocky Mountains of the USA emphasised the importance of cloud cover and edge effects in the establishment of different species (Malanson and Cairns, 1995). In general, much work is needed to refine all of the various computer-based models before they can be used as predictive tools. The latter study, particularly, shows the importance of small scale spatial structure, as well as broad scale climatic change, when projecting the effects on forest ecosystems.

In Central America, a GIS-based approach has been used to investigate potential changes in life (ecoclimatic) zones in Costa Rica (Halpin, 1994). Starting with a data base of 400 m × 400 m pixels for climatic, soil, topographic, and vegetation data, five topographically-distinct climatic regions were defined. Lapse rate, temperature, and precipitation regimes were interpolated from climate station data for each region. This base climate model was then modified to create two climate change sensitivity scenarios; 1) a moderate change scenario (+2.5 °C, +10 percent precipitation) and 2) a more extreme scenario (+3.5 °C, +10 percent precipitation). The scenarios produced significant changes in the location of life zones: 38 percent for the moderate scenario and 47 percent for the extreme one. Both this study and another by Halpin (1994), using the more extreme scenario applied to a 3900 m 'hypothetical mountain' digitised into GIS of Costa Rica, California, and Alaska, showed that, because ecoclimatic zones each have different temperature/precipitation responses, they do not move symmetrically upslope, so that each zone ends up with a new altitudinal range. While some zones may increase in area, others – not necessarily only at the tops of mountains – may completely disappear (Box 17.3). This type of result is of particular importance for ecosystems with a narrow ecological range, such as many cloud forests, which may be seriously threatened by climate change (Hamilton, 1995; Chapter 13).

In general, studies based on the distribution of individual plant species are more realistic than those using plant communities or life zones. This is because paleoecological research shows that each species responds individually to a changing climate, so that animal and plant communities tend to disassemble (Graham and Grimm, 1990; Tallis, 1991). However, while detailed geo-referenced information on the distribution of individual species is available for some mountain ranges, such as the Alps (Guisan *et al.*, 1995) and Rocky Mountains (Morain *et al.*, 1993), this is not true in many mountain ranges so that, as in Costa Rica, a coarser approach has to be used. Nevertheless, the results are valuable for defining research agendas and priorities for nature conservation (Box 17.3).

The assessment of possible changes in populations of animals presents even greater challenges than for plants because of their dependence on other species and their greater mobility. A first

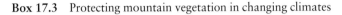

Box 17.3   Protecting mountain vegetation in changing climates

The prospect of climatic change raises many difficult questions concerning the vulnerability of mountain protected areas to future environmental disruption. Likely changes in global temperatures and local precipitation patterns could significantly alter the altitudinal ranges of important species within existing mountain protected areas, and create additional environmental stresses on already fragile mountain ecosystems.

A common interpretation of future climatic change suggests that increases in local temperature would act to move the climatic niches of species upslope, in a linear manner, so that each altitudinal zone would be replaced by the niches of the zone directly below. This conceptual model implies that the boundaries of present climatic ranges for individual species will respond symmetrically to changes in temperature for a particular mountain

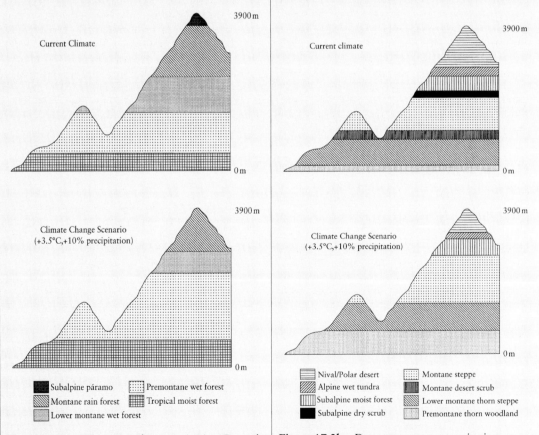

Figure 17.2a   Wet tropical mountain site (Central Cosa Rica: 9°N, 84°W)

Figure 17.2b   Dry temperate mountain site (Sierra Nevada, CA: 38°N, 119°W)

site. Thus, the expected impacts of climate change in mountain protected areas would include the loss of the coolest climatic habitats at the peaks of the mountains, and the linear 'migration' of all remaining habitats upslope.

Both predictive modelling of mountainous landscapes and paleo-ecological records of past climatic changes suggest that changes in vegetation zones are generally not symmetrical along altitudinal gradients. Figure 17.2a and b depicts the area and location of potential vegetation change for a tropical and temperate mountain subjected to an identical magnitude of climatic change (Halpin, 1994). In this simulation, the altitudinal range of the ecoclimatic zone now associated with premontane wet forest for the tropical site expands dramatically, and climatic habitats, including the highest elevation páramo forests, are lost. The temperate example depicts a markedly different pattern of change: the highest altitude ecoclimatic zone is retained, while ecoclimatic zones in the mid-elevations are lost.

These simulation findings raise important questions with regard to the long-term viability of mountain protected areas. Concern about potential impacts of climate change has typically focused on the loss of high alpine sites due to the obvious lack of habitat uphill. This focus has tended to draw attention away from the possibility of dislocations and possible loss of ecoclimatic zones at lower and middle elevations. Clearly, the conservation management solutions required to optimise habitat protection under changing climatic conditions will be very different across latitudes and climatic regions. Researchers and conservation managers will need to define more rigorously the potential habitats that mountain landscapes will provide under likely future climates.

Estimating the potential impacts of changing climatic conditions on the species of mountain protected areas will only result from a significantly better understanding of the highly complex cascade of environmental and ecological feedbacks which control species distributions. More dynamic approaches to the acquisition and maintenance of mountain protected areas will be required to meet the challenges of changing environmental conditions and land use pressures. Techniques for systematically identifying mountain sites which will provide optimal habitat configurations under changing conditions will be an increasingly important research and management issue in the future.

Source: Patrick N. Halpin,
Nicholas School of Environment,
Duke University, USA

approach is to ignore such complications by assuming that climatic factors are the most significant determinants of distribution. For instance, physical and bioclimatic assumptions (e.g. the adiabatic lapse rate and species–area curves) can be used to define new climatic zones. This avoids the need for detailed information on the spatial distribution of climatic variables, which is unavailable for many mountain regions. One example is Nevada's Great Basin, where this approach has been used to examine potential changes for animal species, assuming that a warming of 3 °C would cause potential habitats to shift upward by 500 m. The study suggested that numbers of mammal species would decrease most, followed by butterflies and resident birds (Murphy and Weiss, 1992).

This single species approach can be made more realistic, and therefore more complex, by incorporating a range of climatic variables, together with actual climatic data sets, into a GIS. In the Bavarian Alps, d'Oleire-Oltmanns *et al.* (1995) used habitat maps of 21 bird species, combined with vegetation maps in a GIS, to investigate possible future distributions of these species. The probability of occurrence of a species was defined into three classes: sure/probable, questionable, and no. Assuming a temperature increase of +2.8 °C, two species lost more than 90 percent of their sure/probable habitat, while others gained much potential habitat. However, this type of research is still at a very early stage.

The realities of dispersal and survival

All of the research described above assumes that there are no barriers to dispersal except at the tops of mountains. However, the ability of a plant or animal species to follow its ecological niche depends on a great variety of factors, most of which have not been incorporated into models of any type.

A species' ability to disperse depends on its reproductive capability, its dispersal strategies and, for animals, its mobility (Gadgil, 1971). Every species interacts with others in diverse relationships – including many related to dispersal – to which must be added the multifarious intentional and unintentional human actions which can limit or assist species dispersal. Also, physiological, behavioural, and even genetic adaptations to changing climate are likely to occur in certain species, particularly in response to small-scale alterations to the microclimate mosaic. Such adaptations have been demonstrated, for instance, in grasshoppers responding to microclimatic changes caused by skiing on subalpine meadows in Austria (Illich and Haslett, 1994).

Individuals of a species not only have to be able to disperse successfully, they must also become established in large enough numbers to reproduce and persist in their new environment, where they may encounter new competitors, predators, pests, or pathogens against which they have few, if any, defences. Furthermore, obligate species may become unavailable because of responses to different components of climate change. One example is the subalpine plant *Delphinium nelsonii*, which occurs in the Colorado Rocky Mountains. It is pollinated by hummingbirds, whose migration is determined by photoperiod. However, the annual growth and development of the plant depend on temperature and snowmelt. If climate change led to decreased snowfalls, the hummingbirds might not arrive at the right stage in the plants' annual life cycle to pollinate them, which could lead to their extinction (Harte *et al.*, 1992).

Another factor to consider is the life history strategies of the organisms that compose any ecosystem (Rubinstein, 1992). For instance, many insects and other small organisms have much shorter generation times than the plants and animals they utilise as sources of food. This means they could adapt to new climatic conditions more quickly, leading to increased pressure on their hosts (Fanta, 1992; Franklin *et al.*, 1992). In addition, the mortality rates of many insects are affected by temperature, so that population sizes might be changed directly (Elias, 1991). This is likely to be of major importance for many phytophagous insect species and of particular relevance to the long lived trees of many mountain forests, which tolerate a current climate that is far from optimal for regeneration and growth and therefore could be especially sensitive to the direct and indirect effects of climate change, as noted, for instance, for the *Juniperus polycarpos* forests of northern Oman (Fisher and Gardner, 1995).

In some mountain ecosystems, another critical ecological factor to be considered is fire, whose frequency appears to be influenced by large-scale climatic variability (Swetnam and Betancourt, 1990) and has often been modified considerably by human actions. Both of these factors further complicate the projection of frequencies under new climates. Since photosynthesis is often limited by current temperature and CO_2 concentration, increases in these variables could lead to increased growth of vegetation and therefore to increased fuel loading. This problem could be exacerbated if mortality increases because trees are living in less optimal conditions, there is additional damage from pests or diseases and, particularly, evapotranspiration increases (Franklin *et al.*, 1992). Over time, the distribution of certain species could be significantly changed as a result of new fire frequencies. This could be true not only for forest species, but also for the species of small isolated wetlands with complementary habitats in the surrounding landscape (Pearson, 1994).

In addition to the diverse factors discussed above, there are other physical constraints on the ability of organisms to disperse in response to changing climates and survive in their new environments:

(1) changes in the frequency of other disturbances, such as avalanches, heavy snowfalls, major storms, floods, and droughts (Franklin *et al.*, 1992);

(2) changes in the depth and duration of snow cover and in the length of the snow-free (growing) season (Körner, 1994; Musselmann, 1994);

(3) changes in the hydrochemistry of alpine streams as a result of changes in the snowpack and snowmelt regime (Williams *et al.*, 1996);

(4) changes in cloud cover, which could alter the sunlight available for photosynthesis, balancing or even overcompensating for otherwise positive effects of increased rainfall (Körner, 1995), and also have other effects, such as changing rates of snow melt;

(5) the availability of potential habitat: a major issue for species which would have to move to higher altitudes in response to a warmer climate, since mountains narrow towards their summits, and potential habitats might not be available for species currently restricted to summits (Peters, 1992);

(6) the availability of migration corridors: a crucial issue for species in the complex topography of mountains, which is exacerbated by the east–west orientation of many chains (e.g. Alps, Pyrenees, Himalaya).

Both of the last two constraints could be ameliorated or exacerbated by human actions. For instance, while corridors for migration could be kept open by appropriate land-use management practices and policies, the expansion of agricultural land use to higher altitudes could decrease possibilities for dispersal.

Overall, many mountain species are likely to be particularly sensitive to changes in climate because they are at the edge of their range; geographically localised; genetically impoverished; poor dispersers; slow reproducers; annual; highly specialised; or migratory (McNeely, 1990). Many mountain species have one or more of these characteristics; and many are relicts, having been isolated by past changes in climate. Declines in alpine plant biodiversity are likely to be particularly marked in low island mountains whose floras are rich in endemic species (Grabherr *et al.*, 1995). At the same time, many mountain ecosystems and species may be affected far more in coming decades by changes in land-use practices than by climate change; and the combined effects of land-use change and climate change may be beneficial, neutral, or detrimental.

Recommendations for research

A detailed research agenda has been presented by Guisan and Holten (1995), and only key items are indicated below. One of these should be the monitoring of changes in the distribution of plant and animal species, which should be correlated with changes in climatic parameters both at meteorological height and at the ground surface. Two particular foci for such research should be transects along steep environmental gradients, and ecoclines, such as timberline, where it has been hypothesised that there are likely to be changes in the distribution of plants with different life-forms, as well as the upward movement of heaths, shrubs, and trees at varying rates (Theurillat *et al.*, 1997). Similarly, attention should be focused at the upper limit of the alpine zone, although it should be recognised that some species occupy microsites, such as rock fissures, which already have very extreme conditions. Monitoring of phenology would also be of great benefit, as changes in developmental stages may well be more rapid than changes in distribution (Bucher and Jeanneret, 1994). This type of monitoring also does not require significant resources, and can be done by local inhabitants. Monitoring of both species distribution and phenology requires the establishment of spatially-referenced data bases.

Monitoring is valuable not only to identify and evaluate trends, but also to test the results of research on the potential impacts of climate change, especially with regard to species or ecosystems which such research identifies as being particularly sensitive. This research may use spatial or temporal computer-based models, or experimental approaches. Existing computer-based models have begun to suggest likely directions and patterns of change, but need much further development, taking into account a greater diversity of ecosystem processes than to date. Experimental studies should preferably be conducted in natural ecosystems, focusing on critical ecological stages and processes (Skre, 1995). For plants, these include winter dormancy, carbon balance, flowering and reproduction, and rates of adaptation, migration, and seed or propagule dispersal. Sensitivity studies, in which one or more factors are varied, are also desirable, and could include manipulation (CO_2, temperature, humidity, etc.) and laboratory experiments, as well as transplantation.

A primary goal of all future research should be to establish greater understanding of the possible distribution of plant and animal species under likely scenarios of climate change, taking into consideration the intervening forces of human uses of mountain landscapes and the possibilities for human actions to assist or limit changes in the distribution of individual species and assemblages of species. Nevertheless, it should be recognised that it is likely to prove easier to assess likely changes in the distribution of plant and tree species than of animal species. It seems inevitable that many mountain ecosystems will be considerably different in coming centuries as a result of climate change, in association with other forces deriving from human activities both within and outside mountain areas.

POTENTIAL IMPACTS ON MOUNTAIN SOCIETIES, AND POSSIBLE RESPONSES

As previous chapters in this book have noted, mountain regions provide vital resources for a significant proportion of the world's population: water, minerals, soil nutrients, agricultural products, timber and fuelwood, hydroelectric power, and resources for tourism and recreation. Yet, although changes in mountain climates have the potential to affect hundreds of millions, if not billions, of people, few assessments of the economic and societal impacts of climate change have been conducted for mountain regions. Some of the reasons for this have been mentioned earlier in this chapter: the difficulty of using existing climate models for impact assessment because of their inadequate spatial resolution, the lack of comprehensive climatological data, and the complexity of their biophysical systems. These problems are compounded when considering socio-economic systems in mountain areas, which are also highly complex and for which comprehensive data are also very heterogeneous in their length and frequency of record, spatial coverage, and availability. Two further, related, reasons for the lack of studies in mountain regions are their perceived economic and political marginality, and the fact that many of their most valuable 'products' are not easily quantified in monetary terms (Price, 1990).

The first published discussion of the potential impacts of climate change on mountain societies (Price, 1994) provides the basis for the IPCC assessment (Beniston and Fox, 1996), as relatively little additional research has been conducted in the interim; the work by Cebon et al. (1997) for the Swiss Alps is an exception, but was not finalised when this chapter was completed. However, it must be emphasised that although climate change should certainly be taken into consideration by those concerned with the future of mountain societies, it is only one of many factors influencing their future and, in particular, that the degree of impact – positive or negative – will depend significantly on the type of society experiencing an altered climate.

Types of mountain regions

Although the biological and physical processes characterising mountain regions around the world may exhibit considerable similarities, the inhabitants of these regions differ greatly, interacting with mountain climates in very different ways. As discussed in Chapter 2, in most mountain regions, the dominant human population can trace their ancestry back some centuries, or even millennia. In others – notably those settled by Europeans who displaced indigenous populations – the history of today's predominant population is more recent. Grötzbach (1988; Chapter 2) describes these two broad categories as, respectively, 'old' and 'young' mountains. The latter, in North America, Australia, and New Zealand, are further characterised by sparse settlement, extensive market-oriented pastoral agriculture and forestry, and the recent rise of tourism as a major economic force.

Most 'old' mountain regions are relatively densely settled. They can be divided into three broad categories. The first, including many of the mountains of western Europe and Japan, is characterised by the decline of traditional agriculture and forestry despite support through subsidies. During the post-war period, tourism has grown rapidly, often becoming the basis of the economy. Areas without tourism often experience depopulation (Bätzing et al., 1996; Chapters 4 and 12).

The second category includes most mountains in developing nations, together with many of the mountains surrounding the Mediterranean, in countries that are otherwise industrialised to a greater or lesser extent. Most people are mountain peasants practising traditional subsistence agriculture and, in some cases, transhumance. In some regions, nomads occur in addition to these more settled people. There is a tendency toward over-population, ameliorated somewhat by migration. In all these regions, tourism is becoming locally important, often with considerable effects up to the national scale, as discussed in Chapter 12. However, its initial or continued development may be limited by inaccessibility, political instability, or warfare.

The third category has, until recently, largely been characterised by collectivised or nationalised agriculture and forestry: the mountains of China and the former Soviet Union, and the Balkans and Carpathians. With the fall of communist regimes in all these countries except China, their social, economic, and political structures are now in a state of flux. Parts of the Balkans, Carpathians, and Caucasus appear to be moving towards the first category, while other regions are tending toward the second category; being far from centres of power, the degree of collectivisation or nationalisation was often not as great as in more productive lowland regions. Furthermore, many of these areas are in a state of armed conflict (Chapters 5 and 6).

These categories of mountain populations are inevitably not clear-cut; the immense diversity of mountains rarely allows simplicity! In addition, one should recognise that mountain regions are increasingly being integrated into large economic and political systems, whose influences further complicate such typologies. Nevertheless, recognition of the diversity of mountain societies is important when assessing the potential impacts of climate change on these societies.

Mountain climates and human populations

As discussed above, mountain climates are characterised by marked diurnal and seasonal cycles and high spatial variability. Many resources on which mountain communities depend are climate-dependent, that is, their availability may be affected in the short- and long-term by variability, extremes, and long-term changes in means of climatic parameters. In traditional mountain communities, these resources include water and both domesticated and wild plant, tree, and animal species. Many of these resources are also important in areas where tourism has become important. Other climate-dependent resources in such areas include landscapes of natural and anthropogenic ecosystems, weather conditions that are suitable for specific activities and aid the dispersion of pollutants, and snow for skiing.

As discussed in Chapter 7, the components of the hydrological cycle are particularly critical for those living in mountain communities and downstream. Water is essential for many activities, including agriculture, sewage disposal, food preparation, power generation, and snow-making. At ski resorts, natural snowfall must occur when it is needed. The timing of snowmelt is also crucial for crop or pasture growth, and power generation. Similarly, the timing and volume of rainfall critically constrain many activities. Thus, intra-annual variations in precipitation are at least as important as inter-annual variations.

To understand the interactions of climate and human activities requires detailed long-term knowledge of both sides of these very complicated relationships. Yet, as discussed above, the availability of long-term climatological data varies greatly both within and between mountain regions; and such data are generally more readily available from settlements close to the mountains or in valleys, than from mountain sides and peaks where many activities take place (Barry, 1992a). Even when mountain areas contain quite dense networks of meteorological stations at many altitudes, these often only record data for part of the year, and measurement at individual sites may not have been taken for long enough to provide statistically valid – or useful – descriptions of local climates. Thus, because relationships between climatic variables in mountains and neighbouring areas are highly complex, even when data are available, they often do not adequately describe the climate of an entire mountain region.

In summary, in nearly all regions, assessment of the potential impacts of climate change is constrained by the availability of appropriate models and information and, even where this exists, existing impact assessment methods (Carter *et al.*, 1994; Riebsame, 1989) may not provide suitable outputs for decision-making. Consequently, the conclusions of the following section should be regarded as preliminary, providing an initial overview of issues which need to be considered in future research and in the development of policies, as discussed in the final section.

Potential impacts of climate change

While this section is organised thematically, it must be recognised that, because of the interactions of the resources and sectors discussed below, none should be considered in isolation. Recommendations for research and policy are presented at the end of each theme, with general, cross-cutting conclusions and recommendations in the final section.

Agriculture

As discussed in Chapters 14 and 15, mountain farmers not only make important direct contributions to mountain economies, they also play important roles as managers of the upper parts of watersheds which are vital for downstream populations, and also in maintaining landscapes for tourism (Chapter 12). However, there has been relatively little research on the potential impacts of climate change on mountain agriculture. Similarly to natural ecosystems, predictions have been made that climate change will lead to higher climatic limits to cultivation, in the Alps (Balteanu *et al.*, 1987), China (Li *et al.*, 1993), Japan (Yoshino *et al.*, 1988), Kenya (Downing, 1992), and New Zealand (Salinger *et al.*, 1989). However, as discussed above, even if the climatic conditions are suitable for crop growth, soil conditions may not be; and soil characteristics are likely to change slowly in the absence of significant human intevention.

When considering such broad-scale predictions, the results of detailed studies of the effects of climatic variations on agriculture should be borne in mind. Such studies, in Ecuador (Bravo *et al.*, 1988), Japan (Yoshino *et al.*, 1998), and Papua New Guinea (Allen *et al.*, 1989), show that crop growth and yield are controlled by complex interactions between solar radiation, temperature, precipitation, and frost, and that specific methods of cultivation may permit crop survival in sites with otherwise unsuitable microclimates. Furthermore, while optimal temperature ranges can be identified for each crop species, there are also temperature thresholds to consider. For instance, many tree fruits require winter chilling for buds to set, but late frosts which damage flowers can cause the loss of the fruit crop (Reilly, 1996). Conversely, temperatures which exceed well-recognised thresholds can have severe effects on the production of maize (Decker *et al.*, 1986), potatoes (Prange *et al.*, 1990), and rice (Yoshida, 1981), for example.

It has only been possible to include a few of these complex interactions between crops and their environments in GCM-based impact assessments, and none have considered the direct results of elevated CO_2 on crop growth, although most experiments show some increase in productivity (Reilly, 1996). Assessments conducted to date have suggested both positive and negative impacts, such as decreasing frost risks in the Mexican highlands (Liverman and O'Brien, 1991), increased agricultural production in the Chinese mountains (Hulme *et al.*, 1992), and less productive upland agriculture in Indonesia and Thailand, where impacts would depend on various factors, particularly the availability of irrigation and types of cultivars (Parry *et al.*, 1992).

In many mountain regions, a great variety of cultivars of many crops has been developed over generations. When planted and harvested according to detailed local knowledge in a wide range of microsites, these provide harvests that are adequate for food supplies, and sometimes produce surpluses, under diverse climatic conditions. While some local social movements are emphasising the cultivation of native crops (e.g. Zimmerer, 1992), many remain under utilised even in their native regions in spite of their adaptability to mountain environments and their high nutritional value (National Research Council, 1989). Others have been displaced by varieties that only give high yields under restricted climatic conditions, often with high inputs of fertiliser; or by crops grown for urban or export markets.

As long as a range of cultivars adapted to diverse conditions is maintained in a mountain region, the wide range of microclimates which may be exploited may limit the direct negative effects of climate change on crop yields. On suitable soils, increases in upper limits for cultivation may compensate for losses in yields, which may even

increase if moisture is not limiting. However, this optimistic assessment might be upset by other atmospheric and biological changes (Reilly, 1996). These include increases in cloudiness, numbers of extreme events, and levels of tropospheric ozone and ultraviolet-B radiation; and decreases in available moisture, through either decreased precipitation or increased evaporation and evapotranspiration. Competition from weed species may increase or decrease; these species are likely to respond to changes in CO_2, temperature, and precipitation in a similar way to the crops with which they compete; and changes in climate will also affect the efficacy of herbicides and other means of control. Higher crop yields may also be counterbalanced, during both growth and storage, by increased populations of pests and pathogens, many of which have climatically-controlled distributions. Interspecific interactions between pests and their predators and parasites may also change significantly.

Grazing animals are also likely to be affected by climate change. Changes in weather and, especially, the frequency of severe weather events, may directly affect animal health, growth, and reproduction (Rath *et al.*, 1994). Indirect effects may be through changes in the availability of suitable pasture and fodder, and populations of pathogens and pests. Thus, both sedentary farmers and nomads might have to change seasonal patterns of pasturage and stall feeding, adopt new breeds or species of domesticated animals, or even give up this part of their livelihoods.

There are clearly considerable difficulties in evaluating both the direction and degree of the likely impacts of climate change on mountain agriculture and, in many regions, other factors in changing land use will probably be of similar or greater importance. However, there is an urgent need for increased understanding of the possible impacts of climate change on the potential distribution of different crops; both those that are currently widespread, and also those that offer potential for increased production. Such work, using GIS and sophisticated crop production models, and incorporating scenarios of likely future change, should consider not only the implications of a changing climate, but also the impacts on agriculture of changes in land use likely to result from both climate change and economic and other policies.

In both industrialised and developing countries, there is great need for work on the maintenance of existing cultivars and the breeding of new cultivars which have the potential for ensuring food supplies under a changing climate. Comparable work is also desirable in animal breeding. In developing countries, agricultural production is vital for vast, and often rapidly increasing, numbers of people (Jodha, 1996). In order to increase food security, much work is needed not only in maintaining and distributing adaptable genotypes, but also on identifying and implementing optimum methods of cultivation (especially intensification) and seed and food storage; and identifying and minimising the negative aspects of interactions between domesticated plants and animals, and pests and diseases. Thus, climate change provides a further impetus for greater involvement of mountain farmers in land management; the implementation of development strategies emphasising the cultivation and marketing of indigenous crops; and the general diversification of economic activities in order to enhance the options for life and survival. While many of these activities need to be implemented at the scale of individual mountain regions or countries, there is also a great need for transfer of information and appropriate cultivars through international organisations, such as FAO and CGIAR.

Forestry

Agriculture and forestry have long been inseparable within the economies of traditional communities in 'old' mountains. Trees provide fodder for livestock, shade for crops and livestock, and wood for a multitude of agricultural and domestic purposes. Livestock, in turn, is the principal source of fertiliser for subsistence, and even much cash crop agriculture. Particularly vital, and often dominant in terms of volume, is the use of wood for fuel. This pattern persists in most of these regions, except in Western Europe where wood harvests have declined greatly since the 1950s. This pattern also applies in many 'young' mountain regions, although large-scale wood harvests continue where this is economically and environmentally permissible. Thus, in mountains now mainly inhabited by people of European ancestry, forests tend to be more important for regulating the quality and quantity of water flows, providing protection against 'natural hazards', and as part of the landscape for tourism (Price, 1990).

As described in the previous section, computer-based models and GIS have been used to evaluate potential changes in the distribution of forest species in both the Swiss Alps and western North America. Other studies, with a focus on the productivity rather than the species distribution of forests, have been conducted for eastern and western North America (Botkin and Nisbet, 1992; Cook and Cole, 1991; Winjum and Neilson, 1989). There have also been assessments of the potential impacts of climate change on forest productivity for individual nations and for the tropical, temperate, and boreal zones (Solomon, 1996). These zonal models also incorporate human population growth and agricultural land use, recognising that these may be at least as important as climate change in determining the future distribution of forests (Zuidema et al., 1994).

At present, it is only possible to speculate on possible long-term trends in the productivity of mountain forests. Increasing levels of CO_2 may be beneficial for growth, both directly and through increases in water use efficiency, permitting greater productivity in areas where water supplies are limited (Schlaepfer, 1993). However, as shown by research on Pinus ponderosa in the Sierra Nevada of California and Nevada, USA, changes in climate may result in significant reallocation of biomass among different components of trees (e.g. leaves:wood; primary:secondary branches) (Callaway et al., 1994; DeLucia et al., 1994). Thus, even if total growth rates and productivity stay more or less constant, the availability of different forest resources for mountain and lowland populations may change.

Equally, such research only focuses on one aspect of climate change, ignoring possible changes in the productivity of individual trees and forests resulting from induced changes in frost damage, competitive ability of tree and other plant species, and populations of pests and pathogens. Unusual damage by, or behaviour of, pests may be an early indicator of climate change (Schlaepfer, 1993). Changes in fire frequency are likely to represent the interaction of climate change and human actions; in many mountain areas in North America and Europe, climate change could exacerbate the existing problem of increased fuel loading resulting from decades of fire suppression and decreasing harvests. In addition to land-use change, another complicating factor in industrialised countries may be the effects of acid precipitation through forest dieback. Overall, in industrialised nations, where forest cover and density have often been increasing over recent decades, substantial synergetic negative effects may result from the addition of climate change to other sources of change.

In developing countries, considerable attention has been focused on decreases in forest cover and density, in spite of limited and often contradictory data (Ives and Messerli, 1989; Hamilton, 1992). Climate change is yet another factor to be considered in these complex and ill-defined relationships. As with crops, there is a clear need to maintain genotypes adapted to a wide variety of environmental conditions (Hattemer and Gregorius, 1990); and to develop and use appropriate methods for planting, maintaining, and harvesting trees in both forests and other landscapes with varying densities of trees and crops.

In any mountain region, the importance of the potential impacts of climate change on forests might be expected to be proportional to the density of human population and infrastructure, and further complicated by other diverse environmental and societal factors, including government policies and silvicultural practices. To ameliorate the effects of climate change, applied research should focus on the following responses: increased sanitation logging to remove trees endangered, damaged, or killed by pests, diseases, or fire; changes in harvesting methods; shortened rotations to reduce exposure to changing conditions, modify genetic diversity, and meet local needs for wood; and increased thinning. Combined with these changes in harvesting practices, research is also needed to identify the best options to produce desired outputs in different mountain regions, through planting mixtures of species, including those which may be better suited to likely future conditions; and assisting migration of species by natural mechanisms or transplanting (Solomon, 1996). In addition, the potentially significant role of trees and wood products in sequestering carbon must be recognised (Brown, 1996). Mountain communities may be able to benefit from this through planting programmes which not only provide much-needed resources but also allow the global population to benefit from this 'value' of mountain forests. Proactive government policies are essential to support the types of research described above, and to act on their findings.

In addition to such changes in forestry practice and policy, research is required to develop better understanding for predicting the behaviour of forest ecosystems, species, pests, and diseases in changing environments and land-use systems; appropriate inventory methodologies, which consider regeneration as well as variables necessary to predict future harvests; and appropriate management approaches, including gene conservation. Governments and international organisations, such as FAO and the International Union of Forestry Research Organizations (IUFRO) should support such work, which is vital not only for mountain people but to the billions of people who live downstream, whatever the mix of fuel, timber, protection against hazards, landscape, and other 'products' that mountain forests provide now and will provide in the future.

Water resources

As discussed in Chapters 7 and 15, forests are integral to the hydrological cycle. Thus, as human activities change the area and structure of forests, they also affect the quality and quantity of water supplies which, in turn, affect and are influenced by other aspects of human life from agriculture to health and recreation. In many mountain regions, the quality and quantity of water supplies are already limiting factors, especially in winter. In an era of climate change, conditions may improve or deteriorate through any of the resulting changes in: seasonal or annual precipitation; proportions of solid to liquid precipitation; or frequencies of extreme events. Unfortunately, GCMs vary considerably in their projections of precipitation, though it seems likely that extreme events will increase in frequency (Kattenberg et al., 1996), which could place increased pressure on insurers and governments responsible for compensating or repairing the resulting damage.

Assessments of the potential impacts of climate change on water resources, including snowfall and storage, have been conducted at spatial scales from small watersheds to entire mountain ranges, in Australia (Whetton et al., 1996), France (Martin et al., 1994), Greece (Mimikou et al., 1991), New Zealand (Garr and Fitzharris, 1994), Switzerland (Bultot et al., 1992), and the western USA (Duell, 1994; McCabe and Hay, 1995; Skiles and Hanson, 1994). Research has also been conducted in developing countries, but only at the scale of entire river systems, including some which rise in mountain regions: the Brahmaputra and Ganges (Van Deursen and Kwadijk, 1994), the Indus and Mekong (Riebsame et al., 1995), and the Nile (Strzepek et al., 1995). These studies provide useful information, but their results will not be summarised here, given the limitations in data and models and the numerous uncertainties identified by the IPCC with regard to: GCMs and their regional specificities; knowledge of future climate variability; estimates of changes in water budgets due to altered vegetation and climate; future demands; and socio-economic and environmental impacts of response measures (Kaczmarek, 1996).

As discussed in Chapter 7, demands on most highland water resources systems are likely to continue to increase. The IPCC concludes that arid and semi-arid watersheds and river basins are inherently the most sensitive to climate change, and that this sensitivity will be particularly crucial for irrigated agriculture (Kaczmarek, 1996). With regard to infrastructure for water resources management, integrated multiple reservoir systems should allow greater adaptability than isolated reservoirs, and increased streamflow regulation and water management may be required to meet demands. There is also a great need for institutional adaptation and flexibility in water allocation policies and contingency plans in order to cope with uncertainty.

The IPCC concludes that, though 'climate change will impact the water resource systems of the world . . . we will be able to adapt – though at some cost economically, socially, and ecologically' (Kaczmarek, 1996: p. 473). If this is to be true, governments will have to put considerable resources into resolving the uncertainties described above, as well as institutional change, long-term planning, and the construction and maintenance of appropriate infrastructure. Since many rivers which rise in mountain regions flow into other countries, such initiatives often will have to be multilateral.

Energy

As discussed in Chapter 8, demands for energy are rising in mountain regions world-wide. Climate change will affect these trends; for instance, while warmer winters would decrease energy requirements for domestic heating, warmer summers could cause energy demands for air conditioning. Energy supplies have been considered tangentially above with respect to fuelwood (see also Chapter 8). This is a vital energy source for a significant proportion of mountain people, but its collection and combustion add to emissions of greenhouse gases. Also, as discussed in Chapter 13, available supplies are decreasing in many mountain regions. Another change in the availability of energy supplies resulting from climate change – though it is a policy response, rather than a biophysical force –

is the imposition of 'carbon taxes' on fossil fuels (Goldemberg *et al.*, 1996). Higher costs for fossil fuels will affect a wide range of economic activities, and will be especially important in mountain regions where access is difficult. Thus, as demands for energy increase, and supplies of carbon-based energy sources become less available or more expensive, opportunities will increase for the use of other energy sources with considerable potential in mountain regions: solar, wind, and hydroelectric (Chapter 8).

One of the most likely pressures for the development of new energy sources will probably focus on hydroelectric power, at scales from micro-generators for individual villages to the massive projects described in Chapter 7. The scope for future large-scale hydroelectric projects in Western Europe is limited because most available dam sites have already been developed, while environmental protection policies greatly limit future development in North America (Stone, 1992). In contrast, many large projects are under development or planned in Asia (Badenkov, 1992; Bandyopadhyhay and Gyawali, 1994; Dhakal, 1990). As shown by research in New Zealand (Garr and Fitzharris, 1994), hydrological changes resulting from climate change would undoubtedly influence the viability of these projects, whose impacts on mountain communities may vary from the positive (e.g. availability of irrigation water and power) to the negative (e.g. loss of agricultural land, forests, settlements, and transport infrastructure).

Much research is needed to identify the relative benefits and disadvantages of hydroelectric projects at different scales, and climate change provides additional reasons to move forward on this. At the same time, it provides an impetus for governments, private industry, and other institutions to proactively develop and implement new solar and wind technologies, and to encourage the more efficient use of energy in building design and construction, cooking, and other areas, especially with the aim of decreasing consumption of fuelwood and fossil fuels.

Tourism

As discussed in Chapter 12, tourism has become a major agent of change in mountain regions around the world. Climatic factors are critical in determining the types of activities, and the seasons when they occur. Climate change is likely to affect mountain tourism both directly and indirectly. Direct changes refer to changes in the atmospheric resources necessary for specific activities. Indirect changes may result from both changes in mountain landscapes and wider scale socio-economic changes: for example, in patterns of demand for specific activities or destinations and for fuel prices.

Using scenarios derived from GCMs, the possible implications of climate change for skiing have been evaluated in Australia (Galloway, 1988), Austria (Breiling and Charamza, 1994); eastern Canada (Lamothe and Périard, 1988; McBoyle and Wall, 1987), and Switzerland (Abegg and Froesch, 1994; Abegg et al., 1994). These studies show that even quite small climatic changes could lead to considerable socio-economic disruption in communities that depend on skiing. These impacts might be partially offset by new opportunities in the summer season and also by investment in new technology, such as snow-making equipment, as long as climatic conditions remain within appropriate bounds. Ski resorts are also found in many other mountain regions, in both industrialised and developing countries, and similar impacts might be expected. In the Alps and many other regions, the disappearance of a significant fraction of glaciers could have implications for summer tourism, by shifting the locations visited by mountaineers and other tourists. In South Asia, changes in the timing and intensity of the monsoon (Kattenberg et al., 1996) could greatly affect those who depend on tourism in mountain areas.

In addition to these potential direct impacts of climate change on tourism, increases in the costs of fossil fuels, a major component of the cost of tourism to many mountain regions, may be a critical indirect impact. Other indirect impacts might include decreasing attractiveness of landscapes, new competition from other tourist locations as climates change (particularly at the seasonal scale and in relation to holiday periods), and concerns about the health risks of high levels of ultraviolet radiation at high altitudes.

Tourism is an economic sector which is often characterised by seasonality and uncertainty, and is closely tied to other economic activities. Consequently, the implications of climate change for tourism must be regarded in conjunction with the implications for other economic sectors (Breiling and Charamza, 1994). The extent to which climate change is likely to affect tourism will depend, among other things, on the distance of a mountain community or region from its main market(s), and the degree to which tourism has influenced its

environment and economy. There are numerous examples of the decline of tourism in mountain regions and reversion to an economy based on agriculture and forestry. However, the ability to accomplish this successfully depends on many factors, including the availability of an appropriate labour force, the extent to which essential resources – such as irrigation and terracing systems – have deteriorated, and climate.

In most mountain regions, tourism's development has been characterised by rapid decisions based on far more limited information than decisions relating to other activities on which these communities have relied. Many mountain people recognise that they need to maintain or recover control over the forces of the tourist economy. In order to do this, applied research is essential to provide a clearer understanding of the forces that drive the development of the diverse types of tourism, their relationships to other economic activities, and the various implications – both direct and indirect – of climate change.

Transport

Climate change may influence transport in mountain regions both indirectly, by altering the availability of energy sources, and directly, especially through increases in the frequency of natural hazards, as discussed above and in Chapter 16. Consequently, the maintenance of the stability of both forests and agricultural land is critical, showing the indirect, but crucial importance of these activities. However, the likelihood of increased numbers of extreme precipitation events means that communications – as well as many other activities – could be greatly disrupted. At the same time, a longer snow-free season would ease transport and decrease costs of snow removal. Governments and local authorities should not assume that climate variability will be similar to recent history, and should build scenarios of the likely positive and negative impacts of climate change into long-term land-use and transportation planning.

Human health

Human health could be affected by climate change through changes in temperatures, air quality, and hydrology (McMichael, 1996). Extreme temperatures, both high and low, could cause increased mortality from cardiovascular and respiratory stress. Higher temperatures could also allow populations of disease-carrying organisms to increase significantly and to move to higher altitudes. For instance, incidence of malaria increased in recent exceptionally hot seasons in Pakistan (Bouma *et al.*, 1994) and Rwanda (Loevinsohn, 1994), and could move into large malaria-free populations in Kenya and Zimbabwe with only a small increase in winter temperature (Haines *et al.*, 1993). In general, higher temperatures decrease the reproduction time of disease vectors, increasing the incidence of disease.

Increases in precipitation could also be important in providing additional habitat, in residual puddles and pools, for disease vectors such as mosquitoes and the many micro-organisms which cause diarrhoea and dysentery. Conversely, drier conditions could limit the availability of water for drinking, food preparation, and sewage disposal. A further range of indirect impacts of climate change on human health could derive from changes in the availability of food, as described above. Considerable research is needed to provide better understanding of all of the relationships described above; this should be combined with monitoring of vectors and the incidence of disease, especially just beyond the boundaries of the current incidence of particular diseases, in order to provide information essential for preventative measures. In general, better water and sewage management, as well as reliable food distribution, will be increasingly important. As both permanent and tourist populations increase, the maintenance of adequate health standards will be a growing problem in many mountain regions that climate change is likely to exacerbate.

GENERAL CONCLUSIONS AND RECOMMENDATIONS

Climate change is an issue for careful consideration by people living in mountain regions, visiting them, or living downstream, as well as the authorities responsible for these regions. Their degree of concern will vary considerably in relation to a vast number of interacting factors; and climate change should be seen in terms of not only negative impacts but also positive opportunities (Glantz *et al.*, 1990). Given the complexity of mountain environments at all scales and the lack of detailed scientific knowledge of these environments and how mountain populations utilise them, definitive conclusions are hard to make. The word 'scientific' is important; members of mountain communities

whose history in a given region stretches back many generations are likely to have local knowledge that is at least as relevant for the long-term future of their descendants and their neighbours as information collected according to objective criteria by scientists and their instruments. Such knowledge should be included in activities considering likely changes in climate, their impacts, and responses to these.

Given the high variability of mountain climates, people living in 'old' mountain regions, whose lifestyles continue to be based, to some extent, on the opportunities and constraints of diverse micro-environments may be able to respond or adapt to changes in climate with relatively little difficulty – unless rates of change are very rapid, infrastructure has been allowed to deteriorate significantly, or suitable crop and tree cultivars are unavailable. However, actions based on knowledge of these complex environments must be linked to the existence or development of suitable societal structures. The extensive networks of community and mutual co-operation that characterise many mountain regions (Beaver and Purrington, 1984; Guillet, 1983; Viazzo, 1989) may already be under stress as populations grow and change in structure, and increasing numbers of tourists arrive. Such networks will become more important. Additionally, the central role of women, together with the position of children dependent upon them, must be recognised (Byers and Sainju, 1994; Ives, 1996; Chapter 4).

In Western Europe, tourism's rise to dominance has required massive investments in infrastructure and has been associated with great changes in agriculture and forestry. Similar patterns seem likely in other parts of Europe. Yet, tourism is generally an unreliable long-term basis for mountain economies (Bätzing *et al.*, 1996), to an extent that may well increase with climate change. In all 'old' mountain regions, governments and NGOs should make increased efforts to maintain and optimise the use of traditional knowledge and practices, which are generally declining. They need to be better linked to scientific research on both mountain environments and innovative management techniques, in order to limit the negative effects of climate change and benefit from its positive aspects.

In 'young' mountain regions, both traditional knowledge and long-term climate records tend to be limited, although other aspects of their environments may be well documented. Densities of settlement tend to be quite low, except in tourist areas, and even in these, high population densities often occur only for relatively short periods. Transportation networks are also relatively sparse. As in other regions, communities depending largely on tourism may have to consider other alternatives as demand decreases or climatic conditions become unsuitable. One problem which may increase in severity is the loss of forest cover through epidemics, dieback, or fire, creating unstable and unattractive environments from which emigration appears the best alternative. Substantial loss of forest cover would also be a critical issue for lowland populations through effects on their water supplies; an issue of critical importance world-wide.

In addition to the many types of sectoral or thematic research that are necessary to increase understanding of climate change and its implications and to respond or adapt to them, it will be important to undertake integrated research at regional scales. Such research should involve members of local communities and other stakeholder groups, as well as natural and social scientists, and be designed to have direct application to policy-making. Such research requires considerable resources and personnel, and the clear commitment of participants to work together over a period of years. Some of the methodologies have been outlined by Carter *et al.* (1994); a valuable example of such a regional study is provided by the recently completed Mackenzie Basin Impact Study, in Canada (Cohen, 1995); and many lessons can be learned from past applied interdisciplinary studies, such as those carried out within UNESCO's Man and the Biosphere (MAB) Programme in the mountain regions of Europe (Price, 1995).

Many of the issues for research and policy action can, and should, be addressed at scales from individual mountain communities to nations, there are also needs for international co-operation in research. These have been recognised in the initial definition of a project by the IGBP, involving four of its core projects, on 'Global Change Impacts on Mountain Hydrology and Ecology' (Becker and Bugmann, 1996). The use of the phrase 'global change' recognises that climate change interacts with many other processes of change acting at global scales. The need to take a cross-cutting approach derives from the recognition that mountain hydrology – both natural and human-influenced – and ecology – including human populations and their land uses – are inextricably linked. Two main research themes have been proposed: i) use of ecological and hydrological characteristics in mountain regions as sensitive indicators of global change; ii) integrated studies of river basins, along altitudinal gradients, to ensure sustainable development in mountain regions and downstream under the influence of global change. Initial activities should include a synthesis of current knowledge on the impacts of global change in mountain regions; inventory of current research in mountain environments to support better co-ordination and interpretation; and the planning and establishment of a global network of research sites and river basins for integrated studies and the testing and development of models at different scales. This project is only at the earliest stage of planning; it would be desirable for governments to support it, recognising the need to take a holistic approach to a major challenge for the twenty-first century and beyond.

References

Abegg, B. and Frösch, R. (1994). Climate change and tourism: impact on transport companies in the Swiss Canton of Graubünden. In Beniston, M. (ed.), *Mountain Environments in Changing Climates*. Routledge: London, pp. 328–40

Abegg, B., König, U. and Maisch, M. (1994). Klimaänderung und Gletscherskitourismus. *Geographica Helvetica*, **49**(3): 103–14

Aizen, V., Aizen, E., Melack, J. and Martma, T. (1996). Isotopic measurements of precipitation on Central Asian glaciers (southeastern Tibet, northern Himalayas, central Tien Shan). *Journal of Geophysical Research*, **101**(D4): 185–96

Allan, J.D. (1995). *Stream Ecology: Structure and Function of Running Waters*. Chapman and Hall: London

Allen, B., Brookfield, H. and Byron, Y. (eds.) (1989). Frost and drought in the highlands of Papua New Guinea. *Mountain Research and Development*, **9**: 199–334

Auer, I., Böhm, R. and Mohnl, H. (1990). Die troposphärische Erwarmungsphase die 20 Jahrhunderts in Spiegel der 100 jahrigen Messreihe des alpinen Gipfelobservatoriums auf dem Sonnblick. *CIMA 88, Proceedings, 20th Internationale Tagung für Alpine Meteorologie.* Italian Meterological Service: Rome

Badenkov, Y. (1992). Mountains of the former Soviet Union: value, diversity, uncertainty. In Stone, P. B. (ed.), *The State of the World's Mountains.* Zed Books: London, pp. 257–97

Balteanu, D., Ozenda, P., Huhn, M., Kerschner, H., Tranquillini, W. and Bortenschlager, S. (1987). Impact analysis of climatic change in the central European mountain ranges. In *European workshop on interrelated bioclimatic and land use changes*, Vol. G. Noordwijkerhout: The Netherlands

Bandyopadhyay, J. and Gyawali, D. (1994). Himalayan water resources: Ecological and political aspects of management. *Mountain Research and Development*, **14**: 1–24

Baron, J. S., Ojima, D. S., Holland, E. A. and Parton, W. J. (1994). Analysis of nitrogen saturation potential in Rocky Mountains tundra and forests: implications for aquatic systems. *Biogeochemistry*, **27**: 61–82

Barros, A. and Lettenmeier, D. P. (1993). Dynamic modeling of the spatial distribution of precipitation in remote mountainous areas. *Monthly Weather Review*, **121**: 195–214

Barry, R. G. (1990). Changes in mountain climate and glacio-hydrological responses. *Mountain Research and Development*, **10**: 161–70

Barry, R. G. (1992a). Climate change in the mountains. In Stone, P. B. (ed) *The State of the World's Mountains.* Zed Books: London, pp. 359–380

Barry, R. G. (1992b). Mountain climatology and past and potential future climatic changes in mountain regions: a review. *Mountain Research and Development*, **12**: 71–86

Barry, R. G. (1992c). *Mountain Weather and Climate*, Third edition. Methuen: New York

Barry, R. G. (1995). The importance of mountains in the climate, system climate processes and climate change. *Preprints, Fifth Conference on Mountain Meteorology.* American Meteorological Society: Boston, pp. 1–5

Barry, R. G., Fallot, J.-M. and Armstrong, R. L. (1995). Twentieth-century variability in snow-cover conditions and approaches to detecting and monitoring changes: status and prospects. *Progress in Physical Geography*, **19**(4): 520–32

Barry, R. G. and Van Wie, C. C. (1974). Topo- and microclimatology in alpine areas. In Ives, J. D. and Barry R. G. (eds.), *Arctic and Alpine Environments.* Methuen: London, pp. 73–83

Basist, A., Bell, G. D. and Meentemeyer, V. (1994). Statistical relationships between topography and precipitation patterns. *Journal of Climate*, 7: 305–15

Bätzing, W., Perlik, M. and Dekleva, M. (1996). Urbanization and depopulation in the Alps (with 3 colored maps), *Mountain Research and Development*, **16**(4):335–50

Baumgartner, M. F., Martin, E. and Borel, J.-L. (1995). Climate change impacts on snow cover: Modelling case studies in Switzerland and the French Western Alps. In Guisan, A., *et al.* (eds.), *Potential Ecological Impacts of Climate Change in the Alps and Fennoscandian Mountains.* Conservatoire et Jardin botaniques de la Ville de Genève: Geneva, pp. 105–12

Beaver, P. D. and Purrington, B. L. (eds.) (1984). *Cultural Adaptation to Mountain Environments.* University of Georgia Press: Athens, GA

Becker, A. and Bugmann, H. (eds.) (1996). *Predicting Global Change Impacts on Mountain Hydrology and Ecology: Integrated Catchment Hydrology/Altitudinal Gradient Studies.* IGBP–BAHC Core Project Office: Potsdam

Begon, M., Harper, J.L. and Townsend, C.R. (1990). *Ecology: Individuals, Populations and Communities.* Blackwell: Cambridge

Beniston, M. (ed.) (1994). *Mountain Environments in Changing Climates.* Routledge: London

Beniston, M. and Fox, D. G. (eds.) (1996). Impacts of climate change on mountain regions. In Watson, R. T., Zinyowera, M. C. and Moss, R. H. (eds.), *Climate Change 1995: Impacts, Adaptations and Mitigation of Climate Change: Scientific–Technical Analyses.* Cambridge University Press: Cambridge, pp. 191–213

Beniston, M., Rotach, M., Tschuck, P., Wild, M. and Ohmura, A. (1996). *Simulation of Climate Trends over the Alpine Region – Development of a Physically-based Modelling System for Application to Regional Studies of Current and Future Climate.* Final Scientific

Report to Swiss National Science Foundation, Department of Geography, ETH: Zurich

Billings, W. D. (1974). Adaptation and origins of alpine plants. *Arctic and Alpine Research*, **6**: 129–42

Blumer, F. P. (1994). *Höhenabhängigkeit des Niederschlages im Alpenraum*. Dissertation ETH Nr. 10784, Eidgenössischen Technischen Hochschule, Zürich

Böhm, R. (1986). *Der Sonnblick. Die 100-jahrige Geschichte der Observatoriums und seiner Forschungstätigkeit*. Osterreichischer Bundesverlag: Vienna

Botkin, D. B. and Nisbet, R. A. (1992). Forest response to climatic change: effects of parameter estimation and choice of weather patterns on the reliability of projections. *Climatic Change*, **20**: 87–111

Bouma, M. J., Sondorp, H. E. and van der Kaay, H. J. (1994). Health and climate change. *The Lancet*, **343**: 302

Bowman, W. D., Thodose, T. A., Schardt, J. C. and Conant, R. T. (1993). Constraints of nutrient availability on primary productivity in two alpine tundra communities. *Ecology*, **74**: 2085–97

Braun, J. (1913). Die Vegetationsverhältnisse in der Schneestufe der Rätisch-Lepontischen Alpen. *Neue Denkschriften der Schweizerischen Naturforschenden Gesellschaft*, **48**: 156–307

Braun-Blanquet, J. (1958). Über die obersten Grenzen pflanzlichen Lebens im Gipfelbereich des Schweizerischen Nationalparks. *Kommission der Schweizerischen Naturforschenden Gesellschaft zur Wissenschaftliche Erforschung des Nationalparks*, **6**: 119–42

Bravo, R. E. *et al.* (1988). The effects of climatic variations on agriculture in the Central Sierra of Ecuador. In Parry, M. L., Carter, T. R. and Konijn, N. T. (eds), *The Impact of Climate Variations on Agriculture: Volume 2, Assessments in Semi-arid Regions*. Kluwer: Dordrecht, pp. 381–493

Breiling, M. and Charamza, P. (1994). Localizing the threats due to climate change in mountain environments. In Beniston, M. (ed.), *Mountain Environments in Changing Climates*. Routledge: London, pp. 341–65

Brown, S. (ed.) (1996). Management of forests for mitigation of greenhouse gas emissions. In Watson, R. T., Zinyowera, M. C. and Moss, R. H. (eds.), *Climate Change 1995: Impacts, Adaptations and Mitigation of Climate Change: Scientific–Technical Analyses*. Cambridge University Press: Cambridge, pp. 773–97

Brzeziecki, B., Kienast, F. and Wildi, O. (1994). Potential impacts of a changing climate on the vegetation cover of Switzerland: a simulation experiment using GIS technology. In Price, M. F. and Heywood, D. I. (eds.), *Mountain Environments and Geographic Information Systems*. Taylor and Francis: London, pp. 263–79

Bücher, A. and Dessens, J. (1991). Secular trend of surface temperature at an elevated observatory in the Pyrenees. *Journal of Climate*, **4**: 859–68

Bucher, F. and Jeanneret, F. (1994). Phenology as a tool in topoclimatology. In Beniston, M. (ed.), *Mountain Environments in Changing Climates*. London: Routledge, pp. 270–80

Bultot, F., Gellens, D., Spreafico, M. and Schädler, B. (1992). Repercussions of a CO_2 doubling on the water balance – a case study in Switzerland. *Journal of Hydrology*, **137**: 199–208

Byers, E. and Sainju, M. (1994). Mountain ecosystems and women: Opportunities for sustainable development and conservation. *Mountain Research and Development*, **14**: 213–28

Byron, E. R. and Goldman, C. R. (1990). The potential effects of global warming on the primary productivity of a subalpine lake. *Water Resources Bulletin*, **26**: 983–9

Callaway, R. M., DeLucia, E. H. and Schlesinger, W. H. (1994). Biomass allocation of montane and desert ponderosa pine: an analog for response to climate change. *Ecology*, **75**: 1474–81

Carter, T. R., Parry, M. L., Harasawa, H. and Nishioka, S. (1994). *IPCC Technical Guidelines for Assessing Climate Change Impacts and Adaptations*. University College, London and Center for Global Environmental Research: Tsukuba

Cebon, P., Dahinden, U., Davies, H., Imboden, D. and Jaeger, C. (eds.) (1997). *A View from the Alps: Regional Perspectives on Climate Change*. MIT Press: Boston

Changnon, D. (1992). *Hydroclimate Variability in the Rocky Mountain Region*. Unpublished PhD dissertation, Colorado State University

Chen, J.-Y. and Ohmura, A. (1991). On the influence of alpine glaciers on runoff. In Lang, H. and Musy, A.

(eds.), *Hydrology in Mountainous Regions. I. Hydrological Measurements. The Water Cycle.* International Association of Hydrology Publication 193. IAHS Press: Wallingford, pp. 117–26

Cohen, S. J. (1995). An interdisciplinary assessment of climate change on northern ecosystems: The Mackenzie Basin Impact Study. In Peterson, D. L. and Johnson, D. R. (eds.), *Human Ecology and Climate Change.* Taylor and Francis: Washington DC, pp. 301–15

Collins, D. N. (1987). Climatic fluctuations and run off from glacierised alpine basins. In Solomon, S.L., Beran, M. and Hogg, W. (eds.), *The Influence of Climate Change and Climatic Variability on the Hydrological Regime and Water Resources.* International Association of Hydrology Publication 168. IAHS Press: Wallingford, pp. 77–89

Cook, E. R. and Cole, J. (1991). On predicting the response of forests in eastern North America to future climatic change. *Climatic Change*, 19: 271–82

Cramer, W. P. and Leemans, R. (1993). Assessing impacts of climate change on vegetation using climate classification systems. In Solomon, A. M. and Shugart, H. H. (eds.), *Vegetation Dynamics and Global Change.* Chapman and Hall: New York, pp. 190–217

d'Oleire-Oltmanns, W., Mingozzi, T. and Brendel, U. (1995). In Guisan, A., *et al.* (eds.), *Potential Ecological Impacts of Climate Change in the Alps and Fennoscandian Mountains.* Conservatoire et Jardin botaniques de la Ville de Genève: Geneva, pp. 173–5

Daly, C., Neilson, R. P. and Phillips, D. L. (1994). A statistical–topographic model for mapping climatological precipitation over mountainous terrain. *Journal of Applied Meteorology*, 33: 140–58

Davis, M. B. and Zabinski, C. (1992). Changes in geographical range resulting from greenhouse warming – effects on biodiversity in forests. In Peters, R. L. and Lovejoy, T. J. (eds.), *Global Warming and Biological Diversity.* Yale University Press: New Haven, pp. 297–308

Decker, W. L., Jones, V. K. and Achutuni, R. (1986). *The Impact of Climate Change for Increased Atmospheric Carbon Dioxide on American Agriculture.* Carbon Dioxide Research Division, US Department of Energy: Washington DC

DeLucia, E. H., Callaway, R. M., Schlesinger, W. H. and Scarasia-Mugnozza, G. E. (1994). Offsetting changes in biomass allocation and photosynthesis in ponderosa pine (*Pinus ponderosa*) in response to climate change. *Tree Physiology*, 14: 669–77

Dhakal, D. N. S. (1990). Hydropower in Bhutan: a long-term development perspective. *Mountain Research and Development*, 10: 291–300

Diaz, H. F. and Graham, N. E. (1996). Recent changes in freezing heights and the role of sea surface temperature. *Nature*, 383: 152–5

Downing, T. E. (1992). *Climate Change and Vulnerable Places: Global Food Security and Country Studies in Zimbabwe, Kenya, Senegal, and Chile.* Environmental Change Unit: Oxford

Duell, L. F. W. (1994). The sensitivity of northern Sierra Nevada streamflow to climate change. *Water Resources Bulletin*, 30: 841–59

Duguay, C. R. (1993). Radiation modeling in mountainous terrain: Review and status. *Mountain Research and Development*, 13: 339–57

Duus, A. L. (1992). Estimation and analysis of snow cover in the Snowy Mountains between 1900 and 1991. *Australian Meteorological Magazine*, 40: 195–204

Elias, S. A. (1991). Insects and climate change. *Bioscience*, 41: 552–9

Fanta, J. (1992). Possible impact of climatic change on forested landscapes in central Europe: a review. *Catena* supplement, 22: 133–51

Fischlin, A. (1995). Assessing sensitivities of forests to climate change: Experiences from modelling case studies. In Guisan, A., *et al.* (eds.), *Potential Ecological Impacts of Climate Change in the Alps and Fennoscandian Mountains.* Conservatoire et Jardin botaniques de la Ville de Genève: Geneva, pp. 145–7

Fisher, M. and Gardner, A. S. (1995). The status and ecology of a *Juniperus excelsa* subsp. *polycarpos* woodland in the northern mountains of Oman. *Vegetatio*, 119: 33–51

Fitzharris, B. B. (ed.) (1996). The cryosphere: Changes and their impacts. In Watson, R. T., Zinyowera, M. C. and Moss, R. H. (eds.), *Climate Change 1995: Impacts, Adaptations and Mitigation of Climate Change: Scientific–Technical Analyses.* Cambridge University Press: Cambridge, pp. 240–65

Föhn, P. (1990). Schnee und Lawinen. In Vishcher, D. D. (ed.), *Schnee, Eis und Wasser der Alpen in einer wärmeren Atmosphäre*. Mitteilungen 108, Versuchsanstalt für Wasserbau, Hydrologie und Glaziologie, Eidgenössische Technische Hochschule: Zürich, pp. 38–48

Franklin, J. F. *et al.* (1992). Effects of global climatic change of forests on northwestern North America. In Peters, R. L. and Lovejoy, T. J. (eds.) *Global Warming and Biological Diversity*. Yale University Press: New Haven, pp. 244–57

Frey-Bunes, F., Heimann, D. and Sausen, R. (1995). A statistical–dynamical downscaling procedure for global climate simulations. *Theoretical and Applied Climatology*, **50**: 117–32

Gadgil, M. (1971). Dispersal: population consequences and evolution. *Ecology*, **52**: 253–61

Galloway, R. W. (1988). The potential impact of climate changes on Australian ski fields. In Pearman, G. I. (ed), *Greenhouse: Planning for Climate Change*. CSIRO: East Melbourne, pp. 428–37

Gao, Sh-H. and Peng, J. W. (1994). The climatic features in the Gongga Mountains. In *Glaciers and Environment in the Qinghai-Xizang (Tibet) Plateau (1) The Gongga Mountains*. Science Press: Beijing and New York, pp. 29–38

Garr, C. E. and Fitzharris, B. B. (1994). Sensitivity of mountain runoff and hydro-electricity to changing climate. In Beniston, M. (ed.), *Mountain Environments in Changing Climates*. London: Routledge, pp. 366–81

Giorgi, F. (1990). On the simulation of regional climate using a limited area model nested in a general circulation model. *Journal of Climate*, **3**: 941–63

Giorgi, F., Marinucci, M. R. and Visconti, G. (1992). A $2 \times CO_2$ climate change scenario over Europe generated using a Limited Area Model nested in a General Circulation Model. II. Climate change scenario. *Journal of Geophysical Research*, **97**: 11–28

Giorgi, F. and Mearns, L. O. (1991). Approaches to the simulation of regional climate change: a review. *Reviews of Geophysics*, **29**: 191–216

Giorgi, F., Shields-Brodeur, C. and Bates, G. T. (1994). Regional climate change scenarios over the United States produced with a nested regional climate model: Spatial and seasonal characteristics. *Journal of Climate*, **7**: 375–99

Glantz, M. H., Price, M. F. and Krenz, M. E. (eds.) (1990). *On Assessing Winners and Losers in the Context of Global Warming*. National Center for Atmospheric Research: Boulder

Glassy, J. M. and Running, S. W. (1994). Validating diurnal climatology logic of the MT-CLIM model across a climatic gradient in Oregon. *Ecological Applications*, **4**: 248–57

Gleick, P. H. (1989). Climate change, hydrology and water resources. *Reviews of Geophysics*, **27**: 329–44

Goldemberg, J., *et al.* (1996). Introduction: scope of the assessment. In Bruce, J. P., Lee, H. and Haites, E. F. (eds.), *Climate Change 1995: Economic and Social Dimensions of Climate Change*. Cambridge University Press: Cambridge, pp. 17–51

Grabherr, G., Gottfriend, M. and Pauli, H. (1994). Climate effects on mountain plants. *Nature*, **369**: 448

Grabherr, G., *et al.* (1995). Patterns and current changes in alpine plant diversity. In Chapin, F. S. and Körner, C. (eds.), *Arctic and Alpine Biodiversity: Patterns, Causes and Ecosystem Consequences*. Springer: Berlin, pp. 167–81

Graham, R. W. and Grimm, E. C. (1990). Effects of global climate change on the patterns of terrestrial biological communities. *Trends in Ecology and Evolution*, **5**: 289–92

Grötzbach, E. F. (1988). High mountains as human habitat. In Allan, N. J. R., Knapp, G. W. and Stadel, C. (eds.), *Human Impact on Mountains*. Rowman and Littlefield: Totowa, NJ, pp. 24–35

Guillet, D. (1983). Toward a cultural ecology of mountains: The Central Andes and the Himalaya compared. *Current Anthropology*, **24**: 561–74

Guisan, A. and Holten, J. I. (1995). Impacts of climate change on mountain ecosystems: Future research and monitoring needs. In Guisan, A., *et al.* (eds.), *Potential Ecological Impacts of Climate Change in the Alps and Fennoscandian Mountains*. Conservatoire et Jardin botaniques de la Ville de Genève: Geneva, pp. 179–84

Guisan, A., Tessier, L., Holten, J. I., Haeberli, W. and Baumgartner, M. (eds.) (1995). *Potential Ecological Impacts of Climate Change in the Alps and Fennoscandian Mountains*. Conservatoire et Jardin botaniques de la Ville de Genève: Geneva

Gyalistras, D., Schaer, C., Davies, H. C. and Wanner, H. (1997). Future Alpine climate. In Cebon, P., *et al.* (eds.), *A View from the Alps: Regional Perspectives on Climate Change*. MIT Press: Boston

Haeberli, W. (1994). Accelerated glacier and permafrost changes in the Alps. In Beniston, M. (ed.), *Mountain Environments in Changing Climates*. Routledge: London, pp. 91–107

Haeberli, W. (1995). Climate change impacts on glaciers and permafrost. In Guisan, A., *et al.* (eds.), *Potential Ecological Impacts of Climate Change in the Alps and Fennoscandian Mountains*. Conservatoire et Jardin botaniques de la Ville de Genève: Geneva, pp. 97–103

Haeberli, W. and Hoelzle, M. (1995). Application of inventory data for estimating characteristics of regional climate change effects on mountain glaciers – a pilot with the European Alps. *Annals of Glaciology*, **21**: 206–12

Haines, A., Epstein, P. R. and McMichael, M. J. (1993). Global Health Watch: monitoring the impacts of environmental change. *The Lancet*, **342**: 1464–9

Halpin, P. N. (1994). GIS analysis of the potential impacts of climate change on mountain ecosystems and protected areas. In Price, M. F. and Heywood, D. I. (eds), *Mountain Environments and Geographic Information Systems*. Taylor and Francis: London, pp. 281–301

Hamilton, L. S. (1992). The protective role of mountain forests. *GeoJournal*, **27**: 13–22

Hamilton, L. S. (1995). Mountain cloud forest conservation and research: a synopsis. *Mountain Research and Development*, **15**: 259–66

Harte, J. *et al.* (1992). The nature and consequences of indirect linkages between climate change and biological diversity. In Peters, R. L. and Lovejoy, T. J. (eds.), *Global Warming and Biological Diversity*. Yale University Press: New Haven, pp. 325–43

Hastenrath, S. and Rostom, R. (1990). Variations of the Lewis and Gregory glaciers, Mount Kenya, 1978–86–90. *Erdkunde*, **44**: 313–7

Hattemer, H. H. and Gregorius, H.-R. (1990). Is gene conservation under global climate change meaningful? In Jackson, M. T., Ford-Lloyd, B. V. and Parry, M. L. (eds), *Climatic Change and Plant Genetic Resources*. Belhaven: London, pp. 158–66

Hättenschwiler, S. and Körner, C. (1995). Responses to recent climate warming of *Pinus sylvestris* and *Pinus cembra* within their montane transition zone in the Swiss Alps. *Journal of Vegetation Science*, **6**: 357–68

Hay, L. E. *et al.* (1991). Simulation of precipitation by weather-type analysis. *Water Resources Research*, **27**: 493–501

Hoelzle, M. and Haeberli, W. (1995). Simulating the effects of mean annual air-temperature changes on permafrost distribution and glacier size: an example from the Upper Eingadin, Swiss Alps. *Annals of Glaciology*, **21**: 399–405

Holten, J. I. (1993). Potential effects of climatic change on distribution of plant species, with emphasis on Norway. In Holten, J. I., Paulsen G. and Oechel W. C. (eds.), *Impacts of Climatic Change on Natural Ecosystems*. Norwegian Institute for Nature Research and the Directorate for Nature Management: Trondheim, pp. 84–104

Holten, J. I., Paulsen, G. and Oechel, W. C. (eds.) (1993). *Impacts of Climatic Change on Natural Ecosystems*. Norwegian Institute for Nature Research and the Directorate for Nature Management: Trondheim

Holtmeier, F.-K. (1994). Ecological aspects of climatically-caused timberline fluctuations. In Beniston, M. (ed.), *Mountain Environments in Changing Climates*. Routledge: London, pp. 220–33

Horton, E. B. (1995). Geographical distribution of changes in maximum and minimum temperatures. *Atmospheric Research*, **37**: 102–17

Houghton, J. T., Meira Filho, L. G., Callander, B. A., Harris, N., Kattenberg, A. and Maskell, K. (eds.) (1996). *Climate Change 1995: The Science of Climate Change*. Cambridge University Press: Cambridge

Hulme, M., *et al.* (1992). *Climate Change due to the Greenhouse Effect and its Implications for China*. WorldWide Fund for Nature: Gland

Illich, I. P. and Haslett, J. R. (1994). Responses of assemblages of Orthoptera to management and use of ski slopes on upper subalpine meadows in the Austrian Alps. *Oecologia*, **97**: 470–4

Innes, J. L. (1991). High-altitude and high-latitude tree growth in relation to past, present and future global climate change. *The Holocene*, **1**: 168–73

Ives, J. D. (1996). *Children, Women and Poverty in Mountain Ecosystems*. UNICEF: New York

Ives, J. D. and Messerli, B. (1989). *The Himalayan Dilemma: Reconciling Development and Conservation*. Routledge: London and New York

Järvinen, A. (1995). Effects of climate change on mountain bird populations. In Guisan, A., *et al.* (eds.), *Potential Ecological Impacts of Climate Change in the Alps and Fennoscandian Mountains*. Conservatoire et Jardin botaniques de la Ville de Genève: Geneva, pp. 73–4

Jodha, N. S. (1996). Enhancing food security in a warmer and more crowded world: Factors and processes in fragile zones. In Downing, T. E. (ed.), *Climate Change and World Food Security*. Springer: Berlin, pp. 381–419

Kaczmarek, Z. (ed.) (1996). Water resources management. In Watson, R. T., Zinyowera, M. C. and Moss, R. H. (eds.), *Climate Change 1995: Impacts, Adaptations and Mitigation of Climate Change: Scientific–Technical Analyses*. Cambridge University Press: Cambridge, pp. 469–86

Karl, T. R., Wang, W.-C., Schlesinger, M. E., Knight, R. W. and Portman, D. A. (1990). A method of relating general circulation model simulated climate to the observed local climate. Part I: Seasonal statistics. *Journal of Climate*, 3: 53–79

Kattenberg, A., *et al.* (1996). Climate models – Projections of future climate. In Houghton, J. T., *et al.* (eds.), *Climate Change 1995 – The Science of Climate Change*. Cambridge University Press: Cambridge, pp. 285–357

Katzfey, J. J. (1995). Simulation of extreme New Zealand precipitation events. Part I. Sensitivity to orography and resolution. *Monthly Weather Review*, 123: 737–54

Kimmins, J. P. and Lavender, D. P. (1992). Ecosystem-level changes that may be expected in a changing global climate: a British Columbia perspective. *Environmental Toxicology and Chemistry*, 11: 1061–8

Kirschbaum, M. U. F. and Fischlin, A. (eds.) (1996). Climate change impacts on forests. In Watson, R. T., Zinyowera, M. C. and Moss, R. H. (eds.), *Climate Change 1995: Impacts, Adaptations and Mitigation of Climate Change: Scientific–Technical Analyses*. Cambridge University Press: Cambridge, pp. 95–129

Koch, E. and Rüdel, E. (1990). Mögliche Auswirkungen eines verstärkte Treibhauseffekte auf die Schneeverhältnisse in Österreich. *Wetter und Leben*, 45: 137–3

Konovalov, V. G. and Shchetinnicov, A. S. (1994). Evolution of glaciation in the Pamiro-Alai mountains and its effects on river runoff. *Journal of Glaciology*, 40: 149–57

Körner, C. (1989). The nutritional status of plants from high altitudes. A worldwide comparison. *Oecologia*, 81: 379–91

Körner, C. (1994). Impact of atmospheric changes on high altitude vegetation. In Beniston, M. (ed.), *Mountain Environments in Changing Climates*. Routledge: London: pp. 155–66

Körner, C. (1995). Impact of atmospheric changes on alpine vegetation: The ecophysiological perspective. In Guisan, A., *et al.* (eds.), *Potential Ecological Impacts of Climate Change in the Alps and Fennoscandian Mountains*. Conservatoire et Jardin botaniques de la Ville de Genève: Geneva, pp. 113–20

Kotlyakov, V. M., Serebryanny, J. R. and Solomina, O. N. (1991). Climate and glacier fluctuations in the southern mountains of the USSR during the last 1000 years. *Mountain Research and Development*, 11: 1–12

Kräuchi, N. and Kienast, F. (1993). Modelling subalpine forest dynamics as influenced by a changing environment. *Water, Air, and Soil Pollution*, 68: 185–97

Lamothe and Périard, (1988). *Implications of Climate Change for Downhill Skiing in Quebec*. Climate Change Digest 88-03, Atmospheric Environment Service: Downsview

Laumann, T. and Reeh, N. (1993). Sensitivity to climatic change of the mass balance of glaciers in southern Norway. *Journal of Glaciology*, 39: 656–65

Lertzman, K. P. (1995). Forest dynamics, differential mortality and variable recruitment probabilities. *Journal of Vegetation Science*, 6: 191–204

Lettenmaier, E. P. and Gan, T. Y. (1990). Hydrologic sensitivities of the Sacramento–San Joaquin River basin, California, to global warming. *Water Resources Research*, 26: 69–86

Leung, L. R. and Ghan, S. J. (1995). A subgrid parameterization of orographic precipitation. *Theoretical and Applied Climatology*, 52: 95–118

Leung, L. R., Wigmosta, M. S., Ghan, S. J., Epstein, D. J. and Vail, L. W. (1996). Application of a subgrid orographic precipitation/surface hydrology scheme to a mountain watershed. *Journal of Geophysical Research*, 101(D8): 803–17

Li, Y., Jiang, J., Long, G. and Cheng, Y. (1993). The influence of climate warming on rice production in China. In *The Impact of Climatic Variations on Agriculture and its Strategic Impact*. Beijing University Press: Beijing, pp. 54–130

Liverman, D. M. and O'Brien, K. L. (1991). Global warming and climate change in Mexico. *Global Environmental Change: Human and Policy Dimensions*, 1: 351–64

Loevinsohn, M. (1994). Climatic warming and increased malaria incidence in Rwanda. *The Lancet*, 343: 714–8

Luxmoore, R. J., Wullschleger, S. D. and Hanson, P. J. (1993). Forest responses to CO_2 enrichment and climate warming. *Water, Air, and Soil Pollution*, 70: 309–23

Maisch, M. (1990). Die Gletscher von '1850' und 'Heute' in Bundnerland und in den angrenzenden Gebieten. *Geographica Helvetica*, 42: 127–45

Makarevich, K. G. L. and Rototaeva, O. V. (1986). Present-day fluctuations of mountain glaciers in the northern hemisphere. In Soviet Geophysical Committee, *Data of Glaciological Studies. 57*. Academy of Sciences of the USSR: Moscow, pp. 157–63

Malanson, G. P. and Cairns, D. M. (1995). Effects of cloud cover on a montane forest landscape. *Ecoscience*, 2: 75–82

Mani, M. S. (1968). *Ecology and Biogeography of High Altitude Insects*. Junk: The Hague

Marinucci, M. R., Giorgi, F., Benitson, M., Wild, M., Tschuck, P. and Bernasconi, A. (1995). High resolution simulations of January and July climate over the western Alpine region with a nested regional modeling system. *Theoretical and Applied Climatology*, 51: 119–38

Martin, E., Brun, E. and Durand, Y. (1994). Sensitivity of the French Alps snow cover to the variation of climatic variables. *Annales Geophysicae*, 12: 469–77

McBoyle, G. R. and Wall, G. (1987). The impact of CO_2-induced warming on downhill skiing in the Laurentians. *Cahiers de Géographie de Québec*, 31(82): 39–50

McCabe, G. J. and Hay, L. E. (1995). Hydrological effects of hypothetical climate change in the East River Basin, Colorado, USA. *Hydrological Sciences Journal*, 40: 303–18

McGinnis, D. L. (1995). Downscaling techniques for snowfall prediction in global change studies. In *Sixth International Meeting on Statistical Climatology*, University College, Galway: Ireland, pp. 335–8

McMichael, A. J. (ed.) (1996). Human population health. In Watson, R. T., Zinyowera, M. C. and Moss, R. H. (eds.), *Climate Change 1995: Impacts, Adaptations and Mitigation of Climate Change: Scientific–Technical Analyses*. Cambridge University Press: Cambridge, pp. 561–84

McNeely, J. A. (1990). Climate change and biological diversity: policy implications. In Boer, M. M. and De Groot, R. S. (eds.), *Landscape–ecological impact of climatic change*. IOS Press: Amsterdam, pp. 406–29

Miehe, S., Cramer, T., Jacobsen, J. P. and Winiger, M. (1996). Humidity conditions in the western Karakorum as indicated by climate data and corresponding distribution patterns of the montane and alpine vegetation. *Erdkunde*, 50: 190–204

Mimikou, M. A., Hadjisavva, P. S. and Kouvopoulos, Y. S. (1991). Regional effects of climate change on water resources systems. In van de Ven, F. H. M. *et al.*, (eds.), *Hydrology for the Water Management of Large River Basins*. International Association of Hydrological Sciences: Wallingford, pp. 173–82

Morain, S. A. *et al.* (1993). Design and test of an object-oriented GIS to map plant species in the Southern Rockies. *Geocarto International*, 8(4): 33044

Müller-Lehmans, H., Funk, M., Aellen, G. and Kappenberger, G. (1996). Langjährige Massenbilanzreihen von Gletschern in der Schweiz. *Zeitschrift für Gletscherkunde und Glazialgeologie*, 30: 141–60

Murphy, D. D. and Weiss, S. B. (1992). Effects of climate change on biological diversity in western North America: species losses and mechanisms. In Peters, R. L. and Lovejoy, T. J. (eds.), *Global Warming and Biological Diversity*. Yale University Press: New Haven, pp. 355–68

Musselmann, R. C. (ed.) (1994). *The Glacier Lakes Ecosystem Experiment Site*. General Technical Report RM-249, USDA Forest Service: Fort Collins

National Research Council (1989). *Lost Crops of the Incas*. National Academy Press: Washington DC

Nicholls, N., Gruza, G. V., Jouzel, J., Karl, T. R., Ogallo, L. A. and Parker, D. A. (eds.) (1996). Observed climate variability and change. In Houghton,

J. T., *et al.* (eds.), *Climate Change 1995. The Science of Climate Change*. Cambridge University Press: Cambridge, pp. 132–92

Parry, M. L., Blantan de Rozari, M., Chong, A. L. and Panich, S. (1992). *The Potential Socio-economic Effects of Climate Change in South-east Asia*. United Nations Environment Programme: Nairobi

Pauli, H., Gottfried, M. and Grabherr, G. (1996). Effects of climate change on mountain ecosystems. *World Resource Review*, Woodridge, pp. 382–91

Pearson, S. M. (1994). Landscape-level processes and wetland conservation in the southern Appalachian Mountains. *Water, Air, and Soil Pollution*, **77**: 321–32

Peters, R. L. (1992). Conservation of biological diversity in the face of climate change. In Peters, R. L. and Lovejoy, T. J. (eds.), *Global Warming and Biological Diversity*. Yale University Press: New Haven, pp. 15–30

Pitlick, J. (1994). Relation between peak flows, precipitation and physiography for five mountainous regions in the western USA. *Journal of Hydrology*, **158**: 219–40

Prange, R. K., McRae, K. B., Midmore, D. J. and Deng, R. (1990). Reduction in potato growth at high temperature: Role of photosynthesis and dark respiration. *American Potato Journal*, **67**: 357–69

Price, M. F. (1990). Temperate mountain forests: common-pool resources with changing, multiple outputs for changing communities. *Natural Resources Journal*, **30**: 685–707

Price, M. F. (1994). Should mountain communities be concerned about climate change? In Beniston, M. (ed.), *Mountain Environments in Changing Climates*. Routledge: London, pp. 431–51

Price, M. F. (1995). *Mountain Research in Europe: An Overview of MAB Research from the Pyrenees to Siberia*. Parthenon/UNESCO: Carnforth/Paris

Price, M. F. and Haslett, J. R. (1995). Climate change and mountain ecosystems. In Allan, N. J. R. (ed.), *Mountains at Risk: Current Issues in Environmental Studies*. Manohar: New Delhi, pp. 73–97

Rango, A. and van Katwijk, K. (1990). Climate change effects on the snowmelt hydrology of western North American mountain basins. *IEEE Transactions of Geoscience and Remote Sensing*, GE-**38**: 970–4

Rath, D., McRae, K. B., Midmore, D. J. and Deng, R. (1994). Einfluss von Klimafaktoren auf die Tierproduktion. In Brunnert, H. and Dämmgen (eds.), Klimaveränderungen und Landwirtschaft, Part II, Landbauforschung. *Völkenrode*, **148**: 341–75

Reilly, J. (ed.) (1996). Agriculture in a changing climate: Impacts and adaptation. In Watson, R. T., Zinyowera, M. C. and Moss, R. H. (eds.), *Climate Change 1995: Impacts, Adaptations and Mitigation of Climate Change: Scientific–Technical Analyses*. Cambridge University Press: Cambridge, pp. 427–67

Reisigl, H. and Pitschmann, H. (1958). Obere Grenzen von Flora und Vegetation in der Nivalstufe der zentraler Oetztaler Alpen (Tirol), *Vegetatio*, **8**: 93–129

Richter, M. (1996). Klimatologische und pflanzenmorphologische Vertikalgradienten in Hochgebirgen. *Erdkunde*, **50**: 205–37

Rickenmann, D. and Zimmermann, M. (1993). The 1987 debris flows in Switzerland: documentation and analysis. *Geomorphology* **8**: 175–89

Riebsame, W. E. (1989). *Assessing the Social Implications of Climate Fluctuations*. United Nations Environment Programme: Nairobi

Riebsame, W. E. *et al.* (1995). Complex river basins. In Strezepek, K. M. and Smith, J. B. (eds.), *As Climate Changes: International Impacts and Implications*. Cambridge University Press: Cambridge, pp. 57–91.

Roberts, L. (1989). How fast can trees migrate? *Science*, **243**: 735–7

Rubinstein, D. I. (1992). The greenhouse effect and changes in animal behavior: effects on social structure and life-history strategies. In Peters, R. L. and Lovejoy, T. J. (eds.), *Global Warming and Biological Diversity*. Yale University Press: New Haven, pp. 180–92

Running, S. W., Nemani, R. R. and Hungerford, R. D. (1987). Extrapolation of synoptic meteorological data in mountainous terrain and its use for simulating forest evapotranspiration and photosynthesis. *Canadian Journal of Forest Research*, **17**: 472–83

Salinger, M. J., Williams, J. M. and Williams, W. M. (1989). *CO_2 and Climate Change: Impacts on Agriculture*. New Zealand Meteorological Service: Wellington

Sarmiento, G. (1986). Ecologically crucial features of climate in high tropical mountains. In Vuilleumier, F.

and Monasterio, M. (eds.), *High Altitude Tropical Biogeography*. Oxford University Press: Oxford, pp. 111–45

Schimel, D. *et al.* (eds.) (1996). Radiative forcing of climate change. In Houghton, J. T. *et al.* (eds.), *Climate Change 1995 – The Science of Climate Change*. Cambridge University Press: Cambridge, pp. 65–131

Schlaepfer, R. (ed.) (1993). *Long-term Implications of Climate Change and Air Pollution on Forest Ecosystems*. World Series 4. International Union of Forestry Research Organizations: Vienna

Schoenenberger, W., Frey, W. and Leuenberger, F. (1990). *Ökologie und Technik der Aufforstung im Gebirge – Anregungen für die Praxis*. Bericht 325. EAFV: Birmensdorf

Schubert, C. (1992). The glaciers of the Sierra Nevada de Mérida (Venezuela): a photographic comparison of recent deglaciation. *Erdkunde*, **46**: 58–63

Serebryanny, L. R. and Solomina, O. N. (1996). Glaciers and climate of the mountains of the former USSR during the Neoglacial. *Mountain Research and Development*, **16**: 157–66

Skiles, J. W. and Hanson, J. D. (1994). Responses of arid and semiarid watersheds to increasing carbon dioxide and climate change as shown by simulation studies. *Climatic Change*, **26**: 377–97

Skre, O. (1995). Sensitivity of alpine plants to potential climate changes. In Guisan, A., *et al.* (eds.), *Potential Ecological Impacts of Climate Change in the Alps and Fennoscandian Mountains*. Conservatoire et Jardin botaniques de la Ville de Genève: Geneva, pp. 63–66

Sokolik, I. N. and Toon, O. B. (1996). Direct radiative forcing by anthropogenic mineral aerosols. *Nature*, **381**(6584): 681–3

Solomon, A. M. (ed.), 1996: Wood production under changing climate and land use. In Watson, R. T., Zinyowera, M. C. and Moss, R. H. (eds.), *Climate Change 1995: Impacts, Adaptations and Mitigation of Climate Change: Scientific–Technical Analyses*. Cambridge University Press: Cambridge, pp. 487–510

Starkel, L. (1972). The role of catastrophic rainfall in shaping the relief of the Lower Himalaya (Darjeeling Hills). *Geographica Polonica*, **23**:151–73

Stone, P. B. (ed.) (1992). *The State of the World's Mountains*. Zed Books: London

Strzepek, K., Niemann, J., Somlyody, L. and Kulshrestha, S. (1995). *Global Assessment of the Use of National Water Resources Vulnerabilities*. IIASA: Laxenburg

Sveinbjornsson, B., Nordell, O. and Kauhanen, H. (1992). Nutrient relations of mountain birch growth at and below elevational tree-line in Swedish Lapland. *Functional Ecology*, **6**: 213–20

Swetnam, T. W. and Betancourt, J. L. (1990). Fire–Southern Oscillation relations in the southwestern United States. *Science*, **249**: 1017–20

Tallis, J. H. (1991). *Plant Community History*. Chapman and Hall: London

Taylor, A. H. (1995). Forest expansion and climate change in the mountain hemlock (*Tsuga mertensiana*) zone, Lassen National Park, California, USA. *Arctic and Alpine Research*, **27**: 207–16

Theurillat, J.-P., *et al.* (1997). Sensitivity of plant and soil ecosystems of the Alps to climate change. In Cebon, P., *et al.* (eds.), *A View from the Alps: Regional Perspectives on Climate Change*. MIT Press: Boston (in press)

Thompson, L. G. *et al.* (1993). Recent warming: ice core evidence from tropical ice cores with emphasis on Central Asia. *Global Planetary Change*, **7**: 145–56

Van Deursen, W. P. A. and Kwadijk, J. C. J. (1994). *The Impacts of Climate Change on the Water Balance of the Ganges–Brahmaputra and Yangtze Basin*. Resource Analysis Report RA/94/160, Netherlands Centre for Geo-Ecological Research. Utrecht University: Utrecht

Viazzo, P. P. (1989). *Upland Communities: Environment, Population and Social Structure in the Alps since the Sixteenth Century*. Cambridge University Press: Cambridge

Von Storch, H., Zorita, E. and Cubasch, U. (1993). Downscaling of climate changes to regional scales: An application to winter rainfall in the Iberian Peninsula. *Journal of Climate*, **6**: 1161–71

Wardle, P. and Coleman, M. C. (1992). Evidence for rising limits of four native New Zealand trees. *New Zealand Journal of Botany*, **30**: 303–14

Watanabe, T., Kameyama, S. and Sato, T. (1995). Imja Glacier dead-ice melt rates and changes in a supra-glacial lake, 1989–1994, Khumbu Himal, Napal: Danger of lake drainage. *Mountain Research and Development*, **15**: 293–300

Weber, R. O., Talkner, P. and Stefanicki, G. (1994). Asymmetric diurnal temperature change in the Alpine region. *Geophysical Research Letters*, **21**: 673–6

Weisberg, P. J. and Baker, W. L. (1995). Spatial variation in tree regeneration in the forest–tundra ecotone, Rocky Mountain National Park, Colorado. *Canadian Journal of Forest Research*, **25**: 1326–39

Whetton, P. H., Haylock, M. R. and Galloway, R. (1996). Climate change and snow cover duration in the Australian Alps. *Climatic Change*, **32**: 447–79

Wild, M., Ohmura, A., Gilgen, H. and Roeckner, E. (1995). Regional climate simulation with a high resolution GCM: surface radiative fluxes. *Climate Dynamics*, **11**: 469–86

Williams, M. S., Losleben, M., Caine, N. and Greenland, D. (1996). Changes in climate and hydrochemical responses in a high-elevation catchment, Rocky Mountains. *Limnology and Oceanography*, **41**(5): 939–46

Winjum, J. K. and Neilson, R. P. (1989). Forests. In Smith, J. B. and Tirpak, D. (eds.), *The Potential Effects of Global Climate Change on the United States*. US Environmental Protection Agency: Washington DC, pp. 71–92

Woodward, F. I. (1992). A review of the effects of climate on vegetation: ranges, competition, and composition. In Peters, R. L. and Lovejoy, T. J. (eds.), *Global Warming and Biological Diversity*. Yale University Press: New Haven, pp. 105–23

Wratt, D. S., *et al.* (1996). The New Zealand Southern Alps Experiment. *Bulletin of the American Meteorological Society*, **77**: 683–2

Yoshida, S. (1981). *Fundamentals of Rice Crop Science*. International Rice Research Institute: Los Banos

Yoshino, M. M., *et al.* (1988). The effect of climatic variations on agriculture in Japan. In Parry, M. L., Carter, T. R. and Konijn, N. T. (eds), *The Impact of Climate Variations on Agriculture: Volume 1, Assessments in Cool, Temperate and Cold regions*. Kluwer: Dordrecht, pp. 723–868

Zimmerer, K. S. (1992). The loss and maintenance of native crops in mountain agriculture. *GeoJournal*, **27**: 61–72

Zimmermann, M. and Haeberli, W. (1992). Climatic change and debris flow activity in High Mountain areas – A case study in the Swiss Alps. In Boer, M. and Koster, E. (eds.), Catena Supplement 22, *Greenhouse-Impact on Cold-Climate Ecosystems and Landscapes*. pp. 57–72

Zuidema, G., van der Born, J., Alcamo, J., Kreileman, G. J. J. (1994). Simulating changes in global land cover as affected by economic and climatic factors. *Water, Air, and Soil Pollution*, **76**: 163–98

Sustainable mountain development – Chapter 13 in action

18

El Hadji Sène and Douglas McGuire

INTRODUCTION

In the five years since the United Nations Conference on Environment and Development (UNCED) was held in 1992, the concept of sustainable mountain development has taken on new meaning. Prior to what is now known as the Rio Earth Summit, there was little recognition that mountain regions may require special consideration, despite their importance in terms of biological or landscape diversity, provision of water and other renewable resources, and protection of vital downstream interests, to name but a few. In the wake of Rio, this is no longer the case. Awareness has increased, thanks mainly to the process set in motion by UNCED. Chapter 13 of Agenda 21 of this conference, 'Managing fragile ecosystems: sustainable mountain development', has provided the opportunity for attention to be focused more directly on mountain populations and ecosystems. The fact that mountains were the subject of a separate chapter of Agenda 21 is altogether significant in that it has served to underline the importance of addressing directly issues pertaining to mountain development and conservation. Often in the past, mountain areas, their problems and needs were dealt with mainly in a sectoral context (e.g. agriculture or forestry), and often within a national planning framework. This has tended to marginalise these areas and provide inadequate emphasis to mountain-specific issues which may require special attention. A multi-sectoral, more comprehensive approach to addressing mountain development issues is a relatively new concept, but one whose time has come.

Since the Rio summit in 1992, the UN Food and Agriculture Organization (FAO) has been carrying out the role of Task Manager for Chapter 13. The overall objective of this role is to ensure collaboration and co-operation in the follow up to, including reporting on, the implementation of Agenda 21 by the United Nations system. Major responsibilities as Task Manager include reporting on implementation; strengthening information exchange; promoting inter-agency consultation; catalysing joint activities and programmes; and developing common strategies.

With the aim of enhancing co-operation and collaboration in these areas, an *ad hoc* inter-agency group for Chapter 13 was established in early 1994. The group is made up of representatives of various UN agencies as well as a number of international non-governmental organisations involved in mountain development and conservation issues. The inclusion and active participation of organisations from outside the UN system has provided the opportunity for a wide range of views and perspectives to be considered in the on-going implementation of Chapter 13 and reporting on its progress. This has enabled a more balanced and equitable approach. The group has met on four different occasions since it was established and has proven to be an effective means of ensuring close consultation and interaction among interested parties and a fully collaborative approach to implementation of the Mountain Agenda.

PROGRESS ON THE MOUNTAIN AGENDA

Chapter 13[1], within the framework of its two programme areas[2], has focused mainly on the following objectives:

---

[1]This section is adapted from the Chapter 13 Task Manager report to CSD5.

[2]There are two Programme Areas of Chapter 13: **A**: Generating and strengthening knowledge about the ecology and sustainable development of mountain ecosystems; and **B**: Promoting integrated watershed development and alternative livelihood opportunities. The two areas are often dealt with together in the framework of an integrated ecosystems approach to sustainable mountain development.

(1) Raising awareness of the importance of, and improving understanding of, sustainable mountain development issues at global, regional, and national levels;

(2) Protecting natural resources and developing technical and institutional arrangements for natural disaster reduction;

(3) Strengthening a global information network and database for organisations, governments, and individuals concerned with mountain issues;

(4) Strengthening country capacity to improve planning, implementation, and monitoring of sustainable mountain development programmes and activities;

(5) Combating poverty through the promotion of sustainable income-generating activities and improvement of infrastructure and social services, in particular to protect the livelihood of local communities and indigenous people; and

(6) Formulating and negotiating regional or sub-regional mountain conventions and possibly developing a global mountain charter.

As thinking about sustainable mountain development has evolved over the five years since Rio, the issues raised in Chapter 13 have expanded beyond the original themes of the chapter to include, for example, new emphasis on conservation, culture, sacred values and landscape diversity. The element of spirituality surrounding mountains and the often special relationship between mountain peoples and the land they live on and care for has gained greater recognition over recent years and has become an essential driving force in the movement to conserve mountain environments and cultures. Recently, especially in Africa, the contribution of mountain ranges to water resource flows, and more specifically in the chain linking water to food security, has been highlighted. There is also increasing recognition of the importance of mountain areas in terms of biodiversity conservation, economic potential and protection of downstream interests. This has led to a greater willingness to address development and conservation needs through investment programmes in areas that have traditionally been neglected in national level development planning.

Progress has been made in creating greater awareness of the 'Mountain Agenda' and improving co-ordination of efforts to protect fragile mountain ecosystems and promote sustainable mountain development. This has been achieved largely as a result of the international and regional intergovernmental and non-governmental organisation (NGO) consultations which have provided focus on key mountain issues and led to recommendations relevant at both global and regional levels. Regional intergovernmental consultations have been held for Asia (1994), Latin America (1995), Europe (two sessions, 1996) and Africa (1996); a regional NGO consultation was also held for Europe (1996). Other important meetings include the International NGO Consultation on the Mountain Agenda held in Lima, Peru in February 1995, and other regional, sub-regional, and national initiatives.

New institutional arrangements at global and regional levels have also contributed to progress in achieving and improving communications networking and in providing an information clearing-house function. This has been achieved in part through the creation in 1995 of the Mountain Forum, a network of organisations and institutions with a shared interest in sustainable mountain development. With major financial support provided from the Swiss Government, the Mountain Forum operates primarily as a decentralised structure with regional focal points established to co-ordinate networking activities. The Consortium for the Sustainable Development of the Andean Ecoregion (CONDESAN) convened by the International Potato Centre (CIP), a member of the Consultative Group on International Agricultural Research (CGIAR), has been selected as the focal point for Latin America and the Caribbean. The International Centre for Integrated Mountain Development (ICIMOD) is co-ordinating networking in Asia and the Pacific, both at the regional level, through the Asia Pacific Mountain Network, and at the sub-regional level since establishment of the North and Central Asia Mountain Network. Together with The Mountain Institute, these organisations make up the Mountain Forum's Interim Facilitating Committee, which has been set up at the global level with the specific purpose of assisting in building the network. Regional networks are also being established in Europe and Africa, mainly through NGO efforts.

Land-use planning and management tools are being developed for mountain watershed areas through the preparation of principles and best practices for sustainable mountain development programmes. They are intended to serve as an incentive for national and sub-national planning and are expected to build on existing guidelines for mountain conservation and development. They should also be complementary to guidelines and planning tools which have been developed in other sectors (e.g. national forestry action programmes, national conservation strategies) and consistent with land-use planning tools being developed under Chapter 10 of Agenda 21. This work is expected to be field tested, refined and in (limited) use by the year 2000.

In a growing number of countries, livelihoods of local communities and indigenous populations are being protected and improved through mountain development programmes and projects that include a variety of income-generating activities and improvement of infrastructure and social services. A significant change has been observed in many development assistance projects which presently tend to include greater support to human development and poverty alleviation in addition to more traditional technical concerns of watershed management and protection. The FAO/Italy Interregional Project for Participatory Upland Conservation and Development, which has been operating in five countries since 1991, provides a promising example of an approach which focuses simultaneously on developing new livelihood opportunities while carrying out watershed management and natural resource conservation, through planning efforts led by local stakeholders. Requests for assistance in mountainous developing countries are being received by donors and development co-operation agencies to improve planning and implementation of programmes which aim to achieve a good balance between local development needs and resource protection and conservation.

Progress has been made in the area of biodiversity conservation in mountain areas through action taken by many countries in establishing new protected areas, in undertaking transfrontier collaboration in protected area management, and in the development of several recent initiatives to link mountain protected areas through corridors in mountain ranges so that large protected bioregions are established. Additionally, more attention has been devoted to the important contributions that have been made by farmers and indigenous peoples to the conservation of biological diversity, as demonstrated, for example, by the case of potatoes as a staple food crop in the Andes. As the issue of mountain biodiversity conservation receives increasing attention, linkages are being reinforced with the Convention on Biological Diversity (CBD). For example, the conclusions of the European intergovernmental consultation on

mountains related to this issue were presented and discussed at the Conference of the Parties to the CBD, held in November 1996 in Buenos Aires.

There has also been progress in the development of criteria and indicators for sustainable mountain development, based on a collaborative effort involving United Nations agencies and a limited number of other organisations and institutions. This work has focused on the condition of natural resources, human welfare, and population dynamics. If the indicators currently being developed are to be of practical use and effective, however, further elaboration will be required, including input from planners and practitioners, especially at country level. Once refined and field tested, the indicators being developed under Chapter 13 are expected to provide planners and executing agencies with a basis for monitoring and assessing the impact and sustainability of mountain development programmes.

PROMISING CHANGES

New forms of partnership are coming about as a result of increased collaboration and interaction between the public and private sectors in which governments, non-governmental, and intergovernmental organisations have found common ground and engaged in constructive dialogue. There has been a concerted and participatory effort, particularly over the past three years, to amplify the original aims and objectives of Chapter 13 by forging consensus on priorities, and identifying the differing strengths and capacities of public, private, and NGO sectors to contribute to implementation. In addition, there has been greater emphasis on including local populations and community groups in identifying and planning mountain development activities. New partnerships are also being realised between mountain research and development interests. International research institutes such as CIP, International Centre for Integrated Mountain Development (ICIMOD), the International Centre for Research in Agroforestry (ICRAF) (in particular through the African Highlands Initiative), and the joint United Nations University (UNU)/ International Mountain Society (IMS) project, which have either a specific mountain focus or include strong mountain research programmes, have been active members of the interagency network on mountains. This has served to strengthen the linkages and interaction between mountain research and development interests. Organisations, including the World Conservation Union (IUCN), the International Centre for Alpine Environments (ICALPE), and the International Livestock Research Institute (ILRI) have increased their involvement in mountain conservation and development, and formed new partnerships through their roles in planning and implementing the regional inter-governmental consultations in Europe and Africa.

Although there are still relatively few activities which have been initiated directly as a result of Chapter 13, the willingness of most regions of the world to engage in inter-governmental and non-governmental consultations on the topic of sustainable mountain development has been an important step in implementation since UNCED. This process has allowed for clearer definition of the challenges facing each region and possible responses to them, and a stronger sense of involvement and commitment on the part of governments and other concerned stakeholders as the discussion has focused on issues of greater direct relevance at the regional and national level.

This consultation process has been carried out over a relatively short period, only having begun in late 1994, with the majority of meetings having taken place during 1996. Although so far there has been insufficient interest by governments of North America to hold an inter-governmental consultation there, an NGO-organised meeting, to which governments would also be invited, is currently under discussion. As a result of these regional meetings, greater activity is expected at national and regional levels in terms of specific mountain development and conservation programmes, and reinforced institutional arrangements. Regional networks, in addition to those in Asia and Latin America, are also being established in Europe and Africa, following the consultations held in 1996. In accordance with the recommendations of the third session of the Commission on Sustainable Development (CSD), upon completion of the regional consultations a wider international meeting on sustainable mountain development is expected to be held. Several governments have already expressed their support for holding such a meeting, the objectives of which, however, would need to be clearly defined and perceived as timely and relevant to on-going implemenation to the countries and organisations involved.

The maintenance and generation of data base and information systems to facilitate the integrated management and environmental assessment of mountain ecosystems has seen little progress to date. Nevertheless, this is expected to change in the near future given the new and evolving institutional arrangements and new technologies becoming available for this type of work. In the area of conventional dissemination of scientific and developmental information on mountains, the quarterly journal *Mountain Research and Development*, co-published by UNU and IMS, continues as the only scholarly publication of its kind world-wide.

Improvements to the ecological knowledge base regarding technologies and agricultural and conservation practices in the mountain regions of the world have been slow but recently increasing. Most conservation and development programmes in mountain areas now contain specific components aimed at improving data bases on biological resources. On the other hand, economic, sociological, and cultural information are still largely unavailable. Greater emphasis is being placed on the importance of understanding and promoting appropriate indigenous and locally adapted technologies.

Transnational co-operation has also been improving. There is an increased willingness to work across national borders marked by mountain chains. This is especially so for scientific purposes, such as to study the behaviour of migratory species, for environmental conservation purposes, and for ecological/economic reasons among countries sharing major watercourses. Transborder collaboration among biosphere reserves is a new emerging topic in UNESCO's Programme on Man and the Biosphere, in particular in Eastern and Central Europe. There is also a need for greater co-operation among countries sharing water resources, especially in many regions of the world where there is growing demand for water.

Work has progressed on the better understanding and dissemination of information regarding environmental risks and natural disasters in mountain ecosystems. At the twentieth session of the FAO/European Forestry Commission Working Party on the Management of Mountain Watersheds, held in Lillehammer, Norway in July 1996, the topic of mitigation of natural disasters in mountain areas received significant attention, both in the European and global context. This topic was also dealt with in depth at the June 1996 meeting of INTERPRAEVENT, a group of European experts meeting in Garmisch, Germany. Primary attention was given to co-operation in research, technology transfer, planning and development relating to natural resources management in mountain areas, where protection of alpine areas against natural disasters (e.g. floods, avalanches and mudflows) was emphasised. ICIMOD has also been very active in this field and has recently launched a training programme for mountain risk engineering in the Himalaya. Also, the International Union of Forest Research Organisations (IUFRO) Working Group on Natural Disasters plans to hold a meeting within the framework of the 11th World Forestry Congress (Antalya, Turkey, 13–22 October 1997). The meeting will consider, *inter alia*, hazard prevention measures in mountain regions. This cross-cutting issue is expected to receive increasing attention in the near future as a result of greater awareness created through international initiatives, such as the 1994 World Conference on Natural Disaster Reduction, and new and on-going efforts by international bodies such as those mentioned above.

Perhaps most importantly, and underlying most other factors which have contributed to the relatively successful implementation of Chapter 13, is the notion of commitment. Many of those who advocate the sustainable development and conservation of mountain regions and peoples speak with conviction and deep emotional commitment to most of the causes espoused by the mountain chapter. The various stakeholders working on mountain-related issues, be they from local communities, NGOs, governments, or international organisations, often share a common view of the importance and uniqueness of mountain areas and cultures and carry out their work with a strong sense of dedication. Their conviction, and the partnerships that were forged prior to Rio, were instrumental in gaining chapter status for mountains under Agenda 21 and in maintaining close collaboration throughout the post-Rio period.

EMERGING PRIORITIES

Upon completion of the current consultation process for Chapter 13 follow-up in late 1996, numerous issues concerning sustainable mountain development, which have so far been discussed in rather general terms at the global level, will have also benefited from discussion in a regional context. This has allowed for further elaboration, refinement, and regionally specific proposals for action. This process has achieved an increased focus on specific objectives of Chapter 13 and better understanding of how they can best be attained. Some key priorities which have emerged

during UNCED and the five-year review period, and for which it now seems that activities for conservation and development can increase, include the following:

(1) **Special status of mountain areas** One of the important achievements of the Chapter 13 review process has been greater recognition of mountain areas as special and distinct from lowland areas and, therefore, worthy of special attention.

(2) **Legal and institutional mechanisms** There is also greater recognition of the need for new or reinforced legal mechanisms (charters, conventions, national legislation) to protect fragile mountain ecosystems and promote sustainable and equitable development in mountain regions.

(3) **Investment in mountain development and conservation** More mountain-specific investment programmes and greater mobilisation of financial resources for mountain development and conservation programmes will be required in order for real progress to take place on the 'Mountain Agenda'. In some cases this may mean formulating and financing programmes which are focused exclusively on mountain areas; other situations may lend themselves to expanding conservation and development programmes that have been designed for lowland areas to include mountain areas as well. There are promising signs of greater willingness on the part of governments to increase investment levels in mountain areas, which historically have been neglected.

(4) **Resource flows** Related to (3) above, there remains the need for clearer understanding of resource flows to and from mountain areas. This will lead to increased income to mountain communities and a fairer distribution of earnings from natural resources exploitation and services provided. The electronic conference hosted by the Mountain Forum in 1996, entitled 'Paying for Mountains: Innovative Mechanisms and Promising Examples for Financing Conservation and Sustainable Development' has attracted increased attention to this issue. A number of innovative mechanisms, allowing a greater share of the proceeds from mountain-based economic activities to reach mountain peoples has been presented.

(5) **Status of women and children** The need for greater empowerment, equity and equality of mountain women has been expressed as an important concern throughout the period of UNCED and Chapter 13 follow-up; it will receive even greater attention. A recent publication by UNICEF (*Children, Women and Poverty in Mountain Ecosystems*, Ives, 1996), produced in collaboration with UNU and IMS, is already drawing attention to the basic causes of poverty in mountain areas and the need for policy revisions.

(6) **Cultural integrity and biological diversity** There is growing recognition of mountain areas as valuable sites for preserving cultural integrity and conserving biological diversity. This has come about through increased awareness of specific mountain cultures and their important role in maintaining mountain ecosystems, and the greater recognition of endemism of plant and animal species. More concerted action is needed to address these increasingly important areas of concern.

(7) **Monitoring progress** As the 'Mountain Agenda' moves forward and mountain development and conservation efforts grow, it will be necessary to gauge progress and the extent to which it is sustainable. An enhanced ability to carry out assessment and monitoring of mountain development and conservation activities will be required. The development of a set of criteria and indicators for sustainable development has been underway since late 1995. This work is currently being refined with the aim of providing a practical tool for use at the national or sub-national level.

(8) **Exchange of experience and information collection and dissemination** More direct exchange of experience and information, both among mountain people themselves and among other resource management practitioners is needed. Exchange visits between groups of farmers from neighbouring upland communities have proved to be among the most effective means of stimulating new ideas and transferring knowledge in many conservation and development projects throughout the world in recent years. The networks which have been created over the past few years, especially through activities of the Mountain Forum and from regional inter-governmental and NGO consultations, should be fully ex-

ploited to stimulate and enhance this type of direct exchange at all levels. Greater accessibility of existing information to end-users, both at policy and operational levels has been identified as a priority area.

(9) **Food security** As highlighted at the World Food Summit, held by FAO in Rome in November, 1996, the issue of food security has become a top priority for the global community. Accordingly, efforts to eradicate hunger and malnutrition are expected to receive greater attention under the mountain chapter, along with the overall objective of poverty alleviation in mountain regions.

(10) **Mountain forests** As new interest and attention is being devoted to the forest policy debate, especially since the establishment by the CSD of the Intergovernmental Panel on Forests, new opportunities are available to further discuss the important role played by forests in mountain areas with respect to issues such as hazard prevention, biodiversity conservation, and livelihood opportunities. New institutions and fora, including the European Observatory on Mountain Forests and the IUFRO Task Force on Forests in Sustainable Mountain Development, have recently been established to focus specifically on these types of issues. FAO is also giving greater recognition to the importance of forests in mountain regions and is currently in the process of preparing a publication on the management of mountain forests.

THE ROAD AHEAD

Overall, the Mountain Agenda has made important progress over the relatively short period since work began on its implementation. Nevertheless, there remain certain unfulfilled expectations which will require a more concerted effort in the future. For example, better institutional arrangements are required at the national level to adequately address the multi-dimensional aspect of many mountain issues; greater attention must be paid to the special concerns of mountainous island countries; and more time is required to gain a more thorough understanding, through integrated survey work, of the knowledge base on mountain ecosystems at a global level.

Institutional arrangements for the implementation of Chapter 13 are expected to continue to evolve over the next few years, at all levels.

Exchange of information on mountain issues through new networks, such as the Mountain Forum, is becoming more decentralised as regional and sub-regional networks are established. Greater attention is being paid at the national level to address mountain development issues in a more comprehensive manner through improved inter-ministerial co-operation. UN agencies, such as FAO, are also adapting their programmes and operational methods to be able to more adequately respond to the new challenges and demands of the Mountain Agenda. For example, as a result of its task manager responsibilities for Chapter 13, FAO has revised its traditional approach to watershed management and protection to embrace a much broader vision of conservation and development in mountain regions of the world. Additionally, an inter-departmental Mountain Group has been established within FAO to address mountain issues in a more multi-disciplinary manner in support of the organisation's normative activities and the Chapter 13 Task Manager role.

The action to be undertaken over the next few years within the framework of Chapter 13, and especially towards implementing the recommendations and proposals that have resulted from the many global and regional mountain meetings which have taken place since Rio, will be an important indicator of just how far Chapter 13 has progressed. There is a commonly shared view and sense of accomplishment by those who have been involved in the process so far, that we have achieved a new level of awareness and understanding of the importance of mountain issues, as a result of the dedicated efforts of many over the past few years. But increased awareness must now be followed by action programmes in and for the benefit of mountain regions. These must bring about real change through new legislation, institutional reform, planning tools, networking, and other such activities.

What is clear is that the spirit of co-operation, partnership, and commitment that has led to the progress made so far will be a determining factor in future advances on implementation of the Mountain Agenda. Although steady and solid headway has been made so far, difficult work still lies ahead. Even though consensus has been reached on many important issues in terms of what needs to be done, the real challenge will be finding the sustained political will and the financial means to actually make it happen.

Agenda for sustainable mountain development

19

Jack D. Ives, Bruno Messerli, and Robert E. Rhoades

INTRODUCTION

Any brief synthesis of the foregoing chapters would emphasise a number of common features concerning the *mountain problematique*, to which virtually every contributing author, directly or indirectly, refers. There is the hitherto inadequately appreciated complexity of both the cultural and biophysical components of the mountains. Moreover, there is not only an unacceptably sparse data base and scant theoretical construct, but much of what is known tends to be confined amongst a scholarly elite. Knowledge has not been successfully disseminated to the public and politic at large, or at least not in easily digestible form; much of what has been disseminated is popularised to the point of becoming misinformation. Nevertheless, the mountains and their peoples can no longer be pushed to the margins of our consciousness. In terms of their areal extent and contained populations, their natural resources, and their commanding position over the upper reaches of most of the world's watersheds, they are fast moving to centre stage with regard to the prospects for a humane and sustainable world.

Another set of emphases displayed in the foregoing pages includes the mountains' particular distinction from the other major biomes, or ecological–cultural land assemblages. This distinction, or differentiation, not only embraces their endless and oftentimes perplexing variety, but especially their overpowering verticality – their three-dimensional architecture – that renders them obdurate to conventional processes of modern world development. The consistent and prevailing failure to perceive these 'mountain imperatives', to use Narpat Jodha's incisive phrase (Chapter 14), has been a decisive factor in the high proportion of 'development' and bilateral and multilateral 'aid' disappointments over the last half century. To this must be added the uncertainty factor in all its forms, as epitomised by

Michael Thompson's slogan (Thompson and Warburton, 1985): 'uncertainty on a Himalayan scale' – or even Andean, Pyrenean, or Pamirian scale.

Additional characteristics include: reduced accessibility and susceptibility to natural and human-induced hazards, which lead to unconventionally expensive infrastructure and its maintenance costs. The minority status of most mountain peoples in the developing world, added to the other mountain specifics, have been part of the explanation for the historical and twentieth century neglect of mountain regions. This neglect has increasingly taken the form of resource exploitation, and in the extreme case, military assault, destruction, and ultimately, ethnic genocide (such as in Rwanda, Chechnya, Bosnia, and Iraq–Iran–Turkey).

This seemingly pessimistic introduction must be counterbalanced by acknowledgement of the enormous material and cultural resource base that the world's mountains possess. To this must be added the considerable number of sustainable successes that have been obtained. There is also the impressive spread of mountain awareness that has occurred during these brief five years since the Rio de Janeiro Earth Summit. Nevertheless, it must also be admitted that little material progress has been made since Rio. Despite a global trend toward democratisation and decentralisation, spurred by the liberalisation of world trade (e.g. GATT, NAFTA, WTO), and which may give more voice to mountain communities, these same forces are increasingly placing mountain regions in severe comparative economic disadvantage to the non-mountainous regions.

Concurrent with the growth in mountain awareness, moreover, is the appalling increase in armed conflict and open warfare that is disproportionately destabilising the mountains and their peoples. Even when governments, such as Peru, have seemingly controlled by force such radical mountain-based groups as *Sendero Luminoso,* unbridled economic and political repression of the dispossessed populations generates similar anti-state reactions from yet other groups (*Moviemiento Revolucionario de Tupac Amaru*). When the World Mountain Balance Sheet is reviewed as a whole, military assault on mountain peoples, legitimised by the state, surely greatly offsets all progress that has been made. If we are to produce an agenda for sustainable mountain development, therefore, that can find formal endorsement by the UNCSD and the UN General Assembly, a series of seven over-arching prerequisites must obtain substantial support.

SEVEN PREREQUISITES FOR A TWENTY-FIRST CENTURY MOUNTAIN AGENDA

(1) Mountain perspective

Most of what has been written about mountains, or performed upon them in the guise of aid and development, has been undertaken from outside: by flatlanders and mainstream institutions; whether governmental or private, business or academic, conservationist or political. Very frequently a mountain perspective has been absent. Not only has there been a failure to appreciate *verticality* but the mountain people have often been stereotyped through the eyes of outsiders. This has been very costly to the 'insiders' and, should this situation continue unabated, in the long-term, society-at-large will miss a huge opportunity. A good example is the confusion that has spread concerning efforts to unravel the 'population growth – poverty – environmental degradation' debate, and the development projects mounted to solve 'the problem'. Perhaps it is too trite to argue that poverty is in the eye of the beholder – whose perception of poverty should be counted? We also recognise that poverty is not unique to mountains; here, however, while it does not necessarily compare with some elements of lowland urban poverty (e.g. Calcutta, Bangkok, and Lima), mountain poverty is structurally reinforced by the mountain specifics. But, most emphatically, new technologies, advances in resource management, development of specialist and intensive food production, and birth control measures, will all fail unless the mountain perspective is

interwoven with development processes. More particularly, the female and male farmer must be able to enter the process toward sustainable mountain development, not simply as complete equals, but as recognised possessors of cultural and physical resources that could provide the springboard for real progress.

Change, struggle, hardship, crisis, and adaptation have been integral to mountain living. They are not new, nor even post-1950 phenomena; they originated with the first human penetrations of the mountain regions hundreds, if not thousands of years ago. The current *rate* of change, however, is unprecedented. Herein lies the challenge.

(2) Mountain reciprocity

The lowlands, that is, the non-montane ninety percent of humankind, have a vital need for the gifts that the mountains have to offer – material, spiritual, and aesthetic – water alone is enough to establish this point. If mountain resources, in this increasingly resource-stressed world, are to be exploited in a sustainable manner, the upper watersheds must be utilised as sensitively as possible. The most efficacious way of reaching and maintaining this large goal is to recognise the abilities of the mountain peoples and to work with them. This implies not only equality but compensation – direct payments for resources extracted and services rendered, and indirect payments for beneficial stewardship. At the theoretical level, this becomes simple common sense, although it is rarely seen in practice. At the specific level, of course, it is much more difficult to identify, let alone put into effect. This issue highlights particularly well a problem of mountain perspective especially in developing countries. The extent that poorer mountain farmers will voluntarily adopt sustainable practices will rarely satisfy the externally-defined standards desired by society-at-large. This is due not only to the economic constraints on individual farmers, but to the dearth of easily adaptable technologies that are simultaneously environmentally friendly. The farmer's concern will always have to be short term – for survival, from harvest to family food – while sustainable mountain development will be a long-term process. This gap in chronological scale will only be closed to the extent that the farmer can be compensated for having to bear considerable costs for the benefit of society (see section 'mountain policy').

(3) Mountain devastation

The singlemost obdurate obstacle to sustainable mountain development is warfare, in any or all of its forms. In terms of the world-at-large, the disproportionate burden that mountains and their peoples are obliged to carry, as victims of inhuman treatment, will surely rebound on society unless the current situation can be alleviated and reversed quickly. Both the United Nations Organisation and the so-called rich and powerful nations who consider themselves the world leaders for freedom and security, must work, both together and individually, to end this shame on humanity. This is possibly a utopian call, because the necessary instruments are missing at the UN level. Yet we believe that nothing less than a major restructuring in world affairs will be required. Unless there is a strong measure of success in this arena, the long-term costs will likely exceed the ability of society to pay; the moral costs will be even greater. The losses, in terms of nothing less than cultural and biological extinctions, will be beyond recovery.

(4) Mountain hazards

Closely associated with warfare and its related tentacles of death, disease, poverty, and environmental degradation, is the parallel vulnerability of mountain peoples to natural and human-induced hazards. The wide discrepancies in susceptibility to disaster that relate to age, gender, ethnicity, affluence, and political influence need immediate attention. As Kenneth Hewitt has pointed out (Chapter 16), both with passion and searing statistics, the risks associated with mountain hazards – themselves a component of verticality – are controlled as much or more by socio-economic as by geophysical factors. The geophysicist's approaches to natural hazards require accelerated input from the human scientist.

Again, such hazards are not unique to mountains. Nevertheless, the 'verticality' component of the mountain environment ensures a greater frequency and magnitude of disastrous event, and many lowland catastrophes originate in the highlands.

(5) Mountain awareness

The extent to which the Rio Earth Summit has proved itself a watershed in growth of mountain awareness is literally breathtaking. After spending a quarter century in the intellectual wilderness, this apparent success is still ringing in our ears. United Nations and national governmental agencies, private foundations, universities, non-governmental organisations world-wide, are proclaiming the importance of mountains as vital to world security. But caution is needed. To what extent is this a band-wagon happenstance? Even NGOs need funds to pay the salaries of their rapidly increasing numbers of employees; innumerable instant mountain experts, with their own 'sustainable' agendas, are appearing on the horizon. (This, of course, does not apply in those instances, where local mountain NGOs have developed spontaneously.) The increase in awareness is laudatory; the trick will be to ensure that its benefits outweigh obvious potential costs. Certainly, some organisations (such as the Mountain Forum, CIP, and ICIMOD), in relation to the three avenues of approach to the post-Rio mountain remit outlined in the Preface, warrant strong support to facilitate their further progress. More is required. But equally, critical review is needed to provide guidance and to ensure an adequate degree of altruism and balance – a full incorporation of the mountain perspective.

(6) Mountain knowledge and research

This broad area requires many approaches: basic and applied research in conventional form; participatory research to achieve integration of the mountain perspective; analysis of information and dissemination of results via both traditional and electronic media.

Box 19.1   Information system for multicultural mountain regions

Exchange of information between the different parts of a mountainous region is generally more difficult than in the lowlands. Mountain ridges often form a division between countries and cultural differences are very pronounced here, for topographic reasons. These various borders make it more difficult to share	expertise and scientific knowledge. In many mountainous regions, there is thus a great need for a common system for sharing information. Many other factors point to the urgent need for such a system, as symbolised in the figure below (Figure 19.1). Since several autonomous countries are often involved,

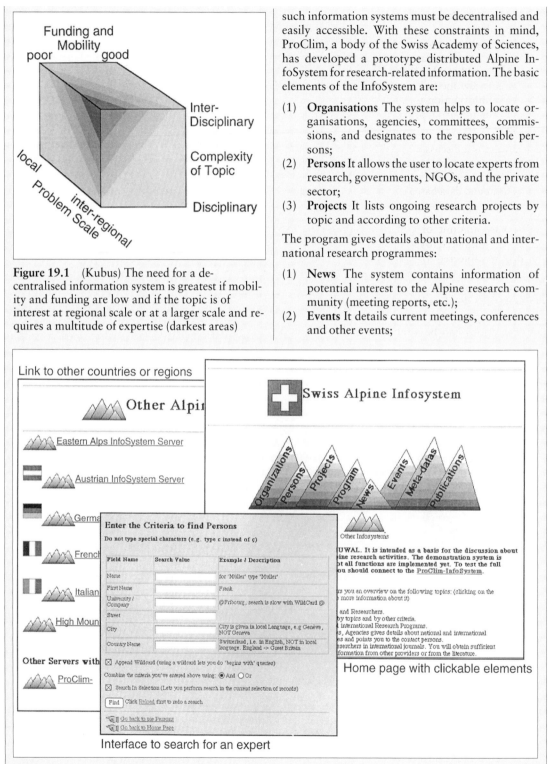

Figure 19.1 (Kubus) The need for a decentralised information system is greatest if mobility and funding are low and if the topic is of interest at regional scale or at a larger scale and requires a multitude of expertise (darkest areas)

such information systems must be decentralised and easily accessible. With these constraints in mind, ProClim, a body of the Swiss Academy of Sciences, has developed a prototype distributed Alpine InfoSystem for research-related information. The basic elements of the InfoSystem are:

(1) **Organisations** The system helps to locate organisations, agencies, committees, commissions, and designates to the responsible persons;

(2) **Persons** It allows the user to locate experts from research, governments, NGOs, and the private sector;

(3) **Projects** It lists ongoing research projects by topic and according to other criteria.

The program gives details about national and international research programmes:

(1) **News** The system contains information of potential interest to the Alpine research community (meeting reports, etc.);

(2) **Events** It details current meetings, conferences and other events;

Figure 19.2 The Alpine InfoSystem serves as a platform for sharing the expertise of a region with other alpine regions. Three windows are shown as examples

(3) **Meta-Data** It identifies the providers of the actual data;
(4) **Publications** It provides listings of publications and reports of researchers in international journals as well as other key articles. The reader can either view the full paper or can obtain sufficient information to quickly retrieve detailed information from providers or from the literature.

The Forum is a discussion platform where the user can post information and questions to others. The distributed information system is based on web servers as front end data bases which are maintained by the information providers (e.g. research centres). Thanks to a common graphical user-interface, the user becomes easily familiar with the system. The underlying database permits a flexible and precise definition of the information categories, which quickly leads the user to the desired material. The prototype system may be tested under the URL: http://www.proclim.unibe.ch/alpine.html

Who benefits from such a system? It can be used by science and policy makers, governmental and non-governmental institutions.

Source: Christoph Ritz, Director ProClim, Swiss Academy of Sciences, Bern, Switzerland

The vast amount of completed mountain research, accumulated over more than a century, needs to be subjected to search, classification, and compilation in accessible data banks. This would involve incorporation of the 'gray' literature. But to be effective the data base needs to be organised systematically so that defensible generalisations and predictions can be attempted and additional research objectives systematically identified and attained. Much progress in this area can be made by furthering the on-going work of the existing regional institutions, such as ICIMOD and CIP. But additional regional mountain centres, precisely linked with the existing ones, need to be established. The entire framework should be subjected to critical and open review.

Box 19.2 A data bank for the Hindu Kush Himalaya

The Hindu Kush–Himalayan region (HKH) is the world's highest mountain region, extending over 3500 km from Afghanistan in the west to Myanmar in the east, and ranging from the Tibetan Plateau in the north, to the Ganges Basin in the south (Figure 19.3). The region represents a formidable range of

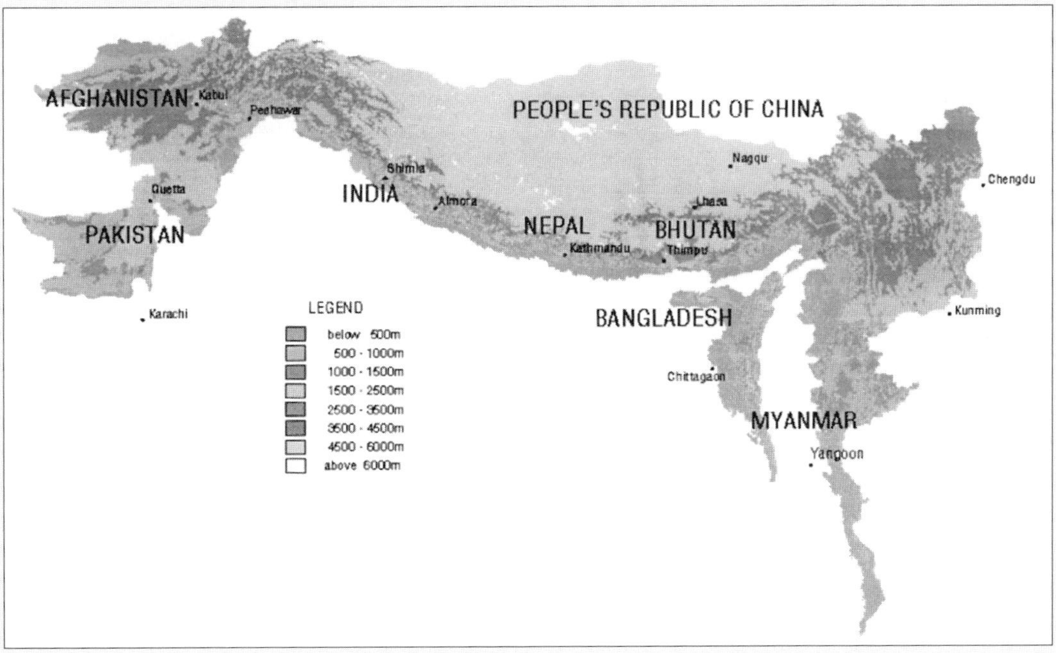

Figure 19.3 The Hindu Kush–Himalayan region

both ecological and developmental problems that have physical as well as socio-economic dimensions. Due to inaccessibility throughout most of this region, information on physical and socio-economic features has in the past been scattered, sectoral and often not comparable among the eight countries to whose territory this region belongs

Today, considerable amounts of data on the HKH region's natural resources are available via satellite. Advances in image processing and computer analysis have made it possible to present information on natural resources on a reliable and uniform data base. With the help of Geographic Information Systems (GIS) it is also possible to integrate biophysical with socio-economic data. GIS has become a powerful and dynamic tool for decision makers in preparing alternative policies and strategies for sustainable and integrated mountain development.

Collecting information and developing a digital data base for the HKH are major functions of ICIMOD's Mountain Environment and Natural Resources Information Service (MENRIS), which started operations in 1989. All information is procured from official government sources and through interpretation of satellite imagery. A comprehensive GIS data base on a scale of 1:250 000 has been completed and is available in both hard copy and digital format. A programme has been introduced to

eventually cover the whole HKH range on this scale. A number of important features are already available on a scale of 1:1 million.

In addition to sharing information, MENRIS has an active programme on introducing remote sensing and GIS technologies in the HKH region. As such, it has collaborating institutions in each of ICIMOD's Regional Member Countries. All these institutions have been provided with relevant hardware and software and more than 1000 persons have benefited from interactions with MENRIS. This may have been in the form of one-day awareness raising policy workshops, to one-month training courses. As a result, national capabilities for data handling in GIS format for sustainable mountain development have increased considerably.

Through this process MENRIS has established a network of institutions that use compatible technologies and methodologies for data collection and analysis, thereby facilitating information sharing on this important mountain range. More recently, improved, cost-effective, and user-friendly communications systems, and local and wide-area networks have begun to provide two-way access for sharing data and information. ICIMOD, in general, and MENRIS, in particular, aspire to play a catalytic role towards this end.

Source: ICIMOD, Kathmandu,
Nepal

None of the above recommendations precludes the encouragement of accelerated traditional or academic research. Basic research will remain fundamental to progress. The continued challenge of unproven assumptions is vital if resource-use decisions are to be based upon real knowledge rather than myths and preconceived opinions. Success in this area will depend to a considerable degree on the existence of rigorously peer reviewed publication. Above all, the chasm between the natural and the human sciences still awaits a bridge. It is perhaps time that mountain scholars and planners alike, from both the natural and social-policy sciences, resurrect and build an interdisciplinery and intersectoral mountain discipline – *montology* (Rhoades, 1997). A policy-sensitive science of montology for the twenty-first century should rely less on shopworn theories, outdated models, and business-as-usual approaches to mountain research and development which, as we have learned, are all too often provided by perspectives born outside the mountain reality. Low-yielding theories, such as neo-Malthusian economics which blame mountain farmers for their breeding habits, or dependency

theories which see the mountain 'crisis' as a result of post-colonial exploitation, need to be replaced with fresh perspectives. These should reflect the global–local realities of the post-modern age, and account for the dynamism of the mountain peoples themselves in shaping their destinies and their highland landscapes.

Regardless of the fact that much mountain research will continue to be, and should be, driven by the idiosyncrasies of the individual scholar, there is a pressing need for specification of a systematic and targeted research programme. This should be the objective of a special international workshop that includes scholars, developers, mountain people, and editors. Undoubtedly, the attainment of sustainable mountain development will be a long-term venture. Development of a targeted research programme is viewed as an objective for the next post-Rio five years.

(7) Mountain policy

Given the disadvantaged economic and political position that most mountain regions occupy

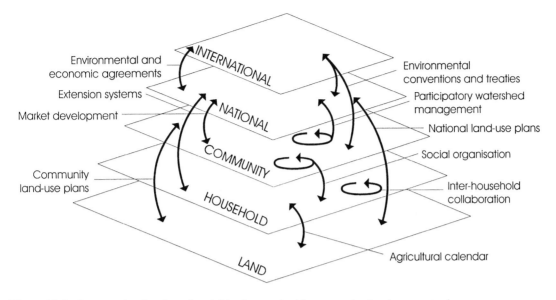

Figure 19.4 Intervention levels and activities in sustainable mountain development and resource management (Hans Hurni, Centre of Environment and Development, University of Bern)

vis-à-vis the lowlands, it is absolutely critical that mountain scholars and planners turn their attention to the formulation of workable policies which are informed by the best possible mountain science. Policy will need to be sensitive to the complex trade-offs that will inevitably occur in the process of sustainable development. We have already alluded to the fact that global societal interests in ecosystem functioning (hydrological cycles, biodiversity maintenance, clean air) cannot be addressed and paid for by the smallholder mountain farmers who are often on the brink of starvation, or at least face severe seasonal food shortages. The global ecosystem solutions are long-term (decades) while food shortage and hunger is a short-run matter (days and weeks) that demands immediate action. The policy implication of this is that the larger society will have to compensate the local mountain people for their efforts to save the mountain environment. But also the mountain communities must be open to innovations and initiatives for improved management of mountain resources and ecosystems. This is where creativity and the *mountain perspective* are needed. Policies applied in the lowlands, or in the developed countries, will be difficult to implement in places such as the Andes or Himalaya (for example, regulations are difficult to enforce, taxes and subsidies are costly to administer, hand-outs can lead to dependency, and so on).

At least two policy directions could show promise: one 'grassroots-focused' and the other 'macro-focused'. These reflect *conservation* with a small 'c' (locally executed) and Conservation with a large 'C' (global, transnational, national) (Janice Alcorn, 1996, personal communication). The first is to design action programmes buttressed by policy instruments and tools (credit, technologies, infrastructure, and so on). These should be provided to complement the indigenous resources (human and biological) since all are required for sustainable mountain development. A mountain perspective approach would appreciate local knowledge, social capital, and biological resources, which need to be complemented by outside resources and knowledge. A second policy approach is to directly compensate mountain people for preserving and enhancing mountain resources that benefit humankind. This approach must be based on an equitably-oriented ecological economics. One radical solution would be to implement a financial 'reverse debt swop' in which policy favours direct compensation to mountain nations and communities at a rate equivalent to the real economic value of the mountains for global ecosystem functioning and society. We do not deny the enormous problems in implementing such policies, but until we begin to at least think in this direction, the mountains will continue to be fed upon by the outside world until they are indeed impoverished.

461

But in this respect there are many extant policies that have been, and are being, implemented in the mountainous regions of Europe that can be reviewed for further development and possible application to all the world's mountain regions (the Alpine Convention, especially, should be explored in this context). These include: water taxes, paid back to mountain communities when their land is used for development of hydropower infrastructure; fair payment for medicinal plants, their collection, and local intellectual knowledge, especially in collaboration with the large pharmaceutical companies involved in modern drug technology and its commercial marketing; and payments and fees in the context of tourism and national park access. Of course, there are already the beginnings of progress in several of these areas, but there is wide scope for much more extensive development and application. A Himalayan or an Andean Convention could indeed help to strengthen the free exchange of data and information, to solve transboundary processes (such as water, traffic, trade) and to create increased confidence between different countries of the same mountain system.

TRAVERSING THE DIVIDE – PATHWAYS TOWARD A MOUNTAIN AGENDA FOR THE TWENTY-FIRST CENTURY

As we approach the twenty-first century, it remains unclear what lies ahead for the world's mountains. Given that we are entering into a global transition never before witnessed by humankind, nor by its predictive sciences, our pathways (maps) to sustainable development for the mountains are indeed crude and imperfect. However, the journey must continue because the negative consequences of inaction are too great. This volume has given us many pathway markers which allow us to see far enough so that we can sketch out the general direction of our map and identify the potentialities and pitfalls which lie along the way. For many reasons the old maps (both intellectual and practical) of the past are no longer adequate to modern times and global change. The indigenous adapted cultures of the mountain peoples of yesteryear, however much remains for the future, have already undergone dramatic change, to varying degrees. Global penetration of market forces increasingly drives local decision making. As external monetised economies replace subsistence–barter economies, mountain peoples find themselves on a treadmill which does not allow them to catch up with their basic – and expanding – needs. Simultaneously, as mountain cultures change and adjust to the external world, the conservation ethic embedded in the local, largely self-sufficient community, is being replaced by individualistic, profit-based values. In these final years of the millennium we still have a chance to throw a bridge between the past and the future in a creative, sustainable way. If we fail, much of the mountains and their treasures will be lost forever; if we succeed, future generations of highlanders and lowlanders alike will continue to enjoy the greatest wonders of planet Earth – the majestic mountains.

In order to preserve the mountains, all stakeholders with access to and interest in mountains – from the smallest landholder to the biggest multinational corporation – must help construct the new map that will lead towards a sustainable future. This premise, and this premise alone, must jointly underscore our Mountain Agenda for the twenty-first century. At the risk of repeating some thoughts introduced earlier in this book, among the many components that will be required to light the path as we struggle toward our goals, are the following (see also, Alcorn, 1991):

(1) political will and public awareness;
(2) guarantees of human rights, ancestral claims, and basic needs of mountain populations;
(3) appreciation of, and support for, indigenous mountain knowledge and management systems;
(4) rejuvenation of a policy-relevant science of montology;
(5) monetary compensation and ethical commitments;
(6) open and continuing dialogue between stakeholder groups.

But the overall goal is a sound balance between self-determination and autonomy on one side, and external influence and support, on the other side.

Political will and public awareness

The progress that has sprung from the Rio Earth Summit will only maintain value if public awareness reaches far and deep enough to catalyse a political will to halt both cultural and physical degradation in the world's mountain regions. Much of the recent rush to defend the mountains came as the result of a suggestion that 1.3 billion dollars in funding would be needed over a 7-year period (1993–2000) for promoting integrated watershed development and alternative livelihood opportunities (Agenda 21, Chapter 13; Article 19). The question now is, how many of the organisations and individuals will still be around when this promise falls short, as well it might, at least in the near-term?

Much of the attempt to raise public awareness in the past has been through the 'scare crisis' approach of painting a dismal picture of how the mountains are washing into the oceans due to the irrational behaviour of an out-of-control mountain peasantry. This worked for a while, but both the lay public, and especially the intellectual mountain research community, have grown weary of what is now seen as a partly self-serving development discourse. Instead, a more positive campaign needs to be launched in which it is shown that the mountain ecosystem *in its own right* needs to be protected and developed for the 'natural and economic' services it provides to humanity. In this regard, the mountain community has much to learn from the multimedia advertising approaches toward raising public awareness of the importance of the tropical rain forests, wetlands, and oceans. The same excitement that we apply to our intellectual studies, or development businesses, needs to be extended to helping politicians, planners, lay public, and anyone who can help make a difference, *understand and appreciate* the contribution mountains make to the spiritual and physical well-being of all peoples of Mother Earth.

It might be worthwhile to design public-friendly non-profit corporations for investing in the protection and the sustainable development of mountain resources. The investors will provide funds to protect mountain qualities, and in return they will receive the excitement of scientific study, other intellectual rewards, such as seminars and courses *in situ*, or simply the moral satisfaction that such a vital part of the world is preserved for future generations. The millions of mountaineering and hiking enthusiasts should be approached to adopt such a protectionist stance, backed up with real monetary support for mountain regions. (While we know this kind of involvement is now beginning to take place, it can be greatly accelerated.) Finally, only by developing a vocal and well-endowed constituency will the world's mountains be protected by political action. We are realistic that, even in the rich North, just compensation to mountains for their contribution to society is far from a reality. But unless the issue is raised again and again with those in power, the mountains will continue to suffer neglect until it is simply too late. Fundamental policy and legislative reforms at all levels are absolutely crucial to sustainable mountain development. To use an American saying about fighting city hall (local government), 'they will not see the light until they feel the heat'. We believe that it is time to turn up the 'heat'.

Guarantees of human rights and basic needs of mountain peoples

We have earlier referred to the issues of civil and political violence against marginal mountain populations. The direct legitimised attacks by state governments against ethnic minorities in the Southeast Asian highlands, New Guinea, Hindu Kush–Karakorum–Himalaya, Zagro–Taurus Arc, Balkans, East African highlands, and the Andean region are all to often overlooked or excused by the political bodies at all levels. It follows nevertheless, that the mountain peoples themselves need to review the destructiveness of their own tendency toward conflict.

Over the past centuries, mountain peoples have often seen their land rights and access to basic resources stripped away in the name of the common good of the larger body politic. Nature reserves, water resources, culturally relevant plants, indigenous knowledge, have all been 'legally' stolen from local control and placed under alien laws and institutions, far removed from local realities. These rights, especially those of ancestral lands, must be re-examined and mechanisms sought to guarantee the return of legal access (within the bounds of a responsible conservation ethic) if mountain peoples have been alienated from their places of origin. If tenure rights have been removed, or the state has redefined them, such rights need to be returned to the historic guardians (we are reminded of the famous case of the nationalisation of Nepal's forests in the 1950s which led to a great deal of ecological and social displacement). Along with this should be a respect for traditional management institutions (which also have been replaced, or circumvented, in the name of conservation or development) and these institutions should be consulted, not discarded, by outsiders. Finally, attention to basic social and physical needs (education, health, and welfare) should be a priority in our twenty-first century Mountain Agenda (see, for example, the UNICEF initiative for Primary Environmental Care – PEC), for without these basic rights we will have little leeway in asking mountain dwellers to help the global community by ensuring an effective stewardship of the mountain lands.

Appreciation of, and support for, indigenous mountain knowledge and management systems

Although rarely recognised by outside development agencies, one of the global attractions of the mountains is the uniqueness and richness of the existing cultural traditions. Along with the natural beauty and grandeur, this cultural ethnoscape is a prime attraction for tourists, sport enthusiasts, philosophers, artists, and naturalists from afar. Yet unless mountain peoples are to be reduced to weekend hotel performers, the real patterns that underlie this cultural diversity (kinship, community organisation, ritual, language, religion, technology) should be appreciated, not deprecated. Unfortunately, in much of the Hindu Kush–Himalayan region and the Andes, traditional mountain ways are looked down on, scorned, and treated as inferior ways of life. Even mountain peoples themselves, as Bista (1991) has so convincingly demonstrated in the case of Nepal, come to believe in their own inferiority *vis-à-vis* the larger society. Without romanticising the adaptability of survival strategies, it can be argued that a long-term intimate relationship with the mountain environment has created a material culture, a highly specialised folk wisdom, and practical management principles, equal in sophistication to any culture on Earth. Tested by time, such local knowledge can form the basis for future planning. For example, instead of rushing to establish nature parks which excluded peoples such as the Sherpas, the Ethiopian Highlanders or the Andean indigenous peoples, a more sensible approach would have been to carefully study how these groups exploited their environment for centuries without degrading their resources significantly. In fact, we would argue that such knowledge is an asset that can be brought to the planning table and placed alongside funds and alien technologies as major 'matching' resources in the execution of sustainable mountain projects.

Creation of a montology – a science that is sensitive to mountain policy

In this post-modern world, there has been a growing suspicion of the ability of our science to ask the right questions, much less provide the right answers. This is due, in part, to the tremendous variety of local responses to articulation with a growing global system which does not give much credence to the simple theorising of days gone by. Similarly, the political rise of cultural enclaves and gendered claims against male power structures has further eroded simplistic social theory and explanation. One result of this has been a retreat by mountain scholars from comparison, extrapolation, and the construction of gendered models to guide our thinking. One outcome of this more particularistic focus has been the infamous problem of 'uncertainty on a Himalayan scale' wherein everyone is right and everyone is wrong. We argue here, however, that it is time to pursue with great vigour and enthusiasm a science of the mountains that will provide a framework for global discussion about sustainability issues. The new science must be capable of walking the tightrope between the particular and the general, the local and the global, the past and the future. We need to identify patterns, however messy the clustering, so as to be able to speak to each other across geographical boundaries and to plan sanely for sustainable development.

This montology, by its very nature, will be part science, part humanities, part social science, part natural science, part policy science, and part folk science. It will be interdisciplinary, intercontinental, intersectoral. We can envision a stronger push in the universities of both the North and the South in the pursuit of this multifaceted approach and a special effort exerted in the education of *in situ* mountain scholars. Such persons are truly lacking today in the mountain research and development community at large. The programme at Bukidnon State University, Philippines, is a leading initiative wherein tribal mountain youth are brought to campus, given an education based on their own cultural traditions, but with a new skills component, and then asked to return with their degree in hand to work with their home communities. Instead of learning American literature, as in the past, they study their tribal legends and laws, along with the science and laws of the larger society. With some novel thinking this model can be duplicated at little cost throughout the mountain regions of the world.

Monetary compensation and ethical commitments

Much as the romantics would like to turn back to a time of subsistence survival and non-monetary economies in the mountains, this is no longer feasible and perhaps, depending on your viewpoint, even desirable. It is foolish to turn our back on the forces of contemporary change. But we need only to remind ourselves that one of the qualities of mountain peoples is their ability to adapt to change through their many strategies of diversification, expansion, regulation, and intensification. While the specific content may change through time, we have every expectation that mountain folk will continue to apply these principles to the introduction of a monetarised economy. One of the problems, however, is how local initiative will meet up with the onslaught of external private enterprise as it continues to penetrate isolated mountain communities. Private enterprise, if it wishes to be sustainable itself, will have to 'internalise' environmental externalities and pay the cost, because someone will eventually have to pay them (Alcorn, 1996). There clearly needs to be a serious examination of the alternatives for income generation that we see today – such as ecotourism, floraculture, micro-enterprises, park-generated revenues, off-season vegetables, specialised vegetable and livestock products. Each of these brings special problems, from cultural dilution to the depletion of soil fertility, which need to be considered and countered. Furthermore, capitalism rewards its practitioners with quick returns, not by saving the environment for future generations. Only by arguing for, and building societal rewards, cultural ethics, and appropriate economic models into the thinking among private enterprises, will the abuses of capitalism in the mountains be controlled. This will require altruistic action on the part of the business community itself, which must come to understand that it too can 'despoil its own nest', or 'kill the goose that lays the golden egg', unless progress can be turned in the direction of the long-term and the sustainable. The unforgivable destruction of Kathmandu by the tourist industry, and the connected national institutions, development agencies, and merchants, in the name of 'progress' is a sad case in point. Today, it is not unusual to see groups of hunkered tourists protected by pollution masks as they attempt to enjoy their 'vacation'. For these, and other reasons, the new montology needs thinkers who will add the tenets of ethics and religious understanding to the twenty-first century mountain perspective. Serious thought must be given to how we can achieve the ethical commitment in the 'hearts and minds of corporate and government decision-makers'

(Alcorn, 1991) which will guarantee that they will be positive agents of change and not selfish forces of destruction.

Open and continuing dialogue between stakeholder groups

Stakeholder analysis argues that each group with an interest in the mountains will have its own understanding and diagnosis of the problem, based on its own perception of the issues. Each, in turn, will believe its own position true and justifiable. The key, therefore, is to make explicit these different positions and perceptions in order to bring the array of actors to the negotiating table in an effort to resolve differences. We have already listed some of the main players in the mountain drama – farmers and herders, researchers, policy makers and planners, development practitioners, tourists, and private enterprise. Today, the institutional landscape of interest groups has become almost as varied as the natural landscape, with GOs, NGOs, international agencies, universities, development groups, and so on, all vying for a place to determine the future of the mountains.

A sustainable mountain development for the next millennium will require 'negotiation round tables', both formal and informal, that will allow for the varied voices to be heard and acted upon. But it must be clear that now we are only hearing the voices of the elite and powerful (mainly flatlander outsiders). The voices of the mountains themselves, along with those speaking the ethical language of moral conservation, must be able to assume a place of equality in the debate. Nevertheless, we might point to the power of the indigenous communities of highland Ecuador which, when denied a voice in matters that affect them, resort to civil disobedience and resistance that stuns the wider non-Indian society into listening. These communities are closed corporate, characterised by a strong sense of place and ethnic identity. They are now becoming aware of the political clout that they have acquired. Similar movements are to be found in the Hindu Kush–Himalayan region, in the Tibesti mountains of the central Sahara, and in Ethiopia, where there are contemporary stirrings which may well lead to mountain power blocs representing the dispossessed. But these movements too, if they are not given a place in the debate, can entrench and resort to violence if the more powerful stakeholders do not care to listen. This is why a sustained debate between all groups is necessary; it will require a long-term commitment over several generations and it must be capable of adapting to the times as they change.

CONCLUSIONS

A great deal of almost unbelievable progress has been made in the five short years since the Mountain Agenda was pushed through at the Rio Earth Summit. Today, there is widespread societal awareness of the global importance of mountains; the 'mountain perspective' is accepted currency by policy makers and planners in many quarters; the global mountain initiative is well on its way. In a sense, these are dreams come true for those of us who have followed the fate of the mountains. But we have only surmounted the first easy slopes in our climb towards the top. Now the steeper paths are immediately ahead. We must bear down, systematically uncover the hard data, prioritise our efforts, gain support from the responsible bodies, and win the confidence of the mountain folk. And we need to talk with each other. Only by working toward these goals will we be able to construct the pathway map that will lead to the peak.

Immediate organisational recommendations

From the foregoing discussion we believe that two carefully interrelated consultations are now required. They would open the next *Rio Plus Ten* 'planning for action' phase:

(1) An international technical meeting with representatives of regional conferences is expected to be implemented by the Task Manager (FAO) in 1998. A major focus of this meeting should be to build a broader constituency for mountains and mountain people, leading to mountain specific policy.

(2) An international conference should be planned on 'Connecting Science with Community Action; this should involve discussion of the formulation of a *montology* discipline'.

Both consultations should include representatives of mountain peoples, scholars from many different disciplines, planners, developers, private foundations, practitioners of international law, and news media representatives. There should be a large overlap in the participants of the two consultations. A full year should be allowed for planning, preparation of position papers, and pre-meeting communication between different groups of participants. Each consultation should be orchestrated to ensure further major progress in augmenting international mountain awareness at all levels.

References

Alcorn, J. B. (1991). Epilogue, In Oldfield, M.L. and Alcorn, J.B. (eds.), *Biodiversity, Culture, Conservation, and Ecodevelopment.* Westview Press: Boulder, CO

Alcorn, J. B. (1996). Discussant Comments. *Global-Articulation of Plant Genetic Resources.* A Symposium organised for the 1996 Society of Applied Anthropology Annual Meeting, Baltimore

Bista, D. B. (1991). *Fatalism and Development – Nepal's Struggle for Modernisation.* Orient Longman Ltd.: Calcutta

Rhoades, R. E. (1997). *Pathways Toward a Sustainable Mountain Agriculture for the 21st Century: The Hindu Kush–Himalayan Experience.* Consultancy report prepared for ICIMOD, Kathmandu, p. 120

Thompson, M. and Warburton, M. (1985). Uncertainty on a Himalayan scale. *Mountain Research and Development,* 5(2): 115–35

Biodata of authors and co-authors

Günther Bächler
Director, Swiss Peace Foundation
Wasserwerkgasse 7
PO Box 75
3000 Bern 13, Switzerland
chfried@dial.eunet.ch

*Political Scientist. Practical experience in
Latin America, Eastern Europe, and the Horn
of Africa. He has published on peace and
conflict research, European security,
environmental security and regional conflicts.*

Jayanta Bandyopadhyay
Visiting Professor, International Academy of
    the Environment
4, chemin de Conches
1231 Geneva
Switzerland
Jayanta.bandyopadhyay@iae.unige.ch

*Head of publications at the International
Academy of Environment. Interest in
environmental conflicts and policy,
collaborated with the Chipko movement for
forest protection in the Indian Himalaya*

Roger G. Barry
Director of the World Data Centre-A
WDC-A for Glaciology/National Snow & Ice Data
University of Colorado
Campus Box 449
Boulder, CO 80309
USA
rbarry@kryos.colorado.edu

*Secretary of the International Mountain
Society, has served on the Polar Research
Board and published in the areas of mountain
and polar climate, snow and ice, and climatic
change.*

Edwin Bernbaum
Research Associate, Center for South Asia Studies
1846 Capistrano Ave.
Berkeley, CA 94707
USA
bernbaum@violet.berkeley.edu

*A director of the American Himalayan
Foundation; member of the Commission on
National Parks and Protected Areas of the
World Conservation Union (IUCN). Works
on the implications of sacred mountains for
environmental and cultural preservation.*

Leendert A. (Sampurno) Bruijnzeel
Research Scientist / Lecturer
Faculty of Earth Sciences
Free University
De Boelelaan 1085
1081 HV Amsterdam
The Netherlands
brul@geo.vu.nl

*Reviews on the Ganges Brahmaputra River
Basin and the effects of conversion of tropical
moist forests led UNESCO's International
Hydrological Programme to commission him
to prepare science-based semi-popular
documents. Recent work and publications on
hydrology of tropical montane cloud forest
and plantations, erosion and sediment
transport processes.*

David S. Cassells
Environmental Specialist, Forest Resources
Environment Division
The World Bank
1818 H Street NW
Washington, DC 20433
USA
dcassells@worldbank.org

*Forest policies and production, watershed
management and conservation practices in
developing countries.Officer, International
Tropical Timber Organisation, NAPAN;
forester, city of Townsville; research officer,
Queensland Forest Service; Teaching,
University of New England.*

David John Fox
Honorary Fellow at University of Manchester, and
   Consul for Chile in Manchester
22 Bollin Hill
Wilmslow
Cheshire SK9 4AW
United Kingdom

Special interest in contemporary economic and social change in Latin America, non-renewable resources and commodity problems in the developing world. Early interest in mining stimulated by a year spent in a mining camp in Ungava (Canada). Subsequent work focused on tin mining, including first hand interest in the mining economy of Bolivia.

Donald A. Gilmour
Head of Forest Conservation Programme, IUCN
Ecosystem Management Group
IUCN, Rue Mauverney 28
1196 Gland
dag@hq.iucn.ch

Co-ordination of IUCN forest conservation activities. Experience in forest policy, community forestry, and watershed management. Also worked with community forestry projects in Nepal, and in watershed management in Australia.

Erwin F. Grötzbach
Professor of Cultural Geography (retired)
Alpspitzstrasse 10
82340 Feldafing
Germany

Cultural geography of high mountains, geography of tourism and religions. Experience in the Alps, Afghanistan, Pakistan and India (Himalaya/Karakorum), Turkey, Jordan. Publications in tourism and environment, religion and environment, and development problems of mountain areas.

Lawrence S. Hamilton
Emeritus Professor of Forest Conservation
Islands and Highlands Environmental Consultancy
RR 3 Box 3010
Charlotte, VT 05445
USA
lsx2_hamilton@together.org

Vice-Chair (Mountains) of IUCN Commission on National Parks and Protected Areas. Senior Fellow, East–West Center (Hawai'i) for 14 years, professor at Cornell University (New York) for 29 years and zone forester, Ontario Government (Canada) for 3 years.

Kenneth Hewitt
Professor of Geography, Cold Regions Research Centre
Dept. of Geography, Wilfrid Laurier University
75 University Avenue West
Waterloo
Ontario
Canada N2L 3C5
Khewitt@mach1.wlu.ca

Geomorphologist, peace, hazards and disasters research. Experience in high mountain environments, especially Karakorum Himalaya, and field studies in the Himalaya (12), Canadian Rockies (6), East Mediterranean earthquake risk (4). Award for scholarly distinction – Canadian Association of Geographers.

Jack D. Ives
Professor of Mountain Geoecology
Department of Geography
Carleton University
1125 Colonel by Drive
Ottawa
Canada K1S 5B6

Co-ordinator, Mountain Ecology and Sustainable Development, UNU; chair, International Working Group, UNESCO MAB-Project 6, 1973–75; President of the International Mountain Society; Founder and co-editor of the quarterly journal 'Mountain Research and Development'. Field research in Iceland, Canadian Arctic, Alps, Rocky Mountains, Himalaya, Pamirs, Yunnan/China, Ecuadorian Andes.

Jan Jeník
Professor
Department of Botany
Charles University
Benátská 2
128 01 Praha 2
Czech Republic

Lecturer in botany and ecology at Prague, Accra, Dar es Salaam, Kabul and Vienna universities. Senior research officer, leader of scientific expeditions, member of UNESCO's MAB Programme. Articles on topoclimate and biodiversity in Hercynian and Carpathian mountains, monographs on root systems and savanna vegetation.

Narpat S. Jodha
Natural Resource Management specialist, ENVSP
The World Bank
1818 H Street NW
Washington DC, 20433
USA
njodha@worldbank.org@internet

Policy and field operations work on community management of natural resources and agricultural systems in fragile resource zones. Research and advisory work in Africa and Asia, especially in dry tropics and mountains. Work includes traditional and modern farming systems, sustainable agriculture, and common property resources.

Richard Kattelmann
Hydrologist with University of California
Sierra Nevada Aquatic Research Lab.
Star Route 1, Box 198
Mammoth Lakes, CA 93546
USA
rick@icess.ucsb.edu

Studies in mountain hydrology and water management to improve streamflow forecasting; evaluations of water resources and riparian systems in the Sierra Nevada. Other research in the Rocky Mountains, Himalaya, Tien Shan, and Andes. Worked for the US Forest Service and US Geological Survey and is advisor to the Imjy Valley Committee in Nepal.

Frank Klötzli
Professor, Geobotanical Institute
Swiss Federal Institute of Technology
Zürichbergstrasse 38
8044 Zürich
Switzerland

Group leader synecology and conservation. Research on ecological limits to dominating organism in forests, wetland, alpine areas and savanna, and influence of mammalian consumers. Worked in north and west Europe, Tanzania, Nepal, Ethiopia, Rwanda, Ecuador, Costa Rica (especially in mountain areas).

Dieter Kraemer
World Meteorological Organisation
41 av Giuseppe Motta
c.p. 2300
1211 Geneva 2
Switzerland
kraemer@www.wmo.ch

Director of Hydrology and Water Resources department, WMO. Civil engineer, experience in water resources development and management; international co-operation including activities with the Mexican Federal Ministry of Water Resources from 1969 to 1979.

Zbigniew W. Kundzewicz
Research Centre for Agricultural and Forest
  Environment Studies
Polish Academy of Sciences
Bukowska 19
60-809 Poznan
Poland
skundze@pd.pl

Secretary to International Commission of Water Resources Systems (International Association of Hydrological Sciences). Main interests in sustainable development, water resources, hydrology, climate and drought.

Stephan Libiszewski
Research Associate, Centre for Security Studies and
   Conflict Research
Swiss Federal Institute of Technology (SEU)
8092 Zürich
Switzerland
libiszewski@sipo.reok.ethz.ch

*Political Scientist (Free University Berlin),
conflict research, security policy,
environmental causes of conflicts (e.g. water
disputes in the Middle East). Research
Associate in the Environment and Conflict
Project (ENCOP).*

Douglas McGuire
Officer - Forest Conservation, Research and
   Education Service (FORC)
FAO Rome
Viale delle Terme di Caracalla
00100 Rome
Italy
douglas.mcguire@fao.org

*Responsible for FAO's Watershed
Management and Sustainable Mountain
Development programme and Task Manager
of Chapter 13 of UNCED Agenda 21; also
provides technical support to field projects in
this area. Worked in natural resources
conservation and international forestry, with
particular focus on Cameroon, Madagascar,
and Rwanda.*

Bruno Messerli
Professor, Department of Geography
Physical Geography
University of Bern
Hallerstrasse 12
3012 Bern
Switzerland
messerli@giub.unibe.ch

*President of IGU, co-ordinator of the UNU
Programme on Mountain Ecology and
Sustainable Development, chairman of the
National Committee of the Swiss
UNESCO-MAB Programme; director of
PAGES (Past Global Changes Programme),
Rector of the University of Bern,
vice-president of the Swiss Academy of
Sciences, member of the Research Council of
the Swiss National Science Foundation
Research in mountain ecology (resource
problems, natural processes, and human
impacts), climate change, geomorphology.
Publications from field work in the Alps,
mountains of the Mediterranean, Africa, the
Himalaya, and the central Andes.*

Laurence A.G. Moss
President and Executive Director of Moss & Associates,
International Cultural Resources Institute
Mestsky rad
381-18 Cesky Krumlov
Czech Republic

*Advisor on strategic planning to the Mayor of
Cesky Krumlov (UNESCO World Heritage
Site), UNEP, The World Bank, and the US
Environmental Protection Agency. Visiting
Professor of Strategic Environmental
Planning at Charles University, Prague.
Founder and director of the International
Cultural Resources Institute.*

Klaus Preiser
Head of Group on Rural Electrification
Frauhofer ISE
Niederdorfstrasse 22
79238 Ehrenkirchen
Germany
preiser@ise.fhg.de

*Research into introduction of photovoltaic
technology for rural electrification schemes;
implementation of stand-alone power supplies
in rural areas (photovoltaic, wind, biogas,
hydropower); scientific and technical
co-operation with Universidad Nacional de
San Juan, Argentina.*

Martin F. Price
Programme Leader, Mountain Regions Programme
University of Oxford
Environmental Change Unit
1A Mansfield Road
Oxford OX1 3TB
United Kingdom
martin.price@ecu.ox.ac.uk

Focal point for Mountain Activities of the European Programme of IUCN; secretary of the standing committee on Human Dimensions of Global Environmental Change of the International Science Council. Research, mainly in the Alps and Rocky Mountains, on interactions of people and environments in mountain areas, and in relation to climate change.

Robert E. Rhoades
Professor of Anthropology
University of Georgia
Athens
Georgia 30602
USA
rrhoades@uga.cc.uga.edu

Former chair of the Department of Anthropology at the University of Georgia. Currently working as a consultant on sustainable mountain agriculture at ICIMOD, Kathmandu. Experience in mountain cultural anthropology, domestic crops and indigenous peoples, especially in the Andes. Ten years field research; former senior scientist with CIP in Lima, and with the International Rice Research Institute in Manila.

Peter Rieder
Professor of Agricultural Economics
ETH Zentrum
8092 Zürich
Switzerland
rieder@iaw.agrl.ethz.ch

Research on effects of liberalisation of marketing orders and direct payments for mountain regions; analysis of the sustainability for agricultural systems; socio-economic impacts of new GATT/WTO regulations; biotechnology effects on developing countries. Worked in Asia, Africa and Latin America; and as consultant to government and development agencies in Switzerland.

John C. Rodda
President, International Association of
    Hydrological Sciences
Institute of Hydrology
Wallingford
Oxfordshire
United Kingdom
106201.1774@compuserve.com

Research into measurements of hydrological variables and the errors associated with them. Director Hydrology and Water Resources, World Meteorological Organisation, Geneva, 1988-1995.

Petra Schweizer
Social scientist
Frauhofer Institut ISE
Eschholzstrasse 36
79106 Freiburg
Germany
petra@ise.fhg.de

Environmental psychologist working on environmental consciousness and energy uses in rural areas. Field research; project management; interdisciplinary work; extension and adaptation of European solar activities to rural areas outside Europe, e.g. Nepal and Argentina.

El-Hadji Sène
Chef, Forest Conservation, Research and
    Education Service (FORC)
FAO Rome
Viale delle Terme di Caracalla
00100 Rome
Italy
elhadji.sene@fao.org

Focal Point for FAO Task Manager role of Chapter 13 of UNCED Agenda 21. Co-ordinator of the Forest Resources Division on agroforestry and land use, dryland forest management, watershed management, wildlife and protected areas, education and extension. Member of CIFOR board and former board member of UNESCO MAB, IUFRO, and IUCN councillor.

Ernst Spiess
Professor of Cartography
ETH Zürich
Langacherstrasse 4B
8127 Forch
Switzerland

Editor-in-Chief Swiss school atlases, map design and production, surveyor, topographic engineer and designer, digital cartographer.

Christoph Stadel
Chair, Human Geography
Institute of Geography
University of Salzburg
Hellbrunnerstrasse 34
5020 Salzburg
Austria
christoph.stadel@sbg.ac.at

Rural development research in the Andes on periodic markets; participation in UNU Highland–Lowland Interactions programme; NGO programmes; research in small towns of the Alps (Austria) trans-alpine transportation, and national parks.

Jim Thorsell
Senior Advisor, Natural Heritage
IUCN
14, Résidence du Golf
1196 Gland
Switzerland
mail@hq.iucn.ch

Responsible for co-ordination of IUCN's work for the UNESCO World Heritage Convention. Also contact point for IUCN Mountain Protected Areas task force. Worked on park projects in Caribbean and Africa before joining IUCN in 1984. Field experience in over 400 national parks in 80 countries.

Peter W. Williams
Director of the Centre for Tourism Policy and Research
Simon Fraser University
Burnaby, BC
Canada
peter_williams@sfu.ca

Research in tourism behaviour in natural environments; growth management in tourism environments; environmental management in commercial enterprises; ski area development and management.

Jörg Wyder
Head of the Swiss Mountain Organisation
Schweiz. Arbeitsgemeinschaft für Berggebiete
Laurstrasse 10
5200 Brugg
Switzerland
brendt@agri.ch

National policy for mountain regions; local/regional initiatives and development; mountain farming/mountain forests. Rural development in South America and Africa; tropical agriculture and husbandry; local/regional development in Eastern Europe; education in rural areas; 10 years chairman of EUROMONTANA.

Index

Page numbers in **bold** type refer to figures or plates; those in *italics*, to tables or boxes.

Abies spp., firs 287, *302*
absolute cost advantage, definition of 86
Acacia, umbrellas, in vegetation belts 233
accessibility
 and amenity migration 270
 limited, and agriculture 313, 314, *315, 317, 321*
 and mining 172–173, 176
 restriction of, on sacred mountains 49
 and tourism 250, 251, 253–254
accessibility *see also* roads
acid drainage from mining 187, 195n
acid precipitation
 and forests 283, 301
 and mining emissions 184
 problems of 135
acid soils, and climate change 417
Aconitum, endangered species of 223
acquired immunodeficiency syndrome, AIDS, global
 problem of 389–391, 398
Adirondack Mountains, United States, forests in
 hazard protection 295
Afghanistan
 conflict in *28*, 253
 disasters in 371, *385, 387, 390, 393, 401,*
  407–408
 drug production in 116
 handicrafts in *28*
 land mines in *388*
 mountain refuge in 27, *28*, 29
 refugees from *384, 385*
Afghanistan *see also* Aral Sea; Himalaya; Pamir, Hindu
 Kush
Africa
 biodiversity in *234*
 climatic records on 412–413
 commercial hunting in 298
 conflicts in 112, *387*
 forest clearance in 282
 glaciation and species distribution in 207
 infectious diseases in *389, 390*, 398
 mountain transfrontier reserves in *240*
 natural disasters in *380, 407–408*
 plant adaptations to adverse conditions 211
 refugees in *385*
 species distribution in 206
 vegetation belts in 232, 233
 vulnerable peoples in *401*
Africa *see also* individual countries
Aga Khan Rural Support Programme (AKRSP), in
 Pakistan 22, 80
Agarvaceae, in vegetation belts 233
Agenda 21 *see* Rio Earth Summit

agricultural crops
 and altitudinal zonation 218–219
 cash crops 110, *111*, 357
 drying, by solar radiation 164
 historical domestication of 62–63, *92*
 rotation, and expoitation of resources 12
agriculture
 and climate change 427, 428–430
 and conflict *110–111*
 cultural ecology of 119, 318–320
 decline of 64
 definition of 313
 economic importance of 87–88
 effects of global conferences on 323–326
 employment in *94*
 and forest clearance 303, **304**
 future planning for 326–333
 and grassland ecosystems 223–224
 in mountain conditions 4–5, 23–27, 121–122,
  314–320
 and over-population 17–18
 regeneration of 337–339
 subsidies for *32, 33, 94*, 263
 in sustainability management 126, 127
 water utilisation in **31**, *149*
 and watershed land use *338, 352*, 355–358, 361
agriculture *see also* fertilisation; livestock;
 transhumance and individual countries
agroforestry 282–283, 288, *289–290*
 and climate change 430
 watershed land use 352, 358–359
AIDS, acquired immunodeficiency syndrome, global
 problem of 389–391, 398
air pollution
 and biodiversity 213–214, *215–216*
 in forests 283, 301, *302–303*
 from mining 177–178, 180, 184, 188
 and modern traffic 33, 97, 188
 permitted levels of 191
air pressure reduction, and altitude 174
air quality, and mining in mountains 173
AKRSP (Aga Khan Rural Support Programme), in
 Pakistan *22*, 80
Alai mountains, Central Asia, potential effects of
 global warming 416
Alaska, United States 46, 112, 174, 210, *235*
Albania
 agriculture in **90**, *98–100*
 disasters in *390, 407–408*
 economic and political framework of **89, 90**,
  98–100
 timber harvesting in 290

Algeria, disasters in *385, 387, 390, 407–408*
alien species 204, 219
    in wetlands and peatlands 225
    in woody ecosystems 225, 299
Alnus mandshurica 210
Aloë, in vegetation belts 233
Alpine Infosystem, Switzerland, information
    dissemination *458*
Alps
    agriculture in **26**, 64
    altitudinal zonation in **26**, 218
    biodiversity in 200, *234*, *419–420*, 421–422
    climatic records on 412
    cultural influences in *29–30*
    endangered ecosystems in 222, 223, 224
    energy supply in **168**, *169*
    fertilisation studies *220*
    forests in hazard protection 295
    game management for hunting in 297
    and global warming, potential impact of 414,
        416, *419–420*, **420**, 421–422, 433
    information database for *458*
    national parks in 243
    population pressure in 12
    sociopolitical conditions in mountains of *29–30,
        62–65*
    spiritual qualities of mountains 47
    tourism in 222, 223, 253, 263, 433
    transport systems in 253
Alps *see also* Austria; France; Germany; Switzerland
Altai mountains, Central Asia, cryosphere and global
    warming 414
Altiplano, South America, solar energy in 162–163,
    166–167
altitude
    and ecosystem diversity 203, 215–219, 232, *235*
    and human adaptation 23–27, *92, 93*
    and mining conditions 172, 174–175, **183**
    and mountain zone definition 3–8
    and precipitation 134
aluminium mining 171
amenity migration 249, 264–276
America *see* North America; South America; United
    States
amphibians, decrease in global populations of 143,
    145–146
Amu Darya River, and the Aral Sea *145*
Anaphalis 210
ancestors and the dead, and sacred mountains 51–52,
    45–46
ancient civilisations, importance of 85, *92, 149*
Andes
    agriculture in *20, 92–93, 149*, 356
    altitudinal belts in *7, 92, 93*, 217, 218, *219*, 287
    biodiversity in 202, 209, *225*, **234**, 234, 242
    economic and political factors in *20*, **21**, **23**, 27,
        92–93, 115, 116
    ecosystems in 92, 203, 225
    endangered species in 297, 298
    fuel source 225
    historical mine labour force 178

    indigenous peoples of 19, *20–21*, 29, 110, *149*
    mining, impact of 188
    national parks in 243, *245–246*
    road construction in, environmental impact of
        188, 350, **351**, 359
    sacred mountains in 43, 44, 46, 50
    treeline forest communities 288
    vegetation belts in 232, 233, 234, 284, **286**
    water pollution in sacred Lake Titicaca *144*
    watershed management programme in *342–344*
    wildlife in 50, 212
Andes *see also* individual countries
Andina mine, Chile **183**
Andropogon, tussock grass, in vegetation belts 233
Androsace alpina 218, **420**
animal dung, uses of 174, 224, 289, 355
animal power, as energy source 157, *158*
animals
    in adverse conditions 212
    in climate change 418–426, *424–426*
    species distribution in 200–202, 207, 208
animals *see also* birds; insects; mammals
Anitiaris, in biodiversity conservation on Holy Hills 45
Annapurna Region, Nepal *259–261*, **260**, 297
Aoraki (Mount Cook), New Zealand, sacred mountain
    46
Appalachian Mountains, biodiversity in 204
Appenines, Italy, dam failure in *394*
aquatic ecology, assessment of 143–147
Arabian Peninsula
    conflicts in 103, 108, 116–117, 122
    disasters in *385, 387, 390*
    national parks in 240, *241*
    water resources in *137*
Aral Sea, Central Asia, destruction of ecosystem of *145*
Ararat, Mount, Turkey, sacred mountain 46
Arctic Circle **5**, 203
Argentina *137*, 240, *241*, *394*
aridity 3, 4, 61
    and biodiversity 226, 227, 233–235
    and land use 27, *31*
    and water loss 344
Arizona 46, 217
armed conflict, definition of 103n
armed conflict *see also* conflict
Armenia
    disasters in *376*, 384, *385*, 388, *401*
    sacred mountain in 46
Armenia *see also* Turkey
arsenic, in mineral extraction 184
Artemisia spp., in the Pamir 201
artistic inspiration, in mountains 47
Arun III hydroelectricity project, Nepal, funding of 150
Ashkhabad Kopper Dagh, Turkmenistan and Iran,
    earthquake disaster in *375*
Asia
    conflicts in *387*
    drug production in 116
    forest clearance in 282
    hydroelectric schemes in 432
    infectious diseases in *389, 390*

inspirational mountains in 47
natural disasters in *380, 408–409*
refugees in *385*
religious groups in **19**
transfrontier reserves in 239, *240*
transport systems and tourism 253
vegetation belts in 287
vulnerable peoples in *401*
Asia *see also* individual countries
Athos, Mount, Greece 44, 47, 51, 54, 57–58, 251
Atlas Mountains, Morocco, agriculture and aridity **31**
atmospheric change, in climate change 418
atmospheric gases, changes in 409
atmospheric pollution *see* air pollution
attractiveness of mountains, and tourism 252–253
Ausangate, Peru, sacred mountain 43, 46
Australia
    and climate change 416, 431, 433
    glaciation in 207, 431
    national parks in 238, 239, *240*
    natural disasters in *380*
    sacred mountains in 46
    skiing and climate change 433
    vegetation ecosystems in 207, 221, 286, 287
    water resources in *137*, 431
Austria
    climatic records on 412
    disasters in 10, **373**, *385, 390, 407–408*
    ecosystems in 204, 223
    game management for hunting in 297
    modern development in *32*
    national parks in *240*
    pollution-impacted forests in 301, 303
    population change in 62, 64
    skiing and climate change 433
    timber harvesting in 290
avalanche
    forests in protection against 295
    impact of 222–223, *373, 380, 407–408*
Ayers Rock (Uluru), Australia, sacred mountain 46
Azerbaijan, disasters in *384, 385, 387, 388, 401*
Azorella, in vegetation belts 233

Bachhni Devi, activist of Chipko movement *128*
bacterial contamination, of water 346
bacterial leaching, in mineral ore processing 177, 184
Badrinath, Indian Himalayas, Hindu shrine 40, **41**, 54–56, 251
Bagrot Valley, Gilgit, Pakistan, agriculture and altitude 25, **26**
Baguio, Luzon, Philippines, amenity migration in 265, 266, 269–270, 272
Bahuguna, Sunderlal, Indian social activist *128, 148*
Bahutu people, Rwanda, and conflict *110–111*
Baikal Lake
    influence on clouds *219*
    Siberia, altitudinal zonation in 218, *219*
Bale National Park, Ethiopia *237*
Bali, sacred mountain in 42
Balkan Peninsula *see* Albania

Baltistan
    Pakistan *68–71*, 79–80; Central Karakorum National Park
bamboo, as forest product *293–294*
Banff National Park, Canada 243
Bangladesh
    disasters in **138**, *141–142, 387, 401*
    ethnic conflict in *117–118*
    hydroelectric potential of *148*
    land tenure, and watershed management 366
    water resources and climate change 431
Bangladesh *see also* Brahmaputra and Ganges Rivers
banking, income from in Switzerland *94*
Batutsi people, Rwanda, and conflict *110–111*
bauxite mining 171
Belize, disasters in *385*
Best Available Technology (BAT) 192
Best Available Technology Not Entailing Excessive Cost (BATNEEC) 191, 192
Best Practical Environmental Option (BPEO) 191
Betula spp., birch *210*, 417–418
Bhagirathi River, Tehri high dam, protest against *128, 148*
Bhatt, activist of Chipko movement *128*
Bhutan
    disasters in *390, 401*
    ecologically over-used landscape in **90**
    forest conservation in 282, *283*, 289
    glacial lakes and climate change *415–416*
    national parks in *240*
    socio-economic factors in *73–79, 90–92*
    tourism in 57, 262
biodiversity
    conservation of 221–226, 227, 238–247, 449, 452
        in sacred mountains 44, *45*
    and ecological interactions 212–221
    in forests 284–288
    in mountains 59, 200
    and pollution 301–303
    and vegetation belts 232–235
biogas, as energy source 161
biogeography, foundations of 203
biomass, as energy source 157, *158, 159*–161
biomass productivity, in mountain zone definition 3–4
biomes, and ecosystem diversity 205
biosphere reserves 238–239
birds
    adaptability of 212, 421
    and endangered ecosystems 222
birth control, contraceptive use 78, 81
birth rates *74*, 91
Black Forest, Germany, forest clearing 297
Black Hills, North America, sacred mountains in 42
blessings, mountains as source of 46–47
Bohem Puyuik (Mount Shasta), California, sacred mountain 46
Bohemian Forest, Austria/Czechoslovakia/Germany, ecosystem changes in 204
Bolivia
    altitudinal belts in **4, 7, 286**
    disasters in *387, 390, 401, 407–408*

Bolivia *contd.*
    drug production in 116
    energy resources in 162–163, *166–167*, **167**,
        167–168, 174
    environmental pollution in *144*
    infectious diseases in *389, 390*
    mining in 174, 175, *178–179*, **181**, 192
    national parks in 234, 237, 240, 241, **246**
    sacred places in 46, *144*
    socio-economic indicators for *74–78, 79*
    tourism in *178–179*, **181**
    watershed management programme in *342–344*
Bolivia *see also* Titicaca Lake
Bombacaceae, bottle trees, in vegetation belts 233
Borneo, biodiversity in *234*
Bosnia, disasters in *385, 386, 387, 388, 401*
Bougainville, Papua New Guinea, conflict over mining
    in 112
BPEO (Best Practical Environmental Option) 191
Brahmaputra River, Bangladesh *141–142, 148*, 431
Brazil *234*, 240
Brianonnais, French Alps, state influence on culture
    29–30
Britain, environmental legislation and mining 189
Bromeliaceae, in vegetation belts 234
Buddhism
    and conflict *117*
    sacred groves and forests of 298–299
    sacred mountains of 39–41, 42, 45, 47, 54
        tourism to 250–251
    temples and sacred places of 44
Buffalo Creek, United States, flood *394, 395*
building construction and styles 66, 256, *374*
    traditional passive heating systems 164
Bukonzo tribe, Uganda, gender discrimination in *71–72*
Bulgaria, national parks in 238
Burundi, disasters in *385, 387, 390, 401*
butterfly farming 239
Bwindi Impenetrable National Park, Uganda 237
Cactaceae, in vegetation belts 233

Calabria, Italy, earthquake disaster in *375*
Calamagrostietum villosae tatricum and soil acidity *216*
Calamagrostis spp., tussock grass 211, 233, *302*
calcium-rich substrates 214
California, United States 46, 147, *376*
Cambodia, land mines in *388*
Cameron Highlands, Malaysia, forest clearing for
    recreation in 297
Cameroon, disasters in *385, 387, 390, 407–408*
Campanula, endangered species of 223
Canada
    amenity migration in 265
    climate change in *415–416*, 433
    glacial lakes *415–416*
    mining and environmental protection issues 112,
        189, 191
    national parks in 237, 238, 240, 241, 243
    sacred mountains in 51
    tourism in 5, 264, 433

Canarium, in biodiversity conservation on Holy Hills
    45
Canary Islands 345, *407–408*
carabid beetles, and altitudinal zonation 203
carbon dioxide (CO_2), in global warming 409,
    413–416, 418
Carex spp., in altitudinal zonation *218*, 233
Carpathian Mountains, human activity in endangered
    ecosystems 224
cash crops (agricultural) 110, *111*, 357
Caspian Sea area, Iran, earthquake disaster in *376*
catastrophic events *see* natural disasters
Cathedral Peak, Sierra Nevada, United States, spiritual
    significance of 44
Catskill Mountains, United States, watershed
    protection in 354
Caucasus
    agriculture in 96
    as barrier 27
    conflict in 9, 98, *113–114*
    cryosphere and global warming 414
    disasters in 371
    economic and political indicators in *96–98*
    ethnic diversity in 96
    health resorts in 96
    indigenous peoples of 109
    industry in 96, 97
    road construction in, environmental impact of
        351, 359
caves, endangered ecosystems in 222
Central America, species diversity in 232, 297
Central Asia
    ecosystem destruction in the Aral Sea *145*
    water resources in *137*
    woody ecosystems as fuel source 225
Central Asia *see also* individual countries
Central Cordillera, Philippine, terracing in farming
    system 356
Central Karakoram National Park, Pakistan *244–245*
centre of the cosmos, mountains in religion 42, 48–49
Cerastium uniflorum **421**
Cerro Rico de Potos, Bolivia, industrial archaeology
    and tourism in *178–179*, **181**
Cesky Krumlov, Czech Republic, amenity migration to
    268
Changbai Mountains, Korea, biodiversity on *210*
Chapter 13, Agenda 21 *see* Rio Earth Summit
Charazani region, Callawaya, Bolivia, agro-ecological
    altitudinal belts 7
charcoal, as energy source *158*, 161, 177, 292, 293
Chechen Ingush, land mines in *388*
Chiang Mai, Thailand, amenity migration in 271, 265,
    266, 272
children
    mortality rates of 68, 71, *74*
    and poverty in developing countries *67–82*
    status of, and Chapter 13 452
    vulnerability of 372, *395, 396, 399*
        in conflict 386
        due to disease 371, 391
    workload of *81–82*

Chile
  disasters in *375*, **383**, *387*, *390*, *401*, *407–408*
  environmental protection legislation in 190, 191
  mines in 171–175, 177, 178, 180–181, **183**
  national parks in *237*, *240*, **246**
  population decline in **20**
  solar energy 162–163, *166–167*
  tourism in 253
  vegetation belts in 284, **286**
  water resources in *137*
China
  biosphere reserve in *239*
  climatic records on 412–413
  conservation practices in **44**, *45*
  disasters in *375*, *380–382*, *387*, *394*, *401*, *407–408*
  drug production in 116
  land tenure, and watershed management 366
  national parks in *237*, *238*, *239*, *240*, *241*, 243
  non-wood forest products in 293
  poverty in mountain regions 61, *330*
  sacred mountains in 40, 43, 44, 45, 46, 47, 54
    tourism to 250
  traditional costumes in **35**
  vegetation belts in 287
  women's status in 80–81
China *see also* Pamir
Chipko movement, Himalayan region *127–128*, *148*, 291
Chittagong Hill Tracts, Bangladesh, ethnic conflict in *117–118*
Chlamydomonas nivalis, alga in red snow 215
chlorofluorocarbons (CFCs), impact of 409–410
cholera, global pandemic of 388
Christianity, churches and sacred places of 44, 47, 54, 57
Chusquea bamboo 226
Chutrun, Baltistan, Pakistan **69**, **70**
cities
  high altitude 30
  mountains as water source for 339
civilians, victims in wars 371, 386–388
cliffs, endangered ecosystems in 222
climate 410–412
  and amenity migration 266
  and biodiversity *201*, 203, 213–215
  dendroclimatological information from trees 285, 286
  and tourism 258, 259
  variation of in a watershed 340
  and vegetation belts 287
climate change 140, 409–410, 412–417
  national parks as research areas *237*
  potential impacts
    on ecosystems 417–426
    on societies 426–434
  and species distribution 206–208
climate change *see also* global warming
climax, in succession in biodiversity 210–211
climbing, and endangered ecosystems 222, 255
clothing, tourism as cause of change in 255

cloud forests
  biodiversity in 225, 232, 284–285
  and pollution-laden precipitation 301
  and water capture 345, 365
clouds 134, *219*
  and climate 410, 411, 413
coal mining 171, 181–182
coastal settlements, and earthquake damage *377*, *378*, *379*, *407–408*
coca, use of by miners 175
Cochamba, Bolivia, watershed management programme in *342–344*
Coleoptera, beetles, adaptations to adverse conditions 212
Collembola, springtails, in adverse conditions 212, *215*, *216*
Colombia
  disasters in *385*, *387*, *390*, *401*, *407–408*
  drug production in 116
  national parks in *240*
  socio-economic indicators in 73–79
  vegetation belts in 284, **286**
Commission for Sustainable Development (CSD), establishment of 1
communications, in mountain areas 27, 31–33, 266, 270–271
comparative cost advantage, definition of 86
conflict
  and human welfare 61, 384–386, 397–400, *401*, 403
  impact of *28*, 371, *373*, 384–388, *407–408*
    on tourism 253, 254
  management of 123–128
  migration and 116–117
  over dam-building and hydropower 83, *152*
  and problems for sustainable development 103–106, 455–456, 457
  and spread of AIDS 389–390
conflict *see also* ethnic conflict
conifers, hydrologic effect of 354
conservation
  of endangered ecosystems 226–228, *239*
  of water resources 152–153
contraceptive use, birth control 78, 81
Cook, Mount(Aoraki), New Zealand, sacred mountain 46
cooking, with solar power *163–164*, *164–165*
cool mountains, management and conservation of biodiversity in 226–228
copper mining 171, 176–177, 180, 182
Cordillera Blanca, Peru, tourism in 253
Cordillera de los Andes *245–246*
corridors, in ecosystems 204–205, *245–246*
corruption, and environmental protection 196
cost advantage, in mountains 86, 101
Costa Rica
  biodiversity in *234*
  disasters in *385*, *387*, *390*, *401*, *407–408*
  forest disturbance and land slope 288
  national parks in *238*, *240*
  stall-feeding of livestock 355

Côte d'Ivoire, Mont Nimba Nature Reserves 237
Croatia, disasters in *385, 387, 388, 401*
crops *see* agricultural crops
Cross Island Highway, Taiwan, environmental impact of 359
Crow people, of North America, sacred mountains of 47
cryosphere, in climate change 414–417
Cryptosporidium, in water contamination, in watersheds 365
CSD (Commission for Sustainable Development), establishment of 1
culture
    and amenity migration 264–265, 266–270, 273
    and community identity 11–12, 341
    and conflicts 117–123
    definition of 119
    diversity of 9, 59, *246–247*
    and environment 22–27, 29–30, 39–42
    ethnic groups 18–22
    and modernisation 31–35
    protection of *237*, 448, 451, 464
    and research on sustainable development 47–54
    and sacred mountains 39–59
    and tourism 249, 253, 255–256, 262–270
Cusco, Peruvian Andes, water management in *149*, 337–339
cyanide, in gold ore processing 177–178, 184, 187
Cyperaceae, in vegetation belts 234
Cyprus, disasters in *387*
Czech Republic
    amenity migration in 265, 266, 267–268
    pollution-impacted forests in 301, *302–303*
Czechoslovakia, Bohemian Forest, ecosystem changes in 204
Czechoslovakia *see also* Giant Mountains
Dacrycarpus, in vegetation belts 287

Dai, Buddhists, conservation practices of 44, *45*
dams
    as cause of conflicts *107–108*, 112
    construction, effects of 83, 109, *117*
    failure of, disasters caused by *381–382, 393, 394*
    and flood protection 143
    and mining 182–183, 188
    and sedimentation, effects on 10, 13
Danthonia, tussocsk grass, in vegetation belts 233
Darjeeling, Himalaya, India, catastrophic events in 10
data
    collection of on water resources 136–141
    historical, on species distribution 200–202
    need for, on traditional conflict management 123–124
    unreliability of 71, 79, 82
data *see also* information
deforestation 8–9, 272, 303–304
deities, present in mountains 43–44, 50–51
Delphi, Greece, sacred mountain 42
Delphinium spp. 223, 425

demand pressure, management of in agriculture planning 329, 330
demographic changes *94, 95, 97*
    and status of women 72
demographic changes *see also* population
dependence, of mountain regions 33–34
desertification, and global warming 413
developing countries, poverty and women and children 67–82
Developing Women's Entrepreneurship in Tourism (DWET), in Nepal 259–261, **260**
development aid *see* foreign aid
development planning, and sacred sites 49
Devil's Tower, North America, sacred mountain 57
Dhampus, Annapurna Region, Nepal, DWET programme **260**
Dharan-Dhankuta road, Nepal, environmental impact of 359
Diamond Mountains (East Asia), sacred mountains 47
Diné people, of North America, sacred mountains of 40
Diptera, flies, adaptations to adverse conditions 212
disasters 371–383, *393–403, 407–408*
    avalanche 222–223, 295, *373, 380, 407–408*
    earthquake 10, 270, 373–374, *374–379, 380–382, 407–408*
    and ecosystem disturbance 204, 209
    fire *373, 408–409, 415, 425, 430*
    floods 141–143, 296, 349, 351, 373–374, *380–382, 383–384, 407–408*
    and human welfare 61
    and hydropower installations 157
    research on 10–11
disasters *see also* hazards
disease hazards 371, *388–391, 407–408*
dispersal, of animals and plants in climate change 424–426
diversification, in agriculture 126, 328, *329, 330*
diversity, of mountain areas, and agriculture 313, 314, *315, 317, 319, 321*
Dolomites, Italy, biodiversity in 201–202, *203*
Doronicum, endangered species of 223
Draba 210
drainage, and biodiversity in wetlands and peatlands 225
Drogue Mafia, Colombia, and drug production 116
drought
    and conflict 116–117, 122
    impact of 143, *373, 407–408*
drug production 115–116, 243, 386
dry cell batteries, as energy source 157, 158, 170
dung, uses of 174, 224, 289, 355
dust aerosols, and global warming 413
dust pollution, in mines 188
DWET, Developing Women's Entrepreneurship in Tourism, in Nepal 259–261, **260**
Dzong Tongsa, Bhutan, ecologically over-used landscape in **90**

Earth Summit *see* Rio Earth Summit
earthquake

impact of 10, 373–374, *374–379*, *380–382*,
407–408
and landslides, in the Philippines *270*
earthquake *see also* seismic activity
ecoclimatic zones, in climate change 422–423
ecological indicators 88–89, *92*, *95*, *97–98*, *100*
ecology
and hazard risk 372
interactions and biodiversity 212–221
problems in *92*
protection of and Chapter 13 447–453
in riverine habitats, changes to 143–147
vulnerability of 400–403
of a watershed 339–341
economic *see also* financial
economic indicators 88–89, *92*, *94–95*, 96, 97,
99–100
economics
autonomy in transport modernisation 35–36
and displaced persons 384–386
of forests **292**, 307
and migrants in hill resorts *269*, 272
of mining in mountains 178–180
in national parks *242–243*
and politics of sustainable development 85–101
of tourism 251, 255–256, *259–261*
in water resource management 150–151
of a watershed 341, 352, 366
ecosystems
and climate change 417–426
distribution of 206–212
diversity in 202–206
endangered 221–226
in forest management 291–292
ecotourism 263
Ecuador
agriculture and climate change 428–429
alien woody species in 225–226
disasters in *387*, *390*, *401*, *407–408*
infectious diseases in *389*, *390*
national parks in *237*, *238*, *241*
socio-economic indicators for *74–78*
vegetation belts in **286**
education 72, 73, 74, 77, 82, 92, 95
and gender discrimination *22*, 69, 79, 80
educational institutions, and amenity migration 270,
272, 273, 275
Egmont, Mount, New Zealand *see* Tongariro
Egremont, Mount (Tongariro), New Zealand, sacred
mountain 40, 43, 49, 52
Egypt 40, 57, 240, 250
EIS (Environmental Impact Statement) 191
El Hierro Island, Canary Islands, trees in water
harvesting 345
El Salvador, disasters in *385*, *387*, *388*, *390*, *401*,
407–408
elderly people, vulnerability of 372, 386, *396*
electricity
hybrid systems 168
for small industries, in management for
sustainability 127

solar powered generation of 162, 164–167,
167–168, 432
electricity *see also* hydropower
'elfin woodlands' *see* cloud forest
Elgon, Mount, Uganda, national park *293–294*, *305*
Elyna myosuroides, and altitudinal zonation in the
Swiss Alps *218*
Emei Shan, China, sacred mountain 54
employment, non-agricultural *94*, *259–261*, 271–272
Encañada, Peruvian Andes, economic and political
indicators in *92–93*
endangered ecosystems, management and conservation
of 226–228, *241*
endangered species, in montane forests *241*, 284–285,
297
energy minerals, mining of 171
energy resources 157–169
and climate change 432
new technology in 168
sustainable development of 158–170, 168
in water, potential for 134, 339
environment
and culture 22–27
in disasters 374, 395
revaluation of 34–35
and the Winter Olympics *252*
environmental adaptation, in pasturalism 126
environmental degradation 18, *97*, *100*, 115
and conflict 103–106, 110–117, 386–388
and highland-lowland interaction 122
and logging techniques 283
and water quality 135–136, 141
environmental functions, of forests **292**, 307
environmental impact
of hydropower installations 157
of mining 112, 172–186, 180–185
of tourism 256–257, 262, 275, 296–297
of transport modernisation 33
environmental impact assessment
and mining 189, 191, 192, *193*
sacred mountains and communal identity 52
in watersheds 365
Environmental Impact Statement (EIS) 191
environmental protection 8, 66–67, *95*
by amenity migrants 273
and Chapter 13 447–453
and mining 172, 186–196
and pilgrimage 54–56
and riverine ecology 150–151
in sacred mountains 39–41, 54
and tourism 56–59
Environmental Protection Agency (EPA), in the United
States 189
environmental resources
and amenity migrants 271–273
global treaties on 325
and tourism 249, 264–265, 266–270
EPA *see* Environmental Protection Agency
Ericaceae, in vegetation belts 217, 234
Eritrea, land mines in *388*
Erzincan, Turkey, earthquake disaster in *375*

Espeletia sp., adaptations of 211
ethical factors
    in future planning 464–465
    in national parks *242–243*
Ethiopia
    agriculture in 119, 126, 356
    ancient civilisation in 119
    disasters in *385, 387, 390, 401, 407–408*
    infectious diseases in *389, 390*
    land mines in *388*
    national parks in *237, 238*
    socio-economic indicators for *74–78, 79,* 119
    soil erosion in 349
    water resources in *137*
ethnic cleansing 386
ethnic conflict 88, 98, *107–108,* 109–113, *113–114,*
    117–118; conflict
ethnic diversity 9, *21,* **25**, *96*
    and culture 18–22
    and mountain refuges 27, 29
ethnic minorities
    in future planning 465–466
    vulnerability of *372, 384, 395, 396,* 399
Eucalyptus spp. 217, 225, 287
Euphorbiaceae, in vegetation belts 233
Euphrates-Tigris basin, and conflict *107–108*
Europe
    agricultural subsidies *32,* 33
    amenity migration, need for comfort in 271
    biodiversity in 200–202, 206, 207, **223**
    climatic records on 412
    conflicts in *387*
    endangered species in 297
    glaciation and species distribution in 207, 208
    human activity and species diversity **223**
    natural disasters in *380, 408–409*
    pollution in precipitation 346
    reforestation in 281
    refugees in *385*
    rocks and soil types 214, 215
    social conditions in mountains of 62–65
    spiritual qualities of mountains in 47
    subsidy for mountain communities 61–62
    transfrontier reserves in 239, *240*
    vulnerable peoples in *401*
    water resources in *134,* 138
Europe *see also* individual countries
European Community, environmental legislation and
    mining 189
evaporation, and the hydrologic cycle 134, 140, *338,*
    344
Everest, Mount 39, 212, 247, 297
external influences, impact of on mountain people
    11–122
'eyebrows of the mountain', *bosque de ceja montaña*
    284

Fagus sylvatica, beech 217, *218, 302, 303,* 422
famine, in disasters 371, *373,* 398, *408–409*

Farfui, Bagrot Valley, Gilgit, Pakistan, agriculture and
    altitude *25,* **26**
fertilisation, use of *289* 220, 224, 355
fertilisation *see also* biomass
fertility, sacred mountains as source of 46
Festuca pilgeri ssp. pilgeri, adaptations of 211
Festucetum versicoloris, and soil acidity 215
field systems, terracing 80, 125, 127
Fiji, cloud forests in 284
financial *see also* economic
financial arrangements, for micro-hydropower systems
    163
financial aspects
    of amenity migration 265–266, *269*
    of dependence of mountain regions 33
    of hydroelectricity *151–152*
    of tourism 57, 251, *259–261*
financial incentives
    for agroforestry *289*
    for development 95
    for environmental protection in mining 191, 192
fire disasters *373, 407–408, 415, 425, 430*
fireplace, social significance of 160
firewood, as energy source 157, *158,* 159–162, *164*
firewood *see also* biomass
fish 143, 146, 225, 419
floods
    forests in protection against 296
    impact of *373–374, 380–382, 408–409*
    occurrence of 141–143, 349, 351
    planning for 383–384
fog, as precipitation 345–346
Folsomia spp., and soil acidity *215, 216*
food
    and amenity migrants 271
    as forest products *293,* 293–294
    and improved transport systems 32–33
    security of, and Chapter 13 448, 452
    shortages of *22,* 33
footpaths and pack routes, stability of 359–360
foreign aid 1–2, 12–13, 67–68, *90–91, 99, 100*
    gender discrimination in 73
foreign aid *see also* World Bank
forestry 66–67, *99–100*
    agroforestry 288, 289–290
    Chipko in protest against *127–128*
    and climate change 427, 430, 430–431
    logging, and environmental damage 283, 291,
        353–354, 360
    road systems and tourism 253
    watershed land use 352, 353–354
forestry *see also* agroforestry
forests *90, 91, 98*
    in altitudinal belt classification 4
    amenity value of *281,* 296–297
    biodiversity in 284–288
    and climate change *423–424,* 435
    degradation of 8–9, *272,* 303–308
    exploitation of 34, 97, 294
    management of 226, *227,* 281–283, 291–292,
        304–309

ownership of 290–291
pollution-impacted 301–303
precipitation influenced by 135
products of *160–161, 281, 282,* 288–297,
    308–309, *353, 355*
as protection from natural hazards 27, 223
protection of 290, 452
restoration of 281–282, 293, 299–301, 361
sacred groves 298–299
wildlife habitat 297–298
as woody ecosystems 225–226, *227*
forests *see also* deforestation; trees
fossil fuels, as energy source 157, *158*
fragility, of mountain areas, and agriculture 313, 314,
    315, 317, 321
France
    climatic records on 412
    culture in 29–30
    dam failure in *394*
    demographic change in 64
    national parks in *240, 241*
    tourism planning 263–264
    water resources **137,** *137,* 431
Frias, Argentina, dam failure in *394*
fuel
    consumption of 18, 174, 177, 271, 287
    fuelwood, effects of use of 225, 292–293, 353, 432
Fuji, Mount, Japan, sacred mountain 39, 43, 47, 250
fungal dieback disease, in Australia *221*

Ganges River 56, *141–142,* 148, 431
Gangotri, India, environmental degradation by
    pilgrimage to 54–55
gardens, spiritual view of mountains as paradise
    44–45, 51
Garibaldi Provincial Park, Canada *237*
gender inequality *see* women, and poverty
genocide, potential for *111,* 113, 386
Gentiana sp., endangered species 223
Genting Highlands, Malaysia, forest clearing for
    recreation in 297
geoecology, foundations of 203
Geographic Information Systems (GIS), database
    421–424, 460
geological evolution, and species distribution 206–208
geomorphic processes, impact of climate change
    415–416
Georgia, disasters in 384, *387, 401*
geothermal energy, use of 168
Germany
    agriculture in 66–67
    dam failure in *394*
    energy supply in *169*
    environmental protection in 66–67
    forests in 66–67, 204, 297, *302, 303*
    national parks in *240*
    tourism in 66–67
    water resources in **137,** *138*
Ghir, Iran, earthquake disaster in *376*
Giant Mountains

air pollution in 215, **216**
biodiversity in 200, 202, 205, 225
Giardia, water-born infection caused by 360–361, 365
Gilgit Agency, Pakistan 25, **26,** 79
glacial cirques, and endangered ecosystems 222–223
Glacial Lake Melbern, Canada, and climate change
    415
glacial lakes, and climate change 143, *415–416*
glaciation, in mountain zone definition 3, 4, **10**
Glacier Bay, Alaska, biodiversity in 210
glaciers
    assessment of 140
    hazards of 382, 383, 414–417
    and species distribution 207
    as water reservoirs 134
Gleno, Italy, dam failure in *394*
Global Biodiversity map 232, *234–235*
global conferences, discourses following 326
global treaties, on environmental resources 325
global warming
    effects of 283, 287–288, 410, 413–417
    mechanism of 409
    records of 412–413
global warming *see also* climate change; temperature
GNP, national statistics 73, *74*
goitre, national statistics 76
Golan Heights, conflict over 103, 108
gold mining 171, 173, 177–178
golf courses, environmental impact of 297, 360
government agencies, and forest restoration 299–301
government interventions, and agriculture 320, *321,*
    323
government ownership, of forests 290–291, 296, *305,*
    305
government policies
    and natural resources in climate change 431,
        432–433
    on socio-economic inequalities 66
    and watershed degradation 357–358
Grasberg, Indonesia, transport in mining in 176
grassland, and watershed protection 223–224, *350,*
    351, 354
grassland *see also* vegetation
graveyards, and environmental protection 51
Great Basin, Nevada, United States, in potential
    climate change 424
Greece
    disasters in *387, 390, 401, 407–408*
    sacred sites 42–43, 44, 47, 51, 54, 57–58, 251
    water resources and climate change 431
Greece *see also* Olympus, Mount
Green Mountains, Vermont, United States, fuelwood
    from 293
greenhouse gases (GHGs) 409–410, 413–414, 432
Greenland, national parks in 238
Gross Kar, Eastern Alps, skiing and endangered
    ecosystems 223
groundwater
    in the hydrologic cycle *338,* 344
    pollution from mining *185,* 187
    reserves, over-utilisation of 188

Guatemala
  disasters in *376, 385, 387, 390, 401, 407–408*
  infectious diseases in *389, 390*
  socio-economic indicators for *74–78*
Guiana Highlands, *tepuis*, microclimate on 209
Guinea, Mont Nimba Nature Reserves 237
Günedogu Anadolu Projesi (GAP), Turkey *107–108*
Gunung Agung, Bali, sacred mountain 42
Haastia pulvinaris 212

handicrafts, as income source 28, 127, 255
Hanumanchatti, India, tree planting by pilgrims 55, 56
Hawai'i 43, 250, 284, **285**, 353
hazards 9–11, 27, 341
  in climate change *415–416*, 425
  disease 371, 388–391, *407–408*
  forests in protection against *281*, 295
  and future planning 451, 457
hazards *see also* disasters
health, safety and environment (HSE) policies, and
  mining 194
health indicators, national statistics *75*
health resorts, and migration to mountains 96, 116
health resorts *see also* sanatoria
health services 22, *75*, 91, 92, 95, 391
heat wave, impact of *373*, *408–409*
Hengduan Mountains, China, montane vegetation
  belts in 287
Herzegovina, land mines in *388*
Hetch Hetchy Valley, Sierra Nevada, protection of 51
hibernation, animal adaptation to adverse conditions
  212
Hieracium spp., distribution of 207
High Atlas Mountains, Morocco **152**, 355
High Tatras, Slovakia **199**, 204, *215–216*
highland-lowland interaction 120–123
  in cultural ecology 119–120
  and dam projects 109, *117*
  and erosion control 351–352
  and future planning 456, 461, 464–465
  and vulnerability *396*, 400
  in watershed management 352, 366
hill stations, in amenity migration 266, 269
hillside erosion, and economic migration 272
Himachal Pradesh, India 251, 252, *330*
Himalaya
  agroforestry in 288, *289–290*
  altitudinal zonation in 219
  biodiversity in 202, 203, *234*
  catastrophic events in 10
  Chipko movement *127–128*, *148*, 291
  conflicts in 103, **104–105**
  endangered species in 297
  energy resources and physiographic regions 159
  forest restoration in 299
  human activity in endangered ecosystems 224, 225
  information database on *459–460*
  poverty in 115
  sacred mountains in 39–41, 44, 46, 54

tourism and endangered ecosystems 222,
  256–257, 297
vegetation belts in 232, 287
water from, and lowland flooding *141–142*,
  383–384
Himalaya *see also* Everest; Karakorum, individual
  countries
Hindu Kush, Afghanistan 28, 29, 203, *459–460*
Hinduism
  sacred mountains of 39–41, 42, 54–56, 250–251
  temples and sacred groves of 44, 298–299
Hira, Mount, sacred mountain 47
HIV, human immunodeficiency virus, global problem
  of 389–391
Hochgebirge, in mountain classification 4, 8, 10, 12
Hohe Tauern National Park, Alps 243
Hohe Wilde, Mount, upward movement of plants on
  419
Honduras, disasters in *385, 387, 390, 401, 407–408*
Hopar, Gilgit Agency, Pakistan, gender discrimination
  in 79
Hopi people, of North America, sacred places of 49, 54
horizontal dimension, in cultural ecology 118–119
Huang Shan, China, sacred mountain 47
Huangshan National Park, China 243
Huangshan Scenic National Area, China 237, *238*
Huascaran, Cordillera Blanca, Peru, landslide in 10
Huascaran National Park, Peru, tourism in 253
human activity 11–12
  and biodiversity 217, 219, *220*
  and climate 213–214, 425–426
  control of in national parks 237
  and disasters 393–400
  and endangered ecosystems 221–228
  hydrological cycle influenced by 135–136
  and sedimentation 364
human activity *see also* agriculture; tourism etc
human adaptation, to mountain environments 23–27,
  314, 316–320
human health, and climate change 433–434
human immunodeficiency virus, HIV, global problem
  of 389–391
human needs, and the mountain environment 2, **3**
human power, as energy source 157, *158*
human rights
  and political will 462, 463–464
  violations against indigenous peoples 399–400,
    401
human settlements 4, 17–18, *24*, 34
Humboldt, Alexander von (1769–1859), altitudinal
  belt classification of 4, **6**
humid tropics, vegetation belts in 232–234
humidity changes, and geological evolution 207
Hungary, pollution-impacted forests in 301, 303
hunting
  in endangered ecosystems 223
  as a forest product *293*, 295, 297–298
Hunza, Pakistan *21–22*, 29, 263, 337
Hurricane Bola (1988), New Zealand, landslips
  following 351, 358

hurricane *see* tropical cyclone
Hushe Valley, Baltistan, Pakistan, women of *68–71*
hydroelectricity *see* hydropower
hydrologic cycle, in watersheds 337, **338**, 344, *346*, **347**
hydrological research, methods 136–141
hydrology, in climate change 414
hydropower
    and climate change 416, 432
    design of schemes 13, 149, 157
    as energy source 157, 158–159, 161–164, 168
    and mining 176, 177
    political aspects of *107–108, 151–152*
    for sustainable development 88
    turbines for 162, *163*
    and water management *107–108*, 134, 149
Hyokiri, Korea, dam failure in *394*

ice, as water reservoir 131, 134
ice cores, information on global warming 412
ice cover, assessment of 140
Iceland, disasters in 10, *407–408*
Ilam District, Nepal, niche advantages in agriculture 330, *331*
Ilex, in cloud forests 232
image, and tourist destinations 254
immunization
    national statistics *75*
    poor hygiene and spread of HIV 391
impenetrability, in highland-lowland migration 120–121, 122–123
Inca people, Peru, ancient civilisation 43, *149*, 337–339
incentive schemes, for agriculture 263
income
    and gender discrimination *68–71, 72–73*
    opportunities, and Chapter 13 447–453
    supplementation of 17
India
    agriculture, diversification in *289, 330*
    amenity migration in 266
    childhood mortality rates 68, *74*
    Chipko movement *127–128*, 291
    conflict over development in 109, 112
    disasters in 10, *385, 387, 401, 407–408*
    environmental degradation by human activity 54–55, 359
    gender discrimination in 68
    national parks in 237, *238, 240*
    pilgrimage in 54–55, **55**, 56
    poverty in mountain areas of 115
    sacred places in 40, **41**, 44, 54–56, 251, 299
    socio-economic indicators for *74–78*
    tourism 251, 252
    water resources in *137*
India *see also* Brahmaputra, Ganges and Indus rivers; Himalaya; Tehri dam
indigenous peoples
    conflict management by 123–124
    and impacts of tourism 275

protest movements of 34, 35, 36, 109–113, *127–128*, 291
    responsibility of, in sustainable development strategies 125
    traditional knowledge of 121–122
    vulnerability of 395, 396, 399–400, *401*
indigenous peoples *see also* ethnic minorities; local communities
Indonesia
    amenity migration in 265
    disasters in *385, 387, 390, 394, 401, 407–408*
    mining in 176, 182, 191
    national parks in 237, *241*
    river sediment load of 351, 359
    road construction 253, 359
    transport in mining in 176
Indus River *148*, 383–384, 431
industrial archaeology, and tourism *178–179*
industrialised countries, social conditions in mountains of 62–65
industry, in North Ossetia-Alania (Caucasus) 96, 97
infant nutrition, national statistics 76
infectious diseases
    and climate change 433–434
    as disaster 388–391, 398
    vulnerability of indigenous people to 399, 400
    and water contamination 346, 360–361, 365
information
    dissemination of, for future planning 447–453, *457–460*
    for tourism 254, 258
information *see also* data
Ingush-Ossetian conflict *113–114*
insects, adaptations to adverse conditions 212
international borders, in watershed management 366
international cooperation, and Chapter 13 447–453
international organisations, on management and conservation of ecosystems 227
Inugsuin Mountains, Baffin Island, Canada, tourism in 5
invertebrates 143, 145, *214–215, 221, 222, 425*
Iran
    disasters in *375, 376, 385, 387, 390, 401, 407–408*
    land mines in *388*
Iraq
    disasters in *107–108, 387, 401*
    land mines in *388*
Ireland, pilgrimage in 251
iron mining 171
irrigation systems 12, 146, *149*, 337–339
Islam
    and conflict *117*
    sacred mountains of 47, 57
    and women's status 69–71, 80
Israel, Golan Heights, conflict over 103, 108
Italy
    biodiversity in 201–202, 203
    demographic change in 64
    disasters in *375, 385, 390, 394, 407–408*

Italy *contd.*
modern development in *32*
national parks in *240*

Jamaica, national parks in *241*
Japan
agriculture and climate change 428–429
disasters in *375*, *376*, *407–408*
national parks in *238*
reforestation in 281
sacred groves and forests of 298
sacred mountains in 39, 42, 43, 45, 46, 47,
51–52, 250
tourism to 250
sacred nature of 42
transport systems and tourism 253
vegetation belts in 287
Jebel Marra, Sudan, agricultural traditions and conflict
116–117, 122
Jordan River, and conflict **106**, 108
Judaism, sacred mountains of 47, 39, 40, 57
Juglans, in cloud forests 232
Juncaceae, in vegetation belts in wetlands 233
Juniperus spp., juniper *201*, *210*, 425

K2, Mount, Central Karakoram National Park,
Pakistan *244–245*
Kaata, Mount, Bolivia, sacred mountain 46
Kabale Highlands, Uganda, soil degradation in
357–358
Kafirs, refuge in the Hindu Kush 29
Kailas, Mount, Tibet, sacred mountain 39, **40**, 42, 44,
54, 56, 250
Kalam, Pakistan, social effects of tourism 263
Kanchenjunga, sacred mountain 57
Kansu, China, earthquake disaster in *375*
Kanto Plain, Japan, earthquake disaster in *375*
Karakorum *see also* Central Karakorum National Park
Karakorum Highway, Pakistan 22, 253
Karakorum Region, Pakistan, women of *68–71*, 79
Karnatika, India, sacred forests in 299
Kasbegh, Mount, Caucasus, conflict around **9**
Kashmir 202, 253, 371
Kasungu National Park, Malawi *237*
Kenya
disasters in *385*, *387*, *390*, *401*
ethnic conflict in 110–112
hydrologic cycle *346*, **347**
infectious diseases in 398
Mount *see* Kere-Nyaga
sacred mountains in 43, 45, 46
socio-economic indicators for *74–78*
vegetation belts 233, 284
water and soil profiles in *346*, *347*
Kenya *see also* Kilimanjaro, Mount
Kere-Nyaga (Mount Kenya), sacred mountain 45, 46,
284
kerosene, as energy source 157, 158
Khumbila, Mount, Nepal, sacred mountain 42, 44, 50

Khumbu region, Nepal 51, 81
Kikuyu people, Kenya, sacred mountains of 43, 46
Kilauea, Hawai'i, sacred mountain 43, 250
Kilimanjaro, Mount, Kenya/Tanzania 12, 45, 212, 284
Kilimanjaro National Park, Tanzania *244*
Kinabalu, Mount, Sabah, Malaysia, forest clearing for
recreation in 297
knowledge-intensive entities *see* educational institutions
Kobe, Japan, earthquake disaster in *376*
Konto River, Indonesia, sediment load of 351, 359
Korab, Mount, Albania **98**
Korea
biodiversity in *210*, 217
dam failure in *394*
sacred mountains in 46
Koya, Mount, Japan, sacred mountain 42, 45, 51–52
Krkonose Mountains *see also* Giant Mountains, Czech
Republic 301, *302–303*
Krössbach, Austria, avalanche protection dam **373**
'krummholz' vegetation 284, 285
Kulekhani Reservoir, Himalaya, effects of
sedimentation on 10
Kurdistan, disasters in 371
Kurds, in ethnic conflict *107–108*, 120–121

labour intensive agriculture systems 125, 126, 127
labour resources, consumption by amenity migrants
271
Lagarostrobus franklinii, huon pine, longevity of 286
lakes, as water reservoirs 134
Lakota people, North America, sacred mountains of
42, 47
land clearance, for agriculture 224, 356–357, *357–358*
land clearance *see also* deforestation; slash-and-burn
agriculture
land contamination, in conflict 388
land holdings
and agroforestry 288
and amenity migrants 268, *269*, 271
fragmentation of 17, *32*, 64
and highland-lowland interaction 122
in pasturalism 126
reform of, in management for sustainability 127
and soil degradation *358*
and watershed management 355, 365–366
land mines 'plague' *388*, 388
land use
agroforestry in 288, *289–290*
and conflict 110–112
and ecosystem patterns 203, 204
impact of amenity migrants 271
management of *319*, *322*
patterns of 4–8, 23–27, *92*, *93*, *94*, *98*
planning, and Chapter 13 449
and soil erosion 349–351
and streamflow 348–349, **350**
in watersheds 352–361
landscape characteristics, in mountain regions 3, 9, 263
landslides
and climate change *415–416*

and earthquake disasters *270*, *380–381*
impact of 10, *72*, *373*, 374–379, *380–382*,
  408–409
mining as cause of 175, 188
road construction as cause of 359–361
Langtang Himal, Nepal 10, 203, 209–210, 211–212
Langtang National Park, Nepal *247*
language
  and conflict *117*
  and culture 18–22
  diversity of, of indigenous people *21–22*, 109–110
  and gender discrimination *69*, 79
  and mountain refuges 27, 29
Laos, land mines in *388*
Larix spp., larch, deciduous conifer *210*, 287
Lascar, Mount, Chile, volcanic eruption of **383**
Latin America *see* South America
latitude, and ecosystem diversity 3, 4, 203
Lauca National Park, Chile *237*, **246**
lead mining 171
Lebanon, disasters in *385*, *387*, *390*, *401*, *407–408*
legal controls
  on hydropower projects *151–152*
  in mining 172, 173–174, 182, 186, 189–193
leisure facilities, and soil erosion 360
leisure time, and amenity migration 270, 271, 266
Lepidoptera, butterflies, adaptations to adverse
  conditions 212
life expectancy rates *74*, 78
Liliaceae, in vegetation belts 234
Lima, Peru, cholera pandemic (1991) 388
limestone bedrock, ecological effect of 214, 225, 345,
  346
linguistic groups *see* language
literacy rates
  national statistics *72*, *74*, *77*, *78*, *91*
  in Nepal *91*
litter (rubbish), and pilgrimage 54, 56
livelihood opportunities, and Chapter 13 447–453
livestock
  in agriculture *22*, *28*, *68–71*, 79
  in agroforestry *289*
  and altitude 25, **26**, 218–219
  and climate change 429
  historical domestication of *92*
  in national parks 244–245, **246**
  on sacred mountains 50
  and water contamination 346, *365*
  as women's work *68–71*, 79
livestock *see also* pastoralism
livestock grazing
  and conflict *110–111*, 116–117
  in endangered ecosystems 223–224, 349, 350, *358*
  as forest product *293*
  watershed land use 352, 354–355
llamas, as pack animals, advantages of *297*, *355*, 360
Lobelia sp., in vegetation belts 211, 233
local communities
  agriculture, traditional knowledge of 332
  in climate change 434
  in disaster aid 394–395

and energy resources, planning for *148*, *152*, 170
in forest management 293, 304–309
and forest restoration 299–301
and forest rights *245*, *247*, 290, 291
in mining planning 194
and mountains in identity 46, 52
protest movements *127–128*, *148*, 291
in resource management planning 325, 449, 456,
  461, 464
and sacred places, local knowledge of *45*, 49
social life of *29–30*
and sustainable development strategies 125, 294
and tourism 261–264
in water management *149*
watershed management by *338–339*, 363, 364,
  365–366
and wildlife park developments 298
logging, and environmental damage 283, 291,
  353–354, 360
Lomatia tasmania, King's holly, longevity of 286
longitude, and ecosystem diversity 203
Lore Lindu National Park, Indonesia 237
lowlands
  characteristics of 118
  favourable development in 65
  resources from mountain regions 106–109,
    131–135, *137–138*
Luzon, Philippines, amenity migration in 265, 266,
  269–270

Macedonia, disasters in *385*
machinery, for mining in mountains 175, 182
Machu Pichu National Park, Andes 243
Makalu–Barun National Park, Nepal *247*
malaria, global problem of *389*, 433–434
Malawi
  infectious diseases in 398
  national parks in *237*, *240*
Malaysia
  agroforestry problems in 290
  amenity migration in 266, 297
  disasters in *385*, *387*, *390*
  national parks in *241*
malnutrition, in mountain communities 62, 63, 92
Malpasset, France, dam failure in *394*
mammals
  adaptations to adverse conditions 212
  in endangered ecosystems 222, 223
  endangered species in forest habitats 297
  new protected areas for *241*
  species diversity, in Australia *221*
  tree damage caused by *302*
mandalas, and mountains 42
Manu National Park, Peru *237*, *238*, *239*
Maori people, New Zealand, sacred mountains and
  communal identity 46, *52–53*
marginalisation, of mountain communities 120, 123
market forces
  and agriculture 320, *321*, *330*, *331*
  liberalisation of, post-UNCD 325

marriage, and women's status in northern Pakistan 69
marriage patterns, and population control 62
Marsabit, Mount, Kenya, vegetation belts on 233
maternal mortality rates 68, *78*
Mato Tipila, North America, sacred mountain *57*
Mauna Kea, Hawai'i, topographic ecosystem profile of 284, **285**
Mauna Loa, Hawai'i, logging on 353
medicinal plants
    in biodiversity conservation *45*, 223
    as forest products *293*, 293–294
    from sacred mountains 46
    local knowledge of 53, *239*
    natural resource *92*, *100*, 101
meditation, role of mountains in 40, 47, 53–54
Mediterranean Vendetta, and impenetrability of mountains 120
Meghalaya Hills, Bangladesh, flood relevance in **138**
Meghna River, and flooding in Bangladesh *141–142*
Mekong River, water resources and climate change 431
men, and gender inequality in northern Pakistan 68–71
mercury, in gold ore processing 177, 184
meteorological factors, in climate 213
Mexico
    disasters in *376*, *385*, *390*, *401*, *407–408*
    sacred mountains of 46
Middle Mountains, Europe 4, 8, 65–67, 222
migration
    animal adaptation to adverse conditions 212
    of people
        and conflict *114*, 116–117
        in cultural ecology 119
        economic refugees 384–386
        and impenetrability 121
        patterns of 86–88
    of plants and animals in climate change 418–426
    of species in biodiversity 204, 208–209
migration *see also* amenity migration; nomadism;
    refugees; transhumance
migration for employment *see* out-migration
military force *see* conflict
military personnel, and hunting 298
military roads, and access for tourism 253
mineral resources *96*, *99*
mines, anti-personnel (land mines) *388*, 388
mining
    animal dung as fuel 174
    as cause of conflicts 112
    closure costs *99–100*, 180, 181, 185, 192
    costs of in mountains 172, 195
    disasters caused by 393
    environmental impact of 112, 147, 172–186, 222
    and environmental protection 186–189, 191, 193–196
    labour for 172, 174, 175, 178, 180
    mineral extraction 174–176, 180
    in national parks 186, *242–243*
        proposal for *242–243*
    open-pit, conditions in 175, **183**
    ore-processing 176–180
    regions of 171–172

road systems and tourism 253
    and sustainable development 186–196
    underground, conditions in 175, 181
    waste from 175–179, 180–185
    water pollution from 135, 187
minority people, vulnerability of 372
Mittelgebirge, Germany 4, 8, 65–67
modernisation, consequences of 31–35, 158
modernisation versus marginalisation 120
Möhne Rhur, Germany, dam failure in *394*
moisture, as aspect of climate 213–214
moisture fluctuations, plant adaptations against 211
molybdenum mining 171
monsoon precipitation *141–142*, *148*, 233, 345, 355–356
Mont Nimba Nature Reserves, Guinea 237
Montenegro, disasters in *385*, *388*
montology, creation of 464
Morocco
    agriculture and aridity **31**, **152**, 355
    disasters in *387*, *390*, *401*, *407–408*
    land mines in *388*
    national parks in *241*
Mount-Ewaso-Ng'iro, Kenya, water and soil profiles *346*, **347**
Mountain Agenda, Earth Summit, Rio de Janeiro (1992) 323–326, 447–453
mountain areas
    agenda for development in 35–36, 455–466
    status of and Chapter 13 447–453
Mountain Forum, creation of 448
mountain inhabitants *see* indigenous people, local communities
Mountain Research and Development magazine 450
mountain zones, definition of 2–8
mountaineering, impact of *246–247*, 255, 256–258
mountains
    biodiversity on *234–235*
    classification of 4, 8–10, 12
    climate in *137*, 410–412
    features of 61, 313–314, *315*
    human dimension in 11–12
    mining in 171–196
    physical processes in 9–11
    protection of 237–247
    spiritual and cultural significance of 39–59
Muktinath, Nepal 55, *163–164*
Muztagh Ata, China, sacred mountain 44
Myanmar, disasters in 371, *387*, *388*, *390*, *401*
Myanmar *see also* Himalaya
Myrtaceae, in vegetation belts 234

Nahuel Huapi National Park, Andes 243
Nanda Devi National Park, India *237*, *238*
national parks
    global inventory of 237–247
    management of 240–247
    and mining 186, *242–243*
    on national borders 237, 238, 239, 243, *245–246*, *247*

nature reserves 204, 237–247
and sacred mountains *42–43*, 49
natural disasters *see* disasters
natural hazards *see* hazards
natural resources 2, **3**, 8, 96, 99, 101
agricultural use of 127, 314, 318, *319*, 322,
323–333
and conflict 103–106, *110–111*, 113–116
and highland-lowland interaction 82–83, 456,
461–462
management of 124, 125, 127–128
and over-population 17–18
protection of and Chapter 13 447–453
renewable, for energy 158–170
and water resource management 151–152
and watershed management *342*
nature reserves 204, 237–247
Navajo people, North America 40, 46, 112
Naxi, China, women's status in 80–81
Nepal
agricultural practices in 12, 62, *289*, 330, *331*, 356
livestock rearing *25*, 25, 50, 355
biodiversity on 203, 209–210, 211–212
disasters in 10, *385*, *390*, *407–408*
DWET programme *259–261*, **260**, **261**
energy resources 150, 159, *160–161*, *163–164*,
256–257, 297
expoitation of resources in 12
forest management in 282, 293, **300**, 300, 305,
306–307
growth of villages in **24**
land tenure, and watershed management 365–366
literacy rates *91*
medical services in *91*
national parks in *238*, *240*, 246–247
out-migration in 68–71
road construction, environmental impact of 359
sacred places in 42, 44, 50, 51, 57, 299
socio-economic indicators for *74–78*, *79*
tourism in 34, 253–257, 297
women's status in 81
Nepal *see also* Himalaya; Muktinath
Netherlands, water resources from the Rhine River
137, *138*
Nevada, United States, in potential climate change 424
Nevado Sajama, Bolivia, world's highest forests **4**
New Guinea 28, 203, 225
New Mexico, tourism in 262
New Zealand
biodiversity in 209, 212, 287, 419
climate change, effects of 212, 419, 431
climatic records on 412–413
disasters in 351, 358, *407–408*
environmental protection legislation in 190
hydroelectric schemes in 432
indigenous people, sacred mountains and
communal identity 46, *52–53*
national parks in *52–53*, 238, *241*
reforestation in 281
sacred mountains in 40, 43, 46, 49, *52–53*
soil erosion, long term effects of 363

water resources 431
Ngati Tuwharetoa, New Zealand, sacred mountains of
49
Nicaragua
disasters in *385*, *387*, *390*, *401*, *407–408*
land mines in *388*
niche potential, and agriculture 314, *315*, *317*, *319*,
321, 330, *331*
Niger, national parks in *238*
Nile River, water resources and climate change 431
Nimba, Mount, Nature Reserves, Côte d'Ivoire *237*
Ningnan County, China, poverty in *330*
Ningxia, China, earthquake disaster *380*
noise, and modern traffic 33
nomadism 4–5, 25–26, 121
Nong, Buddhist Holy Hill *45*
North America
climatic records on 412–413
environmental protection in 189–193, 239, *240*,
281
indigenous peoples, sacred places of 40, 42, 46,
47, 49, 54
mining in 112, 174, 178, 189–193
national parks in 239, *240*, 243
pollution in precipitation 346
transport systems and tourism 253
vegetation belts in 287
water resources in *137*
North America *see also* individual countries
North Ossetia–Alania, Caucasus
agriculture in *96*
economic and political indicators in 96–98
ethnic diversity in 96, 98, 113–114
Norway *240*, *241*, *385*, *390*, *407–408*
Nothofagus sp., beech, in vegetation belts 217, 285,
287
nutritional status, in mountain communities 62, 63,
76, 92
Nuvatukya'ovi (San Francisco Peaks), North America
40, 46, 49, 54
Ochotona spp., hare, adaptations to lack of oxygen 212

Oetzal Alps, Austria, landslide in 10
Ok Tedi mine, Papua New Guinea 112, 182, 183, 191
Olympic Games *see* Winter Olympics
Olympus, Mount, Greece *42–43*, 57
Omine, Mount, Japan, sacred mountain 42
Ontake, Mount, Japan, sacred mountain 47
Oreochloetum distichae and soil acidity *215*
Orinoca, Bolivia, solar energy in **167**
Ossetian–Ingush, Caucasus, inter-ethnic conflict
113–114
Ötztal, Austrian Tyrol, authoritarian control of
population growth 62
out-migration for employment *22*, 28, 62–65, 99, 121
and labour shortage in mountain communities 18,
68–71, 79, *290*
and spread of AIDS 391
over-population, and restricted resources 17–18
Ovis ammonpolii, argali, in the Pamir *201*

oxygen reduction at altitude, adaptations to 174, 212, 213

Oxyria digyna, mountain sorrel, distribution of 206

ozone, in air pollution in forests 301, *302*

Paekdu, Korea, sacred mountains in 46

Paektu, Mount, Korea, tree species diversity 217

Paektu-san, Korea, biodiversity on *210*

Paha Sapa, North America, sacred mountain 42

Pakistan
    agriculture in 22, 25, **26**, *68–71*, 79–80, 356
        irrigation canals 337
        livestock rearing 68–71, 79, 355
    disasters in *375*, *385*, *387*, *390*, *401*, *407–408*
    drug production in 116
    independence of peripheral mountain regions in 29
    indigenous mountain peoples of 19, *21–22*
    Karakorum Highway 22, 253
    linguistic groups in *21*, 29
    malaria in 434
    national parks in *240*, *244–245*, 298
    tourism in 253, 255, 263
    water resources of *137*
    women of *68–71*, 79

Pakistan *see also* Himalaya; Indus river

Pamir 44, *145*, *201*, *234*, 416

Panama, national parks in *238*, *240*

Papua New Guinea
    agriculture and climate change 428–429
    biodiversity in 232, *234*
    disasters in *385*, *390*, *401*, *407–408*
    logging in 353
    mining in 112, 178, 182, 183, 191, 222
    socio-economic indicators for *74–78*, 79

páramo (hot forests), management and conservation in 226, 227

Parc National des Volcans, Rwanda, wildlife in 298

Parinacota, Chilean Andes, population decline in **20**

Parnassus, Mount, Greece, sacred mountain 42

pastoralism, in sustainable development strategies *110–111*, 125–126

Patacancha Valley, Peru, watershed management in *339*

Patagonia, Chile, tourism in 253

patch-corridor-matrix model of ecosystems 204

Pays d'Enhaut, Switzerland, economic and political indicators in *93–95*

Pechoro–Ilychsky Reserve, Russia *237*

peripheral nature, of mountain regions 29–30

personal identity, role of mountains in 46, 52

Peru
    biodiversity in 50, 217, **286**
    disasters in *376*, *385*, *387*, *390*, *401*, *407–408*
        cholera pandemic (1991) 388
        dam failure in *394*
        landslide in 10, *395*
    drug production in 116
    economic and political indicators in *74–78*, *92–93*
    environmental restoration in 50, 192, 337–339
    Inca people, ancient civilisation 43, *149*, 337–339
    mining in 174, 192

national parks in *237*, *238*, *239*, 253

pilgrimage in 251

poverty and conflict in 115

regeneration of agriculture in 337–339

road construction in 253

sacred mountains of 43, 46, 50

tourism in 253

water management in *149*, 337–339

water resources in *137*, *144*

watershed management in *144*, 337–339

pests and diseases, of plants 221, 417, 425, 429, 430

Philippines
    amenity migration in 265, 266, *269–270*, 272
    conflict in 109, 115
    disasters in *270*, *385*, *387*, *401*
    land ownership and tree planting 288
    precipitation and flooding in 345, **349**
    terracing in farming system 356

photovoltaics (PV), conversion of sunlight to electricity 165–167, *169*

photovoltaics (PV) *see also* solar energy

physical well-being, and mountains 46, 53

physiological adaptations, in mountain conditions 174

Phytophthora cinnamoni, fungal dieback disease *221*

Pic du Midi, France, climatic records on 412

Picea spp., Norway spruce 218, 226, 287, *302*, 422

pilgrimage 40–41, 53–56, 250–251

Pinus spp., pines *210*, 218, 226, 286, 287, 430

Pitjantjara, of Australia, sacred mountains of 46

planning
    for environmental preservation 67
    for sustainable development 35–36, 88
    for tourism 257–264, 274
    for water resource management 153

plant collectors, and endangered ecosystems 222, 223

plant growth forms, in species diversity 211–212

plant nutrition, and climate change 417–418

plant sociology, application of 205

plants
    and climate change 418–426
    species diversity in 200–202, 206–208, 215

plate tectonics, effects of 11, 206–207

poaching, of endangered species 298

Podocarpus, conifer, in vegetation belts 217, 232, 287

Poland 224, 225, *240*, 301, *302*; Giant Mountains; High Tatras

politics
    boundaries in mountains 27–28, 173–174
    and conflict 103–106, *114*, 384–386
    and culture 18, *29–30*
    in dam construction *148*
    and economics 85–101
    in ecosystem management and conservation 85–101, 226–227
    of hydroelectricity *151–152*
    liberalisation of post-UNCD 325
    national agrarian policies, effects of 33
    of planning in climate change 434–435
    and power of lowlands 42, 61–62, 67, 82–83
    and protest movements *20–21*
    in regional development 123, 447–453

in vulnerability 396
in watershed management 173–174, 365–366
pollution
energy sources as cause of 157, 158
international issue 301–303
mining as cause of 112, 194
and precipitation 346
in wetlands and peatlands 224, 225
pollution *see also* air pollution; sewage; water pollution
Polylepis sp., in vegetation belts 217, 234
Polytrichum juniperum, distribution of 206
population displacement 17, 96, 103–106, 371, *373*, 384–386, *408–409*, 427
population growth 62, 96, 97, 98–99
in amenity migration 250
policies on 62, 72
and resource exploitation 12, 152–153, 288, 289
population pressure
and agriculture 17, 320–326, 356, *357–358*
and climate change 430
and conflict *110–111*, 110, *114*
and hazard risks *380*, 388
Populus davidiana 210
Portugal, disasters in *390*
potato, impact of 62–63, 92
Pouteria, in biodiversity conservation on Holy Hills *45*
poverty
and conflict 103–106, 113–116
and environmental damage in mining 196
in mountain regions 61–62
and vulnerability of women and children 67–82
power, of sacred mountains 42–43, 49
precipitation
and climate change 413, 414, *415*, 421–422, *423*, 427–429
as climatic feature 214, 410–412
in cloud forests 284
and disasters 10, *380*
and the hydrologic cycle *338*, 344
influence of Baikal Lake on *219*
measurement of 136–139
patterns of 61, *94*, 98, 99, 131–135
and pollution 301, 346
and soil in climate change 418
and species diversity in forests 287, 288
and streamflow 344–349, *348*
and vegetation belts 232, 234, *235*, 284
primary and secondary succession, of species in biodiversity 209–211
prospecting for minerals, environmental impact of 172–174
protection, for mountain parks 240–241
protest movements, of indigenous people 34, 35, 36, *127–128*, 148, 291
Prunus, in cloud forests 232
Puka, Albania **89**, **90**, *98–100*
Puya, in vegetation belts 233
PV (photovoltaics), conversion of sunlight to electricity 165–167, *169*
Pyrenees 27, 47, 253

Qomolangma Nature Preserve, Tibet 247
quality of life indicators, national statistics *74*, *95*
Quercus sp., oak *218*, 232
Quetta, Pakistan, earthquake disaster in *375*

race *see* ethnic groups
Racomitrium lanuginosum, distribution of 206
radioactive radiation, in mines 188
railways, accidents associated with *380–381*, 393
rainfall *see* precipitation
Rainier, Mount, United Sates, national park 44
Ramechhap District, Nepal, energy sources from the forest *160–161*
Read, Mount, Australia, longevity of trees 286
recreational benefits of mountains 252, *281*, 295, 296–297
Red Dog mine, Alaska, ownership of 174
reforestation 281–282, 293, 361; tree planting
refuge or retreat, in mountain areas 22, 27–29, 109, 249
refugee camps, children's mortality rates in 399
refugees *see* population displacement
regeneration programmes, for woody ecosystems 225–226
religion
and conflict *114*, *117*, 386
diversity in 96
and gender discrimination 69, 79
and mountain population groups 18, **19**, 22
and mountain refuges 27, 29, 120–121
mountains in 39–4, 42, 48–49
and research on sustainable development 47–54
and sacred groves and forests 298–299
water in 131
religion *see also* individual religions; sacred mountains
research
in biosphere reserves, need for 243
on climate 409–414
on climate change, need for 417, 426, 435
information on, and Chapter 13 447–453
on physical processes in mountains 9–13
religious and cultural implications of sustainable development 47–54
and water resource management 151–152
resources *see* natural resources
respiratory diseases, and mining 175, 188
Rhine River, seasonal flow in **137**, *138*
Rhizocarpon geographicum, lichen 210
Rhodiola sp., endangered species 210, 223
Rhododendron sp., in vegetation belts 204, 287
rills, development of in soil erosion 349–350
Rimac River, Peru, dam failure in *394*
Rio Earth Summit 1–2
Agenda 21, Chapter 13 1–2, 447–453
Agenda 21 Chapter 13
on European mountain convention 62
international discourses following 326
impact of 323–326, 455–466
and sustainable development 88–89, 101
riparian buffer strips 353, 355, 365

risks, in mountain areas 371–372
rivers
    ecology of and environmental protection 150–151
    pollution of, by mining 182–183, *185*
    as sacred sites 40
    transport, and landslides in China *381*
roads
    and disaster risk 13, 188, 222, 350, 371,
        380–381, 392
    in socioeconomic importance of *96, 98*, 119
    water quality problems caused by 135
    as watershed land use 352, 353, 359–361
rock faces, endangered ecosystems in 222
rock type, and streamflow 346
rocks, and vegetation belts 234
Rocky Mountains, North America 51, 224, *234, 246*,
    354
Rubus chamaemorus, cloudberry, distribution of 200
runoff, in water resources 135–136, 138, 344, 414, 416
Russia
    disasters in *387, 388, 401*
    national parks in *237, 238*
    transplantation of mine labour force 178
Russian Federation
    North Ossetia–Alania, economic and political
        conditions in *96–98*
    Ossetian–Ingush conflict *113–114*
Ruwenzoris, cloud forests on 284
Rwanda
    disasters in *385, 387, 390, 401*
        land mines in *388*
    endangered species in 297, 298
    ethnic conflict in *110–111*, 110, 121
    infectious diseases in 398, 434
    national parks in *240*
    pastoralism in *110–111*
    socio-economic indicators for 73–79
Rwenzori, Uganda *71–72*

Sabah, Malaysia, forest clearing for recreation in 297
sacred groves and forests 298–299
sacred lake, Titicaca Lake, pollution in *144*
sacred mountains
    biodiversity conservation in 44, *45*
    and gender divisions *69*
    and mining 174
    protection of by national parks *237*
    spiritual and cultural significance of 39–59
    tourism in 57, 250–251
sacred sites, on mountains 39–41
sacrifice, on sacred mountains 40, 44
Sagarmatha *see* Mount Everest
Sagarmatha National Park, Nepal 246–247
Sahel, Sudan, agricultural traditions and conflict
    116–117, 122
St Elias Mount (Washetaca), Alaska, sacred mountain
    46
St Francis, United States, dam failure in *394*
St Helen, Mount, volcanic eruption of 382
Sajama, Mount, Bolivia, national park 234, *237*, **246**

Salang Tunnel, Afghanistan (1982), disaster in 393
San Francisco Peaks (Nuvatukya'ovi), North America,
    sacred mountains 40, 46, 49, 54
San Juan Mountains, Colorado, United States, mining
    in *184–185*, 192
sanatoria, on mountains 46, 53
sanatoria *see also* health resorts
sanctuary theory, and mountain cultures 118–120
Sangay National Park, Ecuador *237, 238*
sanitation
    national statistics 75
    and travellers 55, 251, *257, 258*
sanitation *see also* sewage
Santa Fe, United States, amenity migration in 265, 266,
    272
Satureja, in vegetation belts 234
Saussurea 210
Saxifraga sp. 210, *218*
Scandinavian Mountains, ecosystems in 203
scenery, new protected areas for *241*
schooling *see* education
Scotland 222, 223, 224, 297
seasonal agricultural migration *see* transhumance
sediment
    assessment of 136–138, 140
    impact of 10, 13, 146–147, *148*
    process definition of 351, **352**
    in watersheds 349–352
seeds, for regeneration of ecosystems, problems of 226
seismic activity, and construction projects *148*, 359
seismic activity *see also* earthquake disasters
Sempor, Indonesia, dam failure in *394*
Sendero Luminoso (Shining Path), Peru, rebellion by
    115
Senecio sp., giant groundsel, in vegetation belts 211,
    233
Serbia *385, 388, 401*
sericulture, in agroforestry *289*
settlement patterns
    and altitude 23–27
    in disaster regions 377, 383–384, 395
sewage
    and climate change 434
    disposal problem 97, *100*
    impact of in wetlands and peatlands 225
    and water pollution 346, 365
sewage *see also* sanitation
Sewang, China, flood disaster *380*
sexually transmitted diseases 384, 389–391, 398
Shailani, poet of Chipko movement *128*
Shanxi Province, China, dam failure in *394*
Shasta, Mount (Bohem Puyuik), California, sacred
    mountain 46
Sherpa people, Nepal 50, 57, 62, 81
shifting cultivation, impact of 356–357
Shining Path (*Sendero Luminoso*), Peru, rebellion by
    115
Shintoism, sacred groves and forests of 298
Sibbaldia 210
Siberia, altitudinal zonation in 218, *219*
Siberia *see also* Baikal Lake

Sicily, Italy, earthquake disaster in *375*
Sierra Club 47, 186
Sierra Madre Occidental, open-cast mining in 181
Sierra Nevada, California, United States 39, 44, 51, 147, 298–299
Sierra San Luis, Venezuela, water harvesting by trees 345
Sikhachi Alin Mountains, Siberia, mining in *195*
Sikkim, sacred places in 57
Silene acaulis 211
silica-rich rocks, and soil type 214
silver mining 171, *178–179*
Simien National Park, Ethiopia *237, 238*
Sinai, Mount, Egypt, sacred mountain 49, 47, 57, 39, 40, 250
Skaftafell National Park, Iceland, landslide in 10
skiing
    and climate change 433
    development of *42–43*, 262
    impact of 34, 135, 223, 255, 256, 296, 360
    seasonality of 258–259
    technical requirements for skiing events *252*
skiing *see also* Winter Olympics
slash-and-burn agriculture 224, **303, 304, 316,** 356–357
slopes
    instability of
        and earthquake damage *374*
        and road construction 359–360
    and mining conditions 172, 175–176
    in mountain zone definition 3, 61
    and soil erosion 115
    steepness of
        as environmental hazard *379*
        and streamflow 348
        and timber harvesting 288, 290
    windward, and precipitation 134
Slovakia
    air pollution in *215–216*, 301, 303
    biodiversity in 204, 301, 303
    national parks in *240*
    timber harvesting in 290
    white mountains **199**
Slovenia 64, *385*
snow
    assessment of 140
    hazards of 382, 383
    and pollutants 303
    potential impact of global warming 414–417
    as water reservoir 106–107, 131, 134, 303, 354
snow fields, biodiversity in 215
Snowdonia, Wales, *Rhododendron ponticum* in 204
snowlines, in mountain zone definition 4
snowpack, human-induced 213
social aspects
    of management of national parks 243
    of state influence on culture *29–30*
    of a watershed 341
social conflicts, migration as cause of 87
social facilities, in mining proposals 178–180, 194
social factors, in conflict 103–106

Social Impact Assessment (SIA), in mining proposals 194
social impacts
    of amenity migration 265, 271–273
    of hydropower installations 157
    of modernisation 31–36
    of tourism 255–256, 263
social indicators, national statistics 88–89, 92, 95, 97, *98, 100*
socio-economic conditions
    and agriculture 320–323
    and climate change 426–434
    and disasters *374–379, 393–400*
    inequalities in mountain regions 61–67
    in road construction 392–393
    in vulnerability *374–379, 396, 397, 400–403*
socio-economic indicators, national statistics 73–79
soil
    biota, factors affecting *215–216*, 417
    and climate change 417–418
    conditions
        and biodiversity 203, 209–211, 214–215, *215–216*
        and vegetation belts in mountain forests 232, *235, 284, 287*
    conservation, in watershed management *343*
    fertility, and exploitation of resources 5–6, 12
    pollution, effects of *215–216*, 301, 302
    trees in modification of 135
    water storage, and streamflow 344–349
soil erosion 9, 11, **90**, *97–98*, 100
    agroforestry in protection against 226, 358–359
    in land clearance 356–357, *357–358*
    riparian buffer strips in prevention of 353
    roads and paths as cause of 359–360
    and slope gradient 61, 115
    and watershed management 337, 349–352
solar energy 162, *164–167*, 169, 432
solar radiation, effects of 26–27, 211, 212, 284, *411–412*, 413
Sonnblick, Austrian Alps, climatic records on 412
Sorbus spp., mountain ash, distribution of 207, **208**, *210, 302, 303*
South America
    biodiversity in 207, 208, 211, *234*
    climatic records on 412–413
    conflicts in *387*
    disasters in *380, 394, 408–409*
    infectious diseases in *389, 390*
    mining in 172, 178
    national parks in 240, 241
    precipitation from fog in 346
    vulnerable peoples in *385, 401*
    water resources in *137*
South America *see also* individual countries
South Fork, United States, dam failure in *394*
Southeast Asia, vegetation belts in 232
Southern Appalachian Biosphere Reserve, United States *239*
souvenirs, and tourism 255–256
Soviet Union *see* Armenia; Russia; Russian Federation

Spain 39, 40, *240*, 265
species diversity
    assessment of 202–202
    and distribution 206–212
    importance of in climate change 429–430, 431
    remarkable growth forms in 211–212
Sphagno–Empetretum hermaphroditi, and soil acidity *215*
Sphagnum, in vegetation belts in wetlands 233
spiders, adaptations to adverse conditions 212
spiritual significance of mountains 39–59
    and amenity migration 266–268
    and pilgrimage 53–56
sports activities, and tourism 255
sports activities *see also* skiing
Star Mountains, Papua New Guinea, impact of mining 222
State of the World's Children Report (1994) 73–79
Stava, Italy, dam failure in *394*
Stipa sp., tussock grass 211, 233
storm events, impact of 344–345, 348, *373*, 374, *380*, *407–408*
storm events *see also* tropical cyclone
streamflow, factors affecting 344–349
subsidies, for mountain communities *32*, 33, 61, 64, 66–67, 263
Sudan 116–117, 122, *240*, *241*
Sudetes, Central Europe, human activity and species diversity **223**
sulfate aerosols, and global warming 413
Sumava Mountains, Czech Republic, amenity migration in 265, 266, *267–268*
Summitville mine, Colorado, United States *184–185*, 192
sun *see* solar radiation
survival, of animals and plants in climate change 424–426
sustainability
    Chipko in protest against forestry *127–128*
    definition of 313
sustainable development
    and Agenda 21, Chapter 13 88–89, 447–453, 455–466
    in agriculture 126, 320, *322*, 323, 326–333
    in biosphere reserves 239
    and conflict management 103–106, 123–128
    economics and politics of 85–101
    in endangered ecosystems 226–228
    of energy resources 168–170
    of forests 291, 308–309
    and mining 186–196
    in national parks *244–245*, 247
    national problems of 12, 36, 65–67, *94*, *95*, 98
    religious and cultural implications of research on 47–54
    and sacred mountains 39–41, 54–56, 58–59
    and tourism 54–57, 263
    and vulnerability reduction 400–403
    of water resources 143–151
    watershed management in 340–341, *342*, 365–366
Sweden, national parks in *240*

Switzerland
    agricultural protection in *94*
    agriculture **26**, *94*
    altitudinal zonation in *94*, 218
    climate change 431, 433
    cultural features in 29–30, 62, *94*, 152
    demographic change in 64, *94*, *95*
    disasters in *385*, *390*, *407–408*
    economic and political indicators in 62, *93–95*
    environmental protection *95*, 152
    fertilisation studies *220*
    forest restoration in 281, **282**, 300
    forests in hazard protection 295
    glaciation and frost weathering in **10**
    health services in *95*
    hydropower in *151–152*
    national parks in *240*, *241*
    sustainable development in *94*, *95*
    water resources **134**, **136**, 431
Syr Darya River, and the Aral Sea 145
Syria 103, *107–108*, *401*

table mountains (tepuis), ecosystem diversity in 203
Tabor, Mount, sacred mountain 47
T'ai Shan, China, sacred mountain 46, 40, 43, 45, 250
Taiwan 359, *387*, *407–408*
Tajikstan *74–78*, 79, *113*, *241*, 371, *387*; Aral Sea; Pamir
Talamanca ranges, in corridor of nature reserves *245*
Tanzania 12, *238*, *244*, 398; Kilimanjaro, Mount
Taoism, sacred mountains of 47
Taranaki, New Zealand, and communal identity *52–53*
Task Managers, for Chapter 13 1, 447
Tatra National Park, alien species in 225
Tatras, ski development in, environmental impact of **351**, 359
Tatshenshini, Canada, World Heritage status of *242–243*
Tavil Dara region, Tajikstan, conflict in *114*
technological disasters 371, *373*, *391–393*, *407–408*
Tehri Dam, India, controversy over *128*, 148
temperate mountains, management and conservation of biodiversity in 226–228
temperature
    adaptations to variations in 23–27, 211, 212, 418–426
    and altitude 23–27, 174
    in climate, characteristics of 213–214
    and climate change 418–426
    diurnal variations 173, 175
    and mining conditions 173, 174
    in mountain zone definition 3
    and vegetation belts 232, 284
temperature *see also* global warming
temples, and sacred places 44, 51, 44–45
Terai, Nepal, energy resources and physiographic regions 159
terracing, in indigenous farming systems 356
Tesi Lapcha Pass, Nepal 50
Tetracanthella fjellbergi, and soil acidity *215*

Thailand
    amenity migration in 265, 266, 271, 272
    drug production in 116
    infectious diseases in *389, 390*, 398
    landslips in 295
    road construction in, environmental impact of 359
    sacred groves and forests of 298
    shifting cultivation in 356
Thymus, in vegetation belts 234
Tibet 39–41, 42, 44, 54, 56, 224, *247*, 250; Himalaya
Ticino, Switzerland, conflict over hydropower project
    in *152*
Tien Shan mountains, and the Aral Sea *145*
timber, forests as source of 288, 290
timberlines 3–4, 62, **63**
tin mining 171, 176–177
Titicaca Lake, Andes, pollution in *144*
Tlaloc, Mexico, sacred mountain 46
Tongariro, Mount (Egremont), New Zealand, sacred
    mountain 40, 43, 49, 52
topography of mountains, diversity of 199–200
Törbel, Valais, Switzerland, socio-economic balance in
    62
tourism
    activities for 255
    and AIDS 391
    and climate change 427, 432–433
    cultural impact of 11–12, 34–35, 36
    and economic dependency 63, 64, 85–86, 88
    and endangered ecosystems 222, 223, 296–297
    and energy demands 157, *169*
    and environmental conservation 56–57
    income from *32*, 33–34, 63, 68, 70, 82, *94*, 255
    and industrial archaeology *178–179*
    and modernisation of traffic and communications
        33
    in mountain environments 66–67, 249–264,
        274–276
    in national parks **242**, *243*, 243, 244, *247*
    and pilgrimage 250–251
    seasonality of 258–259
    sensitive issues on *42–43*
    and spiritual well-being 53–54, 56–57
    visitor management 257–258
tourism *see also* trekking
tourists, characteristics of 254–255
trade
    international
        effects of 87–88, 101
        restrictions on *94*
trade routes, robbery on *22*, 28
trans-frontier location, national parks in 238, 239,
    243, *245–246, 247*
transhumance 4–5, 22, 25–26
    and conflict *110–111*, 116–117
    and highland-lowland interaction 121–122
    and watershed protection 354
transnational borders, and future planning *457*
transpiration, and the hydrologic cycle *338*, 344
transport, and climate change 433
transport costs, in mining 172, 176

transport systems
    and amenity migration 266, 270–271
    and expoitation of resources 11, 12
    and hazards 27, 371, *380–381*, 383, 392
    modernisation of 31–33
    and mountain barriers 27
treeline belts 285–287, **287**
trees
    altitude and species diversity 217
    in climate change 214, 419
    climate modification by 135
    dendroclimatological information from 214, 285,
        286
    as fuel for tourism 256
    in the hydrologic cycle **338**, 344–348, 354
    planting by pilgrims in India *55–56*, 58
    pollution damage to *302*
    and streamflow 348
    water harvesting by 345
trees *see also* forests
trekking, impact of *68–71*, 255, 256–258, 259, 297
Troll, Carl, altitudinal belt classification of 4
tropical cyclone, impact of *373, 382, 408–409*
tropical mountains, biodiversity on 226–228, 232–235
Tsuga, hemlock, in vegetation belts 287
tsunami damage *377, 378, 379, 408–409*
tundra (montane), conservation of biodiversity in
    223–224, 226, 227
tungsten mining 171
tunnels, accidents associated with 393
Turkey
    conflict over water supply *107–108*
    disasters in *375, 385, 387, 390, 401, 407–408*
    infectious diseases in *389, 390*
    sacred mountains in 46
Turkmenistan, Ashkhabad Kopper Dagh, earthquake
    disaster in *375*
Turkmenistan *see also* Aral Sea
Tyrol, Austria/Italy, modern development in *32*

U.S.A. *see* United States
Uganda
    agriculture in 72
    AIDS in 389, *390*
    disasters in *72, 385, 387, 390, 401*
    education in 72
    endangered species in 297
    forest management systems, degradation of *305*
    gender discrimination in *71–72*
    national parks in *237, 238, 240, 293–294, 305*
    non-wood forest products *293–294*
    soil degradation in *357–358*
    women's life in *71–72*
Uluru (Ayers Rock), Australia, sacred mountain 46
United Nations Conference on Environment and
    Development (UNCED) *see* Rio Earth Summit
United States
    amenity migration in 265, 266
    biodiversity corridor in 242, *245–246*
    biodiversity in 200, 204, 210, 217, *235*

United States *contd.*
    climate change, potential impact of 413–414, 424, 431
    disasters in *376, 394, 395*
    endangered species in 298
    energy resources in 293, 432
    environmental damage, from mining 147, 192
    forests in hazard protection 295
    llamas as pack animals 297
    mining in 112, 147, 172, 174, 181, *184–185*, 192
    national parks in 44, 237, 238, *239 240, 241*
    poverty in mountain regions 62
    sacred places in 40, 44, 46, 57, 298–299
    snow accumulation manipulation 354
    treeline in 286, **287**
    watershed management in 354, 361
    Yellowstone to Yukon Biodiversity Strategy *246*
United States *see also* North America; Rocky Mountains; Sierra Nevada
Unteraargletscher, near Grimselpass, Switzerland, physical processes in **10**
Urals 203, 218, *234*
Uttar Pradesh, India, tourism to 251, 252

Vaiont, Italy, dam failure in 393, *394*
Val d'Aniviers, Switzerland, agriculture and altitude **26**
vegetation
    and altitudinal belts 4, **6, 7,** 203
    and climate change 417–426
    colours of as mountain feature 199–200
    erosion control by 349–351
    in the hydrologic cycle 134, 337, **338**
    new protected areas for *241*
    species diversity *221*
    and streamflow 344–349, 348–349
    in tundra ecosystems 223–224
    and watershed protection 361–362
    in wetlands and peatlands 224–225
vegetation *see also* agriculture; biomass productivity; crops; forest; grassland; trees
vegetation belts 232–235, 284–288
Venezuela
    disasters in *407–408*
    national parks in *238, 240*
    road construction and slope failure 350, **351,** 359
    sacred forest 299
    watershed system of 345–346
Vermont, United States 293, 361
vertical dimension, in cultural ecology 119–120
vertical nomadism *see* transhumance
Victorian Alps, Australia, cryosphere and global warming 416
Vietnam 266, *387, 388, 390, 401*
village councils, and gender discrimination 73
Virunga National Park, Zaire *237, 238*
volcanoes
    eruption of **383**
        and biodiversity *210*
        impact of *373, 380, 382, 407–408*

and mountain zone classification 8
as sacred mountains 42, 43
soil fertility of 12
vegetation belts on 232, 234
vulnerability
    in disasters *374–379, 393–400*
    and highland-lowland interactions 400
    of persons 372, 386
    reduction of 400–403

Wales, Snowdonia, *Rhododendron ponticum* in 204
war, definition of 103n
war *see also* conflict
Washetaca (St Elias Mount), Alaska, sacred mountain 46
waste and rubbish
    disposal of *95*
    and pollution *97, 100,* 360–361
    and tourism/pilgrimage 55, 251, 257
    water quality problems caused by 135
water
    sacred mountains as source of 46
    and streamflow 140, 344–349
    and weathering of rock 173
water *see also* moisture
water pollution 97, *144, 145,* 191, 340, 346, 365
    by livestock 354
    in conflict 388
    and forest decline 301, *302*
    from human activity 135–136, *144, 145,* 146–147
    from mining 135, 180, 182–184, 186, 187
water resources 3–4, 6, 8, 127, 131–136
    assessment of 136–141
    and climate change 418, 428, 431–432
    conflict over 106–109, 116–117
    conservation of 152–153
    and disease spread 434
    migrants/tourists use of 257, *269–270,* 271, 272
    quality of 143–147, 295. *75*
    sustainable development of 147–152
water resources *see also* glaciers; precipitation; snow
water sports, in amenity migration 267
water storage, and the hydrologic cycle *338,* 344–349
watersheds
    catchment, definition 340
    importance of 337–340
    land uses in 352–361
    management of 340–341, *342–344, 345,* 361–366
        failure of programmes 363–364
    and political boundaries 107, 173
    preserves, forests in 296
    protection of *237,* 447–453
    rehabilitation of 361–365
    settlement on *269*
    soil erosion and sediment in 349–352
    spiritual aspects of pure water 53
    streamflow in 344–349
    in water supply management 147, 150–151
wealth, and amenity migration 265, 266, 268, 270, 271, 274

wetlands and peatlands, ecosystems in 224–225, 233, 340
Whistler, Canada, tourism planning in 264
White Mountains 200, 204, 286, **287**, 298–299
wildlife
 habitat for 297–298, 353
 protection for 50, *241*
 water source for 339–340
wind
 as aspect of climate 213–214, 410–412
 as energy resource 168, 432
Winter Olympics, environmental aspects of *252*
women
 energy resources gathered by *160–161*
 and financial opportunities of tourism *259–261*
 and poverty in developing countries 67–81
 status of **35**, *68–71*, 80–81
  in agroforestry *289–290*
  and authoritarian population control 62
  and Chapter 13 452
  and education 22, *69*, 79, 80
  indicators for, national statistics *78*
 vulnerability of 372, 386, 395, *396*, 397–398
  in spread of AIDS 391
woody ecosystems 225–226
World Bank
 mining investment and environmental standards 190

policies on water resource management 150
World Heritage List, of mountain parks 238–239
world market development, effects of 87–88
Wutai Shan, China, sacred mountain 54

Xishuangbanna region, China **44**, *45*, *239*, 293

Yangtze River, China, flood disasters *381*
Yankunjatjara, of Australia, sacred mountains of 46
yareta, resinous shrub used as fuel 174
Yauricocha mine, Peru 174
Yellowstone to Yukon Biodiversity Strategy *246*
Yemen, disasters in *385*, *387*, *390*
Yi people, Yunnan, China, women's status in **35**, 80–81
Yosemite National Park, United States *237*, *238*, 243
Yosemite Valley, Sierra Nevada, protection of 51
Yucca, in vegetation belts 233
Yugoslavia, disasters in 371, 384, *385*, *390*, *401*, *407–408*
Yungay, Peru, landslide, impact of 395

Zaire, ecological protection in *237*, *238*, *240*, 297
Zambia, national parks in *240*
zinc mining 171, 174
Zuni, in North America, sacred sites of 48